HIGH-DENSITY LIPOPROTEINS

HIGH-DENSITY LIPOPROTEINS

Structure, Metabolism, Function, and Therapeutics

ANATOL KONTUSH
M. JOHN CHAPMAN

WILEY

A JOHN WILEY & SONS, INC., PUBLICATION

Published by John Wiley & Sons, Inc., Hoboken, New Jersey
Published simultaneously in Canada

Library of Congress Cataloging-in-Publication Data:

High-density lipoproteins : structure, metabolism, function, and therapeutics /
by Anatol Kontush, M. John Chapman; illustrations by Alexander Marienko.
p. ; cm.
Includes bibliographical references.
ISBN 978-0-470-40821-6 (cloth)
1. High density lipoproteins—Metabolism. 2. Cardiovascular system—Diseases—Treatment.
I. Kontush, Anatol, 1960-. II. Chapman, M. John, 1945-
[DNLM: 1. Lipoproteins, HDL—metabolism. 2. Atherosclerosis—physiopathology.
3. Cardiovascular Diseases—therapy. QU 85]
QP552.L5.H55 2011
572′.68–dc23

2011026194

Printed in the United States of America

10 9 8 7 6 5 4 3 2 1

To my parents, Professor Sergey Kontush and Professor Svetlana Schekatolina; they are the reason why I do science.

Anatol Kontush

To my parents, Madeleine and Richard, who provided every opportunity to pursue biological sciences, to Jane, and to friends and colleagues near and far.

M. John Chapman

CONTENTS

PREFACE

Cardiovascular disease is the leading cause of death among adults worldwide. According to the World Health Organization (WHO), 29.3% of all deaths (approximately 16.7 million) around the globe resulted from cardiovascular disease in 2002 [1]. By comparison, infectious and parasitic diseases, the second major cause of death according to the WHO, were responsible for 23.0% of all deaths. Strikingly, cardiovascular disease kills more people than cancer (12.5%), human immunodeficiency virus (HIV; 4.9%), and road accidents (2.1%) combined [1]. The 2002 global death rate from cardiovascular disease was 268.8 per 100,000. The contribution of cardiovascular mortality is even higher in developed countries; indeed, in the United States, the overall death rate from cardiovascular disease was 262.5 per 100,000 in 2006, accounting for 34.3% (831,272) of all 2,426,264 deaths, or 1 of every 2.9 deaths [2]. On the basis of 2006 mortality rate data, nearly 2300 Americans die of cardiovascular disease each day, an average of 1 death every 38 seconds. Cardiovascular disease equally is the leading cause of mortality in Europe, accounting for over 4 million deaths each year. Nearly half (49%) of all mortality in Europe is from cardiovascular disease (55% in women and 43% in men) [3].

Myocardial infarction and stroke constitute two major clinical manifestations of cardiovascular disease, accounting for 12.6% and 9.7% of all deaths, respectively in the WHO 2002 report [1]. Accordingly, about half of all deaths from cardiovascular disease in Europe result from coronary heart disease and nearly one-third from stroke [3]. The major impact of cardiovascular disease on human health is directly linked to the worldwide pandemic of metabolic diseases, such as type 2 diabetes and metabolic syndrome, which are closely associated with overweight, obesity, and elevated cardiovascular risk. Indeed, the prevalence of diabetes in adults globally was estimated to be 4.6% in 2000 and was projected to rise to 6.4% by the year 2030. The number of adults with diabetes in the world is estimated to rise from 170 million in 2002 to 300 million in 2025 [1].

Atherosclerosis represents the pathological process that underlies cardiovascular morbidity and mortality; in association with plaque rupture, thrombosis, and vessel occlusion, atherosclerosis occurs preferentially at sites of endothelial dysfunction and leads to the formation of atheromatous plaques in the arterial wall. Atherosclerotic plaques result from the progressive accumulation of cholesterol and diverse lipids in native and oxidized forms, extracellular matrix material, inflammatory cells and cell debris in the arterial intima and media. The key role of plasma-derived **cholesterol** in the initiation of atherosclerosis was suggested by Nikolai Anitschkow as early as 1913 [4–6]. Almost 40 years later, this concept was refined on the basis of the analytical ultracentrifugal quantitation and identification of plasma lipoproteins developed by Howard Eder and colleagues at the New York Hospital—Cornell Medical Center in New York [7, 8] and by John Gofman and colleagues at the University of California in Berkeley [9]. Notably, Eder and colleagues demonstrated a prevalence of cholesterol-rich, **low-density lipoprotein (LDL)** and a paucity of protein-rich, **high-density lipoprotein (HDL)** in atherosclerosis and related conditions [10]; the hypothesis proposing the protective role of HDL in cardiovascular disease was born (Fig. 1).

Atherogenic dyslipidemia, recognized as a highly prominent cardiovascular risk factor in subsequent studies, is intimately associated with premature atherosclerosis and involves an imbalance between excess circulating levels of cholesterol in the form of proatherogenic **apolipoprotein B (apoB)**-containing lipoproteins relative to subnormal levels of antiatherogenic **apoA-I**-containing lipoproteins. ApoB is the predominant protein component of proatherogenic, cholesterol-rich LDL, triglyceride-rich very-low density lipoprotein (VLDL), VLDL remnants, and intermediate-density lipoprotein (IDL), whereas apoA-I is the major protein component of antiatherogenic HDL. Indeed, elevated circulating concentrations of **LDL-cholesterol (LDL-C)** occur frequently as hypercholesterolemia, a common form of atherogenic dyslipidemia in which levels of HDL-C are subnormal [46]. LDL is the major vehicle for transport of cholesterol not only to peripheral tissues but also to the arterial wall [47]; preferential ionic interaction of positively charged domains of apoB with negatively charged proteins of the extracellular matrix, including proteoglycans, collagen, and fibronectin, leads to intimal retention of apoB-containing lipoproteins with their subsequent accumulation in arterial wall cells, primarily in macrophages [48]. According to the widely accepted **response-to-retention hypothesis** of atherosclerosis formulated by Kevin Williams and Ira Tabas in the 1990s [49, 50], this process constitutes a major initiating factor in atherosclerotic disease as originally proposed by Dawn Schwenke and Thomas Carew in the 1980s [51–53].

Consistent with the key role of LDL in atherogenesis, inhibitors of the key enzyme of cholesterol biosynthetic pathway, 3-hydroxy-3-methyl-glutaryl-CoA (HMG-CoA) reductase, better known as statins, facilitate marked decrease in

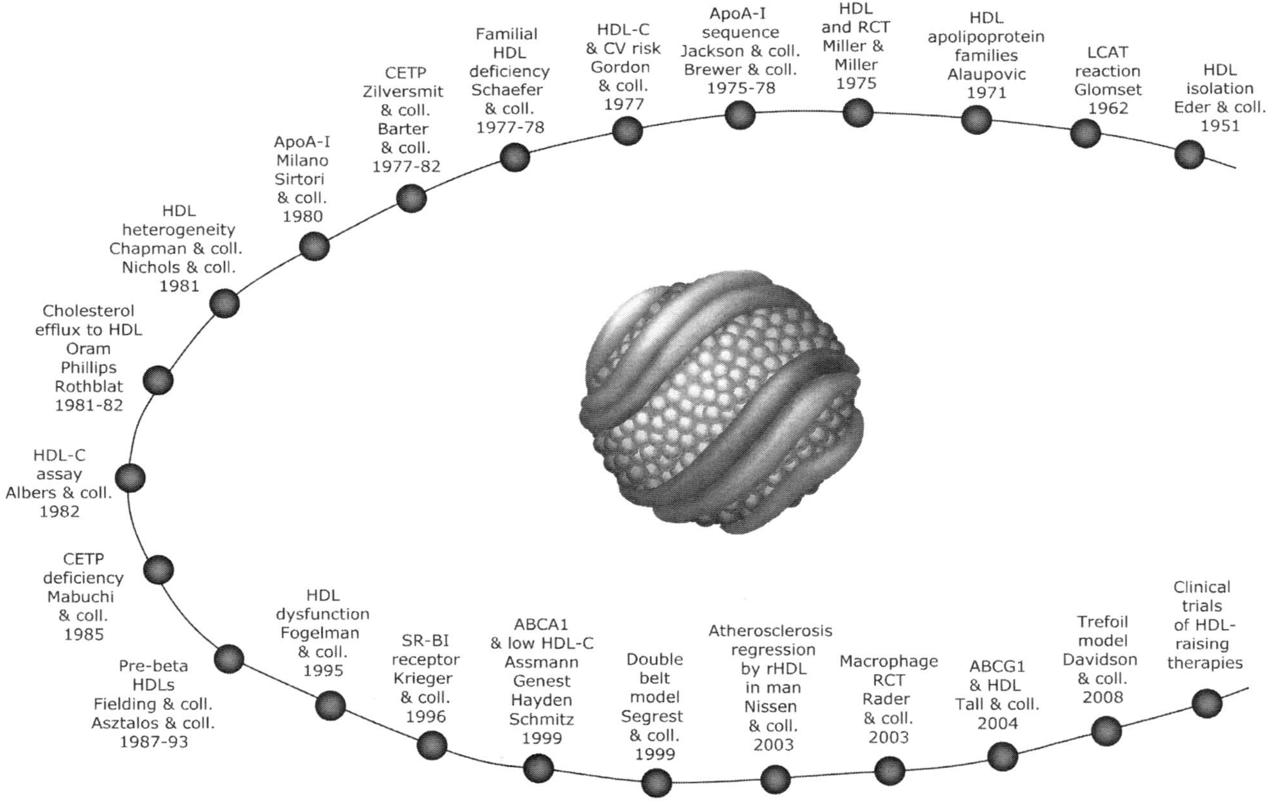

Figure 1 **Timeline of the development of our understanding of human HDL**. Major breakthroughs are presented in chronological order as keywords. It is of note that the list of authors' names is by no means exhaustive; many colleagues and coworkers, whose names are not listed only because of lack of space, provided critical contributions to the discoveries made. The names of such individuals can be found in full references to the studies included in the timeline [7, 8, 11–45].

cardiovascular risk of approximately 30 to 40% [54]; nonetheless, significant residual risk persists. Among factors other than LDL-C that are associated with dyslipidemia, a low level of **HDL-cholesterol (HDL-C)** is most recognized [55]. Several prospective epidemiologic studies, starting with the pioneering investigations of John Gofman in 1966 [56] and Tavia Gordon in 1977 [11] (Table 1) in the Framingham Heart Study and now extending to the US Physicians' Health Study, Prospective Cardiovascular Münster (PROCAM) Study, and Atherosclerosis Risk in Communities (ARIC) Study, have found that low serum HDL-C concentrations (defined as <40 mg/dl in both sexes, or as <40 mg/dl in men and <50 mg/dl in women [57])[1] constitute a strong, independent risk factor for premature coronary heart disease in both nondiabetic and diabetic subjects [55, 58, 59]. Moreover, low HDL-C is characteristic of atherogenic dyslipidemia and increased cardiovascular risk in patients with metabolic diseases such as type 2 diabetes and metabolic syndrome, particularly in those presenting with an inflammatory state.

Consistent with this concept, dyslipidemia defined as a plasma apoB/apoA-I ratio of ≥5:1 was identified as the most strongly predictive cardiovascular risk factor for first myocardial infarction in the standardized case-control INTERHEART study in 52 countries worldwide, representing every inhabited continent [68]. The study, which enrolled 29,972 subjects, revealed that the strongest predictive cardiovascular risk factors for myocardial infarction were dyslipidemia, smoking, hypertension, diabetes, abdominal obesity, psychosocial factors, consumption of fruits, vegetables, and alcohol, and lack of regular physical activity [68]. Collectively, these factors accounted for most (≥90%) of the population attributable risk for the first myocardial infarction in both sexes and at all ages in all regions.

Prospective studies have revealed that the risk for coronary heart disease is elevated by 3% in women and 2% in men for each decrement of 1 mg/dl in HDL-C [46, 61], resulting in the hypothesis that HDL represents a key cardioprotective entity as formulated by David Gordon and Basil Rifkind in 1989 [63] (Table 1). Reliable determination of plasma HDL-C concentrations based on the **dextran sulfate-Mg^{2+} precipitation method** developed by John Albers and associates in 1982 was a prerequisite for this breakthrough [13] (Table 1). Genetic conditions of **familial HDL deficiency** originally described by Ernst Schaefer and colleagues in association with elevated cardiovascular risk provided further contribution to establishing links between HDL and cardiovascular disease [17–19]. The prevalence of low HDL-C levels can vary from 20% in a general population to up to 60% in patients with established coronary heart disease [69]. Not only are low HDL-C levels associated with an increased incidence of coronary heart disease but equally with a greater risk for carotid atherosclerosis and ischemic stroke mortality, and with a more aggressive progression of angiographically defined coronary artery disease [55, 58]. Conversely, decreased risk of cardiovascular events has been frequently observed in subjects with elevated HDL-C

---

[1] To convert mg/dl to mmol/l, multiply by 0.0259. To convert mmol/l to mg/dl, multiply by 38.7.

TABLE 1 Most Cited Papers on HDL

Total Citations	Cites per Year	Authors	Title	Year	Source	References
2986	**87.82**	Gordon, T. et al.	High-density lipoprotein as a protective factor against coronary heart disease: The Framingham study	1977	*The American Journal of Medicine*	[11]
2059	**171.58**	Rubins, H.B. et al.	Gemfibrozil for the secondary prevention of coronary heart disease in men with low levels of high-density lipoprotein cholesterol	1999	*The New England Journal of Medicine*	[60]
2000	55.56	Miller, G.J. and Miller, N.E.	Plasma high-density lipoprotein concentration and development of ischemic heart disease	1975	*The Lancet*	[12]
1713	**77.86**	Gordon, D.J. et al	High-density lipoprotein cholesterol and cardiovascular disease. Four prospective American studies	1989	*Circulation*	[61]
1538	55.03	Warnick, G.R. et al.	Dextran sulfate-Mg^{2+} precipitation procedure for quantitation of high-density lipoprotein cholesterol	1982	*Clinical Chemistry*	[13]
1405	**78.06**	Acton, S. et al.	Identification of scavenger receptor SR-BI as a high-density lipoprotein receptor	1996	*Science*	[14]
1270	**84.67**	Hokanson, J.E. et al.	Plasma triglyceride level is a risk factor for cardiovascular disease independent of high-density lipoprotein cholesterol level: a meta-analysis of population-based prospective studies	1996	*Journal of Cardiovascular Risk*	[62]

(continued)

TABLE 1 ***(Continued)***

Total Citations	Cites per Year	Authors	Title	Year	Source	References
1160	**96.67**	Brooks-Wilson, A. et al.	Mutations in ABC1 in Tangier disease and familial high-density lipoprotein deficiency	1999	*Nature Genetics*	[15]
1089	49.50	Gordon, D.J. and Rifkind, B.M.	High-density lipoprotein—the clinical implications of recent studies	1989	*The New England Journal of Medicine*	[63]
959	35.32	Eisenberg, S.	High-density lipoprotein metabolism	1984	*Journal of Lipid Research*	[64]
948	**118.50**	Nissen, S.E. et al.	Effect of recombinant ApoA-I Milano on coronary atherosclerosis in patients with acute coronary syndromes: a randomized controlled trial	2003	*JAMA*	[16]
545	**77.86**	Brousseau, M.E. et al.	Effects of an inhibitor of cholesteryl ester transfer protein on HDL-cholesterol	2004	*The New England Journal of Medicine*	[65]
529	**75.57**	Cohen, J.C. et al.	Multiple rare alleles contribute to low plasma levels of HDL-cholesterol	2004	*Science*	[66]
324	**108.00**	Kathiresan, S. et al.	Six new loci associated with blood low-density lipoprotein cholesterol, high-density lipoprotein cholesterol, or triglycerides in humans	2008	*Nature Genetics*	[67]

Source: Publish or Perish, www.harzing.com/pop.htm, search of 16.12.2010 for the terms "high-density lipoprotein," "HDL," "apolipoprotein A-I" or "apoA-I" in the title only. Citation numbers for the top 10 most cited papers, selected according to either the total number of citations (first column, which defines the paper ranking in the table) or the number of citations per year (second column) are shown in bold.

levels [55, 58, 70]; in addition, high concentrations of HDL-C (>60 mg/dl) can be associated with longevity [71, 72].

The imbalance between circulating levels of cholesterol transported in HDL relative to that in apoB-containing particles is intimately associated with the induction of both endothelial dysfunction and oxidative stress in the arterial wall, which are in turn closely related to inflammation [47, 73]; as a result, dyslipidemia, oxidative stress, and inflammation are intimately interrelated in the development of atherosclerosis. **Inflammation** is a systemic body response aimed at decreasing the toxicity of harmful agents and repairing damaged tissue. Local inflammation in the arterial wall can be triggered in response to the retention of apoB-containing lipoproteins, primarily LDL, in the arterial intima [49, 50]; modifications of LDL under the action of oxidative stress and other proatherogenic factors can further propagate such a local inflammatory response [74]. Chronic systemic inflammation measured as circulating levels of an acute phase protein biomarker, such as C-reactive protein (CRP), represents a marker of cardiovascular risk [75–78]. Atherosclerosis can therefore be regarded as a chronic inflammatory disease of the arterial wall mediated by modified LDL acting in concert with a spectrum of additional proinflammatory agents on arterial wall cells.

HDL particles are distinguished from atherogenic apoB-containing lipoproteins by their capacity to exert a wide spectrum of **antiatherogenic functional biological activities**. Such activities primarily include the capacity to mediate **cellular cholesterol efflux** by acting as primary acceptors, thereby facilitating **reverse cholesterol transport (RCT)**, a process in which cholesterol is transferred from the arterial wall and peripheral tissues to the liver for excretion into the bile [79]. Thus, RCT involves cellular efflux of free cholesterol as an initial step whose mechanisms were first detailed by John Oram, Michael Phillips, George Rothblat, and colleagues in the early 1980s [20, 21]. As a corollary, reduction in the capacity of HDL to remove cellular cholesterol may result in cholesterol accumulation in arterial wall cells, thereby favoring the development of atherosclerotic plaques [12] (Table 1).

Macrophages represent a major cell type which accumulates cholesterol in atherosclerotic plaques; RCT from macrophages is therefore of a key clinical importance as shown by Dan Rader and colleagues who developed an innovative approach to assess RCT *in vivo* [22]. Cellular receptors and transporters are intimately implicated in cholesterol efflux from macrophages; these primarily are **scavenger receptor class B type I (SR-BI)** as discovered by Monty Krieger [14], **ATP-binding cassette transporter A1 (ABCA1)** as described by Gerd Assmann, Jacques Genest, Michael Hayden, Gerd Schmitz, and colleagues [15, 23, 24] and **ATP-binding cassette transporter G1 (ABCG1)** as observed by Alan Tall and coworkers [25]. Esterification of cell-derived free cholesterol in HDL ensues which is catalyzed by the enzyme **lecithin:cholesterol acyltransferase (LCAT)** as originally reported by John Glomset [26]; hepatic cholesterol removal constitutes a final step of the plasma RCT pathway [12]. Transfer of esterified cholesterol from HDL to apoB-containing lipoproteins represents an important deviation from such "direct" RCT pathway, which is ensured by **cholesteryl ester transfer protein (CETP)** as described by Donald Zilversmit [27, 28], Philip Barter [29, 30], and colleagues in the 1970s.

Other key biological activities of HDL involve (i) the protection of LDL against oxidative stress; (ii) anti-inflammatory actions on arterial wall cells, and (iii) cytoprotective, (iv) anti-infectious, (v) vasodilatory, (vi) antithrombotic, and (vii) antidiabetic activities [80]. Such a wide spectrum of biological activities reflect high heterogeneity of plasma HDL particles, which considerably differ in shape, size, structure, and composition; John Chapman [31], Alex Nichols [32], Christopher Fielding [33], Bela Asztalos [34], and colleagues were among the first to document heterogeneous properties of HDL. Intriguingly, HDL particles can lose their protective activities and even acquire proatherogenic properties under some pathological conditions that primarily include inflammation and induction of the acute phase response. The concept of such **HDL dysfunction** was originally developed by Alan Fogelman, Mohamad Navab, and coworkers in the mid-1990s [35, 36].

The potent atheroprotective properties of HDL particles originate from their unique composition and structure, which are distinguished by the presence of specific (apolipo)proteins and lipids as proposed by Petar Alaupovic in the early 1970s [37]. α-Helical amphipathic structure of apoA-I sequenced in the 1970s by the laboratories of Richard Jackson, Bryan Brewer, and colleagues [38–40] ensures a unique **double-belt configuration** as revealed by studies of Jere Segrest and colleagues in small discoid HDLs [41]. Subsequently, the applicability of the double-belt-like structure to spherical HDL particles was demonstrated by Sean Davidson and colleagues who developed the trefoil model of apoA-I organization in plasma HDL and its major subpopulations [42, 43].

Potent HDL-mediated atheroprotection together with the strong epidemiologic association between low plasma HDL-C concentrations and elevated cardiovascular risk suggest that HDL may represent a promising therapeutic target in the field of cardiovascular disease and atherosclerosis; hence, the concept of **HDL(-C) raising** to treat residual cardiovascular risk remaining after statin treatment. Direct evidence to support this hypothesis was provided in 2003 by Steve Nissen and colleagues who demonstrated that only four-weekly injections of **reconstituted HDL (rHDL)** consisting of apoA-I and phospholipid produced regression of coronary atherosclerosis in patients with acute coronary syndromes [16]. In this study, a genetic variant of apoA-I termed **apoA-I Milano**, which may possess enhanced atheroprotective properties, discovered in 1980 by Cesare Sirtori, Guido Franceschini, and colleagues [44], was used.

Another perspective strategy to target plasma HDL is exemplified by the inhibition of CETP; this approach was prompted by the original observation of Hiroshi Mabuchi and colleagues who reported elevated levels of HDL-C in individuals with genetic CETP deficiency [45]. Clinical trials of HDL-raising therapies, including **CETP inhibitors, nicotinic acid**, and other agents, have therefore attracted particular attention over the last years. Such trials, however, received a major blow when treatment with torcetrapib, the first CETP inhibitor

that entered clinical trials, resulted in excess mortality relative to placebo in patients at high risk for cardiovascular disease [81]. Importantly, other CETP inhibitors under development do not seem to reveal such deleterious effects [82–85], suggesting that it was the failure of the torcetrapib molecule itself rather than the failure of the HDL-raising approach as a whole. Ongoing large-scale trials of novel HDL-raising agents from this class, notably dalcetrapib and anacetrapib, may, if successful, open a new era in the prevention and treatment of cardiovascular disease.

In this book, we will review current knowledge of plasma HDL particles, focussing on their composition, heterogeneity, structure, metabolism, epidemiology, genetics, and biological activities; these topics are covered in Section 1. Furthermore, new findings on functionally defective HDL will be discussed in the context of metabolic diseases associated with elevated cardiovascular risk (Section 2). Finally, in Section 3 we will critically appraise innovative therapeutic strategies to normalize circulating levels and defective functionality of HDL; these exciting developments open new horizons for the treatment of atherogenic dyslipidemia and reduction of cardiovascular risk remaining after statin treatment [54, 86].

The announced quest for HDL as the next therapeutic target to reduce cardiovascular risk has since evolved into a great scientific enterprise full of tough challenges, great advances, unintended errors, and controversial discussions. The major aim of this book is to present a snapshot of this rapidly developing, extremely interesting, and highly challenging field taken at end 2010/beginning 2011, and in turn to relate this picture to the history of the field—a project that looks particularly relevant in view of the 60th anniversary of the discovery that low HDL-C levels are associated with the premature atherosclerosis [10].

Thanks to such a long history, a rich present and a promising future, HDL-targeted therapy has become a well-known concept in biomedicine and beyond. A Google search for the term "high-density lipoprotein" produces some 1,340,000 results; a PubMed search discovers over 27,000 articles related to "high-density lipoprotein," and over 50,000 articles related to HDL (state December 2010), with a clear trend to an increasing number of articles per year (Fig. 2). Despite such strong performance, the level of awareness and knowledge about HDL (commonly termed "good cholesterol") in the general population remains low. Thus, only 14% of adults are familiar with the term HDL-C in Austria, with particularly low knowledge observed for small towns and people with low income [87]. Physicians remain the most important source of information about HDL-C for the general public, emphasizing the insufficiency of available information [87]. In addition, according to a recent questionnaire survey performed in Korea, high cholesterol was identified as a cause of cardiovascular disease by 54.4% of respondents [88]; however, a vast majority of 95.4% did not know their own values and only a tiny 4.1% of respondents were aware of desirable levels of total cholesterol. Regarding HDL-C, a small fraction of 8% of respondents correctly perceived the meaning of this measure as "good cholesterol" [88]. Another aim of our book is therefore to draw more attention to the importance of cholesterol

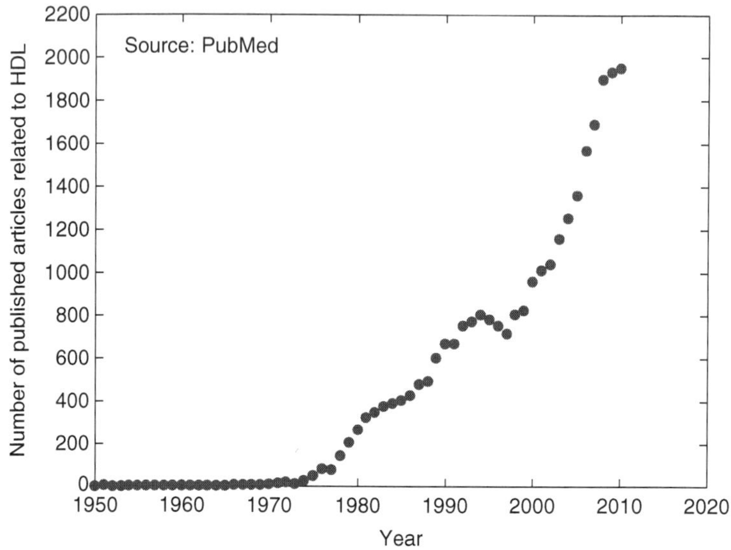

Figure 2 Evolution of the number of HDL-related publications from 1951 to 2010. *Source*: PubMed (December 2010).

metabolism as a whole, and HDL metabolism in particular, for cardiovascular disease. It is important to mention in this regard that, as a result of its focus on cardiovascular disease, this book is essentially devoted to HDL from human plasma. Other types of HDL particles, such as those from human brain [89–91], or from other biological species, largely remain beyond the scope of this edition.

Recently, Sniderman and Furberg have insightfully compared the natural history of coronary disease with a three-act tragedy [92]: "The first act introduces and develops the main characters—namely, atherogenic dyslipoproteinemia, hypertension, and smoking—that appear as we mature and unless something is done, persist during our lifetime. During the second act, which also takes place over decades, these villains incessantly attack and progressively deform the innocent arterial wall. Finally, the third act, which can be tragically brief: in an instant the plaque ruptures, the artery thromboses, and the hero or heroine dies, all too frequently unaware of the drama that was enacted within their arteries. What is the difference, you ask? In the drama of coronary disease, the ending is not fixed; if some of the characters are edited out of the play as soon as they appear, the third act need never take place" [92]. We do believe that this book will be helpful to the scientific community in providing an ultimate answer regarding the role—positive or neutral, major or negligible, causal or bystander—played by HDL in this major human tragedy.

REFERENCES

1. World-Health-Organization. The atlas of heart disease and stroke. World Health Organization; Geneva ; 2004, 112 pp.

2. Lloyd-Jones D, Adams RJ, Brown TM, Carnethon M, Dai S, De Simone G, Ferguson TB, Ford E, Furie K, Gillespie C, Go A, Greenlund K, Haase N, Hailpern S, Ho PM, Howard V, Kissela B, Kittner S, Lackland D, Lisabeth L, Marelli A, McDermott MM, Meigs J, Mozaffarian D, Mussolino M, Nichol G, Roger VL, Rosamond W, Sacco R, Sorlie P, Roger VL, Thom T, Wasserthiel-Smoller S, Wong ND, Wylie-Rosett J. Heart disease and stroke statistics—2010 update: a report from the American Heart Association. *Circulation* 2010; 121: e46–e215.
3. British Heart Foundation. European cardiovascular disease statistics, 2000 ed. British Heart Foundation; London: 2000.
4. Anitschkow NN. Ueber die Veranderungen der Kaninchenaorta bei experimenteller Cholesterinsteatose. *Beitr Pathol Anat* 1913; 56: 379–404.
5. Anitschkow NN, Chalatov S Ueber experimentelle Choleserinsteatose und ihre Bedeutung fur die Entstehung einiger pathologischer Prozesse. *Zentralbl Allg Pathol* 1913; 24: 1–9.
6. Anitschkow NN. Experimental atherosclerosis in animals. In: Cowdry EV, editor. Arteriosclerosis: a survey of the problem. New York: Macmillan; 1933. p 271–322.
7. Russ EM, Eder HA, Barr DP. Protein-lipid relationships in human plasma. I. In normal individuals. *Am J Med* 1951; 11: 468–479.
8. Havel RJ, Eder HA, Bragdon JH. The distribution and chemical composition of ultracentrifugally separated lipoproteins in human serum. *J Clin Invest* 1955; 34: 1345–1353.
9. Lindgren FT, Elliott HA, Gofman JW. The ultracentrifugal characterization and isolation of human blood lipids and lipoproteins, with applications to the study of atherosclerosis. *J Phys Colloid Chem* 1951; 55: 80–93.
10. Barr DP, Russ EM, Eder HA. Protein-lipid relationships in human plasma: II. In atherosclerosis and related conditions. *Am J Med* 1951; 11: 480.
11. Gordon T, Castelli WP, Hjortland MC, Kannel WB, Dawber TR. High density lipoprotein as a protective factor against coronary heart disease. The Framingham Study. *Am J Med* 1977; 62: 707–714.
12. Miller GJ, Miller NE. Plasma-high-density-lipoprotein concentration and development of ischaemic heart-disease. *Lancet* 1975; 1: 16–19.
13. Warnick GR, Benderson J, Albers JJ. Dextran sulfate-Mg2+ precipitation procedure for quantitation of high-density-lipoprotein cholesterol. *Clin Chem* 1982; 28: 1379–1388.
14. Acton S, Rigotti A, Landschulz KT, Xu S, Hobbs HH, Krieger M. Identification of scavenger receptor SR-BI as a high density lipoprotein receptor. *Science* 1996; 271: 518–520.
15. Brooks-Wilson A, Marcil M, Clee SM, Zhang LH, Roomp K, van Dam M, Yu L, Brewer C, Collins JA, Molhuizen HO, Loubser O, Ouelette BF, Fichter K, Ashbourne-Excoffon KJ, Sensen CW, Scherer S, Mott S, Denis M, Martindale D, Frohlich J, Morgan K, Koop B, Pimstone S, Kastelein JJ, Genest J Jr, Hayden MR. Mutations in ABC1 in Tangier disease and familial high-density lipoprotein deficiency. *Nat Genet* 1999; 22: 336–345.
16. Nissen SE, Tsunoda T, Tuzcu EM, Schoenhagen P, Cooper CJ, Yasin M, Eaton GM, Lauer MA, Sheldon WS, Grines CL, Halpern S, Crowe T, Blankenship JC, Kerensky R. Effect of recombinant ApoA-I Milano on coronary atherosclerosis in patients with acute coronary syndromes: a randomized controlled trial. *JAMA* 2003; 290: 2292–2300.

17. Assmann G, Simantke O, Schaefer HE, Smootz E. Characterization of high density lipoproteins in patients heterozygous for Tangier disease. *J Clin Invest* 1977; 60: 1025–1035.
18. Schaefer EJ, Blum CB, Levy RI, Jenkins LL, Alaupovic P, Foster DM, Brewer HB Jr. Metabolism of high-density lipoprotein apolipoproteins in Tangier disease. *N Engl J Med* 1978; 299: 905–910.
19. Schaefer EJ, Santos RD, Asztalos BF. Marked HDL deficiency and premature coronary heart disease. *Curr Opin Lipidol* 2010; 21: 289–297.
20. Oram JF, Albers JJ, Cheung MC, Bierman EL. The effects of subfractions of high density lipoprotein on cholesterol efflux from cultured fibroblasts. Regulation of low density lipoprotein receptor activity. *J Biol Chem* 1981; 256: 8348–8356.
21. Rothblat GH, Phillips MC. Mechanism of cholesterol efflux from cells. Effects of acceptor structure and concentration. *J Biol Chem* 1982; 257: 4775–4782.
22. Zhang Y, Zanotti I, Reilly MP, Glick JM, Rothblat GH, Rader DJ. Overexpression of apolipoprotein A-I promotes reverse transport of cholesterol from macrophages to feces in vivo. *Circulation* 2003; 108: 661–663.
23. Bodzioch M, Orso E, Klucken J, Langmann T, Bottcher A, Diederich W, Drobnik W, Barlage S, Buchler C, Porsch-Ozcurumez M, Kaminski WE, Hahmann HW, Oette K, Rothe G, Aslanidis C, Lackner KJ, Schmitz G. The gene encoding ATP-binding cassette transporter 1 is mutated in Tangier disease. *Nat Genet* 1999; 22: 347–351.
24. Rust S, Rosier M, Funke H, Real J, Amoura Z, Piette JC, Deleuze JF, Brewer HB, Duverger N, Denefle P, Assmann G. Tangier disease is caused by mutations in the gene encoding ATP-binding cassette transporter 1. *Nat Genet* 1999; 22: 352–355.
25. Wang N, Lan D, Chen W, Matsuura F, Tall AR. ATP-binding cassette transporters G1 and G4 mediate cellular cholesterol efflux to high-density lipoproteins. *Proc Natl Acad Sci U S A* 2004; 101: 9774–9779.
26. Glomset JA, Parker F, Tjaden M, Williams RH. The esterification in vitro of free cholesterol in human and rat plasma. *Biochim Biophys Acta* 1962; 58: 398–406.
27. Pattnaik NM, Montes A, Hughes LB, Zilversmit DB. Cholesteryl ester exchange protein in human plasma isolation and characterization. *Biochim Biophys Acta* 1978; 530: 428–438.
28. Pattnaik NM, Zilversmit DB. Interaction of cholesteryl ester exchange protein with human plasma lipoproteins and phospholipid vesicles. *J Biol Chem* 1979; 254: 2782–2786.
29. Barter PJ. Origin of esterified cholesterol transported in the very low density lipoproteins of human plasma. *J Lipid Res* 1974; 15: 11–19.
30. Barter PJ, Hopkins GJ, Calvert GD. Transfers and exchanges of esterified cholesterol between plasma lipoproteins. *Biochem J* 1982; 208: 1–7.
31. Chapman MJ, Goldstein S, Lagrange D, Laplaud PM. A density gradient ultracentrifugal procedure for the isolation of the major lipoprotein classes from human serum. *J Lipid Res* 1981; 22: 339–358.
32. Blanche PJ, Gong EL, Forte TM, Nichols AV. Characterization of human high-density lipoproteins by gradient gel electrophoresis. *Biochim Biophys Acta* 1981; 665: 408–419.
33. Ishida BY, Frolich J, Fielding CJ. Prebeta-migrating high density lipoprotein: quantitation in normal and hyperlipidemic plasma by solid phase radioimmunoassay following electrophoretic transfer. *J Lipid Res* 1987; 28: 778–786.

34. Asztalos BF, Sloop CH, Wong L, Roheim PS. Two-dimensional electrophoresis of plasma lipoproteins: recognition of new apo A-I-containing subpopulations. *Biochim Biophys Acta* 1993; 1169: 291–300.
35. Van Lenten BJ, Hama SY, de Beer FC, Stafforini DM, McIntyre TM, Prescott SM, La Du BN, Fogelman AM, Navab M. Anti-inflammatory HDL becomes pro-inflammatory during the acute phase response. Loss of protective effect of HDL against LDL oxidation in aortic wall cell cocultures. *J Clin Invest* 1995; 96: 2758–2767.
36. Navab M, Berliner JA, Watson AD, Hama SY, Territo MC, Lusis AJ, Shih DM, Van Lenten BJ, Frank JS, Demer LL, Edwards PA, Fogelman AM. The Yin and Yang of oxidation in the development of the fatty streak. A review based on the 1994 George Lyman Duff Memorial Lecture. *Arterioscler Thromb Vasc Biol* 1996; 16: 831–842.
37. Alaupovic P. Apoliproproteins and lipoproteins. *Atherosclerosis* 1971; 13: 141–146.
38. Delahunty T, Baker HN, Gotto AM Jr, Jackson RL. The primary structure of human plasma high density apolipoprotein glutamine I (ApoA-I). I. The amino acid sequence of cyanogen bromide fragment II. *J Biol Chem* 1975; 250: 2718–2724.
39. Baker HN, Gotto AM Jr, Jackson RL. The primary structure of human plasma high density apolipoprotein glutamine I (ApoA-I). II. The amino acid sequence and alignment of cyanogen bromide fragments IV, III, and I. *J Biol Chem* 1975; 250: 2725–2738.
40. Brewer HB Jr, Fairwell T, LaRue A, Ronan R, Houser A, Bronzert TJ. The amino acid sequence of human APOA-I, an apolipoprotein isolated from high density lipoproteins. *Biochem Biophys Res Commun* 1978; 80: 623–630.
41. Segrest JP, Jones MK, Klon AE, Sheldahl CJ, Hellinger M, De Loof H, Harvey SC. A detailed molecular belt model for apolipoprotein A-I in discoidal high density lipoprotein. *J Biol Chem* 1999; 274: 31755–31758.
42. Silva RA, Huang R, Morris J, Fang J, Gracheva EO, Ren G, Kontush A, Jerome WG, Rye KA, Davidson WS. Structure of apolipoprotein A-I in spherical high density lipoproteins of different sizes. *Proc Natl Acad Sci U S A* 2008; 105: 12176–12181.
43. Huang R, Silva RA, Jerome WG, Kontush A, Chapman MJ, Curtiss LK, Hodges TJ, Davidson WS. Apolipoprotein A-I structural organization in high-density lipoproteins isolated from human plasma. *Nat Struct Mol Biol* 2011; 18: 416–422.
44. Franceschini G, Sirtori CR, Capurso A II, Weisgraber KH, Mahley RW. A-I Milano apoprotein. Decreased high density lipoprotein cholesterol levels with significant lipoprotein modifications and without clinical atherosclerosis in an Italian family. *J Clin Invest* 1980; 66: 892–900.
45. Koizumi J, Mabuchi H, Yoshimura A, Michishita I, Takeda M, Itoh H, Sakai Y, Sakai T, Ueda K, Takeda R. Deficiency of serum cholesteryl-ester transfer activity in patients with familial hyperalphalipoproteinaemia. *Atherosclerosis* 1985; 58: 175–186.
46. Wilson PW. High-density lipoprotein, low-density lipoprotein and coronary artery disease. *Am J Cardiol* 1990; 66: 7A–10A.
47. Lusis AJ. Atherosclerosis. *Nature* 2000; 407: 233–241.
48. Khalil MF, Wagner WD, Goldberg IJ. Molecular interactions leading to lipoprotein retention and the initiation of atherosclerosis. *Arterioscler Thromb Vasc Biol* 2004; 24: 2211–2218.
49. Williams KJ, Tabas I. The response-to-retention hypothesis of early atherogenesis. *Arterioscler Thromb Vasc Biol* 1995; 15: 551–561.

50. Tabas I, Williams KJ, Boren J. Subendothelial lipoprotein retention as the initiating process in atherosclerosis: update and therapeutic implications. *Circulation* 2007; 116: 1832–1844.
51. Carew TE, Pittman RC, Marchand ER, Steinberg D. Measurement in vivo of irreversible degradation of low density lipoprotein in the rabbit aorta. Predominance of intimal degradation. *Arteriosclerosis* 1984; 4: 214–224.
52. Schwenke DC, Carew TE. Initiation of atherosclerotic lesions in cholesterol-fed rabbits. I. Focal increases in arterial LDL concentration precede development of fatty streak lesions. *Arteriosclerosis* 1989; 9: 895–907.
53. Schwenke DC, Carew TE. Initiation of atherosclerotic lesions in cholesterol-fed rabbits. II. Selective retention of LDL vs. selective increases in LDL permeability in susceptible sites of arteries. *Arteriosclerosis* 1989; 9: 908–918.
54. Chapman MJ. Beyond the statins: new therapeutic perspectives in cardiovascular disease prevention. *Cardiovasc Drugs Ther* 2005; 19: 135–139.
55. Gotto AM Jr, Brinton EA. Assessing low levels of high-density lipoprotein cholesterol as a risk factor in coronary heart disease: a working group report and update. *J Am Coll Cardiol* 2004; 43: 717–724.
56. Gofman JW, Young W, Tandy R. Ischemic heart disease, atherosclerosis, and longevity. *Circulation* 1966; 34: 679–697.
57. Chapman MJ, Assmann G, Fruchart JC, Shepherd J, Sirtori C. Raising high-density lipoprotein cholesterol with reduction of cardiovascular risk: the role of nicotinic acid—a position paper developed by the European Consensus Panel on HDL-C. *Curr Med Res Opin* 2004; 20: 1253–1268.
58. Maron DJ. The epidemiology of low levels of high-density lipoprotein cholesterol in patients with and without coronary artery disease. *Am J Cardiol* 2000; 86: 11L–14L.
59. Sharrett AR, Ballantyne CM, Coady SA, Heiss G, Sorlie PD, Catellier D, Patsch W. Coronary heart disease prediction from lipoprotein cholesterol levels, triglycerides, lipoprotein(a), apolipoproteins A-I and B, and HDL density subfractions: The Atherosclerosis Risk in Communities (ARIC) Study. *Circulation* 2001; 104: 1108–1113.
60. Rubins HB, Robins SJ, Collins D, Fye CL, Anderson JW, Elam MB, Faas FH, Linares E, Schaefer EJ, Schectman G, Wilt TJ, Wittes J, Veterans Affairs High-Density Lipoprotein Cholesterol Intervention Trial Study Group. Gemfibrozil for the secondary prevention of coronary heart disease in men with low levels of high-density lipoprotein cholesterol. *N Engl J Med* 1999; 341: 410–418.
61. Gordon DJ, Probstfield JL, Garrison RJ, Neaton JD, Castelli WP, Knoke JD, Jacobs DR Jr, Bangdiwala S, Tyroler HA. High-density lipoprotein cholesterol and cardiovascular disease. Four prospective American studies. *Circulation* 1989; 79: 8–15.
62. Hokanson JE, Austin MA. Plasma triglyceride level is a risk factor for cardiovascular disease independent of high-density lipoprotein cholesterol level: a meta-analysis of population-based prospective studies. *J Cardiovasc Risk* 1996; 3: 213–219.
63. Gordon DJ, Rifkind BM. High-density lipoprotein–the clinical implications of recent studies. *N Engl J Med* 1989; 321: 1311–1316.
64. Eisenberg S. High density lipoprotein metabolism. *J Lipid Res* 1984; 25: 1017–1058.
65. Brousseau ME, Schaefer EJ, Wolfe ML, Bloedon LT, Digenio AG, Clark RW, Mancuso JP, Rader DJ. Effects of an inhibitor of cholesteryl ester transfer protein on HDL cholesterol. *N Engl J Med* 2004; 350: 1505–1515.

66. Cohen JC, Kiss RS, Pertsemlidis A, Marcel YL, McPherson R, Hobbs HH. Multiple rare alleles contribute to low plasma levels of HDL cholesterol. *Science* 2004; 305: 869–872.

67. Kathiresan S, Melander O, Guiducci C, Surti A, Burtt NP, Rieder MJ, Cooper GM, Roos C, Voight BF, Havulinna AS, Wahlstrand B, Hedner T, Corella D, Tai ES, Ordovas JM, Berglund G, Vartiainen E, Jousilahti P, Hedblad B, Taskinen MR, Newton-Cheh C, Salomaa V, Peltonen L, Groop L, Altshuler DM, Orho-Melander M. Six new loci associated with blood low-density lipoprotein cholesterol, high-density lipoprotein cholesterol or triglycerides in humans. *Nat Genet* 2008; 40: 189–197.

68. Yusuf S, Hawken S, Ounpuu S, Dans T, Avezum A, Lanas F, McQueen M, Budaj A, Pais P, Varigos J, Lisheng L. Effect of potentially modifiable risk factors associated with myocardial infarction in 52 countries (the INTERHEART study): case-control study. *Lancet* 2004; 364: 937–952.

69. Franceschini G. Epidemiologic evidence for high-density lipoprotein cholesterol as a risk factor for coronary artery disease. *Am J Cardiol* 2001; 88: 9N–13N.

70. Doggen CJ, Smith NL, Lemaitre RN, Heckbert SR, Rosendaal FR, Psaty BM. Serum lipid levels and the risk of venous thrombosis. *Arterioscler Thromb Vasc Biol* 2004; 24: 1970–1975.

71. Barzilai N, Atzmon G, Schechter C, Schaefer EJ, Cupples AL, Lipton R, Cheng S, Shuldiner AR. Unique lipoprotein phenotype and genotype associated with exceptional longevity. *JAMA* 2003; 290: 2030–2040.

72. Barter P. HDL: a recipe for longevity. *Atheroscler Suppl* 2004; 5: 25–31.

73. Chisolm GM, Steinberg D. The oxidative modification hypothesis of atherogenesis: an overview. *Free Radic Biol Med* 2000; 28: 1815–1826.

74. Stocker R, Keaney JF Jr. Role of oxidative modifications in atherosclerosis. *Physiol Rev* 2004; 84: 1381–1478.

75. Ridker PM, Wilson PWF, Grundy SM. Should C-reactive protein be added to metabolic syndrome and to assessment of global cardiovascular risk? *Circulation* 2004; 109: 2818–2825.

76. Willerson JT, Ridker PM. Inflammation as a cardiovascular risk factor. *Circulation* 2004; 109: II–2–10.

77. Verma S, Szmitko PE, Ridker PM. C-reactive protein comes of age. *Nature Clin Pract Cardiovasc Med* 2005; 2: 29–36.

78. C Reactive Protein Coronary Heart Disease Genetics Collaboration (CCGC). Association between C reactive protein and coronary heart disease: mendelian randomisation analysis based on individual participant data. *BMJ* 2011; 342: d548.

79. Glomset JA. The plasma lecithins:cholesterol acyltransferase reaction. *J Lipid Res* 1968; 9: 155–167.

80. Kontush A, Chapman MJ. Functionally defective HDL: A new therapeutic target at the crossroads of dyslipidemia, inflammation and atherosclerosis. *Pharmacol Rev* 2006; 3: 342–374.

81. Barter PJ, Caulfield M, Eriksson M, Grundy SM, Kastelein JJ, Komajda M, Lopez-Sendon J, Mosca L, Tardif JC, Waters DD, Shear CL, Revkin JH, Buhr KA, Fisher MR, Tall AR, Brewer B. Effects of torcetrapib in patients at high risk for coronary events. *N Engl J Med* 2007; 357: 2109–2122.

82. Cannon CP, Shah S, Dansky HM, Davidson M, Brinton EA, Gotto AM, Stepanavage M, Liu SX, Gibbons P, Ashraf TB, Zafarino J, Mitchel Y, Barter P. Safety of anacetrapib in patients with or at high risk for coronary heart disease. *N Engl J Med* 2010; 363: 2406–2415.
83. Fayad ZA, Mani V, Woodward M, Kallend D, Abt M, Burgess T, Fuster V, Ballantyne CM, Stein EA, Tardif JC, Rudd JH, Farkouh ME. Tawakol A, Safety and efficacy of dalcetrapib on atherosclerotic disease using novel non-invasive multimodality imaging (dal-PLAQUE): a randomised clinical trial. Lancet, 2011, in press.
84. Yvan-Charvet L, Kling J, Pagler T, Li H, Hubbard B, Fisher T, Sparrow CP, Taggart AK, Tall AR. Cholesterol efflux potential and antiinflammatory properties of high-density lipoprotein after treatment with niacin or anacetrapib. *Arterioscler Thromb Vasc Biol* 2010; 30: 1430–1438.
85. Niesor EJ, Magg C, Ogawa N, Okamoto H, von der Mark E, Matile H, Schmid G, Clerc RG, Chaput E, Blum-Kaelin D, Huber W, Thoma R, Pflieger P, Kakutani M, Takahashi D, Dernick G, Maugeais C. Modulating cholesteryl ester transfer protein activity maintains efficient pre-beta-HDL formation and increases reverse cholesterol transport. *J Lipid Res* 2010; 51: 3443–3454.
86. Chapman MJ, Redfern JS, McGovern ME, Giral P. Niacin and fibrates in atherogenic dyslipidemia: pharmacotherapy to reduce cardiovascular risk. *Pharmacol Ther* 2010; 126: 314–345.
87. Dorner T, Fodor JG, Lawrence K, Ludvik B, Rieder A. HDL-knowledge in the lay public: results of a representative population survey. *Atherosclerosis* 2007; 195: 195–198.
88. Choi EJ, Jekal Y, Kim S, Yoo JS, Kim HS, Oh EG, Jeon JY, Jang Y, Chu SH, Kim MJ. Middle-aged women's awareness of cholesterol as a risk factor: results from a national survey of Korean Middle-aged Women's Health Awareness (KomWHA) study. *Int J Nurs Stud* 2010; 47: 452–460.
89. Kontush A. Apolipoprotein Abeta: black sheep in a good family. *Brain Pathol* 2004; 14: 433–447.
90. Kontush A, Chapman MJ. HDL: close to our memories? *Arterioscler Thromb Vasc Biol* 2008; 28: 1418–1420.
91. Yu C, Youmans KL, LaDu MJ. Proposed mechanism for lipoprotein remodelling in the brain. *Biochim Biophys Acta* 2010; 1801: 819–823.
92. Sniderman AD, Furberg CD. Age as a modifiable risk factor for cardiovascular disease. *Lancet* 2008; 371: 1547–1549.

ACKNOWLEDGMENTS

It is a pleasure to acknowledge stimulating discussions with our many collaborators and colleagues, including Professors Sean Davidson, Edward Dennis, Wendy Jessup, Kerry-Anne Rye and Alan Tall, both in our INSERM Research Unit at Hospital La Pitié—Salpêtrière and around the world, over the past decade.

We are indebted to the National Institute for Health and Medical Research (INSERM), University of Pierre and Marie Curie (UPMC)—Paris 6, CODDIM Ile-de-France, the Fondation pour la Recherche Médicale, the Fondation de France and the Association for Research on Lipoproteins and Atherogenesis (ARLA, Paris, France) for the continued support of our studies. We gratefully acknowledge the awards of "Contrat d'Interface" from Assistance Publique-Hôpitaux de Paris and INSERM.

ABBREVIATIONS

AAPH, 2,2′-azobis-(2-amidinopropane) hydrochloride
ABC, ATP-binding cassette
ABCA1, ATP-binding cassette transporter A1
ABCG1, ATP-binding cassette transporter G1
AMPK, adenosine monophosphate-activated protein kinase
ANGPTL3, angiopoietin-like protein 3
apo, apolipoprotein
BMI, body-mass index
BMP-1, bone morphogenetic protein 1
BPI, bactericidal permeability-increasing protein
CB1, cannabinoid receptor type 1
CCL, chemokine (C-C motif) ligand
CEOOH, cholesteryl ester hydroperoxide
CETP, cholesteryl ester transfer protein
CI, confidence interval
CRP, C-reactive protein
CX3CL, chemokine (C-X3-C motif) ligand
DGAT, diacylglycerol acyltransferase
DHCR24, 3-beta-hydroxysteroid-delta 24-reductase
eNOS, endothelial nitric oxide synthase
EPR, electron paramagnetic resonance
ERK, extracellular signal-regulated kinase

ESR1, estrogen receptor alpha gene
FADS, fatty acid desaturase
FLIP, FLICE-like inhibitory protein
FPLC, fast protein liquid chromatography
FRET, fluorescence resonance energy transfer
GALNT2, UDP-N-acetyl-alpha-D-galactosamine:polypeptide N-acetylgalactosaminyltransferase 2
GSPx, glutathione selenoperoxidase
HDL, high-density lipoprotein
HDL-C, HDL-cholesterol
HIV, human immunodeficiency virus
HMG-CoA, 3-hydroxy-3-methyl-glutaryl-CoA
HOMA, homeostatic model assessment
HPLC, high-performance liquid chromatography
Hrp, haptoglobin-related protein
hsCRP, high-sensitivity CRP
IC_{50}, half maximal inhibitory concentration
ICAM-1, intercellular adhesion molecule-1
IDL, intermediate-density lipoprotein
IL, interleukin
JAK2, Janus kinase 2
JNK, c-Jun N-terminal kinase
LBP, lipopolysaccharide-binding protein
LCAT, lecithin:cholesterol acyltransferase
LC/MS, liquid chromatography/mass spectrometry
LDL, low-density lipoprotein
LDL-C, LDL-cholesterol
LOOH, lipid hydroperoxide
LOX-1, lectin-like oxidized LDL receptor 1
Lp(a), lipoprotein (a)
LpA-I, HDL particles containing only apoA-I
LpA-I:A-II, HDL particles containing both apoA-I and apoA-II
$LpPLA_2$, lipoprotein-associated phospholipase A_2
LPS, lipopolysaccharide
LTIP, lipid transfer inhibitor protein
LXR, liver X receptor
MALDI, matrix-assisted laser desorption/ionisation

MAPK, mitogen-activated protein kinase
MCP-1, monocyte chemotactic protein-1
MEK, mitogen-activated protein kinase kinase
MMAB, methylmalonic aciduria [cobalamin deficiency] cblB type
MVK, mevalonate kinase
NFκB, nuclear factor kappa B
NMR, nuclear magnetic resonance
NO, nitric oxide
PAF, platelet-activating factor
PAF-AH, platelet-activating factor-acetyl hydrolase
PAI-1, plasminogen activator inhibitor type 1
PCPE2, procollagen C-proteinase enhancer-2
PCSK9, proprotein convertase subtilisin kexin type 9
PI3K, phosphoinositide 3-kinase
PLOOH, phospholipid hydroperoxide
PLPC, 1-palmitoyl-2-linoleoyl phosphatidylcholine
PLTP, phospholipid transfer protein
PON, paraoxonase
POPC, 1-palmitoyl-2-oleoyl phosphatidylcholine
PPAR, peroxisome proliferator-activated receptor
PUFA, polyunsaturated fatty acid
RCT, reverse cholesterol transport
rHDL, reconstituted HDL
ROS, reactive oxygen species
RXR, retinoid X receptor
S1P, sphingosine-1-phosphate
S1P1, 2 and 3, sphingosine-1-phosphate receptor 1, 2 and 3
SAA, serum amyloid A
SELDI, surface-enhanced laser desorption/ionisation
SD, standard deviation
SNP, single-nucleotide polymorphism
sPLA$_2$, secretory phospholipase A_2
SR-BI, scavenger receptor class B type I
SREBP, sterol-regulatory element binding protein
STAT3, signal transducer and activator of transcription 3
TF, tissue factor
TFPI, tissue factor pathway inhibitor

TNF-alpha, tumour necrosis factor-alpha
TOF, time-of-flight
tPA, tissue plasminogen activator
TTC, tetratricopeptide repeat domain
VCAM-1, vascular cell adhesion molecule-1
VLDL, very-low density lipoprotein
WHO, World Health Organization

SECTION 1

NORMAL FUNCTIONAL HIGH-DENSITY LIPOPROTEIN

Lipoproteins are plurimolecular, typically quasi-spherical, pseudomicellar complexes composed of polar and non-polar lipids solubilized by proteins with specialized structure and function, the **apolipoproteins.** Lipids and apolipoproteins thus constitute major building blocks of lipoproteins. The principal lipid-binding structural motifs of apolipoproteins include amphipathic alpha-helixes and beta-sheets. Major lipoproteins in human plasma are, in the order of increasing density and decreasing size, chylomicrons, very low-density lipoproteins (VLDL), intermediate-density lipoproteins (IDL), low-density lipoprotein (LDL), lipoprotein (a) [Lp(a)] and high-density lipoprotein (HDL). Whereas chylomicrons, VLDL, IDL, LDL and Lp(a) are commonly regarded as large, light, lipid-rich particles, **HDL** represents a small, dense, protein-rich lipoprotein with a mean size of 8–10 nm and density of 1.063-1.21 g/ml (Fig. 1.1).

High-Density Lipoproteins: Structure, Metabolism, Function, and Therapeutics, First Edition.
Anatol Kontush and M. John Chapman.

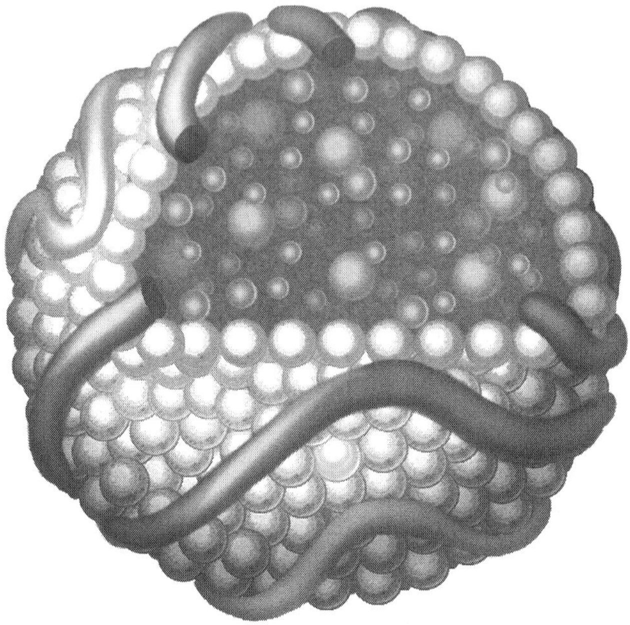

Figure 1.1 Plurimolecular, quasi-spherical, pseudomicellar HDL particle with major structural components. Apolipoprotein A-I (apoA-I) molecules are shown as ribbons, surface phospholipid molecules as light-blue spheres, surface free cholesterol molecules as yellow spheres, core cholesteryl ester molecules as dark-blue spheres and core triglyceride molecules as green spheres. The average number of protein and lipid molecules per HDL particle are as follows: apoA-I, 3–4; phospholipid, 80–120; free cholesterol, 20–40; cholesteryl ester, 100–160; triglyceride, 15–25 (see color plate section; see also Table 1.1).

CHAPTER 1

COMPOSITION

As a result of protein enrichment, proteins form the major building blocks of HDL particles. HDL carries a particularly large number of proteins as compared to other lipoprotein species (Table 1.1). Heterogeneity of the HDL proteome was originally demonstrated as early as 1968–1969 [1–3], with most of the major proteins being discovered in the 1970s.

HDL proteins have traditionally been divided into four major subgroups: apolipoproteins, enzymes, lipid transfer proteins and minor proteins (<5% of total HDL protein; Table 1.1). Whereas apolipoproteins and enzymes are now recognized as key HDL components whose biologic importance is beyond doubt, the role of minor proteins, primarily those involved in complement regulation, protection from infections and the acute-phase response, has received increasing attention in recent years, mainly as a result of advances in proteomic technologies. Indeed, HDL particles have been long thought to contain only apolipoproteins and enzymes directly involved in lipid metabolism. The subsequent discovery of minor amounts of serum amyloid A (SAA), a major positive acute phase reactant, as a component of normal plasma HDL [4] was, at first, considered to be an exception. However, the recent development of proteomic technologies has significantly enhanced the sensitivity of protein detection, revealing that the protein cargo of HDL is much more diverse than previously realized [5–7] (Table 1.2). These studies have allowed the identification of more than 50 proteins in human HDL isolated by ultracentrifugation [5–7] (Table 1.2). Numerous proteins involved in the acute-phase response were unexpectedly found as

High-Density Lipoproteins: Structure, Metabolism, Function, and Therapeutics, First Edition.
Anatol Kontush and M. John Chapman.

TABLE 1.1 Major Components of the HDL Proteome

Protein	Chromosome	Number of Amino Acid Residues	M_r, kDa	Major Function
Apolipoproteins				
ApoA-I	11	243	28	Major structural and functional apolipoprotein, LCAT activator
ApoA-II	1	77	17	Structural and functional apolipoprotein
ApoA-IV	11	376	46	Structural and functional apolipoprotein
ApoA-V	11	343	39	Activator of LPL, inhibitor of TG secretion
ApoC-I	19	57	6.6	Modulator of CETP activity
ApoC-II	19	79	8.8	Activator of LPL
ApoC-III	19	79	8.8	Inhibitor of LPL
ApoC-IV	19	101	11	Regulates TG metabolism
ApoD	3	169	19	Binding of small hydrophobic molecules
ApoE	19	299	34	Structural and functional apolipoprotein, ligand for LDL-R and LRP
ApoF	12	162	29	Inhibitor of CETP
ApoH	17	326	38	Binding of negatively charged molecules
ApoJ	8	427	70	Binding of hydrophobic molecules, interaction with cell receptors

ApoL-I	22	371	44/46	Trypanolytic factor of human serum
ApoM	6	188	25	Binding of small hydrophobic molecules
Enzymes				
LCAT	16	416	63	Esterification of cholesterol to cholesteryl esters
PON1	7	354	43	Calcium-dependent lactonase
PAF-AH ($LpPLA_2$)	6	420	53	Hydrolysis of short-chain phospholipids
GSPx-3	5	206	22	Reduction of hydroperoxides by glutathione
Lipid transfer proteins				
PLTP	20	476	78	Conversion of HDL into larger and smaller particles, transport of LPS
CETP	16	476	74	Heteroexchange of CE and TG and homoexchange of PL between HDL and apoB-containing lipoproteins
Acute-phase proteins				
SAA1	11	104	12	Acute-phase reactant
Complement components				
C3	19	1663	187	Complement activation
Proteinase inhibitors				
Alpha-1-antitrypsin	14	418	52	Inhibitor of serine proteinases

CE, cholesteryl ester; CETP, cholesteryl ester transfer protein; GSPx, glutathione selenoperoxidase; LCAT, lecithin:cholesterol acyltransferase; LDL-R, LDL receptor; LPL, lipoprotein lipase; LpPLA2, lipoprotein-associated phospholipase A2; LRP, LDL receptor-related protein; LPS, lipopolysaccharide; PAF-AH, platelet-activating factor-acetyl hydrolase; PL, phospholipid; PLTP, phospholipid transfer protein; PON1, paraoxonase 1; SAA, serum amyloid A; TG, triglyceride.

TABLE 1.2 Proteins Detected in HDL by Mass Spectrometry

Protein	Reference
Albumin	[8–10]
Alpha-1-acid glycoprotein 2	[8, 9]
Alpha-1-antitrypsin	[8–12]
Alpha-1B-glycoprotein	[8, 11]
Alpha-2-antiplasmin	[8]
Alpha-2-HS-glycoprotein	[8, 9]
Alpha-2-macroglobulin	[9]
Alpha-amylase (salivary)	[12]
Angiotensinogen	[8]
Apo(a)	[10]
ApoA-I	[8–14]
ApoA-II	[8–10, 12–14]
ApoA-IV	[8,10–12] [13, 14]
ApoB	[8, 10]
ApoC-I	[8–10, 12]
ApoC-II	[8, 10, 12, 13]
ApoC-III	[8–14]
ApoC-IV	[8, 10]
ApoD	[8–10, 13, 14]
ApoE	[8–10, 12–14]
ApoF	[8–10]
ApoH	[8]
ApoJ	[8–10, 13]
ApoL-I	[8,10–12] [13]
ApoM	[8, 10, 12, 13]
Bikunin	[8]
C4b binding protein	[8]
CETP	[8, 14]
Complement C3	[8, 13]
Complement C4	[8]
Complement C4b	[8]
Complement C9	[8]
Fibrinogen	[8, 10, 11, 14]
Hrp	[8–10]
Haptoglobin	[13, 14]
Hemopexin	[8]
Inter-alpha-trypsin inhibitor chain H4	[8, 11]
Kininogen-1	[8]
LBP	[11]
LCAT	[8, 14]
PAF-AH (LpPLA$_2$)	[10]
Platelet basic protein	[10]
PLTP	[8, 10]
PON1	[8–10, 13]
PON3	[8, 10]
Prenylcysteine oxidase	[8]
Prothrombin	[10]
Retinol-binding protein	[8]

TABLE 1.2 (*Continued*)

Protein	Reference
SAA1/2	[8, 10, 12]
SAA4	[8–10, 12]
Serpin F1	[8]
Transferrin	[8, 9, 13]
Transthyreitin	[8, 10, 11, 13]
Vitamin D-binding globulin	[8, 9]
Vitronectin	[8]

Modified from Reference 7. HDL was isolated by density gradient or sequential ultracentrigufation in salt solutions in all these studies except [14]. Hrp, haptoglobin-related protein; LBP, lipopolysaccharide-binding protein.

components of normal human plasma HDL. Furthermore, two other large families of HDL-associated proteins, notably those involved in complement regulation and protease inhibition, were discovered [8], raising the possibility that HDL may play a previously unsuspected role in host defence mechanisms and inflammation [7]. It is important to keep in mind, however, that the content of all these proteins in HDL is much lower as compared to that of major HDL apolipoproteins, i.e., apoA-I and apoA-II.

In addition to proteins, HDL contains multiple molecular species of lipids (Table 1.3) [15, 16]. As is the case in other plasma lipoproteins, the HDL lipidome contains phospholipids, unesterified sterols (predominantly cholesterol), cholesteryl esters and triglycerides as its four major classes of lipid (Fig. 1.1). Phospholipids build the surface lipid monolayer of HDL, whereas cholesteryl esters and triglycerides form the hydrophobic lipid core (Fig. 1.1); unesterified (free) sterols are predominantly located in the surface monolayer, partially penetrating the core. Recent lipidomic analyses allowed the identification of more than 200 individual molecular species of lipids in the HDL lipidome [17, 18], the number of which is limited only by the sensitivity of available technologies [19]. Finally, HDL contains multiple sugar moieties as components of glycosylated proteins and has recently been shown to transport microRNA [20].

DEFINITIONS

Lipoprotein: plurimolecular, typically quasi-spherical, pseudomicellar complex composed of polar and non-polar lipids solubilized by proteins possessing specialized structure and function, the apolipoproteins.

Apolipoprotein: a specialized protein that binds and transports lipids in the form of lipoproteins in the circulatory and lymphatic systems; frequently targets lipids to receptors at sites of degradation, storage, transformation and recycling.

Lipid transfer protein: protein involved in the mass transfer and exchange of lipids between lipoprotein particles.

TABLE 1.3 Major Components of the HDL Lipidome

Lipid class	HDL Content, % Total HDL Mass[a]	Major Subspecies	References
Phospholipids	20–30	Phosphatidylcholine, sphingomyelin	[15–17]
Phosphatidylcholine	15–25	18:2/16:0, 18:2/18:0, 20:4/16:0, 18:1/16:0	[15–17]
Sphingomyelin	3–6	18:1/16:0, 18:2/24:0, 18:1/24:1, 18:2/22:0, 18:1/22:1	[15–17]
Phosphatidylinositol	0.6–1.2	18:0/20:3, 18:0/20:4	[15, 21]
Lysophosphatidylcholine	0.2–6	16:0, 18:0, 18:1, 18:2	[15-17]
Phosphatidylethanolamine	0.2–0.6	18:2/16:0, 18:2/18:0, 18:0/20:4	[15–17]
Plasmalogens	0.2–0.6	18:0/20:4, 16:0/20:4, 18:1/20:4, 18:0/18:2, 16:0/22:6	[17]
Ceramide	0.1	24:0, 24:1, 23:0, 22:0, 16:0	[17]
Phosphatidylserine	0.01–0.02	16:0/18:0	[15, 18, 22]
Phosphatidylglycerol	n.e.	18:1/20:2	[21]
Phosphatidic acid	n.e.	20:4/20:0, 20:4/20:2, 18:2/20:1	[21]
Cardiolipin	0.10–0.15		[22]
Sterols	3–5	Cholesterol	[15, 23]
Triacylglycerides	3–6	18:1/16:0/18:1, 18:2/16:0/18:1	[15, 18]
Cholesteryl ester	14–18	Cholesteryl linoleate	[15, 16]

[a] Depending on HDL subpopulation, lipids make 35% to 65% of total HDL mass [24]; n.e., not established.

1.1 PROTEOME

Apolipoproteins

Apolipoprotein A-I. Discovered at the end of the 1960s, **apolipoprotein A-I** (apoA-I; M_r 28 kDa) is the major structural and functional HDL apolipoprotein (see Chapter 3) accounting for approximately 70% of total HDL protein [25]. Almost all HDL particles are believed to contain apoA-I [26, 27]. Major functions of apoA-I involve the activation of lecithin:cholesterol acyltransferase (LCAT), and interaction with cellular receptors; equally apoA-I endows HDL with multiple anti-atherogenic activities (Table 1.1). ApoA-I is synthesized as a 267-residue proprotein which is processed to release an 18-residue signal peptide and a 6-residue propeptide. Mature, circulating apoA-I contains 243 amino acid residues within a single polypeptide that lacks glycosylation or disulfide linkages and is encoded by exon 3 (residues 1–43) and exon 4 (residues 44–243) of a gene located on the long arm of chromosome 11. The apoA-I gene is a part of the APOA1/APOC3/APOA4 gene cluster. The N-terminal region of apoA-I is more highly conserved compared with the C-terminus, suggesting key biologic properties [28, 29].

ApoA-I represents a typical amphipathic protein that contains 8 alpha-helical amphipathic domains of 22 amino acids and two repeats of 11 amino acids (Figs 1.2 and 1.3); as a consequence, apoA-I binds avidly to lipids. Such properties render apoA-I a potent detergent which forms stable micellar complexes with phospholipids, cholesterol, triglycerides, and cholesteryl esters. The elevated amphipacity of apoA-I underlies its capacity to move between lipoprotein particles. As a consequence, apoA-I is also found in chylomicrons and VLDL.

As for many plasma apolipoproteins, the main sites for apoA-I synthesis and secretion are the liver and small intestine, with the liver being the major contributor to the plasma apoA-I pool.

ApoA-II. **ApoA-II** (M_r 17 kDa) is the second major HDL apolipoprotein and represents approximately 15%–20% of total HDL protein. About half of HDL particles may contain apoA-II [43]. Synthesis of apoA-II resembles that of apoA-I, with a 100-residue proprotein processed to release an 18-residue signal peptide and a 5-residue propeptide. ApoA-II circulates as a homodimer composed of two identical polypeptide chains, each containing 77 amino acids [44, 45] (Table 1.1). The two polypeptide chains are connected by a single disulfide bridge at position 6 in the sequence [46] (Figs 1.2 and 1.3). The presence of a Cys residue allows apoA-II to form heterodimers with other cysteine-containing apolipoproteins, such as apoE and apoD [47].

The human apoA-II gene is located on chromosome 1. Despite displaying amphipathic properties, apoA-II is more hydrophobic than apoA-I. Similar to apoA-I, apoA-II is predominantly synthesized in the liver but also in the intestine [48].

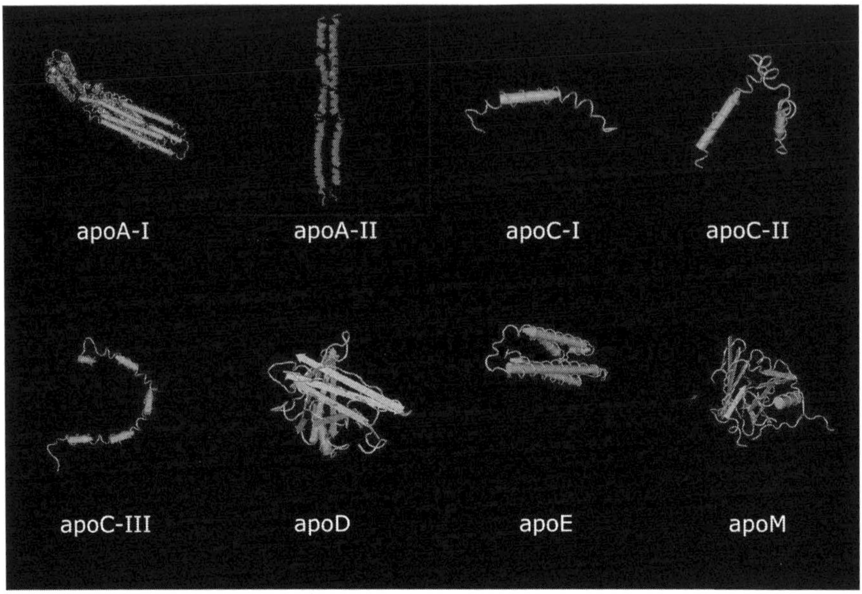

Figure 1.2 Experimentally-obtained structures of major HDL apolipoproteins. Crystal structures of apolipoproteins A-I, A-II, C-I, C-II, C-III, D, E and M are shown as visualized with Cn3D 4.1 (available at: http://www.ncbi.nlm.nih.gov/Structure/CN3D/cn3d.shtml; see the link for the colors) using data from [30–41] (see color plate section).

ApoA-IV. **ApoA-IV** is a 46 kDa O-linked glycoprotein, 376 amino acids in length [49] (Table 1.1). ApoA-IV is the most hydrophilic apolipoprotein; consequently, it readily exchanges between lipoproteins, primarily between chylomicrons, VLDL and HDL, and may also circulate in a free form. ApoA-IV contains thirteen 22-amino acid tandem repeats, nine of which are highly alpha-helical; many of these helices are amphipathic (Fig. 1.3). Such repeats may serve as lipid-binding domains.

In most mammals, including humans, apoA-IV is synthesized in the intestine; hepatic synthesis also occurs in rodents. ApoA-IV is secreted into the circulation on the surface of newly synthesized chylomicron particles and is particularly abundant in follicular fluid [50]. The APOA4 gene resides on chromosome 11 in close linkage to APOA1 and APOC3. The primary translation product of the gene is a 396-residue proprotein which undergoes proteolytic processing to cleave a 20-residue signal peptide and to yield mature apoA-IV.

ApoA-V. The **APOA5** gene is located on chromosome 11 adjacent to the APOA1/APOC3/APOA4 gene cluster [51]. The human APOA5 gene is only expressed in the liver where it encodes a 366-amino acid protein containing a 23-residue signal sequence (Table 1.1). ApoA-V circulates as a 39 kDa protein which is predominantly located on triglyceride-rich particles, chylomicrons and VLDL, but also on HDL; the carboxyl-terminal segment of apoA-V ensures its

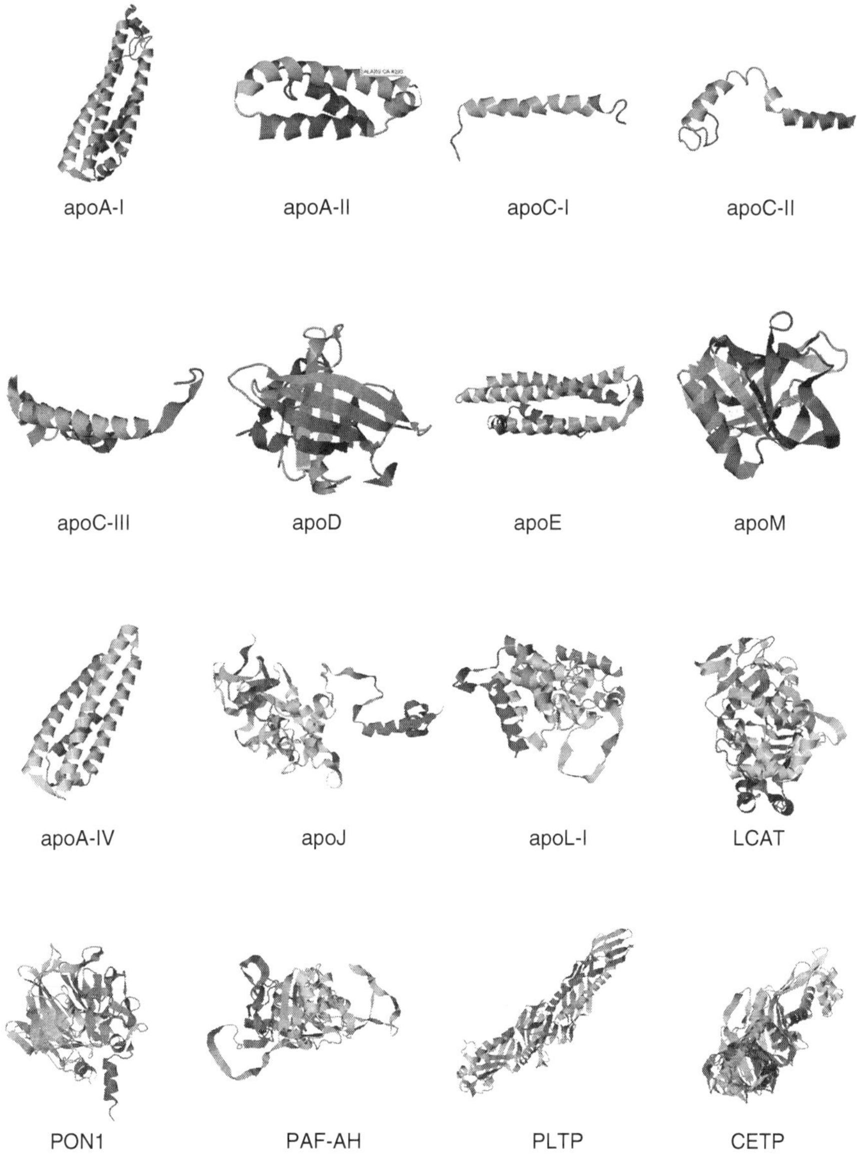

Figure 1.3 Model structures of major HDL proteins. Structures of apolipoproteins A-I, A-II, C-I, C-II, C-III, D, E and M (**upper panel**), as well as of apoA-IV, apoJ, apoL-I, LCAT, PON1, PAF-AH, PLTP and CETP (**lower panel**) calculated by The Protein Model Portal (www.proteinmodelportal.org) are shown [42]. The structures are generated using Molscript; colours describe protein structures according to Jmol viewer (see color plate section).

binding to lipids [52]. ApoA-V functions as an activator of lipoprotein lipase (LPL) and as an inhibitor of hepatic production and secretion of triglyceride.

ApoC-I, Apoc-II, Apoc-III, Apoc-IV. ApoC-I, apoC-II, apoC-III and apoC-IV form a family of small, exchangeable apolipoproteins (Table 1.1). Genes coding for apoC-I, apoC-II and apoC-IV are located close to each other on chromosome 19, forming a gene cluster together with the apoE gene. ApoC proteins are primarily synthesized in the liver and secreted in the circulation.

ApoC-I is the smallest apolipoprotein (M_r 6.6 kDa), containing 57 amino acids [53] (Figs 1.2 and 1.3). ApoC-I associates with both HDL and VLDL and can readily exchange between them. ApoC-I carries a strong positive charge and can thereby bind free fatty acids and modulate activities of several proteins involved in HDL metabolism. In addition, apoC-I modulates the interaction of apoE with VLDL and inhibits binding of VLDL to the LDL receptor-related protein. It is mainly synthesized in the liver and, to a minor degree, in the intestine. ApoC-I synthesis starts with an 83-residue proprotein which is processed to cleave a 26-residue signal peptide. ApoC-I is involved in the activation of LCAT and inhibition of hepatic lipase and cholesteryl ester transfer protein (CETP).

ApoC-II is an 8.8 kDa protein, 79 residues in length, which functions as an activator of several triacylglycerol lipases [54] (Figs 1.2 and 1.3). Similar to apoC-I, apoC-II is associated with HDL and VLDL and can exchange between their surfaces. Region 43–51 of the protein is involved in lipid binding, whereas region 55–78 ensures lipase activation. ApoC-II is produced as a 101-amino acid proprotein cleaved to release a 22-amino acid signal peptide.

ApoC-III is an 8.8 kDa protein containing 79 amino acids [55] (Figs 1.2 and 1.3). As a major VLDL protein, apoC-III is predominantly present in VLDL, with small amounts found in HDL. The protein inhibits LPL and hepatic lipase and decreases the uptake of lymph chylomicrons by hepatic cells. Region 68–99 of apoC-III accounts for lipid binding. The apoC-III gene is a part of the APOA1/APOC3/APOA4 gene cluster on chromosome 11. The proprotein of apoC-III contains 99 amino acid residues which include a 20-residue signal peptide. ApoC-III exists in three different isoforms according to the degree of sialylation of the protein, which differ in their structure, metabolism and effects on triglyceride hydrolysis [56].

Recently discovered in 1996 [57], **apoC-IV** is an 11 kDa protein containing 101 amino acids produced from a 127-residue proprotein. ApoC-IV is expressed in the liver and displays a predicted protein structure characteristic of the other genes in this family. The function of apoC-IV is largely unknown; when over-expressed, it induces hypertriglyceridemia in mice [58, 59]. In normolipidemic plasma, more than 80% of the protein resides in VLDL, with most of the remainder residing in HDL particles. The HDL content of apoC-IV is much lower compared with other apolipoproteins of this family, making apoC-IV a low prevalence HDL apolipoprotein.

ApoD. **ApoD** is a 19 kDa glycoprotein, 169 amino acids in length, mainly associated with HDL [60] (Table 1.1). The protein does not possess a typical apolipoprotein structure and belongs to the lipocalin family, which also includes retinol-binding protein, lactoglobulin and uteroglobulin. Lipocalins form a large, multifunctional family of small lipid-transfer proteins (15–25 kDa) with very limited amino acid sequence identity (often below 20%), but with a common tertiary structure. Lipocalins share a structurally conserved beta-barrel fold which, in many lipocalins, bind hydrophobic ligands (Figs 1.2 and 1.3). The lipocalin fold is followed by an alpha-helix at the C-terminus and surrounds a central cavity lined with hydrophobic aromatic residues that enable binding of small hydrophobic molecules [61]. As a result, apoD transports small hydrophobic ligands, with a high affinity for arachidonic acid [62]. In plasma, apoD forms disulfide-linked homodimers and heterodimers with apoA-II.

The protein is expressed in many tissues, including liver, intestine, pancreas, kidney, placenta, adrenal, spleen, fetal brain and tears. Synthesis of apoD involves the production of a 189-residue precursor and cleavage of a 20-residue signal peptide. The apoD gene is located on chromosome 3.

ApoE. ApoE, a 34 kDa glycoprotein, is a key structural and functional component of HDL despite its much lower content in HDL particles compared with apoA-I [63] (Table 1.1). The major fraction of circulating apoE is carried by triglyceride-containing lipoproteins whose apoE mediates their receptor binding, internalization and catabolism. As apoE possesses the LDL receptor-binding and heparin-binding domains, the apolipoprotein serves as a ligand for the LDL receptor, the LDL receptor-related protein and VLDL receptor, and ensures lipoprotein binding to cell-surface glycosaminoglycans.

ApoE is produced as a 317-residue molecule which is cleaved to release an 18-residue signal peptide. The mature apoE molecule contains 299 amino acids and is extensively glycosylated and sialylated at multiple sites, including Thr194 and Ser290 [64]. Similar to apoA-I and apoA-II, apoE contains eight amphipathic alpha-helical repeats and displays detergent-like properties towards vesicular phospholipids [65] (Figs 1.2 and 1.3).

ApoE is synthesized in multiple tissues and cell types, including liver, endocrine tissues, central nervous system (mainly in astrocytes) and macrophages. The apoE gene resides on chromosome 19. Interestingly, apoE is the evolutionary precursor of mammalian apoA-I; the latter appeared after the divergence of the tetrapod and teleost lineages [66]. ApoE possesses a long, highly conserved region in the amino terminal two-thirds of the protein between residues 22 and 182. Another region of highly conserved sequence is located at the carboxyl terminus [28].

Three common APOE alleles have been identified, APOE2, APOE3, and APOE4, which differ by amino acid substitutions at positions 130 and 176. Specifically, the E2 allele contains Cys residues at positions 130 and 176, the E3 allele displays Cys130 and Arg76, and the E4 allele possesses Arg at both positions. The presence of cysteine residues allows apoE2 and apoE3 to form heterodimers with apoA-II. Human apoE3 preferentially binds to HDL, while apoE4 preferentially binds to VLDL [67]. The stronger lipid-binding ability of apoE4 relative to that of apoE3, together with differences in the nature of the surfaces of VLDL and HDL particles, the former being largely covered with phospholipid and the latter with protein, account for such differential binding properties [67].

ApoF. **ApoF** is a 29 kDa sialoglycoprotein present in human HDL and LDL [68] (Table 1.1). ApoF is also known as lipid transfer inhibitor protein (LTIP) because of its ability to inhibit CETP. ApoF is synthesized in the liver from a gene located on chromosome 12 as a 308-amino acid precursor protein, which contains a signal peptide of 18 amino acids followed by a large proprotein of 290 amino acids. The proprotein is further cleaved to release the 162-amino acid C-terminal fragment which makes up the mature secreted form of the protein.

ApoF is heavily glycosylated with both O- and N-linked sugar groups. Such glycosylation renders the protein highly acidic with an isoelectric point of 4.5, resulting in a molecular mass some 40% greater than predicted [69].

ApoH. **ApoH**, also known as beta-2-glycoprotein 1, is a multifunctional cardiolipin-binding, N- and O-glycosylated protein of M_r 38 kDa (Table 1.1). In addition to cardiolipin, apoH binds to various types of negatively charged substances, such as heparin and dextran sulfate and may prevent activation of the intrinsic blood coagulation cascade by binding to phospholipids on the surface of damaged cells. Such binding properties are related to the presence of a positively charged domain at position 282–287. ApoH regulates platelet aggregation, inhibiting the generation of factor Xa and the activation of both factor XIIa and protein C. ApoH is expressed by the liver as a 345-residue proprotein and is secreted in plasma following cleavage of a signal peptide as a 326-residue protein. The apoH gene is located on chromosome 17.

ApoJ. **ApoJ** (also called **clusterin** and complement-associated protein SP-40,40) is a 70 kDa antiparallel disulfide-linked heterodimeric glycoprotein (Table 1.1). Human apoJ consists of two subunits designated alpha (34–36 kDa) and beta (36–39 kDa) which share limited homology [70]. The two subunits are linked by five disulfide bonds to form an antiparallel ladder-like structure (Fig. 1.3). In each of the mature subunits, the five cysteines that are involved in disulfide bonds are clustered in domains of about 30 amino acids located in the central part of the subunits. ApoJ contains about 30% of N-linked carbohydrate rich in sialic acid and possesses nonspecific binding activity to hydrophobic domains of various proteins. The distinct structure of apoJ allows binding of both a wide spectrum of hydrophobic molecules on the one hand and of specific cell-surface receptors on the other.

In humans, the gene is encoded on chromosome 8 and is highly conserved between species (70%–80% homology). The secretory form of apoJ is initially synthesized as a 449-residue, 60 kDa precursor protein containing a 22-residue signal peptide. The proprotein is glycosylated and proteolytically cleaved into alpha- and beta-subunits held together by disulfide bonds.

ApoL-I. **ApoL-I**, a key functional component of the trypanolytic factor of human serum, is a 371-residue protein associated with HDL [71] (Table 1.1; see also 7.5 of chapter 7). ApoL-I possesses a glycosylation site and shares structural and functional similarities with intracellular apoptosis-regulating proteins of the Bcl-2 family (Fig. 1.3). On the other hand, apoL-I is a lipid-binding protein with a high affinity for phosphatidic acid and cardiolipin [72].

ApoL-I is expressed in pancreas, lung, prostate, liver, placenta and spleen. Synthesis of apoL-I involves alternative splicing of the apoL-I gene located on chromosome 22 with the production of two isoforms of 44 and 46 kDa, and cleavage of a 27-residue signal peptide.

ApoM. **ApoM** is a novel, 25 kDa, 188-amino acid apolipoprotein that was isolated and characterized in 1999 [73]. It is mainly found in HDL, but also in VLDL and LDL [74] (Table 1.1).

ApoM possesses an eight-stranded anti-parallel beta-barrel lipocalin fold and a hydrophobic pocket that ensures the binding of small hydrophobic molecules, such as retinol and retinoic acid [75] (Figs 1.2 and 1.3). ApoM reveals 19% homology with apoD, another apolipoprotein member of the lipocalin family. The 1.95 Å resolution crystal structure analysis reveals that recombinant human apoM is complexed with fatty acids containing 14, 16 or 18 carbon atoms which can be replaced by sphingosine 1-phosphate (S1P) [30].

The gene for apoM is located on chromosome 6 and is synthesized in the liver and kidney. Expression of the apoM gene in the liver is regulated by transcription factors that control key stages in hepatic lipid and glucose metabolism [76]. Interestingly, apoM is exclusively associated with lipoproteins despite lacking an external amphipathic motif. The binding of apoM to lipoproteins is ensured by its hydrophobic N-terminal signal peptide which is retained on secreted apoM, a phenomenon atypical for plasma apolipoproteins. The retained signal peptide functions to anchor apoM to the lipoprotein surface and prevents rapid loss by filtration in the kidney [77–79].

Other Apolipoproteins. Minor protein components isolated within the density range of HDL are exemplified by **apoB** and **apo(a)**, which reflect the presence of Lp(a) and result from overlap in the hydrated densities and physicochemical properties of large, light HDL2 and Lp(a) [10].

ApoO, another minor HDL component, is a 198-amino acid protein that contains a 23-amino acid signal peptide and is expressed in several human tissues [80]. ApoO belongs to the proteoglycan family, is present in HDL, LDL and VLDL and contains chondroitin sulfate chains, a unique feature distinguishing it from other apolipoproteins. At present, the physiologic function of apoO remains unknown [81].

OPEN QUESTION

By what type of interactions are the multiple protein components of HDL associated with its surface?

Enzymes

Plasma HDL particles equally carry enzymes involved in lipid metabolism and redox reactions (Table 1.1), including LCAT, paraoxonase 1 (PON1), PON3, platelet-activating factor-acetyl hydrolase (PAF-AH, also called lipoprotein-associated phospholipase A_2 [$LpPLA_2$]) and glutathione selenoperoxidase 3 (GSPx-3) [82].

LCAT. Discovered by John Glomset in 1962 [83], **LCAT** (EC 2.3.1.43) catalyzes the esterification of cholesterol to cholesteryl esters in plasma lipoproteins, primarily in HDL, but also in apoB-containing particles (Table 1.1; see also 4.1 of chapter 4). Approximately 75% of plasma LCAT activity is associated with HDL. In plasma, LCAT is closely associated with apoD which frequently co-purify [84].

The LCAT gene, located on chromosome 16, is expressed primarily in the liver and, to a lesser extent, in the brain and testes. The primary amino acid sequence of LCAT is highly conserved between species. Mature LCAT protein is synthesized from a 440-residue precursor following cleavage of a 24-residue signal peptide. The mature protein contains 416 amino acids and is heavily N-glycosylated. As a result, its molecular mass of 63 kDa is higher than predicted. The tertiary structure of LCAT is maintained by two disulfide bridges and resembles that of lipases and other proteins of the alpha/beta hydrolase fold family [85] (Fig. 1.3). The structure involves an active site formed by serine (Ser)181, aspartate (Asp)345 and histidine (His)377 residues and a lid formed by residues 53–71 which covers the active site [86]. In addition, LCAT contains two free cysteine residues at positions 31 and 184.

PON1 and PON3. Human **paraoxonases** are calcium-dependent lactonases and consist of three members: PON1, PON2 and PON3 [87]. The three PON genes are aligned on the long arm of chromosome 7 (Table 1.1).

The name "PON" reflects the ability of **PON1**, the first enzyme of the family (discovered in 1946 [88]), to hydrolyze the organophosphate substrate paraoxon (PON activity, EC 3.1.8.1). PON1 is capable of hydrolyzing a broad spectrum of organophosphate substrates and aromatic carboxylic acid esters, thereby representing the main means of protection of the nervous system against organophosphate toxicity. As the latter compounds do not normally occur *in vivo*, subsequent studies have identified lactonase activity as the probable physiologic function of PON1 [89]; hydrolysis of homocysteine thiolactone has been proposed to represent a major step in this pathway [89] (see 7.2 and 7.3 of chapter 7). Catalytic activities of the enzyme require the binding of two calcium ions and involve reversible binding to the substrate as the first step of hydrolytic cleavage.

PON1 is structurally organized as a six-bladed beta-propeller, with each blade consisting of four beta-sheets [90] (Fig. 1.3). Two calcium atoms needed for the stabilization of the structure and catalytic activity are located in the central tunnel of the enzyme. Three helices, located at the top of the propeller, are involved in the anchoring of PON1 to HDL. The enzyme is N-glycosylated and may contain a disulfide bond.

Human PON1 is largely synthesized in the liver, but also in the kidney and colon [91], as a 355-amino acid protein and secreted into the blood after removal of the N-terminal Met residue. Circulating PON1 is, therefore, a 354-residue protein with a molecular mass of 43 kDa. In the circulation, PON1 is almost exclusively associated with HDL; such association is mediated by HDL surface phospholipids and requires the hydrophobic leader sequence retained in the secreted PON1. Interestingly, PON1 protein is much more widely distributed than its mRNA, possibly indicating the delivery of PON1 by HDL to various tissues [91].

PON3 is a 354-residue protein with potent lactonase, limited arylesterase and no PON activities. PON3 displays properties similar to those of PON1, such as the requirement for calcium, N-glycosylation, secretion in the circulation with retained signal peptide and association with HDL. Similar to PON1, PON3 is expressed in the liver but also, to a lesser extent, in the kidney. Interestingly, PON3 is specifically enriched in human follicular fluid [92].

Another member of the PON family, **PON2** is a 354-residue intracellular enzyme not detectable in serum despite its expression in many tissues, including brain, liver, kidney and testis. As for other members of the family, PON2 hydrolyzes organophosphate substrates and aromatic carboxylic acid esters.

PAF-AH (LpPLA2). Equally termed **LpPLA$_2$**, **PAF-AH** (EC 3.1.1.47) is a calcium-independent hydrolytic enzyme that degrades platelet-activating factor (PAF) by hydrolyzing the sn-2 ester bond to yield biologically inactive lyso-PAF [93] (Table 1.1). Such activity allows PAF-AH to regulate the biologic actions of PAF. The enzyme cleaves phospholipid substrates with a short residue at the sn-2 position and can, therefore, hydrolyze pro-inflammatory oxidized short-chain phospholipids in addition to PAF; however, it is inactive against long-chain non-oxidized phospholipids (see 7.2 and 7.3 of chapter 7).

The mammalian PAF-AH family consists of four enzymes, two of which belong to the phospholipase A$_2$ subfamily designated Group VII; the other two have been classified as Group VIII phospholipases. Plasma PAF-AH belongs to the Group VII enzymes and is also known as PLA2G7; another enzyme of this group is liver type II PAF-AH (PAFAH2). PAF-AH is synthesized throughout the brain, white adipose tissue and placenta as a 441- amino acid precursor secreted in the circulation following the cleavage of a 21-residue signal peptide. The mature, 53 kDa enzyme contains 420 amino acid residues and is N-glycosylated. Macrophages secrete the largest amount of PAF-AH and represent the most important source of the circulating enzyme [94]. Plasma PAF-AH circulates in association with LDL and HDL particles, with the majority of the enzyme bound to particularly atherogenic, small, dense LDL and to Lp(a) [95]. The lipoprotein location of PAF-AH may affect both the catalytic efficiency and the function of the enzyme *in vivo*. The PAF-AH gene is located on chromosome 6.

The crystal structure of this enzyme, characterized by x-ray diffraction at a resolution of 1.5 Å, reveals a typical lipase alpha/beta-hydrolase fold and a catalytic triad of Ser273, His351 and Asp296 [96] (Fig. 1.3). The active site is close to the lipoprotein surface and, at the same time, accessible to the aqueous phase.

Two clusters of hydrophobic residues build a lipid-binding domain ensuring association with lipoproteins. C-terminal residues are essential for the association of human PAF-AH with HDL [97]. On the other hand, an acidic domain containing 10 carboxylate residues and a neighboring basic domain of 3 amino acid residues are probably involved in the enzyme partitioning between LDL and HDL [96].

GSPx-3. Also called **glutathione peroxidase 3** (EC 1.11.1.9), plasma **GSPx-3** was first purified in 1987 [98] (Table 1.1). The enzyme is distinct from two other members of the GSPx family, GSPx-1 and GSPx-2, which represent erythrocyte and liver cytosolic enzymes. All GSPx enzymes protect biomolecules from oxidative damage by catalyzing the reduction of hydrogen peroxide, lipid peroxides and organic hydroperoxide in a reaction involving glutathione.

Human GSPx-3 is a homotetrameric protein consisting of 206-residue subunits of a molecular mass of 22.5 kDa containing selenium as a selenocysteine residue at position 73.

Human GSPx-3 is synthesized in the liver, kidney, heart, lung, breast and placenta from a gene located on chromosome 5. A GSPx-3 proprotein of 226-amino acid residues is processed with a cleavage of a 20-residue signal peptide before secretion in the circulation. In plasma, GSPx-3 is exclusively associated with HDL [99]. The enzyme sequence is highly conserved between species, suggesting a critical role in survival.

METHODOLOGIC NOTE

Prolonged ultracentrifugation in highly concentrated salt solutions can remove some proteins from HDL, whereas other methods of HDL isolation (gel filtration, immunoaffinity chromatography, precipitation) provide HDL extensively contaminated with plasma proteins or subject HDL to unphysiologic conditions capable of modifying its structure and/or composition (e.g., extreme pH and ionic strength involved in immunoaffinity separation).

Lipid Transfer Proteins

Lipid transfer between lipoprotein particles is an essential element in plasma lipoprotein metabolism. Two lipid transfer proteins of key importance for lipoprotein metabolism, CETP and phospholipid transfer protein (PLTP), can be associated with HDL [8, 100].

PLTP. **PLTP** is a 476-amino acid protein belonging to the bactericidal permeability-increasing protein (BPI)/lipopolysaccharide (LPS)-binding protein (LBP)/Plunc superfamily of proteins (Table 1.1). PLTP is widely distributed across different tissues, and is synthesized in placenta, pancreas, lung, kidney, heart, liver, skeletal muscle and the brain. In the circulation, PLTP is present as a 78 kDa protein primarily associated with HDL; part of the plasma PLTP pool may exist as inactive complexes [101]. PLTP converts HDL into larger and smaller particles and plays a role in extracellular phospholipid transport. In addition, PLTP can bind LPS (see 4.1 of chapter 4).

Synthesis of PLTP involves the production of a 493-residue precursor which is processed to a mature protein through the cleavage of a signal peptide of 17 residues. Mature PLTP contains multiple glycosylation sites and is stabilized by a disulfide bond (Fig. 1.3). The PLTP gene is located on chromosome 20. PLTP is a positive acute-phase reactant with a potential role in the innate immune system.

CETP. Similar to PLTP, **CETP** is synthesized as a 493-residue proprotein which is converted into the mature 476-residue protein by the cleavage of a signal peptide before secretion (Table 1.1). As for PLTP, CETP belongs to the BPI/LBP/Plunc superfamily and contains multiple N-glycosylation sites. In contrast, CETP is encoded on chromosome 16 and is primarily expressed by the liver and adipose tissue. In the circulation, a 74 kDa CETP protein shuttles between HDL and apoB-containing lipoproteins and facilitates the bidirectional transfer of cholesteryl esters and triglycerides between HDL and (V)LDL (see 4.1 of chapter 4). The structure of CETP, determined at 2.2 Å resolution, includes a hydrophobic tunnel filled with two cholesteryl ester molecules and is plugged by an amphiphilic phosphatidylcholine molecule at each end [102] (Fig. 1.3). Such interactions additionally endow CETP with phosphatidylcholine transfer activity. CETP may also undergo conformational changes to accommodate lipoprotein particles of different size, such as HDL, LDL and VLDL [102].

CRITICAL CONTRIBUTION

Detection of more than 50 proteins involved in lipid metabolism, acute phase response, protease inhibition and complement regulation in human HDL isolated by ultracentrifugation from healthy normolipidemic humans [8].

Acute-Phase Response Proteins

Acute-phase response proteins, whose plasma concentrations are markedly altered by acute inflammation, form the largest family of HDL-associated proteins. Surprisingly, acute-phase proteins (23 out of 48 detected) outnumber proteins implicated in lipid metabolism (22 out of 48) [5, 8], indicative of the biologic significance of their presence on HDL (Table 1.2).

Serum Amyloid A. **Serum amyloid A (SAA)** proteins are major acute-phase reactants secreted during the acute phase of the inflammatory response. This protein family is encoded by four genes found in many species, including humans. Located on chromosome 11, three of these genes are commonly expressed in humans, resulting in three protein isoforms: SAA1, SAA2 and SAA4. The high level of conservation of SAA proteins throughout the invertebrate and vertebrate lineages suggests that they play an important physiologic role.

SAA proteins are predominantly produced by the liver. Hepatic expression of SAA1 and SAA2 genes is regulated by the pro-inflammatory cytokines interleukin (IL)-1, IL-6 and tumour necrosis factor (TNF)-alpha. Expression of SAA1 and SAA2 in the liver is induced during the acute-phase reaction, resulting in an increase in their circulating levels by as much as 1000-fold from basal concentrations of about 1 to 5 mg/l [103]. In contrast, SAA4 is constitutively expressed in the liver and is, therefore, termed "constitutive SAA".

SAA1, the major member of this family, is a 12 kDa acute-phase protein whose circulating levels can be induced up to 1000-fold [104] (Table 1.1; see also see 8.1 of chapter 8). HDL is a major carrier of SAA1 in human, rabbit and murine plasma [105–107]. In the circulation, SAA1 does not exist in a free form and associates with non-HDL lipoproteins in the absence of HDL [108]. SAA1 is bound to HDL via its N-terminal domain [109] and is synthesized as a 122-residue precursor which is subsequently secreted into the circulation following the removal of an 18-residue signal peptide.

Other Proteins. Other proteins involved in the acute-phase response that are associated with HDL, include LBP, fibrinogen, alpha-1-acid glycoprotein 2, alpha-2-HS-glycoprotein, alpha-1-antitrypsin, inter-alpha-trypsin inhibitor heavy chain H4, bikunin (alpha-1-microglobulin), haptoglobin-related protein (Hrp), kininogen-1, transthyretin, transferrin, hemopexin and retinol-binding protein [8] (Table 1.2).

LBP is a 60-kDa acute-phase glycoprotein capable of binding the lipid A moiety of LPS of Gram-negative bacteria and facilitating LPS diffusion [110]. LBP is produced as a 481-residue protein containing a 25-residue signal peptide encoded on chromosome 20. LBP/LPS complexes appear to interact with the CD14 receptor to enhance cellular responses to LPS. In addition to LPS, LBP binds phospholipids, thereby acting as a lipid exchange protein [111]. Indeed, LBP belongs to the same BPI/LBP/Plunc protein superfamily as PLTP and CETP.

Fibrinogen is a common acute-phase protein synthesized by the liver that is converted by thrombin into fibrin during blood coagulation. Fibrinogen is a disulfide-linked heterohexamer containing two sets of three non-identical chains (alpha, beta and gamma) encoded on chromosome 4. Conversion of fibrinogen to fibrin by thrombin involves the cleavage of fibrinopeptides A and B from alpha and beta chains with the subsequent exposure of N-terminal polymerization sites. Formation of blood clot ensues via cross-linking between gamma chains and between alpha chains of different monomers catalyzed by factor XIIIa. Furthermore, factor XIIIa further stabilizes fibrin by incorporating inhibitors of fibrinolysis, such as alpha-2-antiplasmin. Fibrinogen also acts as a cofactor in platelet aggregation.

Alpha-1-acid glycoprotein 2, also termed orosomucoid-2, belongs to the calycin protein superfamily which also includes lipocalins and fatty acid-binding proteins. The protein is a positive acute-phase reactant whose synthesis increases up to 50-fold upon inflammation under control of glucocorticoids, IL-1 and IL-6. Alpha-1-acid glycoprotein 2 appears to modulate the activity of the immune system during the acute-phase reaction. It is encoded on chromosome 9, expressed by the liver and secreted in plasma as a 183-residue protein following the cleavage of an 18-residue signal peptide. In plasma, the protein is N-glycosylated and stabilized by disulfide bonds.

Alpha-2-HS-glycoprotein promotes endocytosis, possesses opsonic properties and influences the mineral phase of bone. The protein shows affinity for calcium and barium ions and contains two chains, A and B, which are held together by a single disulfide bond. Alpha-2-HS-glycoprotein is encoded on chromosome 3, synthesized in the liver as a 367-residue full-length single-chain and secreted into plasma.

Alpha-1-antitrypsin, inter-alpha-trypsin inhibitor heavy chain H4, bikunin, Hrp and kininogen-1 are better known for their roles in the inhibition of proteolysis, whereas transthyretin (a thyroid hormone-binding protein), transferrin (an iron-transport protein), hemopexin (an iron-binding protein) and retinol-binding protein are highly specialized plasma proteins described below.

Importantly, plasma levels and/or activities of some major HDL proteins, such as apoA-I, apoA-IV and PON1 are equally affected by the acute-phase response [82] (see 8.1 of chapter 8). A more recently discovered example of a HDL protein involved in the acute-phase response is provided by parotid secretory protein, a salivary protein secreted from the parotid glands. In hamsters, this protein represents a positive acute-phase reactant present exclusively in HDL [112]. The C-terminus of hamster parotid secretory protein contains a region homologous to the N-termini of a family of HDL-associated proteins, including LBP, CETP and PLTP. In addition to salivary glands, parotid secretory protein is expressed in lung, testis and ovary where it is synthesized as a 235-residue precursor containing a 20-residue signal peptide [112].

Complement Components

HDL-associated proteins involved in complement regulation are represented by complement components 3, 4a, 4b and 9, C4b binding protein and vitronectin [8] (Table 1.2).

Complement component 3 (C3) plays a central role in the activation of the complement system both through the classical and alternative complement activation pathways. The full-length single-chain C3 precursor is a large glycoprotein of 1663 amino acid residues with a predicted molecular mass of 187 kDa (Table 1.1). The precursor is processed to form two chains, beta and alpha, linked by a disulfide bond. The key initiating event in complement activation is the cleavage of C3 into a small, 77-residue C3a anaphylatoxin and a large, 915-residue C3b fragment by protease C3-convertase. After activation, C3b can covalently bind via its reactive thioester to cell surface carbohydrates or immune aggregates.

C4 is a key component involved in the activation of the classical pathway of the complement system. The full-length, single-chain C4 precursor contains

1744 amino acid residues and has a predicted molecular mass of 193 kDa. Prior to secretion, the single-chain precursor is enzymatically cleaved to yield alpha, beta and gamma chains. As a result, C4 circulates in blood as a disulfide-linked trimer of alpha, beta and gamma chains. During activation, the alpha chain is cleaved by C1 into a small, 77-residue **C4a** anaphylatoxin and a large, 690-residue **C4b** fragment that remains associated with the beta and gamma chains. The C4b fragment is the major activation product and an essential subunit of the C3 convertase (C4b2a) and C5 convertase (C3bC4b2a) enzymes of the classical complement pathway.

C4b binding protein controls the classical pathway of complement activation. It binds as a cofactor to C3b/C4b inactivator, which then hydrolyzes complement fragment C4b. It also accelerates the degradation of C3 convertase by dissociating the complement fragment C2a from C4b. C4b binding protein represents a disulfide-linked complex of alpha and beta chains, with the alpha chain ensuring C4b binding. The protein is synthesized as a 597-residue precursor secreted following the cleavage of a 48-residue signal peptide.

C9 is a pore-forming subunit of the membrane attack complex that provides an essential contribution to the innate and adaptive immune response by forming pores in the plasma membrane of target cells. C9 is secreted as a soluble monomer that subsequently oligomerizes at target membranes, forming a pre-pore and, following a conformational change, a pore. C9 is produced as a 559-residue precursor secreted as a mature protein after the cleavage of a 21-residue signal peptide.

Vitronectin is another HDL-associated protein involved in complement regulation. It belongs to cell-to-substrate adhesion molecules present in serum and tissues that interact with glycosaminoglycans and proteoglycans and can be recognized by certain members of the integrin family. In complement regulation, vitronectin serves as an inhibitor of the membrane-damaging effect of the terminal cytolytic pathway. The full 478-residue vitronectin proprotein is first processed with the release of a 19-residue signal peptide that can subsequently be cleaved into two chains: a 65 kDa subunit and a 10 kDa subunit, the first of which can be further cleaved with the release of somatomedin-B. The latter represents a growth hormone-dependent serum factor with protease-inhibiting activity, thereby linking complement regulation and protease inhibition pathways. The role of vitronectin in protease inhibition is further underlined by its interaction with plasminogen activator inhibitor type 1 (PAI-1). Vitronectin is largely expressed in the liver but is also expressed in visceral tissue and the adrenal glands. The presence of vitronectin in HDL indicates that certain HDL components can be derived from non-cellular sources or cells distinct from those that synthesize apoA-I in the liver and intestine [5].

The biologic role of the association of complement components with HDL presently remains unclear. The multitude of complement components found in HDL suggests that this lipoprotein may serve as a platform for the assembly of proteins involved in the innate immune response [5]. In addition, complement regulatory proteins present in HDL, such as vitronectin, may limit injury to cardiac cells and prevent activation of the coagulation cascade [7].

MAJOR CONTROVERSY

As the abundance of most of the HDL-associated proteins is below 1 mol/mol HDL (i.e., less than 1 copy per HDL particle), it remains unclear as to how they are distributed among HDL subpopulations of potentially distinct origin and function. Furthermore, the high molecular mass of some HDL-associated proteins, such as complement components C3 and C4, is comparable with the average mass of HDL particles, suggesting associations with specific HDL subpopulation(s).

Proteinase Inhibitors and Related Proteins

A family of proteins in HDL contains serine proteinase inhibitor domains [8] (Table 1.2). Serine protease inhibitors, termed serpins, are key regulators of numerous biologic pathways involved in inflammation, coagulation, angiogenesis and matrix degradation. This family is exemplified by **alpha-1-antitrypsin**, a potent inhibitor of serine proteinases [12] (Table 1.1). Its primary target is elastase, but it also has moderate affinity for plasmin and thrombin. Alpha-1-antitrypsin is produced from a gene located on chromosome 14 as a 418-residue propeptide with a predicted molecular mass of 47 kDa that contains a 24-residue signal peptide. In the circulation, alpha-1-antitrypsin is present exclusively in HDL as a 52 kDa protein [113].

Alpha-2-antiplasmin, a serine proteinase inhibitor that inhibits plasmin and trypsin as two major targets and also inactivates chymotrypsin, is another key proteinase inhibitor that circulates, in part, associated with HDL. Alpha-2-antiplasmin is encoded by the SERPINF2 gene on chromosome 17, is expressed by the liver and is secreted in plasma as a 452-residue mature peptide.

Next, HDL carries inter-alpha-trypsin inhibitor heavy chain H4 and bikunin, two components of inter-alpha-trypsin inhibitors consisting of three out of four heavy chains from groups 1 to 4 and one light chain from the alpha-1-microglobulin/bikunin precursor or Kunitz-type protease inhibitor 2 groups. The full complex inhibits trypsin, plasmin and lysosomal granulocytic elastase. **Inter-alpha-trypsin inhibitor heavy chain H4** is produced in the liver as a 930-amino acid proprotein secreted following the cleavage of a 28-residue signal peptide. In plasma, the protein is cleaved by kallikrein to yield 100 kDa and 35 kDa fragments; the resulting 100 kDa fragment is further converted to a 70 kDa fragment. The protein is encoded by chromosome 3 and appears to be both N- and O-glycosylated. **Bikunin** is a light chain of the inter-alpha-trypsin inhibitor produced via the proteolytic cleavage of the alpha-1-microglobulin/bikunin precursor protein together with alpha-1-microglobulin and trypstatin. The precursor is synthesized in the liver from a gene located on chromosome 9 as a 352-residue protein.

Other proteolysis-related proteins detected on HDL include Hrp, kininogen-1, prothrombin, angiotensinogen and procollagen C-proteinase enhancer-2 (PCPE2). **Hrp** contains a crippled catalytic triad residue that may allow it to act as a decoy substrate to prevent proteolysis. Hrp belongs to the peptidase S1 family and contains one peptidase S1 domain but does not possess essential catalytic residues and displays no enzymatic activity. It is synthesized as a 348-residue proprotein containing a 19-residue signal peptide from a gene located on chromosome 16.

Kininogen-1 (alpha-2-thiol proteinase inhibitor) plays an important role in blood coagulation and inhibits the thrombin- and plasmin-induced aggregation of thrombocytes. The active peptide, bradykinin, released from high-molecular weight kininogen shows a variety of physiologic effects, including the induction of hypotension. Kininogen-1 is synthesized as a 644-amino acid proprotein that is further processed into a mature form.

Prothrombin is a 579-residue precursor of thrombin, a key serine protease of the coagulation pathway. Thrombin is produced by the enzymatic cleavage of two sites on prothrombin by activated factor X. The activity of factor X is greatly enhanced by binding to activated factor V, termed the prothrombinase complex. Prothrombin is produced in the liver and encoded on chromosome 11.

Angiotensinogen is a 452-amino acid long member of the serpin family, although it is not known to possess inhibitory activity towards enzymes. Angiotensinogen is an alpha-2-globulin constitutively produced and released into the circulation mainly by the liver. Angiotensinogen represents a substrate for renin whose action forms angiotensin I, a 10-amino acid peptide. Angiotensin I serves as a precursor to angiotensin II, an endocrine, autocrine/paracrine and intracrine hormone produced by the action of angiotensin-converting enzyme.

PCPE2, a 415-residue heparin-binding protein containing a cleavable 23-residue signal peptide, is another example of a HDL-associated protein involved in proteolytic regulation [114]. Indeed, PCPE2 binds to the C-terminal propeptide of type I and II procollagen and enhances the cleavage of the propeptide by bone morphogenetic protein 1 (BMP-1, also termed procollagen C-proteinase). Furthermore, PCPE2 plays a role in apoA-I production, accelerating proteolytic processing of pro-apoA-I by enhancing cleavage of the hexapeptide extension present at the N-terminus of apoA-I [114]. PCPE2 interacts with BMP-1 and pro-apoA-I to form a ternary PCPE2/BMP-1/pro-apoA-I complex. Interestingly, PCPE2 is present in mature HDL particles, indicating that it either remains associated with apoA-I after its maturation or that maturation occurs in HDL particles [115].

Alpha-2-HS-glycoprotein and **serpin peptidase inhibitor [clade F, member 1]** are minor HDL-associated proteins involved in the inhibition of proteolysis. The presence of multiple protease inhibitors in HDL suggests that these lipoproteins may play previously unsuspected roles in enhancing plaque stability via the protection of vascular lesions from excessive proteolysis [5], as well in intravascular lipoprotein remodeling.

Other Protein Components

HDL may also carry proteins involved in the regulation of other biologic functions (Table 1.2), such as **Wnt signaling molecules** which participate in cell-to-cell signaling [116]. Wnt proteins are secreted in the circulation and play a role in development and differentiation. In cultured mammalian cells, Wnt3a protein is associated with both HDL and LDL in the culture medium, whereas only HDL allows Wnt3a release from mouse fibroblasts. Palmitoylation of the protein is essential for its association with HDL [116]. Wnt3a is produced as a 352-residue proprotein in placenta, lung, spleen and prostate, with a signal peptide of 18 residues. It is a ligand for members of the frizzled family of seven transmembrane receptors which play distinct roles in cell-to-cell signaling during morphogenesis of the developing neural tube [116].

Another example of a HDL-associated regulatory protein is **progranulin**, a 593-amino acid, cysteine-rich protein with an estimated molecular weight of 68.5 kDa encoded by a single gene on chromosome 17 [117]. Progranulin is a precursor of seven small peptides, termed granulins, which are produced from it through proteolytic cleavage by extracellular proteases. Granulins are synthesized in epithelial cells, inflammatory cells and bone marrow, display cytokine-like activity and may play a role in inflammation, wound repair and tissue remodeling.

HDL has also been reported to transport distinct proteins displaying highly specialized functions. The metabolic purpose of such association is unclear; it might prolong the residence time of a protein or represent a mechanism for protein conservation in the circulation. For example, plasma **retinol-binding protein**, a small 21 kDa protein of 201 amino acid residues that delivers retinol from the liver stores to peripheral tissues, co-isolates with HDL3 [8]. In plasma, the complex of retinol-binding protein and retinol interacts with **transthyretin**, thereby preventing the loss of retinol-binding protein by filtration through the kidney glomeruli. About 40% of plasma transthyretin circulates in a tight protein-protein complex with plasma retinol-binding protein. As a corollary, transthyretin, a thyroid hormone-binding protein, is equally present on HDL [8, 10, 11]. Transthyretin is a homotetramer with four 127-residue subunits assembled around a central channel that can accommodate two ligand molecules. Each subunit possesses two stranded beta sheets and reveals an ellipsoidal shape. Transthyretin circulates in HDL through binding to apoA-I and is able to cleave the C-terminus of lipid-free apoA-I [118].

Transferrin is an iron-transport glycoprotein encoded by chromosome 3 and largely produced in the liver. Transferrin is associated with ultracentrifugally isolated HDL and represents an 80 kDa polypeptide of 679 amino acids that contains 2 specific high-affinity Fe(III) binding sites and reversibly binds iron. **Hemopexin**, an iron-binding protein that binds heme and transports it to the liver for breakdown and iron recovery, equally co-isolates with HDL3 [8]. The protein is encoded by chromosome 11, expressed in the liver, synthesized as a 462-amino acid precursor and is secreted in plasma as a 439-amino acid mature protein; in the circulation, hemopexin is N- and O-glycosylated.

In addition, HDL transports lysosomal proteins such as **prenylcysteine oxidase**, which is involved in the degradation of prenylated proteins [8]. The enzyme consists of 505 amino acids, is N-glycosylated, cleaves the thioether bond of prenyl-L-cysteines, such as farnesylcysteine and geranylgeranylcysteine, and requires flavin adenine dinucleotide as a cofactor. Other minor abundance proteins reported to be associated with HDL are: **albumin, alpha-1B-glycoprotein, alpha-amylase, vitamin D-binding globulin** and **platelet basic protein** [8, 10].

Interestingly, many proteins traditionally known to be largely, or exclusively, carried by HDL are also present in LDL and VLDL isolated by density gradient ultracentrifugation. These proteins include apoA-I, apoA-II, apoA-IV, apoD, apoF, apoJ, apoL-I, apoM, PON1 and SAA4 [119]. Furthermore, the content of some of these proteins, primarily of apoJ and apoF, is strongly (>15-fold) elevated in the electronegative LDL^{-} subfraction [120].

In addition to proteins, HDL carries a number of small **peptides** of between 1 and 5 kDa in mass [11]. Indeed, more than 100 peptide components can be identified in HDL purified by density gradient ultracentrifugation or fractionated by size-exclusion chromatography. These peptides are present in HDL at low concentrations of about 1% of total HDL protein, with some representing fragments of larger proteins, such as apoB, fibrinogen and transthyretin [11]. The family of HDL-associated peptides remains poorly investigated; a 7 kDa anionic peptide factor represents a rare exception as it has been reported to contribute to the biologic activities of HDL [121, 122]. Another example of a specialized HDL-associated peptide with distinct function is provided by **amyloid-beta**, the principal peptide constituent of senile plaques in Alzheimer's disease and a potent chelator of transition metal ions [123]. The association of small peptides with HDL as a vehicle may represent a pathway for peptide delivery or scavenging in order to slow renal clearance and proteolysis, and a significant reservoir of plasma peptides for diagnostic evaluation [11]. The diversity of molecules that bind to lipoprotein particles suggests that lipoproteins can serve as a versatile adsorptive surface for proteins and peptides.

Importantly, the composition of the HDL proteome strongly depends on the method of HDL isolation. Indeed, proteomic analyses of human plasma HDL obtained by ultracentrifugation in KBr solutions detect some 50 individual proteins [5, 8]. However, it is possible that the high ionic strength used for the isolation of lipoproteins with KBr alters the pattern of HDL-associated exchangeable proteins. Lipoprotein fractionation using buffers of physiologic ionic strength and pH prepared with deuterium oxide and sucrose may, therefore, improve characterization of exchangeable apolipoproteins and proteins in HDL [124]. Remarkably, proteomics of apoA-I-containing fractions isolated from human plasma by a non-denaturating approach of fast protein liquid chromatography (FPLC) reveal the presence of up to 115 individual proteins per fraction, only up to 32 of which were identified as HDL-associated proteins in ultracentrifugally-isolated HDL [125]. Co-elution with HDL of plasma proteins of matching size is inevitable in FPLC-based separation; the presence of a particular protein across a range of HDL-containing fractions of different size would, however, suggest that such a protein is, indeed, associated with HDL [126]. Remarkably, several of the most abundant plasma proteins, including albumin, haptoglobin, transferrin and alpha-2-macroglobulin, are indeed present in all apoA-I-containing fractions [125], suggesting their partial association with HDL by a non-specific, low-affinity binding.

CLINICAL SUMMARY

HDL represents a protein-rich particle including multiple protein and lipid components which are potentially relevant to both the protection of the arterial wall from atherogenesis and to the regulation of innate immunity and protection from infection.

1.2 LIPIDOME

Phospholipids

Phospholipids constitute the major lipid class of HDL, build the surface lipid monolayer and ensure specific particle structure. Phospholipids quantitatively predominate in the HDL lipidome, accounting for approximately half of all lipids on a weight basis (Table 1.3). Typically, total plasma HDL contains 20–30% (wt/wt) phospholipids, 3–5% free cholesterol, 14–18% cholesteryl esters and 3–6% triglycerides.

Phosphatidylcholine and sphingomyelin predominate as major molecular classes of phospholipids in HDL. In addition, HDL contains significant amounts of phosphatidylinositol, lysophosphatidylcholine, phosphatidylethanolamine, phosphatidylcholine- and phosphatidylethanolamine-derived plasmalogens and ceramide [16, 17, 127]. Minor HDL phospholipids are represented by phosphatidylglycerol, phosphatidylserine, phosphatidic acid and cardiolipin [18, 21, 22]. As a result, HDL is a major carrier of phosphatidylcholine, phosphatidylethanolamine, phosphatidylethanolamine-derived plasmalogens and lysophosphatidylcholine in plasma; in contrast, the largest amount of sphingomyelin is carried in the circulation by LDL [17].

Phosphatidylcholine accounts for approximately 70% of HDL phospholipid; major molecular species of phosphatidylcholine are represented by the 18:2/16:0, 18:2/18:0 and 20:4/16:0 species [16]. Compared with other lipoproteins, HDL is enriched in phosphatidylcholine containing polyunsaturated fatty acid (PUFA) moieties [17]. Phosphatidylcholine in HDL can be of both hepatic (via formation of nascent HDL) and extrahepatic (via the action of PLTP on apoB-containing lipoproteins) origin. In contrast, **sphingomyelin** is largely delivered to HDL from triglyceride-rich lipoproteins via PLTP-mediated transfer and only originates from nascent HDL to a minor extent [128]. Sphingomyelin content constitutes a critical factor in determining surface pressure in lipid membranes and lipoproteins, enhancing rigidity [129, 130] (see Section 1.8.2 chapter 7). **Lysophosphatidylcholines**, products of the LCAT reaction, constitute a minor subclass of HDL phospholipids, with fatty acid moieties of predominantly 16 and 18 carbon atoms [17].

Phosphatidylinositol, phosphatidylserine and **phosphatidic acid** are negatively charged, minor phospholipids that may significantly impact on the net surface charge of HDL [21, 131, 132], thereby modulating the interaction of HDL with lipases, extracellular matrix and other protein components.

Steroids

Together with phospholipids, **steroids** are located in the surface lipid monolayer of HDL particles and regulate its fluidity. HDL steroids are dominated by cholesterol (Table 1.3), reflecting the key role of this lipoprotein in cholesterol transport through the body. Other sterols are present in HDL at much lower levels, as exemplified by minor amounts of oxysterols (27-hydroxycholesterol, 24-hydroxycholesterol, cholesterol-5,6-beta-epoxide, 7-ketocholesterol) [23] and estrogens (largely present as esters) [133].

Cholesteryl Esters

Cholesteryl esters are formed in HDL as a result of trans-esterification of phospholipids and cholesterol catalyzed by LCAT. These highly hydrophobic lipids form the lipid core of HDL. Most of HDL cholesteryl ester is accounted for by cholesteryl linoleate [16] (Table 1.3).

Triglycerides

HDL **triglycerides** are derived from apoB-containing triglyceride-rich lipoproteins as a result of CETP-mediated heteroexchange with cholesteryl esters originating from HDL. Similar to cholesteryl esters, triglycerides are hydrophobic and are located in the HDL lipid core. Compared with cholesteryl esters, however, triglycerides form a more fluid phase. Triglycerides are dominated by species containing oleic, palmitic and linoleic acid moieties [17].

Minor Lipids

Minor bioactive HDL lipids include diacylglycerides, monoacylglicerides, free fatty acids, glycosphingolipids, gangliosides, sulfatides, and lysosphingolipids [15, 17, 134–136]. Interestingly, HDL is relatively depleted in ceramides, glycosphingolipids and gangliosides compared with LDL. Significantly, **ceramides** play a key role as signaling molecules involved in cellular survival, growth and differentiation, whereas **gangliosides** determine interactions with protein receptors and signal transduction; the physiologic relevance of their presence in HDL remains indeterminate.

Among lysosphingolipids, **sphingosine 1-phosphate (S1P)** is particularly interesting as this bioactive lipid plays a key role in vascular biology and can function as a ligand for the family of G protein-coupled S1P receptors present on endothelial and smooth muscle cells, which regulate cell proliferation, motility, apoptosis, angiogenesis, wound healing and immune response (see chapter 7). HDL is the major carrier of S1P in the circulation, ensuring its bioavailability [137]. Indeed, more than 90% of sphingoid base phosphates are found in HDL and albumin-containing fractions by liquid chromatography/mass spectrometry (LC/MS) analysis [127]. S1P is produced by phosphorylation of sphingosine by sphingosine kinases which are expressed in platelets, erythrocytes, neutrophils and mononuclear cells. Erythrocytes appear to represent the primary source of S1P in plasma followed by platelets [138, 139]; S1P release from erythrocytes can be triggered by HDL and serum albumin [141]. Other biologically active lysolipids carried by HDL are represented by **sphingosylphosphorylcholine** and **lysosulfatide** [141].

Finally, HDL carries minor amounts of **lipophilic vitamins** and **antioxidants**, such as tocopherols, carotenes and co-enzyme Q10 [142–144].

IMPORTANT READING

Hoofnagle AN, Heinecke JW. Lipoproteomics: using mass spectrometry-based proteomics to explore the assembly, structure, and function of lipoproteins. *J Lipid Res* 2009; 50: 1967–1975.

Davidsson P, Hulthe J, Fagerberg B, Camejo G. Proteomics of apolipoproteins and associated proteins from plasma high-density lipoproteins. *Arterioscler Thromb Vasc Biol* 2010; 30: 156–163.

Kontush A, Chapman MJ. Lipidomics as a tool for the study of lipoprotein metabolism. *Curr Atheroscler Rep* 2010; 12: 194–201.

REFERENCES

1. Shore B, Shore V. Heterogeneity in protein subunits of human serum high-density lipoproteins. *Biochemistry* 1968; 7: 2773–2777.
2. Rudman D, Garcia LA, Abell. L, Akgun S. Observations on the protein components of human plasma high- and low-density lipoproteins. *Biochemistry* 1968; 7: 3136–3148.
3. Koga S, Horwitz DL, Scanu AM. Isolation and properties of lipoproteins from normal rat serum. *J Lipid Res* 1969; 10: 577–588.
4. Benditt EP, Eriksen N, Hanson RH. Amyloid protein SAA is an apoprotein of mouse plasma high density lipoprotein. *Proc Natl Acad Sci U S A* 1979; 76: 4092–4096.
5. Heinecke JW. The HDL proteome: a marker-and perhaps mediator-of coronary artery disease. *J Lipid Res* 2009; 50: S167–171.
6. Davidsson P, Hulthe J, Fagerberg B, Camejo G. Proteomics of apolipoproteins and associated proteins from plasma high-density lipoproteins. *Arterioscler Thromb Vasc Biol* 2010; 30: 156–163.

7. Hoofnagle AN, Heinecke JW. Lipoproteomics: using mass spectrometry-based proteomics to explore the assembly, structure, and function of lipoproteins. *J Lipid Res* 2009; 50: 1967–1975.

8. Vaisar T, Pennathur S, Green PS, Gharib SA, Hoofnagle AN, Cheung MC, Byun J, Vuleti, S, Kassim S, Singh P et al. Shotgun proteomics implicates protease inhibition and complement activation in the antiinflammatory properties of HDL. *J Clin Invest* 2007; 117: 746–756.

9. Heller M, Stalder D, Schlappritzi E, Hayn G, Matter U, Haeberli A. Mass spectrometry-based analytical tools for the molecular protein characterization of human plasma lipoproteins. *Proteomics* 2005; 13: 13.

10. Davidson WS, Silva RA, Chantepie S, Lagor WR, Chapman MJ, Kontush A. Proteomic analysis of defined HDL subpopulations reveals particle-specific protein clusters: relevance to antioxidative function. *Arterioscler Thromb Vasc Biol* 2009; 29: 870–876.

11. Hortin GL, Shen RF, Martin BM, Remaley AT. Diverse range of small peptides associated with high-density lipoprotein. *Biochem Biophys Res Commun* 2006; 340: 909–915.

12. Karlsson H, Leanderson P, Tagesson C, Lindahl M. Lipoproteomics II: mapping of proteins in high-density lipoprotein using two-dimensional gel electrophoresis and mass spectrometry. *Proteomics* 2005; 5: 1431–1445.

13. Rezaee F, Casetta B, Levels JH, Speijer D, Meijers JC. Proteomic analysis of high-density lipoprotein. *Proteomics* 2006; 6: 721–730.

14. Kunitake ST, Carilli CT, Lau K, Protter AA, Naya-Vigne J, Kane JP. Identification of proteins associated with apolipoprotein A-I-containing lipoproteins purified by selected-affinity immunosorption. *Biochemistry* 1994; 33: 1988–1993.

15. Skipski VP, Barclay M, Barclay RK, Fetzer VA, Good JJ, Archibald FM. Lipid composition of human serum lipoproteins. *Biochem J* 1967; 104: 340–352.

16. Kontush A, Therond P, Zerrad A, Couturier M, Negre-Salvayre A, de Souza JA, Chantepie S, Chapman MJ. Preferential sphingosine-1-phosphate enrichment and sphingomyelin depletion are key features of small dense HDL3 particles: relevance to antiapoptotic and antioxidative activities. *Arterioscler Thromb Vasc Biol* 2007; 27: 1843–1849.

17. Wiesner P, Leidl K, Boettcher A, Schmitz G, Liebisch G. Lipid profiling of FPLC-separated lipoprotein fractions by electrospray ionization tandem mass spectrometry. *J Lipid Res* 2009; 50: 574–585.

18. Yetukuri L, Soderlund S, Koivuniemi A, Seppanen-Laakso T, Niemela PS, Hyvonen M, Taskinen M-R, Vattulainen I, Jauhiainen M, Oresic M. Composition and lipid spatial distribution of HDL particles in subjects with low and high HDL-cholesterol. *J Lipid Res* 2010; 51: 2341–2351.

19. Quehenberger O, Armando AM, Brown AH, Milne SB, Myers DS, Merrill AH, Bandyopadhyay S, Jones KN, Kelly S, Shaner RL et al. Lipidomics reveals a remarkable diversity of lipids in human plasma. *J Lipid Res* 2010; 51: 3299–3305.

20. Vickers KC, Palmisano BT, Shoucri BM, Shamburek RD, Remaley AT. MicroRNAs are transported in plasma and delivered to recipient cells by high-density lipoproteins. *Nat Cell Biol* 2011; 13: 423–433.

21. Lee JY, Min HK, Choi D, Moon MH. Profiling of phospholipids in lipoproteins by multiplexed hollow fiber flow field-flow fractionation and nanoflow liquid chromatography-tandem mass spectrometry. *J Chromatogr A* 2010; 1217: 1660–1666.

22. Deguchi H, Fernandez JA, Hackeng TM, Banka CL, Griffin JH. Cardiolipin is a normal component of human plasma lipoproteins. *Proc Natl Acad Sci U S A* 2000; 97: 1743–1748.

23. Burkard I, von Eckardstein A, Waeber G, Vollenweider P, Rentsch KM. Lipoprotein distribution and biological variation of 24S- and 27-hydroxycholesterol in healthy volunteers. *Atherosclerosis* 2007; 194: 71–78.

24. Kontush A, Chantepie S, Chapman MJ. Small, dense HDL particles exert potent protection of atherogenic LDL against oxidative stress. *Arterioscler Thromb Vasc Biol* 2003; 23: 1881–1888.

25. Frank PG, Marcel YL. Apolipoprotein A-I: structure-function relationships. *J Lipid Res* 2000; 41: 853–872.

26. Asztalos BF, Schaefer EJ. HDL in atherosclerosis: actor or bystander? *Atherosclerosis* 2003 (Suppl.1); 4: 21–29.

27. Schaefer EJ, Santos RD, Asztalos BF. Marked HDL deficiency and premature coronary heart disease. *Curr Opin Lipidol* 2010; 21: 289–297.

28. Weinberg RB. Identification of functional domains in the plasma apolipoproteins by analysis of inter-species sequence variability. *J Lipid Res* 1994; 35: 2212–2222.

29. Collet X, Marcel YL, Tremblay N, Lazure C, Milne RW, Perret B, Weech PK. Evolution of mammalian apolipoprotein A-I and conservation of antigenicity: correlation with primary and secondary structure. *J Lipid Res* 1997; 38: 634–644.

30. Sevvana M, Ahnstrom J, Egerer-Sieber C, Lange HA, Dahlback B, Muller YA. Serendipitous fatty acid binding reveals the structural determinants for ligand recognition in apolipoprotein M. *J Mol Biol* 2009; 393: 920–936.

31. Kumar MS, Carson M, Hussain MM, Murthy HM. Structures of apolipoprotein A-II and a lipid-surrogate complex provide insights into apolipoprotein-lipid interactions. *Biochemistry* 2002; 41: 11681–11691.

32. Gangabadage CS, Zdunek J, Tessari M, Nilsson S, Olivecrona G, Wijmenga SS. Structure and dynamics of human apolipoprotein CIII. *J Biol Chem* 2008; 283: 17416–17427.

33. Dong LM, Wilson C, Wardell MR, Simmons T, Mahley RW, Weisgraber KH, Agard DA. Human apolipoprotein E. Role of arginine 61 in mediating the lipoprotein preferences of the E3 and E4 isoforms. *J Biol Chem* 1994; 269: 22358–22365.

34. Ajees AA, Anantharamaiah GM, Mishra VK, Hussain MM, Murthy HM. Crystal structure of human apolipoprotein A-I: insights into its protective effect against cardiovascular diseases. *Proc Natl Acad Sci U S A* 2006; 103: 2126–2131.

35. Eichinger A, Nasreen A, Kim HJ, Skerra A. Structural insight into the dual ligand specificity and mode of high density lipoprotein association of apolipoprotein D. *J Biol Chem* 2007; 282: 31068–31075.

36. Storjohann R, Rozek A, Sparrow JT, Cushley RJ. Structure of a biologically active fragment of human serum apolipoprotein C-II in the presence of sodium dodecyl sulfate and dodecylphosphocholine. *Biochim Biophys Acta* 2000; 1486: 253–264.

37. Rozek A, Buchko GW, Cushley RJ. Conformation of two peptides corresponding to human apolipoprotein C-I residues 7-24 and 35-53 in the presence of sodium dodecyl sulfate by CD and NMR spectroscopy. *Biochemistry* 1995; 34: 7401–7408.

38. Rozek A, Buchko GW, Kanda P, Cushley RJ. Conformational studies of the N-terminal lipid-associating domain of human apolipoprotein C-I by CD and 1H NMR spectroscopy. *Protein Sci* 1997; 6: 1858–1868.

39. Dong LM, Parkin S, Trakhanov SD, Rupp B, Simmons T, Arnold KS, Newhouse YM, Innerarity TL, Weisgraber KH. Novel mechanism for defective receptor binding of apolipoprotein E2 in type III hyperlipoproteinemia. *Nat Struct Biol* 1996; 3: 718–722.

40. MacRaild CA, Howlett GJ, Gooley PR. The structure and interactions of human apolipoprotein C-II in dodecyl phosphocholine. *Biochemistry* 2004; 43: 8084–8093.

41. MacRaild CA, Hatters DM, Howlett GJ, Gooley PR. NMR structure of human apolipoprotein C-II in the presence of sodium dodecyl sulfate. *Biochemistry* 2001; 40: 5414–5421.

42. Arnold K, Kiefer F, Kopp J, Battey JN, Podvinec M, Westbrook JD, Berman HM, Bordoli L, Schwede T. The Protein Model Portal. *J Struct Funct Genomics* 2009; 10: 1–8.

43. Duriez P, Fruchart JC. High-density lipoprotein subclasses and apolipoprotein A-I. *Clin Chim Acta* 1999; 286: 97–114.

44. Shimano H. ApoAII controversy still in rabbit? *Arterioscler Thromb Vasc Biol* 2009; 29: 1984–1985.

45. Puppione DL, Della Donna L, Laganowsky AD, Bassilian S, Souda P, Ryder OA, Whitelegge JP. Mass spectral analyses of the two major apolipoproteins of great ape high density lipoproteins. *Comp Biochem Physiol Part D Genomics Proteomics* 2010; 4: 305–309.

46. Brewer HB Jr, Lux SE, Ronan R, John KM. Amino acid sequence of human apoLp-Gln-II (apoA-II), an apolipoprotein isolated from the high-density lipoprotein complex. *Proc Natl Acad Sci U S A* 1972; 69: 1304–1308.

47. Hennessy LK, Kunitake ST, Jarvis M, Hamilton RL, Endeman G, Protter A, Kane JP. Isolation of subpopulations of high density lipoproteins: three particle species containing apoE and two species devoid of apoE that have affinity for heparin. *J Lipid Res* 1997; 38: 1859–1868.

48. Gordon JI, Budelier KA, Sims HF, Edelstein C, Scanu AM, Strauss AW. Biosynthesis of human preproapolipoprotein A-II. *J Biol Chem* 1983; 258: 14054–14059.

49. Utermann G, Beisiegel U. Apolipoprotein A-IV: a protein occurring in human mesenteric lymph chylomicrons and free in plasma. Isolation and quantification. *Eur J Biochem* 1979; 99: 333–343.

50. Jaspard B, Fournier N, Vieitez G, Atger V, Barbaras R, Vieu C, Manent J, Chap H, Perret B, Collet X. Structural and functional comparison of HDL from homologous human plasma and follicular fluid. A model for extravascular fluid. *Arterioscler Thromb Vasc Biol* 1997; 17: 1605–1613.

51. Calandra S, Priore Oliva C, Tarugi P, Bertolini S. APOA5 and triglyceride metabolism, lesson from human APOA5 deficiency. *Curr Opin Lipidol* 2006; 17: 122–127.

52. Mauldin K, Lee BL, Oleszczuk M, Sykes BD, Ryan RO. The carboxyl-terminal segment of apolipoprotein A-V undergoes a lipid-induced conformational change. *Biochemistry* 2010; 49: 4821–4826.

53. McConathy WJ, Quiroga C, Alaupovic P. Studies of the composition and structure of plasma lipoproteins. C- and N-terminal amino acids of C-I polypeptide ("R-Val") of human plasma apolipoprotein C. *FEBS Lett* 1972; 19: 323–326.

54. LaRosa JC, Levy RI, Herbert P, Lux SE, Fredrickson DS. A specific apoprotein activator for lipoprotein lipase. *Biochem Biophys Res Commun* 1970; 41: 57–62.

55. Morrisett JD, Gallagher JG, Aune KC, Gotto AM Jr. Structure of the major complex formed by interaction of phosphatidylcholine bilamellar vesicles and apolipoprotein-alanine (APO-C-III). *Biochemistry* 1974; 13: 4765–4771.

56. Mauger JF, Couture P, Bergeron N, Lamarche B. Apolipoprotein C-III isoforms: kinetics and relative implication in lipid metabolism. *J Lipid Res* 2006; 47: 1212–1218.

57. Zhang LH, Kotite L, Havel RJ. Identification, characterization, cloning, and expression of apolipoprotein C-IV, a novel sialoglycoprotein of rabbit plasma lipoproteins. *J Biol Chem* 1996; 271: 1776–1783.

58. Allan CM, Taylor JM. Expression of a novel human apolipoprotein (apoC-IV) causes hypertriglyceridemia in transgenic mice. *J Lipid Res* 1996; 37: 1510–1518.

59. Kim E, Li K, Lieu C, Tong S, Kawai S, Fukutomi T, Zhou Y, Wands J, Li J. Expression of apolipoprotein C-IV is regulated by Ku antigen/peroxisome proliferator-activated receptor gamma complex and correlates with liver steatosis. *J Hepatol* 2008; 49: 787–798.

60. McConathy WJ, Alaupovic P. Isolation and partial characterization of apolipoprotein D: a new protein moiety of the human plasma lipoprotein system. *FEBS Lett* 1973; 37: 178–182.

61. Campanacci V, Nurizzo D, Spinelli S, Valencia C, Tegoni M, Cambillau C. The crystal structure of the *Escherichia coli* lipocalin Blc suggests a possible role in phospholipid binding. *FEBS Lett* 2004; 562: 183–188.

62. Rassart E, Bedirian A, Do Carmo S, Guinard O, Sirois J, Terrisse L, Milne R. Apolipoprotein D. *Biochim Biophys Acta* 2000; 1482: 185–198.

63. Utermann G. Isolation and partial characterization of an arginine-rich apolipoprotein from human plasma very-low-density lipoproteins: apolipoprotein E. *Hoppe Seylers Z Physiol Chem* 1975; 356: 1113–1121.

64. Lee Y, Kockx M, Raftery MJ, Jessup W, Griffith R, Kritharides L. Glycosylation and sialylation of macrophage-derived human apolipoprotein E analyzed by SDS-PAGE and mass spectrometry: Evidence for a novel site of glycosylation on Ser290. *Mol Cell Proteomics* 2010; 9: 1968–1981.

65. Lund-Katz S, Phillips MC. High density lipoprotein structure-function and role in reverse cholesterol transport. *Subcell Biochem* 2010; 51: 183–227.

66. Powell R, Higgins DG, Wolff J, Byrnes L, Stack., Sharp PM, Gannon F. The salmon gene encoding apolipoprotein A-I: cDNA sequence, tissue expression and evolution. *Gene* 1991; 104: 155–161.

67. Nguyen D, Dhanasekaran P, Nickel M, Nakatani R, Saito H, Phillips MC, Lund-Katz S. Molecular basis for the differences in lipid and lipoprotein binding properties of human apolipoproteins E3 and E4. *Biochemistry* 2010; 49: 10881–10889.

68. Olofsson SO, McConathy WJ, Alaupovic P. Isolation and partial characterization of a new acidic apolipoprotein (apolipoprotein F) from high density lipoproteins of human plasma. *Biochemistry* 1978; 17: 1032–1036.
69. Lagor WR, Brown RJ, Toh S-A, Millar JS, Fuki IV, de la Llera-Moya M, Yuen T, Rothblat G, Billheimer JT, Rader DJ. Overexpression of apolipoprotein F reduces HDL cholesterol levels *in vivo*. *Arterioscler Thromb Vasc Biol* 2009; 29: 40–46.
70. de Silva HV, Stuart WD, Park YB, Mao SJ, Gil CM, Wetterau JR, Busch SJ, Harmony, JA. Purification and characterization of apolipoprotein J. *J Biol Chem* 1990; 265: 14292–14297.
71. Duchateau PN, Pullinger CR, Orellana RE, Kunitake ST, Naya-Vigne J, O'Connor PM, Malloy MJ, Kane JP. Apolipoprotein L, a new human high density lipoprotein apolipoprotein expressed by the pancreas. Identification, cloning, characterization, and plasma distribution of apolipoprotein L. *J Biol Chem* 1997; 272: 25576–25582.
72. Zhaorigetu S, Wan G, Kaini R, Jiang Z, Hu CA. ApoL1, a BH3-only lipid-binding protein, induces autophagic cell death. *Autophagy* 2008; 4: 1079–1082.
73. Xu N, Dahlback B. A novel human apolipoprotein (apoM). *J Biol Chem* 1999; 274: 31286–31290.
74. Axler O, Ahnstrom J, Dahlback B. An ELISA for apolipoprotein M reveals a strong correlation to total cholesterol in human plasma. *J Lipid Res* 2007; 48: 1772–1780.
75. Ahnstrom J, Faber K, Axler O, Dahlback B. Hydrophobic ligand binding properties of the human lipocalin apolipoprotein M. *J Lipid Res* 2007; 48: 1754–1762.
76. Nielsen LB, Christoffersen C, Ahnstrom J, Dahlback B. ApoM: gene regulation and effects on HDL metabolism. *Trends Endocrinol Metab* 2009; 20: 66–71.
77. Christoffersen C, Ahnstrom J, Axler O, Christensen EI, Dahlback B, Nielsen LB. The signal peptide anchors apolipoprotein M in plasma lipoproteins and prevents rapid clearance of apolipoprotein M from plasma. *J Biol Chem* 2008; 283: 18765–18772.
78. Axler O, Ahnstrom J, Dahlback B. Apolipoprotein M associates to lipoproteins through its retained signal peptide. *FEBS Lett* 2008; 582: 826–828.
79. Dahlback B, Nielsen LB. Apolipoprotein M affecting lipid metabolism or just catching a ride with lipoproteins in the circulation? *Cell Mol Life Sci* 2009; 66: 559–564.
80. Lamant M, Smih F, Harmancey R, Philip-Couderc P, Pathak A, Roncalli J, Galinier M, Collet X, Massabuau P, Senard JM et al. ApoO, a novel apolipoprotein, is an original glycoprotein up-regulated by diabetes in human heart. *J Biol Chem* 2006; 281: 36289–36302.
81. Nijstad N, de Boer JF, Lagor WR, Toelle M, Usher D, Annema W, der Giet M, Rader DJ, Tietge UJ. Overexpression of apolipoprotein O does not impact on plasma HDL levels or functionality in human apolipoprotein A-I transgenic mice. *Biochim Biophys Acta* 2011; 1811: 294–299.
82. Navab M, Ananthramaiah GM, Reddy ST, Van Lenten BJ, Ansell BJ, Fonarow GC, Vahabzadeh K, Hama S, Hough G, Kamranpour N et al. The oxidation hypothesis of atherogenesis: the role of oxidized phospholipids and HDL. *J Lipid Res* 2004; 45: 993–1007.
83. Glomset JA. The mechanism of the plasma cholesterol esterification reaction: plasma fatty acid transferase. *Biochim Biophys Acta* 1962; 65: 128–135.
84. Holmquist L. Selective extraction of lecithin:cholesterol acyltransferase (EC 2.3.1.43) from human plasma. *J Biochem Biophys Methods* 2002; 52: 63–68.

85. Peelman F, Goethals M, Vanloo B, Labeur C, Brasseur R, Vandekerckhove J, Rosseneu M. Structural and functional properties of the 154-171 wild-type and variant peptides of human lecithin-cholesterol acyltransferase. *Eur J Biochem* 1997; 249: 708–715.

86. Rousset X, Vaisman B, Amar M, Sethi AA, Remaley AT. Lecithin: cholesterol acyltransferase–from biochemistry to role in cardiovascular disease. *Curr Opin Endocrinol Diabetes Obes* 2009; 16: 163–171.

87. Goswami B, Tayal D, Gupta N, Mallika V. Paraoxonase: a multifaceted biomolecule. *Clin Chim Acta* 2009; 410: 1–12.

88. Mazur A. An enzyme in animal tissues capable of hydrolyzing the phosphorusfluorine bond of alkyl fluorophospates. *J Biol Chem* 1946; 164: 271–289.

89. Jakubowski H. Calcium-dependent human serum homocysteine thiolactone hydrolase. A protective mechanism against protein N-homocysteinylation. *J Biol Chem* 2000; 275: 3957–3962.

90. Harel M, Aharoni A, Gaidukov L, Brumshtein B, Khersonsky O, Meged R, Dvir H, Ravelli RB, McCarthy A, Toker L et al. Structure and evolution of the serum paraoxonase family of detoxifying and anti-atherosclerotic enzymes. *Nat Struct Mol Biol* 2004; 11: 412–419.

91. Mackness B, Beltran-Debon R, Aragones G, Joven J, Camps J, Mackness M. Human tissue distribution of paraoxonases 1 and 2 mRNA. *IUBMB Life* 2010; 62: 480–482.

92. Browne RW, Shelly WB, Bloom MS, Ocque AJ, Sandler JR, Huddleston HG, Fujimoto VY. Distributions of high-density lipoprotein particle components in human follicular fluid and sera and their associations with embryo morphology parameters during IVF. *Hum Reprod* 2008; 23: 1884–1894.

93. Mallat Z, Lambeau G, Tedgui A. Lipoprotein-associated and secreted phospholipases A in cardiovascular disease: roles as biological effectors and biomarkers. *Circulation* 2010; 122: 2183–2200.

94. McIntyre TM, Prescott SM, Stafforini DM. The emerging roles of PAF acetylhydrolase. *J Lipid Res* 2009; 50: 255–259.

95. Tselepis AD, Dentan C, Karabina SA, Chapman MJ, Ninio E. PAF-degrading acetylhydrolase is preferentially associated with dense LDL and VHDL-1 in human plasma. Catalytic characteristics and relation to the monocyte-derived enzyme. *Arterioscler Thromb Vasc Biol* 1995; 15: 1764–1773.

96. Samanta U, Bahnson BJ. Crystal structure of human plasma platelet-activating factor acetylhydrolase: structural implication to lipoprotein binding and catalysis. *J Biol Chem* 2008; 283: 31617–31624.

97. Gardner AA, Reichert EC, Topham MK, Stafforini DM. Identification of a domain that mediates association of platelet-activating factor acetylhydrolase with high density lipoprotein. *J Biol Chem* 2008; 283: 17099–17106.

98. Maddipati KR, Marnett LJ. Characterization of the major hydroperoxide-reducing activity of human plasma. Purification and properties of a selenium-dependent glutathione peroxidase. *J Biol Chem* 1987; 262: 17398–17403.

99. Chen N, Liu Y, Greiner CD, Holtzman JL. Physiologic concentrations of homocysteine inhibit the human plasma GSH peroxidase that reduces organic hydroperoxides. *J Lab Clin Med* 2000; 136: 58–65.

100. Marcel YL, McPherson R, Hogue M, Czarnecka H, Zawadzki Z, Weech PK, Whitlock ME, Tall AR, Milne RW. Distribution and concentration of cholesteryl ester transfer protein in plasma of normolipemic subjects. *J Clin Invest* 1990; 85: 10–17.

101. Cheung MC, Wolfbauer G, Albers JJ. Different phospholipid transfer protein complexes contribute to the variation in plasma PLTP specific activity. *Biochim Biophys Acta* 2011; 1811: 343–347.

102. Qiu X, Mistry A, Ammirati MJ, Chrunyk BA, Clark RW, Cong Y, Culp JS, Danley DE, Freeman TB, Geoghegan KF et al. Crystal structure of cholesteryl ester transfer protein reveals a long tunnel and four bound lipid molecules. *Nat Struct Mol Biol* 2007; 14: 106–113.

103. Khovidhunkit W, Kim MS, Memon RA, Shigenaga JK, Moser AH, Feingold KR, Grunfeld C. Effects of infection and inflammation on lipid and lipoprotein metabolism: mechanisms and consequences to the host. *J Lipid Res* 2004; 45: 1169–1196.

104. Malle E, Steinmetz A, Raynes JG. Serum amyloid A (SAA): an acute phase protein and apolipoprotein. *Atherosclerosis* 1993; 102: 131–146.

105. Hoffman JS, Benditt EP. Changes in high density lipoprotein content following endotoxin administration in the mouse. Formation of serum amyloid protein-rich subfractions. *J Biol Chem* 1982; 257: 10510–10517.

106. Marhaug G, Sletten K, Husby G. Characterization of amyloid related protein SAA complexed with serum lipoproteins (apoSAA). *Clin Exp Immunol* 1982; 50: 382–389.

107. Cabana VG, Lukens JR, Rice KS, Hawkins TJ, Getz GS. HDL content and composition in acute phase response in three species: triglyceride enrichment of HDL a factor in its decrease. *J Lipid Res* 1996; 37: 2662–2674.

108. Cabana VG, Feng N, Reardon CA, Lukens J, Webb NR, de Beer FC, Getz GS. Influence of apoA-I and apoE on the formation of serum amyloid A-containing lipoproteins *in vivo* and *in vitro*. *J Lipid Res* 2004; 45: 317–325.

109. Liang JS, Schreiber BM, Salmona M, Phillip G, Gonnerman WA, de Beer FC, Sipe JD. Amino terminal region of acute phase, but not constitutive, serum amyloid A (apoSAA) specifically binds and transports cholesterol into aortic smooth muscle and HepG2 cells. *J Lipid Res* 1996; 37: 2109–2116.

110. Wurfel MM, Kunitake ST, Lichenstein H, Kane JP, Wright SD. Lipopolysaccharide (LPS)-binding protein is carried on lipoproteins and acts as a cofactor in the neutralization of LPS. *J Exp Med* 1994; 180: 1025–1035.

111. Yu B, Hailman E, Wright SD. Lipopolysaccharide binding protein and soluble CD14 catalyze exchange of phospholipids. *J Clin Invest* 1997; 99: 315–324.

112. Khovidhunkit W, Hachem JP, Medzihradszky KF, Duchateau PN, Shigenaga JK, Moser AH, Movsesyan I, Naya-Vigne J, Kane JP, Feingold KR et al. Parotid secretory protein is an HDL-associated protein with anticandidal activity. *Am J Physiol Regul Integr Comp Physiol* 2005; 288: R1306–R1315.

113. Ortiz-Munoz G, Houard X, Martin-Ventura JL, Ishida BY, Loyau S, Rossignol P, Moreno JA, Kane JP, Chalkley RJ, Burlingame AL et al. HDL antielastase activity prevents smooth muscle cell anoikis, a potential new antiatherogenic property. *Faseb J* 2009; 23: 3129–3139.

114. Zhu J, Gardner J, Pullinger CR, Kane JP, Thompson JF, Francone OL. Regulation of apoAI processing by procollagen C-proteinase enhancer-2 and bone morphogenetic protein-1. *J Lipid Res* 2009; 50: 1330–1339.
115. Sviridov D. (2009). Maturation of apolipoprotein A-I: unrecognized health benefit or a forgotten rudiment? *J Lipid Res* 2009; 50: 1257–1258.
116. Neumann S, Coudreuse DY, van der Westhuyzen DR, Eckhardt ER, Korswagen HC, Schmitz G, Sprong H. Mammalian Wnt3a is released on lipoprotein particles. *Traffic* 2009; 10: 334–343.
117. Okura H, Yamashita S, Ohama T, Saga A, Yamamoto-Kakuta A, Hamada Y, Sougawa N, Ohyama R, Sawa Y, Matsuyama A. HDL/apolipoprotein A-I binds to macrophage-derived progranulin and suppresses its conversion into proinflammatory granulins. *J Atheroscler Thromb* 2010; 17: 568–577.
118. Liz MA, Gomes CM, Saraiva MJ, Sousa MM. ApoA-I cleaved by transthyretin has reduced ability to promote cholesterol efflux and increased amyloidogenicity. *J Lipid Res* 2007; 48: 2385–2395.
119. Sun HY, Chen SF, Lai MD, Chang TT, Chen TL, Li PY, Shieh DB, Young KC. Comparative proteomic profiling of plasma very-low-density and low-density lipoproteins. *Clin Chim Acta* 2010; 411: 336–344.
120. Bancells C, Canals F, Benitez S, Colome N, Julve J, Ordonez-Llanos J, Sanchez-Quesada JL. Proteomic analysis of electronegative low-density lipoprotein. *J Lipid Res* 2010; 51: 3508–3515.
121. Domingo N, Mastellone I, Gres S, Marin V, Lorec AM, Tosini F, Grosclaude J, Farnarier C, Chanussot F. The endothelial cholesterol efflux is promoted by the high-density lipoprotein anionic peptide factor. *Metabolism* 2005; 54: 1087–1094.
122. Attia N, Lorec AM, Domingo N, Nakbi A, Hammami S, Ben Hamda K, Chaaba R, Masmoudi AS, Moulin P, Portugal H et al. High density lipoprotein-anionic peptide factor effect on reverse cholesterol transport in type 2 diabetic patients with and without coronary artery disease. *Clin Biochem* 2010; 43: 1079–1084.
123. Kontush A. Apolipoprotein Abeta: black sheep in a good family. *Brain Pathol* 2004; 14: 433–447.
124. Stahlman M, Davidsson P, Kanmert I, Rosengren B, Boren J, Fagerberg B, Camejo G. Proteomics and lipids of lipoproteins isolated at low salt concentrations in D2O/sucrose or in KBr. *J Lipid Res* 2008; 49: 481–490.
125. Collins LA, Mirza SP, Kissebah AH, Olivier M. Integrated approach for the comprehensive characterization of lipoproteins from human plasma using FPLC and nano-HPLC-tandem mass spectrometry. *Physiol Genomics* 2010; 40: 208–215.
126. Gordon SM, Deng J, Lu LJ, Davidson WS. Proteomic characterization of human plasma high density lipoprotein fractionated by gel filtration chromatography. *J Proteome Res* 2010; 9: 5239–5249.
127. Scherer M, Bottcher A, Schmitz G, Liebisch G. Sphingolipid profiling of human plasma and FPLC-separated lipoprotein fractions by hydrophilic interaction chromatography tandem mass spectrometry. *Biochim Biophys Acta* 2011; 1811: 68–75.
128. Nilsson A, Duan RD. Absorption and lipoprotein transport of sphingomyelin. *J Lipid Res* 2006; 47: 154–171.
129. Rye KA, Hime NJ, Barter PJ. The influence of sphingomyelin on the structure and function of reconstituted high density lipoproteins. *J Biol Chem* 1996; 271: 4243–4250.

130. Saito H, Arimoto I, Tanaka M, Sasaki T, Tanimoto T, Okada S, Handa T. Inhibition of lipoprotein lipase activity by sphingomyelin: role of membrane surface structure. *Biochim Biophys Acta* 2000; 1486: 312–320.

131. Davidson WS, Sparks DL, Lund-Katz S, Phillips MC. The molecular basis for the difference in charge between pre-beta- and alpha-migrating high density lipoproteins. *J Biol Chem* 1994; 269: 8959–8965.

132. Boucher JG, Nguyen T, Sparks DL. Lipoprotein electrostatic properties regulate hepatic lipase association and activity. *Biochem Cell Biol* 2007; 85: 696–708.

133. Tikkanen MJ, Vihma V, Jauhiainen M, Hockerstedt A, Helisten H, Kaamanen M. Lipoprotein-associated estrogens. *Cardiovasc Res* 2002; 56: 184–188.

134. Dawson G, Kruski AW, Scanu AM. Distribution of glycosphingolipids in the serum lipoproteins of normal human subjects and patients with hypo- and hyperlipidemias. *J Lipid Res* 1976; 17: 125–131.

135. Senn HJ, Orth M, Fitzke E, Wieland H, Gerok W. Gangliosides in normal human serum. Concentration, pattern and transport by lipoproteins. *Eur J Biochem* 1989; 181: 657–662.

136. Lalanne F, Pruneta V, Bernard S, Ponsin G. Distribution of diacylglycerols among plasma lipoproteins in control subjects and in patients with non-insulin-dependent diabetes. *Eur J Clin Invest* 1999; 29: 139–144.

137. Lucke S, Levkau B. Endothelial functions of sphingosine-1-phosphate. *Cell Physiol Biochem* 2010; 26: 87–96.

138. Pappu R, Schwab SR, Cornelissen I, Pereira JP, Regard JB, Xu Y, Camerer E, Zheng YW, Huang Y, Cyster JG et al. Promotion of lymphocyte egress into blood and lymph by distinct sources of sphingosine-1-phosphate. *Science* 2007; 316: 295–298.

139. Yatomi Y, Ruan F, Hakomori S, Igarashi Y. Sphingosine-1-phosphate: a platelet-activating sphingolipid released from agonist-stimulated human platelets. *Blood* 1995; 86: 193–202.

140. Bode C, Sensken SC, Peest U, Beutel G, Thol F, Levkau B, Li Z, Bittman R, Huang T, Tolle M et al. Erythrocytes serve as a reservoir for cellular and extracellular sphingosine 1-phosphate. *J Cell Biochem* 2010; 109: 1232–1243.

141. Nofer J-R, Assmann G. Atheroprotective effects of high-density lipoprotein-associated lysosphingolipids. *Trends Cardiovasc Med* 2005; 15: 265–271.

142. Bowry VW, Stanley KK, Stocker R. High density lipoprotein is the major carrier of lipid hydroperoxides in human blood plasma from fasting donors. *Proc Natl Acad Sci U S A* 1992; 89: 10316–10320.

143. Garner B, Witting PK, Waldeck AR, Christison JK, Raftery M, Stocker R. Oxidation of high density lipoproteins. I. Formation of methionine sulfoxide in apolipoproteins AI and AII is an early event that accompanies lipid peroxidation and can be enhanced by alpha-tocopherol. *J Biol Chem* 1998; 273: 6080–6087.

144. Goulinet S, Chapman MJ. Plasma LDL and HDL subspecies are heterogenous in particle content of tocopherols and oxygenated and hydrocarbon carotenoids. Relevance to oxidative resistance and atherogenesis. *Arterioscler Thromb Vasc Biol* 1997; 17: 786–796.

CHAPTER 2

HETEROGENEITY

Plasma HDL particles are highly heterogeneous in their physicochemical properties, intravascular metabolism and biologic activity [1–4] (Fig. 2.1). The elevated degree of heterogeneity appears to represent a universal feature of HDL particles independently of their origin; indeed, multiple HDL subpopulations are equally present in peripheral lymph, follicular fluid and cerebrospinal fluid, as well as in amniotic fluid [5–8]. On a particle basis, HDL is the most numerous per unit volume of plasma and is present at the highest (micromolar) levels compared with other lipoproteins [3, 9].

The prominent heterogeneity of HDL particles primarily results from the **highly dynamic structure of apoA-I,** which allows different conformations as a function of the amount of bound lipid and particle size [10] (see Chapter 3). ApoA-I can be considered to be a uniquely flexible lipid-scavenging protein capable of incorporating phospholipids into stable particles [11]. By doing so, apoA-I can adapt distinct conformations that allow the formation of discoid vs. spherical particles [10].

Mechanisms by which apoA-I can vary HDL diameter and accommodate different numbers of lipid molecules involve specific motions driven by protein-phospholipid interactions [14]. Such motions include relative sliding of apoA-I molecules along their major axes, relative rotation of the protein chains and out-of-plane deformation around proline hinges [14]. The sliding and out-of-plane deformation allows apoA-I to optimize its interactions with phospholipids, while the rotation is useful in maximizing protein-protein salt bridges. These rules

High-Density Lipoproteins: Structure, Metabolism, Function, and Therapeutics, First Edition.
Anatol Kontush and M. John Chapman.

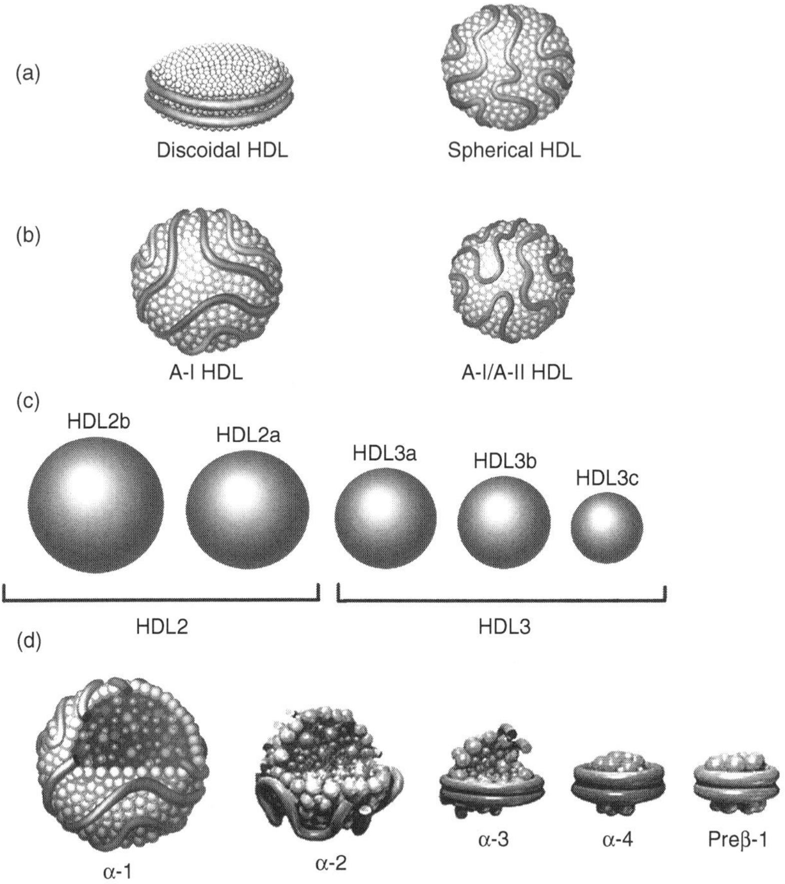

Figure 2.1 Heterogeneity of HDL particles. Major subpopulations of HDL particles which differ in shape (**a**), apolipoprotein composition (**b**), density and size (**c**), and electrophoretic mobility (**c, d**) are shown as revealed by agarose gel electrophoresis (**a**), immunoaffinity chromatography (**b**), nuclear magnetic resonance (NMR) (**c**), polyacrylamide gradient gel electrophoresis (**c, d**) and density gradient ultracentrifugation (**c, d**). Modified from [12] and [13] (see colour plate section).

appear to govern the formation of the smallest discoid HDL particles which differ in size as a function of the amount of bound phospholipid (see Chapter 3). Molecular dynamics simulations on a series of progressively smaller discoid HDLs produced by incremental removal of 1-palmitoyl-2-oleoyl phosphatidylcholine (POPC) reveal that such particles are non-planar, twisted, ellipsoidal saddle-shaped and differ in size secondary to the volume of the lipid pocket able to acquire lipids [11]. In a similar fashion, more than two molecules of apoA-I can co-exist on larger spherical HDL particles in a similar conformation, which allows a gradual increase in the number of apoA-I molecules per HDL with the addition of lipid [15, 16]. Indeed, direct cross-linking measurements have detected the presence of 2–4 mol apoA-I per mol reconstituted HDL (rHDL) as a function of HDL size [17]. These data suggest that structural features of apoA-I determined by its amino acid sequence should represent the primary determinant of the HDL subclass profile. Consistent with this notion, the specific amino acid sequence between helices 7 and 8 of apoA-I markedly influences HDL subclass formation [18]. In addition, the size of nascent HDL particles formed upon the interaction between apoA-I and the adenosine triphosphate (ATP)-binding cassette transporter A1 (ABCA1) is determined by the stability of the N-terminal helix bundle domain and the hydrophobicity of the C-terminal domain of apoA-I [19, 20].

The **capacity to bind and carry distinct proteins** represents another key aspect of HDL heterogeneity. While some HDL proteins possess a high alpha-helical content and may preferentially interact with surface lipids, others may tend to form specific protein-protein complexes [21]. As a result, protein composition may differ significantly between HDL subpopulations, contributing markedly to their heterogeneity [22].

MAJOR CONTROVERSY

Is HDL heterogeneity primarily determined by the interaction of apoA-I and other structural amphipathic proteins with lipids, or by specific protein-protein interactions, or both?

HDL heterogeneity can be assessed using several experimental approaches; the resulting HDL particle profile is critically determined by the method employed (Fig. 2.1). Major techniques used to analyze HDL particle profiles include analytical sequential, zonal, density gradient and vertical ultracentrifugation, gel electrophoresis, nuclear magnetic resonance (NMR), size-exclusion chromatography, gel filtration, ion mobility and affinity chromatography. Despite providing different results, all these approaches agree in distinguishing between five major HDL subpopulations which include **very large, large, medium, small and very small particles,** thereby creating a basis for a unifying nomenclature [23]. The importance of data on HDL heterogeneity obtained using any of these methods should not, however, be overestimated, as all of them provide static, snapshot-like data. Dynamic analyses of fluxes of specific components between distinct HDL subpopulations through the whole HDL metabolic pathway could be more relevant from a physiopathologic point of view; however, the methodology for such analyses remains under-developed at present.

OPEN QUESTION

To what extent can static data provided by the methods to evaluate HDL heterogeneity be translated into dynamic fluxes of components through the HDL metabolic pathway?

2.1 HETEROGENEITY IN PHYSICOCHEMICAL PROPERTIES

The most obvious differences between HDL subpopulations are differences in physical properties, such as size and shape. HDL particles are typically 7–12 nm in size and may be discoid or spherical (Fig. 2.1). **Discoid HDLs** are small (≤8 nm diameter) and lipid-poor (lipid content ≤30%) particles consisting of apolipoproteins (mainly apoA-I) associated with small amounts of lipid (mainly phospholipid and free cholesterol; see 3.2). Discoid HDLs are relatively unstable and short-lived and cannot be easily isolated from plasma; small amounts of discoid HDL can be detected in extravascular compartments, such as peripheral lymph. **Spherical HDL** particles are larger (>8 nm diameter) and additionally contain a hydrophobic core of cholesteryl esters and triglycerides (see 3.3).

Heterogeneity in Density

Historically, ultracentrifugation was the first method applied to lipoprotein isolation and subfractionation. Analytical ultracentrifugation using schlieren optics was originally employed to identify heterogeneity in human plasma HDL in the 1950s by Gofman and colleagues [24–26]. Ultracentrifugal flotation rate in high salt solutions was applied to distinguish between three HDL subpopulations termed **HDL1** [mean density (d) 1.05 g/ml], **HDL2** (d 1.075 g/ml) and **HDL3** (d 1.145 g/ml), which also differed in size [25]; the latter was inversely related to density. Subsequent studies revealed that HDL1 is a minor subfraction present in specific forms of dyslipidemia. Thus, the majority of plasma HDL particles display buoyant density between 1.063 g/ml and 1.21 g/ml. These findings provided a basis for the development of the procedure to fractionate HDL by sequential flotational ultracentrifugation, which allows the isolation of two major subfractions, HDL2 (d 1.063–1.125 g/ml) and HDL3 (d 1.125–1.21 g/ml) [25, 26] (Fig. 2.1). Ultracentrifugally-isolated HDL2 and HDL3 both primarily contain spherical particles as discoid HDL is readily degraded and/or remodeled during repetitive ultracentrifugation steps. Other limitations of this approach include the presence in HDL of other lipoprotein classes, primarily of Lp(a), as a result of the overlap in hydrated densities with light HDL subpopulations [22]. Further development of this approach resulted in the separation of HDL2 and HDL3 within a single ultracentrifugation run using density gradient centrifugation [31]. Moreover, this technique, based on an isopycnic equilibrium, allows reproducible fractionation of five HDL particle subpopulations originally identified by non-denaturing polyacrylamide gradient gel electrophoresis [32], i.e., **large, light HDL2b** and **2a**, and **small, dense HDL3a, 3b** and **3c**, from human plasma in a highly purified form [31, 33–35] (Table 2.1). HDL2b, 2a and 3a subclasses quantitatively predominate in normolipidemic subjects. The size of HDL particles isolated by this approach is progressively reduced with an increase in hydrated density in parallel with a progressive elevation in protein content and in the surface:core ratio, accompanied by diminution in core lipid content consistent with the predominance of surface components in small, dense HDL3 [31,

33–35] (Table 2.1). Critically, these observations were made using fresh plasma. Importantly, ultracentrifugally-isolated HDL is virtually free of other plasma proteins (albumins and globulins) and distinguished by the low content of products of artefactual lipid and protein oxidation that is attenuated during the procedure [30, 36].

DEFINITIONS

HDL2: (d 1.063–1.125 g/ml) a light, large, lipid-rich HDL subfraction isolated by ultracentrifugation with a mean size of 9–10 nm.
HDL3: (d 1.125-1.21 g/ml) a dense, small, lipid-poor HDL subfraction isolated by ultracentrifugation with a mean size of 8–9 nm.
Pre-beta HDL: lipid-poor discoid HDL separated by agarose gel electrophoresis.
Small, dense HDL: generic term to describe lipid-poor and protein-rich discoid and spherical HDL particles of small size ($\leq$9 nm), low molecular mass ($\leq$200 kDa) and high density (1.125–1.21 g/ml), which may include HDL3 separated by ultracentrifugation, as well as pre-beta HDL separated by gel electrophoresis.

The major disadvantage of all methods of HDL fractionation based on ultracentrifugation involves subjecting lipoproteins to high ionic strength and centrifugal shear forces. A shorter centrifugation time and the use of a 'swing-out' rotor therefore represent important advantages of density gradient centrifugation over sequential ultracentrifugation. The replacement of potassium bromide by iodixanol can further diminish lipoprotein modification during the procedure, although separation between individual subfractions may suffer with this approach [37, 38]. Buffers of physiologic ionic strength and pH prepared with deuterium oxide and sucrose can be equally useful in this regard [39]. Centrifugation time can be considerably shortened when HDL is subfractionated using vertical analytical profiling, a technique developed in 1982 [40]. In this method, separation is performed in a vertical rotor and can, therefore, be completed much faster compared with other ultracentrifugation procedures. The method allows the isolation of two major HDL subfractions (HDL2 and HDL3) but suffers from incomplete separation of lipoproteins and major contamination by plasma proteins [41].

Heterogeneity in Electrophoretic Mobility

Non-denaturing polyacrylamide gradient gel electrophoresis has been reproducibly used to separate HDL into five distinct subpopulations of decreasing size: **HDL2b** (9.7–12.9 nm), **HDL2a** (8.8–9.7 nm), **HDL3a** (8.2–8.8 nm), **HDL3b**

TABLE 2.1 Physicochemical Properties of Five Major HDL Subfractions Isolated by Isopycnic Density Gradient Ultracentrifugation from Normolipidemic Human Plasma

	HDL2b	HDL2a	HDL3a	HDL3b	HDL3c
Physical properties					
Diameter, nm	10.2	9.2	8.8	7.8	7.5
Density, g/ml	1.075	1.099	1.120	1.141	1.162
Molecular mass, kDa	450	410	360	200	160
Plasma concentration, μmol/l	1.6	1.8	1.7	1.3	1.0
Surface:core ratio	2.2	2.5	2.7	3.4	3.9
Composition, mol/mol HDL					
ApoA-I	4–5	4	3–4	3	2–3
ApoA-II	0.8	1.0	1.4	1.0	0.4
Phospholipids	130	140	120	45	25
Phosphatidylcholine	85	100	85	30	20
Sphingomyelin	30	25	20	8	3
Phosphatidylinositol	10	10	10	5	2
Phosphatidylethanolamine	2	2	2	1	0.5
Lysophosphatidylcholine	1	1	1	0.5	0.5
Free cholesterol	70	40	25	15	8
Cholesteryl ester	180	160	140	70	40
Triglycerides	30	20	15	10	5

Calculated according to References 27–30.

(7.8–8.2 nm) and **HDL3c** (7.2–7.8 nm) [32] (Fig. 2.1). Importantly, equivalent subpopulations can be quantitatively isolated using isopycnic density gradient ultracentrifugation (see above) [34, 42, 43]. Mean hydrated densities and particle sizes of HDL subpopulations are 1.085 g/ml and 10.57 nm in HDL2b, 1.115 g/ml and 9.16 nm in HDL2a, 1.136 g/ml and 8.44 nm in HDL3a, 1.154 g/ml and 7.97 nm in HDL3b, and 1.171 g/ml and 7.62 nm in HDL3c, respectively [44]. Importantly, levels of the HDL2b subfraction are strongly correlated with those of plasma HDL-cholesterol (HDL-C), indicating that HDL2b particles are the major carrier of cholesterol within the HDL particle spectrum [45] (see also 2.2 of chapter 2). Indeed, HDL size distribution determined by gel electrophoresis is shifted towards large particles when determined by cholesterol compared with protein [46] (see 5.3 of chapter 5). This result suggests that approaches specifically focused on measuring HDL2b as a surrogate marker for HDL-C, such as using a microfluidic lab-on-chip electrophoresis system [47], may provide a biomarker for cardiovascular risk. Further separation of HDL particles using non-denaturing gradient gel electrophoresis allows the identification of up to nine individual apoA-I-containing subpopulations [48].

Another electrophoretic method involves two-dimensional (2-D) electrophoresis that allows reproducible separation by charge and size of more than 10 HDL subspecies in which spherical alpha-HDLs predominate [1, 49]. In this approach, HDL particles are first fractionated by charge in agarose gels into **pre-beta,**

alpha and pre-alpha mobility particles [50] which are subsequently separated by size in polyacrylamide gels and stained using an anti-apoA-I antibody [49]. Five major HDL subpopulations identified by this approach include very small discoid **pre-beta-1 HDL** (about 5.6 nm in size) containing apoA-I and phospholipid, very small discoid **alpha-4 HDL** (about 7.4 nm in size) containing apoA-I, phospholipid and free cholesterol, small spherical **alpha-3 HDL** (about 8.0 nm in size), medium spherical **alpha-2 HDL** (about 9.2 nm in size) containing apoA-I, apoA-II, phospholipid, free cholesterol, cholesteryl ester and triglyceride, and large spherical **alpha-1 HDL** containing apoA-I, phospholipid, free cholesterol, cholesteryl ester and triglyceride (Fig. 2.1). Minor HDL subpopulations observed by 2-D electrophoresis include **pre-alpha particles** that do not contain apoA-II, large **pre-beta-2 HDL**, very large pre-beta HDL containing apoE without apoA-I and small HDL containing apoA-IV without apoA-I [1, 13]. Pre-beta-1 and pre-beta-2 HDL appear to differ in the number of apoA-I molecules (two or three) present per HDL particle. Each subspecies separated by this approach may, however, display further heterogeneity in particle composition, as in the case of ultracentrifugally-isolated subfractions.

More detailed characterization of pre-beta HDL particles reveals that, in addition to apoA-I and phospholipid, they also contain free cholesterol, LCAT, CETP and apoD, but do not contain apoE and apoA-II [6]. The pre-beta-1 particles are small with M_r of 70–80 kDa, whereas pre-beta-2 HDL is much larger with M_r of 325 kDa. By definition, pre-beta HDL possesses a lower negative surface charge than the bulk of plasma HDL, thereby resulting in a slower migration in agarose gels at pH 8.6. The surface potentials of reconstituted pre-beta HDL and alpha-HDL3 are −7.6 and −11.4 mV, respectively [51]. The difference of 3.8 mV corresponds to an apparent 1e greater net negative charge on alpha-migrating HDL3, which can either be provided by phosphatidylinositol or result from the altered conformation of apoA-I in small particles [51]. Interestingly, levels of both pre-beta HDL and small HDL3 are elevated in extravascular fluids, such as peripheral lymph, follicular fluid and amniotic fluid, suggesting a central metabolic role of small HDL particles in these milieux [5, 6, 8].

Heterogeneity in Size

A novel, gas-phase differential electrophoretic macromolecular mobility-based method termed 'ion mobility' allows direct quantification of lipoprotein particle size and concentrations [52]. This rapid procedure allows characterization and quantification of **large, medium** and **small HDL**.

HDL particle size is equally employed as a basis for an alternative approach to the study of HDL heterogeneity which does not involve the isolation of HDL subpopulations. Indeed, NMR spectroscopy takes advantage of the observation that the terminal lipid methyl group proton signal emitted by lipoprotein particles is modulated by lipoprotein size as a result of differences in local magnetic fields

[9, 53]. As a result, HDL subpopulations display distinct NMR signal frequencies (chemical shifts) and can be quantitatively measured following signal deconvolution. Therefore, this approach provides information regarding particle concentrations of HDL subpopulations in plasma. Assumptions concerning the constant number of methyl protons per HDL particle subclass, the arbitrary distinction of HDL subpopulations of large, intermediate and small size, inexact calibration and interferences from variable concentrations of other plasma lipoproteins in different metabolic disorders represent the major limitations of this technique.

Finally, HDL particles can be fractionated using chromatographic approaches. Thus, anion-exchange high-performance liquid chromatography (HPLC) separates HDL into fast- and slowly-eluting fractions that correspond to HDL3 and HDL2, respectively [54]. HDL can be equally subfractionated by size using HPLC with gel permeation columns [55]. Ultrafiltration [56, 57], electrofiltration [58] and size-exclusion chromatography [59] also allow the subfractionation of HDL by size. Such techniques allow the isolation of up to seven HDL subclasses with relative molecular masses of 42–50, 71, 103, 124, 150, 182 and 219 kDa [58]. All the subclasses were reported to contain apoA-I, phosphatidylcholine, sphingomyelin, free cholesterol, cholesteryl ester and triglyceride, whereas some of them also contained apoA-II, C-I, C-II, C-III and E.

Importantly, ultrafiltration permits the isolation of small, protein-rich HDL particles of low molecular mass (40–70 kDa) and of lipoprotein-unassociated apoA-I [56, 57]. Such lipid-free/lipid-poor apoA-I comprises about 5%–10% of human plasma apoA-I and likely derives from direct secretion by the liver and/or intestine or by dissociation from HDL or triglyceride-rich lipoproteins. Currently, there is no strong consensus as to whether this fraction is truly lipid-free or if it contains trace amounts of lipid (lipid-poor). Strictly speaking, **lipid-free apoA-I** does not belong to HDL as its hydrated density of 1.3 g/ml is higher than 1.21 g/ml—the upper limit typically used to define HDL particles. When subjected to ultracentrifugation, lipid-free apoA-I co-isolates with other soluble plasma proteins, primarily with albumin. Interestingly, such co-isolation appears to reflect physiologic interaction between apoA-I and albumin in blood, which may play previously unsuspected roles in HDL metabolism [60]. Indeed, a complex of albumin and apoA-I with M_r of 106 kDa containing small amounts of cholesterol can be isolated from healthy young subjects [58].

METHODOLOGIC NOTE

All existing methods to subfractionate HDL subpopulations (electrophoresis, ultracentrifugation, size-exclusion chromatography, gel filtration, affinity chromatography) suffer from limitations. It is particularly important to emphasize the significance of the use of fresh plasma/serum as starting material in these procedures as freezing/thawing can markedly alter the HDL particle profile.

2.2 HETEROGENEITY IN CHEMICAL COMPOSITION

Heterogeneity in Proteins

The concept of HDL as a family of distinct particles that contain apoA-I but vary in protein composition was originally proposed 40 years ago by Petar Alaupovic [61–63]. Subsequent development has revealed that in all normolipidemic subjects, **HDL particles containing only apoA-I (LpA-I) and both apoA-I and apoA-II (LpA-I:A-II)** can be immunoseparated [64] (Fig. 2.1). ApoA-I is typically distributed approximately equally between LpA-I and LpA-I:A-II, whereas virtually all apoA-II is in LpA-I:A-II [64]. LpA-I and LpA-I:A-II contain approximately 35% and 65% of plasma apoA-I, respectively [65]. On the other hand, approximately half of HDL particles contain apoA-II [66]. In addition to apoA-II, the two subclasses may differ in other protein content, as exemplified by PON1 which preferably associates with LpA-I [67].

LpA-I particles can be further subfractionated according to size. The number of apoA-I molecules in such subpopulations is increased from two to three to four with an increase in the particle size [68]. On the other hand, the entire population of LpA-I:A-II demonstrates the presence of only two apoA-I molecules per particle, while the number of apoA-II molecules varies from one dimeric apoA-II to two and then to three. Upon compositional analyses of individual subpopulations, LpA-I:A-II exhibits comparable proportions for total protein (approximately 58%), phospholipid (approximately 21%), total cholesterol (approximately 16%), triglyceride (approximately 5%) and free cholesterol (approximately 4%) across subfractions; however, LpA-I components show significant variability [68].

Another important subpopulation of HDL particles is formed by apoE-containing HDL. The presence of apoE facilitates the expansion of the lipid core as a result of the accumulation of cholesteryl ester with the formation of large, lipid-rich HDL. These particles represent an excellent ligand for the LDL receptor [69]. Plasma levels of such **apoE-rich HDL** are high at birth but decrease thereafter with reciprocal elevation in LDL-C, consistent with their capacity to deliver cholesterol via apoB/apoE receptors [70]. In addition to apoE, apoA-V also preferentially forms large HDL characterized by enhanced lipid-binding capacity and increased phospholipid content [71].

HDL subpopulations may also markedly differ in the content of minor proteins which can be semi-quantitatively evaluated using proteomic approaches [60]. First, analyses of the HDL proteome reveal that the abundance of most HDL-associated proteins is insufficient to allow one copy per HDL particle, thereby suggesting that specific proteins may be confined to distinct HDL subpopulations of distinct origin and function that are differentially distributed across the HDL particle spectrum [22]. For example, only 5% of HDL particles can be estimated to carry a molecule of apoM [72], suggesting the existence of an apoM-containing subpopulation(s).

Proteomic analysis of five HDL subpopulations isolated from normolipidemic subjects by isopycnic density gradient ultracentrifugation identifies five distinct patterns of distribution of individual protein components across the HDL density subfractions [22]. The most interesting of these identifies small, dense HDL3c as a particle subpopulation in which seven proteins predominantly occur, notably apoJ, apoL-I, apoF, PON1/3, PLTP and PAF-AH. The activities of HDL-associated enzymes (LCAT, PON1, PAF-AH) are equally elevated in small, dense HDL3c [27, 73]. The HDL3c proteome also contains apoA-I, apoA-II, apoD, apoM, SAA 1, 2 and 4, apoC-I, apoC-II and apoE [22]. Consistent with these data, apoL-I [74], apoF [75, 76], apoJ [77, 78], PON1 [27, 78], apoA-IV [79, 80], apoM [81], apoD [82] and SAA1/2 [83, 84] are known to preferentially co-isolate with dense HDL3. Furthermore, small, dense HDL may also represent a preferential carrier for human CETP [85]. On the other hand, however, apoE, apoC-I, apoC-II and apoC-III preferentially localize to large, light HDL2 [22, 86–88]. Importantly, these associations can, in part, be confirmed using an alternative approach of gel filtration subfractionation of HDL particles [89].

CRITICAL CONTRIBUTION

Distinct patterns of distribution and clustering of individual protein components across HDL density subfractions [22].

The low HDL content of the majority of these proteins of less than one copy per HDL particle suggests internal heterogeneity of the HDL3c subfraction. This conclusion is consistent with the isolation of a unique particle containing the trypanosome lytic factor, apoL-I, plus apoA-I and Hrp in the HDL3 density range [90].

In a similar fashion, apoF co-isolates with dense HDL (mean density, 1.134 g/ml) which also contains apoA-I, apoC-II, apoE, apoJ, apoD and PON1 [75]. Complexes formed by PON1 with human phosphate-binding protein and apoJ represent another example of protein-protein interactions occurring within the HDL particle spectrum [21]. In addition, surface plasmon resonance data reveal that interactions with the protein moiety are important for apoE binding to protein-rich HDL3 [91].

Thus, specific protein-protein interactions should drive the formation of such lipoprotein complexes in the circulation. In support of such a mechanism, PLTP in human plasma resides on lipid-poor complexes dominated by apoJ and proteins implicated in host defence and inflammation [92]. Remarkably, 20 out of 24 proteins detected in association with PLTP reveal protein-protein interactions known from online databases. Moreover, apoJ, apoA-I and apoE preserve the lipid-transfer activity of recombinant PLTP in the absence of lipid, indicating that these interactions may have a functional significance [92]. Indeed, apoA-I increases the activity of PLTP secreted by macrophage-derived foam cells [93].

These data show that proteomic analyses can provide deep insights into the heterogeneity of HDL particles. Limitations of such an approach involve its semi-quantitative character, critical dependence of the results on the methodology of

HDL isolation and purification, the nature of the starting biologic material (serum or plasma) and also on the mass spectrometric technology employed for protein analysis and quantitation. For example, HDL particles isolated from serum and plasma samples differ in their content of several proteins, including complement C3, which is elevated in serum-derived HDL [94]. Furthermore, HDL isolated by fast protein liquid chromatography (FPLC) is heavily contaminated by high-molecular weight plasma proteins that co-elute with HDL [95], whereas immunoaffinity chromatography subjects HDL to extreme pH and ionic strength capable of modifying its composition. For this reason, density gradient ultracentrifugation remains the predominant methodology for HDL isolation to study the HDL proteome [22, 96].

Heterogeneity in Lipids

In addition to differences in bound proteins, HDL subpopulations may also differ in their content of lipids. Indeed, particle content of phospholipid, free cholesterol, cholesteryl ester, triglyceride and total fatty acids progressively decreases with an increase in hydrated density from HDL2b to HDL3c [30] (Table 2.1). The same dependency is obtained for phospholipid subclasses, including phosphatidylcholine, phosphatidylethanolamine, phosphatidylinositol, sphingomyelin and lysophosphatidylcholine, as well as for subgroups of fatty acids, including saturated, monounsaturated and polyunsaturated n-6 and n-3 moieties. However, no differences are found between HDL subspecies when data for phosphatidylcholine, phosphatidylethanolamine, phosphatidylinositol, lysophosphatidylcholine, cholesteryl ester and total fatty acids are expressed as a percentage of total lipids, suggesting that their molecular species are in dynamic equilibrium between HDL subpopulations [30].

In contrast, the proportion of sphingomyelin relative to total lipids progressively decreases in parallel with HDL density from 12.8% in HDL2b to 6.2% in HDL3c [30]. Consequently, the sphingomyelin:phosphatidylcholine molar ratio decreases from 0.38 in HDL2b to 0.18 in HDL3c. The distinctly low sphingomyelin content in HDL3c suggests that this pool is not in equilibrium with that of other HDL subpopulations, consistent with the slow rate of exchange of sphingomyelin between lipoproteins and cell membranes [97]. The low sphingomyelin to phosphatidylcholine ratio may reflect a distinct cellular origin(s) of small HDL, as suggested by the low sphingomyelin content of small, nascent HDL particles secreted by J774 macrophages which originate from the exofacial leaflet of the plasma membrane [98]. Similarly to sphingomyelin, the proportion of free cholesterol relative to total lipids decreases twofold from HDL2b to HDL3c [30]. As a result, the cholesteryl ester:free cholesterol ratio significantly increases with HDL density, supporting the argument that small HDL constitutes a major site of cholesterol esterification within the HDL particle spectrum [99]. Elevated LCAT activity and the diminished sphingomyelin:phosphatidylcholine ratio in HDL3c are consistent with this proposal, as sphingomyelin functions as a physiologic inhibitor of LCAT [100, 101]. As both sphingomyelin and free cholesterol decrease the fluidity of liquid-crystal lipids, their low abundance in small, dense HDL suggests elevated fluidity of the surface lipids in this subpopulation [30].

Differences between HDL subpopulations in the abundance of cholesteryl ester and free cholesterol result in the enrichment of total cholesterol in large, light HDL. In this regard, it is essential to emphasize that routine clinical measurement of plasma HDL-C primarily reflects levels of large, cholesterol-rich HDL and frequently lacks the sensitivity to detect small, cholesterol-poor particles. Furthermore, as HDL-C simply quantifies the amount of cholesterol contained within the HDL lipoprotein fraction, this measurement does not necessarily correlate with the number of HDL particles or their particle profile [102]. The marked complexity of the HDL particle profile is, therefore, hardly reflected by a single measurement of HDL-C.

Compositional differences between HDL subpopulations may also include those in minor bioactive lipid components. Thus, the abundance of S1P per HDL particle is asymmetric across the HDL spectrum, with preferential enrichment in HDL3 (40–50 mmol/mol HDL) compared with HDL2 particles (15–20 mmol/mol) [30, 103, 104]. The enrichment of small HDL3 in S1P may be mechanistically related to the potent capacity of such particles to acquire polar lipids of cellular origin [30]. Alternatively, such enrichment can be driven by a specific protein component preferentially present in HDL3 such as apoM [72].

CLINICAL SUMMARY

Plasma HDL particles are highly heterogeneous in their physicochemical properties, intravascular metabolism and biologic activity; such complexity is not revealed by a measurement of HDL-C.

2.3 RELATIONSHIPS BETWEEN HDL SUBFRACTIONS SEPARATED BY DIFFERENT METHODS

Relationships between HDL subfractions separated by different approaches are complex. For example, immuno-isolated LpA-I:A-II is predominantly found in the HDL3 density range, whereas LpA-I is a prominent component of both HDL2 and HDL3 [64]. On the other hand, alpha-migrating HDL predominates in both HDL2 and HDL3 subfractions, whereas pre-beta HDL co-isolates with small, dense HDL particles [1]. Another important example concerns ultracentrifugally-isolated small, dense HDL3c that does not correspond precisely to small HDL subpopulations as determined by other methods. Human HDL3c represents a minor subfraction accounting for about 6% of the total HDL mass and 10% of apoA-I [27–29, 105, 106]. In 2-D electrophoresis, HDL3c reveals further heterogeneity and produces multiple signals corresponding to small alpha-3- and pre-beta-1 HDL (Chantepie S, Kontush A, Chapman MJ, unpublished data).

In contrast, small, alpha-3, pre-beta-3 and pre-beta-1 HDL subfractions measured by 2-D electrophoresis in whole plasma account for about 37%, 4% and 12% of apoA-I, respectively [107]. Moreover, alpha-3, together with alpha-2, represents a major HDL subfraction in normolipidemic subjects. Remarkably, both HDL3c and pre-beta-1 HDL are depleted of sphingomyelin and free cholesterol compared with other HDL subpopulations, strongly suggesting an overlap between these HDL subclasses [6, 30].

IMPORTANT READING

Movva R, Rader DJ. Laboratory assessment of HDL heterogeneity and function. *Clin Chem* 2008; 54: 788–800.

Schaefer EJ, Santos RD, Asztalos BF. Marked HDL deficiency and premature coronary heart disease. *Curr Opin Lipidol* 2010; 21: 289–297.

Rosenson RS, Brewer HB Jr, Chapman MJ, Fazio S, Hussain MM, Kontush A, Krauss RM, Otvos JD, Remaley AT, Schaefer EJ. HDL measures, particle heterogeneity, proposed nomenclature, and relation to atherosclerotic cardiovascular events. *Clin Chem* 2011; 57: 392–410.

REFERENCES

1. Asztalos BF, Schaefer EJ. HDL in atherosclerosis: actor or bystander? *Atheroscler (Suppl.)* 2003; 4: 21–29.
2. Barter P, Kastelein J, Nunn A, Hobbs R. High density lipoproteins (HDLs) and atherosclerosis; the unanswered questions. *Atherosclerosis* 2003; 168: 195–211.
3. Kontush A, Chapman MJ. Functionally defective HDL: A new therapeutic target at the crossroads of dyslipidemia, inflammation and atherosclerosis. *Pharmacol Rev* 2006; 3: 342–374.
4. Kontush A, Chapman MJ. Antiatherogenic small, dense HDL—guardian angel of the arterial wall? *Nat Clin Pract Cardiovasc Med* 2006; 3: 144–153.
5. Vessby B, Gustafson S, Chapman MJ, Hellsing K, Lithell H. Lipoprotein composition of human suction-blister interstitial fluid. *J Lipid Res* 1987; 28: 629–641.
6. Barrans A, Jaspard B, Barbaras R, Chap H, Perret B, Collet X. Pre-beta HDL: structure and metabolism. *Biochim Biophys Acta* 1996; 1300: 73–85.
7. Kontush A, Atwood CS. Amyloid-beta: phylogenesis of a chameleon. *Brain Res Brain Res Rev* 2004; 46: 118–120.
8. Fujita H, Okada T, Inami I, Makimoto M, Hosono S, Minato M, Takahashi S, Mugishima H, Yamamoto T. Heterogeneity of high-density lipoprotein in cord blood and its postnatal change. *Clin Chim Acta* 2008; 389: 93–97.
9. Jeyarajah EJ, Cromwell WC, Otvos JD. Lipoprotein particle analysis by nuclear magnetic resonance spectroscopy. *Clin Lab Med* 2006; 26: 847–870.
10. Davidson WS, Thompson TB. The structure of apolipoprotein A-I in high density lipoproteins. *J Biol Chem* 2007; 282: 22249–22253.

11. Catte A, Patterson JC, Jones MK, Jerome WG, Bashtovyy D, Su Z, Gu F, Chen J, Aliste MP, Harvey SC, et al. Novel changes in discoidal high density lipoprotein morphology: a molecular dynamics study. *Biophys J* 2006; 90: 4345–4360.
12. Tabet F, Rye KA. High-density lipoproteins, inflammation and oxidative stress. *Clin Sci (Lond)* 2009; 116: 87–98.
13. Schaefer EJ, Santos RD, Asztalos BF. Marked HDL deficiency and premature coronary heart disease. *Curr Opin Lipidol* 2010; 21: 289–297.
14. Rocco AG, Gianazza E, Calabresi L, Sensi C, Franceschini G, Sirtori CR, Eberini I. Structural features and dynamics properties of human apolipoprotein A-I in a model of synthetic HDL. *J Mol Graph Model* 2009; 28: 305–312.
15. Silva RA, Huang R, Morris J, Fang J, Gracheva EO, Ren G, Kontush A, Jerome WG, Rye KA, Davidson WS. Structure of apolipoprotein A-I in spherical high density lipoproteins of different sizes. *Proc Natl Acad Sci U S A* 2008; 105: 12176–12181.
16. Huang R, Silva RA, Jerome WG, Kontush A, Chapman MJ, Curtiss LK, Hodges TJ, Davidson WS. Apolipoprotein A-I structural organization in high-density lipoproteins isolated from human plasma. *Nat Struct Mol Biol* 2011; 18: 416–22.
17. Massey JB, Pownall HJ, Macha S, Morris J, Tubb MR, Silva RAGD. Mass spectrometric determination of apolipoprotein molecular stoichiometry in reconstituted high density lipoprotein particles. *J Lipid Res* 2009; 50: 1229–1236.
18. Carnemolla R, Ren X, Biswas TK, Meredith SC, Reardon CA, Wang J, Getz GS. The specific amino acid sequence between helices 7 and 8 influences the binding specificity of human apolipoprotein A-I for high density lipoprotein (HDL) subclasses: a potential for HDL preferential generation. *J Biol Chem* 2008; 283: 15779–15788.
19. Vedhachalam C, Chetty PS, Nickel M, Dhanasekaran P, Lund-Katz S, Rothblat GH, Phillips MC. Influence of apolipoprotein (Apo) A-I structure on nascent high density lipoprotein (HDL) particle size distribution. *J Biol Chem* 2010; 285: 31965–31973.
20. Tanaka M, Dhanasekaran P, Nguyen D, Nickel M, Takechi Y, Lund-Katz S, Phillips MC, Saito H. Influence of N-terminal helix bundle stability on the lipid-binding properties of human apolipoprotein A-I. *Biochim Biophys Acta* 2010; 1811: 25–30.
21. Vaisar T. Thematic Review Series: Proteomics. Proteomic analysis of lipid-protein complexes. *J Lipid Res* 2009; 50: 781–786.
22. Davidson WS, Silva RA, Chantepie S, Lagor WR, Chapman MJ, Kontush A. Proteomic analysis of defined HDL subpopulations reveals particle-specific protein clusters: relevance to antioxidative function. *Arterioscler Thromb Vasc Biol* 2009; 29: 870–876.
23. Rosenson RS, Brewer HB Jr, Chapman MJ, Fazio S, Hussain MM, Kontush A, Krauss RM, Otvos JD, Remaley AT, Schaefer EJ. HDL measures, particle heterogeneity, proposed nomenclature, and relation to atherosclerotic cardiovascular events. *Clin Chem* 2011; 57: 392–410.
24. Lindgren FT, Elliott HA, Gofman JW. The ultracentrifugal characterization and isolation of human blood lipids and lipoproteins, with applications to the study of atherosclerosis. *J Phys Colloid Chem* 1951; 55: 80–93.
25. Delalla OF, Elliott HA, Gofman JW. Ultracentrifugal studies of high density serum lipoproteins in clinically healthy adults. *Am J Physiol* 1954; 179: 333–337.
26. Gofman JW, Young W, Tandy R. Ischemic heart disease, atherosclerosis, and longevity. *Circulation* 1966; 34: 679–697.

27. Kontush A, Chantepie S, Chapman MJ. Small, dense HDL particles exert potent protection of atherogenic LDL against oxidative stress. *Arterioscler Thromb Vasc Biol* 2003; 23: 1881–1888.

28. Hansel B, Giral P, Nobecourt E, Chantepie S, Bruckert E, Chapman MJ, Kontush A. Metabolic syndrome is associated with elevated oxidative stress and dysfunctional dense high-density lipoprotein particles displaying impaired antioxidative activity. *J Clin Endocrinol Metab* 2004; 89: 4963–4971.

29. Nobecourt E, Jacqueminet S, Hansel B, Chantepie S, Grimaldi A, Chapman MJ, Kontush A. Defective antioxidative activity of small, dense HDL particles in type 2 diabetes: Relationship to elevated oxidative stress and hyperglycemia. *Diabetologia* 2005; 48: 529–538.

30. Kontush A, Therond P, Zerrad A, Couturier M, Negre-Salvayre A, de Souza JA, Chantepie S, Chapman MJ. Preferential sphingosine-1-phosphate enrichment and sphingomyelin depletion are key features of small dense HDL3 particles: relevance to antiapoptotic and antioxidative activities. *Arterioscler Thromb Vasc Biol* 2007; 27: 1843–1849.

31. Chapman MJ, Goldstein S, Lagrange D, Laplaud PM. A density gradient ultracentrifugal procedure for the isolation of the major lipoprotein classes from human serum. *J Lipid Res* 1981; 22: 339–358.

32. Anderson DW, Nichols AV, Forte TM, Lindgren FT. Particle distribution of human serum high density lipoproteins. *Biochim Biophys Acta* 1977; 493: 55–68.

33. Guerin M, Bruckert E, Dolphin PJ, Turpin G, Chapman MJ. Fenofibrate reduces plasma cholesteryl ester transfer from HDL to VLDL and normalizes the atherogenic, dense LDL profile in combined hyperlipidemia. *Arterioscler Thromb Vasc Biol* 1996; 16: 763–772.

34. Goulinet S, Chapman MJ. Plasma LDL and HDL subspecies are heterogenous in particle content of tocopherols and oxygenated and hydrocarbon carotenoids. Relevance to oxidative resistance and atherogenesis. *Arterioscler Thromb Vasc Biol* 1997; 17: 786–796.

35. Guerin M, Egger P, Le Goff W, Soudant C, Dupuis R, Chapman MJ. Atorvastatin reduces postprandial accumulation and cholesteryl ester transfer protein-mediated remodeling of triglyceride-rich lipoprotein subspecies in type IIb hyperlipidemia. *J Clin Endocrinol Metab* 2002; 87: 4991–5000.

36. Bowry VW, Stanley KK, Stocker R. High density lipoprotein is the major carrier of lipid hydroperoxides in human blood plasma from fasting donors. *Proc Natl Acad Sci U S A* 1992; 89: 10316–10320.

37. Harman NL, Davies IG, Griffin BA. Separation of the principal HDL subclasses by iodixanol gradient ultracentrifugation. *Atherosclerosis* 2007; 194: 283.

38. Yee MS, Pavitt DV, Tan T, Venkatesan S, Godsland IF, Richmond W, Johnston DG. Lipoprotein separation in a novel iodixanol density gradient, for composition, density, and phenotype analysis. *J Lipid Res* 2008; 49: 1364–1371.

39. Stahlman M, Davidsson P, Kanmert I, Rosengren B, Boren J, Fagerberg B, Camejo G. Proteomics and lipids of lipoproteins isolated at low salt concentrations in D2O/sucrose or in KBr. *J Lipid Res* 2008; 49: 481–490.

40. Cone JT, Segrest JP, Chung BH, Ragland JB, Sabesin SM, Glasscock A. Computerized rapid high resolution quantitative analysis of plasma lipoproteins based upon single vertical spin centrifugation. *J Lipid Res* 1982; 23: 923–935.

41. Kulkarni KR, Marcovina SM, Krauss RM, Garber DW, Glasscock AM, Segrest JP. Quantification of HDL2 and HDL3 cholesterol by the Vertical Auto Profile-II (VAP-II) methodology. *J Lipid Res* 1997; 38: 2353–2364.
42. Tall AR, Blum CB, Forester GP, Nelson CA. Changes in the distribution and composition of plasma high density lipoproteins after ingestion of fat. *J Biol Chem* 1982; 257: 198–207.
43. Guerin M, Lassel TS, Le Goff W, Farnier M, Chapman MJ. Action of atorvastatin in combined hyperlipidemia: preferential reduction of cholesteryl ester transfer from HDL to VLDL1 particles. *Arterioscler Thromb Vasc Biol* 2000; 20: 189–197.
44. Blanche PJ, Gong EL, Forte TM, Nichols AV. Characterization of human high-density lipoproteins by gradient gel electrophoresis. *Biochim Biophys Acta* 1981; 665: 408–419.
45. Anderson DW, Nichols AV, Pan SS, Lindgren FT. High density lipoprotein distribution. Resolution and determination of three major components in a normal population sample. *Atherosclerosis* 1978; 29: 161–179.
46. Toledo-Ibelles P, Garcia-Sanchez C, Avila-Vazzini N, Carreon-Torres E, Posadas-Romero C, Vargas-Alarcon G, Perez-Mendez O. Enzymatic assessment of cholesterol on electrophoresis gels for estimating HDL size distribution and plasma concentrations of HDL subclasses. *J Lipid Res* 2010; 51: 1610–1617.
47. Yang DH, Konigshofer Y, Dike J, Chang E, Mueller O. Performance characteristics of a microfluidic Lab-on-Chip electrophoresis system for high-density lipoprotein (HDL) subfraction separation and measurement. *Clin Chem Lab Med* 2010; 48: 883–889.
48. Atmeh RF, Kasasbeh AZ, Abu Odeh MR. Distribution of apolipoprotein AI-containing lipoprotein subclasses in plasma of normolipidemic subjects. *Acta Biochim Pol* 2010; 57: 285–298.
49. Asztalos BF, Sloop CH, Wong L, Roheim PS. Two-dimensional electrophoresis of plasma lipoproteins: recognition of new apo A-I-containing subpopulations. *Biochim Biophys Acta* 1993; 1169: 291–300.
50. Kunitake ST, La Sala KJ, Kane JP. Apolipoprotein A-I-containing lipoproteins with pre-beta electrophoretic mobility. *J Lipid Res* 1985; 26: 549–555.
51. Davidson WS, Sparks DL, Lund-Katz S, Phillips MC. The molecular basis for the difference in charge between pre-beta- and alpha-migrating high density lipoproteins. *J Biol Chem* 1994; 269: 8959–8965.
52. Caulfield MP, Li S, Lee G, Blanche PJ, Salameh WA, Benner WH, Reitz RE, Krauss RM. Direct determination of lipoprotein particle sizes and concentrations by ion mobility analysis. *Clin Chem* 2008; 54: 1307–1316.
53. Otvos JD. Measurement of lipoprotein subclass profiles by nuclear magnetic resonance spectroscopy. *Clin Lab* 2002; 48: 171–180.
54. Hirowatari Y, Tsunoda Y, Ogura Y, Homma Y. Analyzing of high-density lipoprotein subfractions and low-density lipoprotein subfractions in human serum with anion-exchange chromatography. *Atherosclerosis* 2009; 204: e52–e57.
55. Usui S, Nakamura M, Jitsukata K, Nara M, Hosaki S, Okazaki M. Assessment of between-instrument variations in a HPLC method for serum lipoproteins and its traceability to reference methods for total cholesterol and HDL-cholesterol. *Clin Chem* 2000; 46: 63–72.

56. Atmeh RF. Isolation and identification of HDL particles of low molecular weight. *J Lipid Res* 1990; 31: 1771–1780.
57. Atmeh RF, Abd Elrazeq IO. Small high density lipoprotein subclasses: some of their physico-chemical properties and stability in solution. *Acta Biochim Pol* 2005; 52: 515–525.
58. Atmeh RF, Kana'an BM, Massad TT. Isolation of high density lipoprotein subclasses by electrofiltration and their chemical components. *Prep Biochem Biotechnol* 2009; 39: 248–265.
59. Nanjee MN, Brinton EA. Very small apolipoprotein A-I-containing particles from human plasma: isolation and quantification by high-performance size-exclusion chromatography. *Clin Chem* 2000; 46: 207–223.
60. Hoofnagle AN, Heinecke JW. Lipoproteomics: using mass spectrometry-based proteomics to explore the assembly, structure, and function of lipoproteins. *J Lipid Res* 2009; 50: 1967–1975.
61. Alaupovic P. Apolipoproteins and lipoproteins. *Atherosclerosis* 1971; 13: 141–146.
62. Alaupovic P, Lee DM, McConathy WJ. Studies on the composition and structure of plasma lipoproteins. Distribution of lipoprotein families in major density classes of normal human plasma lipoproteins. *Biochim Biophys Acta* 1972; 260: 689–707.
63. Alaupovic P. The concept of apolipoprotein-defined lipoprotein families and its clinical significance. *Curr Atheroscler Rep* 2003; 5: 459–467.
64. Duriez P, Fruchart JC. High-density lipoprotein subclasses and apolipoprotein A-I. *Clin Chim Acta* 1999; 286: 97–114.
65. James RW, Hochstrasser D, Tissot JD, Funk M, Appel R, Barja F, Pellegrini C, Muller AF, Pometta D. Protein heterogeneity of lipoprotein particles containing apolipoprotein A-I without apolipoprotein A-II and apolipoprotein A-I with apolipoprotein A-II isolated from human plasma. *J Lipid Res* 1988; 29: 1557–1571.
66. Wroblewska M, Kortas-Stempak B, Szutowicz A, Badzio T. Phospholipids mediated conversion of HDLs generates specific apoA-II pre-beta mobility particles. *J Lipid Res* 2009; 50: 667–675.
67. Moren X, Deakin S, Liu ML, Taskinen MR, James RW. HDL subfraction distribution of paraoxonase-1 and its relevance to enzyme activity and resistance to oxidative stress. *J Lipid Res* 2008; 49: 1246–1253.
68. Gauthamadasa K, Rosales C, Pownall HJ, Macha S, Jerome WG, Huang R, Silva RA. Speciated human high-density lipoprotein protein proximity profiles. *Biochemistry* 2010; 49: 10656–10665.
69. Hatters DM, Peters-Libeu CA, Weisgraber KH. Apolipoprotein E structure: insights into function. *Trends Biochem Sci* 2006; 31: 445–454.
70. Nagasaka H, Miida T, Hirano K, Ota A, Yorifuji T, Takatani T, Tsukahara H, Takayanagi M, Hui SP, Kobayashi K, Chiba H. Reduced apolipoprotein E-rich high-density lipoprotein level at birth is restored to the normal range in patients with familial hypercholesterolemia in the first year of life. *J Clin Endocrinol Metab* 2008; 93: 779–783.
71. Zhang X, Wang J, Chen B. Effect *in vitro* of apolipoprotein A-V on the structure and functions of recombinant high density lipoprotein. *Biochem Biophys Res Commun* 2010; 401: 473–479.
72. Dahlback B, Nielsen LB. Apolipoprotein M affecting lipid metabolism or just catching a ride with lipoproteins in the circulation? *Cell Mol Life Sci* 2009; 66: 559–564.

73. Kontush A, Chapman MJ. Antiatherogenic function of HDL particle subpopulations: focus on antioxidative activities. *Curr Opin Lipidol* 2010; 21: 312–318.

74. Hajduk SL, Moore DR, Vasudevacharya J, Siqueira H, Torri AF, Tytler EM, Esko JD. Lysis of Trypanosoma brucei by a toxic subspecies of human high density lipoprotein. *J Biol Chem* 1989; 264: 5210–5217.

75. He Y, Greene DJ, Kinter M, Morton RE. Control of cholesteryl ester transfer protein activity by sequestration of lipid transfer inhibitor protein in an inactive complex. *J Lipid Res* 2008; 49: 1529–1537.

76. Lagor WR, Brown RJ, Toh S-A, Millar JS, Fuki IV, de la Llera-Moya M, Yuen T, Rothblat G, Billheimer JT, Rader DJ. Overexpression of apolipoprotein F reduces HDL cholesterol levels *in vivo*. *Arterioscler Thromb Vasc Biol* 2009; 29: 40–46.

77. de Silva HV, Stuart WD, Duvic CR, Wetterau JR, Ray MJ, Ferguson DG, Albers HW, Smith WR, Harmony JA. A 70-kDa apolipoprotein designated ApoJ is a marker for subclasses of human plasma high density lipoproteins. *J Biol Chem* 1990; 265: 13240–13247.

78. Bergmeier C, Siekmeier R, Gross W. Distribution spectrum of paraoxonase activity in HDL fractions. *Clin Chem* 2004; 50: 2309–2315.

79. Ohta T, Fidge NH, Nestel PJ. Studies on the *in vivo* and *in vitro* distribution of apolipoprotein A-IV in human plasma and lymph. *J Clin Invest* 1985; 76: 1252–1260.

80. Bisgaier CL, Sachdev OP, Megna L, Glickman RM. Distribution of apolipoprotein A-IV in human plasma. *J Lipid Res* 1985; 26: 11–25.

81. Wolfrum C, Poy MN, Stoffel M. Apolipoprotein M is required for prebeta-HDL formation and cholesterol efflux to HDL and protects against atherosclerosis. *Nat Med* 2005; 11: 418–422.

82. Campos E, McConathy WJ. Distribution of lipids and apolipoproteins in human plasma by vertical spin ultracentrifugation. *Arch Biochem Biophys* 1986; 249: 455–463.

83. Benditt EP, Eriksen N. Amyloid protein SAA is associated with high density lipoprotein from human serum. *Proc Natl Acad Sci U S A* 1977; 74: 4025–4028.

84. Coetzee GA, Strachan AF, van der Westhuyzen DR, Hoppe HC, Jeenah MS, de Beer FC. Serum amyloid A-containing human high density lipoprotein 3. Density, size, and apolipoprotein composition. *J Biol Chem* 1986; 261: 9644–9651.

85. Marcel YL, McPherson R, Hogue M, Czarnecka H, Zawadzki Z, Weech PK, Whitlock ME, Tall AR, Milne RW. Distribution and concentration of cholesteryl ester transfer protein in plasma of normolipemic subjects. *J Clin Invest* 1990; 85: 10–17.

86. Schaefer EJ, Foster DM, Jenkins LL, Lindgren FT, Berman M, Levy RI, Brewer HB, Jr. The composition and metabolism of high density lipoprotein subfractions. *Lipids* 1979; 14: 511–522.

87. Cheung MC, Albers JJ. Distribution of high density lipoprotein particles with different apoprotein composition: particles with A-I and A-II and particles with A-I but no A-II. *J Lipid Res* 1982; 23: 747–753.

88. Schaefer EJ, Asztalos BF. Increasing high-density lipoprotein cholesterol, inhibition of cholesteryl ester transfer protein, and heart disease risk reduction. *Am J Cardiol* 2007; 100: S25–S31.

89. Gordon SM, Deng J, Lu LJ, Davidson WS. Proteomic characterization of human plasma high density lipoprotein fractionated by gel filtration chromatography. *J Proteome Res* 2010; 9: 5239–5249.

90. Shiflett AM, Bishop JR, Pahwa A, Hajduk SL. Human high density lipoproteins are platforms for the assembly of multi-component innate immune complexes. *J Biol Chem* 2005; 280: 32578–32585.

91. Nguyen D, Dhanasekaran P, Phillips MC, Lund-Katz S. Molecular mechanism of apolipoprotein E binding to lipoprotein particles. *Biochemistry* 2009; 48: 3025–3032.

92. Cheung MC, Vaisar T, Han X, Heinecke JW, Albers JJ. Phospholipid transfer protein in human plasma associates with proteins linked to immunity and inflammation. *Biochemistry* 2010; 49: 7314–7322.

93. Robciuc MR, Metso J, Sima A, Ehnholm C, Jauhiainen M. Human apoA-I increases macrophage foam cell derived PLTP activity without affecting the PLTP mass. *Lipids Health Dis* 2010; 9: 59.

94. Collins LA, Olivier M. Quantitative comparison of lipoprotein fractions derived from human plasma and serum by liquid chromatography-tandem mass spectrometry. *Proteome Sci* 2010; 8: 42.

95. Collins LA, Mirza SP, Kissebah AH, Olivier M. Integrated approach for the comprehensive characterization of lipoproteins from human plasma using FPLC and nano-HPLC-tandem mass spectrometry. *Physiol Genomics* 2010; 40: 208–215.

96. Vaisar T, Pennathur S, Green PS, Gharib SA, Hoofnagle AN, Cheung MC, Byun J, Vuletic S., Kassim, S., Singh, P. et al. Shotgun proteomics implicates protease inhibition and complement activation in the antiinflammatory properties of HDL. *J Clin Invest* 2007; 117: 746–756.

97. Nilsson A, Duan RD. Absorption and lipoprotein transport of sphingomyelin. *J Lipid Res* 2006; 47: 154–171.

98. Duong PT, Collins HL, Nickel M, Lund-Katz S, Rothblat GH, Phillips MC. Characterization of nascent HDL particles and microparticles formed by ABCA1-mediated efflux of cellular lipids to apoA-I. *J Lipid Res* 2006; 47: 832–843.

99. Nakamura Y, Kotite L, Gan Y, Spencer TA, Fielding CJ, Fielding PE. Molecular mechanism of reverse cholesterol transport: reaction of pre-beta-migrating high-density lipoprotein with plasma lecithin/cholesterol acyltransferase. *Biochemistry* 2004; 43: 14811–14820.

100. Subbaiah PV, Liu M. Role of sphingomyelin in the regulation of cholesterol esterification in the plasma lipoproteins. Inhibition of lecithin-cholesterol acyltransferase reaction. *J Biol Chem* 1993; 268: 20156–20163.

101. Bolin DJ, Jonas A. Sphingomyelin inhibits the lecithin-cholesterol acyltransferase reaction with reconstituted high density lipoproteins by decreasing enzyme binding. *J Biol Chem* 1996; 271: 19152–19158.

102. Joy T, Hegele RA. Is raising HDL a futile strategy for atheroprotection? *Nat Rev Drug Discov* 2008; 7: 143–155.

103. de Souza JA, Vindis C, Negre-Salvayre A, Rye KA, Couturier M, Therond P, Chantepie S, Salvayre R, Chapman MJ, Kontush A. Small, dense HDL 3 particles attenuate apoptosis in endothelial cells: pivotal role of apolipoprotein A-I. *J Cell Mol Med* 2010; 14: 608–620.

104. de Souza JA, Vindis C, Hansel B, Negre-Salvayre A, Therond P, Serrano CV Jr, Chantepie S, Salvayre R, Bruckert, E., Chapman MJ et al. Metabolic syndrome features small, apolipoprotein A-I-poor, triglyceride-rich HDL3 particles with defective anti-apoptotic activity. *Atherosclerosis* 2008; 197: 84–94.

105. Kontush A, de Faria EC, Chantepie S, Chapman MJ. Antioxidative activity of HDL particle subspecies is impaired in hyperalphalipoproteinemia: relevance of enzymatic and physicochemical properties. *Arterioscler Thromb Vasc Biol* 2004; 24: 526–533.

106. Kontush A, de Faria EC, Chantepie S, Chapman MJ. A normotriglyceridemic, low HDL-cholesterol phenotype is characterised by elevated oxidative stress and HDL particles with attenuated antioxidative activity. *Atherosclerosis* 2005; 182: 277–285.

107. Asztalos BF, Cupples LA, Demissie S, Horvath KV, Cox CE, Batista MC, Schaefer EJ. High-density lipoprotein subpopulation profile and coronary heart disease prevalence in male participants of the Framingham Offspring Study. *Arterioscler Thromb Vasc Biol* 2004; 24: 2181–2187.

CHAPTER 3

STRUCTURE

The structure of HDL particles is largely determined by the properties of their exchangeable apolipoprotein components [1]. Apolipoproteins belong to the family of amphipathic proteins whose folding is governed by thermodynamic constraints aimed at the minimization of aqueous exposure of hydrophobic surfaces. ApoA-I is the major structural component of HDL; the predominant part of the apoA-I molecule, with the exception of the N-terminus of 44 amino acids, is organized into 8 alpha-helical amphipathic domains of 22 amino acids and 2 repeats of 11 amino acids. In a similar fashion, apoA-II and apoE, two other key HDL apolipoproteins, also contain a series of amphipathic alpha-helical domains.

The structural organization of HDL varies significantly as a function of HDL composition. From the structural point of view, three major classes of HDL particles can be distinguished: lipid-free apoA-I, discoid lipid-poor HDL and spherical lipid-rich HDL (see also Chapter 2).

3.1 LIPID-FREE APOA-I

Lipid-free apoA-I is highly solvent-exposed. In a lipid-free state, non-polar surfaces of alpha-helical segments of apoA-I are, therefore, sequestered within the protein, while polar surfaces are exposed to the aqueous phase, displaying an array of positively- and negatively-charged areas [2, 3]. Available data reveal that the alpha-helices located in the N-terminal and central regions adopt a

High-Density Lipoproteins: Structure, Metabolism, Function, and Therapeutics, First Edition.
Anatol Kontush and M. John Chapman.

helix bundle structure, whilst the C-terminus forms a separately folded, and less well-organized, domain.

Indeed, the crystal structure of full-length, lipid-free human apoA-I determined at 2.4 Å resolution reveals that the protein crystallized in the presence of chromium tris-acetylacetonate contains an **N-terminal four-helix bundle** [2] (Fig. 3.1). Such intramolecular organization strongly resembles the four-helix bundle structure of the LDL receptor-binding domain of lipid-free apoE [4]; however, the structure of apoA-I is less stable as a result of a lower number of intramolecular hydrogen bonds. Electron paramagnetic resonance (EPR) analysis of the structure of the first 98 N-terminal residues of lipid-free apoA-I also reveals the presence of short, helical fragments with periodic breaks [5] (Fig. 3.1). Intriguingly, EPR data provide evidence for a short beta-strand between amino acid residues 20 and 25; therefore, the resulting structure of the N-terminal domain provided by EPR differs slightly from that suggested by apoA-I crystallization.

X-ray data reveal that the C-terminus of lipid-free apoA-I contains two alpha-helices and forms a hairpin-like domain that interacts with other molecules of apoA-I [2]. The existence of **C-terminal helical fragments** is also supported by EPR; however, the helicity of C-terminal residues is significantly lower when measured by the latter approach [6]. In addition, EPR reveals the presence of another short beta-strand in the C-terminal region which does not appear in the crystal structure [5].

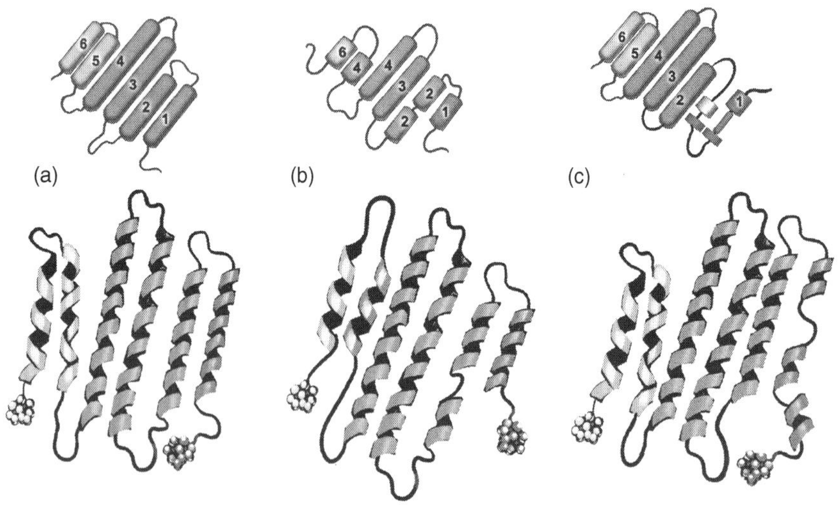

Figure 3.1 Structure of lipid-free apoA-I (modified from Reference [3]). (**a**) Crystal structure of lipid-free apoA-I by x-ray scattering according to Reference [2]. (**b**) Model of lipid-free apoA-I by cross-linking and sequence threading techniques according to Reference [7]. (**c**) Model of lipid-free apoA-I by EPR according to Reference [5]. Red, N-terminal 44 residues; yellow, C-terminal 50 residues; space-filled, extreme terminal amino acids. Insets, elements of secondary structure laid flat with no tertiary organization (see color plate section).

Importantly, the structure of apoA-I obtained using x-ray scattering is consistent with the model of monomeric lipid-free apoA-I obtained using a combined approach of cross-linking chemistry and high-resolution mass spectrometry [7] (Fig. 3.1). The cross-linking model involves a helical bundle structure in which the N- and C-termini are in close proximity, with the N-terminus forming a four-helix bundle. While the C-terminal helices exhibit significant hydrophobic contacts, the N-terminal portion is much less involved in hydrophobic interactions. These properties result in a strong tendency of lipid-free apoA-I to self-aggregate; cross-linking data indicate that the self-association occurs via the interaction of the C- and N-termini of the protein [7]. As a consequence, monomeric lipid-free apoA-I is unstable and readily forms a heterogeneous population of oligomers containing up to five molecules of apoA-I. In such structures, apoA-I may display a molten globule state characterized by the presence of regions with a defined secondary structure but that lack tertiary structure [8].

Interestingly, both x-ray scattering and cross-linking data show that the helices of the bundle are typically, but not always, **punctuated by the proline (Pro) residues**. Consistent with these findings, the crystal structure of a deletion mutant of apoA-I lacking the first N-terminal 43 amino acids determined at 4 Å resolution reveals the presence of an amphipathic alpha-helix punctuated by kinks at regularly spaced Pro residues [9]. However, as the N-terminal domain plays a key role in maintaining the conformation of lipid-free apoA-I, deletion mutants in this region may not be fully relevant for studying the structure of apoA-I. Sequence alignment on apoA-I sequences from 31 species of animals further supports the critical role of Pro residues as three of them (P66, P121 and P220) are completely conserved across the 31 species [10]. Furthermore, there is specific conservation of salt bridge-forming residues in the first 30 residues of apoA-I and general conservation of a variety of residue types in the central domain, helix 2/3 to helix 7/8 [10].

The precise localization of helical segments in the molecule of monomeric lipid-free apoA-I using hydrogen-deuterium exchange and mass spectrometry suggests that alpha-helices are formed by residues 7–44, 54–65, 70–78, 81–115 and 147–178 [11]. Similarly, the crystal structure approach shows that the first N-terminal 44 residues are predominantly helical [2], whereas cross-linking data support the random coil conformation of this region. In a similar fashion, the latter two techniques disagree regarding the helicity of the C-terminus [7]. In general, x-ray scattering estimates helicity of crystallized apoA-I at about 83%, a much higher value than the 50%–60% typically observed for lipid-free apoA-I in solution using other techniques, including circular dichroism [7, 11].

Such discrepancies probably result from the highly dynamic structure of lipid-free apoA-I in solution. Whereas the crystal structure reflects a static conformation of apoA-I corresponding to requirements of minimal energy, conditions of a physiologic solution are characterized by the high conformational dynamics of the protein. Combined, the crystal structure, cross-linking and EPR data provide strong evidence for the presence of the four-helical bundle in the N-terminal part of lipid-free apoA-I; however, such a helical structure is highly dynamic, allowing rapid apoA-I re-organization [11]. In addition, the helices interact with each other, mutually stabilizing the helix bundle.

These data allow us to hypothesize that, as proposed by Davidson and Thompson, "the dynamic nature of apoA-I makes the expectation of a single 'perfect' structure unrealistic" [3]. Therefore, in aqueous solutions, lipid-free monomeric apoA-I most probably exists as a dynamic super-position of multiple interconverting conformations that follow the basic principles of apoA-I organization (Fig. 3.1).

3.2 DISCOID HDL

The major HDL apolipoproteins, apoA-I, apoA-II and apoE, are flexible, lipid-scavenging proteins capable of incorporating phospholipid and cholesterol into stable particles. These apolipoproteins possess detergent-like properties and efficiently solubilize phospholipid to create small discoid HDL [3]. Such particles can be readily reconstituted *in vitro* upon co-incubation of an apolipoprotein with phospholipid. This approach leads to the formation of heterogeneous mixtures of particles that differ in size and protein-to-lipid ratio; the best known example is a 9.6 nm HDL containing two molecules of apoA-I and 150 molecules of phospholipid [12]. Such rHDL can be visualized using electron microscopy or cryotomography techniques which reveal a discoid particle shape [13]. In contrast, the isolation of discoid HDL from plasma is difficult as these particles are present in the circulation at low concentrations and are highly unstable.

In the presence of lipid, apoA-I adopts a conformation that allows hydrophobic surfaces of the protein to contact aliphatic regions of lipid molecules. As a result, apoA-I becomes folded in such a fashion that the hydrophobic surfaces of each amphipathic domain interact with lipid components, while polar faces are exposed to the aqueous environment [3]. Lipid binding is initiated by the hydrophobic C-terminal domain of apoA-I, which induces conformational changes involving an increase in the alpha-helix content and the subsequent opening of the N-terminal helix bundle. The N-terminal hydrophobic patch of apoA-I composed of leucine residues 42, 44, 46, and 47 appears to play an important role in apoA-I binding to lipids [14].

In discoid particles, the existence of a particle 'edge' constrains apoA-I to fold into a distinct conformation. In small discoid HDL, two ring-shaped molecules of apoA-I adopt a 'double belt' orientation with their helices oriented parallel to the plane of the disc and perpendicular to the lipid acyl chains in such a way that they wrap around the lipid bi-layer disc forming two stacked rings in an antiparallel orientation [15–17] (Fig. 3.2). The **double belt model** was proposed by Jere Segrest in the 1970s [18] and has become widely accepted over the last decade. Indeed, the majority of experimental and theoretical data obtained using crystal structure [9], infrared spectroscopy [19] or fluorescence spectroscopy [20] support the original hypothesis of Segrest. The belt formed by apoA-I molecules around reconstituted discoid HDL can be directly observed using cryotomography [13]. Moreover, apoE [21, 22] and apoA-II [23] were also shown to adopt a belt-like conformation in small HDL; these apolipoproteins exist on HDL in the form of dimers.

DEFINITIONS

Double belt model: a widely accepted model of apoA-I organization in discoid HDL, according to which two ring-shaped molecules of apoA-I adopt a 'double belt' orientation with their helices oriented parallel to the plane of the disc and perpendicular to the lipid acyl chains in such a way that they wrap around the lipid bi-layer disc forming two stacked rings in an antiparallel orientation.

Hinge domain: a flexible **domain** of apoA-I that can associate with and dissociate from the HDL surface as the lipid content and diameter of HDL fluctuates.

Trefoil model: a model of apoA-I organization in spherical HDL, according to which all apoA-I molecules are bent to allow their arrangement in identical double belt structures. In the case of 3 mol apoA-I/mol HDL, each molecule is bent 60° with inflection points in helices 5 and 10, forming a trefoil.

The conformation of lipid-bound apoA-I is stabilized by salt bridge interactions between apoA-I molecules; direct intermolecular contacts involving N- and C-termini may be particularly important in this aspect [24, 27]. Detailed assessment of the double belt model suggests that the N- and C-terminal ends are in close contact; such a concept is termed the **belt buckle model**. In addition, both N- and C-terminal fragments can interact with central regions of apoA-I, folding back onto the latter [27]. An alternative view involves the interaction of the N-termini of two molecules of apoA-I forming of a globular domain; this model also implies interactions between two C-termini [28].

Computer analysis predicts that in the double belt configuration, helices 5 of each apoA-I molecule lie in direct opposition (5/5 orientation) [15] (Fig. 3.2). Such registry is supported by fluorescence resonance energy transfer (FRET) [25] and cross-linking [17] studies. Furthermore, apoA-I molecules appear to slide in relation to each other between two stable rotamer conformations, the **5/5 and 5/2 registries** [17] (Fig. 3.2).

Alternative models of apoA-I organization in discoid HDL include the 'picket fence' [29], the 'hairpin' [30] and the 'double super helix' [31]. The **hairpin model** is a modification of the double belt model that suggests that two apoA-I

Figure 3.2 Double belt, hairpin and looped belt models of discoid HDL (modified from Reference [3]). Structures of two molecules of apoA-I in rHDL particles which contain a phospholipid bilayer are shown. (**a**) The 5/5 double belt model built using the 1–43 deletion mutant of apoA-I according to Reference [15]. (**b**) The 5/2 rotamer of the 5/5 double belt model according to Reference [17]. (**c**) The hairpin model according to Reference [24]. Two molecules of apoA-I are shown in different colours. (**d**) Looped belt model of discoid HDL (modified from Reference [25]). Two molecules of apoA-I form two belts aligned antiparallel with a central loop-like structure centered at position 139 and bordered by residues 133 and 146 (see color plate section).

molecules adopt hairpin orientations in such a manner that each molecule interacts with both phospholipid leaflets after a turn [30]. The hairpin turn located in the N-terminus of one apoA-I molecule lies around amino acid residue 44; this domain interacts with the hairpin located in the C-terminal part of the second apoA-I molecule [24] (Fig. 3.2). In contrast to the double belt model, the hairpin model proposes an open-ended structure of apoA-I instead of a closed loop. The hairpins are thought to be stabilized by intramolecular salt bridge interactions; FRET data suggest a random mixture of their head-to-head and head-to-tail orientations [32]. The similarity of the hairpin and the original double belt models makes distinguishing between them challenging.

The **picket fence model** proposes that the 22-amino acid helical repeats wrap around the phospholipid bi-layer in such a way that their axes are parallel to the phospholipid acyl residues. As for the double belt model, the picket fence hypothesis holds that alpha-helices are punctuated by Pro residues [29]. This type of apoA-I organization has, however, been ruled out by data obtained in FRET studies [33].

The **double super helix model** suggests that two molecules of apoA-I form an antiparallel double super helix wrapped around an ellipsoidal lipid phase [31]. The model has been constructed using a 10 Å resolution small angle neutron scattering approach with contrast variation and selective isotopic deuteration applied to 9.6 nm reconstituted discoid HDL. Such low-resolution overall time-averaged conformation of apoA-I reveals an open helical shape; lipids are organized as an ellipsoid in the centre of the helix. An all-atom computational model favors the antiparallel helical orientation of two apoA-I molecules over the conventional double belt model [31]. The major difference between the double super helix and the double belt models is that the N- termini and C-termini of each apoA-I molecule are far apart in the former but close together in the latter. However, cross-linking data demonstrate the existence of N-terminus to N-terminus, as well as C-terminus to C-terminus, apoA-I cross-links [3, 17]. The validity of the double super helix model is a matter of ongoing discussion [34, 35].

Another prominent feature of apoA-I structure in discoid HDL is the presence of a flexible **hinge domain** that can associate and dissociate from the HDL surface as the lipid content and diameter of HDL fluctuate [36]. Fluorescence spectroscopy suggests that helices 5, 6 and 7 (amino acid residues 130 to 174) form such a dynamic loop in the apoA-I molecule [36], whereas EPR data place the hinge domain between residues 133 and 146 [25]. In such a **looped belt model**, one molecule of apoA-I aligns with its counterpart in a 5/5 registry centered at position 139 (Fig. 3.2). The inter-helical segment of apoA-I located in the N-terminus between residues 35 and 49 may also play a role in the adaptation of apoA-I to HDL particle size [12].

METHODOLOGIC NOTE

Structural measurements performed at low spatial resolution should not be employed to build high-resolution models.

The concept of a flexible loop domain of apoA-I extending into the aqueous phase forms the basis for yet another model of HDL structure termed the **solar-flares model**. According to this concept, hydrogen-deuterium exchange mass spectrometry reveals two apoA-I molecules arranged in an antiparallel double belt structure, with residues 159–180 of each molecule 'flaring out' into solution and forming a protruding solvent-exposed loop [28]. The stability of such a solvent-exposed structure may, however, be low as a result of the unfavorable energetic orientation of the hydrophobic protein surface toward the aqueous solvent [37].

An alternative hypothesis to the hinged domain model holds that lipid-free apoA-I molecules capture phospholipid within their highly flexible lipid pockets to produce particles of increasing size [38]. Molecular dynamics simulations on a series of progressively smaller discoid HDL particles produced by the incremental removal of phospholipid results in double belt structures of apoA-I embedded into non-planar, twisted and **ellipsoidal saddle-shaped particles** whose size decreases with a decrease in phospholipid [38]. Discoid HDL assembly begins with the formation of discoid particles which contain apoA-I molecules wrapped around a planar lipid bi-layer [39]. Smaller particles are composed of a minimal surface lipid structure and contain only elements of a lipid bi-layer. Further removal of phospholipid results in a particle family that displays a lipid pocket capable of binding a dynamic range of phospholipid molecules. Such conversion from a discoid to a saddle-shaped particle involves loss of helicity by apoA-I and the formation of loops in opposing antiparallel parts of the double belt with the incremental folding of initially round double belt structures into twisted saddle-shaped apoA-I. Kinetic analyses of reconstituted particles demonstrate that phospholipid-rich particles exist in discrete sizes corresponding to local energetic minima. This theoretical conclusion is supported by experimental data on particle size/shape and apoA-I helicity [39].

This analysis shows that the structure adopted by lipid components of discoid HDL is largely determined by apolipoproteins that build amphipathic protein scaffolding to organize lipids. As a result of the presence of apoA-I, the lipid phase of discoid HDL displays significant structural heterogeneity. Fluorescence studies reveal three regions that differ in packing order: (i) tightly packed lipid close to the disc axis; (ii) intermediately packed lipid; and, (iii) disordered lipid at the disc edge [40].

As can be expected from data obtained in lipid bi-layers, the structure of the lipid phase of HDL is also sensitive to its lipid composition. Indeed, increasing HDL content of sphingomyelin decreases fluidity of the lipid phase; a similar effect is observed upon increasing the level of saturation of HDL fatty acids [41]. Similar to observations made in lipid membranes, cholesterol affects the structure of discoid HDL particles in a dual way, stabilizing relatively unstable HDL complexes by enhancing favorable packing interactions but de-stabilizing more stable particles where such interactions are already highly optimized [42]. In addition, the varying HDL content of free cholesterol exerts similarly dual effects on lipid phase fluidity [42].

Together, the available data indicate that two molecules of apoA-I most probably adopt a conformation described by the 5/5 anti-parallel double belt model in reconstituted discoid HDL (Fig. 3.2). ApoA-I, however, displays considerable conformational dynamics, particularly at the N- and C-termini, allowing transitions between other conformational states, such as the 5/2 registry. In addition, the two apoA-I chains do not wrap around the lipid particle to give a smooth belt, but rather form a flexible hinge domain whose size depends on HDL particle diameter, or twist to create saddle-shaped particles.

MAJOR CONTROVERSY

Details of the anti-parallel organization of two apoA-I molecules in discoid HDL significantly differ between existing models (cf. the classical double belt, double super helix and solar flares models).

3.3 SPHERICAL HDL

Spherical HDLs predominate in human plasma. In such particles, amphipathic helices of apoA-I are located at the HDL surface, with their hydrophilic faces interacting with the aqueous phase and hydrophobic faces interacting with lipids [3]. The structure of apoA-I is less constrained compared with discoid

HDL as there is no particle 'edge'. The structural organization of apoA-I is, however, similar overall between spherical and discoid HDL.

Indeed, apoA-I adopts a **double belt** conformation in both discoid and spherical rHDL of comparable size as determined by cross-linking chemistry and mass spectrometry [26]. Unexpectedly, such conformation of apoA-I is equally independent of the number of apoA-I molecules present in each HDL particle. Moreover, a similar cross-linking pattern is observed in HDL particles isolated from human serum. These findings suggest a general model of apoA-I organization in spherical HDL which assumes that all apoA-I molecules are bent to allow their arrangement in identical double belt structures [26]. In the case of 3 mol apoA-I/mol HDL (the **trefoil model**), each molecule is bent 60° with inflection points in helices 5 and 10, forming a trefoil and breaking the phospholipid surface into three equal lunes of 120° of the sphere surface (Fig. 3.3). In case of more than three (four or five) apoA-I molecules per HDL, such as in large plasma HDL subpopulations, double belts of the molecules can be twisted under different angles to adapt to the available amount of lipid [43]; as a result, apoA-I adopts a symmetrical, cage-like structure. Resulting tetra- and pentafoil structures suggest that apoA-I twisting may account for the decrease in HDL diameter with increasing HDL density at a constant number of apoA-I molecules per HDL, such as observed in small, dense plasma HDL subpopulations [43]. Recently, an alternative model of apoA-I organization in spherical HDL has been proposed on the basis of small angle neutron scattering, which suggests an arrangement of three apoA-I molecules within a low-resolution helical dimeric structure with a hairpin [44].

CRITICAL CONTRIBUTIONS

Double belt model of apoA-I organization in discoid HDL [15–17].

Trefoil model of apoA-I organization in spherical HDL [26].

The trefoil model is fully compatible with dynamic features of HDL, including hinge domains, looped regions and 'solar flares' [26]. Indeed, ApoA-I reveals significant conformational lability in spherical HDL. Molecular dynamics simulations suggest that, during maturation of discoid HDL into spherical particles, N- and C-terminal helices of apoA-I remain flexible, while central regions are stabilized by salt bridges [46]. The N-terminal domain may even dissociate from the lipid surface at high surface concentrations of apoA-I, whereas the C-terminal domain remains attached, as suggested by binding isotherms of apoA-I to small phospholipid vesicles [47].

Similarity in the general structural organization of apoA-I between discoid and spherical HDL is further supported by data on crystal structure, circular dichroism and fluorescence studies [9, 48], as well as by molecular dynamics simulations [46]. However, significant details of apoA-I conformation vary with HDL shape and size. Indeed, interactions between apoA-I molecules are reduced

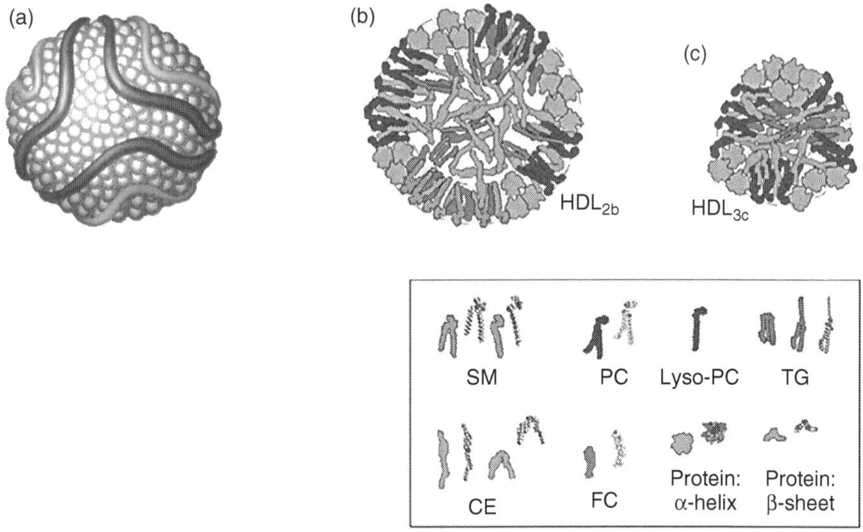

Figure 3.3 Structural organization of spherical HDL. (**a**) Trefoil model of the arrangement of three apoA-I molecules in large spherical HDL (modified from Reference [26]). The model was generated by splitting the right side of two molecules of apoA-I in the double belt 60° out of the plane of the disc; a third molecule, bent the same way, was subsequently inserted. Different colours depict individual molecules of apoA-I. (**b, c**) Arrangement of lipid molecules at the surface and in the core of large, light HDL2b (**b**) and small, dense HDL3c (**c**) particles (modified from Reference [45]) (see color plate section). CE, cholesteryl ester; FC, free cholesterol; lysoPC, lysophosphatidylcholine; PC, phosphatidylcholine; TG, triglyceride; SM, sphingomyelin.

in spherical relative to discoid HDL, presumably as a result of less constrained apoA-I conformation in spherical particles [49]. NMR spectroscopy also reveals that discoid and spherical HDL differ in the conformation of the N-terminal fragment of apoA-I [50]. HDL size also impacts on the reactivity of reconstituted spherical HDL to monoclonal antibodies against apoA-I [51]. Interestingly, such reactivity is strongly affected by the core lipid composition of HDL, implying that apoA-I interacts with core lipids which, in turn, influence its conformation. The influence of surface lipids on apoA-I conformation can be even more pronounced, as exemplified by the induction of the open conformation of the N-terminal helix bundle in the presence of free cholesterol [47]. However, the structural role in spherical HDL of proteins other than apoA-I (primarily apoA-II and apoE) remains largely indeterminate.

OPEN QUESTION

How does the presence of proteins, other than apoA-I, influence the particle structure of HDL?

As in discoid HDL, apoA-I acts as an organizing scaffold that also maintains stable lipid structures in spherical particles. The lipid organization of spherical HDL is commonly viewed as two well-defined regions: a surface monolayer containing phospholipids and free cholesterol, and a hydrophobic core comprised of cholesterol esters and triglycerides (Fig. 1.1). Calculations show that, in spherical particles, the HDL surface is dominated by protein which may cover between 71% of the surface in large, light HDL2b and 87% in small, dense HDL3c [43]. As a result, surface regions of HDL are characterized by a prominent conformational lipid order, whereas the hydrophobic core is much less ordered [52]. The solubility of core lipids in the phospholipid monolayer is low, as shown by molecular dynamics simulations of spherical HDL-sized lipid droplets comprised of POPC and cholesteryl oleate [53]. Cholesteryl ester molecules, therefore, provide only a small contribution to the solvent-accessible surface area of spherical HDL particles [54].

The hydrophobic core of HDL is, however, highly dynamic, as suggested by the broad orientation of sterol moieties of cholesteryl esters, which becomes fully isotropic deep in the core [53]. Therefore, structural defects in the surface monolayer allow the partial penetration of core lipid molecules in the surface monolayer of HDL; these effects are particularly strong in small particles [45] (Fig. 3.3). The penetration of triglyceride and cholesteryl ester molecules into the surface regions occurs mainly via phosphatidylcholine-rich, sphingomyelin- and free cholesterol-poor domains that are loosely packed. Remarkably, in the smallest HDL3c particles which are depleted of both sphingomyelin and free cholesterol [55], virtually all triglyceride and cholesteryl ester molecules penetrate the surface monolayer to some extent. Computational models show that the high propensity of cholesteryl ester molecules to locate in the surface monolayer, together with the high concentration of cholesterol esters in HDL, may result in high surface concentrations of cholesterol esters (from between 13% and 27% of total HDL surface lipids) [45]. Conversely, the low propensity of triglyceride molecules to locate in the surface monolayer, together with a low concentration of triglycerides in HDL, results in low surface concentrations of triglycerides of only up to 5% of total HDL surface lipids [45].

As a result, core lipid molecules reside in contact with surface-located protein components of HDL. Molecular dynamics simulations reveal that cholesteryl esters interact with protein residues mainly through their cholesterol moieties [54]. Such interactions create domains of elevated rigidity, contributing to the decreased mobility of apoA-I and stabilizing the structure of spherical HDL.

IMPORTANT READING

Segrest JP, Harvey SC, Zannis V. Detailed molecular model of apolipoprotein A-I on the surface of high-density lipoproteins and its functional implications. *Trends Cardiovasc Med* 2000; 10: 246–252.

Davidson WS, Thompson TB. The structure of apolipoprotein A-I in high density lipoproteins. *J Biol Chem* 2007, 282, 22249–22253.

Thomas MJ, Bhat S, Sorci-Thomas MG. Three-dimensional models of HDL apoA-I: implications for its assembly and function. *J Lipid Res* 2008, 49, 1875–1883.

REFERENCES

1. Lund-Katz S, Phillips MC. High density lipoprotein structure-function and role in reverse cholesterol transport. *Subcell Biochem* 2010; 51: 183–227.
2. Ajees AA, Anantharamaiah GM, Mishra VK, Hussain MM, Murthy HM. Crystal structure of human apolipoprotein A-I: insights into its protective effect against cardiovascular diseases. *Proc Natl Acad Sci U S A* 2006; 103: 2126–2131.
3. Davidson WS, Thompson TB. The structure of apolipoprotein A-I in high density lipoproteins. *J Biol Chem* 2007; 282: 22249–22253.
4. Wilson C, Wardell MR, Weisgraber KH, Mahley RW, Agard DA. Three-dimensional structure of the LDL receptor-binding domain of human apolipoprotein E. *Science* 1991; 252: 1817–1822.
5. Lagerstedt JO, Budamagunta MS, Oda MN, Voss JC. Electron paramagnetic resonance spectroscopy of site-directed spin labels reveals the structural heterogeneity in the N-terminal domain of apoA-I in solution. *J Biol Chem* 2007; 282: 9143–9149.
6. Oda MN, Forte TM, Ryan RO, Voss JC. The C-terminal domain of apolipoprotein A-I contains a lipid-sensitive conformational trigger. *Nat Struct Biol* 2003; 10: 455–460.
7. Silva RA, Hilliard GM, Fang J, Macha S, Davidson WS. A three-dimensional molecular model of lipid-free apolipoprotein A-I determined by cross-linking/mass spectrometry and sequence threading. *Biochemistry* 2005; 44: 2759–2769.
8. Gursky O, Atkinson D. Thermal unfolding of human high-density apolipoprotein A-1: implications for a lipid-free molten globular state. *Proc Natl Acad Sci U S A* 1996; 93: 2991–2995.
9. Borhani DW, Rogers DP, Engler JA, Brouillette CG. Crystal structure of truncated human apolipoprotein A-I suggests a lipid-bound conformation. *Proc Natl Acad Sci U S A* 1997; 94: 12291–12296.
10. Bashtovyy D, Jones MK, Anantharamaiah GM, Segrest JP. Sequence conservation of apolipoprotein A-I affords novel insights into HDL structure-function. *J Lipid Res* 2011; 52: 435–450.
11. Chetty PS, Mayne L, Lund-Katz S, Stranz D, Englander SW, Phillips MC. Helical structure and stability in human apolipoprotein A-I by hydrogen exchange and mass spectrometry. *Proc Natl Acad Sci U S A* 2009; 106: 19005–19010.
12. Lagerstedt JO, Cavigiolio G, Budamagunta MS, Pagani I, Voss JC, Oda MN. Structure of apolipoprotein A-I N terminus on nascent high density lipoproteins. *J Biol Chem* 2011; 286: 2966–2975.
13. Zhang L, Song J, Cavigiolio G, Ishida BY, Zhang S, Kane JP, Weisgraber KH, Oda MN, Rye K-A, Pownall HJ et al. Morphology and structure of lipoproteins revealed by an optimized negative-staining protocol of electron microscopy. *J Lipid Res* 2011; 52: 175–184.

14. Smith LE, Davidson WS. The role of hydrophobic and negatively charged surface patches of lipid-free apolipoprotein A-I in lipid binding and ABCA1-mediated cholesterol efflux. *Biochim Biophys Acta* 2010; 1801: 64–69.
15. Segrest JP, Jones MK, Klon AE, Sheldahl CJ, Hellinger M, De Loof H, Harvey SC. A detailed molecular belt model for apolipoprotein A-I in discoidal high density lipoprotein. *J Biol Chem* 1999; 274: 31755–31758.
16. Segrest JP, Harvey SC, Zannis V. Detailed molecular model of apolipoprotein A-I on the surface of high-density lipoproteins and its functional implications. *Trends Cardiovasc Med* 2000; 10: 246–252.
17. Silva RA, Hilliard GM, Li L, Segrest JP, Davidson WS. A mass spectrometric determination of the conformation of dimeric apolipoprotein A-I in discoidal high density lipoproteins. *Biochemistry* 2005; 44: 8600–8607.
18. Segrest JP. Amphipathic helices and plasma lipoproteins: thermodynamic and geometric considerations. *Chem Phys Lipids* 1977; 18: 7–22.
19. Koppaka V, Silvestro L, Engler JA, Brouillette CG, Axelsen PH. The structure of human lipoprotein A-I. Evidence for the "belt" model. *J Biol Chem* 1999; 274: 14541–14544.
20. Panagotopulos SE, Horace EM, Maiorano JN, Davidson WS. Apolipoprotein A-I adopts a belt-like orientation in reconstituted high density lipoproteins. *J Biol Chem* 2001; 276: 42965–42970.
21. Narayanaswami V, Maiorano JN, Dhanasekaran P, Ryan RO, Phillips MC, Lund-Katz S, Davidson WS. Helix orientation of the functional domains in apolipoprotein E in discoidal high density lipoprotein particles. *J Biol Chem* 2004; 279: 14273–14279.
22. Schneeweis LA, Koppaka V, Lund-Katz S, Phillips MC, Axelsen PH. Structural analysis of lipoprotein E particles. *Biochemistry* 2005; 44: 12525–12534.
23. Silva RA, Schneeweis LA, Krishnan SC, Zhang X, Axelsen PH, Davidson WS. The structure of apolipoprotein A-II in discoidal high density lipoproteins. *J Biol Chem* 2007; 282: 9713–9721.
24. Bhat S, Sorci-Thomas MG, Alexander ET, Samuel MP, Thomas MJ. Intermolecular contact between globular N-terminal fold and C-terminal domain of ApoA-I stabilizes its lipid-bound conformation: studies employing chemical cross-linking and mass spectrometry. *J Biol Chem* 2005; 280: 33015–33025.
25. Martin DD, Budamagunta MS, Ryan RO, Voss JC, Oda MN. Apolipoprotein A-I assumes a "looped belt" conformation on reconstituted high density lipoprotein. *J Biol Chem* 2006; 281: 20418–20426.
26. Silva RA, Huang R, Morris J, Fang J, Gracheva EO, Ren G, Kontush A, Jerome WG, Rye KA, Davidson WS. Structure of apolipoprotein A-I in spherical high density lipoproteins of different sizes. *Proc Natl Acad Sci U S A* 2008; 105: 12176–12181.
27. Thomas MJ, Bhat S, Sorci-Thomas MG. Three-dimensional models of HDL apoA-I: implications for its assembly and function. *J Lipid Res* 2008; 49: 1875–1883.
28. Wu Z, Wagner MA, Zheng L, Parks JS, Shy JM 3rd, Smith JD, Gogonea V, Hazen SL. The refined structure of nascent HDL reveals a key functional domain for particle maturation and dysfunction. *Nat Struct Mol Biol* 2007; 14: 861–868.
29. Wald JH, Coormaghtigh E, De Meutter J, Ruysschaert JM, Jonas A. Investigation of the lipid domains and apolipoprotein orientation in reconstituted high density lipoproteins by fluorescence and IR methods. *J Biol Chem* 1990; 265: 20044–20050.

30. Brouillette CG, Anantharamaiah GM, Engler JA, Borhani DW. Structural models of human apolipoprotein A-I: a critical analysis and review. *Biochim Biophys Acta* 2001; 1531: 4–46.

31. Wu Z, Gogonea V, Lee X, Wagner MA, Li XM, Huang Y, Undurti A, May RP, Haertlein M, Moulin M et al. Double superhelix model of high density lipoprotein. *J Biol Chem* 2009; 284: 36605–36619.

32. Tricerri MA, Behling Agree AK, Sanchez SA, Bronski J, Jonas A. Arrangement of apolipoprotein A-I in reconstituted high-density lipoprotein disks: an alternative model based on fluorescence resonance energy transfer experiments. *Biochemistry* 2001; 40: 5065–5074.

33. Li H, Lyles DS, Thomas MJ, Pan W, Sorci-Thomas MG. Structural determination of lipid-bound ApoA-I using fluorescence resonance energy transfer. *J Biol Chem* 2000; 275: 37048–37054.

34. Gogonea V, Wu Z, Lee X, Pipich V, Li XM, Ioffe AI, Didonato JA, Hazen SL. Congruency between biophysical data from multiple platforms and molecular dynamics simulation of the double-super helix model of nascent high-density lipoprotein. *Biochemistry* 2010; 49: 7323–7343.

35. Jones MK, Zhang L, Catte A, Li L, Oda MN, Ren G, Segrest JP. Assessment of the validity of the double superhelix model for reconstituted high density lipoproteins: A combined computational-experimental approach. *J Biol Chem* 2010; 285: 41161–41171.

36. Maiorano JN, Jandacek RJ, Horace EM, Davidson WS. Identification and structural ramifications of a hinge domain in apolipoprotein A-I discoidal high-density lipoproteins of different size. *Biochemistry* 2004; 43: 11717–11726.

37. Shih AY, Sligar SG, Schulten K. Molecular models need to be tested: the case of a solar flares discoidal HDL model. *Biophys J* 2008; 94: L87–89.

38. Catte A, Patterson JC, Jones MK, Jerome WG, Bashtovyy D, Su Z, Gu F, Chen J, Aliste MP, Harvey SC et al. Novel changes in discoidal high density lipoprotein morphology: a molecular dynamics study. *Biophys J* 2006; 90: 4345–4360.

39. Gu F, Jones MK, Chen J, Patterson JC, Catte A, Jerome WG, Li L, Segrest JP. Structures of discoidal high density lipoproteins: a combined computational-experimental approach. *J Biol Chem* 2010; 285: 4652–4665.

40. Dergunov AD, Dobretsov GE. Apolipoprotein A-I localization and dipalmitoylphosphatidylcholine dynamics in reconstituted high density lipoproteins. *Chem Phys Lipids* 2000; 104: 161–173.

41. Zerrad-Saadi A, Therond P, Chantepie S, Couturier M, Rye K-A, Chapman MJ, Kontush A. HDL3-mediated inactivation of LDL-associated phospholipid hydroperoxides is determined by the redox status of apolipoprotein A-I and HDL particle surface lipid rigidity: Relevance to inflammation and atherogenesis. *Arterioscler Thromb Vasc Biol* 2009; 29: 2169–2175.

42. Jayaraman S, Benjwal S, Gantz DL, Gursky O. Effects of cholesterol on thermal stability of discoidal high density lipoproteins. *J Lipid Res* 2010; 51: 324–333.

43. Huang R, Silva RA, Jerome WG, Kontush A, Chapman MJ, Curtiss LK, Hodges TJ, Davidson WS. Apolipoprotein A-I structural organization in high-density lipoproteins isolated from human plasma. *Nat Struct Mol Biol* 2011; 18: 416–422.

44. Wu Z, Gogonea V, Lee X, May RP, Pipich V, Wagner MA, Undurti A, Tallant TC, Baleanu-Gogonea C, Charlton F et al. The low resolution structure of ApoA1 in spherical high density lipoprotein revealed by small angle neutron scattering. *J Biol Chem* 2011; 286: 12495–12508.
45. Kumpula LS, Kumpula JM, Taskinen MR, Jauhiainen M, Kaski K, Ala-Korpela M. Reconsideration of hydrophobic lipid distributions in lipoprotein particles. *Chem Phys Lipids* 2008; 155: 57–62.
46. Shih AY, Sligar SG, Schulten K. Maturation of high-density lipoproteins. *J R Soc Interface* 2009; 6: 863–871.
47. Kono M, Okumura Y, Tanaka M, Nguyen D, Dhanasekaran P, Lund-Katz S, Phillips MC, Saito H. Conformational flexibility of the N-terminal domain of apolipoprotein A-I bound to spherical lipid particles. *Biochemistry* 2008; 47: 11340–11347.
48. Jonas A, Wald JH, Toohill KL, Krul ES, Kezdy KE. Apolipoprotein A-I structure and lipid properties in homogeneous, reconstituted spherical and discoidal high density lipoproteins. *J Biol Chem* 1990; 265: 22123–22129.
49. Li HH, Lyles DS, Pan W, Alexander E, Thomas MJ, Sorci-Thomas MG. ApoA-I structure on discs and spheres. Variable helix registry and conformational states. *J Biol Chem* 2002; 277: 39093–39101.
50. Sparks DL, Phillips MC, Lund-Katz S. The conformation of apolipoprotein A-I in discoidal and spherical recombinant high density lipoprotein particles. 13C NMR studies of lysine ionization behavior. *J Biol Chem* 1992; 267: 25830–25838.
51. Curtiss LK, Bonnet DJ, Rye KA. The conformation of apolipoprotein A-I in high-density lipoproteins is influenced by core lipid composition and particle size: a surface plasmon resonance study. *Biochemistry* 2000; 39: 5712–5721.
52. Vuorela T, Catte A, Niemela PS, Hall A, Hyvonen MT, Marrink SJ, Karttunen M, Vattulainen I. Role of lipids in spheroidal high density lipoproteins. *PLoS Comput Biol* 2010; 6: e1000964.
53. Koivuniemi A, Heikela M, Kovanen PT, Vattulainen I, Hyvonen MT. Atomistic simulations of phosphatidylcholines and cholesteryl esters in high-density lipoprotein-sized lipid droplet and trilayer: clues to cholesteryl ester transport and storage. *Biophys J* 2009; 96: 4099–4108.
54. Catte A, Patterson JC, Bashtovyy D, Jones MK, Gu F, Li L, Rampioni A, Sengupta D., Vuorela T, Niemela P et al. Structure of spheroidal HDL particles revealed by combined atomistic and coarse-grained simulations. *Biophys J* 2008; 94: 2306–2319.
55. Kontush A, Therond P, Zerrad A, Couturier M, Negre-Salvayre A, de Souza JA, Chantepie S, Chapman MJ. Preferential sphingosine-1-phosphate enrichment and sphingomyelin depletion are key features of small dense HDL3 particles: relevance to antiapoptotic and antioxidative activities. *Arterioscler Thromb Vasc Biol* 2007; 27: 1843–1849.

CHAPTER 4

METABOLISM

Major pathways of HDL metabolism are based on the highly dynamic nature of the apoA-I molecule, which is capable of adopting an array of conformations in order to bind variable amount of lipids (see Chapter 3)and to form distinct HDL subpopulations (see Chapter 2). Both the N- and C-terminal domains of apoA-I appear to be essential for such HDL remodeling; separation of the N-terminal domains, as HDL particles exceed a critical size, can trigger fusion between HDLs, or between HDLs and membranes, while the C-terminal domains can drive the exchange of polar lipids down concentration gradients [1].

HDL metabolism is thought to primarily ensure the process of reverse cholesterol transport (RCT), which involves the transfer of cholesterol from peripheral cells to the liver for excretion in bile. Metabolic pathways involving HDL are, however, intimately related to those of VLDL and other apoB-containing lipoproteins, reflecting mass exchange and/or the net transfer of both protein and lipid components between apoA-I and apoB-containing particles.

4.1 FORMATION AND INTRAVASCULAR REMODELING

The majority of plasma HDL exists in the form of spherical particles generated by intravascular processes from lipid-free apoA-I or lipid-poor pre-beta HDL [2] (see Chapter 2). These small HDL precursors are produced as nascent HDL by the liver or intestine (Fig. 4.1), and are also released as surface fragments from

High-Density Lipoproteins: Structure, Metabolism, Function, and Therapeutics, First Edition.
Anatol Kontush and M. John Chapman.

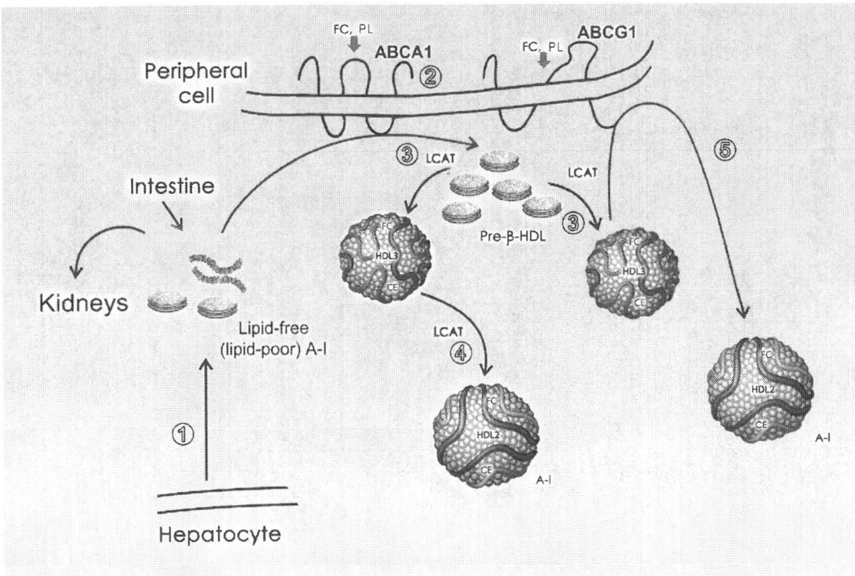

Figure 4.1 Major pathways involved in HDL formation. (1) ApoA-I is secreted by hepatocytes and intestinal cells in a lipid-free and/or lipid-poor (pre-beta) form. (2) ApoA-I lipidation occurs upon interaction with ABCA1, resulting in the formation of lipid-poor, discoid pre-beta HDL particles. (3) Esterification of pre-beta HDL cholesterol catalyzed by LCAT leads to the formation of a HDL lipid core primarily composed of cholesteryl ester; HDL particles grow in size and become small, dense, lipid-poor spherical HDL3. (4) LCAT-mediated cholesterol esterification ensues with a formation of large, light, lipid-rich spherical HDL2. (5) Both HDL3 and, particularly, HDL2 can accept cellular cholesterol and phospholipid via the ABCG1-mediated pathway, increasing HDL lipid load (see color plate section). CE, cholesteryl ester; FC, free cholesterol; PL, phospholipid.

lipolyzed triglyceride-rich lipoproteins (VLDL and chylomicrons); finally, they may be generated during the interconversion of HDL3 and HDL2 as mediated by PLTP [3] (Fig. 4.2). Small nascent HDL is unstable and readily acquires lipids [4]; initial lipidation occurs at cellular membranes via ABCA1-mediated efflux of cholesterol and phospholipid [5] (see 7.1 of chapter 7). It remains unclear as to whether apoA-I is secreted in a lipid-free form or in association with a minor amount of lipid; the presence of lipidated apoA-I in the secretory pathway of hepatic cells has been demonstrated [6, 7].

Small, lipid-poor, discoid HDL particles produced as a result of cellular ABCA1 activity undergo extensive remodeling under the action of multiple factors (Table 4.1; Fig. 4.1). First, such particles further acquire cellular lipids via pathways primarily mediated by ABCA1 and ATP-binding cassette transporter G1 (ABCG1), but also by scavenger receptor class B type I (SR-BI). Second, LCAT-mediated esterification of unesterified cholesterol generates large spherical HDL particles with a neutral lipid core of cholesteryl ester and triglyceride [8]. Such particles may undergo further remodeling via particle

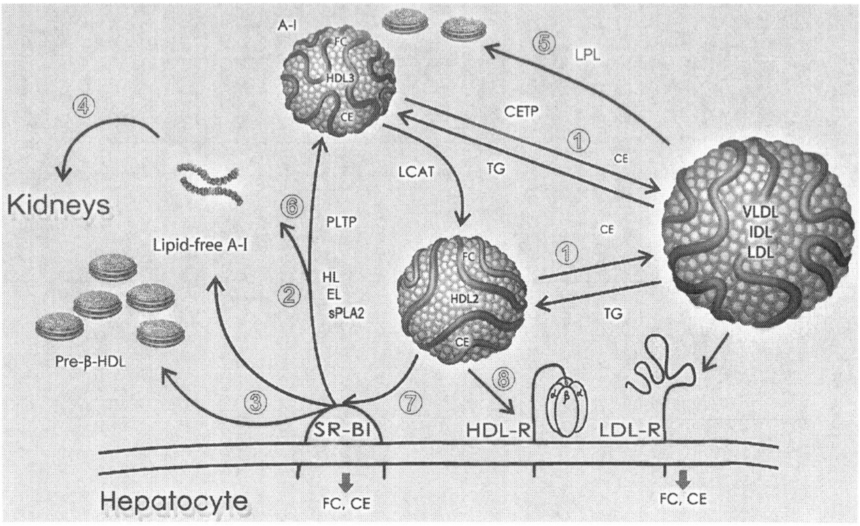

Figure 4.2 Major pathways involved in HDL remodeling. (1) Spherical HDL2 and HDL3 are remodeled by CETP which ensures cholesteryl ester transfer from HDL to triglyceride-rich lipoproteins and reciprocal triglyceride transfer. As a result, HDL particles become enriched in triglyceride. (2) HDL triglyceride can be hydrolyzed by hepatic lipase, whereas HDL phospholipid represents a substrate for endothelial lipase and secreted phospholipase A2. (3) Selective uptake of HDL cholesteryl ester via SR-BI recycles small, lipid-poor HDL subspecies and lipid-free/lipid-poor apoA-I. (4) Lipid-free apoA-I is rapidly removed from the circulation via the renal pathway mediated by cubulin and megalin receptors. (5) Hydrolysis of triglyceride-rich lipoproteins by LPL induces the transfer of surface constituents to the plasma HDL pool. (6) Hydrolysis of HDL triglycerides and phospholipids accelerates apoA-I shedding from HDL particles. HDLs are also remodeled by PLTP with the formation of both larger and smaller particles. (7) HDL cholesteryl ester is selectively taken up via hepatic SR-BI. (8) HDL particles can be also removed from the circulation following interaction with HDL holoparticle receptor(s). CE, cholesteryl ester; EL, endothelial lipase; FC, free cholesterol; HDL-R, holoparticle HDL receptor; HL, hepatic lipase; LDL-R, LDL receptor; TG, triglyceride.

fusion and surface remnant transfer mediated by PLTP [9]. Large HDL2 can, in turn, be converted into small HDL3 upon CETP-mediated transfer of cholesteryl ester from HDL to apoB-containing lipoproteins, upon SR-BI-mediated selective uptake of cholesteryl ester by the liver and steroidogenic organs, and upon hepatic lipase- and endothelial lipase-mediated hydrolysis of core triglyceride [3] (Fig. 4.2). When CETP-mediated transfer of cholesteryl ester occurs between HDL and triglyceride-rich lipoproteins, triglyceride-rich HDL is generated [10] which can be further hydrolyzed by hepatic lipase to small, triglyceride-rich HDL particles [11]. The concerted action of CETP and hepatic lipase promotes a reduction in HDL size, the formation of lipid-poor HDL particles and the shedding of lipid-free/lipid-poor apoA-I from HDL which can interact with ABCA1 in the next lipidation cycle [12, 13]. HDL lipids are either catabolized separately from HDL proteins by selective receptor-mediated uptake or via CETP-mediated transfer to VLDL and LDL, or as holoparticles primarily in the liver via uptake through LDL receptors for apoE-containing HDL and through hitherto unidentified receptors for HDL holoparticles.

ABC Transporters

ABCA1. **ABCA1**, a large membrane multi-pass, cyclic AMP-dependent and sulfonylurea-sensitive anion transporter, is a major player in HDL metabolism, required for both nascent HDL formation and cellular lipid homeostasis [5] (Fig. 4.1). The ABCA1 protein contains 2261 amino acid residues, is heavily glycosylated and is encoded on chromosome 9 (Table 4.1). The sequence of ABCA1 includes two homologous halves, each containing a hydrophobic membrane-anchoring domain and a highly conserved cytoplasmic ABC domain; each of two transmembrane domains of the protein consists of six membrane-spanning segments [14, 15]. ABCA1 is widely expressed, being most abundant in macrophages and hepatocytes.

The structure of ABCA1 is stabilized by disulfide bonds between its N- and C-terminal halves owing to the presence of cysteine residues in the extracellular domains. Specifically, two intramolecular disulfide bonds (between Cys75 and Cys309 in extracellular domain 1 and either Cys1463 or Cys1465 and Cys1477 in extracellular domain 2) are required for ABCA1 to be fully functional in terms of apoA-I binding [16]. It was originally suggested that ABCA1 forms a channel within the plasma membrane through which cholesterol and phospholipid are transferred ('flopped') from the inner to the outer leaflet of the plasma bi-layer membrane [17, 18]. There, the lipids may be picked up by lipid-free apolipoproteins or lipid-poor particles that bind to ABCA1 [19, 20]. Cys1660 appears to be a critical residue for cholesterol translocation by ABCA1 [21].

DEFINITIONS

HDL biogenesis: the formation of nascent, lipid-poor HDL particles as a result of the transfer of plasma membrane cholesterol and phospholipid to lipid-free/lipid-poor apoA-I, apoE and potentially other apolipoproteins during their interaction with the activated ABCA1 transporter.

Indeed, **HDL biogenesis** by ABCA1 depends on the presence of extracellular acceptors, such as lipid-free/lipid-poor apoA-I (Fig. 4.1). In order to efflux cholesterol and phospholipid to the acceptors, ABCA1 generates protruding membrane domains of high curvature, allowing the acceptors to interact with

TABLE 4.1 Major Receptors, Enzymes and Transfer Proteins Involved in HDL Metabolism

	Description	Structure	Number of Amino Acid Residues	Function in HDL Metabolism	Expression Pattern	Regulation
Receptors						
ABCA1	Membrane multi-pass ABC transporter	Two homologous halves, each containing a hydrophobic membrane-anchoring domain and a highly conserved cytoplasmic ABC domain. Heavily glycosylated. Contains disulfide bonds.	2261	Formation of nascent HDL via transfer of cholesterol and phospholipid to lipid-free/lipid-poor apoA-I, other apos and small HDL. Generation of protruding membrane domains of high curvature.	Wide; abundant in macrophages and hepatocytes	Activated by LXRs, RXRs and PPARs. Inhibited by microRNA-33. Degraded by calpain.
ABCG1	Membrane multi-pass half-type ABC transporter	One ABC transmembrane domain and one ABC transporter domain	678	Enrichment of HDL in cellular lipids via their transfer to spherical HDL	In several tissues, including macrophages	Activated by LXRs, RXRs and PPARs. Inhibited by microRNA-33.
SR-BI	Multiligand receptor	Glycosylated integral membrane protein	481	Selective uptake of HDL CE from circulating HDL by hepatocytes and adrenals. Cholesterol efflux to spherical HDL	Primarily in the liver, adrenals, ovaries and testis	Activated by FXR, PPAR alpha and RXR-alpha

Enzymes						
LCAT	Enzyme catalyzing transfer of PC acyl residues to FC	Heavily N-glycosylated. Contains two disulfide bridges	416	Expansion of the HDL lipid core via the formation of CE	Primarily in the liver, also in the brain and testes	Activated by apoA-I and other apos
LPL	Enzyme catalyzing hydrolysis of TG in lipoproteins	Glycosylated. Circulates as a homodimer	448	Generation of surface fragments of TG-rich lipoproteins contributing to HDL formation and growth	Primarily in heart, muscle and adipose tissue	Activated by apoC-II. Inhibited by apoC-III and apoA-II
HL	Enzyme catalyzing hydrolysis of TG, DAG and PL in lipoproteins	Glycosylated	477	Hydrolysis of HDL-associated TG, resulting in decreased circulating levels of HDL	Primarily in hepatocytes	Inhibited by apoC-I, apoC-III and apoA-II
EL	Enzyme catalyzing hydrolysis of PL and TG in HDL	Glycosylated. Stabilized by disulfide bonds. Circulates as a homodimer	480	Hydrolysis of HDL components, promotion of their turnover, acceleration of apoA-I catabolism	Primarily by endothelial cells	Activated by inflammatory cytokines

(*continued*)

TABLE 4.1 (*Continued*)

	Description	Structure	Number of Amino Acid Residues	Function in HDL Metabolism	Expression Pattern	Regulation
Lipid transfer proteins						
CETP	Mediates heteroexchange of CE and TG between HDL and apoB-containing lipoproteins	Glycosylated	476	Transfer of CE from HDL to apoB-containing particles in exchange for TG, recycling of lipid-free/lipid-poor apoA-I	Primarily in the liver, adipose tissue and macrophages	Activated by LXRs and RXRs
PLTP	Mediates transfer and exchange of PL between different lipoproteins	Glycosylated. Stabilized by a disulfide bond	476	Conversion of spherical HDL into larger and smaller particles of modified composition	Wide; includes pancreas, kidney, heart, liver and skeletal muscle	Activated by LXRs and RXRs

DAG, diacylglycerol; EL, endothelial lipase; FC, free cholesterol; FXR, farnesoid X receptor; HL, hepatic lipase; LXR, liver X receptor; lysoPC, lysophosphatidhylcholine; PC, phosphatidylcholine; PL, phospholipid; RXR, retinoid X receptor; TG, triglyceride.

the cell surface. Curvature is critical for apoA-I binding as apoA-I prefers highly curved surfaces. In this pathway, ABCA1 generates loosely packed membrane domains, or non-rafts, decreasing the membrane lipid order [5]. Insertion of the C-terminal domain of apoA-I into structural defects in membrane lipids created in the high curvature domains thereby represents the first step of HDL formation [22].

Interestingly, such membrane ABCA1 activity may lead to lipid efflux even in the absence of apoA-I, releasing large, 24 nm, HDL-like vesicles (microparticles) devoid of apoA-I and apoE [23]. Such microparticles are also an integral component of the classical cholesterol efflux pathway when apoA-I is present and may account for approximately 30% of the total cholesterol released from the cell. The large, spherical microparticles released from J774 cells contain CD14 and ganglioside, consistent with their origin in plasma membrane raft domains [24]. The various HDL particles and microparticles can be created concurrently; there is no precursor-product relationship between them.

Modulation of the physical properties of the plasma membrane may affect HDL production. Specifically, an increase in plasma membrane fluidity promotes HDL and microparticle release, whereas a decrease in the fluidity inhibits the production of the microparticles. ABCA1, therefore, enhances the dynamic movement of the plasma membrane required for apoA-I lipidation and HDL formation [23].

Such functional activity of ABCA1 leads to the formation of a major **high-capacity, lipid-containing site** for the binding and lipidation of apoA-I at the plasma membrane. ABCA1 activity thereby creates two types of high-affinity apoA-I binding sites at the cell surface [25, 26]. The **low capacity site** is formed by direct interaction between apoA-I and ABCA1, and functions in a regulatory role. The much higher capacity lipid-containing site, which accounts for the binding of about 90% of apoA-I, is generated by interactions between apoA-I and lipids, and functions in the assembly of nascent HDL particles [25, 26].

The rate of dissociation of apoA-I from ABCA1 is extremely rapid ($t_{1/2}$ ~8 min), whereas the dissociation of apoA-I from the high-capacity binding site is >10-fold slower ($t_{1/2}$ ~ 85 min) [27]. ApoA-I initially contacts the cell via ABCA1; this association is rapid but transient. Subsequently, ABCA1 mediates the transfer of apoA-I to the high-capacity binding site, thereby allowing phospholipid, and then cholesterol, extraction and dissociation of the lipidated product. The co-operative and productive interaction between apoA-I and the system of the ABCA1/phosphatidylcholine-rich high capacity binding site is consistent with a two-binding site model for nascent HDL biogenesis [27]. ApoA-I dissociation from ABCA1 represents the final step of this pathway, which requires ATP-dependent lipid translocation and apoA-I lipid loading [28].

As a result, **nascent HDL particles** (pre-beta-1 HDL; diameter 7.5 nm) are formed, for example during the incubation of lipid-free apoA-I with fibroblasts in which the expression of ABCA1 is up-regulated [29]. Reportedly, the pre-beta-1 HDL particles contain a single apoA-I molecule, three to four phospholipid molecules and one to two cholesterol molecules, thereby corresponding to M_r of about 30 kDa. The C-terminal alpha-helix of apoA-I, which provides insertion into the plasma membrane, is critical for the formation of such lipid-poor apoA-I-containing particles [29].

Interestingly, the formation of nascent HDL can be influenced by the phospholipid composition of the plasma membrane; specifically, the presence of phosphatidylserine facilitates the interaction of apoA-I with the phospholipid monolayer [30] and accelerates the formation of discoidal HDL [31]. This effect is probably underlain by the capacity of phosphatidylserine to reduce pH at the membrane surface, which may induce conformational changes in apoA-I [31]. Consistent with this mechanism, ABCA1 promotes trans-bi-layer redistribution and surface exposure of phosphatidylserine, thereby stimulating HDL biogenesis [18]. On the other hand, phosphatidylethanolamine increases both the hydration of the membrane interface and the phospholipid acyl chain order, enhancing the binding of amphipathic helices of apoA-I [31].

Importantly, pre-beta-1 HDL represents both a product and a substrate in the ABCA1-mediated reaction to efflux cellular phospholipid and cholesterol to apoA-I. A monomeric apoA-I molecule associated with three to four phospholipid molecules (i.e., lipid-poor apoA-I) has similar properties to the lipid-free apoA-I molecule in terms of cellular cholesterol efflux [29]. Three types of discoid nascent HDL containing two, three, or four molecules of apoA-I per particle are, therefore, formed when J774 mouse macrophages and human skin fibroblasts in which ABCA1 is activated are incubated with human apoA-I [24]. With J774 cells, the predominant species display hydrodynamic diameters of approximately 9 and 12 nm. These discoid HDL particles differ in free cholesterol content and phospholipid composition, and the presence of acidic phospholipids causes them to exhibit alpha-electrophoretic mobility. The mechanisms of the generation of such heterogeneous HDL appear to depend on the location of ABCA1 in more than one membrane microenvironment and, as such, is implicated in the production of different HDL subspecies [24]. Cellular phospholipids and cholesterol incorporated into nascent HDL particles may, therefore, originate from different cell membrane domains [24]. Specifically, lipid composition, size speciation and three-dimensional structures of HDL formed via the activity of ABCA1 are linked to the cholesterol content of the membranes from which they were formed [32]. The heterogeneity of nascent HDL also depends on the C-terminal domain of apoA-I that dictates the distribution of HDL between particle subclasses.

The interaction of apoA-I with ABCA1 also results in the generation of other heterogeneous-sized, nascent pre-beta HDL of discrete chemical composition [33]. These particles are, however, poor substrates for subsequent lipidation by ABCA1 and presumably require additional non-ABCA1-mediated lipidation for further maturation. Therefore, initial interaction of apoA-I with ABCA1 results in a constrained conformation of the apolipoprotein [33]. These mechanisms lead to the formation of small LpA-I particles that contain apoA-I as the unique apolipoprotein [34]. Such particles can subsequently acquire apoA-II secreted by hepatocytes to form LpA-I:A-II. HDL particles can also incorporate additional molecules of apoA-I in a two-step process revealed by surface plasmon resonance techniques. Initially, apoA-I binds to HDL with fast association and dissociation rates, followed by a step exhibiting slower kinetics [34]. Thus, the isolated N-terminal helix-bundle domains of apoA-I interact with proteins resident on the HDL particle surface, resulting in the second slower step involving the opening of the helix bundle domain. Binding is characterized by low affinity with dissociation constants of 10 μM which only apply to a relatively small, labile pool of apoA-I [34].

CRITICAL CONTRIBUTION

Multi-step HDL formation at the plasma membrane mediated by the ABCA1 and ABCG1 transporters with participation of two high-affinity binding sites of high and low capacity [22, 24–26, 29, 35, 36].

In addition to apoA-I and apoA-II, other apolipoproteins are capable of forming nascent HDL following interaction with ABCA1. Thus, apoE participates in a pathway of biogenesis of apoE-containing HDL particles which requires ABCA1 [37]. Specifically, apoE interacts with ABCA1 to extract cholesterol out of the cell and drives the formation of large HDL particles from lipid-loaded macrophages when competing with freely available apo A-I. Furthermore, interaction of ABCA1 with apoC-III promotes the formation of apoC-III-containing HDL-like particles that are distinct from classical apoE- or apoA-I-containing HDL [38]. Metabolically, the formation of apoC-III-containing HDL may shift apoC-III from VLDL to HDL and prevent excess accumulation of plasma apoC-III on VLDL, thereby allowing the efficient lipolysis of VLDL triglyceride by LPL.

Finally, apoM efficiently participates in the formation of small HDL, as shown in apoM-deficient mice which accumulate cholesterol in large HDL1, while the conversion of HDL to pre-beta HDL is impaired [39]. In addition, apoM stimulates the formation of large-sized pre-beta HDL and may, therefore, function intracellularly to transfer lipid onto small pre-beta HDL during, or after, their formation by ABCA1 [40]. Indeed, because of its retained signal peptide, apoM is poorly secreted unless HDL is present in the culture medium [41].

Interestingly, apoA-I, apoA-II and apoE can be also lipidated intracellularly in post-secretory pathways [42]. Intracellular lipidation of apoA-I occurs in two kinetically distinguishable compartments; lipidated apoA-I is rapidly re-secreted in a monomeric form [42, 43]. In contrast, apoA-II can form intracellular dimers before secretion, whereas intracellular apoE is monomeric but can dimerize after secretion and association with HDL [42]. As a corollary, intracellular ABCA1 activity has been proposed to contribute to HDL biogenesis. Indeed, ABCA1 colocalizes with apoA-I in acidic intracellular compartments, including the Golgi and endosome [44]. There, the transporter appears to be able to increase acidity at the membrane surface on the luminal side by translocating negatively-charged phosphatidylserine, thereby causing conformational changes in apoA-I to a rather hydrophobic alpha-helical structure which allows the formation of nascent HDL [44]. However, other data do not support the intracellular route of HDL biogenesis [22, 29, 45]. Thus, only 5.6% of apoA-I that interacts with cells is degraded; internalized apoA-I contributes to approximately 1.4% of total HDL production [46]. Furthermore, blocking endocytosis does not decrease HDL biogenesis. The plasma membrane represents, therefore, the main platform where ABCA1-mediated lipidation of apoA-I occurs. The role of the retro-endocytosis pathway of HDL biogenesis involving ABCA1/apoA-I may, however, be enhanced when cells accumulate excess lipoprotein-derived cholesterol [47].

ABCA1 activity is regulated both at the transcriptional and at the post-translational level via actions on the ABCA1 gene promoter and modulation of protein stability (Table 4.1). **Liver X receptors (LXRs)** play a central role in the regulation of ABCA1 expression in macrophages, ensuring ABCA1 activation following cholesterol load [48, 49]. Activation of LXR results in the enhancement of ABCA1-mediated HDL formation and accelerates cholesterol efflux to apoA-I (see 7.1 of chapter 7). Furthermore, HDL biogenesis via ABCA1 is regulated by retinoid X receptors (RXRs); sterols and fatty acids represent major physiologic modulators of the LXR/RXR pathway [50].

Peroxisome proliferator-activated receptors (PPARs) alpha and gamma are also involved in the activation of ABCA1 expression and associated apoA-I-mediated lipid release in macrophages in a co-operative signal transduction pathway, thereby contributing to HDL biogenesis [51, 52]. The ABCA1 promoter is further stimulated by the ubiquitous factor Sp1 and the hypoxia-induced factor 1, whereas co-repressor complexes involving unliganded LXR, sterol-regulatory element binding protein-2 (SREBP-2), Sp3 and the SCAN-domain zinc finger protein 202 (ZNF202) mediate inhibition of ABCA1 expression in the absence of sterols [50]. In addition, microRNA-33 encoded by SREBP represses the expression of ABCA1 and down-regulates HDL biogenesis in mice [53, 54], acting in concert with the SREBP host genes [55, 56]. Similarly, deficiency of the WW domain-containing oxidoreductase (WWOX), a pro-apoptotic protein and tumor suppressor located in cytoplasm and mitochondria, decreases hepatic expression of ABCA1 and apoA-I and impairs HDL biogenesis through the ABCA1-mediated pathway [57].

Calpain-mediated intracellular degradation is one of the major regulatory factors in the activity of ABCA1 [45]. Surface ABCA1 is internalized and rapidly degraded intracellularly by calpain. In contrast, helical apolipoproteins protect ABCA1 against this degradation and increase the generation of HDL. Notably, apoA-I interferes with the proteolytic degradation and increases recycling of ABCA1 to the surface, resulting in a proportional increase in HDL formation. This pathway provides an autocrine mechanism for the stabilization of hepatic ABCA1, which involves hepatic secretion of apoA-I and results in reduced clearance of the transporter [45]. In addition, ABCA1 can be stabilized by calmodulin which typically binds to calpain substrates to regulate cellular functions. ABCA1 interacts with calmodulin in such a way that the latter directly binds and stabilizes the former in the presence of Ca^{2+} and increases the generation of HDL [58]. Consistent with this observation, cytosolic Ca^{2+} is required for cholesterol transport to apoA-I [59]; the latter can induce Ca^{2+} influx from the medium. Finally, ABCA1 protein degradation can be prevented by the Rho guanine nucleotide exchange factors, PDZ-RhoGEF and LARG, which bind to the C-terminus of ABCA1 by a PDZ-PDZ interaction and activate RhoA [60].

Importantly, HDL biogenesis may contribute significantly to atheroprotection, as suggested by accelerated atherogenesis in ABCA1-deficient mice fed a high-fat diet [61]; impaired cholesterol efflux from macrophages may underlie this observation (see 10.1 of chapter 10). Intriguingly, however, hepatic ABCA1 expression may be more important for atheroprotection than macrophage-specific expression of ABCA1 in mice, with the maintenance of circulating HDL levels as a plausible underlying mechanism [62]. Consistent with this notion, plasma concentrations of HDL-C are directly correlated with sterol markers of intestinal cholesterol absorption [63] which suggests that intestinally-derived cholesterol may make a major contribution to HDL-C levels. The secondary function of macrophage ABCA1 for HDL biogenesis is indirectly supported by the essential role of intestinal-specific, rather than macrophage-specific, LXR agonism for the promotion of macrophage RCT in mice [64].

OPEN QUESTION

How does the transport of cholesterol molecules to HDL particles through the ABCA1 and ABCG1 transporters occur?

ABCG1. **ABCG1**, another membrane multi-pass transporter, plays a key role in HDL metabolism by contributing to HDL formation and remodeling (Fig. 4.1). The ABCG1 protein is much smaller than ABCA1 and contains 678 amino acid residues (Table 4.1). The protein is encoded by chromosome 21 and belongs to half-type ABC transporters which contain one ABC transmembrane domain and one ABC transporter domain. ABCG1 is expressed in several tissues, including macrophages.

In contrast to ABCA1, which interacts with lipid-free apolipoproteins and small, lipid-poor HDL, ABCG1 preferentially binds mature spherical HDL particles [65, 66] (see 7.1 of chapter 7). The particle substrate for ABCG1 can be formed as a result of the interaction of lipid-free apoA-I with ABCA1 [35]; subsequent maturation by LCAT can further improve the ABCG1-binding properties of HDL. The interaction of HDL with ABCG1 is, therefore, regulated by the degree of HDL lipidation. A minor degree of lipidation of apoA-I appears to be sufficient to induce interaction with ABCG1 [67].

As a consequence, ABCG1 efficiently promotes the enrichment of large HDL with cellular lipid, further increasing HDL particle size [65, 66]. Alpha-HDL containing LpA-I and LpA-I:A-II represents a preferential partner for interaction with ABCG1. The receptor-mediated cascade of HDL biogenesis may, therefore, involve the production of small, lipid-poor HDL particles upon interaction of apoA-I with ABCA1 followed by their lipidation by ABCG1 [36] (Fig. 4.1). HDL formation mediated by ABCG1 may also involve the regulation of intracellular movement of cholesterol in the endosomal pathway by the transporter located in intracellular membranes [68]. The capacity of extracellular acceptors to induce ABCG1-mediated efflux is strongly correlated with their total phospholipid content, suggesting that acceptor phospholipids drive ABCG1-mediated efflux [35].

Regulation of ABCG1 activity often parallels that of ABCA1. Thus, LXRs play a central role in the up-regulation of ABCG1 expression in macrophages under cholesterol load [48, 49]. Activation of LXR results in the enhancement of ABCG1-mediated HDL formation and accelerates cholesterol efflux to HDL. RXRs and PPARs are also involved in the activation of ABCG1 expression and in the associated HDL formation by macrophages [50–52]. In addition, transforming growth factor-beta-1 induces, whereas microRNA-33 inhibits, expression of both ABCA1 and ABCG1 in mice [54, 56, 69]. On the other hand, ABCG1 expression can be specifically induced in the absence of effects on ABCA1 expression. Thus, adenosine monophosphate-activated protein kinase (AMPK), which neither influences ABCA1 expression nor acts on LXR, up-regulates ABCG1 in murine macrophages [70]. In addition, fatty acids which differ in their unsaturation may exert differential effects on ABCA1 and ABCG1 expression in macrophages [71].

IMPORTANT READING

Rye KA, Barter PJ. Formation and metabolism of prebeta-migrating, lipid-poor apolipoprotein A-I. *Arterioscler Thromb Vasc Biol* 2004; 24: 421–428.

Rothblat GH, Phillips MC. High-density lipoprotein heterogeneity and function in reverse cholesterol transport. *Curr Opin Lipidol* 2010; 21: 229–238.

Yvan-Charvet L, Wang N, Tall AR. Role of HDL, ABCA1, and ABCG1 transporters in cholesterol efflux and immune responses. *Arterioscler Thromb Vasc Biol* 2010; 30: 139–143.

Enzymes

LCAT. Cholesterol in small, lipid-poor HDL is esterified to cholesteryl ester by LCAT. As a result, small, pre-beta HDL particles acquire alpha mobility (Figs 4.1 and 4.2). LCAT catalyzes trans-esterification of phosphatidylcholine acyl residues and unesterified cholesterol with the formation of cholesteryl ester and lysophosphatidylcholine, thereby possessing lipolytic activity (Table 4.1). This reaction is believed to play a central role in the RCT pathway [72–74];

indeed, the majority of tissue-derived cholesterol is delivered to the liver as cholesteryl ester and only a small amount as free cholesterol [75].

Although some LCAT activity is associated with apoB-containing lipoproteins, the majority of esterified cholesterol is produced within HDL *in vivo* [76]. Nevertheless, apoB-containing lipoproteins still play an important role in this reaction by providing unesterified cholesterol through direct transfer to HDL [76]. From a physicochemical point of view, LCAT traps cholesterol within HDL particles, converting it into a highly hydrophobic, esterified form which locates to the hydrophobic core of HDL. Such cholesteryl ester cannot be spontaneously exchanged with other lipid molecules until it is removed from the circulation by the liver [72]. In parallel, LCAT promotes cholesterol efflux from peripheral cells, acting via the elevation of the concentration gradient for unesterified cholesterol between cell membranes and HDL. Free cholesterol may, however, be rapidly removed from small HDL particles as a result of its lower hydrophobicity compared with cholesteryl ester; the exchange of free cholesterol between small HDL and cell membranes represents an alternative route of cholesterol delivery to the liver.

Cholesterol esterification by LCAT begins with enzyme binding to HDL. Formation of the complex between LCAT and discoid rHDL can be visualized using cryotomography as an outside protrusion at the particle surface as the LCAT molecule is about 5 nm in diameter, comparable to that of rHDL (7–10 nm) [77]. The central region of apoA-I possesses a relatively low mobility relative to the rest of the molecule and therefore provides an optimal location to bind LCAT [78]. A protruding, solvent-exposed loop of apoA-I, including Tyr166, appears to directly interact with, and activate, the enzyme [79]. Cleavage of the phosphatidylcholine fatty acid residue in sn-2 position follows, leading to the transfer of the acyl moiety onto Ser181 from the catalytic center of the enzyme [72]. Subsequently, the fatty acid is trans-esterified to the 3-beta-hydroxyl group on the A-ring of cholesterol to form cholesteryl ester. Phosphatidylcholines containing oleate and linoleate residues form the preferential LCAT substrate in humans; phosphatidylethanolamine can be also hydrolyzed by LCAT, whereas sphingomyelin inhibits the LCAT reaction [72].

Detailed mechanisms of cholesterol esterification by LCAT can be studied by molecular dynamics simulations, which reveal that, after attachment of LCAT to discoid HDL, the helix 5/5 domains in apoA-I form amphipathic 'presentation tunnels' for the migration of hydrophobic acyl chains and amphipathic unesterified cholesterol from the bi-layer to the phospholipase A_2-like and esterification active sites of LCAT, respectively [80]. In contrast, cholesteryl ester, the reaction product of LCAT, partially blocks the amphipathic presentation tunnel [80]. Subsequently, newly synthesized cholesteryl ester molecules should be properly positioned within the HDL particle; formation of a loop on one apoA-I molecule, which weakens interhelical interaction with other molecules, may allow more solvent exposure, resulting in cholesteryl ester insertion between phospholipid leaflets in an expanding particle core [79].

ApoA-I, thereby, plays an essential role in the esterification of plasma cholesterol and the formation of mature, spherical HDL, acting via binding and activation of LCAT in nascent discoid particles. The action of apoA-I may also involve stabilization of a catalytically active conformation of LCAT by virtue of surface-exposed regions of the apolipoprotein [72]. LCAT activation by rHDL is proportional to the number of apoA-I molecules per HDL particle and, as a consequence, to rHDL particle size [81]. Other apolipoproteins, including apoA-IV and apoC-I, are also capable of activating LCAT, although less potently compared with apoA-I. Notably, apoE may play a role in the activation of LCAT on apoB-containing lipoproteins [72]. In addition, the esterification of cholesterol by LCAT is essential for the maturation of nascent apoE-containing HDL [37]. In contrast, apoC-III appears to reduce LCAT activation [82].

MAJOR CONTROVERSY

Is cholesterol esterification by LCAT essential for efficient RCT from peripheral tissues to the liver? What is the relative quantitative contribution of free cholesterol vs. esterified cholesterol to RCT?

Lipases. Intravascular HDL metabolism is strongly influenced by several lipases, including LPL, hepatic lipase and endothelial lipase; other lipases involved in HDL metabolism are LpPLA2 (see 1.1 of chapter 1 and 5.2 of chapter 5) and LCAT (see above).

LPL (EC 3.1.1.34) is encoded by a gene on chromosome 8 which is primarily expressed in heart, muscle and adipose tissue (Table 4.1). The enzyme belongs to the lipase family and is synthesized as a 475-amino acid preprotein which is further processed by cleavage of a 27-residue signal peptide. Human LPL circulates as a homodimer composed of two glycoprotein units of 61 kDa each. The primary function of LPL is the hydrolysis of triglyceride of circulating chylomicrons and VLDL. LPL is an interfacial enzyme that must be liberated from heparin sulfate proteoglycans at the luminal surface of vascular endothelium. Lipolysis of triglyceride-rich lipoproteins by LPL is complex and requires the participation of other proteins, such as glycosylphosphatidylinositol-anchored high density lipoprotein binding protein 1 (GPIHBP1), which binds both LPL and chylomicrons [83].

Plasma LPL may regulate HDL-C levels through several mechanisms (Fig. 4.2). First, phospholipids and apolipoproteins, which are shed from triglyceride-rich lipoproteins during their hydrolysis, can be acquired by HDL particles [84]. Thus, lipolysis of VLDL particles by LPL represents an important source for the formation of pre-beta-1 HDL [85]. Physically, hydrolysis of core triglyceride destabilizes VLDL particles, leading to the release of excessive surface phospholipids and small apolipoproteins. Interestingly, destabilization of VLDL by mild oxidation also results in the generation of HDL-like fragments, highlighting the structural nature of this phenomenon [86].

A decrease in plasma triglyceride concentrations under the action of LPL limits both the reduction of HDL-C and the enrichment of HDL in triglyceride mediated by CETP. As a result of a reduction in HDL triglyceride content, the catabolic rate of HDL apoA-I can be diminished secondary to delayed production of small, lipid-poor apoA-I, which is rapidly cleared from the circulation [87]. Thus, elevated LPL activity is typically associated with increased HDL-C levels.

Several exchangeable apolipoproteins that shuttle between HDL and VLDL can modulate LPL activity. Thus, LPL is activated by apoC-II and inhibited by apoC-III. As a corollary, HDL-C levels are decreased in individuals with increased plasma levels of apoC-III and a hypertriglyceridemic lipid profile [88]. ApoA-IV is required for the efficient activation of LPL by apoC-II. In addition, apoD promotes LPL-mediated hydrolysis of VLDL *in vivo* and *in vitro* [89].

ApoA-II plays a central role in triglyceride catabolism by regulating LPL activity via the displacement of apoCs and apoE from HDL as a result of the higher affinity of apoA-II for the HDL surface compared with other apolipoproteins [90]. Such modifications of the HDL proteome result in delayed triglyceride hydrolysis and prolonged plasma residence time of HDL proteins. Furthermore, the level of hypertriglyceridemia can be up-regulated by the transfer of apoA-II from HDL to VLDL as the accumulation of apoA-II in VLDL inhibits VLDL hydrolysis by LPL [91–93]. Consistent with these data, the expression of human apoA-II in rabbits, a species that does not possess an endogenous apoA-II gene, increases plasma levels of triglycerides, total cholesterol and phospholipids and markedly reduces HDL-C [94]. The action of apoA-II on HDL metabolism can be related to the accelerated catabolism of LpA-I:A-II particles and/or diminished availability of VLDL surface fragments as a result of inefficient triglyceride hydrolysis [94].

Human **hepatic lipase** (EC 3.1.1.3) is produced as a protein of 499 amino acids containing a signal peptide of 22 amino acids (Table 4.1). The human hepatic lipase gene (LIPC) is located on chromosome 15. Mature hepatic lipase is a glycoprotein of approximately 65 kDa, which is primarily synthesized in hepatocytes and secreted in the circulation where it is bound to the surface of hepatocytes and hepatic endothelium. Hepatic lipase specifically catalyzes the hydrolysis of HDL surface lipids (primarily phospholipids but also diacylglycerides), whereas triglycerides are hydrolyzed less efficiently [95–97].

Similar to LPL, hepatic lipase is an interfacial enzyme that must be released from the cell surface. Both HDL and apoA-I can displace hepatic lipase from cell surface proteoglycans [98, 99]. Hepatic lipase displacement by HDL appears to result from a high-affinity association of hepatic lipase and apoA-I. ApoA-II and apoC-I significantly increase hepatic lipase displacement from the cell surface. In contrast, increases in HDL phospholipid, triglyceride and apoE content inhibit hepatic lipase displacement [98, 99].

Hepatic lipase action on HDL involves the hydrolysis of HDL-associated triglycerides, resulting in decreased circulating levels of HDL as a result of enhanced renal clearance (Fig. 4.2). However, hepatic lipase deficiency is associated with elevated HDL-C levels [100] and may compensate for decreases in HDL-C caused by other factors [101].

Similar to LPL, small exchangeable apolipoproteins can regulate the activity of hepatic lipase. Thus, apoC-I, apoC-III and apoA-II all inhibit hepatic lipase [102]. On the other hand, apoC-II, which activates LPL, does not impact hepatic lipase activity.

Endothelial lipase (EC 3.1.1.3) is encoded by the endothelial lipase (LIPG) gene and constitutes another member of the triglyceride lipase gene family [103, 104] (Table 4.1). The human gene is located on chromosome 18 and encodes a 500-residue protein that contains a 20-residue signal peptide. The enzyme is glycosylated, stabilized by disulfide bonds and circulates as a homodimer. Endothelial lipase is expressed in endothelial cells, as well as in the liver, placenta, lung, thyroid, kidney and testis; it hydrolyzes lipid components of lipoproteins, primarily phospholipids but also triglycerides, using a Ser-His-Asp triad [96, 105, 106] (Fig. 4.2). Importantly, HDL represents a preferential substrate for endothelial lipase compared with other lipoproteins.

Similar to hepatic lipase, endothelial lipase is a negative regulator of HDL-C levels; thus, over-expression of the LIPG gene in mice results in a marked reduction of HDL-C [104], whereas LIPG knockout increases HDL-C concentrations [100]. HDL-C raising secondary to decreased endothelial lipase activity is further exemplified by a naturally occurring variant of endothelial lipase associated with elevated HDL-C as a result of impaired secretion and reduced circulating levels of the enzyme [107].

Endothelial lipase expression can be up-regulated by inflammatory cytokines, including tumor necrosis factor (TNF)-alpha and interleukin (IL)-1-beta [108]. Post-translational regulation of endothelial lipase activity involves its inhibition in the presence of angiopoietin-like protein 3 (ANGPTL3) [109] and apoA-II [108].

Endothelial lipase can modulate plasma HDL-C levels through both lipolytic and non-lipolytic functions [110]; the latter may involve enhanced binding to cell-surface proteoglycans [111]. In addition, endothelial lipase promotes the turnover of HDL components and increases the catabolism of apoA-I [106]. The enzyme can also alter the HDL particle profile, converting large HDL particles into small HDL [106]. In support of these data, endothelial lipase expression results in the generation of small pre-beta HDL particles in wild-type mice [112].

Lipid Transfer Proteins

CETP. Plasma **CETP** primarily mediates the heteroexchange of cholesteryl ester and triglyceride between mature spherical HDL and apoB-containing lipoproteins (Table 4.1and Fig. 4.2). Specifically, CETP transfers cholesteryl ester from HDL to apoB-containing particles, particularly VLDL, in exchange for triglyceride [113]. As a result, CETP can act in concert with LCAT, further promoting HDL-mediated cholesterol efflux from peripheral cells via elevation of the concentration gradient for unesterified cholesterol between cell membranes and HDL. Consistent with the role of CETP in RCT, CETP gene expression is activated by LXRs and RXRs concomitantly with that of ABC transporters [114, 115].

HDL-derived cholesteryl esters that accumulate in apoB-containing, triglyceride-rich lipoproteins can be subsequently delivered to the liver through hepatic LDL receptors. In healthy, normolipidemic humans, such indirect cholesterol transport to the liver represents a major pathway for cholesterol removal from peripheral tissues. Indeed, the majority of HDL-derived cholesteryl esters (some 70%) are transported to the liver by apoB-containing lipoproteins in healthy normolipidemic subjects [113]. In parallel, such remodeling of apoB-containing particles, primarily LDL, by CETP is critical to the optimization of LDL particle structure and apoB-100 conformation for high affinity binding to LDL receptors [116].

OPEN QUESTION

What is the net contribution of CETP-mediated cholesteryl ester transfer to RCT under normolipidemic and dyslipidemic conditions?

Although plasma cholesteryl esters are primarily transported by both LDL and HDL, HDL is the preferred lipid donor for CETP compared with LDL. Such specificity originates from the action of plasma apoF (LTIP), predominantly found in LDL. Indeed, apoF predominantly inhibits lipid transfer between VLDL and LDL and targets it to HDL [117, 118]. The inhibitory activity of apoF can be controlled by its reversible incorporation into an inactive 470 kDa HDL complex [119]. ApoC-I is another physiologic inhibitor of CETP activity in human plasma, potentially acting via modification of the HDL surface charge [120]; indeed, apoC-I carries a strong positive charge.

Among apoB-containing, triglyceride-rich particles, those containing apoB and apoCs, as well as apoB and apoE, function as cholesteryl ester acceptors, while those containing only apoB or apoB together with A-II, apoCs and apoE do not [121]. The endogenous plasma activity of CETP is also modulated to a major degree by the magnitude of triglyceridemia [113]. In addition, CETP-mediated cholesteryl ester transfer from HDL to LDL is targeted to LDL particles of intermediate size and density displaying elevated affinity for the LDL receptor, facilitating their rapid removal from plasma [113].

Combined with the subsequent actions of hepatic lipase and SR-BI, the action of CETP ensures the recycling of lipid-free/lipid-poor apoA-I from mature HDL [122]. The generation of lipid-free/lipid-poor apoA-I may represent an essential step in the cycle of HDL maturation [13, 81]. Such CETP-catalyzed HDL speciation can be modulated by apolipoproteins; indeed, it can be accelerated by apoA-IV and inhibited by apoA-II, potentially as a result of their differential affinities for HDL particles [87].

Molecular mechanisms of CETP activity can be deduced from the structure of CETP determined at 2.2 Å resolution [123]. These studies reveal a 60 Å-long tunnel filled with two cholesteryl ester molecules; the tunnel is plugged by a phosphatidylcholine molecule at each end, endowing CETP with phosphatidylcholine transfer activity [123]. The two tunnel openings are large enough to allow lipid access aided by a flexible helix and possibly also by a mobile flap. Point mutations blocking the middle of the tunnel abolish lipid-transfer activities, suggesting that neutral lipids pass through this continuous tunnel. The curvature of the concave surface of CETP matches the radius of curvature of HDL particles [123]. CETP may, however, undergo conformational changes to accommodate larger lipoprotein particles with lower surface curvature, such as LDL and VLDL [123]. In a similar fashion, apoA-I can modulate CETP activity by modulating the properties of the phospholipid surface monolayer of HDL, inducing accumulation of cholesteryl ester molecules close to the surface or by promoting the formation of cholesteryl ester clusters accessible to CETP [124].

In addition to CETP, other proteins may display cholesteryl ester-transfer activity. An interesting example is provided by the serum opacity factor from *Streptococcus pyogenes* which transfers cholesteryl ester from HDL to a cholesteryl ester-rich microemulsion whilst forming neo-HDL, a cholesterol-poor, HDL-like particle [125].

CRITICAL CONTRIBUTION

Crystal structure of CETP at 2.2 Å resolution revealing a 60 Å long tunnel filled with two hydrophobic cholesteryl esters and plugged by an amphiphilic phosphatidylcholine at each end [123].

PLTP. Plasma **PLTP** mediates both net transfer and exchange of phospholipids between different lipoproteins (Table 4.1 and Fig. 4.2). Notably, PLTP is capable of converting spherical, apoA-I-containing HDL into both larger and smaller particles [9]. The net effect of PLTP on HDL may include decreasing HDL size [126] as emphasized by a negative correlation between HDL size and the specific activity of plasma PLTP in healthy humans [127]. The action of PLTP on HDL is accompanied by the dissociation of lipid-free/lipid-poor apoA-I and apoA-II and involves modulation of HDL composition. Furthermore, PLTP interacts with HDL particles and facilitates the transfer of phospholipids from triglyceride-rich lipoproteins to HDL. PLTP can also remodel spherical, apoE-containing rHDL into large and small particles; however, such remodeling does not involve the dissociation of apoE from HDL and is more extensive than that of apoA-I-containing particles [128]. In contrast, the ability of apoE-containing HDL to participate in PLTP-mediated phospholipid transfer is reduced.

In the circulation, PLTP exists in two forms, one of which is catalytically highly active and the other is of low activity. The highly active form is associated with apoE in a 160 kDa complex, while the low activity form exists as a 520 kDa complex containing apoA-I [129]. ApoE can activate the low activity form of

PLTP *in vitro* [130]; in addition, PLTP expression can be induced by LXRs and RXRs [131, 132].

PLTP-mediated lipid transfer plays a crucial role in the maintenance of HDL levels in mouse models [133]. First, PLTP over-expression increases the influx of phospholipid and cholesterol into HDL, leading to an increase in pre-beta HDL particles in mice [133]. Second, elevated plasma PLTP expression may reduce plasma HDL-C by stimulating the plasma turnover and hepatic uptake of HDL cholesteryl ester [101]. In parallel, hepatic VLDL secretion increases, potentially leading to accelerated atherogenesis. In turn, low HDL-C may also impact plasma PLTP as illustrated by low PLTP activity in patients with Tangier disease [134] (see 6.1 of chapter 6); inefficient stabilization of PLTP can account for this observation.

Systemic PLTP activity appears to be pro-atherogenic in animal models; however, low plasma PLTP activity results in a high HDL-C/LDL-C ratio and, combined with low CETP activity, can account for high resistance to atherosclerosis in some animal species [135]. On the other hand, macrophage-derived PLTP can protect from atherosclerosis. Thus, locally produced, macrophage-derived PLTP beneficially alters lipoprotein metabolism, decreasing total plasma cholesterol and increasing HDL-C levels, and reduces lesion progression in LDL receptor-deficient mice [136]. In contrast, macrophages isolated from PLTP-deficient mice display increased cholesterol accumulation, which may result from their reduced alpha-tocopherol content [137].

METHODOLOGIC NOTE

Data obtained in animal models of HDL metabolism should be translated to humans with caution, reflecting basic differences in lipoprotein and cholesterol metabolism between species.

Receptors

SR-BI. **SR-BI**, a multi-ligand receptor member of the CD36 superfamily, is a glycosylated HDL receptor of approximately 80 kDa mass that contains 481 amino acid residues [138] (Table 4.1). The SCARB1 gene coding for SR-BI is located on chromosome 12 and is highly expressed in the liver and steroidogenic tissue, such as adrenals, ovaries and testis [139]. In contrast to ABC transporters, SR-BI promotes ATP-independent bidirectional cholesterol flux between cellular membranes and lipoproteins according to concentration gradients. The major function of SR-BI in HDL remodeling, therefore, involves **selective uptake of HDL cholesteryl ester** from circulating HDL by hepatocytes; this process is coupled to the release of lipid-poor HDL particles into the plasma without the concomitant degradation of the lipoprotein particle [97] (Fig. 4.2). Thus, SR-BI is localized in specialized microvillar channel plasma membrane compartments that retain HDL and are sites for HDL cholesteryl ester selective uptake. Another important role played by SR-BI in HDL metabolism is cholesterol efflux from peripheral cells to mature HDL particles (see 7.1 of chapter 7); this pathway appears to be important in non-lipid-loaded macrophages [140].

At the molecular level, SR-BI mediates the selective translocation of HDL cholesteryl ester by forming a hydrophobic channel through the plasma membrane. In addition, SR-BI facilitates bi-directional flux of cholesterol across the membrane by modifying the phospholipid composition of the latter. The extracellular domain of SR-BI is required for HDL binding and the selective uptake of HDL-derived cholesteryl ester. The hydrophobicity of the N-terminal extracellular regions of the receptor is critical for cholesterol transport, as it may possibly mediate receptor-ligand and/or receptor-membrane interactions [141].

SR-BI primarily shuttles cholesteryl ester between plasma membranes and **large HDL particles**, thereby considerably affecting the HDL plasma profile [142]. Indeed, HDL particles that preferentially interact with SR-BI are larger and contain more lipid compared with those preferentially interacting with ABCA1 or ABCG1 [67]. In addition, hepatic removal of HDL cholesteryl ester accelerates with increasing HDL size [143]. The conformation of apoA-I at the HDL surface is important for the efficient transfer of cholesteryl ester to the cell, mediated by SR-BI [142]. In parallel, HDL interaction with SR-BI results in the release of lipid-poor particles containing apoA-I alone or both apoA-I and apoA-II. The presence of apoA-II improves the ability of apoA-I-containing HDL to re-associate with HDL particles following interaction with SR-BI. As a result, the catabolism of such particles is delayed compared with those that contain only apoA-I [144].

Thus, hepatic levels and the activity of SR-BI can strongly impact HDL metabolism. Elevated SR-BI expression results in enhanced HDL-cholesteryl ester removal from the circulation, reducing plasma HDL-C concentrations as repeatedly demonstrated in mice [145, 146]. In contrast, SR-BI deficiency increases plasma HDL-C and paradoxically enhances atherosclerosis, reflecting disturbed cholesterol transport to the liver [145, 146]. Human data are consistent with these results; thus, the missense P297S mutation in the SCARB1 gene is associated with elevated plasma HDL-C, attenuated adrenal steroidogenesis and diminished adrenal function in humans [147]. In addition, SR-BI protein levels in monocyte-derived macrophages are inversely associated with HDL-C levels and HDL particle size, and positively associated with cholesteryl ester uptake in women with hyperalphalipoproteinemia [148].

CRITICAL CONTRIBUTION

Characterization of SR-BI as a hepatic HDL receptor [138]. Demonstration of the role of SR-BI for HDL metabolism in humans [147].

Importantly, however, SR-BI-mediated removal of cholesterol from plasma is not of major importance in normolipidemic healthy humans. Indeed, as hepatic

SR-BI is expressed at low levels in such subjects, indirect cholesterol transport from HDL to the liver through apoB-containing lipoproteins mediated by CETP represents the major pathway for the clearance of HDL-derived cholesterol [75].

An optimal rate of HDL cholesteryl ester uptake via the SR-BI pathway can, nevertheless, be important for RCT. It is relevant, in this regard, that some compounds may activate SR-BI-mediated selective uptake of HDL cholesteryl ester. SR-BI expression in hepatocytes is activated by such nuclear receptors as farnesoid X receptor [149], PPAR alpha and RXR alpha [150]. Furthermore, aspirin promotes SR-BI expression both *in vitro* in primary human macrophages and *in vivo* in resident peritoneal macrophages of mice and in human carotid atherosclerotic specimens [151]. In addition, probucol up-regulates SR-BI expression and enhances the uptake of HDL cholesteryl esters by hepatocytes in hyperlipidemic rabbits [152]. Transcriptional up-regulation of SR-BI in both HepG2 and RAW 264.7 cells by trichostatin A represents another example of the activation of this pathway [153].

Moreover, over-expression of human and murine apoF accelerates plasma clearance of HDL cholesteryl ester and reduces HDL-C levels in mice by increasing the clearance of HDL-cholesteryl ester, suggesting activation of the interaction between HDL and SR-BI [117]. Selective uptake of cholesteryl ester from HDL can also be enhanced by hepatic lipase [154] and endothelial lipase [112]; the latter phenomenon is essential for the remodeling of large alpha-migrating HDL particles by endothelial lipase.

In addition to cholesteryl ester, SR-BI may also mediate the hepatic uptake of HDL-associated **free cholesterol** and **phosphatidylcholine**. In contrast to cholesteryl ester, however, SR-BI-mediated uptake of free cholesterol results from alterations to membrane structure which provokes enhanced bi-directional flux of this lipid between cells and HDL, and is independent of HDL tethering to the cell surface [155]. On the other hand, HDL-phosphatidylcholine uptake is largely accounted for by HDL internalization [156]. Phosphatidylethanolamine N-methyltransferase, a liver-specific enzyme that converts phosphatidylethanolamine to phosphatidylcholine, links phosphatidylcholine and cholesteryl ester uptake via SR-BI. Mice that lack the enzyme display reduced plasma levels of phosphatidylcholine and cholesterol in HDL [157]. Phosphatidylethanolamine N-methyltransferase deficiency also results in elevated hepatic levels of SR-BI paralleled by accelerated HDL cholesteryl ester uptake.

In addition to HDL, SR-BI binds several other ligands with high affinity, such as apoB-containing lipoproteins and oxidized LDL [144]. Thus, SR-BI plays an important role in the removal of VLDL and/or VLDL remnants [158]. Moreover, hepatic expression of SR-BI is linked to VLDL production. In the liver, a metabolic shunt that delivers HDL-derived cholesterol to a pool from which cholesterol is mobilized for VLDL production may exist [159]. In this regard, it is interesting that HDL uptake inhibits the activation of SREBP-1 in hepatocytes through a cholesterol-dependent mechanism [160]. Furthermore, SR-BI is an essential receptor for hepatitis C virus (HCV) entry into the liver. Hepatitis C virions may initially interact with SR-BI through associated lipoproteins before subsequent direct interactions of the viral glycoproteins with SR-BI occur [161]; PDZK1, a four-PDZ domain-containing adaptor protein that binds to the C-terminus of SR-BI via its first PDZ domain, facilitates virus entry [162].

Intracellularly, SR-BI expression is subject to complex regulation involving PDZK1 [163]. The protein controls the abundance and localization of hepatic SR-BI and is required for normal hepatic SR-BI regulation [164]. In addition to its role as a plasma membrane receptor, SR-BI may also function in cholesterol trafficking from late endosomes/lysosomes [165].

Across the body, SR-BI is responsible for the selective uptake of lipoprotein cholesterol and its ultimate delivery to sites of hormone synthesis in steroidogenic tissues [166]. SR-BI-mediated uptake of serum cholesteryl ester is essential for optimal adrenal glucocorticoid production [167].

IMPORTANT READING

Zannis VI, Chroni A, Krieger M. Role of apoA-I, ABCA1, LCAT, and SR-BI in the biogenesis of HDL. *J Mol Med* 2006; 84: 276–294.

Rousset X, Vaisman B, Amar M, Sethi AA, Remaley AT. Lecithin: cholesterol acyltransferase–from biochemistry to role in cardiovascular disease. *Curr Opin Endocrinol Diabetes Obes* 2009; 16: 163–171.

4.2 CATABOLISM

HDL protein components are catabolized at different rates. The plasma half-life of apoA-I and apoA-II, two major protein components of HDL, is between 4 and 5 days, corresponding to fractional catabolic rates of 0.20–0.25 [168, 169]; apoA-II is catabolized slightly slower than apoA-I. In contrast, other apolipoproteins, such as apoA-IV, apoCs and apoE, which are associated both with HDL and triglyceride-rich lipoproteins, are removed from the circulation at enhanced rates compared with apoA-I [170, 171].

The major sites of catabolism of the protein components of HDL particles are the liver and kidney [172]; HDL-associated cholesterol is primarily removed from the circulation by the liver. Following removal through the liver, HDL-derived cholesterol is secreted into bile [173]. Intriguingly, recent data suggest the existence of a non-biliary pathway that involves direct excretion of cholesterol by the proximal small intestine in mice [174, 175]. Nevertheless, **biliary sterol secretion** appears to be essential for efficient RCT in mice [176]; this pathway is mediated by **hepatic SR-BI** [177]. Such SR-BI-mediated biliary cholesterol secretion is highly potent and does not require transcytosis. This pathway depends largely on ABCB4-mediated phospholipid secretion and mixed micelles as acceptors in bile [177]. Hydrolysis of hepatic cholesteryl esters by neutral cholesteryl ester hydrolase is required to release free cholesterol from lipoprotein-delivered cholesteryl ester [178]. Free cholesterol produced in this reaction is employed for bile acid synthesis or directly secreted into the bile. Hepatic cholesteryl ester hydrolase thereby regulates the last step of RCT by promoting the flux of cholesterol entering the liver via SR-BI and increasing hepatic bile acid output [178]. Biliary cholesterol secretion is also regulated by hepatic ABCG1 which can mediate hepatocyte cholesterol efflux towards plasma from a pool accessible for biliary secretion, resulting in decreased cholesterol output into the bile [179]. Finally, biliary secretion of bile acids, phospholipids and cholesterol is under the control of the ABCB11, ABCB4 and ABCG5/G8 transporters, respectively [173, 180].

HDL particles can be removed from the circulation by **holoparticle HDL receptors** (Fig 4.2). Indeed, hepatic cells rapidly take up holo-HDL particles which are accumulated in endosomal compartments and can be subsequently transferred to multivesicular bodies and, to a minor degree, to lysosomes for degradation [181]. A small proportion of internalized HDL can also be re-secreted by cells [182, 183]. Cubulin, an endocytic receptor for intrinsic factor-vitamin B12 expressed in various absorptive epithelia [184, 185], ectopic beta-chain of membrane-bound ATP synthase [186, 187] and hepatic CD36 [188] represent plausible candidates for the holoparticle HDL receptor. **Cubulin** may be involved in the absorption of HDL in kidney proximal tubules, with megalin functioning as a co-receptor. On the other hand, the **ectopic beta-chain of ATP synthase** may be operative at the surface of hepatocytes [189]. Mechanistically, apoA-I binding to an ectopic F1-ATPase can stimulate extracellular production of adenosine diphosphate (ADP) that activates the HDL endocytosis pathway mediated by a purinergic, G protein-coupled receptor P2Y13 [190]. The key role of P2Y13 in the uptake of holo-HDL particles by hepatocytes is revealed by experiments in P2Y13-deficient mice which exhibit decreased hepatic HDL cholesterol uptake, hepatic cholesterol content and biliary cholesterol output, concomitant with a reduced RCT rate from macrophages to feces [191]. The beta-chain of ATP synthase can also be involved in retro-endocytosis of HDL particles [192]. Finally, hepatic **CD36** may participate in both HDL holoparticle internalization and the selective uptake of cholesteryl ester from HDL [188]. In addition, holo-HDL particle uptake and re-secretion have been proposed to be mediated by SR-BI [193] and may involve a non-identified low-affinity HDL receptor.

ApoE-containing, large HDL can be removed from the circulation via the LDL apoB/E receptor, which represents an alternative route of cholesterol removal from plasma [194]. This pathway may become of particular importance under dyslipidemic conditions, such as in swine and dogs fed a high fat diet [195].

Lipid-free/lipid-poor apolipoproteins, primarily apoA-I, can be rapidly cleared from the circulation by the **kidneys** [196] (Fig. 4.2). Although the glomerular filtration barrier prevents access of mature HDL particles to the proximal tubules, cubulin and megalin receptors may bind filtered lipid-free/lipid-poor HDL-derived apolipoproteins [197]. Consistent with the role of filtration, HDL catabolism by the kidney increases with decreasing pre-beta HDL size as evidenced by experiments in human apoA-I transgenic mice [143]. Renal filtration and receptor-mediated uptake of HDL apolipoproteins may, therefore, influence overall HDL metabolism. Specifically, the rate of apoA-I clearance by kidneys can affect plasma HDL concentrations. Indeed, plasma levels of HDL-C and apoA-I are inversely related to the estimated glomerular filtration rate and creatinine clearance in subjects without severely compromised kidney function, which supports the notion that a high glomerular filtration rate may be an independent determinant of a pro-atherogenic lipoprotein profile [198].

Lipid-free apolipoproteins are filtered by the kidney according to their hydrophobicity. Specifically, more hydrophilic apolipoproteins, such as apoA-I and apoA-IV, can be excreted into urine when renal proximal tubular reabsorption is impaired, whereas more hydrophobic apoA-II cannot [199]. Furthermore, apolipoproteins, such as apoA-I and apoA-II, can be re-absorbed through cubulin/megalin receptors in the kidney proximal tubules [200]. The lipid content of HDL particles can modulate their re-absorption, with cholesteryl ester favoring this process [201].

CLINICAL SUMMARY

HDL metabolism is under the influence of multiple receptors, lipid transfer proteins and enzymes (ABCA1, ABCG1, LCAT, CETP, PLTP, LPL, hepatic lipase, endothelial lipase, SR-BI), each of which is capable of modifying circulating levels of HDL-C.

REFERENCES

1. Jones MK, Gu F, Catte A, Li L, Segrest JP. "Sticky" and "promiscuous", the Yin and Yang of apolipoprotein A-I termini in discoidal high-density lipoproteins: a combined computational-experimental approach. *Biochemistry* 2011; 50: 2249–2263.
2. Rye KA, Barter PJ. Formation and metabolism of prebeta-migrating, lipid-poor apolipoprotein A-I. *Arterioscler Thromb Vasc Biol* 2004; 24: 421–428.
3. von Eckardstein A, Nofer JR, Assmann G. High density lipoproteins and arteriosclerosis. Role of cholesterol efflux and reverse cholesterol transport. *Arterioscler Thromb Vasc Biol* 2001; 21: 13–27.
4. Atmeh RF, Abd Elrazeq IO. Small high density lipoprotein subclasses: some of their physico-chemical properties and stability in solution. *Acta Biochim Pol* 2005; 1: 1.
5. Oram JF. The cholesterol mobilizing transporter ABCA1 as a new therapeutic target for cardiovascular disease. *Trends Cardiovasc Med* 2002; 12: 170–175.
6. Hamilton RL, Guo LS, Felker TE, Chao YS, Havel RJ. Nascent high density lipoproteins from liver perfusates of orotic acid-fed rats. *J Lipid Res* 1986; 27: 967–978.

7. Hamilton RL, Moorehouse A, Havel RJ. Isolation and properties of nascent lipoproteins from highly purified rat hepatocytic Golgi fractions. *J Lipid Res* 1991; 32: 529–543.
8. Jonas A. Lecithin cholesterol acyltransferase. *Biochim Biophys Acta* 2000; 1529: 245–256.
9. van Tol A. Phospholipid transfer protein. *Curr Opin Lipidol* 2002; 13: 135–139.
10. Le Goff W, Guerin M, Chapman MJ. Pharmacological modulation of cholesteryl ester transfer protein, a new therapeutic target in atherogenic dyslipidemia. *Pharmacol Ther* 2004; 101: 17–38.
11. Santamarina-Fojo S, Gonzalez-Navarro H, Freeman L, Wagner E, Nong Z. Hepatic lipase, lipoprotein metabolism, and atherogenesis. *Arterioscler Thromb Vasc Biol* 2004; 24: 1750–1754.
12. Clay MA, Newnham HH, Forte TM, Barter PI. Cholesteryl ester transfer protein and hepatic lipase activity promote shedding of apo A-I from HDL and subsequent formation of discoidal HDL. *Biochim Biophys Acta* 1992; 1124: 52–58.
13. Curtiss LK, Valenta DT, Hime NJ, Rye KA. What is so special about apolipoprotein AI in reverse cholesterol transport? *Arterioscler Thromb Vasc Biol* 2006; 26: 12–19.
14. Langmann T, Klucken J, Reil M, Liebisch G, Luciani MF, Chimini G, Kaminski WE, Schmitz G. Molecular cloning of the human ATP-binding cassette transporter 1 (hABC1): evidence for sterol-dependent regulation in macrophages. *Biochem Biophys Res Commun* 1999; 257: 29–33.
15. Santamarina-Fojo S, Peterson K, Knapper C, Qiu Y, Freeman L, Cheng JF, Osorio J, Remaley A, Yang XP, Haudenschild C et al. Complete genomic sequence of the human ABCA1 gene: analysis of the human and mouse ATP-binding cassette A promoter. *Proc Natl Acad Sci U S A* 2000; 97: 7987–7992.
16. Hozoji M, Kimura Y, Kioka N, Ueda K. Formation of two intramolecular disulfide bonds is necessary for ApoA-I-dependent cholesterol efflux mediated by ABCA1. *J Biol Chem* 2009; 284: 11293–11300.
17. Hamon Y, Luciani MF, Becq F, Verrier B, Rubartelli A, Chimini G. Interleukin-1beta secretion is impaired by inhibitors of the ATP binding cassette transporter, ABC1. *Blood* 1997; 90: 2911–2915.
18. Hamon Y, Broccardo C, Chambenoit O, Luciani MF, Toti F, Chaslin S, Freyssinet JM, Devaux PF, McNeish J, Marguet D et al. ABC1 promotes engulfment of apoptotic cells and transbilayer redistribution of phosphatidylserine. *Nat Cell Biol* 2000; 2: 399–406.
19. Oram JF, Lawn RM, Garvin MR, Wade DP. ABCA1 is the cAMP-inducible apolipoprotein receptor that mediates cholesterol secretion from macrophages. *J Biol Chem* 2000; 275: 34508–34511.
20. Wang N, Silver DL, Costet P, Tall AR. Specific binding of ApoA-I, enhanced cholesterol efflux, and altered plasma membrane morphology in cells expressing ABC1. *J Biol Chem* 2000; 275: 33053–33058.
21. Maekawa M, Kikuchi J, Kotani K, Nagao K, Odgerel T, Ueda K, Kawano M, Furukawa Y, Sakurabayashi I. A novel missense mutation of ABCA1 in transmembrane alpha-helix in a Japanese patient with Tangier disease. *Atherosclerosis* 2009; 206: 216–222.
22. Rothblat GH, Phillips MC. High-density lipoprotein heterogeneity and function in reverse cholesterol transport. *Curr Opin Lipidol* 2010; 21: 229–238.

23. Nandi S, Ma L, Denis M, Karwatsky J, Li Z, Jiang X-C, Zha X. ABCA1-mediated cholesterol efflux generates microparticles in addition to HDL through processes governed by membrane rigidity. *J Lipid Res* 2009; 50: 456–466.
24. Duong PT, Collins HL, Nickel M, Lund-Katz S, Rothblat GH, Phillips MC. Characterization of nascent HDL particles and microparticles formed by ABCA1-mediated efflux of cellular lipids to apoA-I. *J Lipid Res* 2006; 47: 832–843.
25. Vedhachalam C, Ghering AB, Davidson WS, Lund-Katz S, Rothblat GH, Phillips MC. ABCA1-induced cell surface binding sites for apoA-I. *Arterioscler Thromb Vasc Biol* 2007; 27: 1603–1609.
26. Hassan HH, Denis M, Lee D-YD, Iatan I, Nyholt D, Ruel I, Krimbou L, Genest J. Identification of an ABCA1-dependent phospholipid-rich plasma membrane apolipoprotein A-I binding site for nascent HDL formation: implications for current models of HDL biogenesis. *J Lipid Res* 2007; 48: 2428–2442.
27. Bailey D, Iatan I, Hafiane A, Krimbou L, Genest J. Abstract 1315: Evidence for a highly cooperative interaction between apoA-I and ABCA1/phospholipid microdomain binding sites system: Implications for nascent HDL speciation and biogenesis. *Circulation* 2009; 120: S476.
28. Nagao K, Zhao Y, Takahashi K, Kimura Y, Ueda K. Sodium taurocholate-dependent lipid efflux by ABCA1: effects of W590S mutation on lipid translocation and apolipoprotein A-I dissociation. *J Lipid Res* 2009; 50: 1165–1172.
29. Duong PT, Weibel GL, Lund-Katz S, Rothblat GH, Phillips MC. Characterization and properties of pre beta-HDL particles formed by ABCA1-mediated cellular lipid efflux to apoA-I. *J Lipid Res* 2008; 49: 1006–1014.
30. Lecompte MF, Bras AC, Dousset N, Portas I, Salvayre R, Ayrault-Jarrier M. Binding steps of apolipoprotein A-I with phospholipid monolayers: adsorption and penetration. *Biochemistry* 1998; 37: 16165–16171.
31. Nakano M. Elucidation of lipid complex formation mechanisms by static/dynamic structural evaluation. *Yakugaku Zasshi* 2009; 129: 1033–1039 (in Japanese).
32. Massey JB, Pownall HJ. Cholesterol is a determinant of the structures of discoidal high density lipoproteins formed by the solubilization of phospholipid membranes by apolipoprotein A-I. *Biochim Biophys Acta* 2008; 1781: 245–253.
33. Mulya A, Lee J-Y, Gebre AK, Thomas MJ, Colvin PL, Parks JS. Minimal lipidation of pre-{beta} HDL by ABCA1 results in reduced ability to interact with ABCA1. *Arterioscler Thromb Vasc Biol* 2007; 27: 1828–1836.
34. Lund-Katz S, Nguyen D, Dhanasekaran P, Kono M, Nickel M, Saito H, Phillips MC. Surface plasmon resonance analysis of the mechanism of binding of apoA-I to high density lipoprotein particles. *J Lipid Res* 2010; 51: 606–617.
35. Gelissen IC, Harris M, Rye KA, Quinn C, Brown AJ, Kockx M, Cartland S, Packianathan M, Kritharides L, Jessup W. ABCA1 and ABCG1 synergize to mediate cholesterol export to apoA-I. *Arterioscler Thromb Vasc Biol* 2006; 26: 534–540.
36. Jessup W, Gelissen IC, Gaus K, Kritharides L. Roles of ATP binding cassette transporters A1 and G1, scavenger receptor BI and membrane lipid domains in cholesterol export from macrophages. *Curr Opin Lipidol* 2006; 17: 247–257.
37. Zannis VI, Koukos G, Drosatos K, Vezeridis A, Zanni EE, Kypreos KE, Chroni A. Discrete roles of apoA-I and apoE in the biogenesis of HDL species: lessons learned from gene transfer studies in different mouse models. *Ann Med* 2008; 40 (Suppl. 1): 14–28.

38. Kypreos KE. ABCA1 promotes the de novo biogenesis of apolipoprotein CIII-containing HDL particles *in vivo* and modulates the severity of apolipoprotein CIII-induced hypertriglyceridemia. *Biochemistry* 2008; 47: 10491–10502.
39. Wolfrum C, Poy MN, Stoffel M. Apolipoprotein M is required for prebeta-HDL formation and cholesterol efflux to HDL and protects against atherosclerosis. *Nat Med* 2005; 11: 418–422.
40. Mulya A, Seo J, Brown AL, Gebre AK, Boudyguina E, Shelness GS, Parks JS. Apolipoprotein M expression increases the size of nascent prebeta HDL formed by ATP binding cassette transporter A1. *J Lipid Res* 2010; 51: 514–524.
41. Ahnstrom J, Axler O, Dahlback B. HDL stimulates apoM secretion. *Protein Pept Lett* 2010; 17: 1285–1289.
42. Gillard BK, Lin HY, Massey JB, Pownall HJ. Apolipoproteins A-I, A-II and E are independently distributed among intracellular and newly secreted HDL of human hepatoma cells. *Biochim Biophys Acta* 2009; 1791: 1125–1132.
43. Hassan HH, Bailey D, Lee DY, Iatan I, Hafiane A, Ruel I, Krimbou L, Genest J. Quantitative analysis of ABCA1-dependent compartmentalization and trafficking of apolipoprotein A-I: implications for determining cellular kinetics of nascent high density lipoprotein biogenesis. *J Biol Chem* 2008; 283: 11164–11175.
44. Fukuda M, Nakano M, Miyazaki M, Tanaka M, Saito H, Kobayashi S, Ueno M, Handa T. Conformational change of apolipoprotein A-I and HDL formation from model membranes under intracellular acidic conditions. *J Lipid Res* 2008; 49: 2419–2426.
45. Lu R, Arakawa R, Ito-Osumi C, Iwamoto N, Yokoyama S. ApoA-I facilitates ABCA1 recycle/accumulation to cell surface by inhibiting its intracellular degradation and increases HDL generation. *Arterioscler Thromb Vasc Biol* 2008; 28: 1820–1824.
46. Denis M, Landry YD, Zha X. ATP-binding cassette A1-mediated lipidation of apolipoprotein A-I occurs at the plasma membrane and not in the endocytic compartments. *J Biol Chem* 2008; 283: 16178–16186.
47. Azuma Y, Takada M, Shin HW, Kioka N, Nakayama K, Ueda K. Retroendocytosis pathway of ABCA1/apoA-I contributes to HDL formation. *Genes Cells* 2009; 14: 191–204.
48. Laffitte BA, Repa JJ, Joseph SB, Wilpitz DC, Kast HR, Mangelsdorf DJ, Tontonoz P. LXRs control lipid-inducible expression of the apolipoprotein E gene in macrophages and adipocytes. *Proc Natl Acad Sci U S A* 2001; 98: 507–512.
49. Rigamonti E, Chinetti-Gbaguidi G, Staels B. Regulation of macrophage functions by PPAR-alpha, PPAR-gamma, and LXRs in mice and men. *Arterioscler Thromb Vasc Biol* 2008; 28: 1050–1059.
50. Schmitz G, Langmann T. Transcriptional regulatory networks in lipid metabolism control ABCA1 expression. *Biochim Biophys Acta* 2005; 1735: 1–19.
51. Nishimaki-Mogami T, Tamehiro N, Sato Y, Okuhira K, Sai K, Kagechika H, Shudo K, Abe-Dohmae S, Yokoyama S, Ohno Y et al. The RXR agonists PA024 and HX630 have different abilities to activate LXR/RXR and to induce ABCA1 expression in macrophage cell lines. *Biochem Pharmacol* 2008; 76: 1006–1013.
52. Ogata M, Tsujita M, Hossain MA, Akita N, Gonzalez FJ, Staels B, Suzuki S, Fukutomi T, Kimura G, Yokoyama S. On the mechanism for PPAR agonists to enhance ABCA1 gene expression. *Atherosclerosis* 2009; 205: 413–419.

53. Horie T, Ono K, Horiguchi M, Nishi H, Nakamura T, Nagao K, Kinoshita M, Kuwabara Y, Marusawa H, Iwanaga Y et al. MicroRNA-33 encoded by an intron of sterol regulatory element-binding protein 2 (Srebp2) regulates HDL *in vivo*. *Proc Natl Acad Sci U S A* 2010; 107: 17321–17326.

54. Fernandez-Hernando C, Suarez Y, Rayner KJ, Moore KJ. MicroRNAs in lipid metabolism. *Curr Opin Lipidol* 2011; 22: 86–92.

55. Najafi-Shoushtari SH, Kristo F, Li Y, Shioda T, Cohen DE, Gerszten RE, Naar AM. MicroRNA-33 and the SREBP host genes cooperate to control cholesterol homeostasis. *Science* 2010; 328: 1566–1569.

56. Rayner KJ, Suarez Y, Davalos A, Parathath S, Fitzgerald ML, Tamehiro N, Fisher EA, Moore KJ, Fernandez-Hernando C. MiR-33 contributes to the regulation of cholesterol homeostasis. *Science* 2010; 328: 1570–1573.

57. Iatan I, Bailey D, Hafiane A, Zanesi N, Taccioli C, Odeh MA, Pajukanta P, Croce CM, Aqeilan RI, Krimbou L et al. Abstract 1316: Deficiency of the WW domain-containing oxidoreductase (WWOX) impairs the HDL biogenesis pathway. *Circulation* 2009; 120: S476–S477.

58. Iwamoto N, Lu R, Tanaka N, Abe-Dohmae S, Yokoyama S. Calmodulin interacts with ATP binding cassette transporter A1 to protect from calpain-mediated degradation and upregulates high-density lipoprotein generation. *Arterioscler Thromb Vasc Biol* 2010;, 30: 1446–1452.

59. Karwatsky J, Ma L, Dong F, Zha X. Cholesterol efflux to apoA-I in ABCA1-expressing cells is regulated by Ca2+-dependent calcineurin signaling. *J Lipid Res* 2010; 51: 1144–1156.

60. Okuhira K, Fitzgerald ML, Tamehiro N, Ohoka N, Suzuki K, Sawada J, Naito M, Nishimaki-Mogami T. Binding of PDZ-RhoGEF to ATP-binding cassette transporter A1 (ABCA1) induces cholesterol efflux through RhoA activation and prevention of transporter degradation. *J Biol Chem* 2010; 285: 16369–16377.

61. Zhao Y, Van Berkel TJ, Van Eck M. Relative roles of various efflux pathways in net cholesterol efflux from macrophage foam cells in atherosclerotic lesions. *Curr Opin Lipidol* 2010; 21: 441–453.

62. Brunham LR, Singaraja RR, Duong M, Timmins JM, Fievet C, Bissada N, Kang M. H., Samra A, Fruchart J-C, McManus B et al. Tissue-Specific Roles of ABCA1 Influence Susceptibility to Atherosclerosis. *Arterioscler Thromb Vasc Biol* 2009; 29: 548–554.

63. Nunes VS, Leanca CC, Panzoldo NB, Parra E, Cazita PM, Nakandakare ER, de Faria EC, Quintao EC. HDL-C concentration is related to markers of absorption and of cholesterol synthesis: Study in subjects with low vs. high HDL-C. *Clin Chim Acta* 2011; 412: 176–180.

64. Yasuda T, Grillot D, Billheimer JT, Briand F, Delerive P, Huet S, Rader DJ. Tissue-specific liver X receptor activation promotes macrophage reverse cholesterol transport *in vivo*. *Arterioscler Thromb Vasc Biol* 2010; 30: 781–786.

65. Cavelier C, Lorenzi I, Rohrer L, von Eckardstein A. Lipid efflux by the ATP-binding cassette transporters ABCA1 and ABCG1. *Biochim Biophys Acta* 2006; 1761: 655–666.

66. Yvan-Charvet L, Wang N, Tall AR. Role of HDL, ABCA1, and ABCG1 transporters in cholesterol efflux and immune responses. *Arterioscler Thromb Vasc Biol* 2010; 30: 139–143.

67. Lorenzi I, von Eckardstein A, Radosavljevic S, Rohrer L. Lipidation of apolipoprotein A-I by ATP-binding cassette transporter (ABC) A1 generates an interaction partner for ABCG1 but not for scavenger receptor BI. *Biochim Biophys Acta* 2008; 1781: 306–313.

68. Tarr PT, Tarling EJ, Bojanic DD, Edwards PA, Baldan A. Emerging new paradigms for ABCG transporters. *Biochim Biophys Acta* 2009; 1791: 584–593.

69. Hu YW, Wang Q, Ma X, Li XX, Liu XH, Xiao J, Liao DF, Xiang J, Tang CK. TGF-beta1 up-regulates expression of ABCA1, ABCG1 and SR-BI through liver X receptor alpha signaling pathway in THP-1 macrophage-derived foam cells. *J Atheroscler Thromb* 2010; 17: 493–502.

70. Li D, Wang D, Wang Y, Ling W, Feng X, Xia M. Adenosine monophosphate-activated protein kinase induces cholesterol efflux from macrophage-derived foam cells and alleviates atherosclerosis in apolipoprotein E-deficient mice. *J Biol Chem* 2010; 285: 33499–33509.

71. Mauerer R, Ebert S, Langmann T. High glucose, unsaturated and saturated fatty acids differentially regulate expression of ATP-binding cassette transporters ABCA1 and ABCG1 in human macrophages. *Exp Mol Med* 2009; 41: 126–132.

72. Rousset X, Vaisman B, Amar M, Sethi AA, Remaley AT. Lecithin: cholesterol acyltransferase–from biochemistry to role in cardiovascular disease. *Curr Opin Endocrinol Diabetes Obes* 2009; 16: 163–171.

73. Glomset JA. The mechanism of the plasma cholesterol esterification reaction: plasma fatty acid transferase. *Biochim Biophys Acta* 1962; 65: 128–135.

74. Zhou M, Sawyer J, Kelley K, Fordstrom P, Chan J, Tonn G, Carlson T, Retter M, Meininger D, Cheng J et al. Lecithin cholesterol acyltransferase promotes reverse cholesterol transport and attenuates atherosclerosis progression in New Zealand white rabbits. *Circulation* 2009; 120: S1175.

75. Schwartz CC, VandenBroek JM, Cooper PS. Lipoprotein cholesteryl ester production, transfer, and output *in vivo* in humans. *J Lipid Res* 2004; 45: 1594–1607.

76. Bailey D, Ruel I, Hafiane A, Cochrane H, Iatan I, Jauhiainen M, Ehnholm C, Krimbou L, Genest J. Analysis of lipid transfer activity between model nascent HDL particles and plasma lipoproteins: implications for current concepts of nascent HDL maturation and genesis. *J Lipid Res* 2010; 51: 785–797.

77. Zhang L, Song J, Cavigiolio G, Ishida BY, Zhang S, Kane JP, Weisgraber KH, Oda MN, Rye K-A, Pownall HJ et al. Morphology and structure of lipoproteins revealed by an optimized negative-staining protocol of electron microscopy. *J Lipid Res* 2011; 52: 175–184.

78. Shih AY, Sligar SG, Schulten K. Maturation of high-density lipoproteins. *J R Soc Interface* 2009; 6: 863–871.

79. Wu Z, Wagner MA, Zheng L, Parks JS, Shy JM 3rd, Smith JD, Gogonea V, Hazen SL. The refined structure of nascent HDL reveals a key functional domain for particle maturation and dysfunction. *Nat Struct Mol Biol* 2007; 14: 861–868.

80. Jones MK, Catte A, Li L, Segrest JP. Dynamics of activation of lecithin:cholesterol acyltransferase by apolipoprotein A-I. *Biochemistry* 2009; 48: 11196–11210.

81. Cavigiolio G, Shao B, Geier EG, Ren G, Heinecke JW, Oda MN. The interplay between size, morphology, stability, and functionality of high-density lipoprotein subclasses. *Biochemistry* 2008; 47: 4770–4779.

82. Cho KH. Synthesis of reconstituted high density lipoprotein (rHDL) containing apoA-I and apoC-III: the functional role of apoC-III in rHDL. *Mol Cells* 2009; 27: 291–297.
83. Dallinga-Thie GM, Franssen R, Mooij HL, Visser ME, Hassing HC, Peelman F, Kastelein JJ, Peterfy M, Nieuwdorp M. The metabolism of triglyceride-rich lipoproteins revisited: new players, new insight. *Atherosclerosis* 2010; 211: 1–8.
84. Klos KL, Kullo IJ. Genetic determinants of HDL: monogenic disorders and contributions to variation. *Curr Opin Cardiol* 2007; 22: 344–351.
85. Miyazaki O, Fukamachi I, Mori A, Hashimoto H, Kawashiri MA, Nohara A, Noguchi T, Inazu A, Yamagishi M, Mabuchi H et al. Formation of prebeta1-HDL during lipolysis of triglyceride-rich lipoprotein. *Biochem Biophys Res Commun* 2009; 379: 55–59.
86. Guha M, Gursky O. Effects of oxidation on structural stability and remodeling of human very low density lipoprotein. *Biochemistry* 2010; 49: 9584–9593.
87. Rye K-A, Bursill CA, Lambert G, Tabet F, Barter PJ. The metabolism and anti-atherogenic properties of HDL. *J Lipid Res* 2009; 50: S195–S200.
88. Tian L, Wu J, Fu M, Xu Y, Jia L. Relationship between apolipoprotein C-III concentrations and high-density lipoprotein subclass distribution. *Metabolism* 2009; 58: 668–674.
89. Perdomo G, Kim DH, Zhang T, Qu S, Thomas EA, Toledo FGS, Slusher S, Fan Y, Kelley DE, Dong HH. A role of apolipoprotein D in triglyceride metabolism. *J Lipid Res* 2010; 51: 1298–1311.
90. Julve J, Escola-Gil JC, Rotllan N, Fievet C, Vallez E, de la Torre C, Ribas V, Sloan JH, Blanco-Vaca F. Human apolipoprotein A-II determines plasma triglycerides by regulating lipoprotein lipase activity and high-density lipoprotein proteome. *Arterioscler Thromb Vasc Biol* 2010; 30: 232–238.
91. Boisfer E, Lambert G, Atger V, Tran NQ, Pastier D, Benetollo C, Trottier JF, Beaucamps I, Antonucci M, Laplaud M et al. Overexpression of human apolipoprotein A-II in mice induces hypertriglyceridemia due to defective very low density lipoprotein hydrolysis. *J Biol Chem* 1999; 274: 11564–11572.
92. Dugue-Pujol S, Rousset X, Pastier D, Quang NT, Pautre V, Chambaz J, Chabert M, Kalopissis A-D. Human apolipoprotein A-II associates with triglyceride-rich lipoproteins in plasma and impairs their catabolism. *J Lipid Res* 2006; 47: 2631–2639.
93. Castellani LW, Nguyen CN, Charugundla S, Weinstein MM, Doan CX, Blaner WS, Wongsiriroj N, Lusis AJ. Apolipoprotein AII is a regulator of very low density lipoprotein metabolism and insulin resistance. *J Biol Chem* 2008; 283: 11633–11644.
94. Koike T, Kitajima S, Yu Y, Li Y, Nishijima K, Liu E, Sun H, Waqar AB, Shibata N, Inoue T, et al. Expression of human apoAII in transgenic rabbits leads to dyslipidemia: A new model for combined hyperlipidemia. *Arterioscler Thromb Vasc Biol* 2009; 29: 2047–2053.
95. Coffill CR, Ramsamy TA, Hutt DM, Schultz JR, Sparks DL. Diacylglycerol is the preferred substrate in high density lipoproteins for human hepatic lipase. *J Lipid Res* 1997; 38: 2224–2231.
96. Duong M, Psaltis M, Rader DJ, Marchadier D, Barter PJ, Rye KA. Evidence that hepatic lipase and endothelial lipase have different substrate specificities for high-density lipoprotein phospholipids. *Biochemistry* 2003; 42: 13778–13785.

97. Miller M, Rhyne J, Hamlette S, Birnbaum J, Rodriguez A. Genetics of HDL regulation in humans. *Curr Opin Lipidol* 2003; 14: 273–279.

98. Rouhani N, Young E, Chatterjee C, Sparks DL. HDL composition regulates displacement of cell surface-bound hepatic lipase. *Lipids* 2008; 43: 793–804.

99. Young EK, Chatterjee C, Sparks DL. HDL-ApoE content regulates the displacement of hepatic lipase from cell surface proteoglycans. *Am J Pathol* 2009; 175: 448–457.

100. Brown RJ, Lagor WR, Sankaranaravanan S, Yasuda T, Quertermous T, Rothblat GH, Rader DJ. Impact of combined deficiency of hepatic lipase and endothelial lipase on the metabolism of both high-density lipoproteins and apolipoprotein B-containing lipoproteins. *Circ Res* 2010; 107: 357–364.

101. van Haperen R, Samyn H, van Gent T, Zonneveld AJ, Moerland M, Grosveld F, Jansen H, Dallinga-Thie GM, van Tol A, de Crom R. Novel roles of hepatic lipase and phospholipid transfer protein in VLDL as well as HDL metabolism. *Biochim Biophys Acta* 2009; 1791: 1031–1036.

102. Shimano H. ApoAII controversy still in rabbit? *Arterioscler Thromb Vasc Biol* 2009; 29: 1984–1985.

103. Hirata K, Dichek HL, Cioffi JA, Choi SY, Leeper NJ, Quintana L, Kronmal GS, Cooper AD, Quertermous T. Cloning of a unique lipase from endothelial cells extends the lipase gene family. *J Biol Chem* 1999; 274: 14170–14175.

104. Jaye M, Lynch KJ, Krawiec J, Marchadier D, Maugeais C, Doan K, South V, Amin D, Perrone M, Rader DJ. A novel endothelial-derived lipase that modulates HDL metabolism. *Nat Genet* 1999; 21: 424–428.

105. Keller PM, Rust T, Murphy DJ, Matico R, Trill JJ, Krawiec JA, Jurewicz A, Jaye M, Harpel M. Thrall S et al. A high-throughput screen for endothelial lipase using HDL as substrate. *J Biomol Screen* 2008; 13: 468–475.

106. Jaye M, Krawiec J. Endothelial lipase and HDL metabolism. *Curr Opin Lipidol* 2004; 15: 183–189.

107. Brown RJ, Edmondson AC, Griffon N, Hill TB, Fuki IV, Badellino KO, Li M, Wolfe ML, Reilly MP, Rader DJ. A naturally occurring variant of endothelial lipase associated with elevated HDL exhibits impaired synthesis. *J Lipid Res* 2009; 50: 1910–1916.

108. Yasuda T, Ishida T, Rader DJ. Update on the role of endothelial lipase in high-density lipoprotein metabolism, reverse cholesterol transport, and atherosclerosis. *Circ J* 2010; 74: 2263–2270.

109. Shimamura M, Matsuda M, Yasumo H, Okazaki M, Fujimoto K, Kono K, Shimizugawa T, Ando Y, Koishi R, Kohama T et al. Angiopoietin-like protein3 regulates plasma HDL cholesterol through suppression of endothelial lipase. *Arterioscler Thromb Vasc Biol* 2007; 27: 366–372.

110. Tanaka H, Ishida T, Johnston TP, Yasuda T, Ueyama T, Kojima Y, Kundu RK, Quertermous T, Ishikawa Y, Hirata K. Role of endothelial lipase in plasma HDL levels in a murine model of hypertriglyceridemia. *J Atheroscler Thromb* 2009; 16: 327–338.

111. Broedl UC, Maugeais C, Marchadier D, Glick JM, Rader DJ. Effects of nonlipolytic ligand function of endothelial lipase on high density lipoprotein metabolism *in vivo*. *J Biol Chem* 2003; 278: 40688–40693.

112. Nijstad N, Wiersma H, Gautier T, van der Giet M, Maugeais C, Tietge UJ. Scavenger receptor BI-mediated selective uptake is required for the remodeling of high density lipoprotein by endothelial lipase. *J Biol Chem* 2009; 284: 6093–6100.

113. Chapman MJ, Le Goff W, Guerin M, Kontush A. Cholesteryl ester transfer protein: at the heart of the action of lipid-modulating therapy with statins, fibrates, niacin, and cholesteryl ester transfer protein inhibitors. *Eur Heart J* 2010; 31: 149–164.

114. Luo Y, Liang CP, Tall AR. The orphan nuclear receptor LRH-1 potentiates the sterol-mediated induction of the human CETP gene by liver X receptor. *J Biol Chem* 2001; 276: 24767–24773.

115. Honzumi S, Shima A, Hiroshima A, Koieyama T, Ubukata N, Terasaka N. LXR alpha regulates human CETP expression *in vitro* and in transgenic mice. *Atherosclerosis* 2010; 212: 139.

116. Lund-Katz S, Laplaud PM, Phillips MC, Chapman MJ. Apolipoprotein B-100 conformation and particle surface charge in human LDL subspecies: implication for LDL receptor interaction. *Biochemistry* 1998; 37: 12867–12874.

117. Lagor WR, Brown RJ, Toh S-A, Millar JS, Fuki IV, de la Llera-Moya M, Yuen T, Rothbla, G, Billheimer JT, Rader DJ. Overexpression of apolipoprotein F reduces HDL cholesterol levels *in vivo*. *Arterioscler Thromb Vasc Biol* 2009; 29: 40–46.

118. Sun HY, Chen SF, Lai MD, Chang TT, Chen TL, Li PY, Shieh DB, Young KC. Comparative proteomic profiling of plasma very-low-density and low-density lipoproteins. *Clin Chim Acta* 2010; 411: 336–344.

119. He Y, Greene DJ, Kinter M, Morton RE. Control of cholesteryl ester transfer protein activity by sequestration of lipid transfer inhibitor protein in an inactive complex. *J Lipid Res* 2008; 49: 1529–1537.

120. de Barros J-PP, Boualam A, Gautier T, Dumont L, Verges B, Masson D, Lagrost L. Apolipoprotein CI is a physiological regulator of cholesteryl ester transfer protein activity in human plasma but not in rabbit plasma. *J Lipid Res* 2009; 50: 1842–1851.

121. Lee DM, Alaupovic P, Knight-Gibson C, Bagdade JD. Apolipoprotein-B subclasses as acceptors of cholesteryl esters transferred by CETP. *Eur J Clin Invest* 2008; 38: 734–742.

122. Joy T, Hegele RA. Is raising HDL a futile strategy for atheroprotection? *Nat Rev Drug Discov* 2008; 7: 143–155.

123. Qiu X, Mistry A, Ammirati MJ, Chrunyk BA, Clark RW, Cong Y, Culp JS, Danley DE, Freeman TB, Geoghegan KF et al. Crystal structure of cholesteryl ester transfer protein reveals a long tunnel and four bound lipid molecules. *Nat Struct Mol Biol* 2007; 14: 106–113.

124. Dergunov AD, Shabrova EV, Dobretsov GE. Composition, structure and substrate properties of reconstituted discoidal HDL with apolipoprotein A-I and cholesteryl ester. *Spectrochim Acta A Mol Biomol Spectrosc* 2010; 75: 1100–1107.

125. Pownall HJ, Courtney HS, Gillard BK, Massey JB. Properties of the products formed by the activity of serum opacity factor against human plasma high-density lipoproteins. *Chem Phys Lipids* 2008; 156: 45–51.

126. El Harchaoui K, Arsenault BJ, Franssen R, Despres JP, Hovingh GK, Stroes ES, Otvos JD, Wareham NJ, Kastelein JJ, Khaw KT et al. High-density lipoprotein particle size and concentration and coronary risk. *Ann Intern Med* 2009; 150: 84–93.

127. Cheung MC, Wolfbauer G, Deguchi H, Fernandez JA, Griffin JH, Albers JJ. Human plasma phospholipid transfer protein specific activity is correlated with HDL size: implications for lipoprotein physiology. *Biochim Biophys Acta* 2009; 1791: 206–211.
128. Settasatian N, Barter PJ, Rye K-A. Remodeling of apolipoprotein E-containing spherical reconstituted high density lipoproteins by phospholipid transfer protein. *J Lipid Res* 2008; 49: 115–126.
129. Karkkainen M, Oka T, Olkkonen VM, Metso J, Hattori H, Jauhiainen M, Ehnholm C. Isolation and partial characterization of the inactive and active forms of human plasma phospholipid transfer protein (PLTP). *J Biol Chem* 2002; 277: 15413–15418.
130. Janis MT, Metso J, Lankinen H, Strandin T, Olkkonen VM, Rye KA, Jauhiainen M, Ehnholm C. Apolipoprotein E activates the low-activity form of human phospholipid transfer protein. *Biochem Biophys Res Commun* 2005; 331: 333–340.
131. Cao G, Beyer TP, Yang XP, Schmidt RJ, Zhang Y, Bensch WR, Kauffman RF, Gao H, Ryan TP, Liang Y et al. Phospholipid transfer protein is regulated by liver X receptors *in vivo*. *J Biol Chem* 2002; 277: 39561–39565.
132. Laffitte BA, Joseph SB, Chen M, Castrillo A, Repa J, Wilpitz D, Mangelsdorf D, Tontonoz P. The phospholipid transfer protein gene is a liver X receptor target expressed by macrophages in atherosclerotic lesions. *Mol Cell Biol* 2003; 23: 2182–2191.
133. Tzotzas T, Desrumaux C, Lagrost L. Plasma phospholipid transfer protein (PLTP): review of an emerging cardiometabolic risk factor. *Obes Rev* 2009; 10: 403–411.
134. Samyn H, Moerland M, van Gent T, van Haperen R, van Tol A, de Crom R. Reduction of HDL levels lowers plasma PLTP and affects its distribution among lipoproteins in mice. *Biochim Biophys Acta* 2009; 1791: 790–796.
135. Liu HR, Wu G, Zhou B, Chen BS. Low cholesteryl ester transfer protein and phospholipid transfer protein activities are the factors making tree shrew and Beijing duck resistant to atherosclerosis. *Lipids Health Dis* 2010; 9: 114.
136. Valenta DT, Bulgrien JJ, Bonnet DJ, Curtiss LK. Macrophage PLTP is atheroprotective in LDLr-deficient mice with systemic PLTP deficiency. *J Lipid Res* 2008; 49: 24–32.
137. Ogier N, Klein A, Deckert V, Athias A, Bessede G, Le Guern N, Lagrost L, Desrumaux C. Cholesterol accumulation is increased in macrophages of phospholipid transfer protein-deficient mice: Normalization by dietary alpha-tocopherol supplementation. *Arterioscler Thromb Vasc Biol* 2007; 27: 2407–2412.
138. Acton S, Rigotti A, Landschulz KT, Xu S, Hobbs HH, Krieger M. Identification of scavenger receptor SR-BI as a high density lipoprotein receptor. *Science* 1996; 271: 518–520.
139. Cao G, Garcia CK, Wyne KL, Schultz RA, Parker KL, Hobbs HH. Structure and localization of the human gene encoding SR-BI/CLA-1. Evidence for transcriptional control by steroidogenic factor 1. *J Biol Chem* 1997; 272: 33068–33076.
140. Larrede S, Quinn CM, Jessup W, Frisdal E, Olivier M, Hsieh V, Kim M-J, Van Eck M, Couvert P, Carrie A et al. Stimulation of cholesterol efflux by LXR agonists in cholesterol-loaded human macrophages is ABCA1-dependent but ABCG1-independent. *Arterioscler Thromb Vasc Biol* 2009; 29: 1930–1936.
141. Papale GA, Nicholson K, Hanson PJ, Pavlovic M, Drover VA, Sahoo D. Extracellular hydrophobic regions in scavenger receptor BI play a key role in mediating HDL-cholesterol transport. *Arch Biochem Biophys* 2010; 496: 132–139.

142. Temel RE, Parks JS, Williams DL. Enhancement of scavenger receptor class B type I-mediated selective cholesteryl ester uptake from apoA-I(-/-) high density lipoprotein (HDL) by apolipoprotein A-I requires HDL reorganization by lecithin cholesterol acyltransferase. *J Biol Chem* 2003; 278: 4792–4799.

143. Mulya A, Lee J-Y, Gebre AK, Boudyguina EY, Chung S-K, Smith TL, Colvin PL, Jiang X-C, Parks JS. Initial interaction of apoA-I with ABCA1 impacts *in vivo* metabolic fate of nascent HDL. *J Lipid Res* 2008; 49: 2390–2401.

144. Van Eck M, Pennings M, Hoekstra M, Out R, Van Berkel TJ. Scavenger receptor BI and ATP-binding cassette transporter A1 in reverse cholesterol transport and atherosclerosis. *Curr Opin Lipidol* 2005; 16: 307–315.

145. Huby T, Doucet C, Dachet C, Ouzilleau B, Ueda Y, Afzal V, Rubin E, Chapman MJ, Lesnik P. Knockdown expression and hepatic deficiency reveal an atheroprotective role for SR-BI in liver and peripheral tissues. *J Clin Invest* 2006; 116: 2767–2776.

146. Hildebrand RB, Lammers B, Meurs I, Korporaal SJA, De Haan W, Zhao Y, Kruijt JK, Pratico D, Schimmel AWM, Holleboom AG, et al. Restoration of high-density lipoprotein levels by cholesteryl ester transfer protein expression in scavenger receptor class B type I (SR-BI) knockout mice does not normalize pathologies associated with SR-BI deficiency. *Arterioscler Thromb Vasc Biol* 2010; 30: 1439–1445.

147. Vergeer M, Korporaal SJ, Franssen R, Meurs I, Out R, Hovingh GK, Hoekstra M, Sierts JA, Dallinga-Thie GM, Motazacker MM et al. Genetic variant of the scavenger receptor BI in humans. *N Engl J Med* 2011; 364: 136–145.

148. West M, Greason E, Kolmakova A, Jahangiri A, Asztalos B, Pollin TI, Rodriguez A Scavenger receptor class B type I protein as an independent predictor of high-density lipoprotein cholesterol levels in subjects with hyperalphalipoproteinemia. *J Clin Endocrinol Metab* 2009; 94: 1451–1457.

149. Chao F, Gong W, Zheng Y, Li Y, Huang G, Gao M, Li J, Kuruba R, Gao X, Li S, He F. Upregulation of scavenger receptor class B type I expression by activation of FXR in hepatocyte. *Atherosclerosis* 2010; 213: 443–448.

150. Lopez D, McLean MP. Activation of the rat scavenger receptor class B type I gene by PPARalpha. *Mol Cell Endocrinol* 2006; 251: 67–77.

151. Wehinger A, Tancevski I, Seiler R, Frotschnig SM, Frantz S, Huber J, Eller P, Schgoer W, Foeger B, Patsch JR et al. Influence of aspirin on SR-BI expression in human carotid plaques. *Atherosclerosis* 2009; 206: 234–238.

152. Hong SC, Zhao SP, Wu ZH. Effect of probucol on HDL metabolism and class B type I scavenger receptor (SR-BI) expression in the liver of hypercholesterolemic rabbits. *Int J Cardiol* 2007; 115: 29–35.

153. Bao Y, Yang Y, Wang L, Gao L, Jiang W, Wang L, Si S, Hong B. Identification of trichostatin A as a novel transcriptional up-regulator of scavenger receptor BI both in HepG2 and RAW 264.7 cells. *Atherosclerosis* 2009; 204: 127–135.

154. Lambert G, Amar MJ, Martin P, Fruchart-Najib J, Foger B, Shamburek RD, Brewer HB Jr, Santamarina-Fojo S. Hepatic lipase deficiency decreases the selective uptake of HDL-cholesteryl esters *in vivo*. *J Lipid Res* 2000; 41: 667–672.

155. de la Llera-Moya M, Rothblat GH, Connelly MA, Kellner-Weibel G, Sakr SW, Phillips MC, Williams DL. Scavenger receptor BI (SR-BI) mediates free cholesterol flux independently of HDL tethering to the cell surface. *J Lipid Res* 1999; 40: 575–580.

156. Robichaud JC, van der Veen JN, Yao Z, Trigatti B, Vance DE. Hepatic uptake and metabolism of phosphatidylcholine associated with high density lipoproteins. *Biochim Biophys Acta* 2009; 1790: 538–551.
157. Robichaud JC, Francis GA, Vance DE. A role for hepatic scavenger receptor class B, type I in decreasing high density lipoprotein levels in mice that lack phosphatidylethanolamine N-methyltransferase. *J Biol Chem* 2008; 283: 35496–35506.
158. Van Eck M, Hoekstra M, Out R, Bos IST, Kruijt JK, Hildebrand RB, Van Berkel TJC. Scavenger receptor BI facilitates the metabolism of VLDL lipoproteins *in vivo*. *J Lipid Res* 2008; 49: 136–146.
159. Wiersma H, Nijstad N, Gautier T, Iqbal J, Kuipers F, Hussain MM, Tietge UJF. Scavenger receptor BI facilitates hepatic very low density lipoprotein production in mice. *J Lipid Res* 2010; 51: 544–553.
160. Yoshida M, Harada N, Yoshida K, Nakagawa T, Shimohata T, Mawatari K, Takahashi A, Sakaue H, Nakaya Y. High density lipoprotein inhibits the activation of sterol regulatory element-binding protein-1 in cultured cells. *FEBS Lett* 2010; 584: 1217–1222.
161. Catanese MT, Ansuini H, Graziani R, Huby T, Moreau M, Ball JK, Paonessa G, Rice CM, Cortese R, Vitelli A et al. Role of scavenger receptor class B type I in hepatitis C virus entry: kinetics and molecular determinants. *J Virol* 2010; 84: 34–43.
162. Eyre NS, Drummer HE, Beard MR. The SR-BI partner PDZK1 facilitates hepatitis C virus entry. *PLoS Pathog* 2010; 6.
163. Kocher O, Birrane G, Tsukamoto K, Fenske S, Yesilaltay A, Pal R, Daniels K, Ladias JA, Krieger M. *In vitro* and *in vivo* analysis of the binding of the C terminus of the HDL receptor scavenger receptor class B, type I (SR-BI), to the PDZ1 domain of its adaptor protein PDZK1. *J Biol Chem* 2010; 285: 34999–35010.
164. Fenske SA, Yesilaltay A, Pal R, Daniels K, Rigotti A, Krieger M, Kocher O. Overexpression of the PDZ1 domain of PDZK1 blocks the activity of hepatic scavenger receptor, class B, type I by altering its abundance and cellular localization. *J Biol Chem* 2008; 283: 22097–22104.
165. Ahras M, Naing T, McPherson R. Scavenger receptor class B type I localizes to a late endosomal compartment. *J Lipid Res* 2008; 49: 1569–1576.
166. Connelly MA. SR-BI-mediated HDL cholesteryl ester delivery in the adrenal gland. *Mol Cell Endocrinol* 2009; 300: 83–88.
167. Hoekstra M, Ye D, Hildebrand RB, Zhao Y, Lammers B, Stitzinger M, Kuiper J, Van Berkel TJC, Van Eck M. Scavenger receptor class B type I-mediated uptake of serum cholesterol is essential for optimal adrenal glucocorticoid production. *J Lipid Res* 2009; 50: 1039–1046.
168. Schaefer EJ, Zech LA, Jenkins LL, Bronzert TJ, Rubalcaba EA, Lindgren FT, Aamodt RL, Brewer HB Jr. Human apolipoprotein A-I and A-II metabolism. *J Lipid Res* 1982; 23: 850–862.
169. Thompson PD, Cullinane EM, Sady SP, Flynn MM, Bernier DN, Kantor MA, Saritelli AL, Herbert PN. Modest changes in high-density lipoprotein concentration and metabolism with prolonged exercise training. *Circulation* 1988; 78: 25–34.
170. Sun Z, Welty FK, Dolnikowski GG, Lichtenstein AH, Schaefer EJ. Effects of a National Cholesterol Education Program Step II Diet on apolipoprotein A-IV metabolism within triacylglycerol-rich lipoproteins and plasma. *Am J Clin Nutr* 2001; 74: 308–314.

171. Cohn JS, Batal R, Tremblay M, Jacques H, Veilleux L, Rodriguez C, Mamer O, Davignon J. Plasma turnover of HDL apoC-I, apoC-III, and apoE in humans: *in vivo* evidence for a link between HDL apoC-III and apoA-I metabolism. *J Lipid Res* 2003; 44: 1976–1983.

172. Sriram R, Lagerstedt JO, Petrlova J, Samardzic H, Kreutzer U, Xie H, Kaysen GA, Desreux JF, Thonon D, Jacques V et al. Imaging apolipoprotein AI *in vivo*. *NMR Biomed* 2011; 24: 916–924.

173. Dikkers A, Tietge UJ. Biliary cholesterol secretion: More than a simple ABC. *World J Gastroenterol* 2010; 16: 5936–5945.

174. Brown JM, Bell TA 3rd, Alger HM, Sawyer JK, Smith TL, Kelley K, Shah R, Wilson MD, Davis MA, Lee RG et al. Targeted depletion of hepatic ACAT2-driven cholesterol esterification reveals a non-biliary route for fecal neutral sterol loss. *J Biol Chem* 2008; 283: 10522–10534.

175. Temel RE, Brown JM. A new framework for reverse cholesterol transport: Non-biliary contributions to reverse cholesterol transport. *World J Gastroenterol* 2010; 16: 5946–5952.

176. Nijstad N, Gautier T, Briand F, Rader DJ, Tietge UJ. Biliary sterol secretion is required for functional *in vivo* reverse cholesterol transport in mice. *Gastroenterology* 2011; 140: 1043–1051.

177. Wiersma H, Gatti A, Nijstad N, Oude Elferink RP, Kuipers F, Tietge UJ. Scavenger receptor class B type I mediates biliary cholesterol secretion independent of ATP-binding cassette transporter g5/g8 in mice. *Hepatology* 2009; 50: 1263–1272.

178. Zhao B, Song J, Ghosh S. Hepatic overexpression of cholesteryl ester hydrolase enhances cholesterol elimination and *in vivo* reverse cholesterol transport. *J Lipid Res* 2008; 49: 2212–2217.

179. Wiersma H, Nijstad N, de Boer JF, Out R, Hogewerf W, Van Berkel TJ, Kuipers F, Tietge UJ. Lack of Abcg1 results in decreased plasma HDL cholesterol levels and increased biliary cholesterol secretion in mice fed a high cholesterol diet. *Atherosclerosis* 2009; 206: 141–147.

180. Fitzgerald ML, Mujawar Z, Tamehiro N. ABC transporters, atherosclerosis and inflammation. *Atherosclerosis* 2010; 211: 361.

181. Rohrl C, Pagler TA, Strobl W, Ellinger A, Neumuller J, Pavelka M, Stangl H, Meisslitzer-Ruppitsch C. Characterization of endocytic compartments after holo-high density lipoprotein particle uptake in HepG2 cells. *Histochem Cell Biol* 2010; 133: 261–272.

182. Heeren J, Beisiegel U, Grewal T. Apolipoprotein E recycling: implications for dyslipidemia and atherosclerosis. *Arterioscler Thromb Vasc Biol* 2006; 26: 442–448.

183. Faulkner LE, Panagotopulos SE, Johnson JD, Woollett LA, Hui DY, Witting SR, Maiorano JN, Davidson WS. An analysis of the role of a retroendocytosis pathway in ABCA1-mediated cholesterol efflux from macrophages. *J Lipid Res* 2008; 49: 1322–1332.

184. Moestrup SK, Kozyraki R. Cubilin, a high-density lipoprotein receptor. *Curr Opin Lipidol* 2000; 11: 133–140.

185. Hammad SM, Stefansson S, Twal WO, Drake CJ, Fleming P, Remaley A, Brewer HB Jr, Argraves WS. Cubilin, the endocytic receptor for intrinsic factor-vitamin B(12) complex, mediates high-density lipoprotein holoparticle endocytosis. *Proc Natl Acad Sci U S A* 1999; 96: 10158–10163.

186. Martinez LO, Jacquet S, Esteve JP, Rolland C, Cabezon E, Champagne E, Pineau T, Georgeaud V, Walker JE, Terce F et al. Ectopic beta-chain of ATP synthase is an apolipoprotein A-I receptor in hepatic HDL endocytosis. *Nature* 2003; 421: 75–79.

187. Vantourout P, Radojkovic C, Lichtenstein L, Pons V, Champagne E, Martinez LO. Ecto-F(1)-ATPase: A moonlighting protein complex and an unexpected apoA-I receptor. *World J Gastroenterol* 2010; 16: 5925–5935.

188. Brundert M, Heeren J, Merkel M, Carambia A, Herkel J, Groitl P, Dobner T, Ramakrishnan R, Moore KJ, Rinninger F. Scavenger receptor CD36 mediates uptake of high density lipoproteins in mice and by cultured cells. *J Lipid Res* 2011; 52: 745–758.

189. Zhang LH, Kamanna VS, Zhang MC, Kashyap ML. Niacin inhibits surface expression of ATP synthase beta chain in HepG2 cells: implications for raising HDL. *J Lipid Res* 2008; 49: 1195–1201.

190. Malaval C, Laffargue M, Barbaras R, Rolland C, Peres C, Champagne E, Perret B, Terce F, Collet X. Martinez LO. RhoA/ROCK I signalling downstream of the P2Y13 ADP-receptor controls HDL endocytosis in human hepatocytes. *Cell Signal* 2009; 21: 120–127.

191. Fabre AC, Malaval C, Ben Addi A, Verdier C, Pons V, Serhan N, Lichtenstein L, Combes G, Huby T, Briand F et al. P2Y13 receptor is critical for reverse cholesterol transport. *Hepatology* 2010; 52: 1477–1483.

192. Howard AD, Verghese PB, Arrese EL, Soulages JL. The beta-subunit of ATP synthase is involved in cellular uptake and resecretion of apoA-I but does not control apoA-I-induced lipid efflux in adipocytes. *Mol Cell Biochem* 2011; 348: 155–164.

193. Pagler TA, Neuhofer A, Laggner H, Strobl W, Stangl H. Cholesterol efflux via HDL resecretion occurs when cholesterol transport out of the lysosome is impaired. *J Lipid Res* 2007; 48: 2141–2150.

194. Funke H, Boyles J, Weisgraber KH, Ludwig EH, Hui DY, Mahley RW. Uptake of apolipoprotein E-containing high density lipoproteins by hepatic parenchymal cells. *Arteriosclerosis* 1984; 4: 452–461.

195. Mahley RW, Hui DY, Innerarity TL, Weisgraber KH. Two independent lipoprotein receptors on hepatic membranes of dog, swine, and man. Apo-B,E and apo-E receptors. *J Clin Invest* 1981; 68: 1197–1206.

196. Lee JY, Timmins JM, Mulya A, Smith TL, Zhu Y, Rubin EM, Chisholm JW, Colvin PL, Parks JS. HDLs in apoA-I transgenic Abca1 knockout mice are remodeled normally in plasma but are hypercatabolized by the kidney. *J Lipid Res* 2005; 46: 2233–2245.

197. Moestrup SK, Nielsen LB. The role of the kidney in lipid metabolism. *Curr Opin Lipidol* 2005; 16: 301–306.

198. Krikken JA, Gansevoort RT, Dullaart RPF, on behalf of the PSG. Lower HDL-C and apolipoprotein A-I are related to higher glomerular filtration rate in subjects without kidney disease. *J Lipid Res* 2010; 51: 1982–1990.

199. Graversen JH, Castro G, Kandoussi A, Nielsen H, Christensen EI, Norden A, Moestrup SK. A pivotal role of the human kidney in catabolism of HDL protein components apolipoprotein A-I and A-IV but not of A-II. *Lipids* 2008; 43: 467–470.

200. Dugue-Pujol S, Rousset X, Chateau D, Pastier D, Klein C, Demeurie J, Cywiner-Golenzer C, Chabert M, Verroust P, Chambaz J, et al. Apolipoprotein A-II is catabolized in the kidney as a function of its plasma concentration. *J Lipid Res* 2007; 48: 2151–2161.

201. Breznan D, Veereswaran V, Viau FJ, Neville TA, Sparks DL. The lipid composition of high-density lipoprotein affects its re-absorption in the kidney by proximal tubule epithelial cells. *Biochem J* 2004; 379: 343–349.

CHAPTER 5

EPIDEMIOLOGY

5.1 EPIDEMIOLOGY OF HDL-C

HDL-C and Cardiovascular Risk

A substantial body of evidence derived from epidemiological studies, clinical data and intervention trials demonstrates that **low circulating levels of HDL-C** constitute a strong, significant, and independent, predictor of cardiovascular disease (Table 5.1).

DEFINITIONS

Low HDL-C levels: circulating concentrations of HDL-C of <40 mg/dl in both sexes, or <40 mg/dl in men and <50 mg/dl in women [1]; these limits reflect the 50th percentiles of HDL-C distribution in a general population [e.g., 44 mg/dl in men and 53 mg/dl in women in the National Health and Nutrition Examination Survey (NHANES) III cohort [2], or 42 mg/dl in men and 54 mg/dl in women in the Framingham Offspring Study [3]].

High-Density Lipoproteins: Structure, Metabolism, Function, and Therapeutics, First Edition.
Anatol Kontush and M. John Chapman.

Hypoalphalipoproteinemia: a dyslipidemic state in which circulating levels of HDL-C are lower than the fifth (less frequently, tenth) percentile of the local sex- and age-matched healthy population.

Hyperalphalipoproteinemia: a dyslipidemic state in which circulating levels of HDL-C are higher than 95th (less frequently, 90th) percentile of local sex- and age-matched healthy population.

A reduction in the content of cholesterol and protein in the HDL fraction obtained from patients with recent coronary occlusion and other complications of atherosclerosis was first reported as early as 1951 [4]. Furthermore, similar alterations were observed in conditions known to predispose to early and extensive atherosclerosis, such as Type 2 diabetes and familial xanthoma tendinosum, suggesting an inverse relationship between plasma HDL-C concentration and coronary heart disease risk [4]. The power of low plasma HDL-C concentration to predict coronary heart disease risk was confirmed several years later in large-scale **prospective studies** [5, 6]. The landmark epidemiologic study that identified low HDL-C levels as an independent risk factor for coronary heart disease was the Framingham Heart Study. In this prospective study, the inverse relationship between HDL-C and the development of myocardial infarction was found in both men and women aged 50 years and older when HDL-C concentrations were measured between 1969 and 1971 [7]. Subsequent studies of large populations provided even more evidence in support of the role of low HDL-C as a cardiovascular risk factor. Thus, simultaneous analysis of four American prospective studies (Framingham Heart Study, Lipid Research Clinics Prevalence Mortality Follow-up Study, Coronary Primary Prevention Trial and Multiple Risk Factor Intervention Trial) revealed that a 1 mg/dl decrease in plasma levels of HDL-C was associated with a 2%–3% increase in the risk of coronary heart disease [8]. The finding was consistent across the studies and was independent of other cardiovascular risk factors, including plasma levels of LDL-C. Further evidence was provided by the large PROCAM study which documented the inverse relationship between HDL-C and coronary heart disease risk [9]. In this study, men with low HDL-C levels ($<$0.90 mmol/l) displayed a fourfold increased risk of coronary heart disease over a six-year period compared with men who had HDL-C levels of $\geq$0.90 mmol/l [9]. Importantly, the association between low HDL-C and elevated coronary heart disease risk remained significant after adjustment for other risk factors. In addition, when HDL-C levels were assessed in the screening phase of the Bezafibrate Infarction Prevention (BIP) Trial, the number of previous infarctions, severity of congestive heart failure and the severity of angina were negatively correlated with mean HDL-C in a dose-response manner [10].

Large multivariable analyses confirm the inverse relationship between HDL-C and both cardiovascular disease and coronary heart disease mortality. Using the Systematic COronary Risk Evaluation (SCORE) dataset which contains data on HDL-C for 104,961 individuals (45% women) without pre-existing coronary heart disease from seven pooled European prospective studies, a strong, graded, independent, inverse relationship between HDL-C and both cardiovascular disease and coronary heart disease mortality was observed [11]. Thus, adjusted hazard ratios per 0.5 mmol/l increment in HDL-C are 0.60 [confidence interval (CI), 0.51, 0.69] and 0.76 (CI, 0.70, 0.83) in women and men, respectively for the cardiovascular mortality endpoint. The corresponding hazard ratios are 0.53 (CI, 0.42, 0.68) and 0.79 (CI, 0.64, 0.98) in elderly women and men, respectively. The relationship is significant at all levels of cardiovascular risk and in all age groups [11].

Overall, among 58 prospective studies that provided multivariate assessments of the association between low HDL-C and cardiovascular risk, 31 studies found a

TABLE 5.1 Summary of Epidemiologic Findings Concerning the Relationship Between HDL and Cardiovascular Risk

- Low circulating levels of HDL-C are a significant, strong and independent predictor of cardiovascular disease.
- Low plasma HDL-C concentrations predict coronary heart disease risk as shown in large-scale cross-sectional and prospective studies, including the Framingham Heart Study and the Prospective Cardiovascular Münster (PROCAM) Study.
- Low HDL-C is a risk factor for multiple vascular diseases, including myocardial infarction, peripheral artery disease, ischemic stroke and deep vein thrombosis, as well as for recurrent cardiovascular events.
- Multivariate-adjusted meta-analyses confirm the inverse relationship between HDL-C and cardiovascular risk.
- The inverse relationship between HDL-C and cardiovascular risk persists across multiple populations, ethnicities and disease states, including the metabolic diseases Type 2 diabetes and Metabolic Syndrome.
- HDL-C concentrations may reflect the severity of coronary heart disease.
- HDL-C levels are included in all major algorithms to estimate cardiovascular risk, such as the Framingham risk prediction tool, the PROCAM score and the SCORE approach.
- Low HDL-C concentrations are related to an elevated risk, not only of cardiovascular disease, but also of other major disorders, including cancer and cognitive impairment.
- Epidemiologic studies of HDL are limited by extensive confounding. Indeed, low HDL-C levels are strongly linked to obesity, insulin resistance, exercise and alcohol consumption, all of which frequently feature elevated plasma triglyceride levels.
- Low HDL-C levels are frequently observed in general populations, especially in patients with coronary heart disease, and persist, despite statin treatment.
- The inverse relationship between HDL-C and coronary risk persists among coronary heart disease patients treated with statins and displaying low LDL-C levels.
- Measurement of apoA-I constitutes an acceptable alternative to the use of HDL-C in assessing cardiovascular risk. Evidence supporting the preferential use of apoA-I over HDL-C is, however, far from conclusive. At present, measurements of HDL-C should represent the first choice in assessing cardiovascular risk associated with apoA-I-containing lipoproteins.
- The clinical relevance of circulating levels of individual HDL particle subpopulations to atherosclerosis and cardiovascular disease remains unclear.

significant inverse association for all cardiovascular outcomes and subpopulations studied, whereas 17 studies found a significant association for some cardiovascular outcomes and/or subpopulations assessed [12]. These data leave no doubt regarding the role of low HDL-C as a major cardiovascular risk factor.

CRITICAL CONTRIBUTION

Reduction in the content of cholesterol and protein in the HDL fraction obtained from patients with recent coronary occlusion and other complications of atherosclerosis [4].

Identification of low HDL-C levels as an independent risk factor for coronary heart disease in large-scale epidemiologic studies [7–9, 13].

Relevance Across Multiple Populations and Disease States. Importantly, the inverse relationship between HDL-C and cardiovascular risk persists across multiple populations and disease states. First, such a relationship is observed both in **men** and **women**, with women showing stronger associations than men in large prospective studies [7, 11, 14]. The strong association between low HDL-C and cardiovascular disease in women may, however, diminish around the age of menopause [15, 16]. Second, low HDL-C levels remain strong predictors of cardiovascular disease over **extended follow-up** of up to 30 years [17]. The role of HDL-C as a cardiovascular risk factor can, however, be modified by lifestyle. For example, smoking exacerbates the effects of low HDL-C on coronary heart disease risk [18].

Low levels of HDL-C also represent a risk factor for cardiovascular disease in the **elderly population**. Thus, low levels of HDL-C were associated with cardiovascular mortality in elderly subjects, followed for 12 years, from Southern Finland [19]. In a similar fashion, persistently low HDL-C levels over eight years of follow-up are a risk factor for the development of cardiovascular events in an elderly population with a mean age of 70 years [20].

In addition, low levels of HDL-C are associated with major cardiovascular events and mortality in elderly men, but not in women, after adjustment for cardiovascular risk factors [16], consistent with a weakened association between HDL-C and cardiovascular risk after menopause [15]. Low HDL-C levels also represent a risk factor for cardiovascular mortality in elderly Japanese-American men [21]. Finally, low HDL-C and/or abnormal triglyceride levels, when added to abnormal LDL-C, are associated with a higher prevalence of cardiovascular disease in the elderly [22].

The original finding that HDL-C is a major cardiovascular risk factor in **Caucasians** was later confirmed by studies performed in **other populations**, such as in the Middle East [23] and Asia [24], which were both characterized by lower cardiovascular mortality compared with the USA. For example, when 7175 Japanese patients, initially free of cardiovascular disease, were followed for 9.6

years, HDL-C concentrations were inversely correlated with total mortality [24]. In addition, low HDL-C concentrations represented a significant, and independent, risk for coronary artery disease in a general population of Japanese men using long-term (12 years) follow-up data [25].

Low HDL-C is an important cardiovascular risk factor in individuals with Type 2 diabetes [26], hypertensive patients [27], patients with renal transplantation [28], coronary artery disease patients undergoing percutaneous coronary intervention and coronary angiography [29–31], patients with left ventricular dysfunction, regardless of the presence of coronary atherosclerosis [32], patients with carotid artery stenting [33], patients with non-ST-elevation acute coronary syndromes [34] and human immunodeficiency virus (HIV)-1-infected subjects [35]. In addition, low levels of HDL-C are an independent determinant for variant angina pectoris [36]. Moreover, the association between low HDL-C and atherosclerosis can largely underlie the power of metabolic syndrome to predict cardiovascular risk [37].

In addition, low HDL-C may reflect the presence of **multiple vascular diseases**. Most importantly, available data support an inverse association between HDL-C level and **stroke** and/or **carotid atherosclerosis**. Indeed, 8 out of 10 prospective cohort studies on the relationship between HDL-C and stroke risk ($n = 238739$), and 3 out of 8 case-control studies ($n = 3604$ cases, 8220 controls), revealed an association between elevated HDL-C level and decreased risk of stroke [38]. Prospective cohort studies reporting on relative risk per unit increase in HDL-C showed -11% to -15% decreased stroke risk per $+10$ mg/dl increase in HDL-C. Out of 37 studies on carotid intima-media thickness, 31 reported cross-sectional, one longitudinal, and five both cross-sectional and longitudinal associations between HDL-C level and carotid intima-media thickness. Out of 36 cross-sectional studies ($n = 51288$), 20 showed an inverse association between HDL-C level and carotid intima-media thickness. Out of six longitudinal studies ($n = 20065$), three showed no association, one showed a weak association in a subgroup of white women, and two showed a significant inverse relationship between HDL-C level and carotid intima-media thickness [38].

The association between low HDL-C and stroke appears to primarily reflect that between low HDL-C and **ischemic stroke**, which typically occurs more frequently compared with other stroke subtypes. Thus, levels of HDL-C are significantly lower in patients with acute ischemic stroke than in controls [39]. Furthermore, dyslipidemia, defined as an elevated apoB/apoA-I ratio, is associated with an increased risk of ischemic stroke [40]. The relationship between HDL-C and hemorrhagic stroke is, however, much less clear [40]. For example, no association between death owing to intraparenchymal hemorrhage and HDL-C was observed in the Ibaraki Prefectural Health Study [41]. Consistent with this notion, patients with ischemic stroke display lower HDL-C levels compared with those with hemorrhagic stroke [42]. As the latter stroke subtype appears only infrequently in Caucasians, data on its association with plasma lipid profile are limited.

Among other vascular diseases, **venous thrombosis** is associated with dyslipoproteinemia involving subnormal levels of HDL particles, elevated levels of small LDL particles and an elevated ratio of apoB to apoA-I in men aged <55 years [43]. In addition, patients with elevated levels of HDL-C and apoA-I display a decreased risk of recurrent venous thrombo-embolism [44]. Furthermore, the total cholesterol/HDL-C ratio and C-reactive protein (CRP) level are the strongest independent predictors of the development of peripheral arterial disease among 11 atherothrombotic biomarkers assessed at baseline in initially healthy male physicians from the USA [45]. Similarly, levels of HDL-C (adjusted hazard ratio 0.4 for extreme tertiles; 95% CI, 0.3–0.8), CRP, soluble intercellular adhesion molecule-1 (ICAM-1) and the total cholesterol/HDL-C ratio predict the development of peripheral arterial disease in a prospective cohort study with a median follow-up of 12.3 years involving 27,935 initially healthy female health professionals from the USA after multivariable adjustment [46]. In contrast, the risk of venous thromboembolism was not related to HDL-C levels in 19,049 participants of the Longitudinal Investigation of Thromboembolism Etiology (LITE) trial [47]. Finally, low HDL-C dyslipidemia is associated with the risk of heart failure [48, 49], left ventricular systolic dysfunction [50], left ventricular hypertrophy [51] and abdominal aortic aneurysm [52].

The relationship between low HDL-C and cardiovascular risk also holds true for surrogate markers of cardiovascular disease, such as **coronary calcification** [53] and **carotid intima-media thickness** [38]. Thus, low HDL-C levels and advanced age are independently associated with changes in coronary artery calcium scores after two years of follow-up in women with asymptomatic coronary artery disease [53]. Furthermore, carotid intima-media thickness is significantly associated with low HDL-C on multivariate analysis in healthy 8-year-old children from a community-based sample from Sydney, Australia [54]. HDL-C concentrations may, therefore, reflect the **severity of coronary heart disease**. Thus, the number of significantly stenosed vessels is inversely correlated with HDL-C levels in patients who underwent coronary angiography [55].

As a consequence of their clinical importance, HDL-C levels are included in all major **algorithms to estimate cardiovascular risk**, such as the Framingham risk prediction tool, the PROCAM score and the SCORE approach. For example, the updated Framingham risk prediction tool, which features an expanded endpoint of general cardiovascular disease and a 'vascular age concept', reveals a curvilinear relationship between cardiovascular risk and HDL-C involving disproportional increase at low HDL-C and linear decrease at high HDL-C [56]. Next, SCORE, the risk estimation system recommended by the European guidelines on cardiovascular disease prevention, estimates a 10-year risk of cardiovascular disease mortality based on age, sex, country of origin, systolic blood pressure, smoking status and either total cholesterol or the total cholesterol/HDL-C ratio. Remarkably, the additional inclusion of HDL-C alone in risk estimation considerably improves estimates in a clinical setting, particularly in women from high-risk countries and individuals with unusually high, or low, HDL-C levels [57].

The association of low HDL-C with elevated cardiovascular risk, as well as that of high HDL-C with reduced cardiovascular risk, is, therefore, firmly established. The relationship of **hyperalphalipoproteinemia** to cardiovascular disease is, however, more controversial. Thus, in patients undergoing angiographic evaluation for coronary artery disease, low and very low HDL-C levels are associated with coronary artery disease, whereas very high HDL-C levels (>70 mg/dl) are protective when compared with normal levels [58]. The coronary artery disease incidence is 69.5% at very low, 66.3% at low, 49.7% at normal and 34.9% at high HDL-C levels. On the other hand, in the Incremental Decrease in End Points through Aggressive Lipid Lowering (IDEAL; n = 8888) study which compared the efficacy of high-dose to usual-dose statin treatment for the secondary prevention of cardiovascular events, higher HDL-C proved a significant negative cardiovascular risk factor in univariate analysis but a significant positive cardiovascular risk factor following adjustment for age, gender, smoking, apoA-I, and apoB [59]. Increased risk estimates were particularly present at the high ends of the distributions (e.g., >70 mg/dl for HDL-C). Furthermore, elevated HDL-C was associated with additional cardiovascular risk in a subgroup of post-infarction patients at high-risk for recurrent coronary events identified by inflammation and hypercholesterolemia, [60, 61]. High HDL-C and high CRP levels also define a subgroup of men at high-risk for incident cardiovascular disease [62], potentially reflecting impaired HDL particle remodeling in the setting of inflammation [63].

OPEN QUESTION

Are very high levels of HDL-C associated with decreased cardiovascular risk?

The clinical importance of HDL-C as a cardiovascular risk factor presumes strict requirements for the **measurement of HDL-C levels**. Several important comments should be made in this regard.

First, HDL-C levels reveal considerable **seasonal variation**; thus, a statistically significant difference ($p < 0.001$) in HDL-C between winter (median, 37 mg/dl) and summer (median, 39 mg/dl) was observed at baseline in patients with acute coronary syndromes treated with statins in the Pravastatin or Atorvastatin Evaluation and Infection Therapy–Thrombolysis in Myocardial Infarction 22 (PROVE IT–TIMI 22) Study [64]. Upon treatment, median HDL-C was 41 mg/dl in winter vs. 43 mg/dl in summer in the pravastatin group and 39 mg/dl in winter vs. 42 mg/dl in summer in the atorvastatin group ($p < 0.001$). In contrast, LDL-C levels were higher in winter than in summer, resulting in a more atherogenic plasma lipid profile in winter than in summer [64]. Consistent with these data, HDL-C concentrations decrease, while LDL-C concentrations increase in parallel with outdoor temperature [65]. At the same time, measurements of HDL-C in a single spot blood sample still adequately represent average blood levels of this biomarker in the study population as the intra-class correlation coefficient for such measurements is 0.83 across seasons, demonstrating reliability and stability [66].

Second, it appears that to assess cardiovascular risk, HDL-C can be measured in both **fasting and non-fasting** samples. Indeed, levels of HDL-C, at most, change minimally in response to normal food intake in individuals from a general population [13, 67]. As a consequence, levels of HDL-C, triglyceride, total/HDL cholesterol ratio and apoA-I predict cardiovascular disease in prospective studies when measured non-fasting [68, 69]. Third, while the original approach to quantitate plasma HDL-C following dextran sulfate-Mg2+ upper index precipitation of apoB-containing lipoproteins provides high accuracy results [70], recently developed methods for directly measuring HDL-C levels may produce **errors in dyslipidemic individuals**. In particular, results provided by such 'direct' HDL-C assays are strongly influenced by the presence of hypertriglyceridemia [71]. As a consequence, none of these methods meets the National Cholesterol Education Program total error goals for dyslipidemic subjects [72, 73].

METHODOLOGIC NOTES

HDL-C levels display considerable seasonal variation; thus, a statistically significant difference ($p < 0.001$) in HDL-C between winter and summer is observed at baseline in acute coronary syndrome patients treated with statins [64].

In addition, recently developed methods for directly measuring HDL-C levels may produce errors in dyslipidemic individuals [71].

HDL-C and Other Cardiovascular Risk Factors. HDL-C levels progressively increase with **age** for both sexes (e.g., from 0.89 mmol/l at the age of <50 years to 0.96 mmol/l at the age of 65 years or older in men and from 1.06 to 1.14 mmol/l for the respective age groups in women [10]). Male **gender** is consistently associated with lower HDL-C compared with females. Thus, values for both HDL-C and apoA-I are higher in females than males independent of ethnic origin, such as, for example, in the USA, Sweden, Iran and Turkey [74, 75]. On the other hand, the effect of **ethnicity** is illustrated by the observation that HDL-C concentrations are lower in Filipino women (mean, 40.8 mg/dl) than in women from the USA [mean, 60.7 mg/dl according to the NHANES data] [76]. Accordingly, the prevalence of the isolated low HDL-C phenotype is 28.8% in the Philippines compared with 2.1% in the NHANES cohort [76]. Another example involves very low HDL-C levels in indigenous Australians [77]. In addition, ethnicity can modify the effects of gender and age on HDL-C levels. Thus, HDL-C increases with age in women from the USA, whereas it declines with age and converges with male HDL levels in Chinese women [78]. Furthermore, the sex gap in HDL-C is significantly smaller in Asian Indians compared with whites, independent of geographic location [79].

It is essential to emphasize that epidemiologic studies of HDL are limited by **extensive confounding**. Indeed, low HDL-C levels are strongly linked to obesity, insulin resistance, exercise and alcohol consumption, all of which feature elevated triglyceride levels. Furthermore, HDL-C concentrations reveal negative correlations with those of triglycerides, reflecting CETP activity [80]. Elevated triglyceride levels are associated with increased risk of myocardial infaction, ischemic heart disease and death in both men and women in a general population cohort after multifactorially adjusted analysis [81, 82]. Such confounding between triglycerides and HDL-C, in part, explains the absence of hypoalphalipoproteinemia from the World Health Organization (WHO) classification of dyslipidemias. The plasma ratio of triglycerides to HDL-C has, therefore, been proposed to assess cardiovascular risk related to both triglycerides and HDL-C levels; indeed, this ratio provides information about the atherogenicity of the plasma lipoprotein profile, reflecting the rate of cholesterol esterification in apoB-depleted plasma [83]. Furthemore, the plasma triglyceride/HDL-C ratio behaves well in epidemiologic studies; thus, it possesses a high predictive value of a first coronary event, regardless of body-mass index (BMI) [84] and is a powerful independent predictor of all-cause mortality and cardiovascular events in women with suspected ischemia [85]. Furthermore, an elevated triglyceride/HDL-C ratio shows a strong association with the extent of coronary artery disease [86] and is associated with increased coronary heart disease risk in males with a high prevalence of Metabolic Syndrome [87] and in Turkish subjects [88]. In addition, the triglyceride/HDL-C ratio is an imperfect surrogate for insulin resistance and associated coronary heart disease risk in the Framingham Offspring Cohort [89].

These data welcome a minimalistic view that low HDL-C represents a marker of elevated triglyceride levels which reflects high concentrations of pro-atherogenic triglyceride-rich remnant lipoproteins [90, 91]. This notion is supported by the absence of elevated cardiovascular risk in monogenic disorders involving extremely low levels of HDL-C, such as the presence of apoA-I Milano, familial LCAT deficiency and some forms of ABCA1 deficiency [91–96]. Furthermore, although both common and rare ABCA1 variants contribute to levels of HDL-C and the risk of ischemic heart disease in the general population, the association between ABCA1 variants and the risk of ischemic heart disease appears to be independent of plasma levels of HDL-C [97]. Further evidence against the causative role of HDL includes accelerated atherosclerosis in SR-BI-deficiency in association with elevated HDL-C in animal models [98].

DEFINITION

ApoA-I Milano: variant of apoA-I with R173C substitution whose carriers display very low HDL-C levels in association with reduced cardiovascular risk [96, 99].

Inconsistent with this notion, however, an adjustment for triglyceride levels does not eliminate the association between low HDL-C and elevated cardiovascular risk observed in large epidemiologic studies [12]. Thus, the hazard ratio for coronary heart disease is 0.78 (95% CI, 0.74–0.82) for one standard deviation (SD) increase in HDL-C equivalent to 15 mg/dl after adjustment for non-lipid risk factors, non–HDL-C and triglycerides in the Emerging Risk Factors Collaboration (ERFC) analysis of individual records from 302,430 subjects without initial vascular disease from 68 long-term prospective studies, mostly in Europe and North America, with 2.79 million person-years of follow-up [13]. Remarkably, the hazard ratio for coronary heart disease equals 0.99 (95% CI, 0.94–1.05) for 1-SD increase in triglyceride levels in the same study. Furthermore, the relationship of low HDL-C with cardiovascular risk is weakened, but remains significant (hazard ratio 1.38; 95% CI 1.10–1.73), when triglyceride levels are included in the multivariate adjusted model of the risk of ischemic heart disease in men followed for 14 years [100]. Elevated triglyceride and reduced HDL-C concentrations are independent risk factors in univariate analysis in this study. Specifically, men with the lowest level of HDL-C (<30 mg/dl) showed a higher risk (hazard ratio, 1.57; 95% CI 1.26–1.95) when compared with men with the highest level (HDL-C ≥60 mg/dl) [100]. Furthermore, low HDL-C and high triglyceride levels synergistically contribute to coronary heart disease risk in a case-control study of patients with low LDL-C of <130 mg/dl [101].

MAJOR CONTROVERSY

Several monogenic disorders involving low levels of HDL-C (apoA-I Milano, familial LCAT deficiency, ABCA1 deficiency) are not necessarily associated with elevated cardiovascular risk.

Prevalence of Low HDL-C. The strong association between low HDL-C dyslipidemia and cardiovascular risk requires estimation of the prevalence of this condition. Low HDL-C dyslipidemia is frequently observed in a general population, especially in patients with coronary heart disease. Thus, the prevalence of low HDL-C levels can vary from 20% in a general population to up to 60% in patients with established coronary heart disease [102].

Specifically, 37.1% men and 38.9% women display low HDL-C levels in the USA [78]; in Hong Kong, the corresponding prevalences are 34.3% in men and 34.5% in women [78]; in France, the prevalence of low HDL-C in a general population is 11.1% and 26.4% in men and women, respectively [103]; in Spain, 17.5% of the elderly population display low HDL-C, which is strongly, independently and inversely associated with established cardiovascular disease [104]; and, in Turkey, the prevalence of low HDL-C defined as <40 mg/dl equals 21.1%, increasing steadily with BMI [75].

Furthermore, among more than 8500 community-living men with coronary artery disease, 38% displayed low HDL-C (<0.90 mmol/l) [105]. In the screening phase of the BIP trial, 52% of patients with normal cholesterol levels exhibited low HDL-C of less than 0.90 mmol/l [10]. HDL-C levels were low (<40 mg/dl) in 54.6% of patients hospitalized with coronary artery disease, whereas <10% had HDL-C of ≥60 mg/dl [106]. Similarly, low HDL-C concentrations were observed in 56.6% of patients with acute coronary syndrome in Spain (defined as <1.04 mmol/l) [107] and in 53% of patients with acute coronary syndrome in Oman (defined as <40 mg/dl in men and <50 mg/dl in women) [108]. In addition, among 93,263 patients with non-ST-segment elevation acute coronary syndromes from the Can Rapid Risk Stratification of Unstable Angina Patients Suppress ADverse Outcomes with Early Implementation of the ACC/AHA Guidelines (CRUSADE) Initiative, 16,854 (18.1%) displayed very low HDL levels (10–29 mg/dl) and 32,185 (34.5%) had low HDL levels (30–39 mg/dl) [109]. In parallel, patients with very low HDL levels possessed the greatest risk of multi-vessel coronary disease on angiography and in-hospital mortality compared with patients who had normal and high HDL levels. The prevalence of low HDL-C among HIV-infected patients can reach 44.7%, increasing with HIV-1 RNA viral load and hypertriglyceridemia [110].

Importantly, plasma concentrations of HDL-C may reveal population-specific long-term trends. Thus, in women examined in Sweden in 1992–1993 and 2004–2005, HDL-C levels showed a significantly higher mean level in 2004–2005 [111]. In contrast, HDL-C concentrations decreased between 1986 and 2001 in parallel with increases in triglyceride concentrations in the Cardiovascular Risk in Young Finns Study, an ongoing multicentre study of atherosclerosis precursors in Finnish children and young adults [112].

HDL-C and Cardiovascular Risk in Patients Receiving Statins. HDL-raising therapy is largely intended to further decrease the residual cardiovascular risk remaining after statin treatment (see section 3); the significant residual cardiovascular risk that persists after statin treatment may stem, at least in part, from low HDL-C levels [113]. It is particularly important, therefore, to ensure that: (i) low HDL-C is prevalent in patients taking statins and at low LDL-C levels; and, (ii) low HDL-C is also associated with high cardiovascular risk in this population.

Indeed, low HDL-C levels remain prevalent despite statin treatment and the achievement of aggressive LDL-C goals in high-risk patients with coronary heart disease or coronary heart disease risk equivalents. In such subjects, low HDL-C (defined as ≤40 mg/dl in men, ≤50 mg/dl in women) was present in 66% of patients and was prevalent across all LDL-C levels, being most frequent in patients with LDL-C≤70 mg/dl (79% vs. 66% in those with LDL-C of 71–100 mg/dl and 64% in patients with LDL-C >100 mg/dl) [114]. Low HDL-C was highly prevalent in both patients taking statins (67%) and those not taking statins (64%) [114]. Furthermore, non-optimal HDL-C levels defined as less than 40 mg/dl for men and less than 50 mg/dl for women were highly prevalent in patients with acute coronary syndrome and reasonably controlled LDL-C levels

[115]. Indeed, 83% of patients with LDL-C levels of less than 100 mg/dl displayed non-optimal HDL-C. In addition, the prevalence of low HDL-C among dyslipidemic patients largely (by 85%) on lipid-modifying treatment was 40% (women) and 33% (men) in the Pan-European Survey on HDL-C [116]. Similarly, low HDL-C was observed in 27% of men (<40 mg/dl) and 32% of women (<50 mg/dl) in dyslipidemic patients on stable lipid-lowering therapy from nine countries [117]. In this study, the proportion of dyslipidemic adult patients on stable lipid-lowering therapy with low HDL-C was more than 20% in Canada and the USA and less than 12% in France, Spain and Brazil [117].

The deleterious effect of low HDL-C can be maintained in patients taking statins at low LDL-C levels [118]. Thus, baseline HDL-C levels make an important contribution to the residual risk of myocardial infarction in statin-treated patients. Weighted meta-regression analysis of 20 randomized controlled trials of statins with 543,210 person-years of follow-up and 7838 cases of myocardial infarction revealed a significant inverse association between HDL-C and cardiovascular risk in both statin-treated cohorts and non-statin-treated controls [119, 120]. After multiple adjustment, this relationship persists such that every -10 mg/dl decrement in HDL-C is associated with 7.1 (95% CI 6.7–7.3) and 8.3 (95% CI 8.1–8.5) more cases of myocardial infarction per 1000 person-years in statin-treated patients and non-statin controls, respectively ($p = 0.57$ in comparing the effect of HDL-C on cardiovascular risk in statin-treated patients vs. controls) [120].

Similarly, in patients with recent stroke or transient ischemic attack and no coronary heart disease treated with 80 mg/d atorvastatin or placebo for 4.9 years in the Stroke Prevention by Aggressive Reduction in Cholesterol Levels (SPARCL) trial, low baseline HDL-C predicted the risk of recurrent stroke, whereas low HDL-C, high triglyceride and a high LDL/HDL ratio were associated with major cardiovascular events [121]. In addition, a cumulative effect of achieving optimal levels of LDL-C, HDL-C, triglyceride and blood pressure on the risk of recurrent stroke and major cardiovascular events is observed in patients with recent stroke or transient ischemic attack and no known coronary heart disease randomized to 80 mg/d atorvastatin or placebo [122]. The pattern of low HDL-C, low apoA-I, small LDL particles and high triglyceride levels is also associated with elevated cardiovascular risk in statin-treated coronary patients, particularly in those with Type 2 diabetes [123]. Inconsistent with this data, no significant relationships are, however, noted between quartiles of HDL-C concentration and vascular risk, either at baseline or on treatment in participants of the Justification for the Use of Statins in Primary Prevention: An Intervention Trial Evaluating Rosuvastatin (JUPITER) trial given 20 mg rosuvastatin who displayed a low median on-treatment LDL-C concentration of 1.42 mmol/l [124].

The inverse relationship between HDL-C and coronary risk persists among coronary heart disease patients with LDL-C below 60 mg/dl, regardless of statin use [125]. Thus, patients with various clinical manifestations of vascular diseases in the highest HDL-C quintile display a lower risk for subsequent vascular events compared with patients in the lowest HDL-C quintile [126]. Higher HDL-C levels confer a lower risk for developing new vascular events irrespective of the localization of vascular disease, the use of lipid-lowering medication and plasma LDL-C concentration in this study [126]. After multiple adjustment, a graded association is observed between progressively lower levels of HDL-C and higher cardiovascular risk within each level of LDL-C (<100, 100–129, 130–159, 160–189 and ≥190 mg/dL) investigated in the Atherosclerosis Risk in Communities (ARIC) study [127]. In addition, a strong inverse association exists between the HDL-C level and all-cause mortality in patients with LDL-C levels of <70 mg/dl who underwent percutaneous coronary intervention [128]. Finally, in 27,748 apparently healthy women (mean age 54.7 years) with 11-year follow-up for cardiovascular events (n = 1017) in the Women's Health Study, there is a strong inverse association between HDL-C, apoA-I and the risk of cardiovascular disease [129]. The fully-adjusted hazard ratios associated with decreasing quintiles of HDL-C are 1.00, 1.18 (0.90–1.53), 1.32 (1.01–1.72), 1.50 (1.15–1.96) and 1.66 (1.26–2.19). Similarly-adjusted hazard ratios for apoA-I quintiles are 1.00, 1.11 (0.88–1.41), 1.08 (0.83–1.39), 1.51 (1.18–1.92), and 1.42 (1.10–1.83). Similarly for apoA-I, the association of HDL-C with cardiovascular disease is consistent across all levels of LDL-C. Remarkably, women with HDL-C >90th percentile (>69 mg/dl) had onset of cardiovascular disease 10 years later than women with HDL-C <10th percentile (<35 mg/dl) [129].

HDL-C and Other Major Diseases

Intriguingly, low plasma HDL-C concentrations are related to an elevated risk of, not only cardiovascular disease, but also **other major disorders**. Thus, meta-analysis of 21 large, randomized controlled trials with 625,477 person-years of follow-up and 8185 incident **cancers** over a median follow-up of 5 years reveals a significant inverse association between baseline HDL-C levels and the rate of incidence of cancer [130]. The inverse association persists after adjusting for baseline LDL-C, age, BMI and smoking status such that, for every +10 mg/dl increment in HDL-C, there is a −36% (95% CI, 24%–47%) relative reduction in the rate of incidence of cancer [130]. In support of these data, greater HDL-C levels at baseline are associated with a decreased risk of cancer (relative risk for high vs. low quintile (>55.3 vs.<36.2 mg/dl), 0.89(95% CI, 0.83–0.97) in the Alpha-Tocopherol, Beta-Carotene Cancer Prevention (ATBC) Study cohort [131]. The inverse association with HDL-C is evident for cancers of lung, prostate, liver and the hematopoietic system. Reverse causation may, however, have moderately contributed to this association as is obviously the case for the relationship between the risk of cancer and plasma total cholesterol [131]. In agreement with existing case-control studies, a modest association of low plasma HDL-C with a greater incidence of lung cancer was observed in the ARIC cohort [132]. Low HDL-C is also linked to an increased risk of colon cancer in the European Prospective Investigation into Cancer and Nutrition (EPIC) study [133]. Furthermore, low HDL-C may be a marker of increased breast cancer risk among premenopausal women [134]. Consistent with these data, normal and high levels of HDL-C

($\geq$50 mg/dl) are associated with reduced breast cancer risk among premenopausal women [135].

Low circulating HDL-C is equally characteristic of **neurologic disorders**. Thus, low levels of HDL-C are associated with poor memory in middle-aged adults [136, 137]. In addition, a decrease in HDL-C over the 5-year follow-up period in this cohort is associated with a decline in memory in the adjusted analysis. Similarly, low HDL-C levels at baseline predispose to elevated risk of cognitive decline in participants of the Three-City Study [138]. Low HDL-C levels are also associated with a higher severity of cerebral white matter changes on magnetic resonance imaging [139], as well as with lower gray matter volume in cognitively healthy adults [140]. Compared with healthy controls, subjects with long-term depression display lower HDL-C levels which may contribute to a higher risk of adverse coronary events in this disorder [141]. Finally, some studies suggest that low HDL-C levels may be a feature of Alzheimer's disease [142].

Patients with genetically disturbed metabolic pathways that are indirectly linked to HDL metabolism also frequently display low levels of HDL-C. For example, patients with lipid storage diseases like Gaucher's disease or Niemann–Pick disease present with low HDL-C [143]. Examples of other disorders that are associated with low levels of HDL-C include obesity [144], apnea/hypopnea syndrome [145], non-alcoholic fatty liver [146], iron overload [147], chronic kidney disease [148], rheumatoid arthritis [149], inflammatory bowel disease [150], periodontal infection [151], hepatitis B infection [152] and celiac disease [153]. In addition, high HDL-C levels predict low mortality in sepsis [154] and pulmonary arterial hypertension [155]. Such a long, and non-specific, list suggests that low HDL-C levels represent a common condition associated with deteriorated health status that may involve inflammation [156]. Consistent with this conclusion, self-rated health is positively correlated with plasma HDL-C in both men and women across all age groups as found in the Oslo Health Study [157]. Furthermore, high HDL-C is associated with longevity and may increase the odds for healthy aging [21, 158–160]. In particular, high HDL-C is significantly related to survival to 85 years of age after adjusting for other factors associated with longevity in the Veterans Affairs (VA) Normative Aging Study [161].

Not all diseases, however. are non-selectively related to reduced HDL-C concentrations; thus, HDL-C is significantly higher in osteoporotic post-menopausal women than in post-menopausal controls, secondary to the association between a pro-atherogenic lipid profile and higher bone mineral density [162].

5.2 EPIDEMIOLOGY OF HDL-ASSOCIATED PROTEINS AND ENZYMES

ApoA-I

Circulating levels of **apoA-I** strongly correlate with those of HDL-C and may, therefore, represent a biomarker of HDL metabolism, which better reflects the number of HDL particles in the circulation rather than their cholesterol load. Whether plasma apoA-I concentrations are more predictive of coronary heart disease than HDL-C concentrations is, however, difficult to address statistically because of the very tight correlation between apoA-I and HDL-C [163].

Several studies have suggested that apoA-I is a superior **predictor of cardiovascular risk** compared with HDL-C [164]. Thus, plasma apoA-I levels are inversely associated with the risk of myocardial infarction in a large, prospective Apolipoprotein-related MOrtality RISk (AMORIS) study performed in Sweden, which recruited 175,553 individuals [165]. The association remains significant after multivariate adjustment for several variables, including plasma triglyceride levels and is observed across all subgroups, including women. Furthermore, the INTERHEART study, a case-control study involving almost 30,000 participants in 52 countries, shows that an elevated apoB/apoA-I ratio accounts for more than half of the population-attributable risk for a first myocardial infarction [166]. In this large study involving 12,461 cases of acute myocardial infarction and 14,637 age- and sex-matched controls, the apoB/apoA-I ratio was superior to any of the cholesterol ratios, including the total cholesterol/HDL-C and LDL/HDL-C ratios, for the estimation of the risk of acute myocardial infarction in all ethnic groups, in both sexes, and at all ages [167]. It should be noted, in this regard, that plasma total cholesterol/HDL-C and apoB/apoA-I ratios are each typically more closely associated with cardiovascular risk than any of the individual lipoprotein parameters [168]. In patients with acute myocardial infarction, apolipoprotein ratios therefore appear to be more informative about cardiovascular risk than lipid fractions [169].

Consistent with these data, on-trial measures of apoA-I provide incremental information about the risk of cardiovascular events in patients treated with statins in the Air Force-Texas Coronary Atherosclerosis Prevention Study (AFCAPS/TexCAPS) [170]. In the IDEAL study, the on-treatment ratio of apoB/apoA-I was the strongest predictor of congestive heart failure in coronary heart disease patients, mostly driven by the strong association with apoA-I [171]. In the Collaborative Atorvastatin Diabetes Study (CARDS), the apoB/apoA-I ratio was marginally superior over the LDL-C/HDL-C ratio in predicting coronary heart disease [172]. Thus, in statin-treated patients with known coronary heart disease, the apoB/apoA-I ratio was positively associated with clinical events, whereas the LDL-C/HDL-C ratio is not [173].

Further results in support of the superiority of apoA-I over HDL-C as a predictor of cardiovascular risk are accumulating. Thus, in descendants (aged 7–18 years) of male survivors of myocardial infarction, the most informative predictors of future coronary attacks during childhood are plasma apoB-100 and apoB/apoA-I ratio [174]. In the Cardiovascular Risk in Young Finns Study (ages 3–18 years at baseline), the apoB/apoA-I ratio measured in adolescence is superior to the LDL-C/HDL-C ratio in predicting increased intima-media thickness in adulthood [175]. The apoB/apoA-I ratio is also a better discriminator of coronary artery disease risk than any of the conventional lipid ratios in Indian subjects [176]. In contrast to HDL-C, apoA-I remains negatively associated with cardiovascular risk across the major part of its distribution in both the IDEAL and the EPIC-Norfolk studies [59]. In the Treating to New Targets (TNT) study of intensive lipid-lowering treatment performed in coronary heart disease patients, the on-treatment apoB/apoA-I ratio provides the strongest association with the incidence of cerebrovascular events followed by the total cholesterol/HDL-C ratio [177].

The issue of the relative predictive ability of apoA-I vs. HDL-C is, however, far from being settled. Direct comparison of the two measures of HDL levels, HDL-C and apoA-I, does not always support the superiority of apoA-I measurements. Thus, in the EPIC-Norfolk Study that included 2349 individuals, the non-lipid-adjusted risk of a major coronary heart disease event per 1-SD change was 0.78 (0.70–0.87) for HDL-C and 0.79 (0.71–0.87) for apoA-I [178]. Furthermore, the Women's Health Study showed that the magnitude of risk for a major cardiovascular event is higher for low levels of HDL-C than for low levels of apoA-I [129]. In addition, there is no difference between HDL-C and apoA-I levels with regards to risk of recurrent events in fibrate-treated coronary heart disease patients enrolled in the Veterans Administration HDL Intervention Trial (VA-HIT) [179].

Models of cardiovascular risk prediction using the apoB/apoA-I ratio demonstrate performance characteristics comparable with, but not better than, that for other lipid ratios in a population-based, prospective cohort from Framingham, Massachusetts [180] and in the Prevention of Renal and Vascular End Stage Disease (PREVEND) cohort from the Netherlands [181]. In the Fenofibrate Intervention and Event Lowering in Diabetes (FIELD) study of fenofibrate in Type 2 diabetes, the apoB/apoA-I ratio does not perform better than non-HDL-C/HDL-C and total cholesterol/HDL-C ratios as a determinant of cardiovascular risk [182]. In addition, HDL-C and apoA-I levels alone are similarly predictive [183]. In a cross-sectional study of asymptomatic Caucasian subjects with and without Type 2 diabetes, HDL-C concentrations are inversely correlated with coronary artery calcium after adjustment for age, gender and medications [184]; in fully adjusted models, including additional cardiovascular risk factors, only HDL-C remains associated with coronary artery calcium, thereby out-performing apoA-I and apoA-II in terms of coronary artery disease prediction.

The on-treatment apoB/apoA-I ratio is similar to the total cholesterol/HDL-C ratio in terms of risk prediction in patients treated with statins in the PROVE IT-TIMI 22 trial [185]. In the EPIC-Norfolk study population, the total cholesterol/HDL-C and apoB/apoA-I ratios were similarly associated with coronary heart disease risk during a 6-year follow-up [186]. These data indicate that the power of the apoB to apoA-I ratio for predicting coronary heart disease is similar to that of more established lipid ratios.

Overall, the data reveal that the measurement of apoA-I undoubtedly constitutes an acceptable **alternative to the use of HDL-C** in assessing cardiovascular risk. Importantly, apoA-I assays, when performed with well-characterized antibodies, deliver high-quality data and are not inferior, in this regard, to assays determining HDL-C. Evidence supporting the preferential use of apoA-I over

HDL-C is, however, far from conclusive [187]. In view of the ongoing controversy, it appears premature to draw any firm conclusion regarding the relative capacities of apoA-I vs. HDL-C in predicting cardiovascular risk and mortality [40, 188]. Until this controversy is resolved, the well-established measurement of HDL-C should represent the first choice in assessing cardiovascular risk associated with apoA-I-containing lipoproteins.

OPEN QUESTION

Can circulating concentrations and/or enzymatic activites of HDL components other than total cholesterol and apoA-I constitute biomarkers of cardiovascular risk?

Other Apolipoproteins

In addition to apoA-I, other HDL-associated protein components may provide biomarkers of cardiovascular risk. In part, this observation reflects positive correlations between plasma concentrations of HDL-C, apoA-I and several HDL protein components, as well as a negative correlation between apoA-I and SAA. However, independent associations with cardiovascular risk have been reported for HDL components other than cholesterol and apoA-I. Thus, circulating concentrations of **apoA-II** are associated with a decreased risk of future coronary artery disease in apparently healthy people in a nested case-control study in the prospective EPIC-Norfolk cohort [189]. After multiple adjustment, patients in the upper quartile of (apoA-II levels>38.1 mg/dl) displayed an odds ratio for coronary artery disease of 0.48 (95% CI, 0.34–0.67) vs. those in the lowest quartile (<31.1 mg/dl). Importantly, additional adjustment for serum apoA-I levels does not affect risk prediction of apoA-II for future coronary artery disease (odds ratio 0.49, 95% CI 0.34–0.68). In addition, apoA-II levels represented a negative predictor for cardiovascular and total mortality in the Ludwigshafen Risk and Cardiovascular Health (LURIC) study [190]. Finally, low serum apoA-II concentrations predicted incident Metabolic Syndrome and Type 2 diabetes in Turks; increased apoA-II values were not associated with prevalent or incident coronary heart disease in this study and tended to be marginally atheroprotective in males [191]. In contrast, plasma apoA-II adjusted for HDL-C was independently associated with cardiovascular disease and future revascularization in stable cardiac patients [192].

Next, HDL-associated **apoC-III** may represent an important determinant of atherosclerosis progression [193]. In contrast, apoM levels in coronary heart disease patients and controls are similar in case-control studies, suggesting that apoM levels do not affect coronary heart disease risk in humans [194]. Thus, no association between plasma levels of apoM and coronary heart disease was found in two separate case-control studies [195]. On the other hand, plasma apoM concentration is modestly (−9%) reduced in Type 2 diabetes [196].

SAA

Elevated plasma levels of **SAA**, an acute phase protein transported by HDL, have been reported to represent a cardiovascular risk factor. In a prospective case-control study, the plasma level of high-sensitivity CRP (hsCRP) was the strongest univariate predictor of cardiovascular risk in apparently healthy post-menopausal women; the relative risk of events for women in the highest compared with the lowest quartile was 4.4 (95% CI, 2.2–8.9) [197]. However, SAA levels also revealed a significant, albeit weaker, association with cardiovascular events (relative risk for the highest vs. lowest quartile, 3.0). Remarkably, the levels of hsCRP and SAA were significant predictors of cardiovascular risk, even in the subgroup of women with low LDL-C levels [197].

In addition, baseline SAA levels were independently associated with angiographic coronary artery disease [198] and were highly predictive of three-year cardiovascular events in women referred for coronary angiography for suspected myocardial ischemia [198], as well as of the progression of carotid atherosclerosis in patients undergoing ultrasound investigations [199]. In comparison, hsCRP was not associated with angiographic coronary artery disease but, like SAA, was strongly, and independently, predictive of adverse cardiovascular outcome. The elevation of hsCRP and SAA levels at the time of hospital admission predicted poor outcome in patients with unstable angina [198]. Elevated SAA levels were associated with increased mortality in transplant coronary artery disease patients [200]. Both CRP and SAA predicted the risk of stroke in patients with acute coronary syndromes treated with atorvastatin in the Myocardial Ischemia Reduction with Aggressive Cholesterol Lowering (MIRACL) study [201]. In addition, serum SAA represents an inflammatory marker of the extent of coronary stenosis [202]. In the Cardiovascular Risk in Young Finns Study, SAA was associated with several metabolic risk factors but did not represent an independent predictor of intima-media thickness or carotid artery compliance [203].

Furthermore, circulating SAA can be elevated in coronary heart disease patients [200, 204], coronary artery disease subjects [200, 205], patients with ischemic stroke [206] and patients with Type 2 diabetes [207] compared with healthy controls. Local concentrations of SAA are elevated in the coronary artery at sites of plaque rupture in patients with acute myocardial infarction compared with concentrations in the aorta [208]. Finally, plasma levels of SAA correlate with the development of atherosclerosis in mouse models of this disease [209, 210].

PON1

The activities and levels of HDL-associated enzymes may also provide measures of cardiovascular risk. Thus, low **PON1** activity toward paraoxon has been reported to represent an independent risk factor for coronary events in men at high cardiovascular risk [211]. Furthermore, low concentration and enzymatic activity of PON1 may constitute an independent predictor of cardiovascular events in Type 2 diabetic patients [212]. By contrast, In middle-aged women, elevated PON1 activity is associated with an increased risk for acute myocardial infarction [213]. In patients on hemodialysis, the combination of reduced PON1 activity and increased CRP may identify subjects at high cardiovascular risk [214].

In contrast, PON1 activity was similar in cases and controls and was correlated with HDL-C levels in a large, prospective case-control study [215]. Using conditional logistic regression, PON1 activity shows a modest inverse relation with coronary artery disease risk; however, this relationship is abolished after adjustment for HDL-C and apoA-I, suggesting that low PON1 activity is not a causal factor in atherogenesis [215]. Therefore, the relationship between PON1 activity and cardiovascular risk appears controversial at present.

Serum PON1 activity decreases with age [216] and is lower in subjects with Metabolic Syndrome [217], low HDL-C dyslipidemia [218], Type 2 and Type 1 diabetes [216, 219, 220], familial hypercholesterolemia [221], coronary artery disease [222], acute coronary syndrome [223], aortic valve stenosis [224], systolic heart failure [225], polycystic ovary syndrome [226], early pregnancy failure [227], hyperemesis gravidarum [228], hemodialysis [214], end-stage renal disease [229], osteoarthritis [230], ovarian cancer [231], systemic lupus erythematosus [232], antiphospholipid syndrome [233], acute viral hepatitis [234], Takayasu's arteritis [235], *Helicobacter pylori* infection [236, 237], malaria [234] and sepsis [234, 238] compared with age-matched healthy controls. Elevated PON1 activity in psoriasis [239] and unchanged PON1 activity in multiple sclerosis [240] provide rare exceptions of the absence of functional PON1 deficiency in metabolic diseases. These data attest to the lack of specificity of low PON1 activity as a measure of elevated cardiovascular risk.

PAF-AH

Another plasma enzyme, **PAF-AH ($LpPLA_2$)**, of which a minor proportion ($<20\%$ of plasma activity) is carried by HDL, represents a more reliable marker of cardiovascular risk [241]. Thus, the majority of $LpPLA_2$ is present on LDL and increases in parallel with LDL-C, thereby suggesting a positive association with cardiovascular disease. Epidemiologic evidence, however, suggests that $LpPLA_2$ is an independent predictor of cardiovascular risk. Indeed, increased plasma $LpPLA_2$ activity was associated with incident fatal and non-fatal cardiovascular disease in the prospective, population-based Bruneck study [242]. In stable coronary artery disease, elevated levels of $LpPLA_2$ constitute a significant predictor of non-fatal adverse cardiovascular outcomes independent of traditional clinical risk factors and CRP [243]. Levels of $LpPLA_2$ activity were significantly associated with incident coronary heart disease among men and women with Type 2 diabetes in the Health Professionals Follow-Up Study and Nurses' Health Study, independent of traditional and inflammatory risk factors [244]; this positive association is strongest for more severe clinical end-points. $LpPLA_2$ mass and activity were associated with incident cardiovascular events in older adults in the Cardiovascular Health Study [245]; $LpPLA_2$ and CRP were independent and additive in the prediction of events in this study. In elderly people, $LpPLA_2$, together with other inflammatory indices, was a potential biomarker for vascular events, particularly coronary heart disease, in the Prospective Study of Pravastatin in the Elderly (PROSPER) [246]. Elevated levels of $LpPLA_2$ activity and mass were related to the incidence of ischemic stroke independent of established risk factors in the Malmo Diet and Cancer Study [247]. Interestingly, both elevated plasma $LpPLA_2$ activity and urinary concentrations of 8-epi-prostaglandin-2-alpha isoprostanes, products of $LpPLA_2$ activity on esterified phospholipids, are associated with greater coronary artery disease risk; odds ratios for the upper apoA-II levels quartiles of $LpPLA_2$ activity and 8-epi-prostaglandin-2-alpha excretion are 2.47 (95% CI: 1.79, 3.40) and 2.19 (1.52, 3.15), respectively, after multiple adjustment [248].

To further strengthen these data, a recent meta-analysis reveals log-linear-like associations of $LpPLA_2$ activity and mass with a risk of coronary heart disease and vascular death [249]. Relative risks adjusted for conventional risk factors are 1.10 (95% CI 1.05–1.16) with $LpPLA_2$ activity and 1.11 (1.07–1.16) with $LpPLA_2$ mass for coronary heart disease; 1.08 (0.97–1.20) and 1.14 (1.02–1.27) for ischemic stroke; 1.16 (1.09–1.24) and 1.13 (1.05–1.22) for vascular mortality; and, 1.10 (1.04–1.17) and 1.10 (1.03–1.18) for non-vascular mortality, respectively. Interestingly, these calculations reveal that associations of $LpPLA_2$ mass and activity are not exclusive to vascular outcomes and that the vascular associations depend, at least partly, on plasma lipid levels [249].

In addition to cardiovascular disease, $LpPLA_2$ can provide information regarding the risk of heart failure and other heart diseases. Thus, a baseline $LpPLA_2$ level is associated with an increased risk of future congestive heart failure in older adults independent of coronary risk factors [250]. Plasma $LpPLA_2$ levels are also associated with mortality in subjects with heart failure [251]. $LpPLA_2$ activity is positively related to left ventricular dysfunction, congestive heart failure and renal disease in elderly subjects [252]; cardiovascular risk factors only minimally explain variation in plasma levels of $LpPLA_2$ in this study. Plasma $LpPLA_2$ is correlated with carotid intima-media thickness in patients with beta-thalassemia [253]. Finally, $LpPLA_2$ levels are independently associated with the progression of cardiac allograft vasculopathy and predict a higher incidence of cardiovascular events and cardiovascular death in heart transplant patients [254].

It is important to emphasize, however, that, despite such consistent findings, associations of $LpPLA_2$ with cardiovascular risk are often moderate and can be attenuated by adjustment for classic risk factors and other inflammatory markers. Furthermore, Mendelian randomization does not support a causal role for $LpPLA_2$ in cardiovascular disease (see chapter 6). Indeed, unlike $LpPLA_2$ activity, PLA2G7 variants of $LpPLA_2$ associated with modest effects on enzyme activity are not related to cardiovascular risk markers, coronary atheroma or coronary heart disease [255]. Other than the presence of cardiovascular disease, several factors can significantly impact circulating $LpPLA_2$ concentrations. For example, race and sex independently influence $LpPLA_2$ activity and mass [256]. $LpPLA_2$ also increases with aging in parallel with HDL-C levels [257]. In multiple regression analysis, $LpPLA_2$ activity is positively associated with adiposity measures in men [258]. Genetic polymorphisms may also contribute to interindividual variation in $LpPLA_2$ activity and mass [259].

Other Enzymes

The plasma levels and activity of **LCAT** do not show an association with cardiovascular disease. Indeed, low plasma levels of LCAT are not associated with risk of coronary artery disease in the general population [260]. Furthermore, in the community-based, prospective, nested case-control PREVEND study, plasma LCAT activity (determined using an exogenous substrate assay) was only +5% higher in cases of cardiovascular disease compared with controls [261]. In this study, high plasma LCAT activity did not predict reduced cardiovascular risk and may even have attenuated cardioprotection associated with high HDL-C [261]. Consistent with these negative results, familial LCAT deficiency and fish-eye disease, two genetic forms of LCAT deficiency associated with low LCAT activity and reduced HDL-C, do not reveal accelerated atherosclerosis [262] (see chapter 6).

Plasma activity of another enzyme involved in HDL metabolism, **type II secretory phospholipase A_2($sPLA_2$)**, is associated with Metabolic Syndrome in young adults [263].

Lipid Transfer Proteins

Epidemiologic data on the relationship between cardiovascular risk and activities of lipid-transfer proteins are not fully consistent. Thus, plasma **CETP** activity is associated with coronary risk in studies of three common CETP genotypes (Taq1B, I405V and -629C >A) [264–267]. In parallel, however, HDL-C levels are reduced and triglyceride levels are elevated in subjects with high CETP activity in these studies, suggesting potential confounding. In addition, baseline CETP levels positively correlated with carotid intima-media thickness measured two years later in patients with familial hypercholesterolemia treated with statins [268]. Furthermore, baseline CETP levels are associated with future coronary artery disease in a subset of hypertriglyceridemic subjects from an apparently healthy population [269]. Inconsistent with these results, low CETP activity was associated with elevated cardiovascular risk in a prospective investigation of participants of the Framingham Heart Study [270].

Moreover, on one hand, high plasma **PLTP** activity is related to fatal and non-fatal cardiovascular events in coronary artery disease patients under statin treatment in multivariate regression analyses [271]. On the other hand, low, rather than high, PLTP activity is a marker of the presence of peripheral arterial disease in non-diabetic, non-smoking subjects [272].

CLINICAL SUMMARY

Low circulating levels of HDL-C and apoA-I are firmly established as significant and independent predictors of cardiovascular disease. The clinical relevance of other HDL-related biomarkers, such as PON1, remains questionable, at best.

5.3 EPIDEMIOLOGY OF HDL PARTICLE SUBPOPULATIONS

Separated by Density

As HDL particles are highly heterogeneous in structure, metabolism and function, distinct HDL subpopulations may differ in their effects on cardiovascular disease [273–275]. The cardioprotective effects of HDL may, therefore, depend on the HDL particle profile [276]. Cholesterol in HDL is mainly carried by large, light, lipid-rich HDL2 subfractions [277] (see 2.2 of chapter 2); HDL2b and HDL2a levels account, in a major way, for the differences in HDL-C between coronary heart disease patients and controls [278]. As a corollary, HDL2 particles may represent critical components in determining the inverse relationship between HDL-C and cardiovascular risk [279]. Concentrations of **HDL2-C** and **HDL3-C** as estimates for plasma levels of the two major HDL subfractions were, therefore, measured in several studies which differed in separation methods (polyanion precipitation vs. ultracentrifugation). However, conflicting results were obtained [163], with evidence that both HDL2-C and HDL3-C can constitute strong predictors of coronary heart disease or cardiovascular risk factors [277, 280–288]. Indeed, plasma levels of **HDL2b**, which are strongly correlated with HDL-C, are inversely related to coronary heart disease risk [280]; in contrast, increased HDL3b concentrations are associated with an atherogenic lipoprotein phenotype characterized by increased levels of triglycerides and small, dense LDL [289]. Furthermore, HDL2-C levels were inversely associated with coronary heart disease risk in the Kuopio Study [290] and Quebec City Suburbs Study [291]. Consistent with these data, elevated levels of HDL2-C are observed in females vs. males aged 20–60 years old, but this difference disappears in older subjects [292].

In contrast, plasma levels of small HDL3-C are a strong and independent predictor of protection against atherosclerosis in the Physician's Health Study [293], the 5-year follow-up of the Caerphilly and Speedwell study groups [294], and the 9-year follow-up of the Caerphilly study [287] and the VA-HIT study [284]. In addition, levels of HDL3-C were consistently elevated in females vs. males aged 20–60 years old [292]. Finally, both HDL2 and HDL3 mass was related to cardiovascular risk in a prospective, 29-year follow-up of Gofman's Livermore Cohort of 1905 men measured for lipoprotein mass concentrations by analytic ultracentrifugation between 1954 and 1957 [295].

Separated by Electrophoretic Mobility

A more consistent picture emerges when HDL particle heterogeneity is assessed by gel electrophoresis. Thus, plasma levels of **large alpha-1-HDL** are consistently associated with protection from atherosclerosis [296–300]. In addition, analysis of HDL subpopulations carried out in the Framingham Offspring Study revealed that reduced levels of large alpha-1 HDL particles were significantly associated with coronary heart disease prevalence [301]. These data allow definition of a threshold above which the risk of developing coronary heart disease is elevated when the level of apoA-I in plasma alpha-1 HDL is less than 14 mg/dl [301]. In parallel, levels of **small discoid HDL particles** (the sum of apoA-I in pre-beta-1 HDL and alpha-4 HDL mg/dl) are often elevated in coronary heart disease patients and large alpha-2 HDL particles are decreased [296–298, 302]. Importantly, the power of large alpha-1 HDL is superior to that of HDL-C in coronary heart disease risk-prediction [297, 302, 303]. On the other hand, the clinical significance of pre-alpha-HDL particles is unclear. A high degree of complexity and limited reproducibility, however, limit the routine application of this assay to assess cardiovascular risk in clinical practice.

> **CRITICAL CONTRIBUTION**
>
> Identification of low plasma levels of large alpha-1 HDL particles as a strong, independent predictor of cardiovascular events [301, 303].

Separated by Composition

When HDL particles are separated by affinity chromatography, elevated plasma levels of **LpA-I** are often [304, 305], but not always [301, 303, 306], associated with protection from atherosclerosis. The clinical significance of LpA-I:A-II levels remains even more controversial [301,303–305, 307]. Animal data also suggest that LpA-I is more atheroprotective than LpA-I:A-II [308]. Such inconsistent findings probably reflect intrinsic heterogeneity in LpA-I and LpA-I:A-II particles, resulting in complex relationships of their subclasses with HDL-C. As a consequence, elevated HDL-C levels associated with reduced cardiovascular risk may be accompanied by increases in large HDL containing both LpA-I and LpA-I:A-II, as well as by decreases in small HDL containing LpA-I [305].

Separated by Size

When particle concentrations of HDL classes are assessed by NMR, levels of **large HDL** (9.4–14 nm) typically display a negative correlation with cardiovascular risk in univariate analyses, whereas those of small HDL particles (7.3–8.2 nm) show positive correlations [59, 178, 179,309–316]. Consistent with data obtained in ultracentrifugally isolated HDL, concentrations of large HDL particles are elevated in women, increase with age [309, 317], are associated with longevity [159] and can be raised by exercise [318]. Importantly however, multiple confounding related to strong inverse correlations between large and small HDL particles, as well as between large HDL particles and total LDL particles, significantly complicates the interpretation of the relative roles of individual HDL subclasses as evaluated by NMR in the protection from cardiovascular disease [178, 309, 311, 316, 319]. In this regard, it is relevant that the evaluation of plasma levels of HDL subpopulations by the ion mobility assay reveals comparable contributions of small (7.7–10.5 nm) and large (10.5–14.5 nm) HDL to cardiovascular risk, with hazard ratios for coronary events per 1 SD of 0.78 and 0.75, respectively [320].

As HDL subpopulations differ in size, and concentrations of large HDL particles are consistently associated with reduced cardiovascular risk, **mean HDL size** may represent an integrative measure of HDL particle profile and heterogeneity. The ability of HDL size to predict cardiovascular risk has, therefore, been assessed, revealing an inverse association as documented by both NMR and ion mobility measurements [320]. Hazard ratios for incident cardiovascular disease over 11 years calculated using HDL size measured by NMR are, however, comparable, and not superior, to those obtained using HDL-C [316] and can be largely explained by traditional cardiovascular risk factors [321]. Interestingly, ion mobility identifies two independent principal components of such association which can be defined as the atherogenic lipoprotein phenotype, including elevated levels of triglycerides and small LDL, and reduced concentrations of both large and small HDL particles; the two components may reflect independent mechanisms of HDL-mediated cardioprotection [320].

The clinical relevance of circulating levels of individual HDL particle subpopulations to atherosclerosis and cardiovascular disease therefore remains unclear [72]. The inverse association between large alpha-1 particles and cardiovascular risk represents the only notable exception repeatedly reported in several large studies. Such association should not, however, be necessarily interpreted as a direct consequence of the protective properties of such a large HDL subclass; indeed, it may also reflect cardioprotection provided by a metabolic process leading to the formation of large alpha-1 HDL, with efficient RCT being the most relevant candidate. More refined analyses of lipoprotein subspecies may, however, lead to improvements in cardiovascular risk evaluation in future studies [72, 275].

IMPORTANT READING

Asztalos BF, Schaefer EJ. HDL in atherosclerosis: actor or bystander? *Atherosclerosis* 2003; 4 (Suppl.): S21–S29.

Cooney MT, Dudina A, De Bacquer D, Wilhelmsen L, Sans S, Menotti A, De Backer G, Jousilahti P, Keil U, Thomsen T et al. HDL cholesterol protects against cardiovascular disease in both genders, at all ages and at all levels of risk. *Atherosclerosis* 2009; 206: 611–616.

Frikke-Schmidt R. Genetic variation in the ABCA1 gene, HDL cholesterol, and risk of ischemic heart disease in the general population. *Atherosclerosis* 2010; 208: 305–316.

Vergeer M, Holleboom AG, Kastelein JJP, Kuivenhoven JA. The HDL hypothesis: does high-density lipoprotein protect from atherosclerosis? *J Lipid Res* 2010; 51: 2058–2073.

REFERENCES

1. Chapman MJ, Assmann G, Fruchart JC, Shepherd J, Sirtori C. Raising high-density lipoprotein cholesterol with reduction of cardiovascular risk: the role of nicotinic acid—a position paper developed by the European Consensus Panel on HDL-C. *Curr Med Res Opin* 2004; 20: 1253–1268.
2. Third Report of the National Cholesterol Education Program (NCEP) Expert Panel on Detection, Evaluation, and Treatment of High Blood Cholesterol in Adults (Adult Treatment Panel III) final report. *Circulation* 2002; 106: 3143–3421.
3. Contois J, McNamara JR, Lammi-Keefe C, Wilson PW, Massov T, Schaefer EJ. Reference intervals for plasma apolipoprotein A-1 determined with a standardized commercial immunoturbidimetric assay: results from the Framingham Offspring Study. *Clin Chem* 1996; 42: 507–514.
4. Barr DP, Russ EM, Eder HA. Protein-lipid relationships in human plasma: II. In atherosclerosis and related conditions. *Am J Med* 1951; 11: 480.
5. Gofman JW, Young W, Tandy R. Ischemic heart disease, atherosclerosis, and longevity. *Circulation* 1966; 34: 679–697.
6. Miller GJ, Miller NE. Plasma-high-density-lipoprotein concentration and development of ischaemic heart-disease. *Lancet* 1975; 1: 16–19.
7. Gordon T, Castelli WP, Hjortland MC, Kannel WB, Dawber TR. High density lipoprotein as a protective factor against coronary heart disease. The Framingham Study. *Am J Med* 1977; 62: 707–714.
8. Gordon DJ, Probstfield JL, Garrison RJ, Neaton JD, Castelli WP, Knoke JD, Jacobs DR Jr, Bangdiwala S, Tyroler HA. High-density lipoprotein cholesterol and cardiovascular disease. Four prospective American studies. *Circulation* 1989; 79: 8–15.
9. Assmann G, Schulte H, von Eckardstein A, Huang Y. High-density lipoprotein cholesterol as a predictor of coronary heart disease risk. The PROCAM experience and pathophysiological implications for reverse cholesterol transport. *Atherosclerosis* 1996; 124 (Suppl.): S11–S20.
10. Moorjani S, Gagne C, Lupien PJ, Brun D. Plasma triglycerides related decrease in high-density lipoprotein cholesterol and its association with myocardial infarction in heterozygous familial hypercholesterolemia. *Metabolism* 1986; 35: 311–316.
11. Cooney MT, Dudina A, De Bacquer D, Wilhelmsen L, Sans S, Menotti A, De Backer G, Jousilahti P, Keil U, Thomsen T et al. HDL cholesterol protects against

cardiovascular disease in both genders, at all ages and at all levels of risk. *Atherosclerosis* 2009; 206: 611–616.
12. Chirovsky DR, Fedirko V, Cui Y, Sazonov V, Barter P Prospective studies on the relationship between high-density lipoprotein cholesterol and cardiovascular risk: a systematic review. *Eur J Cardiovasc Prev Rehabil* 2009; 16: 404–423.
13. Di Angelantonio E, Sarwar N, Perry P, Kaptoge S, Ray KK, Thompson A, Wood AM, Lewington S, Sattar N, Packard CJ et al. Major lipids, apolipoproteins, and risk of vascular disease. *JAMA* 2009; 302: 1993–2000.
14. Sharrett AR, Ballantyne CM, Coady SA, Heiss G, Sorlie PD, Catellier D, Patsch W. Coronary heart disease prediction from lipoprotein cholesterol levels, triglycerides, lipoprotein(a), apolipoproteins A-I and B, and HDL density subfractions: The Atherosclerosis Risk in Communities (ARIC) Study. *Circulation* 2001; 104: 1108–1113.
15. Fan AZ, Dwyer JH. Sex differences in the relation of HDL cholesterol to progression of carotid intima-media thickness: the Los Angeles Atherosclerosis Study. *Atherosclerosis* 2007; 195: e191–196.
16. Nilsson G, Ohrvik J, Lonnberg I, Hedberg P. Ten-year survival in 75-year-old men and women: Predictive ability of total cholesterol, HDL-C, and LDL-C. *Curr Gerontol Geriatr Res* 2009; 2009: 158425.
17. Pencina MJ, D'Agostino RB Sr, Larson MG, Massaro JM, Vasan RS. Predicting the 30-Year Risk of Cardiovascular Disease: The Framingham Heart Study. *Circulation* 2009; 119: 3078–3084.
18. Nakamura K, Barzi F, Huxley R, Lam TH, Suh I, Woo J, Kim HC, Feigin VL, Gu D, Woodward M. Does cigarette smoking exacerbate the effect of total cholesterol and high-density lipoprotein cholesterol on the risk of cardiovascular diseases? *Heart* 2009; 95: 909–916.
19. Upmeier E, Lavonius S, Lehtonen A, Viitanen M, Isoaho H, Arve S. Serum lipids and their association with mortality in the elderly: a prospective cohort study. *Aging Clin Exp Res* 2010; 21: 424–430.
20. de Freitas EV, Brandao AA, Pozzan R, Magalhaes ME, Fonseca F, Pizzi O, Campana E, Brandao AP. Importance of high-density lipoprotein-cholesterol (HDL-C) levels to the incidence of cardiovascular disease (CVD) in the elderly. *Arch Gerontol Geriatr* 2011; 52: 217–222.
21. Koropatnick TA, Kimbell J, Chen R, Grove JS, Donlon TA, Masaki KH, Rodriguez BL, Willcox BJ, Yano K, Curb JD. A prospective study of high-density lipoprotein cholesterol, cholesteryl ester transfer protein gene variants, and healthy aging in very old Japanese-american men. *J Gerontol A Biol Sci Med Sci* 2008; 63: 1235–1240.
22. Lozano JV, Pallares V, Cea-Calvo L, Llisterri JL, Fernandez-Perez C, Marti-Canales JC, Aznar J, Gil-Guillen V, Redon J. Serum lipid profiles and their relationship to cardiovascular disease in the elderly: the PREV-ICTUS study. *Curr Med Res Opin* 2008; 24: 659–670.
23. Goldbourt U, Yaari S, Medalie JH. Isolated low HDL cholesterol as a risk factor for coronary heart disease mortality. A 21-year follow-up of 8000 men. *Arterioscler Thromb Vasc Biol* 1997; 17: 107–113.
24. Okamura T, Hayakawa T, Kadowaki T, Kita Y, Okayama A, Ueshima H. The inverse relationship between serum high-density lipoprotein cholesterol level and all-cause mortality in a 9.6-year follow-up study in the Japanese general population. *Atherosclerosis* 2006; 184: 143–150.

25. Satoh H, Tomita K, Fujii S, Kishi R, Tsutsui H. Lower high-density lipoprotein cholesterol is a significant and independent risk for coronary artery disease in Japanese men. *J Atheroscler Thromb* 2009; 16: 792–798.
26. Hayashi T, Kawashima S, Itoh H, Yamada N, Sone H, Watanabe H, Hattori Y, Ohrui T, Yokote K, Nomura H et al. Low HDL cholesterol is associated with the risk of stroke in elderly diabetic individuals: changes in the risk for atherosclerotic diseases at various ages. *Diabetes Care* 2009; 32: 1221–1223.
27. Coca A, Cea-Calvo L, Lozano JV, Inaraja V, Fernandez-Perez C, Navarro J, Bonet A, Redon J. High-density lipoprotein cholesterol and cardiovascular disease in Spanish hypertensive women. The RIMHA study. *Rev Esp Cardiol* 2009; 62: 1022–1031.
28. Barn K, Laftavi M, Pierce D, Ying C, Boden WE, Pankewycz O. Low levels of high-density lipoprotein cholesterol: an independent risk factor for late adverse cardiovascular events in renal transplant recipients. *Transpl Int* 2010; 23: 574–579.
29. Sattler KJ, Herrmann J, Yun S, Lehmann N, Wang Z, Heusch G, Sack S, Erbel R, Levkau B. High high-density lipoprotein-cholesterol reduces risk and extent of percutaneous coronary intervention-related myocardial infarction and improves long-term outcome in patients undergoing elective percutaneous coronary intervention. *Eur Heart J* 2009; 30: 1894–1902.
30. Ghazzal ZB, Dhawan SS, Sheikh A, Douglas JS, Veledar E, Mavromatis K, Pohlel FK, Vaccarino V. Usefulness of serum high-density lipoprotein cholesterol level as an independent predictor of one-year mortality after percutaneous coronary interventions. *Am J Cardiol* 2009; 103: 902–906.
31. Alber HF, Wanitschek MM, de Waha S, Ladurner A, Suessenbacher A, Dorler J, Dichtl W, Frick M, Ulmer H, Pachinger O et al. High-density lipoprotein cholesterol, C-reactive protein, and prevalence and severity of coronary artery disease in 5641 consecutive patients undergoing coronary angiography. *Eur J Clin Invest* 2008; 38: 372–380.
32. Sampietro T, Bigazzi F, Puntoni M, Minichilli F, Landi P, Dal Pino B, Carpeggiani C, L'Abbate A. Low high-density lipoprotein predicts death in patients with mild left ventricular dysfunction regardless of coronary atherosclerosis. *J Cardiovasc Med (Hagerstown)* 2009; 10: 898–905.
33. Topakian R, Sonnberger M, Nussbaumer K, Haring HP, Trenkler J, Aichner FT. Postprocedural high-density lipoprotein cholesterol predicts carotid stent patency at 1 year. *Eur J Neurol* 2008; 15: 179–184.
34. Correia LC, Rocha MS, Esteves JP. HDL-cholesterol level provides additional prognosis in acute coronary syndromes. *Int J Cardiol* 2009; 136: 307–314.
35. Cotter AG, Satchell CS, O'Halloran JA, Feeney ER, Sabin CA, Mallon PW. High-density lipoprotein levels and 10-year cardiovascular risk in HIV-1-infected patients. *Aids* 2011; 25: 867–869.
36. Sugano Y, Anzai T, Yagi T, Noma S. Impact of high-density lipoprotein cholesterol level in patients with variant angina pectoris. *Int J Cardiol*, 2010; 140, 175–181.
37. Menotti A, Lanti M, Zanchetti A, Botta G, Laurenzi M, Terradura-Vagnarelli O, Mancini M. The role of HDL cholesterol in metabolic syndrome predicting cardiovascular events. The Gubbio population study. *Nutr Metab Cardiovasc Dis* 2011;21: 315–322.
38. Amarenco P, Labreuche J, Touboul PJ. High-density lipoprotein-cholesterol and risk of stroke and carotid atherosclerosis: a systematic review. *Atherosclerosis* 2008; 196: 489–496.

39. Simundic AM, Nikolac N, Topic E, Basic-Kes V, Demarin V. Are serum lipids measured on stroke admission prognostic? *Clin Chem Lab Med* 2008; 46: 1163–1167.
40. Holme I, Aastveit AH, Hammar N, Jungner I, Walldius G. Relationships between lipoprotein components and risk of ischaemic and haemorrhagic stroke in the Apolipoprotein MOrtality RISk study (AMORIS). *J Intern Med* 2009; 265: 275–287.
41. Noda H, Iso H, Irie F, Sairenchi T, Ohtaka E, Doi M, Izumi Y, Ohta H. Low-density lipoprotein cholesterol concentrations and death due to intraparenchymal hemorrhage: the Ibaraki Prefectural Health Study. *Circulation* 2009; 119: 2136–2145.
42. Mahmood A, Sharif MA, Khan MN, Ali UZ. Comparison of serum lipid profile in ischaemic and haemorrhagic stroke. *J Coll Physicians Surg Pak* 2010; 20: 317–320.
43. Deguchi H, Pecheniuk NM, Elias DJ, Averell PM, Griffin JH. High-density lipoprotein deficiency and dyslipoproteinemia associated with venous thrombosis in men. *Circulation* 2005; 112: 893–899.
44. Eichinger S, Pecheniuk NM, Hron G, Deguchi H, Schemper M, Kyrle PA, Griffin JH. High-density lipoprotein and the risk of recurrent venous thromboembolism. *Circulation* 2007; 115: 1609–1614.
45. Ridker PM, Stampfer MJ, Rifai N. Novel risk factors for systemic atherosclerosis: a comparison of C-reactive protein, fibrinogen, homocysteine, lipoprotein(a), and standard cholesterol screening as predictors of peripheral arterial disease. *JAMA* 2001; 285: 2481–2485.
46. Pradhan AD, Shrivastava S, Cook NR, Rifai N, Creager MA, Ridker PM. Symptomatic peripheral arterial disease in women: nontraditional biomarkers of elevated risk. *Circulation* 2008; 117: 823–831.
47. Chamberlain AM, Folsom AR, Heckbert SR, Rosamond WD, Cushman M. High-density lipoprotein cholesterol and venous thromboembolism in the Longitudinal Investigation of Thromboembolism Etiology (LITE). *Blood* 2008; 112: 2675–2680.
48. Mehra MR, Uber PA, Lavie CJ, Milani RV, Park MH, Ventura HO. High-density lipoprotein cholesterol levels and prognosis in advanced heart failure. *J Heart Lung Transplant* 2009; 28: 876–880.
49. Velagaleti RS, Massaro J, Vasan RS, Robins SJ, Kannel WB, Levy D. Relations of lipid concentrations to heart failure incidence: The Framingham Heart Study. *Circulation* 2009; 120: 2345–2351.
50. Kerola T, Nieminen T, Hartikainen S, Sulkava R, Vuolteenaho O, Kettunen R. High-density lipoprotein is superior to B-type natriuretic peptide as a marker of systolic dysfunction in an elderly general population. *Scand J Clin Lab Invest* 2009; 69: 865–872.
51. Ikee R, Hamasaki Y, Oka M, Maesato K, Mano T, Moriya H, Ohtake T, Kobayashi S. High-density lipoprotein cholesterol and left ventricular mass index in peritoneal dialysis. *Perit Dial Int* 2008; 28: 611–616.
52. Takagi H, Manabe H, Kawai N, Goto SN, Umemoto T. Serum high-density and low-density lipoprotein cholesterol is associated with abdominal aortic aneurysm presence: a systematic review and meta-analysis. *Int Angiol* 2010; 29: 371–375.
53. Dew M, Bateman TM, O'Keefe JH, McGhie AI, Bunten L, McCoy M, Stevens TL.. Age and HDL cholesterol predict progression of coronary artery calcium in women: Findings from the Women's Hearts: Ischemic Screening with Proactive Evaluation and Recommendation (WHISPER) Study. *Circulation* 2009; 120: S345.

54. Ayer JG, Harmer JA, Nakhla S, Xuan W, Ng MKC, Raitakari OT, Marks GB, Celermajer DS. HDL-cholesterol, blood pressure, and asymmetric dimethylarginine are significantly associated with arterial wall thickness in children. *Arterioscler Thromb Vasc Biol* 2009; 29: 943–949.
55. Mitsutake R, Miura S, Zhang B, Saku K. HDL-associated factors provide additional prognostic information for coronary artery disease as determined by multi-detector row computed tomography. *Int J Cardiol* 2010; 143: 72–78.
56. Marma AK, Lloyd-Jones DM. Systematic examination of the updated Framingham Heart Study general cardiovascular risk profile. *Circulation* 2009; 120: 384–390.
57. Cooney MT, Dudina A, De Bacquer D, Fitzgerald A, Conroy R, Sans S, Menotti A, De Backer G, Jousilahti P, Keil U et al. How much does HDL cholesterol add to risk estimation? A report from the SCORE Investigators. *Eur J Cardiovasc Prev Rehabil* 2009; 16: 304–314.
58. Bair TL, May HT, Horne BD, Anderson JL, Muhlestein JB. Differences in characteristics and outcomes by high-density lipoprotein cholesterol levels: The Intermountain Heart Collaborative Study. *Circulation* 2009; 120: S470.
59. van der Steeg WA, Holme I, Boekholdt SM, Larsen ML, Lindahl C, Stroes ES, Tikkanen MJ, Wareham NJ, Faergeman O, Olsson AG et al. High-density lipoprotein cholesterol, high-density lipoprotein particle size, and apolipoprotein A-I: significance for cardiovascular risk: the IDEAL and EPIC-Norfolk studies. *J Am Coll Cardiol* 2008; 51: 634–642.
60. Corsetti JP, Ryan D, Moss AJ, Zareba W, Sparks CE. NAD(P)H oxidase polymorphism (C242T) and high HDL cholesterol associate with recurrent coronary events in postinfarction patients. *Atherosclerosis* 2008; 196: 461–468.
61. Corsetti JP, Ryan D, Rainwater DL, Moss AJ, Zareba W, Sparks CE. Cholesteryl ester transfer protein polymorphism (TaqIB) associates with risk in postinfarction patients with high C-reactive protein and high-density lipoprotein cholesterol levels. *Arterioscler Thromb Vasc Biol* 2010; 30: 1657–1664.
62. Corsetti JP, Gansevoort RT, Sparks CE, Dullaart RP. Inflammation reduces HDL protection against primary cardiac risk. *Eur J Clin Invest* 2010; 40: 483–489.
63. Khovidhunkit W, Kim MS, Memon RA, Shigenaga JK, Moser AH, Feingold KR, Grunfeld C. Effects of infection and inflammation on lipid and lipoprotein metabolism: mechanisms and consequences to the host. *J Lipid Res* 2004; 45: 1169–1196.
64. Tung P, Wiviott SD, Cannon CP, Murphy SA, McCabe CH, Gibson CM. Seasonal variation in lipids in patients following acute coronary syndrome on fixed doses of Pravastatin (40mg) or Atorvastatin (80mg) (from the Pravastatin or Atorvastatin Evaluation and Infection Therapy-Thrombolysis In Myocardial Infarction 22 [PROVE IT-TIMI 22] Study). *Am J Cardiol* 2009; 103: 1056–1060.
65. Halonen JI, Zanobetti A, Sparrow D, Vokonas PS, Schwartz J. Outdoor temperature is associated with serum HDL and LDL. *Environ Res* 2011; 11: 281–287.
66. Lee SA, Wen W, Xiang YB, Fazio S, Linton MF, Cai Q, Liu D, Zheng W, Shu XO. Stability and reliability of plasma level of lipid biomarkers and their correlation with dietary fat intake. *Dis Markers* 2008; 24: 73–79.
67. Langsted A, Freiberg JJ, Nordestgaard BG. Fasting and nonfasting lipid levels: influence of normal food intake on lipids, lipoproteins, apolipoproteins, and cardiovascular risk prediction. *Circulation* 2008; 118: 2047–2056.

68. Mora S, Rifai N, Buring JE, Ridker PM. Fasting compared with nonfasting lipids and apolipoproteins for predicting incident cardiovascular events. *Circulation* 2008; 118: 993–1001.
69. van Dieren S, Nothlings U, van der Schouw YT, Spijkerman AM, Rutten GE, van der A DL, Sluik D, Weikert C, Joost HG, Boeing H et al. Non-fasting lipids and risk of cardiovascular disease in patients with diabetes mellitus. *Diabetologia* 2011; 54: 73–77.
70. Warnick GR, Benderson J, Albers JJ. Dextran sulfate-Mg2+ precipitation procedure for quantitation of high-density-lipoprotein cholesterol. *Clin Chem* 1982; 28: 1379–1388.
71. Perera NJ, Burns JC, Perera RS, Lewis B, Sullivan DR. Adjustment of direct high-density lipoprotein cholesterol measurements according to intercurrent triglyceride corrects for interference by triglyceride-rich lipoproteins. *J Clin Lipidol* 2010; 4: 305–309.
72. Rosenson RS, Brewer HB Jr, Chapman MJ, Fazio S, Hussain MM, Kontush A, Krauss RM, Otvos JD, Remaley AT, Schaefer EJ. HDL measures, particle heterogeneity, proposed nomenclature, and relation to atherosclerotic cardiovascular events. *Clin Chem* 2011; 57: 392–410.
73. Miller WG, Myers GL, Sakurabayashi I, Bachmann LM, Caudill SP, Dziekonski A, Edwards S, Kimberly MM, Korzun WJ, Leary ET et al. Seven direct methods for measuring HDL and LDL cholesterol compared with ultracentrifugation reference measurement procedures. *Clin Chem* 2010; 56: 977–986.
74. Solhpour A, Parkhideh S, Sarrafzadegan N, Asgary S, Williams K, Jungner I, Aastveit A, Walldius G, Sniderman A. Levels of lipids and apolipoproteins in three cultures. *Atherosclerosis* 2009; 207: 200–207.
75. Erem C, Hacihasanoglu A, Deger O, Kocak M, Topbas M. Prevalence of dyslipidemia and associated risk factors among Turkish adults: Trabzon lipid study. *Endocrine* 2008; 34: 36–51.
76. Rutherford JN, McDade TW, Feranil AB, Adair LS, Kuzawa CW. High prevalence of low HDL-c in the Philippines compared to the US: population differences in associations with diet and BMI. *Asia Pac J Clin Nutr* 2010; 19: 57–67.
77. O'Neal DN, Piers LS, Iser DM, Rowley KG, Jenkins AJ, Best JD, O'Dea K. Australian Aboriginal people and Torres Strait Islanders have an atherogenic lipid profile that is characterised by low HDL-cholesterol level and small LDL particles. *Atherosclerosis* 2008; 201: 368–377.
78. Cheung BM, Li M, Ong KL, Wat NM, Tam S, Pang RW, Thomas GN, Woo J, Janus ED, Lau CP et al. High density lipoprotein-cholesterol levels increase with age in American women but not in Hong Kong Chinese women. *Clin Endocrinol* 2009; 70: 561–568.
79. Chandalia M, Mohan V, Adams-Huet B, Deepa R, Abate N. Ethnic difference in sex gap in high-density lipoprotein cholesterol between Asian Indians and Whites. *J Investig Med* 2008; 56: 574–580.
80. Chapman MJ, Le Goff W, Guerin M, Kontush A. Cholesteryl ester transfer protein: at the heart of the action of lipid-modulating therapy with statins, fibrates, niacin, and cholesteryl ester transfer protein inhibitors. *Eur Heart J* 2010; 31: 149–164.
81. Hokanson JE, Austin MA. Plasma triglyceride level is a risk factor for cardiovascular disease independent of high-density lipoprotein cholesterol level: a meta-analysis of population-based prospective studies. *J Cardiovasc Risk* 1996; 3: 213–219.

82. Nordestgaard BG, Benn M, Schnohr P, Tybjaerg-Hansen A. Nonfasting triglycerides and risk of myocardial infarction, ischemic heart disease, and death in men and women. *JAMA* 2007; 298: 299–308.

83. Dobiasova M, Frohlich J, Sedova M, Cheung MC, Brown BG. Cholesterol esterification and atherogenic index of plasma correlate with lipoprotein size and findings on coronary angiography. *J Lipid Res* 2011; 52: 566–571.

84. Cordero A, Andres E, Ordonez B, Leon M, Laclaustra M, Grima A, Luengo E, Moreno J, Bes M, Pascual I et al. Usefulness of triglycerides-to-high-density lipoprotein cholesterol ratio for predicting the first coronary event in men. *Am J Cardiol* 2009; 104: 1393–1397.

85. Bittner V, Johnson BD, Zineh I, Rogers WJ, Vido D, Marroquin OC, Bairey-Merz CN, Sopko G. The triglyceride/high-density lipoprotein cholesterol ratio predicts all-cause mortality in women with suspected myocardial ischemia: a report from the Women's Ischemia Syndrome Evaluation (WISE). *Am Heart J* 2009; 157: 548–555.

86. da Luz PL, Favarato D, Faria-Neto JR Jr, Lemos P, Chagas AC. High ratio of triglycerides to HDL-cholesterol predicts extensive coronary disease. *Clinics* 2008; 63: 427–432.

87. Hadaegh F, Khalili D, Ghasemi A, Tohidi M, Sheikholeslami F, Azizi F. Triglyceride/HDL-cholesterol ratio is an independent predictor for coronary heart disease in a population of Iranian men. *Nutr Metab Cardiovasc Dis* 2009; 19: 401–408.

88. Onat A, Can G, Kaya H, Hergenc G. "Atherogenic index of plasma" (log10 triglyceride/high-density lipoprotein-cholesterol) predicts high blood pressure, diabetes, and vascular events. *J Clin Lipidol* 2010; 4: 89–98.

89. Kannel WB, Vasan RS, Keyes MJ, Sullivan LM, Robins SJ. Usefulness of the triglyceride-high-density lipoprotein versus the cholesterol-high-density lipoprotein ratio for predicting insulin resistance and cardiometabolic risk (from the Framingham Offspring Cohort). *Am J Cardiol* 2008; 101: 497–501.

90. Sarwar N, Sandhu MS, Ricketts SL, Butterworth AS, Di Angelantonio E, Boekholdt SM, Ouwehand W, Watkins H, Samani NJ, Saleheen D et al. Triglyceride-mediated pathways and coronary disease: collaborative analysis of 101 studies. *Lancet* 2010; 375: 1634–1639.

91. Nordestgaard BG, Tybjaerg-Hansen A. Genetic determinants of LDL, lipoprotein(a), triglyceride-rich lipoproteins and HDL: concordance and discordance with cardiovascular disease risk. *Curr Opin Lipidol* 2011; 22: 113–122.

92. Oram JF, Lawn RM, Garvin MR, Wade DP. ABCA1 is the cAMP-inducible apolipoprotein receptor that mediates cholesterol secretion from macrophages. *J Biol Chem* 2000; 275: 34508–34511.

93. Mott S, Yu L, Marcil M, Boucher B, Rondeau C, Genest J Jr. Decreased cellular cholesterol efflux is a common cause of familial hypoalphalipoproteinemia: role of the ABCA1 gene mutations. *Atherosclerosis* 2000; 152: 457–468.

94. Sampietro T, Puntoni M, Bigazzi F, Pennato B, Sbrana F, Dal Pino B, Azzara A, Magagnini E, Perossini P, Cei G et al. Tangier disease in severely progressive coronary and peripheral artery disease. *Circulation* 2009; 119: 2741–2742.

95. Rader DJ. Lecithin: cholesterol acyltransferase and atherosclerosis: another high-density lipoprotein story that doesn't quite follow the script. *Circulation* 2009; 120: 549–552.

96. Chiesa G, Sirtori CR. Apolipoprotein A-I Milano: current perspectives. *Curr Opin Lipidol* 2003; 14: 159–163.
97. Frikke-Schmidt R. Genetic variation in the ABCA1 gene, HDL cholesterol, and risk of ischemic heart disease in the general population. *Atherosclerosis* 2010; 208: 305–316.
98. Van Eck M, Pennings M, Hoekstra M, Out R, Van Berkel TJ. Scavenger receptor BI and ATP-binding cassette transporter A1 in reverse cholesterol transport and atherosclerosis. *Curr Opin Lipidol* 2005; 16: 307–315.
99. Sirtori CR, Calabresi L, Franceschini G, Baldassarre D, Amato M, Johansson J, Salvetti M, Monteduro C, Zulli R, Muiesan ML, et al. Cardiovascular status of carriers of the apolipoprotein A-I Milano mutant: the Limone sul Garda study. *Circulation* 2001; 103: 1949–1954.
100. Linton JA, Kimm H, Ohrr H, Park IS, Jee SH. High-density lipoprotein-cholesterol and ischemic heart disease risk in Korean men with cardiac risk. *Circ J* 2009; 73: 1296–1301.
101. Carey VJ, Bishop L, Laranjo N, Harshfield BJ, Kwiat C, Sacks FM. Contribution of high plasma triglycerides and low high-density lipoprotein cholesterol to residual risk of coronary heart disease after establishment of low-density lipoprotein cholesterol control. *Am J Cardiol* 2010; 106: 757–763.
102. Franceschini G. Epidemiologic evidence for high-density lipoprotein cholesterol as a risk factor for coronary artery disease. *Am J Cardiol* 2001; 88: 9N–13N.
103. Rouvre M, Vol S, Gusto G, Born C, Lantieri O, Tichet J, Lecomte P. Low high density lipoprotein cholesterol: prevalence and associated risk-factors in a large French population. *Ann Epidemiol* 2011; 21: 118–127.
104. Cea-Calvo L, Lozano JV, Fernandez-Perez C, Llisterri JL, Marti-Canales JC, Aznar J, Gil-Guillen V, Redon J. Prevalence of low HDL cholesterol, and relationship between serum HDL and cardiovascular disease in elderly Spanish population: the PREV-ICTUS study. *Int J Clin Pract* 2009; 63: 71–81.
105. Rubins HB, Robins SJ, Collins D, Iranmanesh A, Wilt TJ, Mann D, Mayo-Smith M, Faas FH, Elam MB, Rutan GH et al. Distribution of lipids in 8,500 men with coronary artery disease. Department of Veterans Affairs HDL Intervention Trial Study Group. *Am J Cardiol* 1995; 75: 1196–1201.
106. Sachdeva A, Cannon CP, Deedwania PC, Labresh KA, Smith SC Jr, Dai D, Hernandez A, Fonarow GC. Lipid levels in patients hospitalized with coronary artery disease: an analysis of 136,905 hospitalizations in Get With The Guidelines. *Am Heart J* 2009; 157: 111–117.
107. Pinto X, Millan J, Munoz A, Corbella E, Hernandez-Mijares A, Zuniga M, Mangas A, Pedro-Botet J. A very high prevalence of low HDL cholesterol in Spanish patients with acute coronary syndromes. *Clin Cardiol* 2010; 33: 418–423.
108. Al-Zakwani I, Sulaiman K, Al-Rasadi K, Mikhailidis DP. Prevalence of low high-density lipoprotein cholesterol (HDL-C) as a marker of residual cardiovascular risk among acute coronary syndrome patients from Oman. *Curr Med Res Opin* 2011; 27: 879–885.
109. Roe MT, Ou FS, Alexander KP, Newby LK, Foody JM, Gibler WB, Boden WE, Ohman EM, Smith SC Jr, Peterson ED. Patterns and prognostic implications of low high-density lipoprotein levels in patients with non-ST-segment elevation acute coronary syndromes. *Eur Heart J* 2008; 29: 2480–2488.

110. Bernal E, Masia M, Padilla S, Gutierrez F. High-density lipoprotein cholesterol in HIV-infected patients: evidence for an association with HIV-1 viral load, antiretroviral therapy status, and regimen composition. *AIDS Patient Care STDS* 2008; 22: 569–575.

111. Bjorkelund C, Andersson-Hange D, Andersson K, Bengtsson C, Blomstrand A, Bondyr-Carlsson D, Eiben G, Rodstrom K, Sjoberg A, Sundh V et al. Secular trends in cardiovascular risk factors with a 36-year perspective: observations from 38- and 50-year-olds in the Population Study of Women in Gothenburg. *Scand J Prim Health Care* 2008; 26: 140–146.

112. Viikari JS, Juonala M, Raitakari OT. Trends in cardiovascular risk factor levels in Finnish children and young adults from the 1970s: The Cardiovascular Risk in Young Finns Study. *Exp Clin Cardiol* 2006; 11: 83–88.

113. Chapman MJ. Therapeutic elevation of HDL-cholesterol to prevent atherosclerosis and coronary heart disease. *Pharmacol Ther* 2006; 111: 893–908.

114. Alsheikh-Ali AA, Lin JL, Abourjaily P, Ahearn D, Kuvin JT, Karas RH. Prevalence of low high-density lipoprotein cholesterol in patients with documented coronary heart disease or risk equivalent and controlled low-density lipoprotein cholesterol. *Am J Cardiol* 2007; 100: 1499–1501.

115. Felix-Getzik EM, Kuvin JT, Karas RH. Nonoptimal high-density lipoprotein cholesterol levels are highly prevalent in patients presenting with acute coronary syndromes and well-controlled low-density lipoprotein cholesterol levels. *J Clin Lipidol* 2010; 4: 265–271.

116. Bruckert E, Baccara-Dinet M, McCoy F, Chapman J. High prevalence of low HDL-cholesterol in a pan-European survey of 8545 dyslipidaemic patients. *Curr Med Res Opin* 2005; 21: 1927–1934.

117. Waters DD, Brotons C, Chiang C-W, Ferrieres J, Foody J, Jukema JW, Santos RD, Verdejo J, Messig M, McPherson R et al. for the Lipid Treatment Assessment Project, I. Lipid Treatment Assessment Project 2: A multinational survey to evaluate the proportion of patients achieving low-density lipoprotein cholesterol goals. *Circulation* 2009; 120: 28–34.

118. Assmann G, Cullen P, Schulte H. Non-LDL-related dyslipidaemia and coronary risk: a case-control study. *Diab Vasc Dis Res* 2010; 7: 204–212.

119. Jafri H, Alsheikh-Ali AA, Karas RH. Statin therapy does not reduce the increased cardiovascular risk associated with low levels of high-density lipoprotein cholesterol: Evidence from randomized controlled trials. *Circulation* 2009; 120 (Suppl.): S500.

120. Jafri H, Alsheikh-Ali AA, Karas RH. Meta-analysis: statin therapy does not alter the association between low levels of high-density lipoprotein cholesterol and increased cardiovascular risk. *Ann Intern Med* 2010; 153: 800–808.

121. Amarenco P, Goldstein LB, Callahan A 3rd, Sillesen H, Hennerici MG, O'Neill BJ, Rudolph AE, Simunovic L, Zivin JA, Welch KM. Baseline blood pressure, low- and high-density lipoproteins, and triglycerides and the risk of vascular events in the Stroke Prevention by Aggressive Reduction in Cholesterol Levels (SPARCL) trial. *Atherosclerosis* 2009; 204: 515–520.

122. Amarenco P, Goldstein LB, Messig M, O'Neill BJ, Callahan A 3rd, Sillesen H, Hennerici MG, Zivin JA, Welch KM. Relative and cumulative effects of lipid and blood pressure control in the Stroke Prevention by Aggressive Reduction in Cholesterol Levels trial. *Stroke* 2009; 40: 2486–2492.

123. Drexel H, Aczel S, Marte T, Vonbank A, Saely CH. Factors predicting cardiovascular events in statin-treated diabetic and non-diabetic patients with coronary atherosclerosis. *Atherosclerosis* 2010; 208: 484–489 .

124. Ridker PM, Genest J, Boekholdt SM, Libby P, Gotto AM, Nordestgaard BG, Mora S, MacFadyen JG, Glynn RJ, Kastelein JJ. HDL cholesterol and residual risk of first cardiovascular events after treatment with potent statin therapy: an analysis from the JUPITER trial. *Lancet* 2010; 376: 333–339.

125. de Goma EM, Leeper NJ, Heidenreich PA. Clinical significance of high-density lipoprotein cholesterol in patients with low low-density lipoprotein cholesterol. *J Am Coll Cardiol* 2008; 51: 49–55.

126. Hajer GR, van der Graaf Y, Bots ML, Algra A, Visseren FL. Low plasma HDL-c, a vascular risk factor in high risk patients independent of LDL-c. *Eur J Clin Invest* 2009; 39: 680–688.

127. Muntner P, Lee F, Astor BC. Association of high-density lipoprotein cholesterol with coronary heart disease risk across categories of low-density lipoprotein cholesterol: the atherosclerosis risk in communities study. *Am J Med Sci* 2011; 341: 173–180.

128. Kini AS, Muntner P, Moreno PR, Mann D, Krishnan P, Kim MC, Rafael OC, Farkouh ME, Sharma SK. Relation of high-density lipoprotein cholesterol to mortality after percutaneous coronary interventions in patients with low-density lipoprotein <70 mg/dl. *Am J Cardiol* 2009; 103: 350–354.

129. Mora, S., Cui, Y., Ridker, P. M. Abstract 1462: Association of HDL cholesterol and apolipoprotein A-1 with incident CVD across levels of LDL cholesterol in 27,748 women. *Circulation* 2009; 120 (Suppl.): S507–S508.

130. Jafri H, Alsheikh-Ali AA, Karas RH. Baseline and on-treatment high-density lipoprotein cholesterol and the risk of cancer in randomized controlled trials of lipid-altering therapy. *J Am Coll Cardiol* 2010; 55: 2846–2854.

131. Ahn J, Lim U, Weinstein SJ, Schatzkin A, Hayes RB, Virtamo J, Albanes D. Prediagnostic total and high-density lipoprotein cholesterol and risk of cancer. *Cancer Epidemiol Biomarkers Prev* 2009; 18: 2814–2821.

132. Kucharska-Newton AM, Rosamond WD, Schroeder JC, McNeill AM, Coresh J, Folsom AR. HDL-cholesterol and the incidence of lung cancer in the Atherosclerosis Risk in Communities (ARIC) study. *Lung Cancer* 2008; 61: 292–300.

133. van Duijnhoven FJ, Bueno-De-Mesquita HB, Calligaro M, Jenab M, Pischon T, Jansen EH, Frohlich J, Ayyobi A, Overvad K, Toft-Petersen AP et al. Blood lipid and lipoprotein concentrations and colorectal cancer risk in the European Prospective Investigation into Cancer and Nutrition. *Gut* 2011; 60: 1094–1102.

134. Kucharska-Newton AM, Rosamond WD, Mink PJ, Alberg AJ, Shahar E, Folsom AR. HDL-cholesterol and incidence of breast cancer in the ARIC cohort study. *Ann Epidemiol* 2008; 18: 671–677.

135. Kim Y, Park SK, Han W, Kim DH, Hong YC, Ha EH, Ahn SH, Noh DY, Kang D, Yoo KY. Serum high-density lipoprotein cholesterol and breast cancer risk by menopausal status, body mass index, and hormonal receptor in Korea. *Cancer Epidemiol Biomarkers Prev* 2009; 18: 508–515.

136. Singh-Manoux A, Gimeno D, Kivimaki M, Brunner E, Marmot MG. Low HDL cholesterol is a risk factor for deficit and decline in memory in midlife: the Whitehall II study. *Arterioscler Thromb Vasc Biol* 2008; 28: 1556–1562.

137. Kontush A, Chapman MJ. HDL: close to our memories? *Arterioscler Thromb Vasc Biol* 2008; 28: 1418–1420.

138. Raffaitin C, Feart C, Le Goff M, Amieva H, Helmer C, Akbaraly TN, Tzourio C, Gin H, Barberger-Gateau P. Metabolic syndrome and cognitive decline in French elders: the Three-City Study. *Neurology* 2011; 76: 518–525.

139. Crisby M, Bronge L, Wahlund LO. Low levels of high density lipoprotein increase the severity of cerebral white matter changes: implications for prevention and treatment of cerebrovascular diseases. *Curr Alzheimer Res* 2010; 7: 534–539.

140. Ward MA, Bendlin BB, McLaren DG, Hess TM, Gallagher CL, Kastman EK, Rowley HA, Asthana S, Carlsson CM, Sager MA et al. Low HDL cholesterol is associated with lower gray matter volume in cognitively healthy adults. *Front Aging Neurosci* 2011; 2: 29.

141. Lehto SM, Hintikka J, Niskanen L, Tolmunen T, Koivumaa-Honkanen H, Honkalampi K, Viinamaki H. Low HDL cholesterol associates with major depression in a sample with a 7-year history of depressive symptoms. *Prog Neuropsychopharmacol Biol Psychiatry* 2008; 32: 1557–1561.

142. Michikawa M. Cholesterol paradox: is high total or low HDL cholesterol level a risk for Alzheimer's disease? *J Neurosci Res* 2003; 72: 141–146.

143. von Eckardstein A. Differential diagnosis of familial high density lipoprotein deficiency syndromes. *Atherosclerosis* 2006; 186: 231–239.

144. Weissglas-Volkov D, Pajukanta P. Genetic causes of high and low serum HDL-cholesterol. *J Lipid Res* 2010; 51: 2032–2057.

145. Roche F, Sforza E, Pichot V, Maudoux D, Garcin A, Celle S, Picard-Kossovsky M, Gaspoz JM, Barthelemy JC. Obstructive sleep apnoea/hypopnea influences high-density lipoprotein cholesterol in the elderly. *Sleep Med* 2009; 10: 882–886.

146. Papandreou D, Rousso I, Economou I, Makedou A, Moudiou T, Malindretos P, Pidonia I, Pantoleon A, Mavromichalis I. Is there any association between high-density lipoprotein, insulin resistance and non-alcoholic fatty liver disease in obese children? *Int J Food Sci Nutr* 2009; 60: 312–318.

147. Merono T, Rosso LG, Sorroche P, Boero L, Arbelbide J, Brites F. High risk of cardiovascular disease in iron overload patients. *Eur J Clin Invest* 2011; 41: 479–486.

148. Zoppini G, Targher G, Chonchol M, Perrone F, Lippi G, Muggeo M. Higher HDL cholesterol levels are associated with a lower incidence of chronic kidney disease in patients with type 2 diabetes. *Nutr Metab Cardiovasc Dis* 2009; 19: 580–586.

149. Boyer JF, Gourraud PA, Cantagrel A, Davignon JL, Constantin A. Traditional cardiovascular–risk factors in rheumatoid arthritis: A meta-analysis. *Joint Bone Spine* 2011; 78: 179–183.

150. Sappati Biyyani RS, Putka BS, Mullen KD. Dyslipidemia and lipoprotein profiles in patients with inflammatory bowel disease. *J Clin Lipidol* 2010; 4: 478–482.

151. Saxlin T, Suominen-Taipale L, Kattainen A, Marniemi J, Knuuttila M, Ylostalo P. Association between serum lipid levels and periodontal infection. *J Clin Periodontol* 2008; 35: 1040–1047.

152. Mohamadkhani A, Sayemiri K, Ghanbari R, Elahi E, Poustchi H, Montazeri G. The inverse association of serum HBV DNA level with HDL and adiponectin in chronic hepatitis B infection. *Virol J* 2010; 7: 228.

153. Capristo E, Malandrino N, Farnetti S, Mingrone G, Leggio L, Addolorato G, Gasbarrini G. Increased serum high-density lipoprotein-cholesterol concentration in celiac disease after gluten-free diet treatment correlates with body fat stores. *J Clin Gastroenterol* 2009; 43: 946–949.

154. Grion CM, Cardoso LT, Perazolo TF, Garcia AS, Barbosa DS, Morimoto HK, Matsuo T, Carrilho AJ. Lipoproteins and CETP levels as risk factors for severe sepsis in hospitalized patients. *Eur J Clin Invest* 2010; 40: 330–338.

155. Heresi GA, Aytekin M, Newman J, DiDonato J, Dweik RA. Plasma levels of high-density lipoprotein cholesterol and outcomes in pulmonary arterial hypertension. *Am J Respir Crit Care Med* 2010; 182: 661–668.

156. Robinson JG. Low high-density lipoprotein cholesterol and chronic disease risk: marker or causal? *J Am Coll Cardiol* 2010; 55: 2855–2857.

157. Tomten SE, Hostmark AT. Self-rated health showed a consistent association with serum HDL-cholesterol in the cross-sectional Oslo Health Study. *Int J Med Sci* 2007; 4: 278–287.

158. Glueck CJ, Fallat RW, Millett F, Gartside P, Elston RC, Go RC. Familial hyper-alpha-lipoproteinemia: studies in eighteen kindreds. *Metabolism* 1975; 24: 1243–1265.

159. Barzilai N, Atzmon G, Schechter C, Schaefer EJ, Cupples AL, Lipton R, Cheng S, Shuldiner AR. Unique lipoprotein phenotype and genotype associated with exceptional longevity. *JAMA* 2003; 290: 2030–2040.

160. Barzilai N, Atzmon G, Derby CA, Bauman JM, Lipton RB. A genotype of exceptional longevity is associated with preservation of cognitive function. *Neurology* 2006; 67: 2170–2175.

161. Rahilly-Tierney CR, Spiro A 3rd, Vokonas P, Gaziano JM. Relation between high-density lipoprotein cholesterol and survival to age 85 years in men (from the VA Normative Aging Study). *Am J Cardiol* 2011; 107: 1173–1177.

162. D'Amelio P, Di Bella S, Tamone C, Ravazzoli MG, Cristofaro MA, Di Stefano M, Isaia G. HDL cholesterol and bone mineral density in normal-weight postmenopausal women: is there any possible association? *Panminerva Med* 2008; 50: 89–96.

163. Movva R, Rader DJ. Laboratory assessment of HDL heterogeneity and function. *Clin Chem* 2008; 54: 788–800.

164. Walldius G, Jungner I. Is there a better marker of cardiovascular risk than LDL cholesterol? Apolipoproteins B and A-I - new risk factors and targets for therapy. *Nutr Metab Cardiovasc Dis* 2007; 17: 565–571.

165. Walldius G, Jungner I, Holme I, Aastveit AH, Kolar W, Steiner E. High apolipoprotein B, low apolipoprotein A-I, and improvement in the prediction of fatal myocardial infarction (AMORIS study): a prospective study. *Lancet* 2001; 358: 2026–2033.

166. Yusuf S, Hawken S, Ounpuu S, Bautista L, Franzosi MG, Commerford P, Lang CC, Rumboldt Z, Onen CL, Lisheng L et al. Obesity and the risk of myocardial infarction in 27,000 participants from 52 countries: a case-control study. *Lancet* 2005; 366: 1640–1649.

167. McQueen MJ, Hawken S, Wang X, Ounpuu S, Sniderman A, Probstfield J, Steyn K, Sanderson JE, Hasani M, Volkova E, et al. Lipids, lipoproteins, and apolipoproteins as risk markers of myocardial infarction in 52 countries (the INTERHEART study): a case-control study. *Lancet* 2008; 372: 224–233.

168. Kastelein JJP, van der Steeg WA, Holme I, Gaffney M, Cater NB, Barter P, Deedwania P, Olsson AG, Boekholdt SM, Demicco DA et al. Lipids, apolipoproteins, and their ratios in relation to cardiovascular events with statin treatment. *Circulation* 2008; 117: 3002–3009.

169. Parish S, Peto R, Palmer A, Clarke R, Lewington S, Offer A, Whitlock G, Clark S, Youngman L, Sleight P et al. The joint effects of apolipoprotein B, apolipoprotein A1, LDL cholesterol, and HDL cholesterol on risk: 3510 cases of acute myocardial infarction and 9805 controls. *Eur Heart J* 2009; 30: 2137–2146.

170. Gotto AM Jr, Whitney E, Stein EA, Shapiro DR, Clearfield M, Weis S, Jou JY, Langendorfer A, Beere PA, Watson DJ et al. Relation between baseline and on-treatment lipid parameters and first acute major coronary events in the Air Force/Texas Coronary Atherosclerosis Prevention Study (AFCAPS/TexCAPS). *Circulation* 2000; 101: 477–484.

171. Holme I, Strandberg TE, Faergeman O, Kastelein JJ, Olsson AG, Tikkanen MJ, Larsen ML, Lindahl C, Pedersen TR. Congestive heart failure is associated with lipoprotein components in statin-treated patients with coronary heart disease: Insights from the Incremental Decrease in End points Through Aggressive Lipid Lowering Trial (IDEAL). *Atherosclerosis* 2009; 205: 522–527.

172. Charlton-Menys V, Betteridge DJ, Colhoun H, Fuller J, France M, Hitman GA, Livingstone SJ, Neil HA, Newman CB, Szarek M et al.. Apolipoproteins, cardiovascular risk and statin response in type 2 diabetes: the Collaborative Atorvastatin Diabetes Study (CARDS). *Diabetologia* 2009; 52: 218–225.

173. Liem AH, van de Woestijne AP, Roeters van Lennep HW, Zwinderman AH, van der Steeg WA, Jukema JW. ApoB/A1 and LDL-C/HDL-C and the prediction of cardiovascular risk in statin-treated patients. *Curr Med Res Opin* 2008; 24: 359–364.

174. Aasvee K, Kurvinen E, Sundvall J, Jauhiainen M, Tur I.. Aggregation of lipoprotein and inflammatory parameters in families with a history of premature myocardial infarction: the Tallinn myocardial infarction study. *Clin Chem Lab Med* 2008; 46: 1602–1608.

175. Juonala M, Viikari JS, Kahonen M, Solakivi T, Helenius H, Jula A, Marniemi J, Taittonen L, Laitinen T, Nikkari T et al. Childhood levels of serum apolipoproteins B and A-I predict carotid intima-media thickness and brachial endothelial function in adulthood: the cardiovascular risk in young Finns study. *J Am Coll Cardiol* 2008; 52: 293–299.

176. Goswami B, Rajappa M, Mallika V, Kumar S, Shukla DK. Apo-B/apo-AI ratio: a better discriminator of coronary artery disease risk than other conventional lipid ratios in Indian patients with acute myocardial infarction. *Acta Cardiol* 2008; 63: 749–755.

177. van den Bogaard B, van den Born BJ, Fayyad R, Waters DD, DeMicco DA, LaRosa JC, Kastelein JJ, Holme I. On-treatment lipoprotein components and risk of cerebrovascular events in the Treating to New Targets study. *Eur J Clin Invest* 2011; 41: 134–142.

178. El Harchaoui K, Arsenault BJ, Franssen R, Despres JP, Hovingh GK, Stroes ES, Otvos JD, Wareham NJ, Kastelein JJ, Khaw KT et al. High-density lipoprotein particle size and concentration and coronary risk. *Ann Intern Med* 2009; 150: 84–93.

179. Otvos JD, Collins D, Freedman DS, Shalaurova I, Schaefer EJ, McNamara JR, Bloomfield HE, Robins SJ. Low-density lipoprotein and high-density lipoprotein

particle subclasses predict coronary events and are favorably changed by gemfibrozil therapy in the Veterans Affairs High-Density Lipoprotein Intervention Trial. *Circulation* 2006; 113: 1556–1563.

180. Ingelsson E, Schaefer EJ, Contois JH, McNamara JR, Sullivan L, Keyes MJ, Pencina MJ, Schoonmaker C, Wilson PW, D'Agostino RB et al. Clinical utility of different lipid measures for prediction of coronary heart disease in men and women. *JAMA* 2007; 298: 776–785.
181. Kappelle PJ, Gansevoort RT, Hillege JL, Wolffenbuttel BH, Dullaart RP. Apolipoprotein B/A-I and total cholesterol/high-density lipoprotein cholesterol ratios both predict cardiovascular events in the general population independently of non-lipid risk factors, albuminuria and C-reactive protein. *J Intern Med* 2011; 269: 232–242.
182. Taskinen MR, Barter PJ, Ehnholm C, Sullivan DR, Mann K, Simes J, Best JD, Hamwood S, Keech AC. Ability of traditional lipid ratios and apolipoprotein ratios to predict cardiovascular risk in people with type 2 diabetes. *Diabetologia* 2010; 53: 1846–1855.
183. Taskinen MR, Ehnholm C, Sullivan D, Mann K, Best JD, Simes RJ, Barter P, Keech AC. Apo B/apo A-I ratio does not override the traditional lipid ratios as determinants of CVD risk in patients with type 2 diabetes in the FIELD study. *Circulation* 2009; 120 (Suppl.): S547.
184. Martin SS, Rader DJ, Mehta NN, Wolfe M, Terembula K, Schwartz S, Iqbal N, Schutta M, Bagheri R, Reilly MP et al. High-density lipoprotein cholesterol but not apolipoprotein A-I or -II is associated with coronary calcium in type 2 diabetes. *Circulation* 2009; 120 (Suppl.): S531.
185. Ray KK, Cannon CP, Cairns R, Morrow DA, Ridker PM, Braunwald E. Prognostic utility of apoB/AI, total cholesterol/HDL, non-HDL cholesterol, or hs-CRP as predictors of clinical risk in patients receiving statin therapy after acute coronary syndromes: Results from PROVE IT-TIMI 22. *Arterioscler Thromb Vasc Biol* 2009; 29: 424–430.
186. Arsenault BJ, Despres JP, Stroes ES, Wareham NJ, Kastelein JJ, Khaw KT, Boekholdt SM. Lipid assessment, metabolic syndrome and coronary heart disease risk. *Eur J Clin Invest* 2010; 40: 1081–1093.
187. Sierra-Johnson J, Fisher RM. Is it time to discard the apo B:apo A-I ratio as a predictor of cardiovascular disease? *Nat Clin Pract Cardiovasc Med* 2008; 5: 18–19.
188. Benderly M, Boyko V, Goldbourt U. Apolipoproteins and long-term prognosis in coronary heart disease patients. *Am Heart J* 2009; 157: 103–110.
189. Birjmohun RS, Dallinga-Thie GM, Kuivenhoven JA, Stroes ES, Otvos JD, Wareham NJ, Luben R, Kastelein JJ, Khaw KT, Boekholdt SM. Apolipoprotein A-II is inversely associated with risk of future coronary artery disease. *Circulation* 2007; 116: 2029--2035.
190. Winkler K, Hoffmann MM, Seelhorst U, Wellnitz B, Boehm BO, Winkelmann BR, Marz W, Scharnagl H. Apolipoprotein A-II Is a negative risk indicator for cardiovascular and total mortality: findings from the Ludwigshafen Risk and Cardiovascular Health Study. *Clin Chem* 2008; 54: 1405–1406.
191. Onat A, Hergenc G, Ayhan E, Ugur M, Can G. Impaired anti-inflammatory function of apolipoprotein A-II concentrations predicts metabolic syndrome and diabetes at 4 years follow-up in elderly Turks. *Clin Chem Lab Med* 2009; 47: 1389–1394.

192. Tang WHW, Allayee H, Hartiala J, Wu Y, Hazen SL. Elevated apolipoprotein A2 adjusted for high-density lipoprotein is associated with atherosclerotic burden and progression in stable cardiac patients. *Circulation* 2010; 122 (Suppl.): A18732.

193. Blankenhorn DH, Alaupovic P, Wickham E, Chin HP, Azen SP. Prediction of angiographic change in native human coronary arteries and aortocoronary bypass grafts. Lipid and nonlipid factors. *Circulation* 1990; 81: 470–476.

194. Dahlback B, Nielsen LB. Apolipoprotein M affecting lipid metabolism or just catching a ride with lipoproteins in the circulation? *Cell Mol Life Sci* 2009; 66: 559–564.

195. Ahnstrom J, Axler O, Jauhiainen M, Salomaa V, Havulinna AS, Ehnholm C, Frikke-Schmidt R, Tybjaerg-Hansen A, Dahlback B. Levels of apolipoprotein M are not associated with the risk of coronary heart disease in two independent case-control studies. *J Lipid Res* 2008; 49: 1912–1917.

196. Plomgaard P, Dullaart RP, de Vries R, Groen AK, Dahlback B, Nielsen LB. Apolipoprotein M predicts pre-beta-HDL formation: studies in type 2 diabetic and nondiabetic subjects. *J Intern Med* 2009; 266: 258–267.

197. Ridker PM, Hennekens CH, Buring JE, Rifai N. C-reactive protein and other markers of inflammation in the prediction of cardiovascular disease in women. *N Engl J Med* 2000; 342: 836–843.

198. Liuzzo G, Biasucci LM, Gallimore JR, Grillo RL, Rebuzzi AG, Pepys MB, Maseri A. The prognostic value of C-reactive protein and serum amyloid A protein in severe unstable angina. *N Engl J Med* 1994; 331: 417–424.

199. Schillinger M, Exner M, Mlekusch W, Sabeti S, Amighi J, Nikowitsch R, Timmel E, Kickinger B, Minar C, Pones M et al. Inflammation and Carotid Artery—Risk for Atherosclerosis Study (ICARAS). *Circulation* 2005; 111: 2203–2209.

200. Fyfe AI, Rothenberg LS, DeBeer FC, Cantor RM, Rotter JI, Lusis AJ. Association between serum amyloid A proteins and coronary artery disease: Evidence from two distinct arteriosclerotic processes. *Circulation* 1997; 96: 2914–2919.

201. Kinlay S, Schwartz GG, Olsson AG, Rifai N, Szarek M, Waters DD, Libby P, Ganz P. Inflammation, statin therapy, and risk of stroke after an acute coronary syndrome in the MIRACL study. *Arterioscler Thromb Vasc Biol* 2008; 28: 142–147.

202. Chen GL, Liu LW, Xie S, Liu H, Liu YQ, Li YS. High-density lipoprotein associated factors apoA-I and serum amyloid A in Chinese non-diabetic patients with coronary heart disease. *Chin Med J (Engl)* 2010, 123, 658–663.

203. Jylhava J, Haarala A, Eklund C, Pertovaara M, Kahonen M, Hutri-Kahonen N, Levula M, Lehtimaki T, Huupponen R, Jula A et al. Serum amyloid A is independently associated with metabolic risk factors but not with early atherosclerosis: the Cardiovascular Risk in Young Finns Study. *J Intern Med* 2009; 266: 286–295.

204. Delanghe JR, Langlois MR, De Bacquer D, Mak R, Capel P, Van Renterghem L, De Backer G. Discriminative value of serum amyloid A and other acute-phase proteins for coronary heart disease. *Atherosclerosis* 2002; 160: 471–476.

205. Winkler K, Winkelmann BR, Scharnagl H, Hoffmann MM, Grawitz AB, Nauck M, Bohm BO, Marz W. Platelet-activating factor acetylhydrolase activity indicates angiographic coronary artery disease independently of systemic inflammation and other risk factors: The Ludwigshafen Risk and Cardiovascular Health Study. *Circulation* 2005; 111: 980–987.

206. Brea D, Sobrino T, Blanco M, Fraga M, Agulla J, Rodriguez-Yanez M, Rodriguez-Gonzalez R, Perez de la Ossa N, Leira R, Forteza J et al. Usefulness of haptoglobin and serum amyloid A proteins as biomarkers for atherothrombotic ischemic stroke diagnosis confirmation. *Atherosclerosis* 2009; 205: 561–567.

207. Choudhury RP, Leyva F. C-Reactive protein, serum amyloid A protein, and coronary events. *Circulation* 1999; 100: e65–e66.

208. Maier W, Altwegg LA, Corti R, Gay S, Hersberger M, Maly FE, Sutsch G, Roffi M, Neidhart M, Eberli FR et al. Inflammatory markers at the site of ruptured plaque in acute myocardial infarction: Locally increased interleukin-6 and serum amyloid A but decreased C-reactive protein. *Circulation* 2005; 111: 1355–1361.

209. Herrington DM, Parks JS. Estrogen and HDL: all that glitters is not gold. *Arterioscler Thromb Vasc Biol* 2004; 24: 1741–1742.

210. Lewis KE, Kirk EA, McDonald TO, Wang S, Wight TN, O'Brien KD, Chait A. Increase in serum amyloid A evoked by dietary cholesterol is associated with increased atherosclerosis in mice. *Circulation* 2004; 110: 540–545.

211. Mackness B, Durrington P, McElduff P, Yarnell J, Azam N, Watt M, Mackness M. Low paraoxonase activity predicts coronary events in the Caerphilly Prospective Study. *Circulation* 2003; 107: 2775–2779.

212. Ikeda Y, Inoue M, Suehiro T, Arii K, Kumon Y, Hashimoto K. (2009). Low human paraoxonase predicts cardiovascular events in Japanese patients with type 2 diabetes. *Acta Diabetol*, 46, 239–242.

213. van Himbergen TM, van der Schouw YT, Voorbij HAM, van Tits LJH, Stalenhoef AFH, Peeters PHM, Roest M. Paraoxonase (PON1) and the risk for coronary heart disease and myocardial infarction in a general population of Dutch women. *Atherosclerosis* 2008; 199: 408–414.

214. Kannampuzha J, Darling PB, Maguire GF, Donnelly S, McFarlane P, Chan CT, Connelly PW. Paraoxonase 1 arylesterase activity and mass are reduced and inversely related to C-reactive protein in patients on either standard or home nocturnal hemodialysis. *Clin Nephrol* 2010; 73: 131–138.

215. Birjmohun RS Vergeer M, Stroes ES, Sandhu MS, Ricketts SL, Tanck MW, Wareham NJ, Jukema JW, Kastelein JJ, Khaw KT et al. Both paraoxonase-1 genotype and activity do not predict the risk of future coronary artery disease; the EPIC-Norfolk Prospective Population Study. *PLoS One* 2009; 4: e6809.

216. Costa LG, Vitalone A, Cole TB, Furlong CE. Modulation of paraoxonase (PON1) activity. *Biochem Pharmacol* 2005; 69: 541–550.

217. Blatter Garin MC, Kalix B, Morabia A, James RW. Small, dense lipoprotein particles and reduced paraoxonase-1 in patients with the metabolic syndrome. *J Clin Endocrinol Metab* 2005; 90: 2264–2269.

218. Brites FD, Verona J, Schreier LE, Fruchart JC, Castro GR, Wikinski RL. Paraoxonase 1 and platelet-activating factor acetylhydrolase activities in patients with low HDL-cholesterol levels with or without primary hypertriglyceridemia. *Arch Med Res* 2004; 35: 235–240.

219. Boemi M, Leviev I, Sirolla C, Pieri C, Marra M, James RW. Serum paraoxonase is reduced in type 1 diabetic patients compared to non-diabetic, first degree relatives; influence on the ability of HDL to protect LDL from oxidation. *Atherosclerosis* 2001; 155: 229–235.

220. Dullaart RP, de Vries R, Sluiter WJ, Voorbij HA. High plasma C-reactive protein (CRP) is related to low paraoxonase-I (PON-I) activity independently of high leptin and low adiponectin in type 2 diabetes mellitus. *Clin Endocrinol (Oxf)* 2009; 70: 221–226.

221. Mackness MI, Harty D, Bhatnagar D, Winocour PH, Arrol S, Ishola M, Durrington PN. Serum paraoxonase activity in familial hypercholesterolaemia and insulin-dependent diabetes mellitus. *Atherosclerosis* 1991; 86: 193–199.

222. Mohamed RH, Mohamed RH, Karam RA, Abd El-Aziz TA. The relationship between paraoxonase1-192 polymorphism and activity with coronary artery disease. *Clin Biochem* 2010; 43: 553–558.

223. Senturk T, Sarandol E, Gullulu S, Erdinc S, Ozdemir B, Baran I, Aydinlar A. Association between paraoxonase 1 activity and severity of coronary artery disease in patients with acute coronary syndromes. *Acta Cardiol* 2008; 63: 361–367.

224. Cagirci G, Cay S, Karakurt O, Durmaz T, Yazihan N, Canga A, Aydin C, Acikel S, Kilic H, Topaloglu S et al. Paraoxonase activity might be predictive of the severity of aortic valve stenosis. *J Heart Valve Dis* 2010; 19: 453–458.

225. Tang WH, Wu Y, Mann S, Pepoy M, Shrestha K, Borowski AG, Hazen SL. Diminished antioxidant activity of high-density lipoprotein-associated proteins in systolic heart failure. *Circ Heart Fail* 2011; 4: 59–64.

226. Mohamadin AM, Habib FA, Elahi TF. Serum paraoxonase 1 activity and oxidant/antioxidant status in Saudi women with polycystic ovary syndrome. *Pathophysiology* 2010; 17: 189–196.

227. Toy H, Camuzcuoglu H, Celik H, Erel O, Aksoy N. Assessment of serum paraoxonase and arylesterase activities in early pregnancy failure. *Swiss Med Wkly* 2009; 139: 76–81.

228. Verit FF, Erel O, Celik H. Paraoxonase-1 activity in patients with hyperemesis gravidarum. *Redox Rep* 2008; 13: 134–138.

229. Gugliucci A, Kinugasa E, Kotani K, Caccavello R, Kimura S. Serum paraoxonase 1 (PON1) lactonase activity is lower in end-stage renal disease patients than in healthy control subjects and increases after hemodialysis. *Clin Chem Lab Med* 2011; 49: 61–67.

230. Soran N, Altindag O, Cakir H, Celik H, Demirkol A, Aksoy N. Assessment of paraoxonase activities in patients with knee osteoarthritis. *Redox Rep* 2008; 13: 194–198.

231. Camuzcuoglu H, Arioz DT, Toy H, Kurt S, Celik H, Erel O. Serum paraoxonase and arylesterase activities in patients with epithelial ovarian cancer. *Gynecol Oncol* 2009; 112: 481–485.

232. Batuca JR, Ames PR, Amaral M, Favas C, Isenberg DA, Delgado Alves J. Anti-atherogenic and anti-inflammatory properties of high-density lipoprotein are affected by specific antibodies in systemic lupus erythematosus. *Rheumatology* 2009; 48: 26–31.

233. Charakida M, Besler C, Batuca JR, Sangle S, Marques S, Sousa M, Wang G, Tousoulis D, Delgado Alves J, Loukogeorgakis SP, et al. Vascular abnormalities, paraoxonase activity, and dysfunctional HDL in primary antiphospholipid syndrome. *JAMA* 2009; 302: 1210–1217.

234. Kedage V, Muttigi MS, Shetty MS, Suvarna, R, Rao SS, Joshi C, Prakash M. Serum paraoxonase 1 activity status in patients with liver disorders. *Saudi J Gastroenterol* 2010; 16: 79–83.

235. Espinola-Zavaleta N, Soto-Lopez ME, Carreon-Torres E, Gamboa R, Mejia AM, Marquez-Velasco R, Bojalil R, Huesca-Gomez C, Reyes PA, Perez-Mendez O. Altered flow-mediated vasodilatation, low paraoxonase-1 activity, and abnormal high-density lipoprotein subclass distribution in Takayasu's arteritis. *Circ J* 2009; 73: 760–766.

236. Akbas HS, Basyigit S, Suleymanlar I, Kemaloglu D, Koc S, Davran F, Demir I, Suleymanlar G. The assessment of carotid intima media thickness and serum paraoxonase-1 activity in Helicobacter pylori positive subjects. *Lipids Health Dis* 2010; 9: 92.

237. Aslan M, Nazligul Y, Horoz M, Bolukbas C, Bolukbas FF, Gur M, Celik H, Erel O. Serum paraoxonase-1 activity in Helicobacter pylori infected subjects. *Atherosclerosis* 2008; 196: 270–274.

238. Novak F, Vavrova L, Kodydkova J, Novak F Sr, Hynkova M, Zak A, Novakova O. Decreased paraoxonase activity in critically ill patients with sepsis. *Clin Exp Med* 2010; 10: 21–25.

239. Toker A, Kadi M, Yildirim AK, Aksoy H, Akcay F. Serum lipid profile paraoxonase and arylesterase activities in psoriasis. *Cell Biochem Funct* 2009; 27: 176–180.

240. Jamroz-Wisniewska A, Beltowski J, Stelmasiak Z, Bartosik-Psujek H. Paraoxonase 1 activity in different types of multiple sclerosis. *Mult Scler* 2009; 15: 399–402.

241. Mallat Z, Lambeau G, Tedgui A. Lipoprotein-associated and secreted phospholipases A in cardiovascular disease: roles as biological effectors and biomarkers. *Circulation* 2010; 122: 2183–2200.

242. Tsimikas S, Willeit J, Knoflach M, Mayr M, Egger G, Notdurfter M, Witztum JL, Wiedermann CJ, Xu Q, Kiechl S. Lipoprotein-associated phospholipase A2 activity, ferritin levels, metabolic syndrome, and 10-year cardiovascular and non-cardiovascular mortality: results from the Bruneck study. *Eur Heart J* 2009; 30: 107–115.

243. Sabatine MS, Morrow DA, O'Donoghue M, Jablonksi KA, Rice MM, Solomon S, Rosenberg Y, Domanski MJ, Hsia J for the PI. Prognostic utility of lipoprotein-associated phospholipase A2 for cardiovascular outcomes in patients with stable coronary artery disease. *Arterioscler Thromb Vasc Biol* 2007; 27: 2463–2469.

244. Hatoum IJ, Hu FB, Nelson JJ, Rimm EB. Lipoprotein-associated phospholipase A2 activity and incident coronary heart disease among men and women with type 2 diabetes. *Diabetes* 2010; 59: 1239–1243.

245. Jenny NS, Solomon C, Cushman M, Tracy RP, Nelson JJ, Psaty BM, Furberg CD. Lipoprotein-associated phospholipase A2 (Lp-PLA2) and risk of cardiovascular disease in older adults: Results from the Cardiovascular Health Study. *Atherosclerosis* 2010; 209: 528–532.

246. Caslake MJ, Packard CJ, Robertson M, Cooney J, Nelson JJ, Ford I, Gaw A, Jukema JW, Macfarlane PW, Stott DJ et al. Lipoprotein-associated phospholipase A2, inflammatory biomarkers, and risk of cardiovascular disease in the Prospective Study of Pravastatin in the Elderly at Risk (PROSPER). *Atherosclerosis* 2010; 210: 28–34.

247. Persson M, Berglund G, Nelson JJ, Hedblad B. Lp-PLA2 activity and mass are associated with increased incidence of ischemic stroke: A population-based cohort study from Malmö, Sweden. *Atherosclerosis* 2008; 200: 191–198.

248. Kim JY, Hyun YJ, Jang Y, Lee BK, Chae JS, Kim SE, Yeo HY, Jeong TS, Jeon DW, Lee. H. Lipoprotein-associated phospholipase A2 activity is associated with

coronary artery disease and markers of oxidative stress: a case-control study. *Am J Clin Nutr* 2008; 88: 630–637.

249. Thompson A, Gao P, Orfei L, Watson S, Di Angelantonio E, Kaptoge S, Ballantyne C, Cannon CP, Criqui M, Cushman M et al. Lipoprotein-associated phospholipase A(2) and risk of coronary disease, stroke, and mortality: collaborative analysis of 32 prospective studies. *Lancet* 2010; 375: 1536–1544.

250. Suzuki T, Solomon C, Jenny NS, Tracy R, Nelson JJ, Psaty BM, Furberg C, Cushman M. Lipoprotein-associated phospholipase A2 and risk of congestive heart failure in older adults: The Cardiovascular health Study. *Circ Heart Fail* 2009; 2: 429–436.

251. Gerber Y, Dunlay SM, Jaffe AS, McConnell JP, Weston SA, Killian JM, Roger VL. Plasma lipoprotein-associated phospholipase A2 levels in heart failure: association with mortality in the community. *Atherosclerosis* 2009; 203: 593–598.

252. Furberg CD, Nelson JJ, Solomon C, Cushman M, Jenny NS, Psaty BM. Distribution and correlates of lipoprotein-associated phospholipase A2 in an elderly cohort: the Cardiovascular Health Study. *J Am Geriatr Soc* 2008; 56: 792–799.

253. Tselepis AD, Hahalis G, Tellis CC, Papavasiliou EC, Mylona PT, Kourakli A, Alexopoulos DC. Plasma levels of lipoprotein-associated phospholipase A2 are increased in patients with beta-thalassemia. *J Lipid Res* 2010; 51: 3331–3341.

254. Raichlin E, McConnell JP, Bae JH, Kremers WK, Lerman A, Frantz RP. Lipoprotein-associated phospholipase A2 predicts progression of cardiac allograft vasculopathy and increased risk of cardiovascular events in heart transplant patients. *Transplantation* 2008; 85: 963–968.

255. Casas JP, Ninio E, Panayiotou A, Palmen J, Cooper JA, Ricketts SL, Sofat R, Nicolaides AN, Corsetti JP, Fowkes FGR et al. PLA2G7 Genotype, lipoprotein-associated phospholipase A2 activity, and coronary heart disease risk in 10 494 cases and 15 624 controls of European ancestry. *Circulation* 2010; 121: 2284–2293.

256. Brilakis ES, Khera A, McGuire DK, See R, Banerjee S, Murphy SA, de Lemos JA. Influence of race and sex on lipoprotein-associated phospholipase A2 levels: observations from the Dallas Heart Study. *Atherosclerosis* 2008; 199: 110–115.

257. Campo S, Sardo MA, Trimarchi G, Bonaiuto A, Saitta C, Bitto A, Castaldo M, Cinquegrani M, Bonaiuto M, Cristadoro S et al. Platelet activating factor-acetylhydrolase (PAF-AH) activity and HDL levels, but not PAF-AH gene polymorphisms, are associated with successful aging in Sicilian octogenarians. *Aging Clin Exp Res* 2008; 20: 171–177.

258. Detopoulou P, Nomikos T, Fragopoulou E, Panagiotakos DB, Pitsavos C, Stefanadis C, Antonopoulou S. Lipoprotein-associated phospholipase A2 (Lp-PLA2) activity, platelet-activating factor acetylhydrolase (PAF-AH) in leukocytes and body composition in healthy adults. *Lipids Health Dis* 2009; 8: 19.

259. Suchindran S, Rivedal D, Guyton JR, Milledge T, Gao X, Benjamin A, Rowell J, Ginsburg GS, McCarthy JJ. Genome-wide association study of Lp-PLA(2) activity and mass in the Framingham Heart Study. *PLoS Genet* 2010; 6: e1000928.

260. Holleboom AG, Kuivenhoven JA, Vergeer M, Hovingh GK, van Miert JN, Wareham NJ, Kastelein JJ, Khaw KT, Boekholdt SM. Plasma levels of lecithin:cholesterol acyltransferase and risk of future coronary artery disease in apparently healthy men and women: a prospective case-control analysis nested in the EPIC-Norfolk population study. *J Lipid Res* 2010; 51: 416–421.

261. Dullaart RPF, Perton F, van der Klauw MM, Hillege HL, Sluiter WJ. High plasma lecithin:cholesterol acyltransferase activity does not predict low incidence of cardiovascular events: Possible attenuation of cardioprotection associated with high HDL cholesterol. *Atherosclerosis* 2010; 208: 537–542.

262. Calabresi L, Baldassarre D, Castelnuovo S, Conca P, Bocchi L, Candini C, Frigerio B., Amato M, Sirtori CR, Alessandrini P et al. Functional lecithin: Cholesterol acyltransferase is not required for efficient atheroprotection in humans. *Circulation* 2009; 120: 628–635.

263. Mattsson N, Magnussen CG, Ronnemaa T, Mallat Z, Benessiano J, Jula A, Taittonen L, Kahonen M, Juonala M, Viikari JSA et al. Metabolic syndrome and carotid intima-media thickness in young adults: Roles of apolipoprotein B, apolipoprotein A-I, C-reactive protein, and secretory phospholipase A2: The Cardiovascular Risk in Young Finns Study. *Arterioscler Thromb Vasc Biol* 2010; 30: 1861–1866.

264. Thompson A, Di Angelantonio E, Sarwar N, Erqou S, Saleheen D, Dullaart RP, Keavney B, Ye Z, Danesh J. Association of cholesteryl ester transfer protein genotypes with CETP mass and activity, lipid levels, and coronary risk. *JAMA* 2008; 299: 2777–2788.

265. Regieli JJ, Jukema JW, Grobbee DE, Kastelein JJ, Kuivenhoven JA, Zwinderman AH, van der Graaf Y, Bots ML, Doevendans PA. CETP genotype predicts increased mortality in statin-treated men with proven cardiovascular disease: an adverse pharmacogenetic interaction. *Eur Heart J* 2008; 29: 2792–2799.

266. Wilson PW. CETP genes, metabolic effects, and coronary disease risk. *JAMA* 2008; 299: 2795–2796.

267. Boekholdt SM, Sacks FM, Jukema JW, Shepherd J, Freeman DJ, McMahon AD, Cambien F, Nicaud V, de Grooth GJ, Talmud PJ et al. Cholesteryl ester transfer protein TaqIB variant, high-density lipoprotein cholesterol levels, cardiovascular risk, and efficacy of pravastatin treatment: individual patient meta-analysis of 13,677 subjects. *Circulation* 2005; 111: 278–287.

268. de Grooth GJ, Smilde TJ, van Wissen S, Klerkx AHEM, Zwinderman AH, Fruchart J-C, Kastelein JJP, Stalenhoef AFH, Kuivenhoven JA. The relationship between cholesteryl ester transfer protein levels and risk factor profile in patients with familial hypercholesterolemia. *Atherosclerosis* 2004; 173: 261–267.

269. Boekholdt SM, Kuivenhoven JA, Wareham NJ, Peters RJ, Jukema JW, Luben R, Bingham SA, Day NE, Kastelein JJ, Khaw KT. Plasma levels of cholesteryl ester transfer protein and the risk of future coronary artery disease in apparently healthy men and women: the prospective EPIC (European Prospective Investigation into Cancer and nutrition)-Norfolk population study. *Circulation* 2004; 110: 1418–1423.

270. Vasan RS, Pencina MJ, Robins SJ, Zachariah JP, Kaur G, D'Agostino RB, Ordovas JM. Association of circulating cholesteryl ester transfer protein activity with incidence of cardiovascular disease in the community. *Circulation* 2009; 120: 2414–2420.

271. Schlitt A, Blankenberg S, Bickel C, Lackner KJ, Heine GH, Buerke M, Werdan K, Maegdefessel L, Raaz U, Rupprecht HJ et al. PLTP activity is a risk factor for subsequent cardiovascular events in CAD patients under statin therapy: the AtheroGene Study. *J Lipid Res* 2009; 50: 723–729.

272. Schgoer W, Mueller T, Jauhiainen M, Wehinger A, Gander R, Tancevski, I, Salzmann K, Eller P, Ritsch A, Haltmayer M et al. Low phospholipid transfer protein

(PLTP) is a risk factor for peripheral atherosclerosis. *Atherosclerosis* 2008; 196: 219–226.

273. Wierzbicki AS. Intervening on HDL-C: is it possible? Does it work? *Int J Clin Pract* 2007; 61: 1782–1786.
274. Ghali WA, Rodondi N. HDL cholesterol and cardiovascular risk. *BMJ* 2009; 338: a3065.
275. Krauss RM. Lipoprotein subfractions and cardiovascular disease risk. *Curr Opin Lipidol* 2010; 21: 305–311.
276. Heinecke JW. The HDL proteome: a marker-and perhaps mediator-of coronary artery disease. *J Lipid Res* 2009; 50: S167–171.
277. Skinner ER. High-density lipoprotein subclasses. *Curr Opin Lipidol* 1994; 5: 241–247.
278. Kontush A, Chapman MJ. Functionally defective HDL: A new therapeutic target at the crossroads of dyslipidemia, inflammation and atherosclerosis. *Pharmacol Rev* 2006; 3: 342–374.
279. Anderson DW, Nichols AV, Pan SS, Lindgren FT. High density lipoprotein distribution. Resolution and determination of three major components in a normal population sample. *Atherosclerosis* 1978; 29: 161–179.
280. Johansson J, Carlson LA, Landou C, Hamsten A. High density lipoproteins and coronary atherosclerosis. A strong inverse relation with the largest particles is confined to normotriglyceridemic patients. *Arterioscler Thromb* 1991; 11: 174–182.
281. Drexel H, Amann FW, Rentsch K, Neuenschwander C, Luethy A, Khan SI, Follath F. Relation of the level of high-density lipoprotein subfractions to the presence and extent of coronary artery disease. *Am J Cardiol* 1992; 70: 436–440.
282. Drexel H, Amann FW, Beran J, Rentsch K, Candinas R, Muntwyler J, Luethy A, Gasser T, Follath F. Plasma triglycerides and three lipoprotein cholesterol fractions are independent predictors of the extent of coronary atherosclerosis. *Circulation* 1994; 90: 2230–2235.
283. Drexel H, Steurer J, Muntwyler J, Meienberg S, Schmid HR, Schneider E, Grochenig E, Amann FW. Predictors of the presence and extent of peripheral arterial occlusive disease. *Circulation* 1996; 94: 199–205.
284. Robins SJ, Collins D, Wittes JT, Papademetriou V, Deedwania PC, Schaefer EJ, McNamara JR, Kashyap ML, Hershman JM, Wexler LF et al. Relation of gemfibrozil treatment and lipid levels with major coronary events: VA-HIT: a randomized controlled trial. *JAMA* 2001; 285: 1585–1591.
285. Alagona C, Soro A, Ylitalo K, Salonen R, Salonen JT, Taskinen MR. A low high density lipoprotein (HDL) level is associated with carotid artery intima-media thickness in asymptomatic members of low HDL families. *Atherosclerosis* 2002; 165: 309–316.
286. Barter P, Kastelein J, Nunn A, Hobbs R. High density lipoproteins (HDLs) and atherosclerosis; the unanswered questions. *Atherosclerosis* 2003; 168: 195–211.
287. Yu S, Yarnell JW, Sweetnam P, Bolton CH. High density lipoprotein subfractions and the risk of coronary heart disease: 9-years follow-up in the Caerphilly Study. *Atherosclerosis* 2003; 166: 331–338.
288. Desai MY, Rodriguez A, Wasserman BA, Gerstenblith G, Agarwal S, Kennedy M, Bluemke DA, Lima JAC. Association of cholesterol subfractions and carotid lipid core measured by MRI. *Arterioscler Thromb Vasc Biol* 2005; 25: e110–e111.

289. Berneis KK, Krauss RM. Metabolic origins and clinical significance of LDL heterogeneity. *J Lipid Res* 2002; 43: 1363–1379.

290. Salonen JT, Salonen R, Seppanen K, Rauramaa R, Tuomilehto J. HDL, HDL2, and HDL3 subfractions, and the risk of acute myocardial infarction. A prospective population study in eastern Finnish men. *Circulation* 1991; 84: 129–139.

291. Lamarche B, Moorjani S, Cantin B, Dagenais GR, Lupien PJ, Despres JP. Associations of HDL2 and HDL3 subfractions with ischemic heart disease in men. Prospective results from the Quebec Cardiovascular Study. *Arterioscler Thromb Vasc Biol* 1997; 17: 1098–1105.

292. Delalla OF, Elliott HA, Gofman JW. Ultracentrifugal studies of high density serum lipoproteins in clinically healthy adults. *Am J Physiol* 1954; 179: 333–337.

293. Stampfer MJ, Sacks FM, Salvini S, Willett WC, Hennekens CH. A prospective study of cholesterol, apolipoproteins, and the risk of myocardial infarction. *N Engl J Med* 1991; 325: 373–381.

294. Sweetnam PM, Bolton CH, Yarnell JW, Bainton D, Baker IA, Elwood PC, Miller NE. Associations of the HDL2 and HDL3 cholesterol subfractions with the development of ischemic heart disease in British men. The Caerphilly and Speedwell Collaborative Heart Disease Studies. *Circulation* 1994; 90: 769–774.

295. Williams PT, Feldman DE. Prospective study of coronary heart disease vs. HDL2, HDL3, and other lipoproteins in Gofman's Livermore Cohort. *Atherosclerosis* 2011; 214: 196–202.

296. Asztalos BF, Schaefer EJ. HDL in atherosclerosis: actor or bystander? *Atherosclerosis* 2003; 4 (Suppl.): 21–29.

297. Asztalos BF, Batista M, Horvath KV, Cox CE, Dallal GE, Morse JS, Brown GB, Schaefer EJ. Change in alpha1 HDL concentration predicts progression in coronary artery stenosis. *Arterioscler Thromb Vasc Biol* 2003; 23: 847–852.

298. Asztalos BF, Schaefer EJ. High-density lipoprotein subpopulations in pathologic conditions. *Am J Cardiol* 2003; 91: 12E–17E.

299. Schaefer EJ, Asztalos BF. Where are we with high-density lipoprotein raising and inhibition of cholesteryl ester transfer for heart disease risk reduction? *Curr Opin Cardiol* 2007; 22: 373–378.

300. Schaefer EJ, Asztalos BF. Increasing high-density lipoprotein cholesterol, inhibition of cholesteryl ester transfer protein, and heart disease risk reduction. *Am J Cardiol* 2007; 100: S25–S31.

301. Asztalos BF, Cupples LA, Demissie S, Horvath KV, Cox CE, Batista MC, Schaefer EJ. High-density lipoprotein subpopulation profile and coronary heart disease prevalence in male participants of the Framingham Offspring Study. *Arterioscler Thromb Vasc Biol* 2004; 24: 2181–2187.

302. Lamon-Fava S, Herrington DM, Reboussin DM, Sherman M, Horvath KV, Cupples LA, White C, Demissie S, Schaefer EJ, Asztalos BF. Plasma levels of HDL subpopulations and remnant lipoproteins predict the extent of angiographically-defined coronary artery disease in postmenopausal women. *Arterioscler Thromb Vasc Biol* 2008; 28: 575–579.

303. Asztalos BF, Collins D, Cupples LA, Demissie S, Horvath KV, Bloomfield HE, Robins SJ, Schaefer EJ. Value of high-density lipoprotein (HDL) subpopulations in predicting recurrent cardiovascular events in the Veterans Affairs HDL Intervention Trial. *Arterioscler Thromb Vasc Biol* 2005; 25: 2185–2191..

304. Parra HJ, Arveiler D, Evans AE, Cambou JP, Amouyel P, Bingham A, McMaster D, Schaffer P, Douste-Blazy P, Luc G et al. A case-control study of lipoprotein particles in two populations at contrasting risk for coronary heart disease. The ECTIM Study. *Arterioscler Thromb* 1992; 12: 701–707.

305. Asztalos B, Zhang W, Roheim PS, Wong L. Role of free apolipoprotein A-I in cholesterol efflux. Formation of pre-alpha-migrating high-density lipoprotein particles. *Arterioscler Thromb Vasc Biol* 1997; 17: 1630–1636.

306. Montali A, Vega GL, Grundy SM. Concentrations of apolipoprotein A-I-containing particles in patients with hypoalphalipoproteinemia. *Arterioscler Thromb* 1994; 14: 511–517.

307. Duriez P, Fruchart JC. High-density lipoprotein subclasses and apolipoprotein A-I. *Clin Chim Acta* 1999; 286: 97–114.

308. Schultz JR, Verstuyft JG, Gong EL, Nichols AV, Rubin EM. Protein composition determines the anti-atherogenic properties of HDL in transgenic mice. *Nature* 1993: 365: 762–764.

309. Mora S, Szklo M, Otvos JD, Greenland P, Psaty BM, Goff DC Jr, O'Leary DH, Saad MF, Tsai MY, Sharrett AR. LDL particle subclasses, LDL particle size, and carotid atherosclerosis in the Multi-Ethnic Study of Atherosclerosis (MESA). *Atherosclerosis* 2007; 192: 211–217.

310. Garvey WT, Kwon S, Zheng D, Shaughnessy S, Wallace P, Hutto A, Pugh K, Jenkins AJ, Klein RL, Liao Y. Effects of insulin resistance and type 2 diabetes on lipoprotein subclass particle size and concentration determined by nuclear magnetic resonance. *Diabetes* 2003; 52: 453–462.

311. Goff DC Jr, D'Agostino RB Jr, Haffner SM, Otvos JD. Insulin resistance and adiposity influence lipoprotein size and subclass concentrations. Results from the Insulin Resistance Atherosclerosis Study. *Metabolism* 2005; 54: 264–270.

312. Festa A, Williams K, Hanley AJG, Otvos JD, Goff DC, Wagenknecht LE, Haffner SM. Nuclear magnetic resonance lipoprotein abnormalities in prediabetic subjects in the Insulin Resistance Atherosclerosis Study. *Circulation* 2005; 111: 3465–3472.

313. Kathiresan S, Otvos JD, Sullivan LM, Keyes MJ, Schaefer EJ, Wilson PW, D'Agostino RB, Vasan RS, Robins SJ. Increased small low-density lipoprotein particle number: a prominent feature of the metabolic syndrome in the Framingham Heart Study. *Circulation* 2006; 113: 20–29.

314. Rosenson RS, Otvos JD, Freedman DS. Relations of lipoprotein subclass levels and low-density lipoprotein size to progression of coronary artery disease in the Pravastatin Limitation of Atherosclerosis in the Coronary Arteries (PLAC-I) trial. *Am J Cardiol* 2002; 90: 89–94.

315. Kuller L, Arnold A, Tracy R, Otvos J, Burke G, Psaty B, Siscovick D, Freedman DS, Kronmal R. Nuclear magnetic resonance spectroscopy of lipoproteins and risk of coronary heart disease in the Cardiovascular Health Study. *Arterioscler Thromb Vasc Biol* 2002; 22: 1175–1180.

316. Mora S, Otvos JD, Rifai N, Rosenson RS, Buring JE, Ridker PM. Lipoprotein particle profiles by nuclear magnetic resonance compared with standard lipids and apolipoproteins in predicting incident cardiovascular disease in women. *Circulation* 2009; 119: 931–939.

317. Freedman DS, Otvos JD, Jeyarajah EJ, Shalaurova I, Cupples LA, Parise H, D'Agostino RB, Wilson PW, Schaefer EJ. Sex and age differences in lipoprotein

subclasses measured by nuclear magnetic resonance spectroscopy: the Framingham Study. *Clin Chem* 2004; 50: 1189–1200.

318. Kraus WE, Houmard JA, Duscha BD, Knetzger KJ, Wharton MB, McCartney JS, Bales CW, Henes S, Samsa GP, Otvos JD et al. Effects of the amount and intensity of exercise on plasma lipoproteins. *N Engl J Med* 2002; 347: 1483–1492.
319. Jeyarajah EJ, Cromwell WC, Otvos JD. Lipoprotein particle analysis by nuclear magnetic resonance spectroscopy. *Clin Lab Med* 2006; 26: 847–870.
320. Musunuru K, Orho-Melander M, Caulfield MP, Li S, Salameh WA, Reitz RE, Berglund G, Hedblad B, Engstrom G, Williams PT et al. Ion mobility analysis of lipoprotein subfractions identifies three independent axes of cardiovascular risk. *Arterioscler Thromb Vasc Biol* 2009; 29: 1975–1980.
321. Arsenault BJ, Lemieux I, Despres JP, Gagnon P, Wareham NJ, Stroes ES, Kastelein JJ, Khaw KT, Boekholdt SM. HDL particle size and the risk of coronary heart disease in apparently healthy men and women: the EPIC-Norfolk prospective population study. *Atherosclerosis* 2009; 206: 276–281.

CHAPTER 6

GENETICS

Evolutionary genetics assumes that coronary heart disease has not been subject to direct selective pressure because it typically manifests in middle age and beyond, and is unlikely to have an effect on reproductive success [1]. According to the **ancestral-susceptibility model for complex diseases**, such common complex diseases as coronary heart disease may, therefore, result from a mismatch between the ancestral alleles and current environments. The model holds that ancestral alleles that reflect adaptations to the lifestyle of ancient human populations can be deleterious in the modern-life environment. High serum levels of cholesterol represent an example of such adaptation which may have been advantageous during the rapid increase in human size across human evolution because of its role in steroid hormone synthesis.

Similarly, the **'thrifty-gene' hypothesis** posits that certain genetic variants in humans have evolved to maximize metabolic efficiency, lipid storage and food-searching behaviour [1]; the modern environment promotes the development of Metabolic Syndrome in this case. Another example is provided by genes in the coagulation pathway which may have been under selective pressure given the potential survival advantage from a robust clotting mechanism in the ancestral environment, possibly by enhancing hemostasis in the setting of bleeding from injury; this pathway becomes maladaptive in terms of

High-Density Lipoproteins: Structure, Metabolism, Function, and Therapeutics, First Edition.
Anatol Kontush and M. John Chapman.

atherogenesis. Finally, host response to pathogens has been under selective pressure in the past; heightened immune response developed to fight pathogens may increase the risk of coronary heart disease [1].

Whatever their evolutionary role, HDL-C levels are under strong genetic control, with **heritability** estimates typically ranging between 40% and 60% and attaining up to 80% [2–4]. Equally, the HDL subclass profile and mean HDL size are genetically controlled, with heritability estimated from 27% up to 89% for small HDL particles [4–7]. As with many other complex traits, determinants of HDL-C levels can be monogenic, purely environmental, or, in most cases, result from multiple genes, environmental factors and their interactions, revealing a polygenic and multifactorial nature [8]. Indeed, multiple genes contribute to HDL-C levels, with the effects of individual genes typically small and population-dependent. Genes known to affect HDL metabolism primarily include those encoding HDL components (APOA1, APOC3, APOE, LCAT), lipases (LPL, LIPC, LIPG), lipid transfer proteins (CETP) and HDL receptors [ABCA1, scavenger receptor class B member 1 (SCARB1), LDL-R; Fig. 6.1 and Table 6.1] but also many others (Table 6.2).

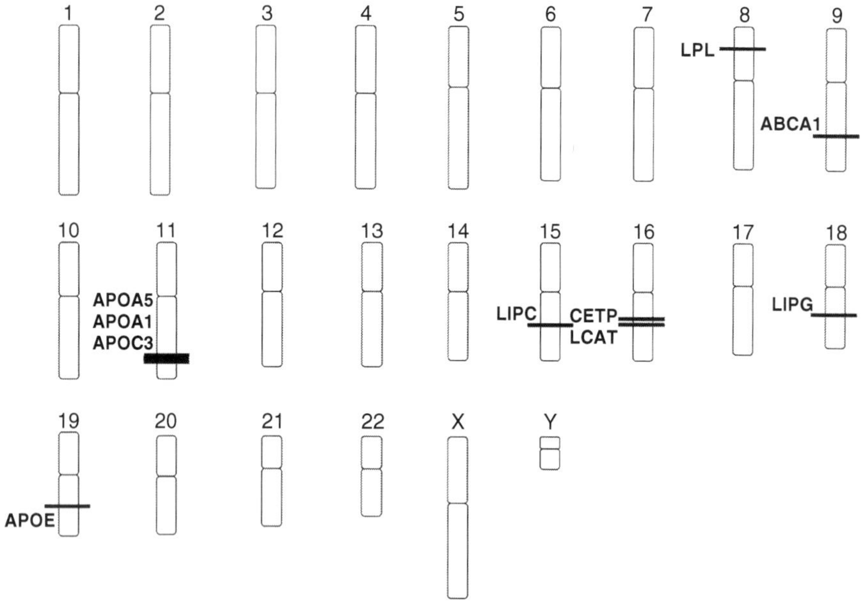

Figure 6.1 Human genome with the positions of major genes influencing HDL-C levels. Positions of the genes are as follows: LPL, 8p21.3; ABCA1, 9q31.1; APOA1/APOC3/APOA4/APOA5, 11q23-24; APOC3, 11q23.1-q23.2; LIPC, 15q21-23; CETP, 16q21; LCAT, 16q22.1; LIPG, 18q21.1; APOE, 19q13.2.

TABLE 6.1 Single Genes Associated with HDL-C Levels (Modified from Reference 8)

Gene	Deficiency	Lipoprotein Phenotype	Elevated Cardiovascular Risk
LPL (8p21.3)	Type I hyperlipoproteinemia (familial chylomicronemia)	Low HDL-C (<20 mg/dl), severe hypertriglyceridemia, chylomicronemia and low LDL-C	Uncommon
ABCA1 (9q31.1)	Tangier disease	Very low HDL-C (<5 mg/dl) and apoA-I, absence of large HDL, low LDL-C and moderately elevated TG	Reported
APOA1 (11q23-q24)	ApoA-I deficiency	Very low HDL-C (<5 mg/dl), absence of apoA-I, normal LDL-C and TG	Reported
LIPC (15q21-23)	Hepatic lipase deficiency	High HDL-C (≥70 mg/dl), moderately elevated apoA-I, large HDL enriched in TG, severe hypertriglyceridemia, elevated LDL-C	Reported
CETP (16q21)	CETP deficiency	Very high HDL-C (> 100 mg/dl), large HDL2 enriched in CE and apoE and depleted of TG, elevated apoA-I, low to normal LDL-C and apoB	Controversial
LCAT (16q22.1)	LCAT deficiency	Very low HDL-C (<5 mg/dl), low apoA-I and LDL-C, small HDL, high TG, abnormal apoB-containing lipoproteins, elevated FC/CE ratio	Reported
	Fish-eye disease (partial LCAT deficiency)	Low HDL-C (<10 mg/dl), low apoA-I, small HDL, low LDL-C, high TG, moderately elevated FC/CE ratio	Reported

CE, cholesteryl ester; FC, free cholesterol; TG, triglyceride.

TABLE 6.2 Genomic loci Associated with HDL-C Levels in Genome-Wide Association Studies (Modified from Reference 8)

Gene(s)	Locus	SNP	Minor Allele Frequency	Effect Size[a]
APOA2	1q23.3	rs4073054	0.38	0.09 μmol/l
GALNT2	1q42.13	rs2144300	0.40	1.11 mg/dl
		rs4846914	0.40	0.05 SD
			0.40	0.07 SD
APOB	2p24.1	rs6754295	0.25	0.07 SD
PCCB	3q22.3	rs3856637	0.28	0.37 μmol/l
LPL	8p21.3	rs10503669	0.10	2.09 mg/dl
		rs12678919	0.10	0.23 SD
		rs2083637	0.26	0.11 SD
		rs328	0.09	0.17 SD
TTC39B	9p22.3	rs471364	0.12	0.08 SD
ABCA1	9q31.1	rs1883025	0.26	0.08 SD
		rs3890182	0.13	0.10 SD
		rs3905000	0.14	0.11 SD
		rs4149268	0.36	0.82 mg/dl
NR1H3 (LXRA)	11p11.2	rs7120118	0.42	1.54 mg/dl
MADD, FOLH1		rs7395662	0.39	0.07 SD
FADS1/2/3	11q12.2	rs174547	0.33	0.09 SD
APOA1/C3/A4/A5	11q23.3	rs964184	0.14	0.17 SD
		rs12225230	0.18	3.2 mg/dl
		rs518181	0.37	0.68 μmol/l
Intergenic, ASCL1, PAH	12q23.2	rs1818702	0.29	0.42 μmol/l
MVK/MMAB	12q24.11	rs2338104	0.45	0.07 SD
			0.45	0.48 mg/dl
CCDC92/DNAH10/ZNF664	12q24.31.B	rs7307277	0.34	0.26 μmol/l
LIPC	15q22.1	rs10468017	0.30	0.10 SD
		rs4775041	0.33	1.38 mg/dl
		rs1532085	0.44	1.93 mg/dl
			0.37	1.83 mg/dl
			0.41	0.13 SD
		rs1800588	0.21	0.14 SD
CETP	16q13	rs1532624	0.43	0.21 SD
		rs1800775	0.49	0.18 SD
		rs173539	0.32	0.25 SD
		rs1864163	0.20	4.12 mg/dl
		rs3764261	0.28	3.47 mg/dl
			0.31	3.96 mg/dl
			0.31	3.47 mg/dl
		rs9989419	0.35	1.72 mg/dl
		rs7499892	0.18	0.76 μmol/l
LCAT	16q22.1	rs2271293	0.13	0.13 SD
			0.11	0.07 SD
		rs255049	0.22	1.93 mg/dl

TABLE 6.2 (*Continued*)

Gene(s)	Locus	SNP	Minor Allele Frequency	Effect Size[a]
WIPI1	17q24.2	rs2909207	0.22	0.09 μmol/l
LIPG	18q21.1	rs2156552	0.16	1.20 mg/dl
		rs4939883	0.17	0.14 SD
			0.17	0.10 SD
ANGPTL4	19p13.2	rs2967605	0.16	0.12 SD
APOE/C1/C4/C2	19q13.32	rs405509	0.48	0.08 μmol/l
PLTP	20q13.12	rs7679	0.19	0.07 SD
HNF4A	20q13.2	rs1800961	0.03	0.19 SD

[a]Change in HDL-C concentrations per one copy of the risk allele.

CLINICAL SUMMARY

Circulating concentrations of HDL-C are under strong genetic control; effects of the majority of common gene variants on HDL-C are, however, small, whereas the effects of some rare variants can be pronounced.

6.1 ABC TRANSPORTERS AND OTHER RECEPTORS

ABCA1

The **ABCA1** gene is highly polymorphic with more than 1000 polymorphic sites and more than 100 known mutations [9]. The most known phenotype resulting from mutations in the gene is known as **Tangier disease**, a rare, monogenic, autosomal recessive disorder with approximately 50 cases identified worldwide. This disorder owes its name to its discovery on Tangier Island, off the coast of Virginia, USA and is characterized by the deposition of cholesterol in the reticulo-endothelial organs which is manifested in orange-coloured tonsils [10–12]. Other important features of the disease are mild hypertriglyceridemia, splenomegaly, hepatomegaly and peripheral neuropathy [13]. A dysfunctional ABCA1 transporter incapable of the efflux of cellular cholesterol to lipid-free and lipid-poor apolipoproteins underlies these symptoms, resulting in marked HDL deficiency in homozygous carriers, with levels of HDL-C of less than 5 mg/dl and those of apoA-I less than 4 mg/dl [14–16] (Table 6.1). In parallel, levels of triglycerides are somewhat elevated (>200 mg/dl) and those of LDL-C levels reduced (about 50% of normal), secondary to the contribution of hepatic ABCA1 to the formation of LDL via CETP. In these subjects, apoA-I is not properly lipidated and becomes rapidly cleared from the circulation, resulting in markedly reduced levels of apoA-I and the presence of only small pre-beta HDL particles. Truncated ABCA1 proteins can be responsible for such dysfunction, resulting from both exonic and intronic mutations of the ABCA1 gene [17]. In heterozygous carriers, HDL-C levels are only moderately reduced, with a decrease in large HDL particles. Greatly reduced levels of HDL-C in patients with Tangier disease frequently result in an elevated cardiovascular risk [14–16]. Similar to LCAT deficiency, however (see this Chapter below), cardiovascular risk can also be significantly modulated by reduced LDL-C levels in this disorder.

> **DEFINITION**
>
> **Tangier disease:** a rare, monogenic, autosomal recessive disorder that owes its name to its discovery on Tangier Island off the coast of Virginia, USA. It results from mutations in the ABCA1 gene, is characterized by markedly reduced HDL-C levels, features deposition of cholesterol in the reticulo-endothelial organs manifested in orange-coloured tonsils, and may enhance cardiovascular risk.

Other mutations in the ABCA1 gene may also account for low HDL-C and lead to **familial hypoalphalipoproteinemia** across different populations [10–12, 18–21]. Such genetically-determined low HDL-C dyslipidemia may result in an elevated cardiovascular risk [22]. On the other hand, polymorphisms in the ABCA1 gene exert moderate or null effects on HDL-C levels in large-scale studies, potentially reflecting a very low frequency of mutations with strong effects on HDL metabolism [23]. For example, HDL-C levels were decreased by -29% in heterozygous carriers of four rare ABCA1 mutations (P1065S, G1216V, N1800H and R2144X) in the Copenhagen City Heart Study [24]. Paradoxically, however, such a decrease in HDL-C levels was not associated with increased cardiovascular risk in this study. Vice versa, single-nucleotide polymorphisms (SNPs) in the ABCA1 gene that are associated with elevated cardiovascular risk do not affect, or increase, HDL-C levels [25]. In addition, ABCA1 and ABCG1 polymorphisms may influence cardiovascular risk in patients with Type 2 diabetes or coronary artery disease [26, 27].

The effects of ABCA1 polymorphisms on HDL-C are also observed in genome-wide association studies [28–34] (Table 6.2). Inter-correlations of ABCA1 polymorphisms are weak, suggesting that the missing associations of individual ABCA1 SNPs with HDL-C levels are independent of each other [23].

> **CRITICAL CONTRIBUTION**
>
> Identification of ABCA1 mutations as a cause of Tangier disease and familial HDL deficiency [10–12].

SR-BI

Genetic variants of **SCARB1** are associated with serum lipid levels [35]. A synonymous variant within the SCARB1 gene, exon 8 rs5888, is associated with altered lipid levels and cardiovascular risk in humans [36]. The rs5888 variant affects SR-BI RNA secondary structure, resulting in reduced SR-BI protein expression and impaired function. Furthermore, rs4238001 and rs2278986 SNPs are independent predictors of SR-BI protein levels [37]. Estrogen and SCARB1 genotype may act synergistically to regulate expression of SCARB1 isoforms and impact serum levels of HDL-C and triglycerides [38].

The effect of polymorphisms in the SCARB1 gene is inconsistent across human studies; the variants present in the HapMap database are not correlated [23]. Thus, variation in the SCARB1 gene at rs10846744 was significantly associated with carotid intima-media thickness in the Multi-Ethnic Study of Atherosclerosis (MESA) independent of lipids and other well-established cardiovascular risk factors [39]. In contrast, the development of carotid atherosclerosis is not influenced by the P297S mutation in the SCARB1 gene in humans [40].

LDL Receptor

Being of pivotal importance for circulating levels of LDL-C, the LDL receptor gene may also be involved in the regulation of HDL-C levels, as suggested by studies of polymorphisms in the **LDLR** gene [23]. Altered catabolism of apoE-containing HDL, as well as changes in circulating LDL particle numbers resulting in the modulation of CETP activity, can account for the link between LDLR polymorphisms and HDL metabolism. However, these data need to be confirmed by larger studies.

6.2 APOLIPOPROTEINS

ApoA-I

Homozygous or compound heterozygous deficiency of apoA-I is a rare condition that results in the complete absence of apoA-I from plasma accompanied by a marked decrease in HDL-C to undetectable levels [41, 42] (Table 6.1). In contrast, concentrations of LDL-C and triglycerides are normal. Patients with apoA-I deficiency exhibit xanthomas, mild-to-moderate corneal opacification and are under increased risk of premature coronary artery disease [43].

Subjects with heterozygous forms of apoA-I deficiency feature plasma HDL-C and apoA-I levels that are about 50% of normal and usually do not present specific clinical symptoms [8]. Structural mutations in the **APOA1** gene, such as missense mutations, are a frequent cause of familial hypoalphalipoproteinemia. Although the impact of these mutations is variable, they appear to cause the greatest elevation in the risk of coronary artery disease when compared with other gene defects [8, 42, 44–47]. For example, patients that are heterozygous for the L178P substitution are characterized by low HDL-C levels and decreased survival [48].

In contrast, individuals that express the **apoA-I Milano** variant with R173C substitution represent a notable exception as they display very low HDL-C levels in association with reduced cardiovascular risk [49, 50] (see 5.1 of chapter 5). Similarly cardioprotective is the apoA-I Paris R151C variant [51]. Both the apoA-I Milano and Paris mutations are distinguished by significant impact on HDL structure [52]. Consistent with these data, common genetic variation in the APOA1 gene associated with increased apoA-I and HDL-C levels does not result in decreased cardiovascular risk in the Copenhagen City Heart Study [53].

Other APOA1 polymorphisms that impact on HDL-C involve a common A-75G substitution in the promoter region [54, 55], as well as several SNPs, such as T84C (rs5070) and C-151T (rs5069) [23]. Two genome-wide association studies document a relationship between polymorphisms in the APOA1/C3/A4/A5 gene cluster and HDL-C [28, 56] (Table 6.2).

ApoA-V

The **APOA5** gene polymorphisms display the most pronounced associations with HDL-C levels compared with genes coding for other major HDL apolipoproteins [9, 23]. Most, but not all, APOA5 polymorphisms, including -1131T>C (rs662799b) and -3A>G (rs651821), are associated with decreased concentrations of HDL-C [9, 23, 57, 58]. Only four APOA5 SNPs are available in HapMap which are weakly associated with each other, except rs662799 and rs651821 which display a strong inter-correlation [23]. The key role of apoA-V in triglyceride metabolism may significantly contribute to the observed association between APOA5 variants and HDL-C [59]. Indeed, the -1131T>C polymorphism is associated with lower HDL-C (mean difference per C allele -3.5%), lower apoAI (-1.3%), higher apoB ($+3.2\%$) and higher triglyceride levels ($+16.0\%$), resulting in an odds ratio for coronary heart disease of 1.18 [60].

ApoC-III

APOC3 mutations can be associated with lower levels of triglycerides and LDL-C, higher levels of HDL-C and decreased atherosclerotic burden [61]. The APOA1/APOC3/APOA4/APOA5 gene locus is highly significantly associated with both triglyceride and HDL-C levels in genome-wide association studies [32] (Table 6.2). Data on the association between the APOC3 gene and genetic variability in HDL-C levels are, however, heterogeneous, suggesting that the role of this gene in the modulation of HDL-C is minor [23].

ApoE

The association between the **APOE** gene and HDL-C levels was investigated in several large-scale population-based studies, the majority of which found elevated

HDL-C in individuals with the **E2** allele and reduced HDL-C in those with the **E4** allele compared with the common **E3** allele [23, 62]. These results are consistent with the well established inverse associations between the same alleles and concentrations of VLDL and LDL, i.e., reduced LDL-C in individuals with the E2 allele and elevated LDL-C in those with the E4 allele compared with the E3 allele [62].

6.3 ENZYMES

LCAT

Homozygous **LCAT** deficiency is an underlying cause of familial LCAT deficiency and fish-eye disease (Table 6.1). **Familial LCAT deficiency** results from the complete loss of LCAT activity. In contrast, **fish-eye disease** is associated with a change in the substrate specificity of LCAT that becomes inactive towards HDL while retaining its activity towards apoB-containing lipoproteins. LCAT deficiency results in free cholesterol and phosphatidylcholine deposition in membranes followed by corneal opacification (fish-eye disease), anemia and renal failure, secondary to the accumulation of lipids in glomeruli [63]. The deficiency may result from functionally impaired truncated LCAT protein with undetectable LCAT activity [64].

In homozygous subjects, both familial LCAT deficiency and fish-eye disease are characterized by low HDL-C (<10 mg/dl) and apoA-I (<50 mg/dl) levels, representing typical forms of familial hypoalphalipoproteinemia [65–67]. Heterozygous carriers of LCAT deficiency typically display low HDL-C levels and are clinically normal. In addition, triglyceride levels are elevated and LDL-C is decreased in LCAT-deficient states. In parallel, free cholesterol is elevated in plasma and peripheral tissues, concomitant with the accumulation of small, lipid-poor HDL.

The association of homozygous LCAT deficiency with premature coronary artery disease remains controversial, with studies showing both accelerated and delayed atherogenesis [67–70] (Table 6.1). Decreased concentrations of LDL-C may significantly impact on cardiovascular risk in this metabolic disorder. Similarly inconsistent are the associations between HDL-C levels and polymorphisms of the LCAT gene [23]. Of several genome-wide association studies, only two report a relationship between LCAT gene polymorphism and HDL-C [29, 32] (Table 6.2). Two LCAT SNPs available in HapMap (rs4986970 and rs2292318) are not correlated with each other, an observation consistent with differences in their association with HDL-C [23].

PON1

The **PON1** gene displays two well known polymorphisms, the Gln192Arg and the Met55Leu, which exist in strong linkage disequilibrium and influence PON1 activity. The PON1 **Q192R** polymorphism is the main determinant of

PON1 activity associated with coronary heart disease risk and progression [71–73]. Specifically, the 192R allele is associated with elevated PON1 activity, diminished levels of biomarkers of oxidative stress, decreased risk of all-cause death and reduced incidence of cardiovascular events, as observed in a large, prospective study in subjects undergoing coronary angiography [74]. Mechanisms accounting for such an association may involve lower stability of the Q192 vs. R192 alloenzyme as a result of weaker binding to HDL; however, reverse causation cannot be excluded [74]. In contrast, the Q192R polymorphism displays a profound impact on PON1 arylesterase and lactonase activities, but does not predict coronary artery disease risk in another study [75].

Furthermore, cardiovascular risk can be modulated by other PON1 variants [76], as well as by polymorphisms in the gene of PON2, a related enzyme from the same family [77]. However, a meta-analysis of 88 case-control studies suggests that an overall association between PON1 polymorphisms and cardiovascular risk is weak, if there is any [78].

Data on the relationships between the PON1 polymorphisms and HDL-C levels are similarly inconsistent [23]. Furthermore, genome-wide association studies typically do not reveal an association between the PON1 gene and HDL-C levels [79], suggesting that the potential action of PON1 on cardiovascular disease is unlikely to involve modulation of HDL-C levels. Regulation of anti-atherogenic activities of HDL by paraoxonases appears a more plausible hypothesis [80]. Consistent with this pathway, genetic components explain more than 60% of the variation in PON1 activity in humans [81].

LPL

Familial **LPL** deficiency (Type I hyperlipoproteinemia, familial chylomicronemia) is a rare, autosomal recessive disorder characterized by the absence of LPL activity and a massive accumulation of chylomicrons. Homozygous deficiency is associated with severe hypertriglyceridemia and marked reductions in HDL-C and LDL-C levels [82], resulting in pancreatitis (Table 6.1). The phenotype is less severe in heterozygotes that display normal-to-elevated triglyceride levels and decreased HDL-C.

LPL polymorphisms may exert significant influence on HDL-C concentrations. Notably, D9N (rs1801177), Gly188Glu and Asn291Ser (rs268) are associated with decreased levels of HDL-C, whereas S447X (rs328) and HindIII (Int8, rs320) tend to result in the elevation of HDL-C [23]. These associations are largely confirmed by meta-analysis [83]. Furthermore, LPL gene variants influence longitudinal variations in HDL-C levels [84]. Mechanistically, mutations resulting in the loss of LPL activity typically lead to the impaired flux of surface fragments from triglyceride-rich lipoproteins to HDL and thereby decrease HDL-C. In addition, accelerated CETP-mediated heteroexchange of cholesteryl ester and triglyceride between triglyceride-rich lipoproteins and HDL may contribute to the reduction of HDL-C levels in LPL deficiency. Cardiovascular risk is not, however, elevated in this metabolic state, potentially as a result of both the anti- and pro-atherogenic properties of LPL [8].

Five LPL polymorphisms reveal strong linkage disequilibrium forming a block, according to HapMap data [85]. Importantly, studies of genetic variants in the non-coding region of the LPL gene with no known direct functional significance reveal an association between a combination of low concentrations of HDL-C and elevated triglycerides with increased cardiovascular risk, suggesting potential pathophysiologic relevance [83, 86]. The relationship between LPL polymorphisms and HDL-C is further consistent with results of genome-wide association studies [28–34, 56, 85, 87] (Table 6.2).

Hepatic Lipase

Complete hepatic lipase deficiency is a rare, autosomal recessive condition resulting in elevated plasma concentrations of cholesterol, triglyceride, HDL-C and apoA-I with the accumulation of large HDL (Table 6.1). The lack of triglyceride hydrolysis by hepatic lipase leads to triglyceride enrichment of both HDL and LDL particles in this condition [88]. As a result, variation in the **LIPC** gene may underlie hyperalphalipoproteinemia.

Two hepatic lipase promoter polymorphisms, C-514T (rs1800588) and G-250A (rs2070895), result in different circulating levels of the enzyme and reveal consistent associations with elevated levels of HDL-C across several studies [23]. For example, one and two copies of the minor allele in the C-514T polymorphism increase HDL-C levels by +1.5 and +3.5 mg/dl, respectively, as shown by meta-analysis on more than 24,000 subjects [89]. Other LIPC polymorphisms are also associated with increased HDL-C [90–93]. Genome-wide association studies confirm the relationship between LIPC polymorphisms and HDL-C levels [28–34, 56, 87] (Table 6.2).

According to the HapMap data, several SNPs located in the promoter region of the LIPC gene are inter-correlated [23]. Paradoxically, however, SNPs in the hepatic lipase gene associated with reduced hepatic lipase activity and increased plasma levels of HDL-C are linked to increased cardiovascular risk [94–96]. Furthermore, an association of HDL-raising CETP Taq1B (B2B2) and LIPC-514C/T (TT) alleles is related to an elevated risk of coronary artery disease, suggesting impaired regeneration of lipid-poor apoA-I by both CETP and hepatic lipase with impaired RCT [95]. The relationship between hepatic lipase deficiency and cardiovascular risk, therefore, appears controversial; both too low and too high hepatic lipase activity might be pro-atherogenic [8]. Interestingly, genome-wide association studies reveal a strong association between the LIPC gene and age-related macular degeneration which is paralleled by a similar association with the CETP gene, suggesting the implication of HDL in this disorder [97].

Endothelial Lipase

Five polymorphisms in the **LIPG** gene are associated with HDL-C levels [23]. The -384A>C (rs3813082) and 584C/T (rs2000813) SNPs result in elevated

HDL-C, whereas C+42T (rs2276269), T+2864C (rs6507931) and 2237G>A (rs3744841) lead to diminished HDL-C concentrations.

Mechanisms underlying the associations between the LIPG polymorphisms and HDL metabolism may involve loss of function [98], as well as impaired secretion of endothelial lipase protein, both resulting in elevated levels of HDL-C [99].

According to the HapMap project, three LIPG SNPs are strongly intercorrelated. The role of LIPG polymorphisms in the variation of HDL-C is consistent with the results of genome-wide association studies [28, 29, 32, 34, 85] (Table 6.2).

6.4 LIPID TRANSFER PROTEINS

CETP

The **CETP** gene is highly polymorphic with several common polymorphisms, as well as rare mutations [100]. CETP deficiency typically results in elevated levels of HDL-C with decreased-to-normal LDL-C concentrations [93, 101] (Table 6.1). Homozygous CETP deficiency can be associated with complete loss of CETP activity, leading to the accumulation of large, cholesteryl ester-rich HDL and elevation of HDL-C levels up to fivefold; this metabolic condition is further characterized by the presence of highly heterogeneous LDL particles. Such complete CETP deficiency is exceedingly rare in Caucasians but frequently observed in patients of Asian origin, where it may account for up to 50% of all cases of hyperalphalipoproteinemia [8, 102]. Heterozygous deficiency of CETP results in less pronounced increases in HDL-C levels of +10 to +30% [82]. Mutations in the CETP gene frequently cause familial hyperalphalipoproteinemia, with a splice-site mutation in intron 14 and a missense D442G mutation in exon 15, representing the most frequent alterations in Japan. On the other hand, substitutions in exons 6 (544C>T) and 9 (c.802C>T) may be responsible for hyperalphalipoproteinemia in the Italian population [103]. Both mutations introduce a premature substitution stop codon and are predicted to cause the production of truncated proteins devoid of function [103]. Impaired secretion of CETP protein may also account for elevated HDL-C [104].

Most CETP polymorphisms are also associated with elevated HDL-C levels. Meta-analyses including more than 10,000 subjects reveal strong associations of common **Taq1B** (+279G>A, B2 allele, rs708272), Ile405Val (+16A>G, rs5882) and −629C>A (rs1800775) polymorphisms with HDL-C [105–108]; in parallel, CETP levels are decreased. In addition, G-2708A (rs12149545), G-971A (rs4783961) and +784CCC>A (rs34145065c) reveal positive relationships with HDL-C. In contrast, two rare polymorphisms, A373P (rs5880) and R451Q (rs1800777), are associated with reduced HDL-C. The relationship between CETP polymorphisms and HDL-C is confirmed by many genome-wide association studies [9,28–34, 56, 85, 87, 109, 110] (Table 6.2).

Importantly, the three common CETP genotypes (Taq1B, I405V and -629C>A) associated with lower CETP activity and higher HDL-C levels are inversely associated with coronary risk [106, 107, 111]. Furthermore, odd ratios for coronary heart disease are compatible with the expected reductions in risk for equivalent increases in HDL-C concentration in available prospective studies [107]. Similar results were reported earlier by other meta-analyses of the relationship between TaqIB variants, HDL-C and cardiovascular risk [112]. Moreover, individuals with longevity syndrome display elevated frequency of the B2 allele of the TaqIB polymorphism and of the Int 14A variant in the CETP gene, both of which are associated with high levels of HDL-C [113, 114].

Consistent with these data, complete CETP deficiency does not predispose to atherosclerosis in the absence of major cardiovascular risk factors [102]. The presence of a CETP mutation does not appear to increase cardiovascular risk in the Japanese population [115, 116]. Furthermore, SNPs near, or in, the CETP gene at 16q13 are associated with both elevated HDL-C and a reduced risk of incident myocardial infarction in a prospective cohort of initially healthy women from the USA [117]. These results support the causal relationship between low CETP activity, high HDL-C and reduced cardiovascular risk, and are consistent with the epidemiologic data (see 5.1 of chapter 5).

The TaqIB genotype is, however, associated not only with reduced cardiovascular risk, but also with step-wise decreases in triglyceride, LDL-C, apoB and remnant cholesterol levels, thereby complicating definitive attribution of the cardioprotective effect to the increase in HDL-C alone [107, 118, 119]. Consistent with these data, the TaqIB polymorphism influences the risk of metabolic syndrome [120]. The causal role of CETP mutations for atherosclerotic disease is further undermined by the results of the INTERHEART study, which reported an association between three CETP SNPs and levels of apoA-I, but no association between these SNPs and the risk of myocardial infarction [121]. These findings are consistent with earlier data showing that increased HDL-C levels caused by the Ile405Val mutation in CETP are associated with an elevated risk of ischemic heart disease in white women [122].

CETP polymorphisms are in strong linkage disequilibrium; indeed, dense genotyping of the CETP gene in more than 2000 individuals provides a complex, nonlinear set of linkage disequilibrium bins associated with HDL-C levels [100]. Despite such a pattern of disequilibrium, CETP polymorphisms are not necessarily strongly inter-correlated [100]. For example, SNPs genotyped in HapMap are, for the most part, only weakly correlated with each other [23]; some of them reveal additive influences on HDL-C concentrations.

MAJOR CONTROVERSY

Do mutations resulting in high (e.g., those in CETP) or low (e.g., those in ABCA1) levels of HDL-C consistently modify cardiovascular risk?

6.5 OTHER GENES

Polymorphisms in some other genes involved in HDL metabolism may impact circulating concentrations of HDL-C (Table 6.2). Thus, HDL-C levels are related to polymorphisms in the APOA2 gene [123, 124]; familial apoA-II deficiency does not, however, strongly impact HDL-C levels or cardiovascular risk [125]. Interestingly, the APOB gene has been recently reported to be associated with HDL-C levels in a genome-wide association study of 22 loci associated with serum lipid levels across 16 population-based cohorts [34]. Association between HDL-C and proprotein convertase subtilisin/kexin type 5 (PCSK5) may reflect the inactivation of endothelial lipase by PCSK5 [126].

The -6G>A (rs11039155) substitution in the NR1H3 gene coding for LXR alpha [127] and the Pro12Ala (rs1805192) in the PPARG2 gene coding for PPAR gamma 2 both result in elevated levels of HDL-C [23]. In contrast, the L162V (rs1800206) polymorphism in the PPARA gene coding for PPAR alpha is associated with a reduction in HDL-C [23]. Next, ABCG5/G8 polymorphisms influence plasma HDL-C in smokers who participated in the Boston Puerto Rican Health Study [128]. Ghrelin, a small, 28-amino acid residue peptide regulator of body weight, may also be involved in HDL metabolism, as suggested by inverse associations of four polymorphisms in the ghrelin gene with HDL-C levels [129, 130]. Plasma levels of ghrelin and HDL-C are positively correlated [131]. The effects of these SNPs on HDL-C may be independent, as most of them are not strongly correlated in the HapMap database [23].

Other genes whose variants are reportedly associated with HDL-C levels include those coding for APOC4 [124], ABCC6 [124], ABCG1, GPR109A/B/81, NFKB1 [79], RXR alpha (RXRA) [124], cubulin (CUBN) [124], 3-hydroxy-3-methyl-glutaryl (HMG)-CoA reductase (HMGCR) [132], acyl CoA:diacylglycerol acyltransferase 1 (DGAT1), selectin P (SELP) [124], chemokine receptor 5 (CCR5) [133], CD36 antigen (CD36), endothelin-1 (EDN1), mast cell chymase (MCC), AMP-kinase alpha 2 subunit (PRKAA2), ZNF202, cytochrome P450, family 3, subfamily A, polypeptide 4 (CYP3A4) [23], cathepsin S (CTSS) [134], aldehyde dehydrogenase 2 (ALDH2) [135], methylene tetrahydrofolate reductase (MTHFR) [136], retinol-binding protein 4 (RBP4) [137], estrogen receptor alpha (ESR1) [138, 139], pre-B-cell colony-enhancing factor 1 (PBEF1; visfatin) [140], uncoupling protein 3 (UCP3) [141], WWOX [142, 143], glutathione S-transferase (GSTM1 and GSTT1) [144] and potassium voltage-gated channel, Shaw-related subfamily, member 2 (KCNC2) [145] genes. Mechanisms of the regulation of HDL-C levels by the majority of the genes from this group are unknown.

In addition, genome-wide association studies identify mevalonate kinase (MVK)/methylmalonic aciduria [cobalamin deficiency] cblB type (MMAB), CCCTC-binding factor (CTCF)/protein arginine methyltransferase 8 (PRMT8), mitogen-activated protein (MAP) kinase-activating death domain protein (MADD)/folate hydrolase 1 (FOLH1), cleft lip and palate associated transmembrane protein 1 (CLPTM1), glutamate receptor, ionotropic, N-methyl-D-aspartate

3A (GRIN3A) [23, 146], UDP-N-acetyl-alpha-D-galactosamine:polypeptide N-acetylgalactosaminyltransferase 2 (GALNT2) [9, 23], ANGPTL3, fatty acid desaturase 1 (FADS1)-FADS2-FADS3, PLTP, tetratricopeptide repeat domain 39B (TTC39B) [32], hepatocyte nuclear factor 4 alpha (HNF4A) [32, 147], nuclear receptor interacting protein 1 (NRIP1), membrane-bound transcription factor peptidase, site 2 (MBTPS2), sterol-C4-methyl oxidase-like protein (SC4MOL), protein tyrosine phosphatase, non-receptor type 11 (PTPN11), adducin 1 alpha (ADD1), SCAN domain containing protein 1 (SCAND1) [147], SAA [148], ribosomal protein L21 pseudogene 98 (RPL21P98) [149], potassium channel tetramerisation domain containing protein 10 (KCTD10) [150] and solute carrier family 39 (zinc transporter), member 8 (SLC39A8) [151] as candidate genes involved in the regulation of HDL metabolism. Polymorphisms in the PLTP, LIPC, FADS1, TTC39B and MMAB genes were associated with transcript levels in human liver [32]. Recent meta-analysis performed in more than 100,000 individuals of European ancestry adds protein phosphatase 1, regulatory (inhibitor) subunit 3B (PPP1R3B), insulin receptor substrate 1 (IRS1), chromosome 6 open reading frame 106 (C6orf106) and Kruppel-like factor 14 (KLF14) genes to this list [152].

Of these genes, neighbouring MVK/MMAB genes appear particularly interesting as MVK catalyzes an early step in cholesterol biosynthesis, and MMAB protein participates in a metabolic pathway that degrades cholesterol. Furthermore, both genes are regulated by SREBP2. MMAB appears to be the most likely gene influencing HDL-C levels at this locus [153]. ANGPTL3 may act via the inhibition of endothelial lipase, LPL and hepatic lipase, providing a key contribution to the regulation of HDL-C levels [154, 155]. Furthermore, mutations in ANGPTL3 can impact on the metabolism of apoB-containing lipoproteins, resulting in familial combined hypolipidemia [156]. The role of the PLTP gene is consistent with an increased frequency of polymorphisms in this gene in patients with hyperalphalipoproteinemia [157, 158]. Furthermore, a gene score based on two PLTP SNPs is associated with lower PLTP transcription and activity, an increased number of HDL particles, smaller HDL particle size and decreased cardiovascular risk [159].

SNPs characterizing common variation in the SAA genes are associated with carotid intima-media thickness, HDL-C levels and cardiovascular events in the participants of the Cardiovascular Health Study [148] and with carotid intima-media thickness in Chinese subjects [160]. The GALNT2 gene encodes a widely expressed glycosyltransferase that can modify a lipoprotein or a lipoprotein receptor. Interestingly, the MADD-FOLH1 locus of chromosome 11 represents a gene desert close to the centromere with no known gene on the 550 kb flanking region; the two genes, MADD and FOLH1, flanking the locus have not been implicated in lipid metabolism [23, 34]. The CTCF-PRMT8 gene encodes a transcriptional regulator potentially involved in hormone-dependent gene silencing [23, 34].

Finally, genome-wide association studies reveal that distinct genes may be specifically associated with HDL size, as well as with plasma levels of individual

HDL subfractions [161]. Specifically, plasma levels of HDL subpopulations are affected by the variation in the ESR1, CETP and LPL genes [139].

Intriguingly, genome-wide association studies observe strong association signals with HDL-C for polymorphisms located in intergenic regions relative to the genes coding for CETP, LIPG and LPL [85]. These findings support the relevance of intergenic regions for the regulation of genes involved in HDL metabolism.

CRITICAL CONTRIBUTION

Novel genes involved in HDL metabolism as discovered in genome-wide association studies [28–34, 56, 85, 87, 109].

6.6 GENE INTERACTIONS

Monogenic disorders of HDL metabolism involving defects in the APOA1, LCAT, ABCA1, LIPG, LIPC, CETP and SCARB1 genes, have, so far, failed to provide definitive insights into the relationship between HDL and cardiovascular disease [47, 162, 163]. However, the identified genes together only explain a small amount of less than 10% of the HDL-C variance, which leaves enormous space for further yet-to-be identified genetic variants [23]. This task may be accomplished by large, population-based, genome-wide, meta-analyses and by deep-sequencing approaches on the identified genes. The resulting findings will probably re-draw and extend the map of metabolic pathways involved in HDL metabolism [23]. In particular, the 15q25 quantitative trait locus, the most significant locus identified to date for the HDL-C trait, provides a framework for the ultimate identification of the underlying HDL-C variant and gene which may provide insights into novel regulatory mechanisms of HDL-C metabolism [164].

In genome-wide association studies of genes known to impact HDL metabolism, the effects that such variants exert on HDL-C concentrations account for <10%, with most accounting for <1%–2% [28, 29, 165]. Moreover, genetic loci associated with HDL-C are likely to harbor multiple, independent causative variants, frequently with opposite effects on the HDL-C phenotype [79]. These data imply that there is no single gene that exerts predominant influence on HDL-C concentration, which instead represents the integral sum of many gene effects on lipoprotein metabolism. When considering the results of these studies, it is important to remember that the effect size of a given variant has little to do with the role of a gene in lipoprotein metabolism. The best example of this kind is provided by the role of HMGCR in LDL metabolism; indeed, an HMGCR variant accounts for only 1 mg/dl difference in LDL-C, whose regulation despite being a key metabolic function of this gene [32, 166].

> **OPEN QUESTION**
>
> How can small effects on HDL-C levels of individual variants, each accounting for <5%, and the majority for <1%–2%, combine to translate into a heritability of HDL-C concentrations ranging between 40% and 60% and attaining up to 80%?

Combining contributions from several common variants, each of which only modestly impacts HDL-C, can improve the prediction of HDL-C levels. Indeed, a genotype score calculated on the basis of the number of unfavorable alleles in nine SNPs located at nine loci, including the CETP, LIPC, LPL and ABCA1 genes, among others, is inversely related to HDL-C levels measured in 5414 subjects from the cardiovascular cohort of the Malmo Diet and Cancer Study [166]. Similarly, the proportion of individuals exceeding clinical cut-points defining low HDL-C increases, whereas mean HDL-C decreases with an increase in the HDL genotype score, reflecting allelic dosage [32]. These results suggest that **cumulative effects of multiple common variants** contribute to low HDL-C dyslipidemia.

Furthermore, rare variants occurring with a frequency of <1% can also significantly contribute to the regulation of HDL-C levels in the general population when combined [167]. Indeed, a population-based study revealed that non-synonymous sequence variants in ABCA1, APOA1 and LCAT genes are eight times more common (16% vs. 2%) in individuals with hypoalphalipoproteinemia (<5th percentile) compared with those with hyperalphalipoproteinemia (>95th percentile) [167]. These data support the hypothesis that **rare genetic variants** may still lead to a common disease, such as atherosclerosis [168]. Effects of numerous polymorphisms on HDL-C are often paralleled by their reciprocal effects on triglyceride levels, consistent with an inverse relationship between the two traits.

Genetic studies offer a rare possibility to assess the causal basis of the association between HDL-C levels and cardiovascular risk based on the concept of **Mendelian randomization** [169], which can be considered as a randomized trial of nature. Such an approach assumes that, as both alleles of a given gene (e.g., CETP or ABCA1) are of lifelong persistence, they will determine, to a certain degree, whether the carrier is exposed to a given trait (e.g., low HDL-C) and, thereby, to a risk of related disease (e.g., cardiovascular disease). In the framework of this concept, the hypothesis regarding the causative role of HDL for cardiovascular disease is supported by the relationship between elevated HDL-C levels and decreased cardiovascular risk associated with CETP polymorphisms [107]. In contrast, the absence of such an association observed for polymorphisms in the ABCA1 [24] and LIPC [94–96] genes is inconsistent with such a role. The latter negative results can be explained by a low number of carriers of a given mutation, by only moderately decreased HDL-C levels in affected subjects, by normal triglyceride levels in low HDL-C subjects, and/or by a dissociation between HDL-C levels and HDL function [23], on the assumption that it is functional deficiency of HDL, rather than low HDL-C levels per se that confers cardiovascular risk (see Section 2).

Overall, only 6 out of 15 common SNPs associated with HDL-C levels reveal associations with the risk of myocardial infarction; remarkably, all of them exert pleiotropic effects on triglyceride and/or LDL-C concentrations, making clear conclusions impossible. As a comparison, 9 out of 10 common polymorphisms associated with LDL-C also show associations with myocardial infarction. Such controversy shows that the role of different genes for metabolic pathways involved in HDL metabolism requires further clarification. The unexplained heritability of the HDL-C trait should be investigated in large, meta- and mega-genome wide association studies, consistent with polygenic inheritance of largely modest effects of individual genes [8]. Future studies should also focus on diverse populations, well-defined phenotypes and environmental factors; indeed, interactions between genetic and environmental factors in their effects on HDL should be uncovered [170]. In addition, different types of variants, such as structural, rare, and low-frequency variants should be investigated in terms of their potential effects on HDL metabolism in order to provide a comprehensive understanding of HDL genetics [8]. Finally, such an approach should involve studies of the effect of genetic variance on HDL metabolism and function in order to provide a mechanistic basis for genetic associations.

IMPORTANT READING

Hegele RA. Plasma lipoproteins: genetic influences and clinical implications. *Nat Rev Genet* 2009; 10: 109–121.

Boes E, Coassin S, Kollerits B, Heid IM, Kronenberg F. Genetic-epidemiological evidence on genes associated with HDL cholesterol levels: a systematic in-depth review. *Exp Gerontol* 2009; 44: 136–160.

Weissglas-Volkov D, Pajukanta P. Genetic causes of high and low serum HDL-cholesterol. *J Lipid Res* 2010; 51: 2032–2057.

Musunuru K, Kathiresan S. Genetics of coronary artery disease. *Annu Rev Genomics Hum Genet* 2010; 11: 91–108.

REFERENCES

1. Ding K, Kullo IJ. Evolutionary genetics of coronary heart disease. *Circulation* 2009; 119: 459–467.
2. Kronenberg F, Coon H, Ellison RC, Borecki I, Arnett DK, Province MA, Eckfeldt JH, Hopkins PN, Hunt SC. Segregation analysis of HDL cholesterol in the NHLBI Family Heart Study and in Utah pedigrees. *Eur J Hum Genet* 2002; 10: 367–374.

3. Perusse L, Rice T, Despres JP, Bergeron J, Province MA, Gagnon J, Leon AS, Rao DC, Skinner JS, Wilmore JH et al. Familial resemblance of plasma lipids, lipoproteins and postheparin lipoprotein and hepatic lipases in the HERITAGE Family Study. *Arterioscler Thromb Vasc Biol* 1997; 17: 3263–3269.
4. Pietilainen KH, Soderlund S, Rissanen A, Nakanishi S, Jauhiainen M, Taskinen MR, Kaprio J. HDL subspecies in young adult twins: heritability and impact of overweight. *Obesity* 2009; 17: 1208–1214.
5. Kaess B, Fischer M, Baessler A, Stark K, Huber F, Kremer W, Kalbitzer HR, Schunkert H, Riegger G, Hengstenberg C. The lipoprotein subfraction profile: heritability and identification of quantitative trait loci. *J Lipid Res* 2008; 49: 715–723.
6. Tejero ME, Voruganti VS, Cai G, Cole SA, Laston S, Wenger CR, Mac Cluer JW, Dyke B, Devereux R, Ebbesson SO et al. Pleiotropic effects on subclasses of HDL, adiposity, and glucose metabolism in adult Alaskan Eskimos. *Am J Hum Biol* 2010; 22: 444–448.
7. Kaess BM, Tomaszewski M, Braund PS, Stark K, Rafelt S, Fischer M, Hardwick R, Nelson CP, Debiec R, Huber F et al. Large-scale candidate gene analysis of HDL particle features. *PLoS One* 2011; 6: e14529.
8. Weissglas-Volkov D, Pajukanta P. Genetic causes of high and low serum HDL-cholesterol. *J Lipid Res* 2010; 51: 2032–2057.
9. Weissglas-Volkov D, Aguilar-Salinas CA, Sinsheimer JS, Riba L, Huertas-Vazquez A, Ordonez-Sanchez ML, Rodriguez-Guillen R, Cantor RM, Tusie-Luna T, Pajukanta P. Investigation of variants identified in Caucasian genome-wide association studies for plasma high-density lipoprotein cholesterol and triglycerides levels in Mexican dyslipidemic study samples. *Circ Cardiovasc Genet* 2010; 3: 31–38.
10. Bodzioch M, Orso E, Klucken J, Langmann T, Bottcher A, Diederich W, Drobnik W, Barlage S, Buchler C, Porsch-Ozcurumez M et al. The gene encoding ATP-binding cassette transporter 1 is mutated in Tangier disease. *Nat Genet* 1999; 22: 347–351.
11. Brooks-Wilson A, Marcil M, Clee SM, Zhang LH, Roomp K, van Dam M, Yu L, Brewer C, Collins JA, Molhuizen HO et al. Mutations in ABC1 in Tangier disease and familial high-density lipoprotein deficiency. *Nat Genet* 1999; 22: 336–345.
12. Rust S, Rosier M, Funke H, Real J, Amoura Z, Piette JC, Deleuze JF, Brewer HB, Duverger N, Denefle P et al. Tangier disease is caused by mutations in the gene encoding ATP-binding cassette transporter 1. *Nat Genet* 1999; 22: 352–355.
13. Oram JF, Vaughan AM. ATP-Binding cassette cholesterol transporters and cardiovascular disease. *Circ Res* 2006; 99: 1031–1043.
14. Oram JF, Lawn RM, Garvin MR, Wade DP. ABCA1 is the cAMP-inducible apolipoprotein receptor that mediates cholesterol secretion from macrophages. *J Biol Chem* 2000; 275: 34508–34511.
15. Mott S, Yu L, Marcil M, Boucher B, Rondeau C, Genest J Jr. Decreased cellular cholesterol efflux is a common cause of familial hypoalphalipoproteinemia: role of the ABCA1 gene mutations. *Atherosclerosis* 2000; 152: 457–468.
16. Sampietro T, Puntoni M, Bigazzi F, Pennato B, Sbrana F, Dal Pino B, Azzara A, Magagnini E, Perossini P, Cei G et al. Tangier disease in severely progressive coronary and peripheral artery disease. *Circulation* 2009; 119: 2741–2742.

17. Bocchi L, Pisciotta L, Fasano T, Candini C, Puntoni MR, Sampietro T, Bertolini S, Calandra S. Multiple abnormally spliced ABCA1 mRNAs caused by a novel splice site mutation of ABCA1 gene in a patient with Tangier disease. *Clin Chim Acta* 2010; 411: 524–530.

18. Slatter TL, Jones GT, Williams MJ, van Rij AM, McCormick SP. Novel rare mutations and promoter haplotypes in ABCA1 contribute to low-HDL-C levels. *Clin Genet* 2008; 73: 179–184.

19. Aguilar-Salinas CA, Canizales-Quinteros S, Rojas-Martinez R, Mehta R, Villarreal-Molina MT, Arellano-Campos O, Riba L, Gomez-Perez FJ, Tusie-Luna MT. Hypoalphalipoproteinemia in populations of Native American ancestry: an opportunity to assess the interaction of genes and the environment. *Curr Opin Lipidol* 2009; 20: 92–97.

20. Guevara-Cruz M, Tovar AR, Larrieta E, Canizales-Quinteros S, Torres N. Increase in HDL-C concentration by a dietary portfolio with soy protein and soluble fiber is associated with the presence of the ABCA1 R230C variant in hyperlipidemic Mexican subjects. *Mol Genet Metab* 2010; 101: 268–272.

21. Berge KE, Leren TP. Mutations in APOA-I and ABCA1 in Norwegians with low levels of HDL cholesterol. *Clin Chim Acta* 2010; 411: 2019–2023.

22. Ma XY, Liu JP, Song ZY. Associations of the ATP-binding cassette transporter A1R219K polymorphism with HDL-C level and coronary artery disease risk: A meta-analysis. *Atherosclerosis* 2011; 215: 428–434.

23. Boes E, Coassin S, Kollerits B, Heid IM, Kronenberg F. Genetic-epidemiological evidence on genes associated with HDL cholesterol levels: a systematic in-depth review. *Exp Gerontol* 2009; 44: 136–160.

24. Frikke-Schmidt R, Nordestgaard BG, Stene MC, Sethi AA, Remaley AT, Schnohr P, Grande P, Tybjaerg-Hansen A. Association of loss-of-function mutations in the ABCA1 gene with high-density lipoprotein cholesterol levels and risk of ischemic heart disease. *JAMA* 2008; 299: 2524–2532.

25. Frikke-Schmidt R, Nordestgaard BG, Jensen GB, Steffensen R, Tybjaerg-Hansen A. Genetic variation in ABCA1 predicts ischemic heart disease in the general population. *Arterioscler Thromb Vasc Biol* 2008; 28: 180–186.

26. Porchay-Balderelli I, Pean F, Emery N, Maimaitiming S, Bellili N, Travert F, Mohammedi K, Roussel R, Marre M, Fumeron F. Relationships between common polymorphisms of adenosine triphosphate-binding cassette transporter A1 and high-density lipoprotein cholesterol and coronary heart disease in a population with type 2 diabetes mellitus. *Metabolism* 2009; 58: 74–79.

27. Furuyama S, Uehara Y, Zhang B, Baba Y, Abe S, Iwamoto T, Miura S, Saku K. Genotypic effect of ABCG1 gene promoter -257T>G polymorphism on coronary artery disease severity in Japanese men. *J Atheroscler Thromb* 2009; 16: 194–200.

28. Kathiresan S, Melander O, Guiducci C, Surti A, Burtt NP, Rieder MJ, Cooper GM, Roos C, Voight BF, Havulinna AS et al. Six new loci associated with blood low-density lipoprotein cholesterol, high-density lipoprotein cholesterol or triglycerides in humans. *Nat Genet* 2008; 40: 189–197.

29. Willer CJ, Sanna S, Jackson AU, Scuteri A, Bonnycastle LL, Clarke R, Heath SC, Timpson NJ, Najjar SS, Stringham HM et al. Newly identified loci that influence lipid concentrations and risk of coronary artery disease. *Nat Genet* 2008; 40: 161–169.

30. Wallace C, Newhouse SJ, Braund P, Zhang F, Tobin M, Falchi M, Ahmadi K, Dobson RJ, Marcano AC, Hajat C et al. Genome-wide association study identifies genes for biomarkers of cardiovascular disease: serum urate and dyslipidemia. *Am J Hum Genet* 2008; 82: 139–149.

31. Kooner JS, Chambers JC, Aguilar-Salinas CA, Hinds DA, Hyde CL, Warnes GR, Gomez Perez FJ, Frazer KA, Elliott P, Scott J et al. Genome-wide scan identifies variation in MLXIPL associated with plasma triglycerides. *Nat Genet* 2008; 40: 149–151.

32. Kathiresan S, Willer CJ, Peloso GM, Demissie S, Musunuru K, Schadt EE, Kaplan L, Bennett D, Li Y, Tanaka T et al. Common variants at 30 loci contribute to polygenic dyslipidemia. *Nat Genet* 2009; 41: 56–65.

33. Talmud PJ, Drenos F, Shah S, Shah T, Palmen J, Verzilli C, Gaunt TR, Pallas J, Lovering R, Li K et al. Gene-centric association signals for lipids and apolipoproteins identified via the HumanCVD BeadChip. *Am J Hum Genet* 2009; 85: 628–642.

34. Aulchenko YS, Ripatti S, Lindqvist I, Boomsma D, Heid IM, Pramstaller PP, Penninx BW, Janssens AC, Wilson JF, Spector T et al. Loci influencing lipid levels and coronary heart disease risk in 16 European population cohorts. *Nat Genet* 2009; 41: 47–55.

35. Acton S, Osgood D, Donoghue M, Corella D, Pocovi M, Cenarro A, Mozas P, Keilty J, Squazzo S, Woolf EA et al. Association of polymorphisms at the SR-BI gene locus with plasma lipid levels and body mass index in a white population. *Arterioscler Thromb Vasc Biol* 1999; 19: 1734–1743.

36. Constantineau J, Greason E, West M, Filbin M, Kieft JS, Carletti MZ, Christenson LK, Rodriguez A. A synonymous variant in scavenger receptor, class B, type I gene is associated with lower SR-BI protein expression and function. *Atherosclerosis* 2010; 210: 177.

37. West M, Greason E, Kolmakova A, Jahangiri A, Asztalos B, Pollin TI, Rodriguez A. Scavenger receptor class B type I protein as an independent predictor of high-density lipoprotein cholesterol levels in subjects with hyperalphalipoproteinemia. *J Clin Endocrinol Metab* 2009; 94: 1451–1457.

38. Chiba-Falek O, Nichols M, Suchindran S, Guyton J, Ginsburg GS, Barrett-Connor E, McCarthy JJ. Impact of gene variants on sex-specific regulation of human scavenger receptor class B type 1 (SR-BI) expression in liver and association with lipid levels in a population-based study. *BMC Med Genet* 2010; 11: 9.

39. Naj AC, West M, Rich SS, Post W, Kao WHL, Wasserman BA, Herrington DM, Rodriguez A. Association of scavenger receptor class B type I polymorphisms with subclinical atherosclerosis: The Multi-Ethnic Study of Atherosclerosis. *Circ Cardiovasc Genet* 2010; 3: 47–52.

40. Vergeer M, Korporaal SJ, Franssen R, Meurs I, Out R, Hovingh GK, Hoekstra M, Sierts JA, Dallinga-Thie GM, Motazacker MM et al. Genetic variant of the scavenger receptor BI in humans. *N Engl J Med* 2011; 364: 136–145.

41. Funke H. Genetic determinants of high density lipoprotein levels. *Curr Opin Lipidol* 1997; 8: 189–196.

42. Wada M, Iso T, Asztalos BF, Takama N, Nakajima T, Seta Y, Kaneko K, Taniguchi Y, Kobayashi H, Nakajima K et al. Marked high density lipoprotein deficiency due to apolipoprotein A-I Tomioka (codon 138 deletion). *Atherosclerosis* 2009; 207: 157–161.

43. von Eckardstein A. Differential diagnosis of familial high density lipoprotein deficiency syndromes. *Atherosclerosis* 2006; 186: 231–239.
44. Ng DS, Vezina C, Wolever TS, Kuksis A, Hegele RA, Connelly PW. Apolipoprotein A-I deficiency. Biochemical and metabolic characteristics. *Arterioscler Thromb Vasc Biol* 1995; 15: 2157–2164.
45. Ikewaki K, Matsunaga A, Han H, Watanabe H, Endo A, Tohyama Ji, Kuno M, Mogi J, Sugimoto K, Tada N. A novel two nucleotide deletion in the apolipoprotein A-I gene, apoA-I Shinbashi, associated with high density lipoprotein deficiency, corneal opacities, planar xanthomas, and premature coronary artery disease. *Atherosclerosis* 2004; 172: 39–45.
46. Dastani Z, Dangoisse C, Boucher B, Desbiens K, Krimbou L, Dufour R, Hegele RA, Pajukanta P, Engert JC, Genest J et al. A novel nonsense apolipoprotein A-I mutation (apoA-I(E136X)) causes low HDL cholesterol in French Canadians. *Atherosclerosis* 2006; 185: 127–136.
47. Schaefer EJ, Santos RD, Asztalos BF. Marked HDL deficiency and premature coronary heart disease. *Curr Opin Lipidol* 2010; 21: 289–297.
48. Hovingh GK, Brownlie A, Bisoendial RJ, Dube MP, Levels JH, Petersen W, Dullaart RP, Stroes ES, Zwinderman AH, de Groot E et al. A novel apoA-I mutation (L178P) leads to endothelial dysfunction, increased arterial wall thickness, and premature coronary artery disease. *J Am Coll Cardiol* 2004; 44: 1429–1435.
49. Sirtori CR, Calabresi L, Franceschini G, Baldassarre D, Amato M, Johansson J, Salvetti M, Monteduro C, Zulli R, Muiesan ML et al. Cardiovascular status of carriers of the apolipoprotein A-I Milano mutant: the Limone sul Garda study. *Circulation* 2001; 103: 1949–1954.
50. Chiesa G, Sirtori CR. Apolipoprotein A-I Milano: current perspectives. *Curr Opin Lipidol* 2003; 14: 159–163.
51. Bruckert E, von Eckardstein A, Funke H, Beucler I, Wiebusch H, Turpin G, Assmann G. The replacement of arginine by cysteine at residue 151 in apolipoprotein A-I produces a phenotype similar to that of apolipoprotein A-IMilano. *Atherosclerosis* 1997; 128: 121–128.
52. Rocco AG, Sensi C, Gianazza E, Calabresi L, Franceschini G, Sirtori CR, Eberini I. Structural and dynamic features of apolipoprotein A-I cysteine mutants, Milano and Paris, in synthetic HDL. *J Mol Graph Model* 2010; 29: 406–414.
53. Haase CL, Tybjaerg-Hansen A, Grande P, Frikke-Schmidt R. Genetically elevated apolipoprotein A-I, high-density lipoprotein cholesterol levels, and risk of ischemic heart disease. *J Clin Endocrinol Metab* 2010; 95: E500–E510.
54. Pagani F, Sidoli A, Giudici GA, Barenghi L, Vergani C, Baralle FE. Human apolipoprotein A-I gene promoter polymorphism: association with hyperalphalipoproteinemia. *J Lipid Res* 1990; 31: 1371–1377.
55. Dawar R, Gurtoo A, Singh R. Apolipoprotein A1 gene polymorphism (G-75A and C+83T) in patients with myocardial infarction: a pilot study in a north Indian population. *Am J Clin Pathol* 2010; 134: 249–255.
56. Chasman DI, Pare G, Zee RYL, Parker AN, Cook NR, Buring JE, Kwiatkowski DJ, Rose LM, Smith JD, Williams PT et al. Genetic loci associated with plasma concentration of low-density lipoprotein cholesterol, high-density lipoprotein cholesterol, triglycerides, apolipoprotein A1, and apolipoprotein B among 6382 white women in genome-wide analysis with replication. *Circ Cardiovasc Genet* 2008; 1: 21–30.

57. Chen ES, Furuya TK, Mazzotti DR, Ota VK, Cendoroglo MS, Ramos LR, Araujo LQ, Burbano RR, de Arruda Cardoso Smith M. APOA1/A5 variants and haplotypes as a risk factor for obesity and better lipid profiles in a Brazilian Elderly Cohort. *Lipids* 2010; 45: 511–517.
58. Zhao T, Zhao J. Association of the apolipoprotein A5 gene -1131 T>C polymorphism with fasting blood lipids: a meta-analysis in 37859 subjects. *BMC Med Genet* 2010; 11: 120.
59. Johansen CT, Wang J, Lanktree MB, Cao H, McIntyre AD, Ban MR, Martins RA, Kennedy BA, Hassell RG, Visser ME et al. Excess of rare variants in genes identified by genome-wide association study of hypertriglyceridemia. *Nat Genet* 2010; 42: 684–687.
60. Sarwar N, Sandhu MS, Ricketts SL, Butterworth AS, Di Angelantonio E, Boekholdt SM, Ouwehand W, Watkins H, Samani NJ, Saleheen D et al. Triglyceride-mediated pathways and coronary disease: collaborative analysis of 101 studies. *Lancet*, 2010; 375, 1634–1639.
61. Pollin TI, Damcott CM, Shen H, Ott SH, Shelton J, Horenstein RB, Post W, McLenithan JC, Bielak LF, Peyser PA et al. A null mutation in human APOC3 confers a favorable plasma lipid profile and apparent cardioprotection. *Science* 2008; 322: 1702–1705.
62. Burman D, Mente A, Hegele RA, Islam S, Yusuf S, Anand SS. Relationship of the ApoE polymorphism to plasma lipid traits among South Asians, Chinese, and Europeans living in Canada. *Atherosclerosis* 2009; 203: 192–200.
63. Miller M, Zhan M. Genetic determinants of low high-density lipoprotein cholesterol. *Curr Opin Cardiol* 2004; 19: 380–384.
64. Park CW, Lim M-H, Youn D-Y, Jung SE, Chung S, Ahn YS, Chang YS, Lee J-H Two novel frame shift mutations in lecithin:cholesterol acyltransferase (LCAT) gene associated with a familial LCAT deficiency phenotype. *Atherosclerosis* 2009; 206: 346.
65. Rousset X, Vaisman B, Amar M, Sethi AA, Remaley AT. Lecithin: cholesterol acyltransferase–from biochemistry to role in cardiovascular disease. *Curr Opin Endocrinol Diabetes Obes* 2009; 16: 163–171.
66. Jonas A. Lecithin cholesterol acyltransferase. *Biochim Biophys Acta* 2000; 1529: 245–256.
67. Calabresi L, Baldassarre D, Castelnuovo S, Conca P, Bocchi L, Candini C, Frigerio B, Amato M, Sirtori CR, Alessandrini P et al. Functional lecithin: cholesterol acyltransferase is not required for efficient atheroprotection in humans. *Circulation* 2009; 120: 628–635.
68. Hovingh GK, Hutten BA, Holleboom AG, Petersen W, Rol P, Stalenhoef A, Zwinderman AH, de Groot E, Kastelein JJ, Kuivenhoven JA. Compromised LCAT function is associated with increased atherosclerosis. *Circulation* 2005; 112: 879–884.
69. Ayyobi AF, McGladdery SH, Chan S, John Mancini GB, Hill JS, Frohlich JJ. Lecithin: cholesterol acyltransferase (LCAT) deficiency and risk of vascular disease: 25 year follow-up. *Atherosclerosis* 2004; 177: 361–366.
70. Holleboom AG, Duivenvoorden R, van den Bogaard B, de Groot E, Nederveen AJ, Lameris JS, Hutten BA, Kastelein JJ, Kuivenhoven JA, Stroes ES. Patients with low plasma high density lipoprotein-cholesterol due to mutations in the gene encoding for lecithin: cholesterol acyl transferase have increased atherosclerosis: A 3.0 Tesla MRI Study. *Circulation* 2010: 122 (Suppl): A17773.

71. Mackness MI, Mackness B, Durrington PN. Paraoxonase and coronary heart disease. *Atherosclerosis* 2002; 3 (Suppl.): 49–55.
72. Aviram M, Rosenblat M. Paraoxonases and cardiovascular diseases: pharmacological and nutritional influences. *Curr Opin Lipidol* 2005; 16: 393–399.
73. Goswami B, Tayal D, Gupta N, Mallika V. Paraoxonase: a multifaceted biomolecule. *Clin Chim Acta* 2009; 410: 1–12.
74. Bhattacharyya T, Nicholls SJ, Topol EJ, Zhang R, Yang X, Schmitt D, Fu X, Shao M, Brennan DM, Ellis SG et al. Relationship of paraoxonase 1 (PON1) gene polymorphisms and functional activity with systemic oxidative stress and cardiovascular risk. *JAMA* 2008; 299: 1265–1276.
75. Bayrak A, Bayrak T, Demirpence E, Kilinc K. Differential hydrolysis of homocysteine thiolactone by purified human serum (192)Q and (192)R PON1 isoenzymes. *J Chromatogr B Analyt Technol Biomed Life Sci* 2011; 879: 49–55.
76. Najafi M, Gohari LH, Firoozrai M. Paraoxonase 1 gene promoter polymorphisms are associated with the extent of stenosis in coronary arteries. *Thromb Res* 2009; 123: 503–510.
77. Marchegiani F, Spazzafumo L, Provinciali M, Cardelli M, Olivieri F, Franceschi C, Lattanzio F, Antonicelli R. Paraoxonase2 C311S polymorphism and low levels of HDL contribute to a higher mortality risk after acute myocardial infarction in elderly patients. *Mol Genet Metab* 2009; 98: 314–318.
78. Wang M, Lang X, Zou L, Huang S, Xu Z. Four genetic polymorphisms of paraoxonase gene and risk of coronary heart disease: a meta-analysis based on 88 case-control studies. *Atherosclerosis* 2011; 214: 377–385.
79. Edmondson AC, Braund PS, Stylianou IM, Khera AV, Nelson CP, Wolfe ML, DerOhannessian SL, Keating BJ, Qu L, He J et al. Dense genotyping of candidate gene loci identifies variants associated with high-density lipoprotein cholesterol. *Circ Cardiovasc Genet* 2011; 4: 145–155.
80. Kontush A, Chapman MJ. Functionally defective HDL: A new therapeutic target at the crossroads of dyslipidemia, inflammation and atherosclerosis. *Pharmacol Rev* 2006; 3: 342–374.
81. Rainwater DL, Rutherford S, Dyer TD, Rainwater ED, Cole SA, Vandeberg JL, Almasy L, Blangero J, Maccluer JW, Mahaney MC. Determinants of variation in human serum paraoxonase activity. *Heredity* 2009; 102: 147–154.
82. Klos KL, Kullo IJ. Genetic determinants of HDL: monogenic disorders and contributions to variation. *Curr Opin Cardiol* 2007; 22: 344–351.
83. Wittrup HH, Tybjaerg-Hansen A, Nordestgaard BG. Lipoprotein lipase mutations, plasma lipids and lipoproteins, and risk of ischemic heart disease. A meta-analysis. *Circulation* 1999; 99: 2901–2907.
84. Tang W, Apostol G, Schreiner PJ, Jacobs DR J., Boerwinkle E, Fornage M. Associations of lipoprotein lipase gene polymorphisms with longitudinal plasma lipid trends in young adults: The Coronary Artery Risk Development in Young Adults (CARDIA) Study. *Circ Cardiovasc Genet* 2010; 3: 179–186.
85. Heid IM, Boes E, Muller M, Kollerits B, Lamina C, Coassin S, Gieger C, Doring A, Klopp N, Frikke-Schmidt R et al. Genome-wide association analysis of high-density lipoprotein cholesterol in the population-based KORA study sheds new light on intergenic regions. *Circ Cardiovasc Genet* 2008; 1: 10–20.

86. Hu Y, Liu W, Huang R, Zhang X. A systematic review and meta-analysis of the relationship between lipoprotein lipase Asn291Ser variant and diseases. *J Lipid Res* 2006; 47: 1908–1914.

87. Chasman DI, Pare G, Zee RY, Parker AN, Cook NR, Buring JE, Kwiatkowski DJ, Rose LM, Smith JD, Williams PT et al. Genetic loci associated with plasma concentration of low-density lipoprotein cholesterol, high-density lipoprotein cholesterol, triglycerides, apolipoprotein A1, and apolipoprotein B among 6382 white women in genome-wide analysis with replication. *Circ Cardiovasc Genet* 2008; 1: 21–30.

88. Cohen JC, Vega GL, Grundy SM. Hepatic lipase: new insights from genetic and metabolic studies. *Curr Opin Lipidol* 1999; 10: 259–267.

89. Isaacs A, Sayed-Tabatabaei FA, Njajou OT, Witteman JC, van Duijn CM. The -514 C->T hepatic lipase promoter region polymorphism and plasma lipids: a meta-analysis. *J Clin Endocrinol Metab* 2004; 89: 3858–3863.

90. Iijima H, Emi M, Wada M, Daimon M, Toriyama S, Koyano S, Sato H, Hopkins PN, Hunt SC, Kubota I et al. Association of an intronic haplotype of the LIPC gene with hyperalphalipoproteinemia in two independent populations. *J Hum Genet* 2008; 53: 193–200.

91. Grarup N, Andreasen CH, Andersen MK, Albrechtsen A, Sandbaek A, Lauritzen T, Borch-Johnsen K, Jorgensen T, Schmitz O, Hansen T et al. The -250G>A promoter variant in hepatic lipase associates with elevated fasting serum high-density lipoprotein cholesterol modulated by interaction with physical activity in a study of 16,156 Danish subjects. *J Clin Endocrinol Metab* 2008; 93: 2294–2299.

92. Feitosa MF, Myers RH, Pankow JS, Province MA, Borecki IB. LIPC variants in the promoter and intron 1 modify HDL-C levels in a sex-specific fashion. *Atherosclerosis* 2009; 204: 171–177.

93. Plengpanich W, Siriwong S, Khovidhunkit W. Two novel mutations and functional analyses of the CETP and LIPC genes underlying severe hyperalphalipoproteinemia. *Metabolism* 2009; 58: 1178–1184.

94. Andersen RV, Wittrup HH, Tybjaerg-Hansen A, Steffensen R, Schnohr P, Nordestgaard BG. Hepatic lipase mutations, elevated high-density lipoprotein cholesterol, and increased risk of ischemic heart disease: the Copenhagen City Heart Study. *J Am Coll Cardiol* 2003; 41: 1972–1982.

95. van Acker BA, Botma GJ, Zwinderman AH, Kuivenhoven JA, Dallinga-Thie GM, Sijbrands EJ, Boer JM, Seidell JC, Jukema JW, Kastelein JJ et al. High HDL cholesterol does not protect against coronary artery disease when associated with combined cholesteryl ester transfer protein and hepatic lipase gene variants. *Atherosclerosis* 2008; 200: 161–167.

96. Johannsen TH, Kamstrup PR, Andersen RV, Jensen GB, Sillesen H, Tybjaerg-Hansen A, Nordestgaard BG. Hepatic lipase, genetically elevated high-density lipoprotein, and risk of ischemic cardiovascular disease. *J Clin Endocrinol Metab* 2009; 94: 1264–1273.

97. Chen W, Stambolian D, Edwards AO, Branham KE, Othman M, Jakobsdottir J, Tosakulwong N, Pericak-Vance MA, Campochiaro PA, Klein ML et al. Genetic variants near TIMP3 and high-density lipoprotein-associated loci influence susceptibility to age-related macular degeneration. *Proc Natl Acad Sci* 2010: 107: 7401–7406.

98. Edmondson AC, Brown RJ, Kathiresan S, Cupples LA, Demissie S, Manning AK, Jensen MK, Rimm EB, Wang J, Rodrigues A et al. Loss-of-function variants in

endothelial lipase are a cause of elevated HDL cholesterol in humans. *J Clin Invest* 2009; 11: 1042–1050.

99. Brow RJ, Edmondson AC, Griffon N, Hill TB, Fuki IV, Badellino KO, Li M, Wolfe ML, Reilly MP, Rader DJ. A naturally occurring variant of endothelial lipase associated with elevated HDL exhibits impaired synthesis. *J Lipid Res* 2009; 50: 1910–1916.
100. Thompson JF, Wood LS, Pickering EH, DeChairo B, Hyde CL. High-density genotyping and functional SNP localization in the CETP gene. *J Lipid Res* 2007; 48: 434–443.
101. Miller M, Rhyne J, Hamlette S, Birnbaum J, Rodriguez A. Genetics of HDL regulation in humans. *Curr Opin Lipidol* 2003; 14: 273–279.
102. Calabresi L, Nilsson P, Pinotti E, Gomaraschi M, Favari E, Adorni MP, Bernini F, Sirtori CR, Calandra S, Franceschini G et al. A novel homozygous mutation in CETP gene as a cause of CETP deficiency in a Caucasian kindred. *Atherosclerosis* 2009; 205: 506–511.
103. Cefalu AB, Noto D, Magnolo L, Pinotti E, Gomaraschi M, Martini S, Vigna GB, Calabresi L, Tarugi P, Averna MR. Novel mutations of CETP gene in Italian subjects with hyperalphalipoproteinemia. *Atherosclerosis* 2009; 204: 202–207.
104. Thompson JF, Reynolds JM, Williams SP, Wood LS, Paciga SA, Lloyd DB. Frequency and function of CETP variants among individuals of Asian ancestry. *Atherosclerosis* 2009; 202: 241–247.
105. Boekholdt SM, Thompson JF. Natural genetic variation as a tool in understanding the role of CETP in lipid levels and disease. *J Lipid Res* 2003; 44: 1080–1093.
106. Boekholdt SM, Sacks FM, Jukema JW, Shepherd J, Freeman DJ, McMahon AD, Cambien F, Nicaud V, de Grooth GJ, Talmud PJ et al. Cholesteryl ester transfer protein TaqIB variant, high-density lipoprotein cholesterol levels, cardiovascular risk, and efficacy of pravastatin treatment: individual patient meta-analysis of 13,677 subjects. *Circulation* 2005; 111: 278–287.
107. Thompson A, Di Angelantonio E, Sarwar N, Erqou S, Saleheen D, Dullaart RP, Keavney B, Ye Z, Danesh J. Association of cholesteryl ester transfer protein genotypes with CETP mass and activity, lipid levels, and coronary risk. *JAMA* 2008; 299: 2777–2788.
108. Tsai MY, Johnson C, Kao WHL, Sharrett AR, Arends VL, Kronmal R, Jenny NS, Jacobs Jr, DR, Arnett D, O'Leary D et al. Cholesteryl ester transfer protein genetic polymorphisms, HDL cholesterol, and subclinical cardiovascular disease in the Multi-Ethnic Study of Atherosclerosis. *Atherosclerosis* 2008; 200: 359–367.
109. Hiura Y, Shen CS, Kokubo Y, Okamura T, Morisaki T, Tomoike H, Yoshida T, Sakamoto H, Goto Y, Nonogi H et al. Identification of genetic markers associated with high-density lipoprotein-cholesterol by genome-wide screening in a Japanese population: the Suita study. *Circ J* 2009; 73: 1119–1126.
110. Smith EN, Chen W, Kahonen M, Kettunen J, Lehtimaki T, Peltonen L, Raitakari OT, Salem RM, Schork NJ, Shaw M et al. Longitudinal genome-wide association of cardiovascular disease risk factors in the Bogalusa Heart Study. *PLoS Genet* 2010; 6: e1001094.
111. Regieli JJ, Jukema JW, Grobbee DE, Kastelein JJ, Kuivenhoven JA, Zwinderman AH, van der Graaf Y, Bots ML, Doevendans PA. CETP genotype predicts increased mortality in statin-treated men with proven cardiovascular disease: an adverse pharmacogenetic interaction. *Eur Heart J* 2008; 29: 2792–2799.

112. Wilson PW. CETP genes, metabolic effects, and coronary disease risk. *JAMA* 2008; 299: 2795–2796.
113. Kolovou G, Stamatelatou M, Anagnostopoulou K, Kostakou P, Kolovou V, Mihas C, Vasiliadis I, Diakoumakou O, Mikhailidis DP, Cokkinos DV. Cholesteryl ester transfer protein gene polymorphisms and longevity syndrome. *Open Cardiovasc Med J* 2010; 4: 14–19.
114. Koropatnick TA, Kimbell J, Chen R, Grove JS, Donlon TA, Masaki KH, Rodriguez BL, Willcox BJ, Yano K, Curb JD. A prospective study of high-density lipoprotein cholesterol, cholesteryl ester transfer protein gene variants, and healthy aging in very old Japanese-american men. *J Gerontol A Biol Sci Med Sci* 2008; 63: 1235–1240.
115. Curb JD, Abbott RD, Rodriguez BL, Masaki K, Chen R, Sharp DS, Tall AR. A prospective study of HDL-C and cholesteryl ester transfer protein gene mutations and the risk of coronary heart disease in the elderly. *J Lipid Res* 2004; 45: 948–953.
116. Moriyama Y, Okamura T, Inazu A, Doi M, Iso H, Mouri Y, Ishikawa Y, Suzuki H, Iida M, Koizumi J et al. A low prevalence of coronary heart disease among subjects with increased high-density lipoprotein cholesterol levels, including those with plasma cholesteryl ester transfer protein deficiency. *Prev Med* 1998; 27: 659–667.
117. Ridker PM, Pare G, Parker AN, Zee RYL, Miletich JP, Chasman DI. Polymorphism in the CETP gene region, HDL cholesterol, and risk of future myocardial Infarction: Genomewide analysis among 18 245 initially healthy women from the Women's Genome Health Study. *Circ Cardiovasc Genet* 2009; 2: 26–33.
118. Frikke-Schmidt R, Johannsen TH, Schou J, Nordestgaard BG, Tybjcrg-Hansen A. cholesteryl ester transfer protein genotype predicts anti-atherogenic lipid profile and decreased risk of ishemic heart disease, ischemic cerebrovascular disease, and death not attributable to HDL cholesterol alone. *Circulation* 2009; 120 (Suppl.): S1175.
119. Tanrikulu-Kucuk S, Ademoglu E, Gurdol F, Bilge AK, Mutlu-Turkoglu U, Nisanci Y. Cholesteryl ester transfer protein Taq1B polymorphism in an angiographically assessed Turkish population: no effects on coronary artery disease risk. *Genet Test Mol Biomarkers* 2010; 14: 637–642.
120. Sandhofer A, Tatarczyk T, Laimer M, Ritsch A, Kaser S, Paulweber B, Ebenbichler CF, Patsch JR. The Taq1B-variant in the cholesteryl ester-transfer protein gene and the risk of metabolic syndrome. *Obesity* 2008; 16: 919–922.
121. Anand SS, Xie C, Pare G, Montpetit A, Rangarajan S, McQueen MJ, Cordell HJ, Keavney B, Yusuf S, Hudson TJ et al. Genetic variants associated with myocardial infarction risk factors in over 8000 individuals from five ethnic groups: The INTERHEART Genetics Study. *Circ Cardiovasc Genet* 2009; 2: 16–25.
122. Agerholm-Larsen B, Nordestgaard BG, Steffensen R, Jensen G, Tybjaerg-Hansen A. Elevated HDL cholesterol is a risk factor for ischemic heart disease in white women when caused by a common mutation in the cholesteryl ester transfer protein gene. *Circulation* 2000; 101: 1907–1912.
123. Xiao J, Zhang F, Wiltshire S, Hung J, Jennens M, Beilby JP, Thompson PL, McQuillan BM, McCaskie PA, Carter KW et al. The apolipoprotein AII rs5082 variant is associated with reduced risk of coronary artery disease in an Australian male population. *Atherosclerosis* 2008; 199: 333–339.
124. Peloso GM, Demissie S, Collins D, Mirel DB, Gabriel SB, Cupples LA, Robins SJ, Schaefer EJ, Brousseau ME. Common genetic variation in multiple metabolic pathways influences susceptibility to low HDL-cholesterol and coronary heart disease. *J Lipid Res* 2010; 51: 3524–3532.

125. Deeb SS, Takata K, Peng RL, Kajiyama G, Albers JJ. A splice-junction mutation responsible for familial apolipoprotein A-II deficiency. *Am J Hum Genet* 1990; 46: 822–827.
126. Iatan I, Dastani Z, Do R, Weissglas-Volkov D, Ruel I, Lee JC, Huertas-Vazquez A, Taskinen MR, Prat A, Seidah NG et al. Genetic variation at the proprotein convertase subtilisin/kexin type 5 gene modulates high-density lipoprotein cholesterol levels. *Circ Cardiovasc Genet* 2009; 2: 467–475.
127. Legry V, Cottel D, Ferrieres J, Chinetti G, Deroide T, Staels B, Amouyel P, Meirhaeghe A. Association between liver X receptor alpha gene polymorphisms and risk of metabolic syndrome in French populations. *Int J Obes* 2008; 32: 421–428.
128. Junyent M, Tucker KL, Smith CE, Garcia-Rios A, Mattei J, Lai C-Q, Parnell LD, Ordovas JM. The effects of ABCG5/G8 polymorphisms on plasma HDL cholesterol concentrations depend on smoking habit in the Boston Puerto Rican Health Study. *J Lipid Res* 2009; 50: 565–573.
129. Cummings DE, Shannon MH. Roles for ghrelin in the regulation of appetite and body weight. *Arch Surg* 2003; 138: 389–396.
130. Hubacek JA, Bohuslavova R, Skodova Z, Adamkova V. Variants within the ghrelin gene–association with HDL-cholesterol, but not with body mass index. *Folia Biol* 2007; 53: 202–206.
131. Nanjo Y, Adachi H, Hirai Y, Enomoto M, Fukami A, Otsuka M, Yoshikawa K, Yokoi K, Ogata K, Tsukagawa E et al. Factors associated with plasma ghrelin level in Japanese general population. *Clin Endocrinol* 2011; 74: 453–458.
132. Chen YC, Chen YD, Li X, Post W, Herrington D, Polak JF, Rotter JI, Taylor KD. The HMG-CoA reductase gene and lipid and lipoprotein levels: the multi-ethnic study of atherosclerosis. *Lipids* 2009; 44: 733–743.
133. Hyde CL, MacInnes A, Sanders FA, Thompson JF, Mazzarella RA, Faergeman O, van Wijk DF, Wood L, Lira M, Paciga SA. Genetic association of the CCR5 region with lipid levels in at-risk cardiovascular patients. *Circ Cardiovasc Genet* 2010; 3: 162–168.
134. Spielmann N, Mutch DM, Rousseau F, Tores F, Hager J, Bertrais S, Basdevant A, Tounian P, Dubern B, Galan P et al. Cathepsin S genotypes are associated with Apo-A1 and HDL-cholesterol in lean and obese French populations. *Clin Genet* 2008; 74: 155–163.
135. Wada M, Daimon M, Emi M, Iijima H, Sato H, Koyano S, Tajima K, Kawanami T, Kurita K, Hunt SC et al. Genetic association between aldehyde dehydrogenase 2 (ALDH2) variation and high-density lipoprotein cholesterol (HDL-C) among non-drinkers in two large population samples in Japan. *J Atheroscler Thromb* 2008; 15: 179–184.
136. Real JT, Martinez-Hervas S, Garcia-Garcia AB, Chaves FJ, Civera M, Ascaso JF, Carmena R. Association of C677T polymorphism in MTHFR gene, high homocysteine and low HDL cholesterol plasma values in heterozygous familial hypercholesterolemia. *J Atheroscler Thromb* 2010; 16: 815–820.
137. Shea JL, Loredo-Osti JC, Sun G. Association of RBP4 gene variants and serum HDL cholesterol levels in the Newfoundland population. *Obesity* 2010; 18: 1393–1397.
138. Figtree GA, Grieve SM, Speller B, Geiger M-J, Robinson BG, Channon KM, Ragoussis J, Collins P, Watkins H. A commonly occurring polymorphism upstream of the estrogen receptor alpha alters transcription and is associated with increased HDL. *Atherosclerosis* 2008; 199: 354–361.

139. Lamon-Fava S, Asztalos BF, Howard TD, Reboussin DM, Horvath KV, Schaefer EJ, Herrington DM. Association of polymorphisms in genes involved in lipoprotein metabolism with plasma concentrations of remnant lipoproteins and HDL subpopulations before and after hormone therapy in postmenopausal women. *Clin Endocrinol*; 2010; 72: 169–175.

140. Johansson LM, Johansson LE, Ridderstrale M. The visfatin (PBEF1) G-948T gene polymorphism is associated with increased high-density lipoprotein cholesterol in obese subjects. *Metabolism* 2008; 57: 1558–1562.

141. Hamada T, Kotani K, Fujiwara S, Sano Y, Domichi M, Tsuzaki K, Sakane N. The common -55 C/T polymorphism in the promoter region of the uncoupling protein 3 gene reduces prevalence of obesity and elevates serum high-density lipoprotein cholesterol levels in the general Japanese population. *Metabolism* 2008; 57: 410–415.

142. Lee JC, Weissglas-Volkov D, Kyttala M, Dastani Z, Cantor RM, Sobel EM, Plaisier CL, Engert JC, van Greevenbroek MM, Kane JP et al. WW-domain-containing oxidoreductase is associated with low plasma HDL-C levels. *Am J Hum Genet* 2008; 83: 180–192.

143. Saez ME, Gonzalez-Perez A, Martinez-Larrad MT, Gayan J, Real LM, Serrano-Rios M, Ruiz A. WWOX gene is associated with HDL cholesterol and triglyceride levels. *BMC Med Genet* 2010; 11: 148.

144. Maciel SS, Pereira Ada C, Silva GJ, Rodrigues MV, Mill JG, Krieger JE. Association between glutathione S-transferase polymorphisms and triglycerides and HDL-cholesterol. *Atherosclerosis* 2009; 206: 204–208.

145. Vaidya D, Yanek LR, Mathias RA, Becker LC, Becker DM. Genomewide analysis finds a variant in the KCNC2 gene associated with sex-dependent differences in HDL cholesterol levels. *Circulation* 2010; 122 (Suppl.): A20876.

146. Keebler ME, Sanders CL, Surti A, Guiducci C, Burtt NP, Kathiresan S. Association of blood lipids with common DNA sequence variants at 19 genetic loci in the Multiethnic United States National Health and Nutrition Examination Survey III. *Circ Cardiovasc Genet* 2009; 2: 238–243.

147. Lu Y, Dolle MET, Imholz S, van 't Slot R, Verschuren WMM, Wijmenga C, Feskens EJM, Boer JMA. Multiple genetic variants along candidate pathways influence plasma high-density lipoprotein cholesterol concentrations. *J Lipid Res* 2008; 49: 2582–2589.

148. Carty CL, Heagerty P, Heckbert SR, Enquobahrie DA, Jarvik GP, Davis S, Tracy RP, Reiner AP. Association of genetic variation in serum amyloid-A with cardiovascular disease and interactions with IL6, IL1RN, IL1beta and TNF genes in the Cardiovascular Health Study. *J Atheroscler Thromb* 2009; 16: 419–430.

149. Ma L, Yang J, Runesha HB, Tanaka T, Ferrucci L, Bandinelli S, Da Y. Genome-wide association analysis of total cholesterol and high-density lipoprotein cholesterol levels using the Framingham Heart Study data. *BMC Med Genet* 2010; 11: 55.

150. Keebler ME, Deo RC, Surti A, Konieczkowski D, Guiducci C, Burtt N, Buxbaum SG, Sarpong DF, Steffes MW, Wilson JG et al. Fine-mapping in African Americans of 8 recently discovered genetic loci for plasma lipids/clinical perspective. *Circ Cardiovasc Genet* 2010; 3: 358–364.

151. Waterworth DM, Ricketts SL, Song K, Chen L, Zhao JH, Ripatti S, Aulchenko YS, Zhang W, Yuan X, Lim N et al. Genetic variants influencing circulating lipid

levels and risk of coronary artery disease. *Arterioscler Thromb Vasc Biol* 2010; 30: 2264–2276.

152. Teslovich TM, Musunuru K, Smith AV, Edmondson AC, Stylianou IM, Koseki M, Pirruccello JP, Ripatti S, Chasman DI, Willer CJ et al. Biological, clinical and population relevance of 95 loci for blood lipids. *Nature* 2010; 466: 707–713.
153. Fogarty MP, Xiao R, Prokunina-Olsson L, Scott LJ, Mohlke KL. Allelic expression imbalance at high-density lipoprotein cholesterol locus MMAB-MVK. *Hum Mol Genet* 2010; 19: 1921–1929.
154. Moon HD, Nakajima K, Kamiyama K, Takanashi K, Sakurabayashi I, Nagamine T. Higher frequency of abnormal serum angiopoietin-like protein 3 than abnormal cholesteryl ester transfer protein in Japanese hyperalphalipoproteinemic subjects. *Clin Chim Acta* 2008; 398: 99–104.
155. Nettleton JA, Volcik KA, Hoogeveen RC, Boerwinkle E. Carbohydrate intake modifies associations between ANGPTL4[E40K] genotype and HDL-cholesterol concentrations in white men from the Atherosclerosis Risk in Communities (ARIC) study. *Atherosclerosis* 2009; 203: 214–220.
156. Musunuru K, Pirruccello JP, Do R, Peloso GM, Guiducci C, Sougnez C, Garimella KV, Fisher S, Abreu J, Barry AJ et al. Exome sequencing, ANGPTL3 mutations, and familial combined hypolipidemia. *N Engl J Med* 2010; 363: 2220–2227.
157. Aouizerat BE, Engler MB, Natanzon Y, Kulkarni M, Song J, Eng C, Huuskonen J, Rivera C, Poon A, Bensley M et al. Genetic variation of PLTP modulates lipoprotein profiles in hypoalphalipoproteinemia. *J Lipid Res* 2006; 47: 787–793.
158. Engler MB, Pullinger CR, Malloy MJ, Natanzon Y, Kulkarni MV, Song J, Eng C, Huuskonen J, Rivera C, Poon A et al. Genetic variation in phospholipid transfer protein modulates lipoprotein profiles in hyperalphalipoproteinemia. *Metabolism* 2008; 57: 1719–1724.
159. Vergeer M, Boekholdt SM, Sandhu MS, Ricketts SL, Wareham NJ, Brown MJ, de Faire U, Leander K, Gigante B, Kavousi M et al. Genetic variation at the phospholipid transfer protein locus affects its activity and high-density lipoprotein size and is a novel marker of cardiovascular disease susceptibility. *Circulation* 2010; 122: 470–477.
160. Xie X, Ma YT, Yang YN, Fu ZY, Li XM, Huang D, Ma X, Chen BD, Liu F. Polymorphisms in the SAA1/2 gene are associated with carotid intima media thickness in healthy Han Chinese subjects: the Cardiovascular Risk Survey. *PLoS One* 2010; 5: e13997.
161. Chasman DI, Pare G, Mora S, Hopewell JC, Peloso G, Clarke R, Cupples LA, Hamsten A, Kathiresan S, Malarstig A et al. Forty-three loci associated with plasma lipoprotein size, concentration, and cholesterol content in genome-wide analysis. *PLoS Genet* 2009; 5: e1000730.
162. Hegele RA. Plasma lipoproteins: genetic influences and clinical implications. *Nat Rev Genet* 2009; 10: 109–121.
163. Vergeer M, Holleboom AG, Kastelein JJP, Kuivenhoven JA. The HDL hypothesis: does high-density lipoprotein protect from atherosclerosis? *J Lipid Res* 2010; 51: 2058–2073.
164. Yang R, Li L, Seidelmann SB, Shen G-Q, Sharma S, Rao S, Abdullah KG, MacKinlay KG, Elston RC, Chen Q et al. A genome-wide linkage scan identifies multiple quantitative trait loci for HDL-cholesterol levels in families with premature CAD and MI. *J Lipid Res* 2010; 51: 1442–1451.

165. Saleheen D, Soranzo N, Rasheed A, Scharnagl H, Gwilliam R, Alexander M, Inouye M, Zaidi M, Potter S, Haycock P et al. Genetic determinants of major blood lipids in Pakistanis compared with Europeans/clinical perspective. *Circ Cardiovasc Genet* 2010; 3: 348–357.

166. Kathiresan S, Melander O, Anevski D, Guiducci C, Burtt NP, Roos C, Hirschhorn JN, Berglund G, Hedblad B, Groop L et al. Polymorphisms associated with cholesterol and risk of cardiovascular events. *N Engl J Med* 2008; 358: 1240–1249.

167. Cohen JC, Kiss RS, Pertsemlidis A, Marcel YL, McPherson R, Hobbs HH. Multiple rare alleles contribute to low plasma levels of HDL cholesterol. *Science* 2004; 305: 869–872.

168. Hu S, Zhong Y, Hao Y, Luo M, Zhou Y, Guo H, Liao W, Wan D, Wei H, Gao Y et al. Novel rare alleles of ABCA1 are exclusively associated with extreme high-density lipoprotein-cholesterol levels among the Han Chinese. *Clin Chem Lab Med* 2009; 47: 1239–1245.

169. Katan MB. Apolipoprotein E isoforms, serum cholesterol, and cancer. *Lancet* 1986; 1: 507–508.

170. Ahmad T, Chasman DI, Buring JE, Lee IM, Ridker PM, Everett BM. Physical activity modifies the effect of LPL, LIPC, and CETP polymorphisms on HDL-C levels and the risk of myocardial infarction in women of European ancestry/clinical perspective. *Circ Cardiovasc Genet* 2011; 4: 74–80.

CHAPTER 7

BIOLOGIC ACTIVITIES

Epidemiologic data unambiguously demonstrate that elevated plasma concentrations of HDL-C are one of the strongest negative risk factors for cardiovascular disease. However, considerable cardiovascular risk may still exist, even at supranormal levels of HDL-C [1]. There is convincing evidence that plasma levels of HDL-C (i.e., **HDL quantity**) do not reflect all facets of the atheroprotective properties of HDL [2–5]. Such biologic activities of HDL (i.e., **HDL quality**), also broadly termed '**HDL functionality**' or '**HDL function**', do not necessarily track with HDL-C concentrations and may be as important, if not more important, in determining the net atheroprotective capacity of HDL.

DEFINITIONS

HDL quantity: circulating plasma levels of HDL-C.
HDL quality (function): atheroprotective properties of HDL.

HDL particles are well known to possess multiple anti-atherogenic and vasoprotective activities which include cellular cholesterol efflux capacity together with anti-inflammatory, antioxidative, cytoprotective, vasodilatory, anti-infectious and antithrombotic activities [6–12] (Fig. 7.1). Moreover, recent

High-Density Lipoproteins: Structure, Metabolism, Function, and Therapeutics, First Edition.
Anatol Kontush and M. John Chapman.

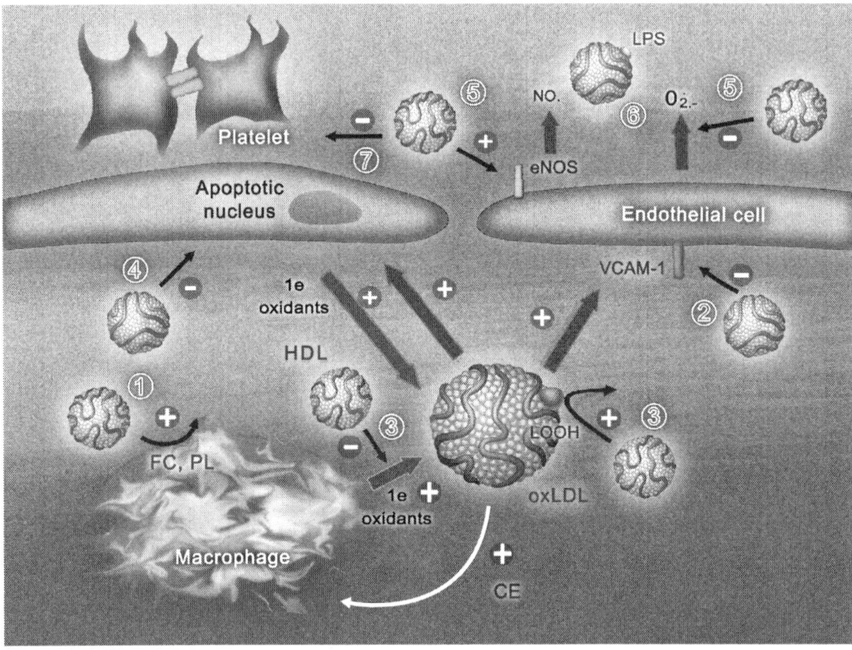

Figure 7.1 Major biologic activities of HDL. (1) Cholesterol efflux capacity (as cholesterol and phospholipid efflux from macrophages). (2) Anti-inflammatory activity (as inhibition of VCAM-1 expression in endothelial cells induced by oxidized LDL). (3) Antioxidative activity (as inhibition of LDL oxidation by cell-derived one-electron oxidants and removal of oxidized lipids from oxidized LDL). (4) Cytoprotective activity (as inhibition of endothelial cell apoptosis induced by oxidized LDL). (5) Vasodilatory activity (as stimulation of cellular NO production and inhibition of cellular superoxide release). (6) Anti-infectious activity (as LPS-binding). (7) Antithrombotic activity (as inhibition of platelet aggregation). 1e oxidants, one-electron oxidants; FC, free cholesterol; PL, phospholipid; oxLDL, oxidized LDL.

data add new facets to the complex pattern of the biologic activities of HDL. Notably, human plasma HDL has been recently shown to improve glucose metabolism by multiple mechanisms, including enhanced insulin secretion by pancreatic beta-cells, improved insulin sensitivity and maintenance of cholesterol homeostasis [13–15]. Another study has recently ascribed a potent anti-elastase activity to HDL, which was observed as the inhibition of extracellular matrix degradation, cell detachment and apoptosis induced by elastase in human vascular smooth muscle cells [16]. Such a novel biologic property of HDL is most probably accounted for by the presence of serpin peptidase inhibitors in HDL, such as alpha-1-antitrypsin [17].

The latter example reveals that the multitude of HDL functions reflects, at least in part, the wide variety of proteins carried by HDL; indeed, specific biologic activities of HDL appear to be mediated by one or more distinct proteins potentially acting synergistically [18, 19]. In turn, such compositional heterogeneity suggests a high level of functional heterogeneity among HDL particles [18, 20, 21]. Indeed, despite the fact that virtually all HDL subpopulations display multiple biologic activities, many of these activities are especially potent in small, dense, protein-rich HDL [11, 22]. The multiple biologic functions of HDL may, therefore, be mediated by distinct particle subspecies defined by specific clusters of bound proteins [18] (Table 7.1). As the lipid composition of HDL may influence associations of different proteins at the HDL surface, HDL particle subsets may serve as platforms for the assembly of certain protein components that perform specific biologic functions. Indeed, HDL subspecies can be thought of as distinct metabolic entities carrying distinct protein cargoes and performing numerous functions, some of which, but not all, can be cardioprotective [19].

The capacity to efflux cholesterol from peripheral cells, which represents the first, and key, step in the RCT pathway, has long been considered to represent the most important and clinically relevant atheroprotective property of HDL, thereby forming the basis for the negative association between circulating levels of HDL-C and cardiovascular disease [23]. Cholesterol efflux from macrophages to HDL has been unambiguously demonstrated in animal models via several mechanisms involving ABCA1 and ABCG1 transporters, the SR-BI receptor and passive diffusion [24–26]. In addition, however, compelling evidence has accumulated over the last two decades in favor of the physiologic and pathologic relevance of other features of the anti-atherogenic action of HDL [7–11] (Fig. 7.1). Recent breakthroughs from experiments involving rHDL infusion have clearly emphasized the role of anti-inflammatory, anti-infectious, antithrombotic and antidiabetic activities of HDL in humans [13, 15, 27–29]. However, it remains indeterminate as to whether some biologic activities of HDL are more relevant for protection of the vascular wall from atherogenesis and cardiovascular events compared with others.

OPEN QUESTIONS

Which biologic activities of HDL are relevant for the protection of the arterial wall from atherogenesis? How should the biologic activities of HDL be evaluated in order to provide information on cardiovascular risk?

The multitude of biologic activities expressed by HDL raises the question of the assessment of HDL functionality *ex vivo*, and also of how to assess the clinical benefit of such activities; indeed, certain activities may be relevant to a specific pathophysiologic context, for example cholesterol accumulation, inflammation, thrombosis or plaque progression. Obviously, the measurement of any selected HDL activity provides only limited information on HDL quality. An integrative assay is desperately required, which would ideally reflect all major facets of the functionality of HDL particles. Such a single common denominator reflecting multiple HDL functions may not, however, be attainable, leaving direct measurements of each biologic activity of HDL as the major route to

TABLE 7.1 Major Protein and Lipid Components Involved in Biologic Activities of HDL

Activity	Component	Role
Cholesterol efflux	ApoA-I, apoA-II, apoA-IV, apoE	Interaction with cellular receptors, including ABCA1, ABCG1 and SR-BI
		Solubilization and transport of lipids
	LCAT	Esterification of cholesterol
	PL	Incorporation of cellular lipids
	PL fatty acid moieties, SM, FC	Modulation of surface lipid fluidity
Protection from oxidative damage	ApoA-I	Reduction of lipid hydroperoxides to redox-inactive hydroxides
		Scavenging of ROS
	ApoA-IV, apoE, apoM	Removal of oxidized lipids from cells and lipoproteins
		Scavenging of ROS
	ApoJ	Removal of oxidation products
	PON1, PAF-AH, LCAT	Hydrolysis of short-chain oxidized PLs
	PL fatty acid moieties, SM, FC	Modulation of surface lipid fluidity
	Lipophilic antioxidants	Scavenging of ROS
Anti-inflammatory activities	ApoA-I	Removal of pro-inflammatory oxidized lipids from cells and lipoproteins
	PON1, PAF-AH	Hydrolysis of pro-inflammatory oxidized PL
	PL fatty acid moieties	Modulation of surface lipid fluidity

Cytoprotective activities	ApoA-I	Interaction with cellular SR-BI and ABCG1, activation of intracellular signaling
		Efflux of toxic oxysterols
	ApoE	Efflux of toxic oxysterols
	ApoJ	Removal of oxidation products
	S1P	Activation of intracellular signaling through S1P receptors
Anti-infectious activities	ApoA-I	Interaction with LPS
	LBP	LPS-binding and transfer to soluble CD14 and to HDL
	ApoL-I, Hbp, Hrp	Trypanosome lytic factor; toxicity mediated by pore formation in endosomal membranes
Vasodilatory activities	ApoA-I	Interaction with endothelial SR-BI, activation of intracellular signaling
		Efflux of cellular lipids, activation of eNOS
	S1P	Activation of intracellular signaling through S1P receptors
Antithrombotic activities	ApoA-I	Interaction with endothelial SR-BI, efflux of cellular lipids, activation of eNOS
	ApoE	Activation of eNOS
	PL	Incorporation of cellular cholesterol

FC, free cholesterol. Hbp, hemoglobin-binding protein. PL, phospholipid. SM, sphingomyelin.

assess HDL quality at present [30]. The identification of analytical biomarkers of multiple HDL functions that can serve as readily measurable surrogates for functional assays therefore remains particularly attractive [31].

7.1 CHOLESTEROL EFFLUX CAPACITY

DEFINITIONS

Reverse cholesterol transport (RCT): lipoprotein-facilitated transfer of cholesterol from peripheral cells to the liver for excretion into the bile
Cholesterol efflux capacity: capacity of HDL to remove cholesterol from cells.

Continuous removal of accumulating cholesterol from the arterial wall for excretion into the bile constitutes a major element of atheroprotection. Such transfer of cholesterol through the body is termed **RCT** and is largely mediated by HDL. Indeed, HDL represents a key factor in ensuring the removal of cellular cholesterol from the arterial wall [32].

CRITICAL CONTRIBUTION

The identification of HDL as a key factor involved in the removal of cellular cholesterol from arterial wall cells and the maintenance of cellular cholesterol homeostatsis [32].

The concept of RCT originally proposed by Glomset in the 1960s [33] is largely based on the observation that, whereas every cell in the body possesses the capacity to synthesize cholesterol, only the liver can excrete this compound in significant amounts, either as biliary cholesterol or bile acids. Indeed, all major cell types present in the arterial wall possess machinery necessary to efflux cholesterol. Since then, our understanding of the role of RCT has evolved to its present recognition as a key biologic pathway involved in atheroprotection. Indeed, the notion that macrophages become foam cells as a result of a disturbed balance between the uptake of cholesterol from atherogenic lipoproteins and the efflux of cholesterol to anti-atherogenic particles is now widely accepted [34, 35]. The **cholesterol efflux capacity** of HDL particles is defined accordingly as their ability to remove cholesterol from the membranes of peripheral cells. In the framework of atherosclerotic disease, cell types of critical importance participating in this pathway are macrophages and macrophage-derived foam cells. Furthermore, HDL is also capable of removing cholesterol from other arterial wall cells as well as extracellular cholesterol located in extracellular, cholesterol-rich microdomains [36]. Thus, cholesterol efflux represents the first, and probably essential, step of the RCT pathway. Indeed, this initial step of RCT has been suggested to be more relevant to the atheroprotective ability of HDL than the final step, i.e., the promotion of hepatic cholesterol excretion (as biliary cholesterol or bile acids) into the feces [37], despite the essential role of the latter [34].

METHODOLOGIC NOTE

RCT can be evaluated *in vivo* in animal models using a method in which macrophages are loaded with cholesterol, labelled with a ^{3}H-cholesterol tracer *ex vivo* and injected intraperitoneally where they efflux ^{3}H-cholesterol to HDL-based acceptor particles. The appearance of the tracer can be followed in plasma and the amount of macrophage-derived tracer excreted in feces can be quantitated as a measure of macrophage-to-feces RCT [38].

Consistent with the central role of HDL-mediated cholesterol efflux in protection from the development of atherosclerosis, rates of cholesterol efflux and RCT from macrophages have recently been suggested to reflect atherogenesis better than HDL-C levels [24]. Indeed, circulating levels of HDL-C frequently, but not always, parallel RCT rates [39]; furthermore, such levels do not necessarily reflect the capacity of plasma to efflux cellular cholesterol [4, 40]. As a consequence, an increase in RCT rates, potentially resulting from enhanced cellular cholesterol efflux to HDL, can be more relevant to protection from atherosclerosis than those in HDL-C levels [41].

OPEN QUESTION

Does cellular cholesterol efflux better reflect cardiovascular risk compared with plasma HDL-C levels?

Strong evidence in support of this suggestion is provided by individuals that express the apoA-I Milano variant [42] (see 6.2 of Chapter 6). These individuals do not display elevated cardiovascular risk despite extremely low HDL-C levels as a result of enhanced catabolism of HDL. RCT is, however, accelerated in these individuals, potentially reflecting the formation of apoA-I Milano homodimers with prolonged plasma residence time [42, 43]. Measurement of the cholesterol efflux capacity of HDL therefore possesses the potential to be developed as a clinical laboratory tool for cardiovascular risk assessment and evaluation of the pharmacologic efficacy of lipid-modulating or anti-atherosclerotic drugs [4, 31].

Interestingly, the capacity to efflux cellular cholesterol appears to constitute a common property of all HDL particles independent of their origin. Indeed, HDL obtained from cerebrospinal fluid efficiently removes cholesterol from neuronal cells [44]. In addition, HDL isolated from follicular fluid is as potent as plasma HDL in promoting cholesterol efflux from hepatic cells in culture [45, 46].

Mechanisms of Cellular Cholesterol Efflux

During initiation of the RCT pathway, HDL must first cross the endothelium to gain access to arterial intimal cells. Importantly, the endothelium does not behave as a passive exchange barrier for lipoproteins, but rather actively controls their trafficking between intravascular and extravascular compartments. Aortic endothelial cells are capable of binding, internalizing and translocating HDL and apoA-I from the apical to the basolateral compartment *in vitro* [47]. Endothelial cells transcytose HDL by a mechanism involving SR-BI and ABCG1 [48], while transcytosis of lipid-free apoA-I occurs by a distinct mechanism involving ABCA1 [49]. **Trans-endothelial trafficking** of lipid-free apoA-I results in its lipidation [50]; in contrast, transcytosis of HDL is accompanied by a decrease in HDL size. Interestingly, HDL transport through aortic endothelial cells is facilitated by endothelial lipase, implying the role of phospholipid hydrolysis [51].

Following transcytosis, both lipid-free apolipoproteins and HDL particles interact with cellular surfaces to induce cholesterol efflux. Key elements of such efflux involve interactions of lipid-free apolipoproteins and HDL with specific receptors on the cell surface and with cellular surface lipids. Cellular HDL receptors primarily include ABCA1 and ABCG1 transporters and the SR-BI receptor, as well as a holoparticle HDL receptor, such as the pathway involving the beta-chain of the surface ATP synthase [52] (see also Chapter 4).

ABCA1-Mediated Efflux. Lipid-free HDL apolipoproteins induce fast, saturable, unidirectional and LCAT-independent efflux of cellular cholesterol and phospholipids [23, 53]. As a result, HDL particles efficiently acquire cholesterol in the extravascular compartment [54]. Apolipoprotein-mediated lipid efflux involves specific interactions with cell membrane proteins, desorption of membrane lipids from caveolae, lipidation of lipid-free apolipoproteins and the production of small, lipid-poor HDL [23, 55]. Lipid-free apolipoproteins efficiently remove cholesterol and phospholipids from macrophages, aortic smooth muscle cells and normal human skin fibroblasts, but not from fibroblasts of patients with Tangier disease [56, 57].

Defective ABCA1 transporter function in Tangier disease provided the initial evidence that **ABCA1** is central to lipid efflux mediated by **lipid-free/lipid-poor apolipoproteins** [58, 59] (see 6.1 of Chapter 6). For example, the Tangier disease mutation W590S, which resides in the extracellular domain and reduces apoA-I-dependent lipid efflux, impairs ATP-dependent translocation of cellular lipids to apoA-I [60]. In support of this mechanism, apoA-I–mediated cholesterol efflux is severely decreased by the inhibition of ABCA1 with either antisense oligonucleotides or pharmacologic compounds, but is increased by ABCA1 over-expression [61]. As ABCA1-mediated HDL formation constitutes a key step in HDL metabolism and RCT, ABCA1 over-expression leads to hyperalphalipoproteinemia and increases biliary cholesterol excretion in transgenic mice [62], while ABCA1 inactivation in macrophages significantly impairs *in vivo* RCT [63].

In a similar fashion, experiments with cholesterol-loaded mouse macrophages demonstrate that ABCA1 may account for at least 80% of cholesterol efflux stimulated by cellular cholesterol loading [64]. Moreover, cholesterol efflux from macrophages to apoB-depleted human sera with HDL-C levels in the range between the 25th and 75th percentiles is correlated with specific ABCA1-mediated efflux [65]. **Small, pre-beta HDL particles**, both initially present in HDL preparations and generated during remodeling of large HDL by macrophages, are key factors in this pathway. Indeed, plasma from healthy men matched for age, BMI, triglyceride, apoA-I and HDL-C levels displays enhanced cholesterol efflux capacity via ABCA1 (+31%) compared with women, secondary to an increase (+240%) in plasma levels of pre-beta HDL particles [40]. Thus, ABCA1 is a pivotal regulator of cellular cholesterol efflux and of the lipidation of apoA-I, a key step in the formation of mature, spherical HDL particles (Fig. 7.2).

Solubilization of cellular phospholipids and cholesterol by apoA-I appears to represent the rate-determining step in ABCA1-mediated cellular lipid efflux [66]. Cellular cholesterol and phospholipid efflux mediated by ABCA1 involves the formation of surface lipid microdomains under the action of ABCA1 followed by the binding of apoA-I to ABCA1 with the subsequent efflux of cholesterol [67]. Indeed, extracellular, cholesterol-rich microdomains can be detected on the surface of monocyte-derived macrophages using a monoclonal antibody raised against organized cholesterol microarrays containing 10–20 molecules of the sterol [68]. Consistent with the role of ABCA1-mediated cholesterol efflux in the RCT pathway from arterial intimal macrophages, such an antibody preferentially stains arterial intima.

From a kinetic point of view, the process of cholesterol efflux from cells occurs in two steps: (1) the first-order desorption of a cholesterol molecule from the hydrophobic environment of the plasma membrane, followed by (2) a second-order collision process involving cholesterol incorporation into the acceptor particle [69]. Formation of small, lipid-poor HDL ensues which may subsequently mature to large, lipid-rich particles. Such HDL maturation involves interaction with other HDL receptors, primarily with ABCG1 [70], and can be facilitated by apoE secreted from macrophages [68] (see below). Importantly, a significant fraction of ABCA1-mediated cholesterol efflux from macrophages can be mediated by large, HDL-like microparticles [71].

ABCA1-mediated cholesterol efflux to apoA-I involves the mobilization of cholesterol from the plasma membrane, endoplasmic reticulum, trans-Golgi network, late endocytic and lysosomal compartments, and cholesteryl ester droplets [72]. The late endosome/lysosome pool of cholesterol is a critical part of the

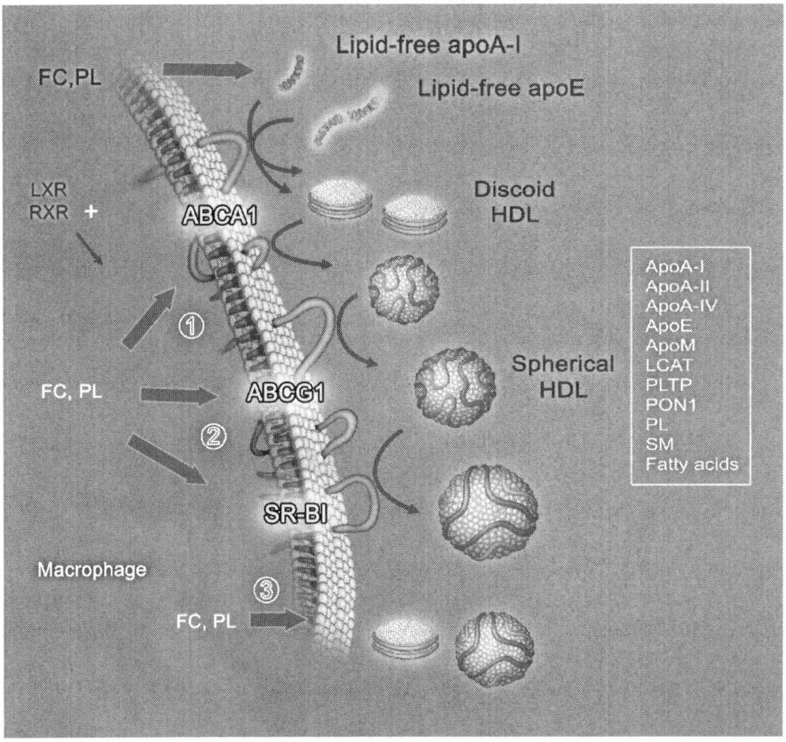

Figure 7.2 Major pathways involved in HDL-mediated cholesterol efflux. (1) Cellular cholesterol and phospholipid are effluxed to lipid-free apolipoproteins, such as apoA-I and apoE, and to lipid-poor discoid HDL via ABCA1. (2) Spherical HDL particles represent acceptors of cell-derived cholesterol effluxed via ABCG1 and SR-BI. (3) All HDL subpopulations appear to be able to accept cellular cholesterol delivered by passive diffusion. Inset lists HDL components potentially involved in HDL-mediated cholesterol efflux (see color plate section). FC, free cholesterol; PL, phospholipid; SM, sphingomyelin.

total cholesterol substrate pool for ABCA1. Neutral cholesteryl ester hydrolase-mediated hydrolysis of cellular cholesteryl esters is required to generate free cholesterol for efflux from macrophages [73]. The caveolin-1-dependent system ensures the transport of cholesterol from intracellular compartments to caveolae for subsequent transfer to extracellular acceptors [74]. Internalization of ABCA1 appears to be a requirement for ABCA1-mediated cholesterol mobilization. Indeed, intracellular trafficking of ABCA1 is essential for cholesterol efflux [75]; the former process involves palmitoylation of ABCA1 with the participation of a subunit of the serine palmitoyltransferase enzyme, serine palmitoyltransferase long chain base subunit 1, which regulates ABCA1 translocation through the endoplasmic reticulum [76]. Importantly, ABCA1-dependent cholesterol efflux to different acceptors from human macrophage-derived foam cells is strongly reduced at acidic pH as a result of attenuated expression of the transporter; such pH-sensitive cholesterol removal may accelerate lipid accumulation in deep areas of advanced atherosclerotic plaques where the intimal fluid is acidic [77].

The **post-translational regulation** of the transporter also includes stimulation by beta-1-syntrophin via delayed ABCA1 degradation [76]. Regulation of the delivery of ABCA1 to the cell surface is initially provided by Rab8 [78]. In addition, the binding of apoA-I to ABCA1 activates Janus kinase 2 (JAK2), which is important for cholesterol efflux [67]. Interestingly, the addition of HDL is capable of increasing ABCA1 expression in endothelial cells, providing a positive feedback mechanism for the up-regulation of cholesterol efflux [79]. Continuous endocytic recycling of ABCA1 mediated by apoA-I may contribute to this pathway [80]. HDL-induced oligomerization and Golgi exit of caveolin-1 may provide an additional contribution to enhanced cholesterol efflux [81]. In intestinal cells, cholesterol movement to ABCA1 can be facilitated by the Niemann-Pick C1-like 1 protein [82].

In contrast, unsaturated fatty acids can inhibit cholesterol efflux from macrophages by accelerating the degradation of ABCA1 [83, 84]. Other lipids can also regulate ABCA1-mediated efflux; indeed, endothelial lipase appears to promote apoA-I-mediated cholesterol efflux through mechanisms involving increased lysophosphatidylcholine production associated with augmented apoA-I binding [85]. Products of human 15-lipoxygenase-1 can also promote ABCA1-mediated cholesterol efflux from macrophages, acting through accelerated hydrolysis of cholesteryl esters [86].

ABCA1-mediated cholesterol efflux is also under tight **transcriptional control**. LXRs efficiently up-regulate ABCA1 expression in, and cholesterol efflux from, macrophages [87]. The stimulation of cholesterol efflux to HDL by LXR agonists in human foam cells involves an ATP-dependent transport mechanism mediated by ABCA1, which is largely independent of other HDL receptors [88]. PPAR gamma [89] and PPAR alpha [87] can also regulate ABCA1 expression. Thus, adiponectin activates apoA-I-mediated cholesterol efflux from macrophages through an ABCA1-dependent pathway by the activation of both LXR alpha and PPAR gamma [90]. Enhancement of hepatic HDL synthesis by adiponectin is also dependent on chicken ovalbumin upstream promoter transcription factor 2 (COUP-TFII), an orphan receptor [91]. This pathway may contribute to a positive correlation between plasma HDL-C and adiponectin concentrations in humans [92, 93].

In addition to ABCA1, other members of the ABCA subfamily of ABC transporters can be involved in cholesterol efflux from macrophages. Indeed, selective disruption of ABCA5 in macrophages leads to increased atherosclerotic lesion development in LDL receptor-null mice [94]. ABCA5, B9, D3, and D4 transporters can also be operative in Kupffer cells [95], while ABCA7 is involved in cholesterol export from mammalian spermatozoa [96].

Thus, ABCA1 plays a central role in cellular cholesterol efflux. However, significant RCT from macrophages is observed, even in the total absence of ABCA1 in mice, highlighting the role of other cholesterol transporters and/or passive cholesterol diffusion [63]. Indeed, several other sterol-regulated ABC transporters belonging to the ABCG subfamily, notably ABCG1 and ABCG4, can contribute to cholesterol efflux from macrophages [97–99].

ABCG1-Mediated Efflux. **ABCG1** is a half-type ABC transporter which effluxes cholesterol to mature spherical HDL2 and HDL3 particles (Fig. 7.2). Acting principally on **large, cholesteryl ester-rich HDL**, ABCG1 may form a sequential pathway of HDL biogenesis with ABCA1 and LCAT. Such sequential growth of HDL particles under the action of ABCA1, LCAT and ABCG1 can be supported by several cell types, such as macrophages and endothelial cells [25, 100]. Within the plasma membrane, ABCG1 redistributes cell cholesterol to domains that interact preferentially with mature HDL, but not with lipid-poor apolipoproteins [70]. More specifically, ABCG1 increases cell membrane cholesterol pools and changes the rate of desorption of cholesterol into the aqueous phase without enhancing its association with the acceptor [101].

MAJOR CONTROVERSY

Does ABCG1 play a major quantitative role in cholesterol efflux to HDL from human macrophages?

The role of the ABCG1-mediated pathway of cholesterol efflux from macrophages appears to be quantitatively less important compared with that of ABCA1. Moreover, the data on the role of ABCG1 in cholesterol efflux are controversial. Indeed, ABCG1 does not significantly contribute to stimulated cholesterol efflux from human macrophages induced by LXR agonists [88] and accounts for not more than 20% of the stimulation of cholesterol efflux from mouse peritoneal macrophages following cholesterol loading [64]. Furthermore, transgenic human ABCG1 does not influence atherosclerosis in apoE-null mice but may participate in the regulation of tissue cholesterol biosynthesis [102]. Plasma cholesterol levels may, however, have been too high and the pathogenesis too robust in this model for ABCG1 to affect atherogenesis. Consistent with these data, ABCG1 deletion in macrophages does not aggravate atherosclerosis in LDL-receptor and apoE knock-out mice [103–105]. Finally, ABCG1 does not appear to significantly contribute to phospholipid efflux [106].

On the other hand, there are data supporting a role for ABCG1 in cholesterol efflux. Thus, total body ABCG1 expression may protect against early atherosclerotic lesion development in a mouse model [35]. In addition, ABCG1 efficiently effluxes cellular phospholipid and sphingomyelin, as shown in two cellular models [101, 107]. Furthermore, cholesterol mobilization from mouse macrophages to rHDL has recently been shown to be largely mediated by ABCG1 [108]. Data indicate that macrophage ABCG1 and apoE exert additive, independent and beneficial effects on both cholesterol efflux and atherosclerosis in mice [109]. Similarly, ABCA1 and ABCG1 are additive in their effects on RCT in mice [110].

A major compensatory role of ABCG1-mediated cholesterol efflux is further consistent with the absence of the effect of ABCA1 deletion on the centripetal movement of cholesterol from peripheral tissues to the liver and intestine in the mouse [111]. Thus, ABCA1 and ABCG1 may act synergistically in enhancing cellular cholesterol efflux [105]. Finally, ABCG1 deletion may moderately accelerate atherosclerosis in LDL-receptor knock-out mice [109, 112]. It appears, therefore, that the net effect of ABCG1 on atherosclerosis is rather subtle and strongly depends on the experimental setting and/or stage of atherogenesis [113].

Similar to ABCA1, the cholesterol-effluxing activity of ABCG1 can be **regulated at the post-translational level**, such as, for example, by fatty acids, and is under transcriptional control. For example oxidized fatty acids generated by 12/15-lipoxygenase may induce serine phosphorylation of ABCG1, thereby increasing the degradation rate of the transporter and causing macrophage cholesterol accumulation [114].

SR-BI-Mediated Efflux. In contrast to lipid-free apolipoproteins, lipid-containing HDL particles induce both specific and non-specific forms of cholesterol efflux [23]. Non-specific cholesterol efflux can be also mediated by phospholipid vesicles, synthetic cyclodextrins, albumin or partially proteolyzed HDL; it is slow, unsaturable and bi-directional and, thus, appears to occur by passive diffusion [23, 55]. Such bi-directional flux between mature HDL and plasma membranes is largely mediated by **SR-BI** (Fig. 7.2); this pathway involves the binding of HDL particles to the receptor with subsequent reorganization of lipids within cholesterol- and caveolae-rich domains in the plasma membrane [115, 116]. Interestingly, HDL binding to SR-BI reveals negative co-operativity caused by either classic allostery or ensemble effects [117]. The phospholipid content of HDL is a principal determinant of such SR-BI-mediated cholesterol efflux which is predominantly mediated by large, (phospho)lipid-rich HDL [118]. Consistent with this observation, plasma from healthy women matched for age, BMI, triglyceride, apoA-I and HDL-C levels displays an enhanced capacity to mediate cellular cholesterol efflux via SR-BI (+14%) compared with men [40]; such enhancement probably results from a significant increment in plasma levels of large, lipid-rich HDL2 particles (+20%) in females.

SR-BI may play an important role in net cellular cholesterol efflux, acting primarily in hepatocytes but also in macrophages, adipocytes and other cell types [119, 120]. Importantly, **Cla-1** (human SR-BI) appears to provide a major contribution to cholesterol efflux from macrophages under basal non-stimulated conditions; such efflux is ATP-independent and does not require an energy supply [88]. In addition, studies in mice reveal that the bulk mobilization of cholesterol observed in the circulation after infusion of rHDL is primarily mediated by SR-BI [108]. SR-BI may also be important for cholesterol efflux from mouse macrophages under high dietary lipid load [121]. It is, however, essential to keep in mind that cholesterol flux via SR-BI is always **bi-directional**. For this reason, SR-BI does not promote net cholesterol efflux from cells to plasma HDL containing active LCAT owing to the reuptake of cholesteryl ester into cells [122].

Findings of enhanced atherosclerosis in SR-BI-deficient mice cannot, therefore, be explained by a defect in macrophage cholesterol efflux [122, 123].

Intracellularly, cholesterol efflux via SR-BI is subject to complex regulation by multiple proteins, including sterol carrier protein-2 and liver fatty acid binding protein, which play a significant role in such efflux, as demonstrated in mouse hepatocytes [124]. On the other hand, cholesterol efflux can be under the repressive influence of CD36, as documented by enhanced cholesterol and phospholipid efflux from CD36-deficient macrophages [125]; increased surface ABCA1 localization in CD36-deficient cells may underlie this observation. As a corollary, CD36 decreases hepatic production of apoA-I and thereby reduces RCT, suggesting potential benefits of CD36 inhibition. Controversially, however, CD36 can also bind HDL and mediate selective cholesteryl ester uptake, although much less efficiently than SR-BI [126].

Other Pathways. **Passive receptor-independent diffusion** of cholesterol from plasma membranes to HDL particles, which involves the direct interaction of HDL with the membrane surface, may also significantly contribute to net cholesterol removal from cells (Fig. 7.2). This pathway may be particularly important under non-stimulated basal conditions, being capable of effluxing cholesterol even more potently than SR-BI [64]; however, the role of passive cholesterol diffusion becomes minor upon cholesterol loading of cells, as this pathway cannot be up-regulated [64]. Importantly, passive diffusion of cellular cholesterol from plasma membranes to HDL does not appear to involve the passage of the hydrophobic cholesterol molecule through the aqueous phase, but rather comprises direct movement of cholesterol between the hydrophobic compartments of the plasma membrane and HDL. The term 'passive diffusion' therefore appears to better describe this process compared to 'aqueous diffusion', which is still frequently used.

Another mechanism potentially implicated in HDL-mediated cholesterol efflux is **retro-endocytosis**, i.e., the uptake of HDL into clathrin-coated endosomes followed by intracellular enrichment with lipids and resecretion [127–129]. The amount of apoA-I resecreted by this pathway is, however, small and unlikely to be sufficient to account for the HDL produced by the cholesterol efflux process [130, 131]. In addition, cholesterol efflux from macrophages mediated by lipid-free apoA-I can be uncoupled from apoA-I internalization into cells, making this pathway of minor importance.

Finally, HDL-mediated cholesterol efflux from macrophages may be facilitated by apoE secretion [132]. Indeed, macrophage-derived apoE can associate with HDL and improve its cholesterol-acceptor properties. As a corollary, the production of apoE by both human and mouse macrophages enhances cholesterol efflux to HDL [133, 134]. Furthermore, apoE secreted by macrophages can inhibit the oxidation of LDL [135], potentially acting via sequestration of lipids. Macrophage apoE secretion can be induced by both HDL and apoA-I, providing a positive feedback mechanism to up-regulate lipid efflux [136–138]. In contrast, the up-regulation of LDL receptors inhibits the secretion of apoE [139].

ApoE-mediated cholesterol efflux can protect from accelerated atherosclerosis in ABCG1-deficient mice [104]. Importantly, as atherosclerotic plaques are only partially permeable to plasma HDL, **apoE-mediated cholesterol efflux** may represent a key mechanism in removing cholesterol from plaque macrophages [140]. Consistent with this notion, local production of apoE in macrophages corrects dyslipidemia and completely protects apoE-null mice from diet-induced atherosclerosis, even in the absence of changes in plasma lipid concentrations [141, 142]. Moreover, macrophage-derived apoE significantly contributes to macrophage RCT *in vivo*, acting via the acceleration of cholesterol efflux [143].

In addition to activating cellular cholesterol efflux, HDL can reduce cholesterol influx. Such an effect can in part be accounted for by the inhibitory action of HDL on LDL oxidation by cells, which otherwise results in the formation of oxidatively-modified LDL avidly taken up by macrophages [144]. Furthermore, HDL can down-regulate receptors involved in the uptake of oxidized LDL, such as CD36 and Fc gamma receptors [145].

Cellular cholesterol efflux is normally paralleled by **phospholipid efflux** (Fig. 7.2). Indeed, ABCA1 ensures the transfer of phosphatidylcholine to apoA-I, both during and after secretion of the apolipoprotein [146]. These studies have introduced the concept of reverse phosphatidylcholine transport in which both HDL and LDL supply phosphatidylcholine to the liver. In this concept, the liver is both a donor of phosphatidylcholine during the assembly and secretion of lipoproteins, as well as a recipient of phosphatidylcholine from plasma lipoproteins [146]. The processes of cholesterol and phosphatidylcholine efflux do not, however, always occur simultaneously as suggested by selective impairment of phospholipid efflux accompanied by normal efflux of cholesterol in subjects of French Canadian ancestry with severe HDL deficiency [147]. In addition, phospholipid efflux can be selectively regulated via Ser2054 phosphorylation in ABCA1 by protein kinase A [148]. Such dissociation between phospholipid and cholesterol efflux is related to defective ABCA1 protein regulation, can be corrected by oxysterols and results in the abnormal formation of large alpha-HDL.

Role of HDL Components

Proteins. Both protein and lipid components of HDL particles contribute to their capacity to efflux cellular lipids (Table 7.1; Fig. 7.2). As the surface of spherical HDL is dominated by protein components [149], they appear to primarily contribute to such biologic activity. Interactions between HDL and external molecules, such as cellular receptors should, therefore, be largely mediated by HDL-associated proteins.

Among HDL proteins, **apoA-I** predominates in terms of its role in lipid efflux, as the major physiologic function of apoA-I is apparently to solubilize and transport cholesterol and phospholipid (Table 7.1). In this context, the central domain of apoA-I, which ensures correct helix orientation at the polar face, is particularly important for cell cholesterol trafficking and solubilization [150].

Conformational lability of the N-terminal helix bundle domain and hydrophobicity of the C-terminal domain also appear to be important for the capacity of apoA-I to induce cellular cholesterol efflux [151]. Furthermore, highly hydrophobic and bulky residues seem to be required at the Trp positions on the non-polar face of apoA-I for its lipid-binding and lipid-acceptor functions [152]. As a consequence, apoA-I adopts a conformation with a high affinity for membranes in small, cholesterol-poor HDL, thereby facilitating cholesterol efflux [153]. Following cholesterol enrichment, a conformational change occurs in apoA-I which decreases its affinity for membranes, allowing lipoprotein particle release. Consistent with its potent cholesterol efflux capacity, apoA-I is thought to play a central role in cholesterol transport from macrophages to the liver, as highlighted by the demonstration of accelerated RCT in mice over-expressing human apoA-I [38].

ApoA-II is also able to act as a primary acceptor and to efficiently remove cholesterol from macrophages *in vivo* [154]. Other apolipoproteins, including **apoA-IV, apoC-I, apoE** and **apoM**, can also function as efficient acceptors of cellular cholesterol [23, 53, 155] (Table 7.1). For example, apoA-IV can significantly contribute to the potent cholesterol efflux properties of HDL from human follicular fluid [45], whereas apoM forms small, pre-beta HDL with elevated cholesterol efflux capacity [156]. As a consequence, the over-expression of apoM in LDL-receptor-null mice may protect against atherosclerosis [156].

The cholesterol efflux capacity of apolipoproteins can be significantly influenced by polymorphisms and mutations. The most prominent example is provided by the apoE2 and apoE3 isoforms of apoE which appear to efflux cellular cholesterol to a superior degree compared with apoE4 [157, 158]. The role of polymorphisms is further highlighted by T347S, a common amino acid substitution in apoA-IV that confers the protein with improved lipid-binding properties and enhanced cholesterol efflux capacity via ABCA1 [159]. In contrast, Q360H, another common isoform, does not modify efflux compared with wild-type apoA-IV. On the other hand, the apoA-I Milano variant is at least as efficient as wild-type apoA-I in promoting macrophage ABCA1-mediated cholesterol efflux and SR-BI-mediated cholesterol uptake [43, 160–163].

Enzymes carried by HDL are also of major relevance for cholesterol efflux and RCT. **LCAT** is a key player in the net transfer of cholesterol from peripheral cells to the liver (Table 7.1). The relationship between LCAT and RCT is, however, complex as the latter is reduced only in complete LCAT deficiency, as observed in animal models [164]. In contrast, neither partial deficiency nor over-expression of LCAT modifies RCT from macrophages in mice despite significant changes in HDL-C [164]. Accelerated cholesterol efflux via ABCA1, mediated by elevated concentrations of small HDL, including pre-beta particles, can account for the absence of the effect of partial LCAT deficiency on RCT, as suggested by data obtained in human carriers of LCAT mutations [165]. A compensatory mechanism may also account for normal RCT in LCAT-over-expressing mice as a result of accelerated cholesterol efflux via ABCG1 mediated by elevated concentrations of large HDL [164, 166]. These data show that in the absence of LCAT, cholesterol can still be efficiently transported to the liver in the non-esterified form via a compensatory pathway.

Another HDL protein implicated in cholesterol efflux is **PON1**, which stimulates the capacity of HDL to promote cholesterol efflux through enhanced HDL binding to macrophages [167] (Table 7.2). Interaction between HDL and ABCA1 stimulated by PON1 may underlie this effect [168], which is potentially related to cellular accumulation of lysophosphatidylcholine and may stimulate cholesterol efflux via the ABCA1 pathway [169]. Cellular cholesterol efflux can also be modulated by lipid transfer proteins, such as **PLTP**. Indeed, systemic overexpression of PLTP impairs macrophage cholesterol efflux and RCT in mice, resulting in accelerated atherogenesis [170].

Lipids. The multitude of the effects of proteins on HDL-mediated cholesterol efflux by no means implies that the role of HDL lipids is negligible. Despite their relatively minor contribution to the total surface of spherical HDL, lipids provide an environment for the acquisition of exogeneous lipid molecules, primarily of cholesterol, thereby assuring cellular lipid efflux. As a direct consequence, properties of HDL surface lipids may strongly impact efflux. First, the cholesterol

TABLE 7.2 Potential Roles of PON1 in Biologic Activities of HDL

Activity	Effect	Mechanism	
Cholesterol efflux	Activation	Enhancement of HDL binding to macrophages	[167]
		Stimulation of ABCA1 by lysoPC	[168, 169]
Protection of LDL from oxidative damage	Enhancement	Hydrolysis of pro-inflammatory short-chain oxidized PLs; weak activity toward LOOH	[11, 22, 171–177]
Anti-inflammatory actions	Enhancement	Hydrolysis of homocysteine thiolactone, protection against protein homocysteinylation	[174, 178–180]
		Hydrolysis of pro-inflammatory short-chain oxidized PLs	
Vasodilatory activity	Enhancement	Stimulation of Akt/eNOS phosphorylation and endothelial NO production	[181]

LysoPC, lysophosphatidylcholine; PL, phospholipid.

efflux capacity of HDL via SR-BI dose-dependently increases in parallel with the increasing HDL content of **phospholipids**, the major surface lipid component [118] (Table 7.1). Recent methodologic developments allowing marked increase in HDL content of phospholipids without loss of apoA-I may, therefore, hold promise as a means of improving HDL functionality and accelerate RCT *in vivo* [182].

Qualitatively, the ability of HDL to accept cellular cholesterol depends on the physical state of surface phospholipids, with liquid crystal unsaturated phospholipids forming the most efficient cholesterol acceptor particles [69] (Table 7.1). In contrast, HDLs prepared from gel-phase saturated phospholipids are much less efficient. Consistent with these data, phase-transition temperatures of molecular species of phosphatidylcholine [183] are inversely related to the capacity of rHDL to induce cellular cholesterol efflux [69]. The physical properties of HDL lipids are particularly important for cholesterol efflux mediated by SR-BI, which strongly depends on the HDL content of phospholipid [118]. Interestingly, the differential conformation of apoA-I observed in the liquid crystal vs. gel state does not contribute to differences in efflux, as demonstrated by data obtained using protein-free vesicles [69].

In a similar fashion, high **sphingomyelin** content decreases cellular cholesterol uptake by HDL containing unsaturated lipids in a liquid crystal state [184] (Table 7.1); indeed, a decrease in the fluidity of surrounding liquid-crystal lipids represents one of the major physiologic effects of sphingomyelin. According to the two-step mechanism of cholesterol efflux involving desorption of cholesterol from cellular membranes with subsequent collision with, and incorporation into, acceptor HDL particles [69], the percentage of effective collisions may be higher for HDL particles containing liquid crystal phospholipids compared with those containing gel-phase lipids. These data reveal **fluidity of HDL surface lipids** as a major structural determinant of cholesterol efflux capacity.

The requirement of the fluid lipid phase for cholesterol efflux is consistent with the notion that HDL structural lability is a prerequisite for this process. Indeed, HDL structure is stabilized by kinetic barriers that decelerate protein dissociation and lipoprotein fusion [185]. As similar barriers modulate metabolic remodeling of plasma HDL, compositional changes that destabilize HDL particles may accelerate their metabolic remodeling, including that which occurs during cellular lipid efflux. Such compositional changes may include reduced chain length and an increased level of **unsaturation of fatty acid residues**, which increases HDL structural lability and accelerates particle fusion and apolipoprotein dissociation. Indeed, reduced length or increased cis-unsaturation of phosphatidylcholine acyl chains destabilize discoid HDL and promote protein dissociation and lipoprotein fusion [185]. Consequently, such changes may be beneficial for cholesterol efflux by facilitating the insertion of cholesterol and promoting the dissociation of lipid-poor apolipoproteins [186].

Specific interactions between lipid molecules may also modulate the cholesterol efflux capacity of HDL. As sphingomyelin possesses a high affinity for free cholesterol, HDL enrichment in sphingomyelin can decrease cholesterol influx from HDL to cells without affecting efflux [118], thereby counteracting the effects of diminished fluidity [184].

> **CLINICAL SUMMARY**
>
> Cholesterol efflux capacity represents the major biologic activity of HDL and is critical for cholesterol removal from the arterial wall and for protection from cardiovascular disease.

Functional Heterogeneity of HDL

It is well established that HDL subpopulations differ in their cholesterol efflux properties. Indeed, lipid-poor and lipid-rich HDL particles display distinct cholesterol efflux capacities indicative of functional heterogeneity of HDLs (Table 7.3).

A decrease in the lipid content of HDL is generally thought to increase its capacity to remove cellular cholesterol [187, 188]; **small, dense, lipid-poor, protein-rich HDL** particles have, therefore, long been thought to represent more

TABLE 7.3 Functional Heterogeneity of HDL Particles

Biologic function	Subpopulation	Feature
Cholesterol efflux	Small, dense HDL	Enhanced efflux from lipid-loaded macrophages
		Enhanced efflux via ABCA1
		Enhanced LCAT activation
	Large, light HDL	Enhanced efflux via ABCG1 and SR-BI
Protection from oxidative damage	Small, dense HDL	Enhanced protection of LDL from oxidative stress
		Enhanced inactivation of LOOH and other oxidized PLs
Anti-inflammatory effects	Small, dense HDL	Potent inhibition of VCAM-1 expression on endothelial cells
Cytoprotective effects	Small, dense HDL	Potent protection of endothelial cells from oxidized LDL-induced apoptosis
Anti-infectious effects	Dense HDL of high M_r	Trypanosome lytic factor; cytotoxicity mediated by pore formation in endosomal membranes; LPS binding and transfer to soluble CD14
Antithrombotic effects	Large, light HDL	Inhibition of platelet aggregation
	Small, dense HDL	Enhanced anticoagulant activity

PL, phospholipid.

efficient cholesterol acceptors compared with their large, light, lipid-rich, protein-poor counterparts [189]. Recent data, however, show that the relative efficiency of different HDL particles in promoting cholesterol efflux critically depends on cellular receptors which mediate this process. Thus, lipid-free and/or lipid-poor HDL apolipoproteins potently, and dose-dependently, induce cholesterol uptake via interaction with **ABCA1** (Table 7.3). Consistent with this observation, plasma levels of small pre-beta-1 HDL particles correlate with serum capacity to induce ABCA1-mediated cholesterol efflux from J774 macrophages [190]. Indeed, ABCA1-mediated cellular cholesterol efflux can be efficiently driven not only by monomolecular lipid-free/lipid-poor apoA-I, but also by small discoid rHDL particles of 7.8 nm diameter resembling plasma pre-beta1 HDL [191]. Furthermore, cholesterol efflux from macrophages to apoB-depleted human sera with similar HDL-C values between the 25th and 75th percentiles is correlated with serum levels of small pre-beta HDL [65]. Next, HDL particles from a fetus with Smith-Lemli-Opitz syndrome involving 7-dehydrocholesterol reductase deficiency and characterized by very low plasma cholesterol levels, are small and efflux approximately 50% more cholesterol from a trophoblast cell line compared with HDL from unaffected fetuses or adults [192]. Finally, cholesterol efflux from spermatozoa mediated by HDL from pre-ovulatory human follicular fluid is primarily accounted for by small pre-beta HDL particles [193]; physiologically, this phenomenon may underlie the capacity of the fluid to promote hyperactivation of the flagellar movement of spermatozoa.

CRITICAL CONTRIBUTION

Identification of ABCA1 as a key cell membrane transporter involved in cholesterol efflux [58, 59, 64, 88]. Demonstration of the primary role of small, lipid-poor HDL in ABCA1-mediated cholesterol efflux [64, 65, 191, 194].

Conversely, **large, lipid-rich HDL** particles represent a more efficient ligand for cellular uptake of cholesteryl ester mediated by **SR-BI** compared with small, lipid-poor HDL [195, 196] (Table 7.3), consistent with the role of these particles in RCT from peripheral cells to the liver [23, 190]. In a similar way, SR-BI-mediated free cholesterol efflux to large HDL2 is greater than that to small HDL3 [118]. Higher phospholipid content in large HDL, which results in greater lipid surface, probably underlies the latter observation. Large, lipid-rich HDL, with a diameter of 9.6 nm or greater, also efficiently promotes cholesterol efflux via **ABCG1** [191] (Table 7.3). Such in part opposite relationships of cholesterol efflux and cholesteryl ester uptake to HDL size suggest a complex relationship between HDL size and RCT. Indeed, combined deficiency of hepatic lipase and endothelial lipase in a mouse model increases HDL levels, decreases HDL size and accelerates HDL capacity to efflux cellular cholesterol but reduces the clearance of HDL cholesteryl ester, resulting in unchanged RCT rate [197].

The key role played by small HDL particles in cholesterol efflux *in vivo* is consistent with the capacity of ABCA1 to account for the greater part of cholesterol efflux from lipid-loaded macrophages compared with other HDL transporters and receptors [64, 65]. Consistent with the role of small, lipid-poor HDL, experiments with RAW 264.7 macrophages and mice reveal that pathways of cholesterol efflux using apoA-I as an acceptor make a predominant contribution to cholesterol export from macrophages *in vivo* compared with pathways that use HDL [198]. However, pathways promoted by large HDL via SR-BI and ABCG1 may also significantly contribute to net cholesterol efflux; their significance should not be underestimated. Indeed, even the largest HDL1 particles are capable of efficiently removing cellular cholesterol from lipid-loaded macrophages [199].

The central role of small HDL in net cellular cholesterol efflux is further supported by a number of other observations, the most compelling of which probably represents the enhanced cholesterol efflux capacity and potent anti-atherogenic activity of *in vitro* de-lipidated HDL [200]. Another piece of evidence in support of the key role of small HDL particles in cholesterol efflux is provided by data obtained in rabbits expressing human apoA-I. In these animals, small, lipid-poor HDL predominates. In parallel, the cholesterol efflux capacity of rabbit serum increases [201]. Consistent with these findings, small, cholesterol-containing rHDL induces the mobilization of intracellular cholesterol to the plasma membrane, whereas large rHDL is inactive [150]. The central domain of apoA-I, whose exposure to the aqueous environment is diminished with an increase in HDL size and/or cholesterol content, appears to underlie this difference. In small, lipid-poor particles, apoA-I may adopt a conformation that results in a high affinity for lipid surfaces, thereby facilitating cholesterol efflux [153].

Physically, the potent capacity of small, lipid-poor HDL to efflux cellular cholesterol appears to result from low lateral pressure in the surface lipid monolayer, which is well below the values obtained for large unilamellar vesicles [202]. Such low lateral pressure is paralleled by the presence of structural defects in the lipid packing of small HDL [203] (Fig. 3). Moreover, such small particles are discoid and can form saddle surface. As a consequence, they are capable of readily acquiring large amounts of phospholipid and cholesterol.

An obvious consequence of the inverse relationship between HDL size and the capacity to efflux cholesterol is the conclusion that lipid-free apoA-I, which can be regarded as an HDL subpopulation of the smallest possible size, should represent the most potent acceptor of cellular cholesterol. Available data indirectly support this notion. Indeed, despite their potent efflux capacity, small pre-beta HDL particles are not essential for cellular cholesterol efflux [204], thereby suggesting that lipid-free, rather than lipid-poor, apolipoproteins function as primary cholesterol acceptors [189].

The distinct conformation of apoA-I observed in small HDL may also contribute to its propensity to activate LCAT. Indeed, LCAT activity is highest in the smallest HDL subpopulation isolated from human plasma [205], indicative of the role of small HDL as a major site of cholesterol esterification (Table 7.3). In support of this notion, the cholesteryl ester/free cholesterol ratio increases with decreasing HDL size [205]. In addition, small HDL particles display a faster cholesterol esterification rate compared with large HDL [31].

When comparing the biologic activities of HDL subpopulations, it is essential to keep in mind the concentration basis employed for such comparisons. Notably, small, dense, spherical HDL more potently induces cholesterol efflux than its large, light, spherical counterpart when compared on the basis of phospholipid. In contrast, if HDL particles are compared on a particle number basis, large HDL that contains more phospholipid per particle may reveal greater cholesterol efflux capacity via SR-BI or ABCG1 compared with small HDL. It remains unknown as to whether the same is true for small discoid HDL particles which *a priori* efflux cholesterol more efficiently than other HDL subpopulations.

As HDL particles differ not only in size and density, but also in apolipoprotein composition, the latter differences may also impact on HDL capacity to induce cholesterol efflux. Indeed, HDL containing only apoA-I appears to remove cellular cholesterol more efficiently than particles containing both apoA-I and apoA-II [206]. In addition, HDL subpopulations with distinct protein composition may participate in cholesterol efflux in different body fluids. Indeed, apoA-I-containing nascent HDL particles play a critical role in cholesterol transport in the circulation, whereas apoE-containing HDL particles mediate cholesterol transport in the extracellular fluid of the brain [207, 208].

IMPORTANT READING

Lund-Katz S, Phillips MC. High density lipoprotein structure-function and role in reverse cholesterol transport. *Subcell Biochem* 2010; 51: 183–227.

Tall AR. Cholesterol efflux pathways and other potential mechanisms involved in the athero-protective effect of high density lipoproteins. *J Intern Med* 2008; 263: 256–273.

7.2 ANTIOXIDATIVE ACTIVITY

According to the response-to-retention hypothesis of atherosclerosis, cholesterol-rich lipoproteins, primarily LDL, retained in the arterial wall can become modified under the action of cells present in the arterial wall intima [209–213]. Such pathophysiologic lipoprotein alterations in the arterial wall include oxidative modifications [214]. Indeed, atherosclerotic plaques contain multiple products of protein and lipid oxidation, such as oxidized, chlorinated and nitrated amino acids, lipid hydroperoxides (LOOHs), short-chain aldehydes, oxidized phospholipids and oxysterols [215]. Various oxidative systems can be involved in lipoprotein oxidation *in vivo*, including myeloperoxidase, NADPH oxidase, nitric oxide synthase and lipoxygenase. Accordingly, reactive chlorine, nitrogen and oxygen species are the most probable lipoprotein oxidants *in vivo* [215].

Lipoprotein oxidation in the arterial intima can be regarded as a consequence of local **oxidative stress**, which represents an imbalance between pro-oxidants and antioxidants in favor of the former. Oxidative stress in the arterial wall intima is closely associated with the development of atherosclerosis and gives rise to the formation of **oxidized LDL** [215] (Fig. 7.3). This modified lipoprotein has a major impact on atherogenesis, as highlighted by a recent finding that selective

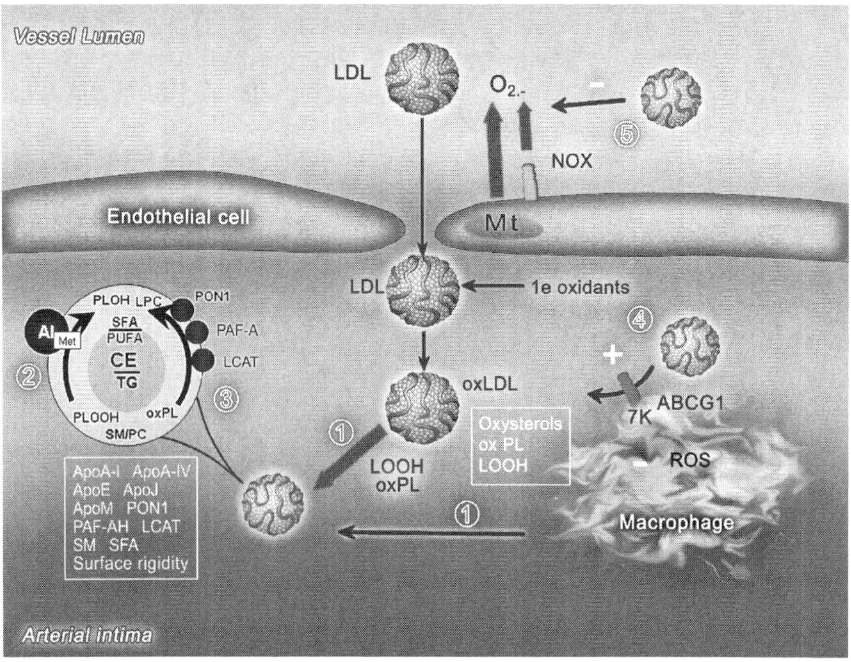

Figure 7.3 Major pathways involved in antioxidative activity of HDL. HDL inhibits LDL oxidation in the arterial intima induced by one-electron oxidants (1e oxidants), acting via a two-step mechanism involving (1) the transfer of oxidized lipids, such as lipid hydroperoxides and short-chain oxidized phospholipids (oxPL) from oxidized LDL (oxLDL) or arterial wall cells to HDL with subsequent inactivation by HDL protein components. (2) ApoA-I, acting by virtue of Met residues, reduces PLOOH to redox-inactive phospholipid hydroxides (PLOH). (3) Hydrolytic enzymes (PON1, PAF-AH, LCAT) catalyse the hydrolysis of oxidized phospholipids to lysophosphatidylcholine (LPC). The efficacy of the transfer of oxidized lipids to HDL depends on the fluidity of HDL surface lipids and decreases with decreasing fluidity, such as with increasing SM/PC ratio or wt % of saturated fatty acids (SFAs). The capacity of apoA-I to react with PLOOH may be influenced by apoA-I conformation which, in turn, depends on the CE/TG ratio in the HDL lipid core. (4) HDL decreases ROS production in macrophages acting via ABCG1-mediated efflux of oxidized sterols, including 7-ketosterol (7K). (5) HDL inhibits superoxide generation in endothelial cells by NADPH oxidase (NOX) and mitochondrial (Mt) electron chain. Inset lists HDL components potentially involved in the antioxidative activity of HDL.

removal of oxidized LDL from the circulation by ectopical over-expression of lectin-like oxidized LDL receptor 1 (LOX-1) in the liver completely prevents progression of the disease in apoE -/- mice, despite the persistence of severe dyslipidemia [216].

> **DEFINITION**
>
> **Antioxidative activity**: the capacity of HDL to protect biomolecules from oxidative damage by one- or two-electron oxidants.

The key pro-atherogenic property of modified LDL involves its capacity to accumulate in arterial wall cells, primarily in macrophages, via scavenger receptor pathways which are not down-regulated as a function of cellular cholesterol content and therefore lead to the formation of lipid-loaded cells [214]. Furthermore, modified lipoproteins provoke biologic responses that can account for all known features of atherosclerosis, including endothelial dysfunction and the development of the lipid-rich, vulnerable plaque [217]. Downstream products of lipoprotein modifications are particularly deleterious as they eventually block the normal efferocytosis of monocyte-derived cells (i.e., macrophages, foam cells and dendritic cells) from the plaque, thereby accelerating disease progression. For example, oxidized phospholipids, such as 1-palmytoyl-2-(5-oxovaleroyl)-sn-glycero-3-phosphocholine and 1-palmitoyl-2-glutaroyl-sn-glycero-3-phosphocholine found in oxidized LDL and apoptotic cells, are concentrated within atherosclerotic lesions and constitute potent pro-inflammatory mediators that induce monocyte adhesion [3]. Oxidized phospholipids also strongly interact with specific endothelial cell proteins which may be responsible for the inflammatory response and monocyte recruitment [218]. In addition, LDL-derived oxidized lipids are capable of migrating into cellular membranes, providing 'seeding' molecules for the propagation of oxidative damage and further promoting oxidative stress [219].

LDL has long been thought to represent the major physiologic target of protection from oxidative damage provided by HDL (Fig. 7.3). HDL **antioxidative activity** is, therefore, typically measured *in vitro* as the inhibition of LDL oxidation by HDL [10, 11, 220]. However, HDL is also able to inhibit the generation of **ROS** (Fig. 7.3) and to decrease intracellular oxidative stress *in vitro* in endothelial cells [221–223] and in macrophages [224, 225], as well as *in vivo* in animal models of acute arterial inflammation [226]. The protective actions of HDL towards LDL oxidation can be observed both *in vitro* upon their co-incubation [144] and *in vivo* upon HDL supplementation [227]. Consistent with the role of HDL in the protection of biomolecules from oxidative stress *in vivo*, circulating concentrations of HDL-C are negatively associated with biomarkers of systemic oxidative stress in dyslipidemic individuals with Type 2 diabetes and Metabolic Syndrome [228, 229]. Furthermore, HDL-C levels are inversely related to leukocyte telomere length which, in turn, reflects long-term cellular exposure to oxidative stress in both cross-sectional and prospective analyses, as recently demonstrated in the Bogalusa Heart Study [230].

The relative importance of antioxidative activity in the overall cardioprotective effect of HDL compared with other biologic actions remains indeterminate but may be especially relevant under metabolic conditions featuring oxidative stress (e.g. Metabolic Syndrome or Type 2 diabetes; see Chapter 10.2). The antioxidative activity of HDL has been proposed to be less important than cholesterol efflux capacity, as suggested by the absence of systemic antioxidative effects of human apoA-I expression in apoE -/- mice accompanied by delayed atherosclerosis [231]. The increase in non-HDL-C levels observed in this animal model, however, renders interpretation of these results complex. In contrast, both the cholesterol efflux capacity and the antioxidative activity of HDL are impaired in parallel to a similar extent in apoA-I -/- mice, as shown in another study [232].

OPEN QUESTION

Can HDL-mediated inhibition of LDL oxidation inhibit the development of atherosclerosis?

Mechanisms of Protection Against Oxidative Stress

HDL can protect LDL and other lipoproteins from oxidative stress induced by various oxidants, including one- and two-electron species. One-electron oxidants, primarily free radicals, oxidize both the lipid and protein moieties of HDL with the formation of lipid and protein radicals which, in turn, further propagate the oxidation of HDL components to stable termination products [214]. HDL particles potently protect both the lipid and protein moieties of LDL from free radical-induced oxidation and inhibit the accumulation of products of both lipid and protein oxidation, including oxidized phospholipids, short-chain aldehydes and oxidized amino acid residues [10, 220, 233]. Such protective capacity is mediated by several HDL components and involves distinct molecular pathways [233]. In contrast, HDL-mediated protection of LDL from oxidation by two-electron oxidants, such as hypochlorite, which mainly modify the protein moiety of LDL, appears unspecific and probably reflects direct oxidant scavenging by HDL [234].

Removal of oxidized lipids from LDL or cells represents the first step of HDL-mediated protection from oxidative damage induced by free radicals (Fig. 7.3). Indeed, phospholipid hydroperoxides (PLOOHs) are rapidly transferred from oxidized LDL to HDL upon their co-incubation [171]. The addition of a hydroperoxyl group to a fatty acid, phospholipid, triglyceride or cholesteryl ester molecule strongly increases hydrophilicity. As a result, linoleic acid, methyl linoleate and trilinolein hydroperoxides are more surface-active than their non-peroxidized counterparts [235]. LOOH molecules are therefore exposed to the aqueous phase at the HDL surface and can escape the lipoprotein more readily than their non-oxidized counterparts. The transfer of LOOHs from LDL to HDL can occur directly between lipoprotein phospholipid monolayers, either spontaneously or mediated by lipid transfer proteins, including CETP, which can accelerate the transfer of both cholesteryl ester hydroperoxides (CEOOHs) and PLOOHs [236, 237]. As HDL and LDL represent the major carriers of phospholipid and cholesteryl ester in human plasma [238], the exchange of LOOH with other lipoproteins (e.g., VLDL) should be of secondary importance.

Consistent with these mechanisms, HDL particles constitute a major transport vehicle of LOOH in human plasma and may therefore function as a 'sink' for oxidized lipids [239] which can accumulate in the particle when the LOOH-inactivating capacity of HDL is overwhelmed. In support of this hypothesis, HDL protects apoB-containing lipoproteins from the accumulation of diet-derived LOOH following a hydroperoxide-rich meal [240]. Furthermore, HDL has recently been demonstrated to serve as a major carrier of plasma F2-isoprostanes, the stable final products of lipid peroxidation [241]. The accumulation of oxidized lipids in HDL most probably results from their transfer from LDL, remnant triglyceride-rich lipoproteins and/or arterial wall cells, which can be mediated, in part, by lipid transfer proteins [236, 237]. Subsequently, CEOOHs and their corresponding hydroxides can be rapidly removed from HDL via selective uptake by the liver mediated by SR-BI [242]; enhanced hydrophilicity of CEOOHs compared with non-oxidized cholesteryl esters may underlie this observation [237]. This pathway may significantly contribute to the removal of toxic oxidized lipids from the body. As CEOOHs are removed from the circulation more rapidly than the corresponding non-oxidized cholesteryl ester, the addition of a hydroperoxyl group may accelerate elimination of cholesterol from the body [243].

The transfer of PLOOH from oxidized LDL to HDL can beneficially impact pro-atherogenic properties of oxidized LDL. Thus, the capacity of oxidized LDL to induce foam cell formation is diminished following the removal of LOOH [244], most probably as a result of attenuated interaction of oxidized LDL with scavenger receptors [245]. In addition to the transfer of LOOH from oxidized LDL to HDL, the transfer of non-oxidized phospholipids in the opposite direction may also contribute to a decrease in the charge of oxidized LDL. The significant cardioprotective potential of this pathway is highlighted by protection from reperfusion injury provided in a rabbit model by a synthetic HDL preparation comprising recombinant apoA-I-Milano complexed with POPC (ETC-216) [246]. As reperfusion leads to the production of oxidized lipids that can damage myocardium, ETC-216 decreases LOOH levels in cardiac left ventricle muscle, prevents the left ventricular end-diastolic pressure elevation and reduces reperfusion injury [246].

The role of the transfer of oxidized lipids to HDL is further consistent with the correlation between HDL capacities to efflux cholesterol and to prevent LDL oxidation observed in obese women [247]. Importantly, both the capacities of HDL to efflux cholesterol [69] and to remove oxidized phospholipids [171] are regulated by the fluidity of the HDL surface lipid monolayer. Indeed, HDL particles with a less rigid lipid surface more efficiently efflux cholesterol [69] and more efficiently acquire oxidized phospholipids, resulting in a more potent protection of LDL from oxidative stress [171].

The capacity of HDL to remove LOOH is not specific for LDL, as similar effects have been reported for erythrocytes and astrocytes [248–250]. Furthermore, HDL efficiently effluxes oxidized sterols from macrophages, monocytes and neutrophils [251–254]. Indeed, CEOOHs and PLOOHs are transferred between ghost erythrocyte membranes and LDL with apparent first-order rate constants approximately 60 and 35 times greater than those of the respective parent native lipids [255]. Similar PLOOH transfer kinetics are observed with unilamellar liposomes as an acceptor, consistent with desorption from the donor membrane as the rate-limiting step. For individual CEOOHs, the rate constant of the transfer inversely correlates with reverse-phase HPLC retention time, suggesting that faster transfer reflects greater hydrophilicity [256].

The inactivation of oxidized lipids associated with HDL particles represents the second step in the capacity of HDL to protect from oxidative damage [233] (Fig. 7.3). Depending on their structure, oxidized lipids can be reduced (LOOHs) or hydrolyzed (short-chain oxidized phospholipids, lysophosphatidylcholine; Fig. 7.3). Indeed, HDL rapidly reduces LOOH to the corresponding hydroxides [171, 236, 257–259]. Furthermore, HDL particles are capable of inactivating the biologic activity of short-chain oxidized phospholipids by their hydrolysis to lysophospholipid and to the free fatty acid or aldehydic moiety [10, 220].

In clear contrast to the mechanisms of the inhibitory action of HDL on the oxidative modification of LDL, those of the antioxidative activity of HDL towards cells are much less well elucidated. HDL-mediated **inhibition of ROS generation** can be observed in endothelial cells challenged with oxidized LDL [221–223] and in macrophages exposed to oxidized phospholipids [224] or glucose [225] (Fig. 7.3). Such antioxidative action of HDL is paralleled by improved cell survival and does not necessitate direct contact between HDL and toxic agents in the extracellular compartment, consistent with their intracellular origin. Nicotinamide adenine dinucleotides phosphate (NADPH) oxidase and the mitochondrial electron transport chain provide major contributions to ROS generated by endothelial cells in the presence of oxidized LDL [260]. Superoxide anion $O_2^{\cdot-}$ and hydrogen peroxide, H_2O_2, are two major ROS produced by NADPH oxidase and the mitochondrial electron transport chain; superoxide anion appears, however, to be rapidly dismuted to hydrogen peroxide, which may participate in oxidized LDL-induced signaling [261]. Diminished production of superoxide and/or hydrogen peroxide may, therefore, be implicated in the antioxidative effect of HDL in endothelial cells [260]. Similarly, HDL attenuates glucose-mediated increases in macrophagic superoxide production via attenuation of NADPH oxidase activity [225]; activation of cellular superoxide dismutases can be also involved. Such cellular antioxidative activity of HDL requires interaction with surface receptors; interaction with SR-BI appears to largely account for the antioxidative effects observed in endothelial cells [260], whereas both SR-BI and ABCG1 are involved in HDL-mediated inhibition of superoxide production in macrophages [225].

METHODOLOGIC NOTE

Plasma levels of F2-alpha isoprostanes should be measured as a gold standard, non-enzymatic biomarker of oxidative stress *in vivo*, which may reflect the antioxidative function of HDL [241, 262].

Role of HDL Components

Apolipoproteins. **ApoA-I** plays a central role in HDL-mediated protection from oxidative damage (Table 7.1); as a corollary, oxidative stress is elevated following the deletion of apoA-I in animal models [232, 263]. ApoA-I underlies the inactivation of LOOHs, as its methionine (**Met) residues** 112 and 148 can reduce LOOHs into redox-inactive lipid hydroxides, thereby terminating chain reactions of lipid peroxidation [171, 243, 259]. Molecular dynamics simulation of human apoA-I structure show that Met112 and Met148 interact with the Tyr115 residue, creating a microenvironment unique to human apoA-I [264], which may be optimal for such a redox reaction. LOOH can also be inactivated by apoA-II, although to a lesser extent. The key role of apoA-I in the HDL-mediated protection of LDL from one-electron oxidants is critically supported by the observation that rHDL containing only purified apoA-I and POPC, but devoid of enzymatic components, and authentic small, dense HDL3b+3c, are comparable in their capacities to delay lipid peroxidation of LDL [171]. Concentrations of redox-active Met residues in apoA-I and of PLOOH suggest a 1:1 reaction stoichiometry [171, 243, 259]. In addition, lipid-free apoA-I attenuates LDL oxidation by the removal of LOOH molecules from LDL [265]. The HDL content of apoA-I and the oxidative status of apoA-I Met residues are, therefore, important determinants of the capacity of HDL to inactivate LOOH and to protect LDL from free radical-induced oxidation [171]. In addition to Met, **Cys residues** may be important for the antioxidative properties of apoA-I, as demonstrated by the elevated antioxidant potency of the N74C mutant of apoA-I toward LDL oxidation *in vitro*, paralleled by remarkable anti-inflammatory and anti-atherogenic activities *in vivo* [266]. Histidine (His) residues may also contribute to the apoA-I-mediated inhibition of LDL oxidation as a result of their metal-chelating properties [267].

The redox reaction between HDL Met residues and LOOH is paralleled by the formation of oxidized forms of apoA-I and apoA-II which contain methionine sulfoxides [243]. Methionine sulfoxide moieties formed in apoA-I can be reduced back to Met by methionine sulfoxide reductases, ubiquitous enzymes present in virtually all mammalian tissues essential for protecting cells against oxidative damage [268]. Indeed, peptide methionine sulfoxide reductase is capable of reducing methionine sulfoxide in oxidized apoA-I in the presence of its co-factor dihydrolipoic acid [243]. Interestingly, only oxidized proteins with methionine sulfoxides located on their surface can be a substrate of methionine sulfoxide reductases unless the oxidation of Met residues provokes protein unfolding [268], suggestive of the probing capacity of this reaction for the conformation of apoA-I.

Specifically, the introduction of a polar sulfoxide may re-orient the direction of the hydrophobic face of the apoA-I molecule in the vicinity of Met112, thereby altering its lipid affinity [243].

Other apolipoprotein components of HDL may also contribute to HDL-mediated protection of LDL from free radical-induced oxidation. Indeed, HDL particles contain multiple apolipoproteins, including apoA-II, apoA-IV, apoA-V, apoE, apoJ and apoM, known to protect LDL from oxidative damage *in vitro* [17, 18, 269] (Table 7.1). The capacity of **apoA-II** to protect LDL from oxidation is, however, questionable, given the fact that over-expression of human apoA-II in dyslipidemic mice accelerates atherosclerosis in parallel with alterations in the HDL proteome, increases in aortic accumulation of oxidized LDL and a reduction of HDL antioxidative activity [270, 271]. Such pro-atherogenic actions of apoA-II may be related to the displacement of anti-atherogenic apoA-I and PON1 by apoA-II from HDL particles [271]. Antioxidative properties have also been reported for **apoA-IV** [272, 273].

ApoE possesses established anti-atherosclerotic activity which is normally ascribed to its lipid-transport properties [274]. However, the action of apoE goes beyond such action. Indeed, apoE possesses distinct antioxidative properties [275] and can promote the regression of atherosclerosis independently of lowering plasma cholesterol levels [276–278]. HDL-associated **apoJ** can attenuate oxidation of LDL by artery wall cells [279]; in addition, apoJ is cytoprotective at low physiologic levels [280]. The beneficial actions of apoJ may be related to its ability to maintain the integrity of membrane and lipoprotein lipids via its hydrophobic-binding domains [281].

Next, experiments in transgenic mice suggest that **apoM** possesses anti-atherogenic properties; possible mechanisms underlying this association may include antioxidative activity [282]. Finally, HDL is able to function as a preventive antioxidant through its capacity to bind transition metal ions [283] which, in the free form, are potent catalyzers of LDL oxidation. Intriguingly, plasma HDL carries **amyloid-beta peptide**, a major component of senile neuritic plaques and a strong chelator of transition metals [208].

Enzymes. Enzymatic components potentially contributing to the antioxidative properties of HDL include PON1, PAF-AH and LCAT, all of which were proposed to hydrolyze pro-inflammatory short-chain oxidized phospholipids [11, 22] (Tables 7.1 and 7.2). In contrast, PON1, PAF-AH and LCAT are weakly reactive towards LOOH [172–175]. In addition, HDL carries GSPx which can reduce LOOH to the corresponding hydroxides and thereby detoxify them [284–286].

PON1 is a lactonase which displays potent anti-atherosclerotic effects *in vivo* [287, 288] (Table 7.2). Genetically determined PON1 activity is inversely related to systemic oxidative stress and prospective cardiovascular risk in coronary heart disease patients [289, 290]. Levels of PON1 expression determine the development of atherosclerosis in mouse models and display inverse relationships to biomarkers of systemic oxidative stress [291, 292]. PON1 gene transfer also decreases local oxidative stress in the vascular wall, resulting in improved endothelial function in a mouse model [293] The capacity of PON1 to diminish oxidative stress *in vivo* is relevant for different cell types, including macrophages and beta-cells, and may result in the delayed development of diabetes [294, 295]. Such protective actions can be mediated by specific binding sites on the cell surface, as demonstrated for macrophages [294].

In the circulation, PON1 is exclusively present in HDL where it is anchored to lipids via its hydrophobic N-terminus [296, 297]. The association of PON1 with HDL is a prerequisite for maintaining the normal serum activity of the enzyme. HDL provides the optimal physiologic acceptor complex for PON1, in terms of both stimulating enzyme secretion and stabilizing the secreted peptide [298]; PON1 interaction with apoA-I is critical for enzyme stability [299]. HDL and, less efficiently, VLDL, but not LDL, promote PON1 secretion from cells; the differences observed between these lipoproteins are related to differences in their lipid composition [300]. Cellular SR-BI plays a major role in the transfer of PON1 to HDL, functioning as a docking molecule for HDL particles [301]. Following association with PON1, HDL may serve as a vehicle for the enzyme for transfer back to cell membranes to improve cellular resistance to oxidative stress [302].

PON3, another member of the PON family carried by HDL, possesses properties similar to those of PON1. Specifically, the enzyme is a lactonase whose activity is selectively stimulated by binding to apoA-I on HDL with a concomitant increase in its stability [303]. PON3 also exhibits potentially anti-atherogenic activities which appear, however, less potent compared with those of PON1 [303]. Interestingly, PON2, another enzyme of the PON gene cluster only found intracellularly and not on HDL, is also protective against elevated oxidative stress, as observed in macrophages in the presence of high concentrations of glucose [304]; this protective effect is paralleled by diminished triglyceride accumulation and can be mediated by the attenuation of NADPH-oxidase activity. Remarkably, PON2 over-expression decreases atherosclerosis in apoE-deficient mice [305]. Such similarities in the effects of different PON enzymes on atherosclerosis and oxidative stress suggest common underlying mechanisms. In support of this hypothesis, over-expression of the whole PON gene cluster consisting of the PON1, PON2, and PON3 genes represses atherogenesis and promotes atherosclerotic plaque stability *in vivo* [306].

MAJOR CONTROVERSY

Does PON1 directly contribute to the antioxidative activities of HDL?

Accumulating data, however, question the ability of PON1 to hydrolyze LOOH [174–177]. The antioxidative activity of PON1 purified from human serum has been ascribed to the presence of detergents or to unidentified proteins [173]. Use of specific enzymatic inhibitors does not support the contention that PON1 contributes significantly to the inactivation of LDL-derived LOOH [171]. We [171] and others [259, 307] demonstrated the existence of PON1-independent antioxidative activity of HDL, indicating that HDL-associated PON1 does not significantly contribute to the inactivation of LDL-derived LOOH.

The distinction between the hydrolytic actions of HDL-associated PAF-AH and of PON1 on LOOH remains, however, difficult as the two activities are determined using substrates, such as PAF (for PAF-AH) and paraoxon (for PON1), whose hydrolysis does not parallel the hydrolysis of LOOH [308]. Indeed, esterase and hydroperoxidase activities of PON1 often dissociate, suggesting distinct active sites [309]. Furthermore, the capacity to protect LDL from oxidation, but not the esterase activity, of PON1 appears to require a cysteine (Cys) residue at position 283 [310].

The major activity of PON1 has been demonstrated to be that of a **calcium-dependent lactonase** rather than a peroxidase [174, 178, 179]. The affinity of PON1 for LOOH is several orders of magnitude lower than its affinity for lactones [174, 178]; LOOH do not, therefore, constitute a significant substrate of PON1. Indeed, the established anti-atherosclerotic properties of PON1 [287, 288] may be unrelated to its capacity to inactivate LOOH, but rather may involve its major activity as a lactonase through a still unknown pathway upstream of the regulation of systemic oxidative stress. In particular, as PON1 catalyzes the hydrolysis of homocysteine thiolactone, this activity could protect against homocysteinylation, a pro-atherogenic post-translational modification of proteins [180]. Indeed, N-homocysteinylated proteins acquire auto-immunogenic, pro-inflammatory and prothrombotic properties accompanied by attenuated biologic activity, potentially contributing to atherogenesis [180]. In support of this pathway, recent evidence suggests that the homocysteine thiolactonase activity of PON1 is a determinant of plasma levels of N-homocysteinylated proteins in humans [180]. Together with the action on the cholesterol efflux properties of HDL [167–169], the lactonase activity may contribute to the anti-atherosclerotic effects of PON1 *in vivo* [287, 288].

It is more probable that **PAF-AH**, rather than PON1, represents the hydrolase for oxidized phospholipids in HDL [176, 177]. Indeed, plasma PAF-AH readily hydrolyzes PAF-like oxidatively-fragmented phospholipids with shortened and/or derivatised *sn*-2 groups and thereby functions as a highly selective phospholipase for oxidized phospholipids [176]. Furthermore, PAF-AH can hydrolyze PLOOH [175], leading to the formation of lysophosphatidylcholine and a fatty acid hydroperoxide.

LCAT may equally hydrolyze short-chain oxidized phosphatidylcholines generated during lipoprotein oxidation [172]. In addition, LCAT can delay LDL oxidation, acting as a chain-breaking antioxidant most likely via its Cys residues [311]. In our studies, however, the inhibition of PAF-AH and LCAT was without effect on the HDL-mediated inactivation of LDL-derived LOOH [171], thereby indicating that both the enzymes are, at most, a minor factor in LOOH inactivation by HDL. In contrast, PON1, PAF-AH and LCAT can all contribute to the inactivation of short-chain aldehydic oxidized phospholipids derived from LOOH

(Table 7.1). In addition, HDL-associated enzymes, such as PAF-AH and/or PON1, may inactivate phospholipid core aldehydes generated upon HDL oxidation by peroxynitrite with the participation of apoA-I [312]. Such a PAF-AH/PON1-coupled protective function of apoA-I can effectively divert pro-atherogenic LOOH to less harmful products [10, 220, 313].

Consistent with this hypothesis, HDL-associated PAF-AH is thought to play an anti-atherogenic role in contrast to the LDL-associated enzyme [314–316]. Indeed, local arterial expression of PAF-AH reduces accumulation of oxidized LDL and inhibits inflammation, shear stress-induced thrombosis and neointima formation in balloon-injured carotid arteries of non-hyperlipidemic rabbits [317]. However, it remains unclear as to whether products of the hydrolysis of short-chain oxidized phospholipids, lysophosphatidylcholine and oxidized free fatty acid, are more or less pro-atherogenic compared with the oxidized phospholipid substrate [318]. Being more hydrophilic than their precursors, oxidized free fatty acid would be more prone to diffuse away from the lipoprotein, thus reducing oxidative pressure on the latter. In addition, free fatty acid hydroperoxides represent excellent substrates for plasma glutathione peroxidises. Together, these findings suggest that hydrolysis of oxidized lipids in HDL should act beneficially on oxidative stress and atherosclerosis.

Finally, the trypanosome lytic factor exclusively present in very dense subpopulations of human HDL (see Chapter 7.5) has been reported to exhibit a peroxidase-like activity in the presence of hydrogen peroxide [319], which may provide a minor contribution to the LOOH-inactivating properties of HDL. Hemopexin, an iron-binding protein, may also contribute to the capacity of HDL to protect LDL from iron-mediated oxidation as the protein inhibits LDL oxidation by hemoglobin [320].

Lipids. HDL lipids can significantly modulate antioxidative activities displayed by the protein components (Table 7.2). Indeed, the **rigidity of the phospholipid monolayer** of HDL particles is a key modulator of the transfer efficiency of PLOOH from LDL to HDL [171]. The strong inverse correlation of the PLOOH transfer rate from oxidized LDL to liposomes with liposomal membrane rigidity emphasizes the importance of both physical and chemical properties in this process. Furthermore, both the PLOOH transfer rate from oxidized LDL and the capacity of HDL to delay LDL oxidation decrease in parallel with surface lipid rigidity in rHDL containing apoA-I and various phospholipid species [171]. Elevated lipid membrane rigidity also inhibits the transfer of LOOH to HDL from cell membranes [255]. The surface monolayer rigidity of HDL is primarily determined by the relative content of lipids such as sphingomyelin (relative to phosphatidylcholine), free cholesterol (relative to phosphatidylcholine) and saturated + monounsaturated fatty acids (relative to polyunsaturated ones); the increasing content of each of them rigidifies the surface monolayer of HDL particles at a given protein/lipid ratio [171].

Finally, HDL particles carry small amounts of lipophilic antioxidants, primarily tocopherols, which may make a minor contribution to the antioxidative properties of HDL [239, 258, 321] (Table 7.1).

Functional Heterogeneity of HDL

Apolipoproteins, enzymes and lipids that determine the antioxidative activities of HDL are non-uniformly distributed across the spectrum of HDL subpopulations. ApoA-I, the major protein component of HDL participating in the reduction of LOOHs, is enriched relative to apoA-II in small, dense HDL3c and in large, light HDL2b compared with HDL2a, 3a and 3b subpopulations [205]. *in vivo*, PON1 is preferentially associated with large HDL but can be displaced to small, dense particles upon ultracentrifugation [322–324]. The size and shape of HDL seems to be critical for PON1 binding [296]. ApoJ is associated with a subset of small HDL which also contains PON1 [325], PLTP and CETP [326]. Similarly, LCAT activity [323], PAF-AH activity [323] and apoA-IV [327] are enriched in small, dense HDL isolated by density gradient ultracentrifugation [323] or by gel filtration [327].

As a consequence, HDL particles are heterogeneous in their capacity to protect LDL from oxidative damage induced by one-electron oxidants. Under mild oxidative stress induced by an azo-initiator 2,2′-azobis-(2-amidinopropane) hydrochloride (AAPH) or copper ions (Cu^{2+}), the antioxidative activity of HDL subfractions isolated by density gradient ultracentrifugation against LDL oxidation increases with increment in density and decrement in size from HDL2b to HDL3c, thereby establishing that **small, dense HDL** acts as a potent protector of LDL from oxidation [323] (Table 7.3). Similarly, the HDL3 subfraction is a more potent protector of LDL from *in vitro* oxidation by one-electron oxidants compared with HDL2 [323, 328, 329]. Such potent antioxidative activity of small, dense HDL is related to the inactivation of LOOHs, which are primary products of LDL lipid peroxidation [323]. As a consequence, the accumulation of secondary lipid peroxidation products, such as aldehydes and short-chain oxidized phospholipids, is also potently inhibited by small, dense HDL3 [144, 323, 328, 329]. Furthermore, as aldehydes interact with amino acid residues of apoB to form protein adducts, the covalent oxidation of the LDL protein moiety is also significantly retarded by HDL3 [144, 323, 328, 329]. Consistent with these data, small, dense HDL3 is more resistant to oxidative modification compared with large, light HDL2 [330].

CRITICAL CONTRIBUTION

Small, dense, protein-rich HDL particles potently protect LDL from oxidative damage [323]. This activity is accounted for by the inactivation of LOOHs via a two-step mechanism involving the transfer of LOOHs to HDL facilitated by enhanced fluidity of the HDL surface lipid monolayer with subsequent reduction to redox-inactive hydroxides by Met residues of apoA-I [171, 233, 236, 257–259].

The efficiency of both of the major steps involved in the antioxidative activity of HDL can differ between HDL subpopulations. Indeed, small, dense HDL3 particles may be superior to large, light HDL2 in terms of their **capacity to remove oxidized lipids** from other lipoproteins and cellular membranes. Structural defects in the packing of surface lipids which allow the insertion of exogenous molecules and become more pronounced with decreasing HDL particle size may account for this property [203, 256]. Consistent with this possibility, small, discoid rHDL complexes appear to remove negatively-charged lipids from oxidized LDL to a greater degree compared with native spherical HDL [245]. In addition, a diminished amount of sphingomyelin and free cholesterol in small, dense HDL may result in the increased fluidity of the surface lipid monolayer, thereby facilitating the incorporation of oxidized lipids of exogenous origin [233].

The inactivation of oxidized lipid molecules following their transfer to HDL may also occur more rapidly in small, dense particles (Table 7.3). First, the reduction of LOOHs to hydroxides is more efficient in HDL3 compared with HDL2 [171]. The relative enrichment of small, dense HDL in apoA-I may underlie this feature [205]. Moreover, the distinctly low lipid content of small, dense HDL can induce conformational changes in apoA-I relative to large, light HDL, resulting in enhanced exposure to the aqueous phase and modified reactivity towards monoclonal antibodies [331, 332] which may, in turn, facilitate the redox reaction between Met residues of apoA-I and LOOHs. The critical role of apoA-I conformation and stability for its antioxidative properties is highlighted by the observation that recombinant apoA-I-Milano is much more potent than wild-type apoA-I in attenuating LDL oxidation by copper ions [246].

Second, hydrolysis of oxidized lipids by HDL-associated hydrolytic enzymes also appears to be predominantly associated with small, dense HDL3 [323]. Preferential localization of enzymatic activities of PON1, LCAT and PAF-AH with HDL3 is probably responsible for such an association. Furthermore, enzymatic activities may be beneficially influenced by the lipidome of small, dense HDL3 which is distinct in displaying a low sphingomyelin-to- phosphatidylcholine ratio [205]. Indeed, sphingomyelin belongs to structural lipids which exert a positive impact on surface rigidity and a negative impact on LCAT activity [333, 334].

IMPORTANT READING

Stocker R, Keaney JF Jr. Role of oxidative modifications in atherosclerosis. *Physiol Rev* 2004; 84: 1381–1478.

Kontush A, Chapman MJ. Antiatherogenic small, dense HDL—guardian angel of the arterial wall? *Nat Clin Pract Cardiovasc Med* 2006; 3: 144–153.

Finally, the unique proteome of HDL3c, in which several proteins occur predominantly [18, 205], may have implications for its antioxidative activity which is highly correlated with the presence of apoJ, apoM, SAA4, apoD, apoL-1,

PON1/3 and PLTP [18]. Consistent with these data, PLTP is present on HDL, together with apoJ, apoL-I, apoD, apoA-I, apoA-II and several other proteins [326], suggesting direct functional interaction between them. Protein-rich, lipid-poor lipoprotein complexes of this type may display distinct biologic activities involving the protection of LDL from oxidative damage. It is relevant in this context that the simultaneous presence of PON1 and apoA-IV on a subset of small HDL particles whose plasma levels are elevated by the consumption of extra virgin olive oil endows these particles with potent antioxidative activity and is associated with decreased atherosclerosis in apoE -/- mice [335]. The potential influence of HDL protein composition on the antioxidative activities of HDL is further highlighted by the enhanced capacity of LpA-I:A-II particles to stabilize PON1 and to protect the enzyme from inactivation by oxidation compared with LpA-I particles [336]. Furthermore, apoE and apoA-II are both capable of binding PON1 and of stimulating its activity, suggesting that PON1-HDL recognition is primarily based on surface properties of the apolipoproteins and that specific protein-protein interactions should play a secondary role in this process [337].

These data implicate both protein and lipid components in the potent antioxidative activities of small, dense HDL. According to the two-step mechanistic hypothesis, the elevated capacity of HDL3 to incorporate and/or to inactivate (P)LOOH and other oxidized lipids may derive from: (i) enrichment in apoA-I; (ii) enrichment in hydrolytic enzymes, such as PON1, LCAT and PAF-AH and other functional protein components; (iii) depletion of sphingomyelin; and/or, (iv) altered conformation of apoA-I compared with HDL2 [22, 205].

CLINICAL SUMMARY

HDL potently protects LDL from oxidative damage *in vitro*, a plausible anti-atherogenic activity. Clinical evidence to support the link between the development of atherosclerosis and the capacity of HDL to protect biomolecules from oxidative damage is, however, lacking in humans.

7.3 ANTI-INFLAMMATORY ACTIVITY

Atherosclerosis can be regarded as a chronic inflammatory disease that evolves in response to LDL-derived cholesterol deposition in arteries. A local **inflammatory reaction** in the arterial wall in response to the accumulation of pro-atherogenic lipoproteins represents an important element of the response-to-retention hypothesis of atherosclerosis [209–213]. Monocytes, lymphocytes, neutrophils and other blood cells act in concert to mediate this reaction, with endothelial cell activation and monocyte adhesion to endothelium representing key early steps [338]. The tethering of monocytes to endothelial cells occurs via the interaction of integrins, such as CD11/CD18, with vascular cell adhesion molecule-1 (VCAM-1),

ICAM-1 and E-selectin on the endothelial cell surface. Monocyte transmigration across the endothelial surface and then into the subendothelial space is dependent on endothelial expression of chemokines, such as monocyte chemotactic protein-1 (MCP-1). In the intima, monocytes are transformed into macrophages, resulting in the accumulation of the latter. This process is accompanied by the accumulation of T-cells which may exert a deleterious effect on atherogenesis through direct cellular contact with monocytes/macrophages, inducing a massive up-regulation of pro-inflammatory cytokine production [339].

DEFINITION

Anti-inflammatory activity: the capacity of HDL to attenuate an inflammatory response.

Circulating HDL particles display multiple **anti-inflammatory actions** which may collectively contribute to the immunosuppresion of atherosclerosis significantly [340–343]. Such a spectrum of activities qualifies HDL as a major player in host defence and as a key element of the innate immune system. First, HDL is capable of **decreasing cytokine-induced adhesion molecule expression** on endothelial cells and of inhibiting monocyte adhesion to the endothelium (Fig. 7.4a). Indeed, intravenous infusion of a single dose of rHDL (CSL-111, which is apoA-I/phosphatidylcholine at 1:150 mol/mol) given over 4 hours at 80 mg/kg significantly decreases VCAM-1 expression in atherosclerotic plaques obtained from patients with peripheral vascular disease in a placebo-controlled trial [27]. Thus, the percentage of plaque cells expressing VCAM-1 comprises 28% in rHDL-treated subjects vs. 50% in the placebo group. In parallel, rHDL infusion reduces both monocyte activation and circulating levels of TNF-alpha. In a similar fashion, infusion of rHDL prior to that of recombinant human CRP completely prevents both the inflammatory response and thrombin generation in healthy humans [28]. Infusion of rHDL also inhibits adhesion molecule expression across arterial endothelium in a rabbit model of acute arterial inflammation [226]; such inhibitory effects of rHDL on adhesion molecules may translate into decreased atherosclerotic lesion formation in mice [344].

In support of these *in vivo* data, *in vitro* experiments reveal that HDL particles potently inhibit the expression of VCAM-1, ICAM-1 and E-selectin induced in endothelial cells by TNF-alpha, IL-1, endotoxin or CRP [28, 345–347] (Fig. 7.4a). HDL also decreases MCP-1 secretion by endothelial cells [348] and by human synoviocytes [349] induced by modified LDL and monosodium urate crystals, respectively. These effects may limit the inflammatory process by diminishing monocyte/macrophage recruitment and adhesion to the endothelial surface.

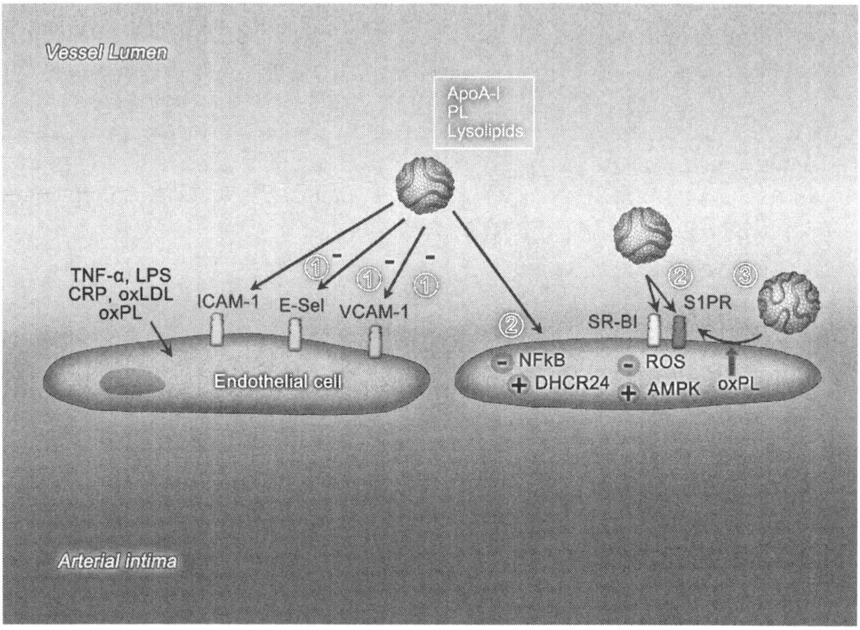

Figure 7.4 Major pathways involved in anti-inflammatory activity of HDL in endothelial cells (a) and leukocytes (b). (1) HDL decreases the expression of adhesion molecules (ICAM-1, VCAM-1, E-selectin [E-Sel]) on endothelial cells activated by TNF-alpha, CRP, oxidized LDL (oxLDL), oxidized phospholipid (oxPL) and LPS. (2) HDL inhibits activation of NFκB in endothelial cells induced by oxidized LDL and TNF-alpha, with subsequent reduction of inflammatory signaling and decrease in intracellular ROS generation. Such intracellular action of rHDL involves interaction with SR-BI and S1P receptors and requires up-regulation of DHCR24 and activation of AMPK. (3) HDL removes pro-inflammatory oxidized lipids from endothelial cells and/or from oxidized LDL followed by hydrolytic inactivation under the action of HDL-associated enzymes. (4) Actions of HDL on monocytes include the inhibition of monocyte adhesion to the endothelium, a decrease in monocyte activation by LPS with subsequent inhibition of monocyte transformation into a macrophage, and reduction of CD11/CD11b expression. These effects are mediated by the efflux of non-oxidized and/or oxidized cholesterol from macrophages via ABCA1 and ABCG1 with concomitant reduction in lipid raft content. (5) HDL acts on T-cells, inhibiting their stimulation by antigen-presenting cells. (6) HDL attenuates activation of monocytes by stimulated T-lymphocytes, acting via diminished production of cytokines and chemokines (TNF-alpha, IL-1beta, IL-6). (7) HDL potently reduces neutrophil activation and infiltration in the arterial wall; this effect involves cholesterol efflux as a major underlying pathway. (8) Finally, HDL restores the emigratory process of monocyte-derived dendritic cells from atherosclerotic plaques. Inset lists HDL components potentially involved in anti-inflammatory activity of HDL. FC, free cholesterol; PL, phospholipid.

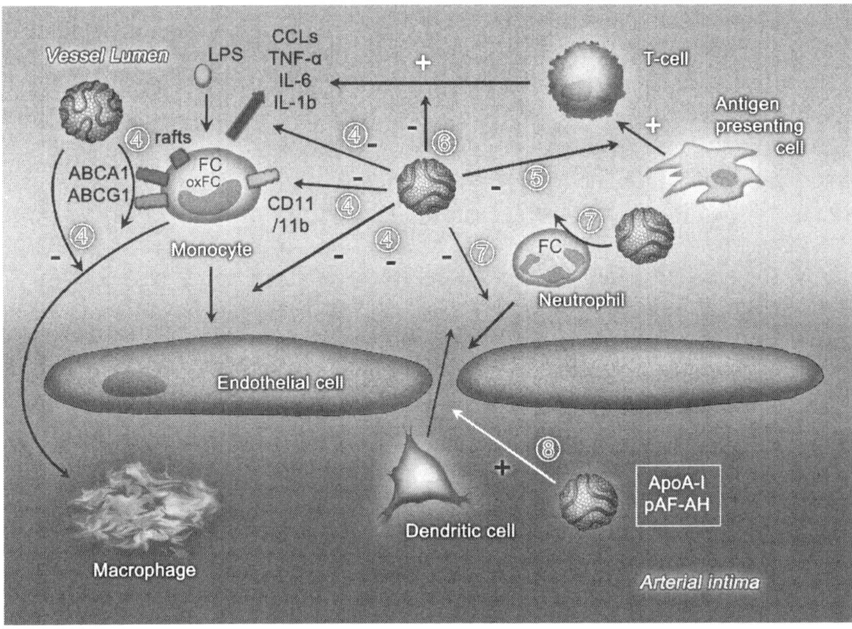

Figure 7.4 (*Continued*)

> **CRITICAL CONTRIBUTION**
>
> Decreased expression of adhesion molecules in atherosclerotic plaques after a single intravenous infusion of rHDL in patients with peripheral vascular disease in a placebo-control trial [27] and across arterial endothelium in a rabbit model of acute arterial inflammation [226].

HDL can also directly act on monocytes and other blood cells participating in the inflammatory response. The **inhibition of monocyte activation** constitutes an important facet of HDL action on the innate immune system (Fig. 7.4b). Indeed, rHDL reduces the expression of chemokines, including chemokine (C-C motif) ligand 2 (CCL2), CCL5 and chemokine (C-X3-C motif) ligand 1 (CX3CL1), as well as chemokine receptors, including CCR2 and CX3CR1 by monocytes *in vivo* and *in vitro*, acting via modulation of nuclear factor kappa B (NFκB) and PPAR gamma [350]. Furthermore, rHDL infusion suppresses CD11b expression on circulating monocytes, consistent with the *in vitro* finding that HDL reduces fatty acid-induced cytokine release from macrophages by suppressing the c-Jun N-terminal kinase (JNK) pathway [13, 15]. In addition, monocyte activation *in vitro* measured as surface expression of CD11b induced by phorbol myristic acid or receptor-dependent agonists, such as LPS, is attenuated by both HDL and apoA-I

[251]. HDL also inhibits a subset of LPS-stimulated macrophage genes that regulate the type I interferon response, acting independently of sterol metabolism [351]. In addition, apoA-I can inhibit chemotaxis of human monocytes [352].

HDL actions on monocytes include **decreased myeloid cell proliferation** and reduced monocytosis, as observed in LDL receptor-deficient mice [353]; these observations may underlie dramatic suppression of atherosclerosis, leukocytosis, monocytosis and myeloid cell infiltration by apoA-I transgene expression in these animals. Consistent with these data, mice deficient in ABCA1 and ABCG1 display enhanced proliferation of hematopoietic stem and multipotential progenitor cells, as well as accelerated atherosclerosis, which can be inhibited by HDL [354]. Another pathway implicated in the anti-inflammatory actions of HDL on monocyte/macrophages includes the diminished conversion of macrophage-secreted factors into pro-inflammatory complexes via direct binding; co-immunoprecipitation of macrophage-produced progranulin with apoA-I supports this notion [355].

HDL particles also inhibit the ability of antigen-presenting cells to stimulate T-cells [356] (Fig. 7.4b). In addition, HDL displays inhibitory activity towards contact-mediated activation of human monocytes by stimulated T-lymphocytes [357], as well as by microparticles generated by stimulated T-lymphocytes [358] (Fig. 7.4b). As cellular contact with stimulated T-cells is a potent inducer of cytokine production in monocytes, HDL specifically **inhibits the production of pro-inflammatory cytokines and chemokines** (TNF-alpha, IL-1beta, IL-6, IL-8, CCL3, and CCL4) induced by T-cell contact [359] (Fig. 7.4b). This pathway may be particularly relevant to the attenuation of atherogenesis in chronic inflammatory diseases. Direct interaction of HDL with T-lymphocytes, which can block subsequent activation of monocytes by lymphocytes, can account for this effect [360]. In support of this pathway, apoA-I expression prevents cholesterol-associated T-lymphocyte activation and proliferation in peripheral lymph nodes of double knockout LDL-receptor -/-, apoA-I -/- mice [361]. Such a reduction in autoimmune response can be of therapeutic relevance as it is associated with decreased severity of aortic atherosclerosis [361]. Furthermore, injection of apoA-I reduces inflammation in the double knockout mice by augmenting the effectiveness of the T-regulatory cell response [362].

Besides pro-inflammatory cytokines, the contact-mediated activation of monocytes induces the production of cytokine inhibitors, such as the secreted form of the IL-1 receptor antagonist. Interestingly, production of this protein is not inhibited in the presence of HDL [363]. HDL can also restore the emigratory process of macrophages and monocyte-derived dendritic cells and thus result in resolution of inflammatory reactions in atherosclerotic plaques [356, 364] (Fig. 7.4b). In addition, apoA-I possesses the capacity to induce differentiation of bone marrow-derived progenitor cells *in vitro* [365].

Next, HDL and apoA-I potently **reduce neutrophil activation** *in vitro* [252, 366] (Fig. 7.4b). Importantly, HDL is equally capable of decreasing neutrophil activation *in vivo*; indeed, infusion of rHDL significantly attenuates the activation of circulating neutrophils in patients with peripheral vascular disease [252]. Similarly, rHDL decreases neutrophil infiltration in the arterial wall in a rabbit model of acute arterial inflammation [226] (Fig. 7.4b). An interesting corollary of these findings is the superiority of HDL over other tested agents in reducing neutrophil activation when used for foreign surface pacification in cardiopulmonary bypass circuits [367]. Furthermore, the potent capacity of HDL to reduce mortality and cerebral infarct volume in a rat model of embolic stroke can be underlaid by decreased neutrophil recruitment in the infarct area with the protection of blood-brain barrier integrity [368].

The anti-inflammatory activity of HDL *in vivo* is consistent with elevated levels of CRP in subjects with hypoalphalipoproteinemia [369], with negative correlations between plasma levels of CRP and HDL-C [93,370–372], between plasma levels of pentraxin 3 and HDL-C [373], but also between plasma levels of ICAM-1 and HDL-C, and particularly small, dense HDL3-C [374]. Furthermore, decreased HDL-C levels are an independent determinant of circulating concentrations of TNF-alpha in the general Japanese population, suggesting that elevation of this pro-inflammatory mediator may contribute to increased cardiovascular risk in low HDL-C subjects [375]. The ability of HDL to inhibit endothelial adhesion molecule expression may, however, be independent of HDL-C concentrations, reflecting substantial interindividual variability in humans [31].

Finally, HDL has been shown to bind LPS and to neutralize its pro-inflammatory effects; this biologic activity is, however, typically considered to belong to the anti-infectious properties of HDL (see 7.5 of Chapter 7).

Mechanisms of Anti-Inflammatory Protection

The multiple anti-inflammatory effects of HDL suggest that several mechanisms of action may be operative. Among them, **cellular lipid efflux** appears to form a mechanistic basis for several key pathways involved, including the capacity of HDL to decrease adhesion molecule expression (Fig. 7.4a). Indeed, reduced expression of VCAM-1 observed in atherosclerotic plaques obtained from patients with peripheral vascular disease following intravenous infusion of rHDL is paralleled by decreased plaque lipid content and diminished macrophage size [27]. Concomitantly, the capacity of apoB-depleted plasma to efflux cholesterol together with circulating levels of HDL-C increases after infusion of rHDL.

Direct inhibitory actions of HDL and apoA-I on monocyte and neutrophil activation also appear to involve cellular lipid efflux as a main mechanism (Fig. 7.4b). Rapid depletion of cholesterol from cell membranes probably represents a key pathway accounting for the anti-inflammatory effects of HDL and apoA-I on monocytes [251]. Consistent with these data, inhibition of neutrophil activation by HDL and apoA-I correlates with their capacity to reduce lipid raft content in cell membranes [252]. Furthermore, a reduction in both membrane cholesterol and lipid raft content reduces pro-inflammatory LPS-induced signaling [376]; impaired signaling via Toll-like receptor 4 (TLR4) may be involved in this effect [377,378].

HDL-mediated reduction in myeloid cell proliferation and monocytosis can be equally related to cholesterol efflux from cells. Indeed, HDL promotes cholesterol efflux and decreases Ras membrane association and extracellular signal-regulated kinase (ERK) signaling in response to IL-3 or granulocyte macrophage colony stimulating factor, thereby inhibiting proliferation of marrow myeloid cells [353].

HDL-mediated cholesterol efflux results in concerted changes in the expression of genes involved in the regulation of the inflammatory response. Indeed, the removal of cellular cholesterol down-regulates the inflammatory phenotype of macrophages with consequent attenuation of signaling via TLRs [379]. Macrophage ABCA1 is pivotal for this effect as it dampens inflammatory signaling by reducing TLR4 trafficking to lipid rafts as a result of a reduction of membrane cholesterol content [380]. In contrast, inflammatory gene expression is elevated in peritoneal macrophages deficient in ABCA1 and/or ABCG1 transporters concomitant with cholesterol accumulation. In a similar fashion, ABCG1 deficiency in aortic endothelial cells activates endothelial IL-6 receptor/signal transducer and activator of transcription 3 (STAT3) signaling, thereby enhancing monocyte-endothelial interactions and vascular inflammation [381]. These data highlight the importance of **ABCA1** and **ABCG1** in regulation of the local inflammatory response and atherosclerotic plaque stabilization, expanding the traditional roles of HDL and ABC transporters in cholesterol efflux and RCT to anti-inflammatory and immunosuppressive functions of HDL [26]. Available data suggest that sterol efflux activities of ABCA1 and ABCG1 modulate macrophage expression of inflammatory cytokines and chemokines, as well as lymphocyte proliferative responses. Modulation of sterol levels and lipid organization in cell membranes, primarily in lipid rafts, may constitute the underlying mechanism of these observations [26].

ApoA-I interaction with ABC transporters may also induce anti-inflammatory signaling independent of cholesterol efflux. Thus, apoA-I interaction with ABCA1 induces JAK2 and STAT3 activation, thereby decreasing cytokine expression in macrophages [382]. This pathway may result in mRNA destabilization of pro-inflammtory cytokines in macrophages activated by LPS [383].

OPEN QUESTIONS

Does cholesterol efflux mechanistically underlie the major anti-inflammatory activities of HDL particles? Is apoA-I sufficient for the anti-inflammatory effects of HDL?

The anti-inflammatory action of HDL may also involve the **removal of oxidized lipids** from endothelial cells, macrophages and/or oxidized LDL followed by hydrolytic inactivation under the action of HDL-associated enzymes (PAF-AH, PON1, LCAT; Fig. 7.4, Tables 7.1 and 7.2). Such a pathway mechanistically resembles those involved in the antioxidative activity of HDL [10, 220, 233] (see 7.2 of Chapter 7). The pathophysiologic relevance of oxidized phospholipids in atherosclerosis is reflected by their potent pro-inflammatory activities which can trigger arterial wall inflammation [384]. Inactivation of such pro-inflammatory agents by HDL may be associated with decreased expression of adhesion molecules and decreased macrophage adhesion to endothelial cells [10, 385]. HDL may also remove non-esterified fatty acids from LDL which contribute to the capacity of a minor electronegative LDL^- subpopulation to induce cytokine expression in monocytes [386].

The capacity of HDL to decrease inflammatory stimulation of endothelial cells induced by oxidized LDL-derived oxidized phospholipids can be measured *in vitro* in aortic cell co-culture [348]. A surrogate assay assessing the inhibitory action of HDL on the generation of pro-inflammatory oxidized phospholipids in LDL has been developed [387] that is mechanistically similar to other assays assessing the antioxidative activity of HDL [233, 323]. The relevance of these assays as prognostic markers of cardiovascular risk remains to be established.

MAJOR CONTROVERSY

Does the *in vitro* assay of anti-inflammatory activity of HDL better reflect cardiovascular risk than determination of HDL-C? [388, 389].

HDL may also limit the propagation of the inflammatory response as suggested by attenuated release of MCP-1, soluble L-selectin, soluble ICAM-1 and soluble VCAM-1 from LPS-stimulated human acute monocytic leukemia (THP-1) cells in the presence of apoA-I [352]. This pathway can account for the HDL-mediated inhibition of monocyte chemotaxis.

The inhibiting action of HDL on adhesion molecule expression in endothelial cells involves the modulation of multiple genes related to protection against inflammation. First, HDL inhibits the activation of **NFκB** induced by oxidized LDL and TNF-alpha with a subsequent reduction in inflammatory signaling [390, 391] (Fig. 7.4a). Indeed, rHDL decreases VCAM-1 promoter activity via the NFκB binding site in human endothelial cells activated with TNF-alpha [392, 393]. Remarkably, this effect is dependent on the up-regulation of 3-beta-hydroxysteroid-delta 24-reductase (DHCR24) [393] (Fig. 7.4a). Such intracellular action of rHDL requires interaction with SR-BI on the cellular surface. **DHCR24** is an enzyme involved in steroid metabolism that also contributes to hydrogen peroxide scavenging and possesses anti-apoptotic activity. Increased expression of DHCR24 is also observed *in vivo* following rHDL infusion in a rabbit model of acute vascular inflammation, consistent with the anti-inflammatory role of this enzyme [394]. NFκB-mediated inhibition of intracellular signaling induced by HDL leads to decreased intracellular production of ROS [390, 391]; links between ROS generation and the NFκB-mediated signaling pathway may involve proteasome activation [390].

HDL-induced inhibition of adhesion molecule expression is dependent on **AMPK** activation (Fig. 7.4a). In turn, HDL-induced activation of AMPK depends on both S1P receptors and SR-BI and necessitates calcium/calmodulin-dependent protein kinase [395]. HDL is also able to inhibit TNF-alpha-stimulated activation of sphingosine kinase resulting in reduced cellular production of S1P [391]. The induction of endothelial adhesion molecule expression by intracellular S1P forms a basis for the anti-inflammatory role of this effect. S1P-mediated signaling with subsequent activation of the lysosphingolipid receptor/phosphoinositide 3-kinase (PI3K)/Akt axis may also be involved in the modulation of the expression of long pentraxin 3 in human endothelial cells in the presence of HDL [396].

Transforming growth factor beta may function as another important mediator of the anti-inflammatory activity of HDL. Indeed, expression of this anti-inflammatory protein is up-regulated by HDL in cultured endothelial cells [397]. Modified production of ILs can be equally involved in the effects of HDL on inflammatory cascades. In fact, HDL attenuates IL-6 production in endothelial cells exposed to pro-inflammatory stimuli, such as TNF-alpha or endotoxin [398]. In addition, HDL increases the production of IL-10, an anti-inflammatory cytokine, by monocytes in the basal state but decreases it under pro-inflammatory conditions in the presence of IL-1-beta [399]. Next, HDL blocks oxidized LDL-induced signaling *via* the epidermal growth factor receptor with a subsequent reduction of matrix metalloproteinase-2 up-regulation [400, 401]. The protective effect of HDL along this pathway is related to the capacity of HDL particles to block intracellular ROS generation and to prevent the interaction of oxidized lipids with the receptor [401].

Other pathways that may potentially contribute to the anti-inflammatory actions of HDL *in vivo* include direct interactions of infused HDL with circulating acute-phase proteins. Experiments with rHDL infusions in humans injected with recombinant CRP [28] and in rabbits developing acute vascular inflammation [394] do not, however, favor this possibility.

CLINICAL SUMMARY

Clinical evidence to support the link between the development of atherosclerosis and the capacity of HDL to decrease inflammation is accumulating both *in vitro* and *in vivo*, but remains insufficient.

Role of HDL Components

Proteome. The ability of HDL to inhibit adhesion molecule expression appears to be related to the content of specific protein and lipid components, including apoA-I, apoA-II, apoA-IV and distinct molecular species of (phospho)lipids, such as S1P and sphingosylphosphorylcholine [347, 402, 403] (Table 7.2).

Experiments with infusion of rHDL containing only apoA-I and phospholipid demonstrate that other components are not essential for the capacity of HDL to diminish the inflammatory response *in vivo* [338]. **ApoA-I** appears to majorly contribute to the biologic activity of such rHDL complexes (Table 7.1). Consistent with the central role of cholesterol efflux from lipid rafts for the anti-inflammatory effects of HDL and apoA-I on monocytes and neutrophils, apoA-I reduces lipid raft content and decreases neutrophil activation more efficiently than total HDL [252]. In addition, apoA-I is as effective as HDL in suppressing the activation of monocytes [251]. ApoA-I, therefore, appears to act through the ABCA1 transporter, whereas HDL may act through several receptors, including SR-BI, ABCG1 and ABCA1.

ApoA-I also determines the capacity of HDL to inhibit contact-mediated activation of monocytes by stimulated T-cells [357]. Furthermore, apoA-I inhibits inflammatory functions in peripheral blood mononuclear cells activated by either specific antigens or lectins [357]. The potent anti-inflammatory properties of apoA-I are further supported by reduced inflammatory responses and decreased renal damage in a rat model of ischemia/reperfusion renal injury [404]. In this model, apoA-I appears to protect the kidney from injury by inhibiting inflammatory cytokine release and neutrophil infiltration and activation. Monocyte recruitment into the vascular wall can be also inhibited by apoA-I, as suggested in studies with the apoA-I-mimetic peptide, D-4F [405]. Interestingly, HDL isolated from carriers of apoA-I Milano is more effective than HDL from control subjects in down-regulating TNF-alpha-induced expression of VCAM-1 in endothelial cells [406], suggesting potent anti-inflammatory properties of this variant.

The role of protein components other than apoA-I in the capacity of HDL to modulate inflammatory reactions is poorly investigated. HDL-associated **PAF-AH** has been shown to determine the capacity of HDL in restoring the emigratory process of monocyte-derived dendritic cells, thereby resolving inflammatory reactions in atherosclerotic plaques [356]. **PON1** may ensure hydrolytic inactivation of pro-inflammatory oxidized phospholipids [10, 220, 407], as well as the hydrolysis of homocysteine thiolactone involved in pro-inflammatory processes [180] (Table 7.2). In contrast, **LCAT** has been proposed to mitigate the anti-inflammatory properties of HDL as circulating levels of CRP are positively correlated with LCAT activity in non-diabetic men without cardiovascular disease [371]; generation of pro-inflammatory lysolipids in the course of the LCAT reaction may provide a mechanistic link for such association.

Lipidome. The **(phospho)lipid** component of HDL may also be involved in the anti-inflammatory actions of HDL (Table 7.1). Indeed, the **lysosphingolipid** cargo of HDL has been shown to inhibit aortic expression of MCP-1 [408]. In addition, 1-palmitoyl-2-linoleoyl phosphatidylcholine (PLPC) is capable of inhibiting pro-inflammatory signaling via the NFκB pathway and may thereby contribute to the capacity of rHDL to inhibit inflammation in cultured endothelial cells [347, 391]. In contrast, S1P may endow HDLs with pro-inflammatory activity as suggested by its chemotactic effect toward macrophages [409]; S1P3 receptors are essential for this effect, which may play a key role in atherosclerosis by promoting inflammatory monocyte/macrophage recruitment [409].

Interestingly, the anti-inflammatory activity of rHDL containing different molecular species of phosphatidylcholine is regulated by the fatty acid composition of this lipid [410]. Specifically, rHDL particles containing PLPC and 1-palmitoyl-2-arachidonoyl phosphatidylcholine inhibit VCAM-1 expression in activated endothelial cells by −95% and −70%, respectively, at an apoA-I concentration of 16 μM. At this concentration of apoA-I, POPC rHDL inhibits VCAM-1 expression by only −16% and dipalmitoyl phosphatidylcholine rHDL does not inhibit at it all. These differences cannot be explained by differential binding of rHDL to endothelial cells [410] and appear to be related to phase transition temperatures of phosphatidylcholines [183], consistent with the role of the efflux of pro-inflammatory lipids as an underlying mechanism whose efficiency is modulated by the lipid phase fluidity.

Functional Heterogeneity of HDL

The potential heterogeneity of HDL anti-inflammatory activity remains poorly characterized. **Small, dense, protein-rich HDL3** has been reported to be superior to large, light, lipid-rich HDL2 in terms of its capacity to inhibit VCAM-1 expression in endothelial cells on both an apoA-I and total cholesterol basis [411], indicative of the enhanced intrinsic anti-inflammatory properties of HDL3 (Table 7.3). This finding, consistent with the potent antioxidative activity of small, dense HDL3 particles [323, 328, 329], may reflect their distinct proteome and/or lipidome [18, 205]. Moreover, spherical HDLs are more effective than discoid HDL in this assay, emphasizing the importance of the particle shape for the anti-inflammatory activity [347].

Interestingly, a distinct particle subpopulation present in small, dense HDL exhibits potent immunomodulatory capacities towards neutrophils [412]. Factor H-related proteins quantitatively predominate in these particles [413].

IMPORTANT READING

Navab M, Yu R, Gharavi N, Huang W, Ezra N, Lotfizadeh A, Anantharamaiah GM, Alipour N, Van Lenten BJ, Reddy ST et al. High-density lipoprotein: antioxidant and anti-inflammatory properties. *Curr Atheroscler Rep* 2007; 9: 244–248.

Murphy AJ, Chin-Dusting JP, Sviridov D, Woollard KJ. The anti inflammatory effects of high density lipoproteins. *Curr Med Chem* 2009; 16: 667–675.

Rye K-A, Bursill CA, Lambert G, Tabet F, Barter PJ. The metabolism and anti-atherogenic properties of HDL. *J Lipid Res* 2009; 50: S195–S200.

7.4 CYTOPROTECTIVE ACTIVITY

Apoptosis and necrosis of arterial wall cells, primarily of macrophages, endothelial cells, smooth muscle cells and lymphocytes, may significantly contribute to the development of atherosclerotic plaque [414, 415]. Macrophage apoptosis is induced by cellular accumulation of free cholesterol which typically originates from modified LDL. Furthermore, oxidized sterols formed during LDL oxidation or consumed in the diet are toxic to macrophages and endothelial cells and may have a central role in promoting atherogenesis.

HDL protects both macrophages and endothelial cells from apoptosis [416] (Fig. 7.5). When macrophage apoptosis is induced by loading with free cholesterol or oxidized LDL, addition of HDL to the cell medium potently inhibits apoptotic cell death [253]. In a similar fashion, HDL preserves the viability of macrophages during phagocytosis of apoptotic cells (efferocytosis) [224].

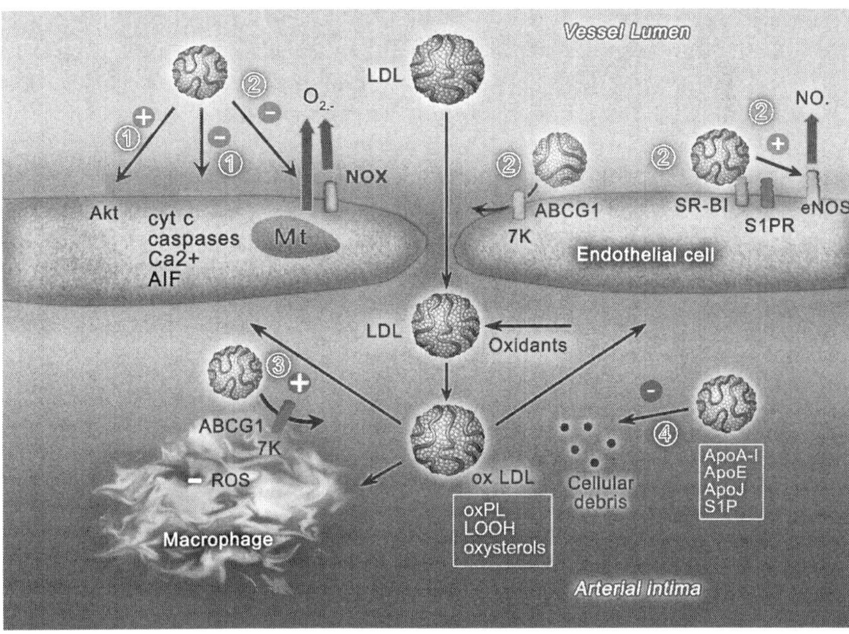

Figure 7.5 Major pathways involved in cytoprotective activity of HDL. (1) HDL protects endothelial cells from apoptosis induced by oxidized LDL (oxLDL), free cholesterol, chylomicron (CM) remnants, TNF-alpha and complement and growth factor deprivation. This action involves the inhibition of: (i) cytochrome c (cyt c) release; (ii) caspase activation; (iii) calcium release; (iv) release of apoptosis-inducing factor (AIF); and, (v) DNA fragmentation. In parallel, Akt is activated. (2) Cellular NO production by eNOS is elevated, whereas superoxide production by NADPH oxidase (NOX) and the mitochondrial (Mt) electron chain is diminished by HDL. HDL-mediated cytoprotection involves the efflux of cellular 7-ketosterol (7K) via ABCG1, and HDL interaction with SR-BI and S1P receptors. (3) HDL exerts an anti-apoptotic action on macrophages that requires 7K efflux via ABCG1 and is accompanied by decreased ROS generation. (4) HDL may also act directly via the scavenging of denatured proteins in the extracellular space, in this way providing clearance of toxic cellular debris. Insets list HDL components potentially involved in the cytoprotective activity of HDL as well as cytotoxic components of oxLDL.

HDL also protects endothelial cells against the toxic effects of several agents, including oxidized LDL [221, 223], chylomicron remnants [417], TNF-alpha [222] and proteins of the complement system [418] (Fig. 7.5). The dose-dependent capacity of HDL to protect cells against LDL-induced cytotoxicity was first demonstrated in 1979 in human aortic smooth muscle cells and umbilical vein endothelial cells [419]. More recent studies have revealed that HDL potently inhibits apoptosis in endothelial cells induced by oxidized LDL [221, 223] or TNF-alpha [222]. HDL also prevents endothelial cell apoptosis elicited by growth factor withdrawal [420]. Together, the cytoprotective properties of HDL towards major types of arterial wall cells may form a basis for its beneficial effect on endothelial function.

The **cytoprotective properties** of HDL are not confined to arterial wall cells and appear to represent a broad physiologic phenomenon. Indeed, HDL improves survival of both human and murine beta-cells in isolated islets, protecting them from basal, as well as IL-1-beta and glucose-induced, apoptosis [421]. Next, apoE-containing HDL synthesized by glial cells protects neurons from apoptosis [422]. In addition, levels of HDL-C and apoA-I in follicular fluid are negative predictors for embryo fragmentation score in women undergoing *in vitro* fertilization, consistent with cytoprotective effects of HDL in this environment [423]. Interestingly, a role for HDL in embryonic development and fertility has been directly demonstrated in animal models; beneficial effects on intra-follicular cholesterol homeostasis that are important for oocyte and embryo development may underlie this finding [46]. Finally, HDL is capable of preventing oxidized LDL-induced cell death of human osteoblastic cells, potentially contributing to protection from osteoporosis [424].

In addition to cytoprotection from toxic agents, HDL is **mitogenic** [425, 426]. Furthermore, HDL can potentiate beneficial effects of growth factors [426]. Such effects may contribute to the net atheroprotective activity of HDL, for example proliferation of smooth muscle cells induced by HDL can strengthen the fibrous cap of the atherosclerotic plaque and improve plaque stability. Depending on the experimental setting however, HDL can also inhibit cell proliferation [427].

The effects of HDL on cell survival are closely related to those on **cell migration**. HDL exhibits differential effects on cell migration depending on the cell type and on the agonist employed. In endothelial cells, both migration and survival are potently stimulated by HDL [428]. Furthermore, HDL may stimulate macrophage emigration from atherosclerotic lesions, acting via cholesterol removal from plasma membrane via ABCA1 and ABCG1 and decreased Rac1 signaling [364]; emigration of dendritic cells can also be induced [356]. Interestingly, HDL also stimulates migration, survival and invasion of prostate cancer cells [429, 430], potentially attesting low HDL-C levels in cancer to a protective response.

Mechanisms of Cytoprotection

Cellular efflux of oxidized cholesterols, primarily of 7-ketosterol, mediated by HDL plays a key role in the HDL-mediated protection of macrophages from apoptosis induced by loading with free cholesterol or oxidized LDL [224, 253] (Fig. 7.5). Interestingly, even when macrophage apoptosis is induced by oxidized phospholipids, HDL-mediated efflux of oxidized sterols, rather than of oxidized phospholipids, is still the major mechanism assuring cytoprotection [224]. This pathway may also contribute to the protection of endothelial cells [431].

Potent inhibition of apoptosis observed upon incubation of HDL with endothelial cells is correlated with **decreased intracellular generation of ROS**, suggesting that cytoprotection can be related to the intracellular antioxidative actions of HDL [221–223, 260] (Fig. 7.5). Reduced cellular generation of superoxide anion and/or hydrogen peroxide may underlie such antioxidative effects of HDL [260]; down-regulation of superoxide production by NADPH oxidase and/or mitochondrial electron transport chain appears to be implicated in this observation. In a similar fashion, HDL dampens the oxidative burst in macrophages following their exposure to oxidized phospholipids and/or apoptotic cells [224]. Intracellular signaling induced by HDL binding to cellular receptors provides a major contribution to such intracellular antioxidative activity of HDL; direct interaction between HDL and toxic oxidized lipids and/or proteins, which allows their inactivation and removal, may also contribute to the effect [10, 432].

Major intracellular mechanisms underlying the anti-apoptotic actions of HDL include the **preservation of mitochondrial integrity** resulting in the inhibition of the release of cytochrome c and of apoptosis-inducing factor from mitochondria followed by the abrogation of caspase cascade activation [260, 416] (Fig. 7.5). In addition, **reduced fragmentation of nuclear DNA** in the presence of HDL blocks cytotoxic calcium signaling pathways triggered by oxidized LDL and other cytotoxic agents [221]. Importantly, HDL efficiently inhibits both caspase-dependent (mediated by calpain/Bid/cytochrome c/caspase-3) and caspase-independent (mediated by apoptosis-inducing factor) apoptotic pathways [260]. The **activation of the survival Akt pathway** in a PI3K-dependent manner represents a key signaling cascade induced by HDL [390, 391] (Fig. 7.5). Indeed, HDL induces activation of Akt, Foxo 3a and ERK 1/2 and decreases pro-apoptotic Bcl-2-like-protein Bim with an increase in the anti-apoptotic myeloid leukemia cell differentiation protein, MCL-1 [433].

CRITICAL CONTRIBUTION

Cellular efflux of toxic oxysterols plays a major role in cytoprotection mediated by HDL [224, 253].

The stimulation of Akt signaling may enhance endothelial nitric oxide (NO) synthesis and promote endothelial cell survival by inhibiting caspases through nitrosylation-based mechanisms [434]. Indeed, the **stimulation of NO synthesis** by HDL, which is central for the vasodilatory effects of the lipoprotein, may significantly contribute to its cytoprotective effects [435] (Fig. 7.5). Regulation of the signaling function of proteins concentrated in caveolae, such as endothelial nitric oxide synthase (eNOS) and SR-BI, may underlie this activity [436, 437]. Such regulation may occur via modulation of cholesterol abundance within caveolae in the presence of HDL [436]. HDL interaction with **SR-BI** is, indeed, essential for the protection of endothelial cells from apoptosis induced by oxidized LDL [260, 438, 439] (Fig. 7.5). In contrast, SR-BI is not critical to the anti-apoptotic activity of HDL towards beta-cells [421].

Another HDL receptor, **ABCG1**, makes a critical contribution to the anti-apoptotic action of HDL in macrophages [224, 253] (Fig. 7.5). As ABCG1 is highly expressed on endothelial cells [440], this pathway can also be relevant for the protection of the endothelium [431]. The role of ABCG1 in such cyto-protection is fully consistent with its major function in promoting cellular efflux of cholesterol and 7-ketocholesterol [431]. In addition, the anti-apoptotic effects of apoA-I on human umbilical vein endothelial cells may require interaction between apoA-I and cell-surface mitochondrial-related F1-ATPase independently of other cellular receptors [441].

Interestingly, mechanisms of HDL-mediated cytoprotection may differ between cell types and toxic agents. Indeed, HDL protects beta-cells from apoptosis by specifically blocking the expression of 4E-binding protein 1 which is up-regulated by serum starvation and by IL-1-beta [442]. Intriguingly, in the same cell type, HDL blocks apoptosis induced by endoplasmic reticulum stress without affecting 4E-binding protein 1, indicative of the existence of multiple protective pathways potentially mediated by distinct HDL components [442].

Mechanisms of accelerated cell growth in the presence of HDL may include elevated supply of lipid nutrients as well as triggering of mitogenic signaling [433, 443]. Indeed, HDL can increase cytosolic concentrations of calcium [444] and activate mitogen-activated protein kinase (MAPK) through a SR-BI/Ras-mediated pathway [445]. F1-ATPase can also participate in the proliferative activities of HDL particles as its activation by apoA-I induces the proliferation of endothelial cells, suggestive of the involvement of extracellular production of ADP [441].

The effects of HDL on endothelial cell migration involve the induction of intra-cellular signaling via **S1P receptors 1 and 3** (S1P1 and S1P3) [428] (Fig. 7.5). On the other hand, S1P2 receptors can mediate inhibitory effects of HDL on the migration of smooth muscle cells [446]. Intracellular mechanisms involved may include modulation of the production of matrix metalloproteases [447] that participate in smooth muscle cell migration [448] by HDL.

OPEN QUESTION

What is the relative role of SR-BI, ABCG1 and S1P receptors in mediating cytoprotection by HDL?

Role of HDL Components

Proteome. Apolipoproteins and minor bioactive lipids represent two major groups of HDL components involved in cytoprotective effects of HDL (Fig. 7.5 and Table 7.1). Indeed, **apoA-I** [221, 260] and **apoE** (as an apoE3/3 isoform) [449] appear to be central for HDL-mediated protection from apoptosis induced by growth factor deprivation or by oxidized LDL. In addition, apoJ may also constitute an important cytoprotective element [432].

The beneficial actions on apoA-I and apoE on cell viability are most probably related to their interactions with cellular receptors, such as SR-BI, ABCG1 and others, with subsequent induction of intracellular survival pathways and the promotion of lipid efflux. In contrast, **apoJ** may act directly via the scavenging of denatured proteins in the extracellular space, in this way providing clearance of toxic cellular debris [432] (Fig. 7.5 and Table 7.1). Indeed, the propensity of apoJ to bind a variety of hydrophobic molecules renders it a major player in the removal of dead cell remnants following apoptosis and necrosis. As a typical example, over-expression of apoJ diminishes TNF-alpha-induced apoptosis of human endothelial cells and restores eNOS expression [450].

Lipidome. HDL-associated **sphingolipids** such as S1P, sphingosylphosphorylcholine and lysosulfatide may significantly contribute to HDL-mediated protection from apoptosis induced by growth factor deprivation [403, 428, 451–453] (Fig. 7.5 and Table 7.1). **S1P** is a potent anti-apoptotic agent that may mediate the anti-apoptotic effect of HDL via increased NO production and the improved survival of endothelial cells [454]. Furthermore, S1P confers survival benefits in cardiomyocytes subjected to oxidative stress, acting through signal activation via both Akt and the mitogen-activated protein kinase kinase (MEK 1/2)-ERK1/2 pathways and stimulation of the phosphorylation of glycogen synthase kinase-3-beta [455]. ERK1/2 activation occurs through an S1P1 receptor-G-protein-dependent pathway, whereas the activation of Akt is mediated by S1P3 receptors. Activation of the S1P2 receptors, tyrosine kinase family Src and transcription factor STAT3 can also be involved in the cytoprotection induced by S1P [456, 457]. Sphingosylphosphorylcholine can also act on S1P3 receptors, resulting in reduced myocardial apoptosis and diminished infarct size in a mouse model of myocardial infarction [458].

Next, HDL contains minor amounts of lipophilic antioxidants, such as tocopherols [321] which may participate in the modulation of redox-dependent cell signaling and gene expression, thereby neutralizing toxic pro-oxidant effects of oxidized LDL [459].

The capacity of the both the protein and lipid HDL component to protect cells from apoptosis is not specific to arterial wall cells. Indeed, apoptosis of beta-cells induced by IL-1-beta is also inhibited in the presence of either de-lipidated protein or de-proteinated lipid moieties of HDL; both apoA-I and S1P appear to contribute to such cytoprotection [421].

The mitogenic effects of HDL can be equally attributed to sphingosylphosphorylcholine and lysosulfatide [460]. In contrast, apoA-I and apoE are apparently not mitogenic per se and rather inhibit the proliferation of smooth muscle cells [427], potentially acting via NO-dependent pathways [461].

The effects of HDL on endothelial cell and smooth muscle cell migration can also be largely mediated by S1P, acting via the S1P receptors [416, 428, 446]. HDL proteins and enzymes are both capable of contributing to this biologic activity via the inactivation and/or removal of chemotactic oxidized lipids. For example, HDL-associated PON1 may degrade oxidized lipids and thereby inhibit ensuing macrophage migration [462], whereas apoA-I-mimetic peptide D-4F binds oxidized lipids and inhibits influenza-induced trafficking of macrophages into the vascular wall [405]. Furthermore, apoJ efficiently decreases the migration of smooth muscle cells through the reduction of TNF-alpha-stimulated expression of adhesion molecules and metalloproteases; this effect may underlie the reduction of neointimal thickening in balloon-injured rat carotid arteries under the action of apoJ [450].

MAJOR CONTROVERSY

Both apoA-I and S1P have been shown to be pivotal for cytoprotective properties of HDL; are their activities independent or synergistic?

Functional Heterogeneity of HDL

Despite the fact that cytoprotective properties of HDL have been known for several decades, their functional heterogeneity remains poorly characterized. Recent studies reveal that **small, dense, protein-rich HDL3** particles fractionated from normolipidemic human plasma by isopycnic density gradient ultracentrifugation potently protect human microvascular endothelial cells against primary apoptosis induced by oxidized LDL [260, 463] (Table 7.3). On a per particle basis, small, dense HDL3c displays twofold superior cytoprotective activity relative to large, light HDL2b. Cytoprotective effects of HDL3 involve reductions of DNA fragmentation, annexin V binding and caspase-3 activation, as well as decreases in cytoplasmic release of cytochrome c and of apoptosis-inducing factor. Equally, all HDL subfractions attenuate intracellular generation of ROS; such antioxidative activity also diminishes from HDL3c to HDL2b and primarily involves the inhibition of cellular superoxide production [260]. Two components with potent cytoprotective properties, S1P and apoA-I, are enriched in small, dense HDL3c [205]. Mechanistic studies suggest that apoA-I, rather than S1P, is pivotal to the potent anti-apoptotic activities of small, dense HDL3 on endothelial cells incubated with oxidized LDL [260].

Finally, HDL3 appears to be more effective than HDL2 in promoting endothelial cell proliferation by decreasing LPS- and TNF alpha-induced expression of a disintegrin and metalloproteinase with thrombospondin motifs 1 (ADAMTS-1) which inhibits endothelial cell proliferation, as well as by enhancing intracellular calcium levels via a pertussis toxin-sensitive G protein-mediated pathway involving phospholipase C [444, 464].

IMPORTANT READING

Negre-Salvayre A, Dousset N, Ferretti G, Bacchetti T, Curatola G, Salvayre R. Antioxidant and cytoprotective properties of high-density lipoproteins in vascular cells. *Free Radic Biol Med* 2006; 41: 1031–1040.

Nofer J-R, Assmann G. Atheroprotective effects of high-density lipoprotein-associated lysosphingolipids. *Trends Cardiovasc Med* 2005; 15: 265–271.

7.5 ANTI-INFECTIOUS ACTIVITY

Infection is thought to play a role in the progression and instability of atherosclerotic plaques [465]. Inflammatory response, endothelial dysfunction and thrombotic activation appear to mediate the association between infection and atherosclerosis. LPS, a key functional component of the outer membrane of Gram-negative bacteria, is an important factor in determining their pathogenicity. LPS is released from bacteria on multiplication or lysis, acts as endotoxin and elicits a strong immune response.

Innate immunity is the first line of defence against invading microorganisms. Plasma HDL displays several key anti-infectious activities and thereby appears to form part of the innate immune system. First, HDL plays a major role in the **binding of circulating LPS** and its hepatic clearance to bile and is thereby capable of inhibiting LPS-induced cellular activation [466–470] (Fig. 7.6). As a result, HDL provides protection against death owing to endotoxic shock in animal models [471, 472] and attenuates organ damage by LPS, as exemplified by reduced acute lung injury in a mouse model following HDL administration [473]. It remains to be shown whether HDL infusion can protect from septic shock in humans [469]. Consistent with these data, circulating levels of LPS are negatively associated with serum concentrations of HDL-C in the multi-ethnic UK Wandsworth Heart and Stroke Study [474] and in coronary heart disease patients [475]. Interestingly, HDL is also capable of binding other functional components of pathogenic bacteria, such as streptococcal collagen-like protein 1, a virulence factor on the surface of the M41-type group A Streptococcus [476]; the physiologic role of such interaction remains to be established.

Next, human plasma HDL is a major carrier of specific **trypanosome lytic activity**, which selectively protects humans from *Trypanosome brucei brucei*, a parasitic species that causes sleeping sickness [477] (Fig. 7.6). Although *T. b. brucei* fatally infects livestock in sub-Saharan Africa, humans are innately resistant to this infection as a result of the capacity of human plasma to lyse this unicellular protozoan parasite. Trypanosome lytic factor constitutes a minor

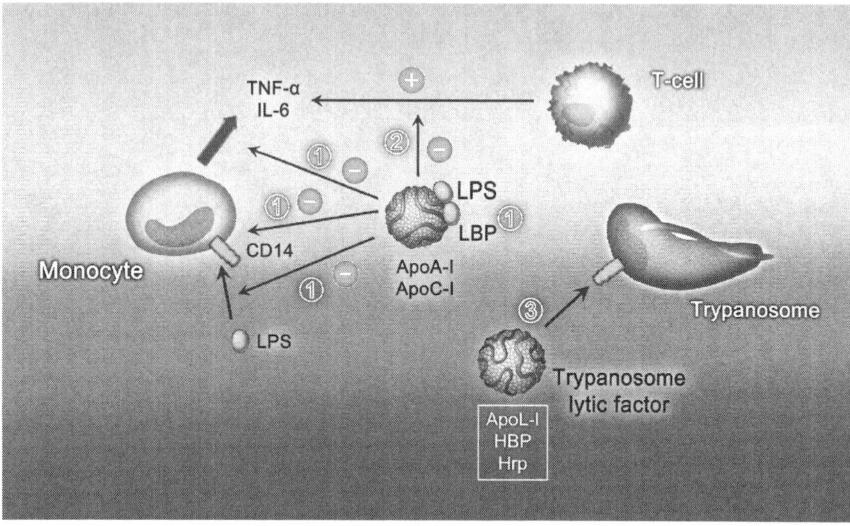

Figure 7.6 Major pathways involved in anti-infectious activity of HDL. (1) HDL binds circulating LPS and provides its clearance to the bile, thereby inhibiting LPS-induced monocyte activation with reduced expression of CD14 and decreased TNF-alpha and IL-6 production; HDL-associated LBP is involved in LPS binding. (2) HDL inhibits T-cell-mediated activation of monocytes. (3) HDL is a major carrier of specific trypanosome lytic activity. Trypanosome lytic factor constitutes a minor subpopulation of HDL particles that is characterized by the ability to kill extracellular protozoon parasites of the genus *Trypanosoma*. Inset lists HDL components which form trypanosome lytic factor.

subpopulation of HDL particles that is characterized by the ability to kill *T. b. brucei* [478]. Iy contrast, two other subspecies of the genus, *T. b. gambiense* and *T. b. rhodesiense*, are resistant to the lysis and can infect humans. In addition, trypanosome lytic factor can ameliorate infection by the intracellular pathogen *Leishmania*, a genus of trypanosome protozoa that replicates in phagolysosomes of macrophages [479, 480]. Trypanosome lytic factor may, therefore, have a broad function as a component of innate immunity that can limit infections by its ability to selectively damage pathogens in phagolysosomes within the reticulo-endothelial system (see below).

HDL may also contribute to protection from leprosy, enhancing the innate immune response to *Mycobacterium leprae*, the causative agent of leprosy [481]. Finally, it is interesting to mention that the antibacterial activity of HDL is not specific to humans and is also observed in other species. For example, HDL and apoA-I isolated from teleostean fishes display potent *in vitro* activity against Gram-positive and Gram-negative bacteria [482].

OPEN QUESTION

Can HDL infusion protect against bacterial and viral infection in man?

Mechanisms of Protection from Infection

The inactivation of LPS by HDL in humans involves a **reduced pro-inflammatory response** *in vivo* with diminished release of TNF-alpha, IL-6 and IL-8, attenuated changes in leukocyte counts and enhanced expression of CD11b/CD18 on granulocytes [466, 483] (Fig. 7.6). Reduced monocyte expression of **CD14**, the major LPS receptor, represents an underlying element of such antibacterial actions of HDL; this event may diminish TNF-alpha production in monocytes upon stimulation with LPS [466]. In support of the importance of this pathway, rHDL potently decreases both LPS-induced CD14 expression and TNF-alpha production in monocytes obtained from patients with cirrhosis and from control subjects [484]. In addition, HDL mediates an enhancement of integrin-mediated cell adhesion of neutrophils in response to LPS [412]; this response requires the presence of CD14 on the neutrophil surface.

HDL action on LPS metabolism may involve macrophagic **ABCA1** with the removal of cell-associated LPS and restoration of normal macrophage responsiveness [485]. **SR-BI** may also be involved in the effects of HDL on LPS, acting via hepatic LPS clearance and glucocorticoid production [486].

Anti-infectious activities of HDL involve **anti-inflammatory pathways** as a major component. Indeed, circulating levels of HDL-associated LPS are negatively correlated with clinical parameters and plasma cytokine concentrations as observed in patients with periodontitis [487]. Consistent with these data, HDL-mediated tissue protection from LPS-induced injury involves the inhibition of oxidative damage and of pro-inflammatory cytokine response by HDL, resulting in decreases in hemorrhage, intra-alveolar edema and neutrophil infiltration in lungs [473].

Similar pathways may underlie HDL-mediated protection from leprosy which is characterized by the accumulation of host-derived oxidized phospholipids in macrophages infected with *M. leprae* [481]. Such accumulation of oxidized phospholipids in lepra lesions is similar to that observed in atherosclerotic plaques and may result in the inhibition of the innate immune response. Interestingly, HDL is capable of reversing such inhibition induced by mycobacterial infection and host-derived oxidized phospholipids, suggesting a common mechanism through which lipid metabolism and innate immunity may beneficially impact the pathogenesis of both atherosclerosis and microbial infection [481].

CRITICAL CONTRIBUTION

Identification of trypanosome lytic factor of human plasma as a minor subpopulation of HDL particles with a buoyant density of 1.21–1.24 g/ml, molecular mass of 490 kDa and particle diameter of 15–21 nm [477, 488, 489], and characterization of the mechanisms of its toxicity to trypanosomes via ionic pore formation in endosomal membranes of the parasite [490].

HDL-mediated killing of *T. b. brucei* occurs through a unique mechanism that is unrelated to that of LPS inactivation and involves ionic pore formation in endosomal membranes of the parasite [490, 491]. Mechanistically, killing is initiated in the lysosome of the target trypanosome where the acidic pH facilitates a membrane-disrupting activity by trypanosome lytic factor. An osmotic effect linked to this ionic flux leads to uncontrolled swelling of the lysosome, ultimately causing the death of the parasite. Trypanosome lytic factor also efficiently binds and permeabilizes unilamellar liposomes at lysosomal pH [492]. **Lysosomal membrane damage in trypanosomes** is initiated by the stable association of the trypanosome lytic factor with the membrane. In a similar manner, trypanosome lytic factor directly damages the *Leishmania* parasite in the acidic parasitophorous vacuole [480].

HDL also possesses nonspecific, broad-spectrum antiviral properties, being active against both DNA and RNA viruses [493]. The protective action of HDL toward viral infections involves attachment to target cells and the inhibition of hemagglutination [494, 495].

Role of HDL Components

Proteome. The inactivation of LPS by HDL is mediated through direct interaction with apoA-I [496] (Fig. 7.6 and Table 7.1). Indeed, **apoA-I** tightly binds LPS and neutralizes its toxicity; the C-terminal domain of apoA-I is responsible for these effects [497]. As a consequence, apoA-I efficiently decreases LPS-induced mortality in a mouse model, diminishes LPS-induced acute organ injury and attenuates cellular LPS toxicity *in vitro* [496, 498]. These effects are mediated by the protective effect of apoA-I against LPS-induced systemic inflammation as documented by the attenuation in LPS-induced production of TNF-alpha, IL-6 and IL-1-beta, and diminished expression of CD14 on monocytes [498]. Furthermore, apoA-I can augment phagocytic function of macrophages, acting via stabilization of the ABCA7 transporter which is critical for phagocytosis [499]. Finally, apoA-I may contribute to the antiviral properties of HDL particles as it inhibits cell fusion induced by herpes simplex virus and blocks penetration of the virus into cells [500]; amphipathic helices of apoA-I appear to account for these activities. In contrast, **apoA-II** may promote, rather than inhibit, the proinflammatory monocyte response to LPS, acting via suppression of the inhibitory activity of high concentrations of LBP which plays an important role in regulating leukocyte responses to LPS [501]. Such dual action of the major HDL components may contribute to finely tuning host responses to LPS.

LBP is another HDL-associated protein involved in the anti-infectious activities of HDL [502] (Table 7.1). LBP transfers LPS to soluble CD14, thereby enhancing neutrophil responses to LPS [471]. Furthermore, the interaction of LBP and LPS enables prompt binding and neutralization of LPS by HDL as LBP can also transfer LPS to HDL [503]. In a similar fashion, LPS can be bound and transferred by **PLTP**, which is, however, unable to transfer LPS to CD14 and thus cannot mediate cellular response to LPS [504].

ApoC-I represents another protein component of HDL that modulates the innate immune response. Indeed, high plasma levels of apoC-I are associated with reduced risk of mortality from infection independently of HDL-C concentrations in 85-year-old participants of the prospective, population-based Leiden 85-Plus Study [505]. Consistent with this observation, high levels of apoC-I protect mice against mortality caused by bacterial sepsis [506]. The molecular mechanisms underlying this association may involve the enhancement of LPS-induced inflammatory response in macrophages by apoC-I [507].

ApoL-I, a minor HDL apolipoprotein, is a major functional component of the trypanosome lytic factor (Table 7.1). Consistent with the key role of apoL-I, lack of the protein because of frameshift mutations in both APOL1 alleles results in trypanosome infection in humans [508]. In addition to apoL-I, the complex contains **hemoglobin-binding protein, Hrp** and apoA-I which act cooperatively to achieve the maximal killing capacity of the pathogen [479, 509] (Fig. 7.6 and Table 7.1). The potent cytotoxic properties of apoL-I are highlighted by its ability to induce autophagic cell death in a variety of cells; apoL-I is further classified as a Bcl-2 homology domain 3 (BH3)-only pro-death protein [510]. In human endothelial cells, apoL-I expression can be induced by interferon-gamma and TNF-alpha, suggesting that the apolipoprotein may play a role in cytokine-induced inflammatory response [511]. Interestingly, two sequence variants of APOL1 common in subjects of African origin, but absent from those of European origin, potently lyse *T. b. rhodesiense in vitro* and are associated with end-stage kidney disease in African Americans, suggesting positive selection [512, 513].

Two other components, hemoglobin-binding protein and Hrp, confer the trypanosome lytic factor with the capacity to specifically bind to the pathogen [514]; apoL-III may also play a role [515]. *T. b. brucei* possesses a specific surface receptor for the haptoglobin-hemoglobin complex as a way to capture heme into hemoproteins that contribute to parasite viability. As this receptor does not discriminate between hemoglobin-binding protein and Hrp, Hrp-containing HDL particles of human serum are efficiently taken up by the parasite, leading to the simultaneous internalization of apoL-I, Hrp and hemoglobin-binding protein-derived heme [516]. Once in the lysosome, apoL-I is targeted to the lysosomal membrane where its colicin-like anionic pore-forming activity triggers an influx of chloride ions from the cytoplasm and subsequent cellular death [514]. In addition to apoL-I, Hpr and apoA-I are also capable of efficiently permeabilizing lipid bilayers at low pH [492]. Trypanosome lytic factor, apoL-I and apoA-I exhibit specificity for anionic membranes, whereas Hrp permeabilizes both anionic and zwitter-ionic membranes.

According to this mechanism, selective resistance of the *T. brucei* subspecies *rhodesiense* and *gambiense* to this lytic activity reflects two distinct pathways. In the case of *T. b. rhodesiense*, resistance to lysis involves the interaction of the serum resistance-associated protein, a truncated version of the major and variable surface antigen of the parasite, with the C-terminal helix of apoL-I, which inhibits the pore-forming activity of the latter [491]. Presumably as a result of its truncated nature, the serum resistance-associated protein is not targeted to the plasma membrane as are regular variant surface glycoproteins, but ends up in the late endosomal compartment. In this location, the protein is thought to neutralize apoL-I through coiled-coil interactions between alpha-helices [514]. On the other hand, the resistance to lysis by apoL-I of *T. b. gambiense*, which lacks the serum resistance-associated protein gene, is caused by both mutations and reduced expression of the surface receptor for the haptoglobin-hemoglobin complex [517].

Intriguingly, not all components of normal plasma HDL beneficially contribute to protection from infection. Specifically, **PLTP** is essential for the association of lipid A, the membrane anchor of LPS, with lipoproteins, leading to the extension of its plasma residence time and to the magnification of its pro-inflammatory properties [518].

Lipids. Specific HDL lipids may also endow HDL with anti-infectious properties; indeed, rHDL containing ganglioside monosialotetrahexosylganglioside (**GM1**), a component of cholera toxin receptor complex, can potently bind the toxin and divert it away from healthy mammalian cells [519]. It is relevant, in this regard, that plasma HDL contains low levels of several biologically active gangliosides [520–522].

IMPORTANT READING

Pays E, Vanhollebeke B, Vanhamme L, Paturiaux-Hanocq F, Nolan DP, Perez-Morga D. The trypanolytic factor of human serum. *Nat Rev Microbiol* 2006; 4: 477–486.

Gordon SM, Hofmann S, Askew DS, Davidson WS. High density lipoprotein: it's not just about lipid transport anymore. *Trends Endocrinol Metab* 2011; 22: 9–15.

Functional Heterogeneity of HDL

Two key anti-infectious activities, the trypanosome-lytic and LPS-binding activities of human plasma, are associated with specific subpopulations of HDL (Table 7.3). The trypanosome-lytic factor is isolated from plasma as a minor subpopulation of dense HDL particles with molecular mass of 490 kDa, particle diameter of 15–21 nm and buoyant density of 1.21–1.24 g/ml [477, 489]. In addition to apoL-I, hemoglobin-binding protein, Hrp and apoA-I, the complex appears to contain apoA-II, apoC-I, apoC-II and apoC-III [477].

Another minor subpopulation of HDL accounts for the capacity of plasma to enhance integrin-mediated cell adhesion of neutrophils in response to LPS mediated by CD14. This HDL subspecies is small and dense, exhibiting a density of 1.219–1.264 g/ml and a molecular mass of 200 kDa [412]. Factor H-related proteins predominate in these particles which also contain apoA-I and LBP;

moreover, all plasma factor H-related protein-1 appears to be associated with this subset of HDL [412].

7.6 VASODILATORY ACTIVITY

Endothelial function is largely determined by endothelial production and the bioavailability of NO. ROS can inactivate NO; increased production of ROS in the vascular wall under such pro-atherogenic conditions as atherogenic dyslipidemia, hypercholesterolemia, Type 2 diabetes and hypertension can, therefore, decrease NO bioavailability and impair endothelial function [523, 524]. The **endothelial dysfunction** that results plays a pivotal role in the development, progression and clinical complications of atherosclerosis and contributes to an increased risk of cardiovascular disease. Major mechanisms underlying endothelial dysfunction include the pro-inflammatory activation of endothelial cells, diminished function of endothelial progenitor cells and endothelial cell apoptosis [523, 524].

HDL particles display potent **vasodilatory activity**, which reflects their capacity to **stimulate NO release** by endothelial cells [414, 525, 526] (Fig. 7.7). HDL can also **stimulate the production of prostacyclin**, which possesses potent vasorelaxing activity [527, 528]. The vasodilatory activity of HDL can be detected as its capacity to induce endothelium-dependent vasodilation in isolated aortic rings, to improve heart rhythm and to correct arrhythmia [529, 530]. Indeed, rHDL shortens duration of ventricular tachycardia or ventricular fibrillation after reperfusion in a rat model [529].

HDL and apoA-I can act beneficially, not only on endothelium, but also on **endothelial progenitor cells**. Indeed, HDL stimulates early progenitor cell-mediated endothelial repair in a mouse model [530]. Moreover, human apoA-I transfer increases the number of circulating endothelial progenitor cells, promotes endothelial regeneration and attenuates neointima formation in a murine model of transplant arteriosclerosis; stimulation of NO production underlies these effects [531]. The beneficial action of HDL on the endothelium probably underlies the attenuation of vein graft atherosclerosis in apoE-null mice by topical administration of HDL-containing gel (50 μl of 200 μg/ml formulation) on the adventitial side of vein grafts [532]. Increased incorporation of circulating progenitor cells in the endothelium, enhanced endothelial regeneration and reduced intimal inflammation are associated with the anti-atherosclerotic action of HDL which acts in an NO-dependent fashion [532, 533].

Observational data are consistent with the vasodilatory function of HDL. Circulating HDL-C concentrations are directly correlated with measures of endothelial function [31, 525, 534] and inversely correlated with arterial stiffness [535]. Furthermore, epicardial endothelial function is correlated with circulating levels of HDL particles, suggesting local impact of HDL on the coronary circulation [536]. Finally, HDL-C concentrations are positively associated with aortic pulse wave velocity and left ventricular diastolic function in subjects without and with metabolic syndrome [537].

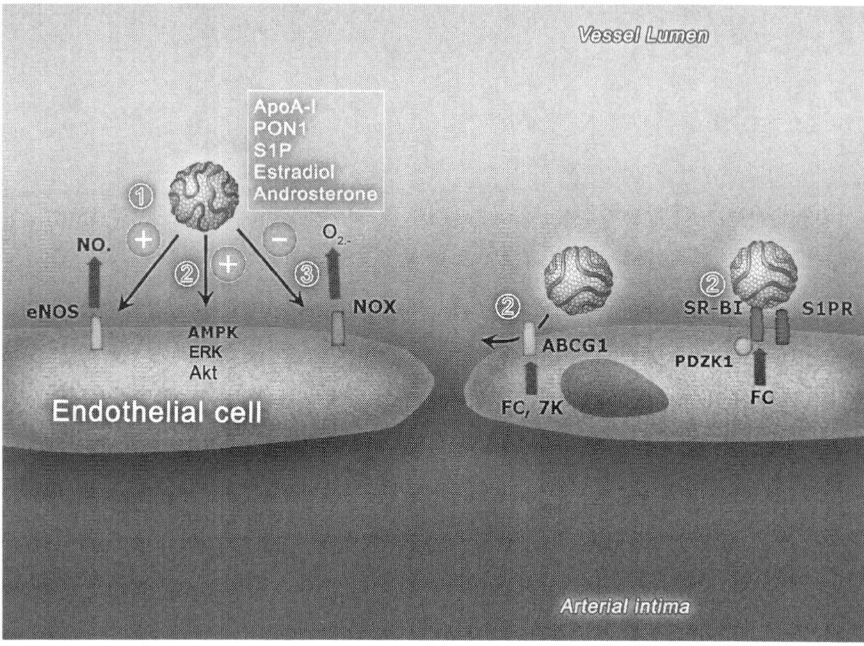

Figure 7.7 Major pathways involved in the vasodilatory activity of HDL. (1) HDL enhances NO production by eNOS in endothelial cells. (2) Such vasodilatory activity involves HDL interaction with SR-BI and S1P receptors followed by such intracellular events as the interaction of PDZK1 with SR-BI, the activation of Akt, AMPK and ERK, and the efflux of free cholesterol (FC) and 7-ketosterol (7K) via ABCG1; in addition, FC can be effluxed via SR-BI. (3) In contrast, cellular superoxide production by NADPH oxidase (NOX) is inhibited in the presence of HDL. Inset lists HDL components potentially involved in vasodilatory activity of HDL.

CRITICAL CONTRIBUTION

HDL particles display potent vasodilatory activity that reflects enhanced NO bioavailability as a result of both the stimulation of NO release by endothelial cells and the reduction in the production of ROS capable of inactivating NO [525, 526].

Mechanisms of Vasodilatory Activity

The activation of NO production involves HDL **binding to SR-BI** as an initiating event (Fig. 7.7). SR-BI is important for HDL action on both vascular endothelium and endothelial progenitor cells [433, 533, 538]. Enhanced NO production is a result of the activation of intracellular signaling cascades that require the binding of membrane cholesterol to SR-BI [433]; such cholesterol-sensing is mediated by the C-terminal transmembrane domains of this receptor. Another essential early step in the process is **cholesterol efflux** to HDL bound to the extracellular loop of SR-BI [31, 539] (Fig. 7.7). Subsequently, SR-BI initiates signaling in the endothelium through the interaction of the C-terminal PDZ-interacting domain of SR-BI with the adaptor PDZ domain-containing protein **PDZK1** [539] (Fig. 7.7). In contrast, endothelial cell responses to other stimuli, including vascular endothelial growth factor, are PDZK1-independent [540]. Src interaction with SR-BI represents another important element of the signaling cascades which is also independent of PDZK1 [540]. Endothelial expression and membrane translocation of the cationic amino acid transporter-1 with subsequent L-arginine uptake are also critical for this pathway [541].

Subsequent intracellular events culminating in the activation of NO production are mediated by endothelial **Akt** and involve intracellular calcium mobilization, an increase in intracellular ceramide levels, eNOS-Ser1177-phosphorylation and eNOS-Thr495-dephosphorylation [525, 526, 542, 543].

HDL-induced activation of eNOS and Akt is dependent on **AMPK** activation which, in turn, depends on both SR-BI and **S1P receptors** (Fig. 7.7). Such signaling involves calcium/calmodulin-dependent protein kinase kinase and, in the case of SR-BI, serine-threonine kinase LKB1 [395]. AMPK thereby functions as a regulator of vascular reactivity via the preservation of endothelial cell function.

Another pathway involved in maintaining eNOS activity and endothelium-dependent vasorelaxation in endothelial cells by HDL is mediated by **ABCG1** and involves cholesterol efflux (Fig. 7.7). As observed in cholesterol-fed mice, HDL-mediated **efflux of cholesterol and 7-oxysterols** via ABCG1 improves the formation of active eNOS dimers and results in decreased ROS production, which can account for the beneficial effects on the endothelium [431]. Endothelial 7-ketocholesterol efflux prevents ROS production in an ABCG1-dependent manner, thus preserving eNOS activity and NO bioavailability and improving vascular reactivity [544]. Such HDL-mediated decrease in ROS production can be observed in atherosclerotic mice at a systemic level as a reduction in circulating levels of 8-isoprostanes [531]. In support of these data, vascular deficiency of ABCG1 impairs endothelial NO synthase-mediated relaxation and accelerates atherosclerosis in LDL receptor-deficient mice [545].

The induction of NO production mediated by ABCG1 may involve down-regulation of the inhibitory influence of caveolin-1 on eNOS [546]. In addition, an **Akt/ERK** pathway can be involved, potentially contributing to the suppression of reperfusion-induced arrhythmias by rHDL [529] (Fig. 7.7). ABCG1 expression can be up-regulated by AMPK activation independently of LXR alpha through a post-transcriptional mechanism that increases mRNA stability.

The increase in NO bioavailability and improved vasodilation under the action of HDL may also directly reflect **diminished cellular production of superoxide**, which inactivates NO (Fig. 7.7). NADPH oxidase is a major vascular source of superoxide. Remarkably, HDL reduces NADPH oxidase activity and expression of its major subunits in endothelial cells; this pathway is associated with beneficial actions of HDL on the renin-angiotensin system, such as decreased responsiveness to angiotensin II [547].

CLINICAL SUMMARY

Vasodilatory activity of HDL may translate into improved endothelial function; clinical evidence to support this hypothesis is accumulating.

Role of HDL Components

Proteins. **ApoA-I** appears to make a major contribution to vasodilatory effects of HDL on the endothelium (Table 7.1). Mechanisms of such actions involve the interaction of apoA-I with endothelial SR-BI and the activation of intracellular signaling cascades [414]. Similarly, apoA-I is a major HDL component underlying the beneficial effects of HDL on endothelial progenitor cells [531]. Indeed, expression of both human apoA-I and apoA-I Milano increases the number of endothelial progenitor cells and improves their function, assessed as incorporation into, and endothelial regeneration in, allografts, thereby inhibiting progression of atherosclerosis [548]. In contrast, lipid-free apoA-I alone is typically incapable of expressing vasoprotective activity [529], consistent with the requirement for its lipidation in the circulation with the formation of mature HDL particles able to interact with SR-BI. Interestingly, HDL isolated from carriers of apoA-I Milano is more effective than HDL from control subjects in stimulating eNOS expression and activation in endothelial cells [406], a finding which may contribute to the potent atheroprotective properties of apoA-I Milano [42].

Another protein component of HDL potentially implicated in vasodilation is represented by **PON1** (Table 7.2). Interestingly, inhibition of PON1 activity prevents the capacity of HDL to stimulate Akt/eNOS phosphorylation, endothelial NO production and NO-dependent vasodilation [181].

IMPORTANT READING

Mineo C, Deguchi H, Griffin JH, Shaul PW. Endothelial and antithrombotic actions of HDL. *Circ Res* 2006; 98: 1352–1364.

Lipids. Among HDL-associated lipids, **S1P** is a key molecule that may mediate the effects of HDL on the endothelium (Table 7.1). S1P interacts with S1P receptors, activating intracellular signaling cascades including the small G-protein Rac, Src kinase, PI3K, Akt and MAPK [414]. In particular, the S1P3 subclass of S1P receptors may mediate vasoactive effects of HDL, such as eNOS activation [525, 549]. The S1P2 and S1P3 subclasses are involved in the stimulating effects of S1P on endothelial cell proliferation and tube formation mediated by Akt/ERK/NO [550]. S1P2 receptors can also be activated by S1P in smooth muscle cells, resulting in suppressed cellular growth in arteries; HDL-associated

S1P can thereby regulate neointimal development [551]. In addition, S1P can be involved in the stimulation of prostacyclin production by HDL, acting via up-regulation of cyclooxygenase-2 in smooth muscle cells mediated by the S1P2 and S1P3 receptors [552]. On the other hand, S1P does not contribute to cholesterol efflux and cannot, therefore, account for the effects of HDL on the endothelium, which are mediated by such efflux [550].

Other HDL-carried lipids capable of regulating endothelial relaxation include estradiol and androsterone (Table 7.1). HDL-associated **estradiol** stimulates endothelial relaxation through the induction of eNOS [553], an effect which can be linked to accelerated cholesterol efflux [554]. **Dehydroepiandrosterone**, a potent vascular relaxant, is present in HDL in the form of fatty acyl esters, its naturally-occurring lipophilic derivatives. Such dehydroepiandrosterone fatty acyl esters can induce endothelial eNOS and thereby lead to vascular relaxation [555].

Functional Heterogeneity of HDL

It remains largely indeterminate as to whether vasodilatory effects of HDL are specifically associated with individual HDL subpopulations. As induction of NO production involves HDL interaction with S1P receptors [525, 526], small, dense HDL3, which is enriched in S1P, may be critical for HDL-dependent vasodilation [205, 556]. Further along this line, HDL3 stimulates the secretion of prostacyclin by endothelial cells more potently compared with HDL2 [557]. In contrast, large, light HDL2, which represents a preferential ligand for ABCG1, may provide a key contribution in vasodilation mediated by cholesterol efflux through this receptor [431].

> **OPEN QUESTION**
>
> Do specific HDL subpopulation(s) in the total pool of HDL particles predominate in exerting vasodilatory properties?

7.7 ANTITHROMBOTIC ACTIVITY

Platelets are involved in the initiation, progression and rupture of atherosclerotic lesions. Arterial thrombus formation is determined by the balance between pro-thrombotic mediators, such as tissue factor (TF) and PAI-1, and antithrombotic factors, such as tissue factor pathway inhibitor (TFPI) and tissue plasminogen activator (tPA) [558]. Furthermore, platelets can modify the properties of endothelial cells so as to facilitate arterial penetration of both monocytes and lymphocytes [559]. In addition, platelets facilitate the conversion of arterial wall cells into foam cells via endothelial activation, the secretion of growth factors and chemokines, and through the recruitement of inflammatory cells [560]. Finally, platelet aggregation represents the key element in the formation of a platelet- and fibrin-rich thrombus formed following the rupture of the atherosclerotic plaque.

HDL exerts multiple **antithrombotic effects** and may thereby beneficially impact the progression of atherothrombotic disease [560] (Fig. 7.8). First, the capacity of HDL to reduce platelet activity *ex vivo* has been directly demonstrated in interventional trials. Indeed, infusion of rHDL (CSL-111; 20 mg/kg/h) significantly **attenuated the *ex vivo* platelet aggregation** response to multiple agonists in individuals with Type 2 diabetes characterized by enhanced platelet aggregation capacity relative to healthy controls; this effect is paralleled by a transient increase in plasma HDL-C levels [29]. The inhibitory action of rHDL on platelet function is time- and dose-dependent and results in the attenuation of hypercoagulant activity and a reduction in thrombus formation under flow. In a similar manner, infusion of rHDL (25–40 mg/kg) transiently inhibits *ex vivo* platelet aggregation induced by arachidonic acid and collagen in healthy subjects [561]. These data are supported by the observation in a population of normo- and dyslipidemic subjects that plasma levels of HDL-C are negative independent predictors of *ex vivo* platelet thrombus formation induced by venous blood in a porcine aorta model [562].

Consistent with the antithrombotic effects of HDL, plasma levels of HDL-C are inversely related to the incidence of venous thrombo-embolism in men [563] and in post-menopausal women [564]; patients with venous thrombo-embolism display reduced levels of HDL-C compared with controls [565, 566]. Furthermore, a meta-analysis of 21 case-control and cohort studies with a total of more than 63,000 patients reveals that HDL-C levels are significantly decreased in patients with venous thrombosis [567]. In contrast, no association between the risk of venous thrombo-embolism and baseline HDL-C concentrations was found in the LITE study [568].

CRITICAL CONTRIBUTION

Infusion of rHDL decreases *ex vivo* platelet aggregation response to multiple agonists in individuals with Type 2 diabetes characterized by enhanced platelet aggregation capacity [29], as well as in healthy subjects [561].

The *in vitro* antithrombotic activities of HDL are expressed as dose-dependent inhibitory actions on agonist-stimulated platelet aggregation, fibrinogen binding, granule secretion and thromboxane A2 and 12-hydroxyeicosatetraenoic acid production [7, 561, 569, 570] (Fig. 7.8). HDL decreases platelet aggregation mediated by glycoprotein IIb/IIIa in response to thrombin, collagen, ADP and adrenaline [569, 571–574] (Fig. 7.8). In addition, HDL **inhibits factors that promote blood coagulation**, including TF and factors X, Va and VIIIa [7, 570] (Fig. 7.8). Data on the antithrombotic activities of HDL are, however, not fully consistent as no effect of the lipoprotein on platelet aggregation and related events was observed

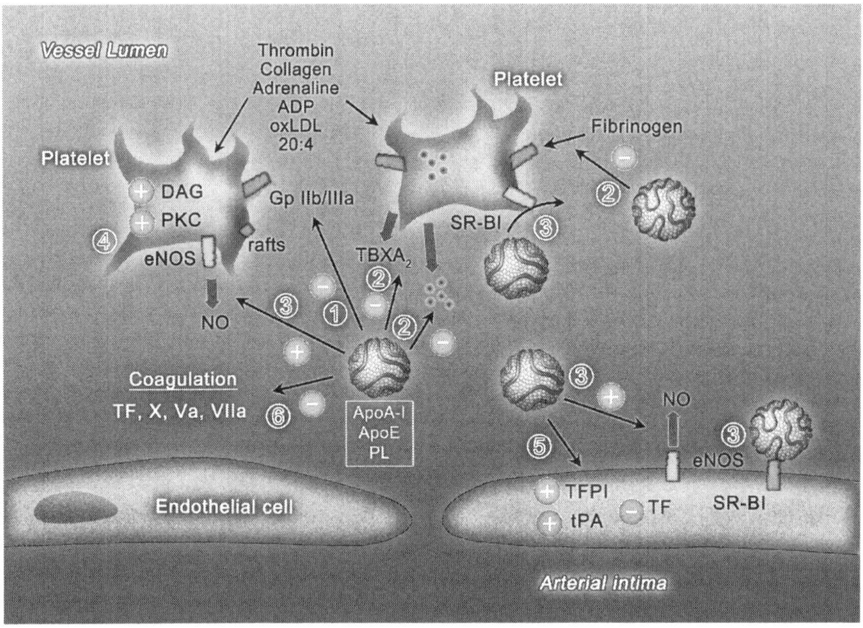

Figure 7.8 Major pathways involved in antithrombotic activity of HDL. (1) HDL effects on platelets primarily involve decreased aggregation response to multiple agonists mediated by glycoprotein (gp) IIb/IIIa; such agonists include thrombin, collagen, adrenaline, ADP, oxidized LDL (oxLDL) and arachidonic acid (20:4). (2) HDL exerts inhibitory actions on fibrinogen binding, granule secretion and thromboxane A2 (TBXA2) and 12-hydroxyeicosatetraenoic acid production. (3) Increased production of NO may form a basis for the inhibitory action of HDL on platelet aggregation, as well as for antithrombotic effects of HDL towards endothelial cells. In both platelets and endothelial cells, the generation of NO can be triggered by the interaction of HDL with SR-BI, which may also ensure cellular cholesterol efflux and diminished lipid raft assembly. (4) Interaction of HDL with the platelet SR-BI receptor may trigger intracellular signaling cascades, including release of diacylglycerol (DAG) and activation of protein kinase C (PKC). (5) In endothelial cells, HDL impairs thrombin-induced TF expression and activity, increases TFPI expression and enhances tPA release. (6) HDL inhibits factors that promote blood coagulation, including TF and factors X, Va and VIIIa. Inset lists HDL components potentially involved in antithrombotic activity of HDL. PL, phospholipid.

in some studies [575]; differences in experimental system (type and concentration of agonist, incubation time, baseline platelet phenotype, etc.) are probably responsible for such discrepancies.

In addition to its effects on platelets, HDL exerts antithrombotic effects on endothelial cells. Specifically, HDL impairs thrombin-induced TF expression and activity, increases TFPI expression and enhances tPA release, but does not affect PAI-1 expression [576] (Fig. 7.8).

Mechanisms of Antithrombotic Effects

Increased production of NO may form the basis for the inhibitory action of HDL on platelet aggregation [577] (Fig. 7.8; see 7.6 of Chapter 7). The antithrombotic effects of HDL towards endothelial cells also depend on the stimulation of cellular NO production [576]. In both platelets and endothelial cells, the generation of NO can be triggered by the interaction of HDL with platelet **SR-BI** [560] (Fig. 7.8). As a consequence, the inhibitory effects of HDL towards platelet aggregation can also be mediated by SR-BI, as suggested by the absence of inhibition in platelets obtained from SR-B1-deficient mice [560] as well as in the presence of ligands for SR-BI [578]. Negative correlation between platelet SR-BI expression and *ex vivo* aggregation further supports the role of this pathway [579]. Furthermore, platelets from carriers of the P297S mutation in the SCARB1 gene feature increased cholesterol content and impaired function [580]. Consistent with these data, platelets from SR-BI-deficient mice show increased *ex vivo* adherence to immobilized fibrinogen and circulate in an activated state [581]. The apoE receptor 2/LDL receptor-related protein 8 is another surface molecule potentially involved in the interaction of platelets with HDL [582]; HDL apoE may be implicated in this pathway.

Mechanistically, **enhanced cholesterol efflux** from platelets, which results in reduced cholesterol content of platelet membranes, diminished lipid raft assembly and stimulation of eNOS, can account for the beneficial action of HDL on NO production and platelet aggregation [29] (Fig. 7.8). Indeed, increased platelet cholesterol content as a result of non-bone marrow-derived SR-BI deficiency and associated dyslipidemia leads to platelet hyper-reactivity [583]. Similar pathways involving the modulation of lipid raft composition with subsequent effects on raft-associated proteins therefore appear to underlie the vasodilatory, cytoprotective and antithrombotic activities of HDL.

Interaction of HDL with the platelet SR-BI receptor may trigger intracellular signaling cascades that encompass the intracellular release of diacylglycerol (DAG) from plasma membrane phosphatidylcholine, activation of protein kinase C (PKC), stimulation of the Na^+/H^+ antiport, alkalization of the cytoplasm and inhibition of calcium release from storage sites [560] (Fig. 7.8). Inhibition of the activity of phosphatidylinositol-specific phospholipase C may also contribute to the effects observed [569].

Another interesting aspect of the modulatory action of HDL on thrombus formation involves anticoagulatory activity. Blood coagulation occurs through the formation of enzymatic protein complexes that assemble on the surface of anionic phospholipids. ApoA-I **abrogates the procoagulant activity of anionic phospholipids**, which lose their capacity to mediate activation of prothrombin by factor Xa in the presence of factor Va and to facilitate binding of factor Va [584]. Such activity of apoA-I appears to involve a rearrangement of anionic phospholipids into surface domains which are not sufficiently large to accommodate the prothrombinase complex. PLTP-mediated transfer of anionic phospholipids may contribute to this pathway [584]. Finally, HDL can stimulate the secretion of PAI-1 from adipocytes, thereby enhancing fibrinolysis [585].

OPEN QUESTION

Can cholesterol efflux capacity mechanistically underlie multiple anti-atherogenic activites of HDL?

Role of HDL Components

The antithrombotic activity of HDL appears to be intimately associated with its **phospholipid** component, which enhances cholesterol efflux from platelets and reduces lipid raft assembly [29] (Table 7.1). Indeed, lipids significantly contribute to the effects of HDL on coagulation; thus, enrichment of **cardiolipin**, a minor anionic phospholipid with potent anticoagulant properties, in the HDL fraction can be particularly relevant in this case [586]. Negatively-charged phosphatidylserine and phosphatidylinositol can also contribute to the capacity of HDL to inhibit platelet aggregation, acting via enhancement of HDL interaction with SR-BI [578]. Furthermore, HDL-mediated stimulation of the secretion of PAI-1 by adipocytes can be accounted for by its **S1P** content [585], thereby highlighting the role of HDL lipids in its anticoagulant properties. On the other hand, the presumed capacity of ultracentrifugally-isolated HDL to function as a cofactor to activated protein C in the degradation of factor Va has been recently reported to result from a contamination by negatively-charged phospholipid membranes [587].

The protein components of HDL may also significantly contribute to its effects on thrombosis and coagulation (Table 7.1). For example, HDL acts via its protein moiety to enhance the anticoagulant activity of protein S [588]. This effect can be accounted for by HDL apolipoprotein components with anticoagulant properties, primarily by **apoA-I** [584]. In contrast, isolated apoA-I displays only minimal effects on platelet function, cholesterol efflux or lipid raft assembly [29].

ApoE is another HDL apolipoprotein that exerts effects on platelet function. Indeed, **apoE** represents a potent inhibitory factor, accounting, in part, for HDL action on platelet aggregation [571]. Inhibitory actions of apoE on platelets can be mediated by the activation of NO production, as suggested by data obtained with apoE-containing liposomes [589]. Rather, **apoA-II** may play a prothrombotic role, as suggested by positive correlations between plasma levels of apoA-II, and coagulant activities of factor VII and factor X in healthy males [590].

The effects of HDL on thrombosis and fibrinolysis can also be related to the recently discovered presence in HDL of multiple **proteins involved in protease inhibition** [17, 18], as exemplified by alpha-1-antitrypsin, a potent inhibitor of serine proteinases which has a moderate affinity for plasmin and thrombin [591]. HDL particles also contain alpha-2-antiplasmin, bikunin and kininogen-1 [17], all of which can potently inhibit proteases involved in coagulation and/or thrombosis, as well as prothrombin, a precursor of thrombin [18]. The role of these proteins in the biologic activity of HDL remains to be elucidated.

Minor HDL components modulating thrombotic and coagulation pathways are also represented by HDL-associated **complement regulatory proteins**, which have been hypothesized to limit injury to cardiac cells and to prevent activation of the coagulation cascade [592]. In contrast, **CETP** may accelerate coagulation, as suggested by the positive correlation between coagulation rates and CETP levels in the plasma of healthy adults [593]. Interestingly, recombinant CETP dose-dependently increases prothrombin activation in reaction mixtures containing purified factors Xa and Va and prothrombin, suggesting enhancement of prothrombinase activity [593]. Finally, HDL-derived haptoglobin may also contribute to the antiplatelet activities of the lipoprotein [594].

IMPORTANT READING

Nofer JR, Brodde MF, Kehrel BE. High-density lipoproteins, platelets and the pathogenesis of atherosclerosis. *Clin Exp Pharmacol Physiol* 2010; 37: 726–735.

Functional Heterogeneity of HDL

The potential heterogeneity of antithrombotic activities across the HDL particle spectrum is largely indeterminate. **Large, light, apoE-containing HDL2** particles appear to be more potent than small, dense HDL3 in terms of their capacity to inhibit platelet aggregation [595] (Table 7.3). Consistent with these data, an important difference exists in the capacity of the two HDL subclasses to modulate endothelial production of thromboxane A2, which is reduced by HDL2 but enhanced by HDL3 [557].

In contrast, TFPI has been reported to be preferentially associated with dense subspecies of HDL in human plasma, resulting in their potent anticoagulant activity [596] (Table 7.3). Furthermore, small, dense HDL3, but not large, light HDL2, specifically activate PAI-1 secretion by adiposytes [585], potentially reflecting enrichment of HDL3 in S1P [205]. Interestingly, post-translational modifications of apoA-I by protease activity are elevated in small, dense HDL, suggesting enrichment of this fraction in proteases [597].

OPEN QUESTIONS

Do low HDL-C levels contribute to a prothrombotic state in dyslipidemic subjects?

Which protein and/or lipid components of HDL are involved in the inhibition of blood coagulation cascades?

7.8 ANTIDIABETIC ACTIVITY

Type 2 diabetes and insulin resistance are well known to significantly increase cardiovascular risk [598]. The abnormal function of beta-cells underlies the development of insulin-resistant states which are characterized by a constellation of cardiovascular risk factors. Insulin-resistant states typically feature low HDL-C levels which have long been considered to reflect elevated levels of triglyceride-rich lipoproteins as a result of cholesteryl ester/triglyceride heteroexchange mediated by CETP [599]. The role of CETP in the association between low HDL-C and impaired glucose homeostasis is illustrated by the influence of Taq1B polymorphism on insulin levels and homeostatic model assessment (HOMA) index. Specifically, carriers of the B1B1 genotype display low HDL-C levels and high insulin and HOMA levels compared with individuals with at least one B2 allele [600].

The relationship between HDL-C levels and glucose metabolism has, therefore, been thought to be associative. Recent data, however, suggest that low HDL-C may play a causative role in the development of insulin resistance [14]. Epidemiologically, low levels of HDL-C may precede the development of Type 2 diabetes in a clinical setting [14, 601]. Furthermore, low in-treatment HDL-C levels predict the development of Type 2 diabetes in hypertensive subjects [602]. The recently discovered capacity of HDL to improve the functionality of beta-cells may provide a mechanistic link between low HDL-C and Type 2 diabetes [603, 604].

OPEN QUESTION

Can reduced antidiabetic activity of HDL predict development of Type 2 diabetes?

HDL acutely modulates glucose metabolism by a number of mechanisms. Indeed, rHDL infusion (80 mg/kg over 4h) **reduces plasma glucose, increases plasma insulin** and improves the HOMA index compared with placebo in patients with Type 2 diabetes [13, 15]. Consistent with these findings, HDL **accelerates insulin secretion** by Min6 beta-cells *in vitro* [13, 605]. Both lipid-free apoA-I and apoA-II are also able to stimulate insulin secretion from beta-cells, indicative of the involvement of **ABCA1** [605]. Animal models confirm the participation of ABCA1 in the pathogenesis of Type 2 diabetes. Mice lacking ABCA1 specifically in beta-cells quickly develop glucose intolerance and display cholesterol accumulation within the beta-cell plasma membrane [606, 607]. The role of ABCA1 in glucose metabolism is further highlighted by the finding of impaired insulin secretion in carriers of loss-of-function mutations in ABCA1, which provide evidence that ABCA1 is important for normal beta-cell function in humans [608]. The **removal of excess cholesterol from beta-cells** via ABCA1 may, therefore, play a role in the HDL-mediated modulation of the insulin secretory pathway and in the maintenance of cholesterol homeostasis in pancreatic beta-cells. Other HDL receptors, such as SR-BI and ABCG1, have also been proposed to contribute to the effects of HDL on beta-cells, but this question remains controversial at present [421, 605]. These data provoke the hypothesis that cholesterol efflux capacity may mechanistically underlie multiple anti-atherogenic activites of HDL, including anti-inflammatory, cytoprotective, antithrombotic and antidiabetic actions.

Another mechanism potentially contributing to improved insulin sensitivity under the action of HDL includes the **protection of pancreatic beta-cells from apoptosis**. Indeed, HDL enhances survival of beta-cells *in vitro*, protecting them from apoptotic death [421, 442]. Activation of intracellular survival pathways forms the basis for this effect and involves altered expression of inducible nitric oxide synthase, Fas and Fas-associated death domain protein (FADD)-like interleukin-1 beta-converting enzyme (FLICE) like inhibitory protein (FLIP), three proteins regulating the process of apoptosis. Inducible nitric oxide synthase, which mediates the pro-apoptotic effects of glucose and cytokines, and the death receptor Fas, are down-regulated by HDL. In contrast, the expression of FLIP, an inhibitory protein of the Fas pathway, is up-regulated [421]. As both apoA-I and S1P protect islet cells from IL-1-beta and glucose-induced apoptosis *in vitro*, multiple mechanisms may be involved in the protective effect of HDL [421].

CRITICAL CONTRIBUTION

Infusion of rHDL reduces plasma glucose, increases plasma insulin secretion and improves insulin sensitivity compared with placebo in patients with Type 2 diabetes [13, 15]. Stimulation of cholesterol efflux from beta-cells via the ABCA1-mediated pathway may play a key role in these effects.

HDL also **improves insulin sensitivity at the level of skeletal muscle**. Specifically, infusion of rHDL activates the AMPK pathway in skeletal muscle biopsies via increased phosphorylation of acetyl-CoA carboxylase 2 (ACC-beta) in humans [13]. *In vitro*, HDL increases glucose uptake by cultured skeletal muscle cells via activation of AMPK; this effect involves the binding of ABCA1 and activation of calcium/calmodulin-dependent protein kinase kinase [13]. ApoA-I is also capable of activating AMPK and stimulating glucose uptake in myocytes [37], suggesting the involvement of ABCA1. Mounting evidence points to a role for islet inflammation in the pathogenesis of Type 2 diabetes; as HDL suppresses both CD11b expression on circulating monocytes *in vivo* and reduces fatty acid-induced cytokine release from macrophages *in vitro* [13, 15], it can enhance insulin sensitivity via **suppression of lipid-induced macrophage inflammation** [609].

The beneficial effects of HDL and apoA-I on glucose metabolism in insulin-resistant states are not limited to their actions on beta-cells, skeletal muscle cells and monocytes but may also target **adipocytes**. Thus, both over-expression of apoA-I and treatment with D-4F, an apoA-I mimetic peptide, reduce white fat mass and improve insulin sensitivity in mice who develop obesity upon feeding on a high fat diet [610]. In this experimental setting, both apoA-I and D-4F act to enhance energy expenditure, increasing expression of uncoupling protein 1 in brown fat tissue and stimulating AMPK phosphorylation in brown adipocytes in culture [610]. AMPK activation in adipose tissue induced by HDL may lead to the inhibition of fatty acid lipolysis and oxidation as observed in patients with Type 2 diabetes infused with rHDL [611]; elevation of plasma insulin concentrations may also contribute to this effect.

IMPORTANT READING

Fryirs M, Barter P.J, Rye K.A. Cholesterol metabolism and pancreatic beta-cell function. *Curr Opin Lipidol* 2009; 20: 159–164.

Getz GS, Reardon CA. High-density lipoprotein function in regulating insulin secretion: possible relevance to metabolic syndrome. *Arterioscler Thromb Vasc Biol* 2010; 30: 1497–1499.

von Eckardstein A, Sibler RA. Possible contributions of lipoproteins and cholesterol to the pathogenesis of diabetes mellitus type 2. *Curr Opin Lipidol* 2011; 22: 26–32.

REFERENCES

1. Chapman MJ. Beyond the statins: new therapeutic perspectives in cardiovascular disease prevention. *Cardiovasc Drugs Ther* 2005; 19: 135–139.
2. Francis GA. The complexity of HDL. *Biochim Biophys Acta* 2010; 1801: 1286–1293.
3. Navab M, Reddy ST, Van Lenten BJ, Anantharamaiah GM, Fogelman AM. The role of dysfunctional HDL in atherosclerosis. *J Lipid Res* 2009; 50: S145–S149.
4. Khera AV, Cuchel M, de la Llera-Moya M, Rodrigues A, Burke MF, Jafri K, French BC, Phillips JA, Mucksavage ML, Wilensky RL et al. Cholesterol efflux capacity, high–density lipoprotein function, and atherosclerosis. *N Engl J Med* 2011; 364: 127–135.
5. Navab M, Reddy ST, Van Lenten BJ, Fogelman AM. HDL and cardiovascular disease: atherogenic and atheroprotective mechanisms. *Nat Rev Cardiol* 2011; 8: 222–232.
6. Stein O, Stein Y. Atheroprotective mechanisms of HDL. *Atherosclerosis* 1999; 144: 285–301.
7. Nofer JR, Kehrel B, Fobker M, Levkau B, Assmann G, von Eckardstein A. HDL and arteriosclerosis: beyond reverse cholesterol transport. *Atherosclerosis* 2002; 161: 1–16.
8. Assmann G, Nofer JR. Atheroprotective effects of high-density lipoproteins. *Annu Rev Med* 2003; 54: 321–341.

9. Assmann G, Gotto AM Jr. HDL cholesterol and protective factors in atherosclerosis. *Circulation* 2004; 109: III–8–14.

10. Navab M, Ananthramaiah GM, Reddy ST, Van Lenten BJ, Ansell BJ, Fonarow GC, Vahabzadeh K, Hama S, Hough G, Kamranpour N et al. The oxidation hypothesis of atherogenesis: the role of oxidized phospholipids and HDL. *J Lipid Res* 2004; 45: 993–1007.

11. Kontush A, Chapman MJ. Functionally defective HDL: A new therapeutic target at the crossroads of dyslipidemia, inflammation and atherosclerosis. *Pharmacol Rev* 2006; 3: 342–374.

12. Gordon SM, Hofmann S, Askew DS, Davidson WS. High density lipoprotein: it's not just about lipid transport anymore. *Trends Endocrinol Metab* 2011; 22: 9–15.

13. Drew BG, Duffy SJ, Forbes JM, Chin-Dusting J, Kaye DM, Kemp BE, Sviridov D, Steinberg GR, Kingwell BA. High density lipoprotein modulates glucose metabolism by multiple mechanisms. *Circulation* 2009; 120 (Suppl.): S468–S469.

14. von Eckardstein A, Sibler RA. Possible contributions of lipoproteins and cholesterol to the pathogenesis of diabetes mellitus type 2. *Curr Opin Lipidol* 2011; 22: 26–32.

15. Drew BG, Duffy SJ, Formosa MF, Natoli AK, Henstridge DC, Penfold SA, Thomas WG, Mukhamedova N, de Courten B, Forbes JM et al. High-density lipoprotein modulates glucose metabolism in patients with type 2 diabetes mellitus. *Circulation* 2009; 119: 2103–2111.

16. Ortiz-Munoz G, Houard X, Martin-Ventura JL, Ishida BY, Loyau S, Rossignol P, Moreno JA, Kane JP, Chalkley RJ, Burlingame AL et al. HDL antielastase activity prevents smooth muscle cell anoikis, a potential new antiatherogenic property. *FASEB J* 2009; 23: 3129–3139.

17. Vaisar T, Pennathur S, Green PS, Gharib SA, Hoofnagle AN, Cheung MC, Byun J, Vuletic S, Kassim S, Singh P et al. Shotgun proteomics implicates protease inhibition and complement activation in the antiinflammatory properties of HDL. *J Clin Invest* 2007; 117: 746–756.

18. Davidson WS, Silva RA, Chantepie S, Lagor WR, Chapman MJ, Kontush A. Proteomic analysis of defined HDL subpopulations reveals particle-specific protein clusters: relevance to antioxidative function. *Arterioscler Thromb Vasc Biol* 2009; 29: 870–876.

19. Heinecke JW. The HDL proteome: a marker-and perhaps mediator-of coronary artery disease. *J Lipid Res* 2009; 50: S167–171.

20. Gordon S, Durairaj A, Lu JL, Davidson WS. High-density lipoprotein proteomics: Identifying new drug targets and biomarkers by understanding functionality. *Curr Cardiovasc Risk Rep* 2010; 4: 1–8.

21. Gordon SM, Deng J, Lu LJ, Davidson WS. Proteomic characterization of human plasma high density lipoprotein fractionated by gel filtration chromatography. *J Proteome Res* 2010; 9: 5239–5249.

22. Kontush A, Chapman MJ. Antiatherogenic small, dense HDL - guardian angel of the arterial wall? *Nat Clin Pract Cardiovasc Med* 2006; 3: 144–153.

23. von Eckardstein A, Nofer JR, Assmann G. High density lipoproteins and arteriosclerosis. Role of cholesterol efflux and reverse cholesterol transport. *Arterioscler Thromb Vasc Biol* 2001; 21: 13–27.

24. Rader DJ, Alexander ET, Weibel GL, Billheimer J, Rothblat GH. The role of reverse cholesterol transport in animals and humans and relationship to atherosclerosis. *J Lipid Res* 2009; 50: S189–S194.

25. Zhao Y, Van Berkel TJ, Van Eck M. Relative roles of various efflux pathways in net cholesterol efflux from macrophage foam cells in atherosclerotic lesions. *Curr Opin Lipidol* 2010; 21: 441–453.

26. Yvan-Charvet L, Wang N, Tall AR. Role of HDL, ABCA1, and ABCG1 transporters in cholesterol efflux and immune responses. *Arterioscler Thromb Vasc Biol* 2010; 30: 139–143.

27. Shaw JA, Bobik A, Murphy A, Kanellakis P, Blombery P, Mukhamedova N, Woollard K, Lyon S, Sviridov D, Dart AM. Infusion of reconstituted high-density lipoprotein leads to acute changes in human atherosclerotic plaque. *Circ Res* 2008; 103: 1084–1091.

28. van Leuven SI, Birjmohun RS, Franssen R, Bisoendial RJ, de Kort H, Levels JH, Basser RL, Meijers JC, Kuivenhoven JA, Kastelein JJ et al. ApoAI-phosphatidylcholine infusion neutralizes the atherothrombotic effects of C-reactive protein in humans. *J Thromb Haemost* 2009; 7: 347–354.

29. Calkin AC, Drew BG, Ono A, Duffy SJ, Gordon MV, Schoenwaelder SM, Sviridov D, Cooper ME, Kingwell BA, Jackson SP. Reconstituted high-density lipoprotein attenuates platelet function in individuals with Type 2 diabetes mellitus by promoting cholesterol efflux. *Circulation* 2009; 120: 2095–2104.

30. Sviridov D, Mukhamedova N, Remaley AT, Chin-Dusting J, Nestel P. Antiatherogenic functionality of high density lipoprotein: how much versus how good. *J Atheroscler Thromb* 2008; 15: 52–62.

31. Movva R, Rader DJ. Laboratory assessment of HDL heterogeneity and function. *Clin Chem* 2008; 54: 788–800.

32. Miller GJ, Miller NE. Plasma-high-density-lipoprotein concentration and development of ischaemic heart-disease. *Lancet* 1975; 1: 16–19.

33. Glomset JA. The plasma lecithins:cholesterol acyltransferase reaction. *J Lipid Res* 1968; 9: 155–167.

34. Nijstad N, Gautier T, Briand F, Rader DJ, Tietge UJ. Biliary sterol secretion is required for functional in vivo reverse cholesterol transport in mice. *Gastroenterology* 2011; 140: 1043–1051.

35. Out R, Hoekstra M, Meurs I, de Vos P, Kuiper J, Van Eck M, Van Berkel TJC. Total body ABCG1 expression protects against early atherosclerotic lesion development in mice. *Arterioscler Thromb Vasc Biol* 2007; 27: 594–599.

36. Ong DS, Anzinger JJ, Leyva FJ, Rubin N, Addadi L, Kruth HS. Extracellular cholesterol-rich microdomains generated by human macrophages and their potential function in reverse cholesterol transport. *J Lipid Res* 2010; 51: 2303–2313.

37. Remaley AT, Amar M, Sviridov D. HDL-replacement therapy: mechanism of action, types of agents and potential clinical indications. *Expert Rev Cardiovasc Ther* 2008; 6: 1203–1215.

38. Zhang Y, Zanotti I, Reilly MP, Glick JM, Rothblat GH, Rader DJ. Overexpression of apolipoprotein A-I promotes reverse transport of cholesterol from macrophages to feces in vivo. *Circulation* 2003; 108: 661–663.

39. Harchaoui KE, Franssen R, Hovingh GK, Bisoendial RJ, Stellaard F, Kuipers F, Kastelein JJP, Kuivenhoven JA, Stroes ESG, Groen AK. Reduced fecal sterol excretion in subjects with familial hypoalphalipoproteinemia. *Atherosclerosis* 2009; 207: 614.

40. Catalano G, Duchene E, Julia Z, Le Goff W, Bruckert E, Chapman MJ, Guerin M. Cellular SR-BI and ABCA1-mediated cholesterol efflux are gender-specific in healthy subjects. *J Lipid Res* 2008; 49: 635–643.

41. Leite JO, Fernandez ML. Should we take high-density lipoprotein cholesterol levels at face value? *Am J Cardiovasc Drugs* 2010; 10: 1–3.

42. Chiesa G, Sirtori CR. Apolipoprotein A-I Milano: current perspectives. *Curr Opin Lipidol* 2003; 14: 159–163.

43. Franceschini G, Calabresi L, Chiesa G, Parolini C, Sirtori CR, Canavesi M, Bernini F. Increased cholesterol efflux potential of sera from ApoA-I Milano carriers and transgenic mice. *Arterioscler Thromb Vasc Biol* 1999; 19: 1257–1262.

44. Demeester N, Castro G, Desrumaux C, De Geitere C, Fruchart JC, Santens P, Mulleners E, Engelborghs S, De Deyn PP, Vandekerckhove J et al. Characterization and functional studies of lipoproteins, lipid transfer proteins, and lecithin:cholesterol acyltransferase in CSF of normal individuals and patients with Alzheimer's disease. *J Lipid Res* 2000; 41: 963–974.

45. Jaspard B, Fournier N, Vieitez G, Atger V, Barbaras R, Vieu C, Manent J, Chap H, Perret B, Collet X. Structural and functional comparison of HDL from homologous human plasma and follicular fluid. A model for extravascular fluid. *Arterioscler Thromb Vasc Biol* 1997; 17: 1605–1613.

46. Fujimoto VY, Kane JP, Ishida BY, Bloom MS, Browne RW. High-density lipoprotein metabolism and the human embryo. *Hum Reprod Update*, 2010; 16, 20–38.

47. von Eckardstein A, Rohrer L. Transendothelial lipoprotein transport and regulation of endothelial permeability and integrity by lipoproteins. *Curr Opin Lipidol* 2009; 20: 197–205.

48. Rohrer L, Ohnsorg PM, Lehner M, Landolt F, Rinninger F, von Eckardstein A. High-density lipoprotein transport through aortic endothelial cells involves scavenger receptor BI and ATP-binding cassette transporter G1. *Circ Res* 2009; 104: 1142–1150.

49. Cavelier C, Rohrer L, von Eckardstein A. ATP-Binding cassette transporter A1 modulates apolipoprotein A-I transcytosis through aortic endothelial cells. *Circ Res* 2006; 99: 1060–1066.

50. Rohrer L, Cavelier C, Fuchs S, Schluter MA, Volker W, von Eckardstein A. Binding, internalization and transport of apolipoprotein A-I by vascular endothelial cells. *Biochim Biophys Acta* 2006; 1761: 186–194.

51. Rohrer L, Ohnsorg PM, von Eckardstein A. Endothelial lipase participates in the high density lipoprotein transport through aortic endothelial cells. *Circulation* 2009; 120 (Suppl.): S1176.

52. Martinez LO, Jacquet S, Esteve JP, Rolland C, Cabezon E, Champagne E, Pineau T, Georgeaud V, Walker JE, Terce F et al. Ectopic beta-chain of ATP synthase is an apolipoprotein A-I receptor in hepatic HDL endocytosis. *Nature* 2003; 421: 75–79.

53. Lewis GF, Rader DJ. New insights into the regulation of HDL metabolism and reverse cholesterol transport. *Circ Res* 2005; 96: 1221–1232.

54. Nanjee MN, Cooke CJ, Wong JS, Hamilton RL, Olszewski WL, Miller NE. Composition and ultrastructure of size subclasses of normal human peripheral lymph lipoproteins: quantification of cholesterol uptake by HDL in tissue fluids. *J Lipid Res* 2001; 42: 639–648.

55. Rothblat GH, de la Llera-Moya M, Atger V, Kellner-Weibel G, Williams DL, Phillips MC. Cell cholesterol efflux: integration of old and new observations provides new insights. *J Lipid Res* 1999; 40: 781–796.

56. Brousseau ME, Eberhart GP, Dupuis J, Asztalos BF, Goldkamp AL, Schaefer EJ, Freeman MW. Cellular cholesterol efflux in heterozygotes for Tangier disease is markedly reduced and correlates with high density lipoprotein cholesterol concentration and particle size. *J Lipid Res* 2000; 41: 1125–1135.

57. Oram JF. Tangier disease and ABCA1. *Biochim Biophys Acta* 2000; 1529: 321–330.

58. Oram JF, Lawn RM, Garvin MR, Wade DP. ABCA1 is the cAMP-inducible apolipoprotein receptor that mediates cholesterol secretion from macrophages. *J Biol Chem* 2000; 275: 34508–34511.

59. Mott S, Yu L, Marcil M, Boucher B, Rondeau C, Genest J Jr. Decreased cellular cholesterol efflux is a common cause of familial hypoalphalipoproteinemia: role of the ABCA1 gene mutations. *Atherosclerosis* 2000; 152: 457–468.

60. Nagao K, Zhao Y, Takahashi K, Kimura Y, Ueda K. Sodium taurocholate-dependent lipid efflux by ABCA1: effects of W590S mutation on lipid translocation and apolipoprotein A-I dissociation. *J Lipid Res* 2009; 50: 1165–1172.

61. Oram JF. The cholesterol mobilizing transporter ABCA1 as a new therapeutic target for cardiovascular disease. *Trends Cardiovasc Med* 2002; 12: 170–175.

62. Vaisman BL, Lambert G, Amar M, Joyce C, Ito T, Shamburek RD, Cain WJ, Fruchart-Najib J, Neufeld ED, Remaley AT et al. ABCA1 over-expression leads to hyperalphalipoproteinemia and increased biliary cholesterol excretion in transgenic mice. *J Clin Invest* 2001; 108: 303–309.

63. Wang M-D, Franklin V, Marcel YL. In vivo reverse cholesterol transport from macrophages lacking ABCA1 expression is impaired. *Arterioscler Thromb Vasc Biol* 2007; 27: 1837–1842.

64. Adorni MP, Zimetti F, Billheimer JT, Wang N, Rader DJ, Phillips MC, Rothblat GH. The roles of different pathways in the release of cholesterol from macrophages. *J Lipid Res* 2007; 48: 2453–2462.

65. de la Llera-Moya M, Drazul-Schrader D, Asztalos BF, Cuchel M, Rader DJ, Rothblat GH. The ability to promote efflux via ABCA1 determines the capacity of serum specimens with similar high-density lipoprotein cholesterol to remove cholesterol from macrophages. *Arterioscler Thromb Vasc Biol* 2010; 30: 796–801.

66. Vedhachalam C, Chetty PS, Nickel M, Dhanasekaran P, Lund-Katz S, Rothblat GH, Phillips MC. Influence of apolipoprotein (Apo) A-I structure on nascent high density lipoprotein (HDL) particle size distribution. *J Biol Chem* 2010; 285: 31965–31973.

67. Vaughan AM, Tang C, Oram JF. ABCA1 mutants reveal an interdependency between lipid export function, apoA-I binding activity, and Janus kinase 2 activation. *J Lipid Res* 2009; 50: 285–292.

68. Ong DS, Anzinger JJ, Leyva FJ, Rubin N, Addadi L, Kruth HS. Extracellular cholesterol-rich microdomains generated by human macrophages and their potential function in reverse cholesterol transport. *J Lipid Res* 2010; 51: 2303–2313.

69. Davidson WS, Gillotte KL, Lund-Katz S, Johnson WJ, Rothblat GH, Phillips MC. The effect of high density lipoprotein phospholipid acyl chain composition on the efflux of cellular free cholesterol. *J Biol Chem* 1995; 270: 5882–5890.

70. Vaughan AM, Oram JF. ABCG1 redistributes cell cholesterol to domains removable by high density lipoprotein but not by lipid-depleted apolipoproteins. *J Biol Chem* 2005; 280: 30150–30157.

71. Duong PT, Collins HL, Nickel M, Lund-Katz S, Rothblat GH, Phillips MC. Characterization of nascent HDL particles and microparticles formed by ABCA1-mediated efflux of cellular lipids to apoA-I. *J Lipid Res* 2006; 47: 832–843.

72. Boadu E, Bilbey NJ, Francis GA. Cellular cholesterol substrate pools for adenosine-triphosphate cassette transporter A1-dependent high-density lipoprotein formation. *Curr Opin Lipidol* 2008; 19: 270–276.

73. Zhao B, Song J, Ghosh S. Hepatic overexpression of cholesteryl ester hydrolase enhances cholesterol elimination and in vivo reverse cholesterol transport. *J Lipid Res* 2008; 49: 2212–2217.

74. Luo DX, Cao DL, Xiong Y, Peng XH, Liao DF. A novel model of cholesterol efflux from lipid-loaded cells. *Acta Pharmacol Sin* 2010; 31: 1243–1257.

75. Singaraja RR, Kang MH, Vaid K, Sanders SS, Vilas GL, Arstikaitis P, Coutinho J, Drisdel RC, El-Husseini AED, Green WN et al. Palmitoylation of ATP-binding cassette transporter A1 is essential for its trafficking and function. *Circ Res* 2009; 105: 138–147.

76. Tamehiro N, Zhou S, Okuhira K, Benita Y, Brown CE, Zhuang DZ, Latz E, Hornemann T, von Eckardstein A, Xavier RJ et al. SPTLC1 binds ABCA1 to negatively regulate trafficking and cholesterol efflux activity of the transporter. *Biochemistry* 2008; 47: 6138–6147.

77. Lee-Rueckert M, Lappalainen J, Leinonen H, Pihlajamaa T, Jauhiainen M, Kovanen PT. Acidic extracellular environments strongly impair ABCA1-mediated cholesterol efflux from human macrophage foam cells. *Arterioscler Thromb Vasc Biol* 2010; 30: 1766–1772.

78. Linder MD, Mayranpaa MI, Peranen J, Pietila TE, Pietiainen VM, Uronen R-L, Olkkonen VM, Kovanen PT, Ikonen E. Rab8 regulates ABCA1 cell surface expression and facilitates cholesterol efflux in primary human macrophages. *Arterioscler Thromb Vasc Biol* 2009; 29: 883–888.

79. Li GX, Liu P, Xu L, Miao F, Liu YF, Zhao H, Zhang ZW. Effects of high-density lipoprotein on the cholesterol efflux from endothelial cells. *Nan Fang Yi Ke Da Xue Xue Bao* 2009; 29: 2191–2193 [in Chinese].

80. Hassan HH, Bailey D, Lee DY, Iatan I, Hafiane A, Ruel I, Krimbou L, Genest J. Quantitative analysis of ABCA1-dependent compartmentalization and trafficking of apolipoprotein A-I: implications for determining cellular kinetics of nascent high density lipoprotein biogenesis. *J Biol Chem* 2008; 283: 11164–11175.

81. Lin YC, Lin CH, Kuo CY, Yang VC. ABCA1 modulates the oligomerization and Golgi exit of caveolin-1 during HDL-mediated cholesterol efflux in aortic endothelial cells. *Biochem Biophys Res Commun* 2009; 382: 189–195.

82. Field FJ, Watt K, Mathur SN. Origins of intestinal ABCA1-mediated HDL-cholesterol. *J Lipid Res* 2008; 49: 2605–2619.

83. Hu YW, Ma X, Li XX, Liu XH, Xiao J, Mo ZC, Xiang J, Liao DF, Tang CK. Eicosapentaenoic acid reduces ABCA1 serine phosphorylation and impairs ABCA1-dependent cholesterol efflux through cyclic AMP/protein kinase A signaling pathway in THP-1 macrophage-derived foam cells. *Atherosclerosis* 2009; 204: e35–e43.

84. Yang Y, Jiang Y, Wang Y, An W. Suppression of ABCA1 by unsaturated fatty acids leads to lipid accumulation in HepG2 cells. *Biochimie* 2010; 92; 958–963.

85. Qiu G, Hill JS. Endothelial lipase promotes apolipoprotein AI-mediated cholesterol efflux in THP-1 macrophages. *Arterioscler Thromb Vasc Biol* 2009; 29: 84–91.

86. Weibel GL, Joshi MR, Alexander ET, Zhu P, Blair IA, Rothblat GH. Overexpression of human 15(S)-lipoxygenase-1 in RAW macrophages leads to increased cholesterol mobilization and reverse cholesterol transport. *Arterioscler Thromb Vasc Biol* 2009; 29: 837–842.

87. Rigamonti E, Chinetti-Gbaguidi G, Staels B. Regulation of macrophage functions by PPAR-alpha, PPAR-gamma, and LXRs in mice and men. *Arterioscler Thromb Vasc Biol* 2008; 28: 1050–1059.

88. Larrede S, Quinn CM, Jessup W, Frisdal E, Olivier M, Hsieh V, Kim M-J, Van Eck M, Couvert P, Carrie A et al. Stimulation of cholesterol efflux by LXR agonists in cholesterol-loaded human macrophages is ABCA1-dependent but ABCG1-independent. *Arterioscler Thromb Vasc Biol* 2009; 29: 1930–1936.

89. Mogilenko DA, Shavva VS, Dizhe EB, Orlov SV, Perevozchikov AP. PPAR gamma activates ABCA1 gene transcription but reduces the level of ABCA1 protein in HepG2 cells. *Biochem Biophys Res Commun* 2010; 402: 477–482.

90. Tsubakio-Yamamoto K, Matsuura F, Koseki M, Oku H, Sandoval JC, Inagaki M, Nakatani K, Nakaoka H, Kawase R, Yuasa-Kawase M et al. Adiponectin prevents atherosclerosis by increasing cholesterol efflux from macrophages. *Biochem Biophys Res Commun* 2008; 375: 390–394.

91. Tsubakio-Yamamoto K, Yamashita S, Ohama T, Masuda D, Sandoval JC, Inagaki M, Nakatani K, Yuasa-Kawase M, Kawase R, Nakaoka H et al. Adiponectin enhances de novo hepatic HDL synthesis through LXR alpha-and COUP-TFII-dependent pathways. *Circulation* 2009; 120: S446.

92. Park SH, Kim JY, Lee JH, Park HY. Association between plasma adiponectin and high-density lipoprotein cholesterol in postmenopausal women. *Clin Biochem* 2010; 43: 1069–1073.

93. Peng YS, Chiu YL, Chen HY, Yang JY, Lai CF, Hsu SP, Pai MF. Decreased high-density lipoprotein cholesterol is associated with inflammation and insulin resistance in non-diabetic haemodialysis patients. *Nephrology* 2010; 15: 692–699.

94. Ye D, Meurs I, Ohigashi M, Calpe-Berdiel L, Habets KL, Zhao Y, Kubo Y, Yamaguchi A, Van Berkel TJ, Nishi T et al. Macrophage ABCA5 deficiency influences cellular cholesterol efflux and increases susceptibility to atherosclerosis in female LDLr knockout mice. *Biochem Biophys Res Commun* 2010; 395: 387–394.

95. Ye D, Hoekstra M, Out R, Meurs I, Kruijt JK, Hildebrand RB, Van Berkel TJ, Van Eck M. Hepatic cell-specific ATP-binding cassette (ABC) transporter profiling identifies putative novel candidates for lipid homeostasis in mice. *Atherosclerosis* 2008; 196: 650–658.

96. Morales CR, Marat AL, Ni X, Yu Y, Oko R, Smith BT, Argraves WS. ATP-binding cassette transporters ABCA1, ABCA7, and ABCG1 in mouse spermatozoa. *Biochem Biophys Res Commun* 2008; 376: 472–477.

97. Nakamura K, Kennedy MA, Baldan A, Bojanic DD, Lyons K, Edwards PA. Expression and regulation of multiple murine ATP-binding cassette transporter G1 mRNAs/isoforms that stimulate cellular cholesterol efflux to high density lipoprotein. *J Biol Chem* 2004; 279: 45980–45989.

98. Wang N, Lan D, Chen W, Matsuura F, Tall AR. ATP-binding cassette transporters G1 and G4 mediate cellular cholesterol efflux to high-density lipoproteins. *Proc Natl Acad Sci U S A* 2004; 101: 9774–9779.

99. Kennedy MA, Barrera GC, Nakamura K, Baldan A, Tarr P, Fishbein MC, Frank J, Francone OL, Edwards PA. ABCG1 has a critical role in mediating cholesterol efflux to HDL and preventing cellular lipid accumulation. *Cell Metab* 2005; 1: 121–131.

100. Stefulj J, Panzenboeck U, Becker T, Hirschmugl B, Schweinzer C, Lang I, Marsche G, Sadjak A, Lang U, Desoye G et al. Human endothelial cells of the placental barrier efficiently deliver cholesterol to the fetal circulation via ABCA1 and ABCG1. *Circ Res* 2009; 104: 600–608.

101. Sankaranarayanan S, Oram JF, Asztalos BF, Vaughan AM, Lund-Katz S, Adorni MP, Phillips MC, Rothblat GH. Effects of acceptor composition and mechanism of ABCG1-mediated cellular free cholesterol efflux. *J Lipid Res* 2009; 50: 275–284.

102. Burgess B, Naus K, Chan J, Hirsch-Reinshagen V, Tansley G, Matzke L, Chan B, Wilkinson A, Fan J, Donkin J et al. Overexpression of human ABCG1 does not affect atherosclerosis in fat-fed ApoE-deficient mice. *Arterioscler Thromb Vasc Biol* 2008; 28: 1731–1737.

103. Baldan A, Pei L, Lee R, Tarr P, Tangirala RK, Weinstein MM, Frank J, Li AC, Tontonoz P, Edwards PA. Impaired development of atherosclerosis in hyperlipidemic Ldlr-/- and ApoE-/- mice transplanted with Abcg1-/- bone marrow. *Arterioscler Thromb Vasc Biol* 2006; 26: 2301–2307.

104. Ranalletta M, Wang N, Han S, Yvan-Charvet L, Welch C, Tall AR. Decreased atherosclerosis in low-density lipoprotein receptor knockout mice transplanted with Abcg1-/- bone marrow. *Arterioscler Thromb Vasc Biol* 2006; 26: 2308–2315.

105. Out R, Hoekstra M, Habets K, Meurs I, de Waard V, Hildebrand RB, Wang Y, Chimini G, Kuiper J, Van Berkel TJ et al. Combined deletion of macrophage ABCA1 and ABCG1 leads to massive lipid accumulation in tissue macrophages and distinct atherosclerosis at relatively low plasma cholesterol levels. *Arterioscler Thromb Vasc Biol* 2008; 28: 258–264.

106. Baldan A, Bojanic DD, Edwards PA. The ABCs of sterol transport. *J Lipid Res* 2009; 50: S80–S85.

107. Sano O, Kobayashi A, Nagao K, Kumagai K, Kioka N, Hanada K, Ueda K, Matsuo M. Sphingomyelin-dependence of cholesterol efflux mediated by ABCG1. *J Lipid Res* 2007; 48: 2377–2384.

108. Cuchel M, Lund-Katz S, de la Llera-Moya M, Millar JS, Chang D, Fuki I, Rothblat GH, Phillips MC, Rader DJ. Pathways by which reconstituted high-density lipoprotein mobilizes free cholesterol from whole body and from macrophages. *Arterioscler Thromb Vasc Biol* 2010; 30: 526–532.

109. Lammers B, Out R, Hildebrand RB, Quinn CM, Williamson D, Hoekstra M, Meurs I, Van Berkel TJ, Jessup W, Van Eck M. Independent protective roles for macrophage ABCG1 and apoE in the atherosclerotic lesion development. *Atherosclerosis* 2009; 205: 420–426.

110. Wang X, Collins HL, Ranalletta M, Fuki IV, Billheimer JT, Rothblat GH, Tall AR, Rader DJ. Macrophage ABCA1 and ABCG1, but not SR-BI, promote macrophage reverse cholesterol transport in vivo. *J Clin Invest* 2007; 117: 2216–2224.

111. Xie C, Turley SD, Dietschy JM. ABCA1 plays no role in the centripetal movement of cholesterol from peripheral tissues to the liver and intestine in the mouse. *J Lipid Res* 2009; 50: 1316–1329.

112. Out R, Hoekstra M, Hildebrand RB, Kruit JK, Meurs I, Li Z, Kuipers F, Van Berkel TJ, Van Eck M. Macrophage ABCG1 deletion disrupts lipid homeostasis in alveolar macrophages and moderately influences atherosclerotic lesion development in LDL receptor-deficient mice. *Arterioscler Thromb Vasc Biol* 2006; 26: 2295–2300.

113. Curtiss LK. Is two out of three enough for ABCG1? *Arterioscler Thromb Vasc Biol* 2006; 26: 2175–2177.

114. Nagelin MH, Srinivasan S, Lee J, Nadler JL, Hedrick CC. 12/15-Lipoxygenase activity increases the degradation of macrophage ATP-binding cassette transporter G1. *Arterioscler Thromb Vasc Biol* 2008; 28: 1811–1819.

115. de la Llera-Moya M, Rothblat GH, Connelly MA, Kellner-Weibel G, Sakr SW, Phillips MC, Williams DL. Scavenger receptor BI (SR-BI) mediates free cholesterol flux independently of HDL tethering to the cell surface. *J Lipid Res* 1999; 40: 575–580.

116. Yancey PG, Asztalos BF, Stettler N, Piccoli D, Williams DL, Connelly MA, Rothblat GH. SR-BI- and ABCA1-mediated cholesterol efflux to serum from patients with Alagille syndrome. *J Lipid Res* 2004; 45: 1724–1732.

117. Nieland TJ, Xu S, Penman M, Krieger M. Negatively cooperative binding of high-density lipoprotein to the HDL receptor SR-BI. *Biochemistry* 2011; 50: 1818–1830.

118. Yancey PG, de la Llera-Moya M, Swarnakar S, Monzo P, Klein SM, Connelly MA, Johnson WJ, Williams DL, Rothblat GH. High density lipoprotein phospholipid composition is a major determinant of the bi-directional flux and net movement of cellular free cholesterol mediated by scavenger receptor BI. *J Biol Chem* 2000; 275: 36596–36604.

119. Van Eck M, Pennings M, Hoekstra M, Out R, Van Berkel TJ. Scavenger receptor BI and ATP-binding cassette transporter A1 in reverse cholesterol transport and atherosclerosis. *Curr Opin Lipidol* 2005; 16: 307–315.

120. Zhang Y, McGillicuddy FC, Hinkle CC, O'Neill S, Glick JM, Rothblat GH, Reilly MP. Adipocyte modulation of high-density lipoprotein cholesterol. *Circulation* 2010; 121: 1347–1355.

121. Zhao Y, Pennings M, Hildebrand RB, Ye D, Calpe-Berdiel L, Out R, Kjerrulf M, Hurt-Camejo E, Groen AK, Hoekstra M et al. Enhanced foam cell formation, atherosclerotic lesion development, and inflammation by combined deletion of ABCA1 and SR-BI in bone marrow-derived cells in LDL receptor knockout mice on Western-type diet. *Circ Res* 2010; 107: e20–e31.

122. Yvan-Charvet L, Pagler TA, Wang N, Senokuchi T, Brundert M, Li H, Rinninger F, Tall AR. SR-BI inhibits ABCG1-stimulated net cholesterol efflux from cells to plasma HDL. *J Lipid Res* 2008; 49: 107–114.

123. Huby T, Doucet C, Dachet C, Ouzilleau B, Ueda Y, Afzal V, Rubin E, Chapman MJ, Lesnik P. Knockdown expression and hepatic deficiency reveal an atheroprotective role for SR-BI in liver and peripheral tissues. *J Clin Invest* 2006; 116: 2767–2776.

124. Storey SM, Atshaves BP, McIntosh AL, Landrock KK, Martin GG, Huang H, Payne HR, Johnson JD, Macfarlane RD, Kier AB et al. Effect of sterol carrier protein-2 gene ablation on HDL-mediated cholesterol efflux from primary cultured mouse hepatocytes. *Am J Physiol Gastrointest Liver Physiol* 2010; 299: G244–G254.

125. Yue P, Chen Z, Nassir F, Bernal-Mizrachi C, Finck B, Azhar S, Abumrad NA. Enhanced hepatic apoA-I secretion and peripheral efflux of cholesterol and phospholipid in CD36 null mice. *PLoS One* 2010; 5: e9906.

126. de Beer MC, Castellani LW, Cai L, Stromberg AJ, de Beer FC, van der Westhuyzen DR. ApoA-II modulates the association of HDL with class B scavenger receptors SR-BI and CD36. *J Lipid Res* 2004; 45: 706–715.

127. Heeren J, Weber W, Beisiegel U. Intracellular processing of endocytosed triglyceride-rich lipoproteins comprises both recycling and degradation. *J Cell Sci* 1999; 112: 349–359.

128. Takahashi Y, Smith JD. Cholesterol efflux to apolipoprotein AI involves endocytosis and resecretion in a calcium-dependent pathway. *Proc Natl Acad Sci U S A* 1999; 96: 11358–11363.

129. Heeren J, Beisiegel U, Grewal T. Apolipoprotein E recycling: implications for dyslipidemia and atherosclerosis. *Arterioscler Thromb Vasc Biol* 2006; 26: 442–448.

130. Faulkner LE, Panagotopulos SE, Johnson JD, Woollett LA, Hui DY, Witting SR, Maiorano JN, Davidson WS. An analysis of the role of a retroendocytosis pathway in ABCA1-mediated cholesterol efflux from macrophages. *J Lipid Res* 2008; 49: 1322–1332.

131. Howard AD, Verghese PB, Arrese EL, Soulages JL. The beta-subunit of ATP synthase is involved in cellular uptake and resecretion of apoA-I but does not control apoA-I-induced lipid efflux in adipocytes. *Mol Cell Biochem* 2011; 348: 155–164.

132. Mazzone T. Apolipoprotein E secretion by macrophages: its potential physiological functions. *Curr Opin Lipidol* 1996; 7: 303–307.

133. Mazzone T, Reardon C. Expression of heterologous human apolipoprotein E by J774 macrophages enhances cholesterol efflux to HDL3. *J Lipid Res* 1994; 35: 1345–1353.

134. Zhang WY, Gaynor PM, Kruth HS. Apolipoprotein E produced by human monocyte-derived macrophages mediates cholesterol efflux that occurs in the absence of added cholesterol acceptors. *J Biol Chem* 1996; 271: 28641–28646.

135. Mabile L, Lefebvre C, Lavigne J, Boulet L, Davignon J, Lussier-Cacan S, Bernier L. Secreted apolipoprotein E reduces macrophage-mediated LDL oxidation in an isoform-dependent way. *J Cell Biochem* 2003; 90: 766–776.

136. Dory L. Regulation of apolipoprotein E secretion by high density lipoprotein 3 in mouse macrophages. *J Lipid Res* 1991; 32: 783–792.

137. Rees D, Sloane T, Jessup W, Dean RT, Kritharides L. Apolipoprotein A-I stimulates secretion of apolipoprotein E by foam cell macrophages. *J Biol Chem* 1999; 274: 27925–27933.

138. Kockx M, Rye KA, Gaus K, Quinn CM, Wright J, Sloane T, Sviridov D, Fu Y, Sullivan D, Burnett JR et al. Apolipoprotein A-I-stimulated apolipoprotein E secretion from human macrophages is independent of cholesterol efflux. *J Biol Chem* 2004; 279: 25966–25977.

139. Lucic D, Huang ZH, Gu de S, Altenburg MK, Maeda N, Mazzone T. Regulation of macrophage apoE secretion and sterol efflux by the LDL receptor. *J Lipid Res* 2007; 48: 366–372.

140. Ishiguro H, Yoshida H, Major AS, Zhu T, Babaev VR, Linton MF, Fazio S. Retrovirus-mediated expression of apolipoprotein A-I in the macrophage protects against atherosclerosis in vivo. *J Biol Chem* 2001; 276: 36742–36748.

141. Linton MF, Atkinson JB, Fazio S. Prevention of atherosclerosis in apolipoprotein E-deficient mice by bone marrow transplantation. *Science* 1995; 267: 1034–1037.

142. Hasty AH, Linton MF, Brandt SJ, Babaev VR, Gleaves LA, Fazio S. Retroviral gene therapy in ApoE-deficient mice: ApoE expression in the artery wall reduces early foam cell lesion formation. *Circulation* 1999; 99: 2571–2576.

143. Zanotti I, Pedrelli M, Poti F, Stomeo G, Gomaraschi M, Calabresi L, Bernini F. Macrophage, but not systemic, apolipoprotein E is necessary for macrophage reverse cholesterol transport in vivo. *Arterioscler Thromb Vasc Biol* 2011; 31: 74–80.

144. Parthasarathy S, Barnett J, Fong LG. High-density lipoprotein inhibits the oxidative modification of low-density lipoprotein. *Biochim Biophys Acta* 1990; 1044: 275–283.

145. Carvalho MD, Vendrame CM, Ketelhuth DF, Yamashiro-Kanashiro EH, Goto H, Gidlund M. High-density lipoprotein inhibits the uptake of modified low-density lipoprotein and the expression of CD36 and Fc gamma RI. *J Atheroscler Thromb* 2010; 17: 844–857.

146. Vance DE. Role of phosphatidylcholine biosynthesis in the regulation of lipoprotein homeostasis. *Curr Opin Lipidol* 2008; 19: 229–234.

147. Rashid S, Marcil M, Ruel I, Genest J. Identification of a novel human cellular HDL biosynthesis defect. *Eur Heart J* 2009; 30: 2204–2212.

148. See RH, Caday-Malcolm RA, Singaraja RR, Zhou S, Silverston A, Huber MT, Moran J, James ER, Janoo R, Savill JM et al. Protein kinase A site-specific phosphorylation regulates ATP-binding cassette A1 (ABCA1)-mediated phospholipid efflux. *J Biol Chem* 2002; 277: 41835–41842.

149. Huang R, Silva RA, Jerome WG, Kontush A, Chapman MJ, Curtiss LK, Hodges TJ, Davidson WS. Apolipoprotein A-I structural organization in high-density lipoproteins isolated from human plasma. *Nat Struct Mol Biol* 2011; 18: 416–422.

150. Gonzalez MC, Toledo JD, Tricerri MA, Garda HA. The central type Y amphipathic alpha-helices of apolipoprotein AI are involved in the mobilization of intracellular cholesterol depots. *Arch Biochem Biophys* 2008; 473: 34–41.

151. Alexander ET, Vedhachalam C, Sankaranarayanan S, de la Llera-Moya M, Rothblat GH, Rader DJ, Phillips MC. Influence of apolipoprotein A-I domain structure on macrophage reverse cholesterol transport in mice. *Arterioscler Thromb Vasc Biol* 2011; 31: 320–327.

152. Peng D-Q, Brubaker G, Wu Z, Zheng L, Willard B, Kinter M, Hazen SL, Smith JD. Apolipoprotein A-I tryptophan substitution leads to resistance to myeloperoxidase-mediated loss of function. *Arterioscler Thromb Vasc Biol* 2008; 28: 2063–2070.

153. Tricerri A, Corsico B, Toledo JD, Garda HA, Brenner RR. Conformation of apolipoprotein AI in reconstituted lipoprotein particles and particle-membrane interaction: effect of cholesterol. *Biochim Biophys Acta* 1998; 1391: 67–78.

154. Rotllan N, Ribas V, Calpe-Berdiel L, Martin-Campos JM, Blanco-Vaca F, Escola-Gil JC. Overexpression of human apolipoprotein A-II in transgenic mice does not impair macrophage-specific reverse cholesterol transport in vivo. *Arterioscler Thromb Vasc Biol* 2005; 25: e128–e132.

155. Westerterp M, Van Eck M, de Haan W, Offerman EH, Van Berkel TJ, Havekes LM, Rensen PC. Apolipoprotein CI aggravates atherosclerosis development in ApoE-knockout mice despite mediating cholesterol efflux from macrophages. *Atherosclerosis* 2007; 195: e9–e16.

156. Wolfrum C, Poy MN, Stoffel M. Apolipoprotein M is required for prebeta-HDL formation and cholesterol efflux to HDL and protects against atherosclerosis. *Nat Med* 2005; 11: 418–422.

157. Hara M, Matsushima T, Satoh H, Iso-o N, Noto H, Togo M, Kimura S, Hashimoto Y, Tsukamoto K. Isoform-dependent cholesterol efflux from macrophages by apolipoprotein E is modulated by cell surface proteoglycans. *Arterioscler Thromb Vasc Biol* 2003; 23: 269–274.

158. Michikawa M, Fan QW, Isobe I, Yanagisawa K. Apolipoprotein E exhibits isoform-specific promotion of lipid efflux from astrocytes and neurons in culture. *J Neurochem* 2000; 74: 1008–1016.

159. Gomaraschi M, Putt WE, Pozzi S, Iametti S, Barbiroli A, Bonomi F, Favari E, Bernini F, Franceschini G, Talmud PJ et al. Structure and function of the apoA-IV T347S and Q360H common variants. *Biochem Biophys Res Commun* 2010; 393: 126–130.

160. Chiesa G, Monteggia E, Marchesi M, Lorenzon P, Laucello M, Lorusso V, Di Mario C, Karvouni E, Newton RS, Bisgaier CL et al. Recombinant apolipoprotein A-I Milano infusion into rabbit carotid artery rapidly removes lipid from fatty streaks. *Circ Res* 2002; 90: 974–980.

161. Weibel GL, Alexander ET, Joshi MR, Rader DJ, Lund-Katz S, Phillips MC, Rothblat GH. Wild-type apoA-I and the Milano variant have similar abilities to stimulate cellular lipid mobilization and efflux. *Arterioscler Thromb Vasc Biol* 2007; 27: 2022–2029.

162. Alexander ET, Weibel GL, Joshi MR, Vedhachalam C, de la Llera-Moya M, Rothblat GH, Phillips MC, Rader D.J. Macrophage reverse cholesterol transport in mice expressing apoA-I Milano. *Arterioscler Thromb Vasc Biol* 2009; 29: 1496–1501.

163. Favari E, Gomaraschi M, Zanotti I, Bernini F, Lee-Rueckert M, Kovanen PT, Sirtori CR, Franceschini G, Calabresi L. A unique protease-sensitive high density lipoprotein particle containing the apolipoprotein A-I Milano dimer effectively promotes ATP-binding Cassette A1-mediated cell cholesterol efflux. *J Biol Chem* 2007; 282: 5125–5132.

164. Tanigawa H, Billheimer JT, Tohyama J-i, Fuki IV, Ng DS, Rothblat GH, Rader DJ. Lecithin: cholesterol acyltransferase expression has minimal effects on macrophage reverse cholesterol transport in vivo. *Circulation* 2009; 120: 160–169.

165. Calabresi L, Favari E, Moleri E, Adorni MP, Pedrelli M, Costa S, Jessup W, Gelissen IC, Kovanen PT, Bernini F et al. Functional LCAT is not required for macrophage cholesterol efflux to human serum. *Atherosclerosis* 2009; 204: 141–146.

166. Hossain MA, Tsujita M, Akita N, Kobayashi F, Yokoyama S. Cholesterol homeostasis in ABCA1/LCAT double-deficient mouse. *Biochim Biophys Acta* 2009; 1791: 1197–1205.

167. Efrat M, Aviram M. Paraoxonase 1 interactions with HDL, antioxidants and macrophages regulate atherogenesis—a protective role for HDL phospholipids. *Adv Exp Med Biol* 2010; 660: 153–166.

168. Rosenblat M, Vaya J, Shih D, Aviram M. Paraoxonase 1 (PON1) enhances HDL-mediated macrophage cholesterol efflux via the ABCA1 transporter in association with increased HDL binding to the cells: a possible role for lysophosphatidylcholine. *Atherosclerosis* 2005; 179: 69–77.

169. Hara S, Shike T, Takasu N, Mizui T. Lysophosphatidylcholine promotes cholesterol efflux from mouse macrophage foam cells. *Arterioscler Thromb Vasc Biol* 1997; 17: 1258–1266.

170. Samyn H, Moerland M, van Gent T, van Haperen R, Grosveld F, van Tol A, de Crom R. Elevation of systemic PLTP, but not macrophage-PLTP, impairs macrophage reverse cholesterol transport in transgenic mice. *Atherosclerosis* 2009; 204: 429–434.

171. Zerrad-Saadi A, Therond P, Chantepie S, Couturier M, Rye K-A, Chapman MJ, Kontush A. HDL3-mediated inactivation of LDL-associated phospholipid hydroperoxides is determined by the redox status of apolipoprotein A-I and HDL particle surface lipid rigidity: Relevance to inflammation and atherogenesis. *Arterioscler Thromb Vasc Biol* 2009; 29: 2169–2175.

172. Goyal J, Wang K, Liu M, Subbaiah PV. Novel function of lecithin-cholesterol acyltransferase. *J Biol Chem* 1997; 272: 16231–16239.

173. Teiber JF, Draganov DI, La Du BN. Purified human serum PON1 does not protect LDL against oxidation in the in vitro assays initiated with copper or AAPH. *J Lipid Res* 2004; 45: 2260–2268.

174. Draganov DI, Teiber JF, Speelman A, Osawa Y, Sunahara R, La Du BN. Human paraoxonases (PON1, PON2, and PON3) are lactonases with overlapping and distinct substrate specificities. *J Lipid Res* 2005; 46: 1239–1247.

175. Kriska T, Marathe GK, Schmidt JC, McIntyre TM, Girotti AW. Phospholipase action of platelet-activating factor acetylhydrolase, but not paraoxonase-1, on long fatty acyl chain phospholipid hydroperoxides. *J Biol Chem* 2007; 282: 100–108.

176. Marathe GK, Zimmerman GA, McIntyre TM. Platelet-activating factor acetylhydrolase, and not paraoxonase-1, is the oxidized phospholipid hydrolase of high density lipoprotein particles. *J Biol Chem* 2003; 278: 3937–3947.

177. Connelly PW, Draganov D, Maguire GF. Paraoxonase-1 does not reduce or modify oxidation of phospholipids by peroxynitrite. *Free Radic Biol Med* 2005; 38: 164–174.

178. Jakubowski H. Calcium-dependent human serum homocysteine thiolactone hydrolase. A protective mechanism against protein N-homocysteinylation. *J Biol Chem* 2000; 275: 3957–3962.

179. Khersonsky O, Tawfik DS. Structure-reactivity studies of serum paraoxonase PON1 suggest that its native activity is lactonase. *Biochemistry* 2005; 44: 6371–6382.

180. Perla-Kajan J, Jakubowski H. Paraoxonase 1 protects against protein N-homocysteinylation in humans. *FASEB J* 2010; 24: 931–936.

181. Besler C, Heinrich K, Rohrer L, Yonekawa K, Doerries C, Manes C, Muller M, Templin C, Corti R, Maier W et al. Mechanisms leading to impaired endothelial-vasoprotective properties of high density lipoprotein (HDL) in patients with coronary disease (CAD). *Circulation* 2009; 120 (Suppl.): S1009.

182. Tchoua U, Gillard BK, Pownall HJ. HDL superphospholipidation enhances key steps in reverse cholesterol transport. *Atherosclerosis* 2010; 209: 430–435.

183. Litman BJ, Lewis EN, Levin IW. Packing characteristics of highly unsaturated bilayer lipids: Raman spectroscopic studies of multilamellar phosphatidylcholine dispersions. *Biochemistry* 1991; 30: 313–319.

184. Marmillot P, Patel S, Lakshman MR. Reverse cholesterol transport is regulated by varying fatty acyl chain saturation and sphingomyelin content in reconstituted high-density lipoproteins. *Metabolism* 2007; 56: 251–259.

185. Guha M, Gantz DL, Gursky O. Effects of acyl chain length, unsaturation, and pH on thermal stability of model discoidal HDLs. *J Lipid Res* 2008; 49: 1752–1761.

186. Guha M, Gao X, Jayaraman S, Gursky O. Correlation of structural stability with functional remodeling of high-density lipoproteins: the importance of being disordered. *Biochemistry* 2008; 47: 11393–11397.

187. Ohta T, Saku K, Takata K, Nakamura R, Ikeda Y, Matsuda I. Different effects of subclasses of HDL containing apoA-I but not apoA-II (LpA-I) on cholesterol esterification in plasma and net cholesterol efflux from foam cells. *Arterioscler Thromb Vasc Biol* 1995: 15: 956–962.

188. Sasahara T, Nestel P, Fidge N, Sviridov D. Cholesterol transport between cells and high density lipoprotein subfractions from obese and lean subjects. *J Lipid Res* 1998; 39: 544–554.

189. Asztalos B, Zhang W, Roheim PS, Wong L. Role of free apolipoprotein A-I in cholesterol efflux. Formation of pre-alpha-migrating high-density lipoprotein particles. *Arterioscler Thromb Vasc Biol* 1997; 17: 1630–1636.

190. Asztalos BF, de la Llera-Moya M, Dallal GE, Horvath KV, Schaefer EJ, Rothblat G.H. Differential effects of HDL subpopulations on cellular ABCA1- and SR-BI-mediated cholesterol efflux. *J Lipid Res* 2005; 46: 2246–2253.

191. Favari E, Calabresi L, Adorni MP, Jessup W, Simonelli S, Franceschini G, Bernini F. Small discoidal pre-beta1 HDL particles are efficient acceptors of cell cholesterol via ABCA1 and ABCG1. *Biochemistry* 2009; 48: 11067–11074.

192. Jenkins KT, Merkens LS, Tubb MR, Myatt L, Davidson WS, Steiner RD, Woollett LA. Enhanced placental cholesterol efflux by fetal HDL in Smith-Lemli-Opitz syndrome. *Mol Genet Metab* 2008; 94: 240–247.

193. Hamdi SM, Vieitez G, Jaspard B, Barbaras R, Perret B, Mieusset R, Parinaud J, Collet X. Effects of human follicular fluid and high-density lipoproteins on early spermatozoa hyperactivation and cholesterol efflux. *J Lipid Res* 2010; 51: 1363–1369.

194. Singaraja RR, Fievet C, Castro G, James ER, Hennuyer N, Clee SM, Bissada N, Choy JC, Fruchart JC, McManus BM et al. Increased ABCA1 activity protects against atherosclerosis. *J Clin Invest* 2002; 110: 35–42.

195. de Beer MC, Durbin DM, Cai L, Jonas A, de Beer FC, van der Westhuyzen DR. Apolipoprotein A-I conformation markedly influences HDL interaction with scavenger receptor BI. *J Lipid Res* 2001; 42: 309–313.

196. Thuahnai ST, Lund-Katz S, Dhanasekaran P, de la Llera-Moya M, Connelly MA, Williams DL, Rothblat GH, Phillips MC. Scavenger receptor class B type I-mediated cholesteryl ester-selective uptake and efflux of unesterified cholesterol. Influence of high density lipoprotein size and structure. *J Biol Chem* 2004; 279: 12448–12455.

197. Brown RJ, Lagor WR, Sankaranarayanan S, Yasuda T, Quertermous T, Rothblat GH, Rader DJ. Impact of combined deficiency of hepatic lipase and endothelial Lipase on the metabolism of both high-density lipoproteins and apolipoprotein B-containing lipoproteins. *Circ Res* 2010; 107: 357–364.
198. Mukhamedova N, Escher G, D'Souza W, Tchoua U, Grant A, Krozowski Z, Bukrinsky M, Sviridov D. Enhancing apolipoprotein A-I-dependent cholesterol efflux elevates cholesterol export from macrophages in vivo. *J Lipid Res* 2008; 49: 2312–2322.
199. Wu P, Liang DH, Liu YF, Liu YY, Zhang XL, Fu Q, Miao F. Effects of high-density lipoprotein 1 on the formation of foam cells from human monocyte-derived macrophages. *Zhonghua Xin Xue Guan Bing Za Zhi* 2009; 37: 156–160 [in Chinese].
200. Sacks FM, Rudel LL, Conner A, Akeefe H, Kostner G, Baki T, Rothblat G, de la Llera-Moya M, Asztalos B, Perlman T et al. Selective delipidation of plasma HDL enhances reverse cholesterol transport in vivo. *J Lipid Res* 2009; 50: 894–907.
201. Duverger N, Viglietta C, Berthou L, Emmanuel F, Tailleux A, Parmentier-Nihoul L, Laine B, Fievet C, Castro G, Fruchart JC et al. Transgenic rabbits expressing human apolipoprotein A-I in the liver. *Arterioscler Thromb Vasc Biol* 1996; 16: 1424–1429.
202. Miyazaki M, Nakano M, Fukuda M, Handa T. Smaller discoidal high-density lipoprotein particles form saddle surfaces, but not planar bilayers. *Biochemistry* 2009; 48: 7756–7763.
203. Kumpula LS, Kumpula JM, Taskinen MR, Jauhiainen M, Kaski K, Ala-Korpela M. Reconsideration of hydrophobic lipid distributions in lipoprotein particles. *Chem Phys Lipids* 2008; 155: 57–62.
204. Sviridov D, Miyazaki O, Theodore K, Hoang A, Fukamachi I, Nestel P. Delineation of the role of pre-beta 1-HDL in cholesterol efflux using isolated pre-beta 1-HDL. *Arterioscler Thromb Vasc Biol* 2002; 22: 1482–1488.
205. Kontush A, Therond P, Zerrad A, Couturier M, Negre-Salvayre A, de Souza JA, Chantepie S, Chapman MJ. Preferential sphingosine-1-phosphate enrichment and sphingomyelin depletion are key features of small dense HDL3 particles: relevance to antiapoptotic and antioxidative activities. *Arterioscler Thromb Vasc Biol* 2007; 27: 1843–1849.
206. Ohta T, Nakamura R, Ikeda Y, Shinohara M, Miyazaki A, Horiuchi S, Matsuda I. Differential effect of subspecies of lipoprotein containing apolipoprotein A-I on cholesterol efflux from cholesterol-loaded macrophages: functional correlation with lecithin: cholesterol acyltransferase. *Biochim Biophys Acta* 1992; 1165: 119–128.
207. Lund-Katz S, Phillips MC. High density lipoprotein structure-function and role in reverse cholesterol transport. *Subcell Biochem* 2010; 51: 183–227.
208. Kontush A. Apolipoprotein Abeta: black sheep in a good family. *Brain Pathol* 2004; 14: 433–447.
209. Williams KJ, Tabas I. The response-to-retention hypothesis of early atherogenesis. *Arterioscler Thromb Vasc Biol* 1995; 15: 551–561.
210. Tabas I, Williams KJ, Boren, J. Subendothelial lipoprotein retention as the initiating process in atherosclerosis: update and therapeutic implications. *Circulation* 2007; 116: 1832–1844.
211. Carew TE, Pittman RC, Marchand ER, Steinberg D. Measurement in vivo of irreversible degradation of low density lipoprotein in the rabbit aorta. Predominance of intimal degradation. *Arteriosclerosis* 1984; 4: 214–224.

212. Schwenke DC, Carew TE. Initiation of atherosclerotic lesions in cholesterol-fed rabbits. I. Focal increases in arterial LDL concentration precede development of fatty streak lesions. *Arteriosclerosis* 1989; 9: 895–907.
213. Schwenke DC, Carew TE. Initiation of atherosclerotic lesions in cholesterol-fed rabbits. II. Selective retention of LDL vs. selective increases in LDL permeability in susceptible sites of arteries. *Arteriosclerosis* 1989; 9: 908–918.
214. Stocker R, Keaney JF Jr. Role of oxidative modifications in atherosclerosis. *Physiol Rev* 2004; 84: 1381–1478.
215. Gaut JP, Heinecke JW. Mechanisms for oxidizing low-density lipoprotein. Insights from patterns of oxidation products in the artery wall and from mouse models of atherosclerosis. *Trends Cardiovasc Med* 2001; 11: 103–112.
216. Ishigaki Y, Katagiri H, Gao J, Yamada T, Imai J, Uno K, Hasegawa Y, Kaneko K, Ogihara T, Ishihara H et al. Impact of plasma oxidized low-density lipoprotein removal on atherosclerosis. *Circulation* 2008; 118: 75–83.
217. Williams KJ, Feig JE, Fisher EA. Rapid regression of atherosclerosis: insights from the clinical and experimental literature. *Nat Clin Pract Cardiovasc Med* 2008; 5: 91–102.
218. Gugiu BG, Mouillesseaux K, Duong V, Herzog T, Hekimian A, Koroniak L, Vondriska TM, Watson AD. Protein targets of oxidized phospholipids in endothelial cells. *J Lipid Res* 2008; 49: 510–520.
219. Hansen-Hagge TE, Baumeister E, Bauer T, Schmiedeke D, Renne T, Wanner C, Galle J. Transmission of oxLDL-derived lipid peroxide radicals into membranes of vascular cells is the main inducer of oxLDL-mediated oxidative stress. *Atherosclerosis* 2008; 197: 602–611.
220. Van Lenten BJ, Navab M, Shih D, Fogelman AM, Lusis AJ. The role of high-density lipoproteins in oxidation and inflammation. *Trends Cardiovasc Med* 2001; 11: 155–161.
221. Suc I, Escargueil-Blanc I, Troly M, Salvayre R, Negre-Salvayre A. HDL and ApoA prevent cell death of endothelial cells induced by oxidized LDL. *Arterioscler Thromb Vasc Biol* 1997; 17: 2158–2166.
222. Sugano M, Tsuchida K, Makino N. High-density lipoproteins protect endothelial cells from tumor necrosis factor-alpha-induced apoptosis. *Biochem Biophys Res Commun* 2000; 272: 872–876.
223. Robbesyn F, Garcia V, Auge N, Vieira O, Frisach MF, Salvayre R, Negre-Salvayre A. HDL counterbalance the proinflammatory effect of oxidized LDL by inhibiting intracellular reactive oxygen species rise, proteasome activation, and subsequent NF-kappaB activation in smooth muscle cells. *FASEB J* 2003; 17: 743–745.
224. Yvan-Charvet L, Pagler TA, Seimon TA, Thorp E, Welch CL, Witztum JL, Tabas I, Tall AR. ABCA1 and ABCG1 protect against oxidative stress-induced macrophage apoptosis during efferocytosis. *Circ Res* 2010; 106: 1861–1869.
225. Tabet F, Lambert G, Cuesta Torres LF, Hou L, Sotirchos I, Touyz RM, Jenkins AJ, Barter PJ, Rye K-A. Lipid-free apolipoprotein A-I and discoidal reconstituted high-density lipoprotein differentially inhibit glucose-induced oxidative stress in human macrophages. *Arterioscler Thromb Vasc Biol* 2011; 31: 1192–1200.
226. Nicholls SJ, Dusting GJ, Cutri B, Bao S, Drummond GR, Rye KA, Barter PJ. Reconstituted high-density lipoproteins inhibit the acute pro-oxidant and proinflammatory vascular changes induced by a periarterial collar in normocholesterolemic rabbits. *Circulation* 2005; 111: 1543–1550.

227. Klimov AN, Gurevich VS, Nikiforova AA, Shatilina LV, Kuzmin AA, Plavinsky SL, Teryukova NP. Antioxidative activity of high density lipoproteins in vivo. *Atherosclerosis* 1993; 100: 13–18.

228. Botelho PB, Fioratti CO, Abdalla DS, Bertolami MC, Castro IA. Classification of individuals with dyslipidaemia controlled by statins according to plasma biomarkers of oxidative stress using cluster analysis. *Br J Nutr* 2010; 103: 256–265.

229. Hansel B, Giral P, Nobecourt E, Chantepie S, Bruckert E, Chapman MJ, Kontush A. Metabolic syndrome is associated with elevated oxidative stress and dysfunctional dense high-density lipoprotein particles displaying impaired antioxidative activity. *J Clin Endocrinol Metab* 2004; 89: 4963–4971.

230. Chen W, Gardner JP, Kimura M, Brimacombe M, Cao X, Srinivasan SR, Berenson GS, Aviv A. Leukocyte telomere length is associated with HDL cholesterol levels: The Bogalusa heart study. *Atherosclerosis* 2009; 205: 620–625.

231. Choudhury RP, Rong JX, Trogan E, Elmalem VI, Dansky HM, Breslow JL, Witztum JL, Fallon JT, Fisher EA. High-density lipoproteins retard the progression of atherosclerosis and favorably remodel lesions without suppressing indices of inflammation or oxidation. *Arterioscler Thromb Vasc Biol* 2004; 24: 1904–1909.

232. Moore RE, Navab M, Millar JS, Zimetti F, Hama S, Rothblat GH, Rader DJ. Increased atherosclerosis in mice lacking apolipoprotein A-I attributable to both impaired reverse cholesterol transport and increased inflammation. *Circ Res* 2005; 97: 763–771.

233. Kontush A, Chapman MJ. Antiatherogenic function of HDL particle subpopulations: focus on antioxidative activities. *Curr Opin Lipidol* 2010; 21: 312–318.

234. Chantepie S, Malle E, Sattler W, Chapman MJ, Kontush A. Distinct HDL subclasses present similar intrinsic susceptibility to oxidation by HOCl. *Arch Biochem Biophys* 2009; 487: 28–35.

235. Nuchi CD, Hernandez P, McClements DJ, Decker EA. Ability of lipid hydroperoxides to partition into surfactant micelles and alter lipid oxidation rates in emulsions. *J Agric Food Chem* 2002; 50: 5445–5449.

236. Christison JK, Rye KA, Stocker R. Exchange of oxidized cholesteryl linoleate between LDL and HDL mediated by cholesteryl ester transfer protein. *J Lipid Res* 1995; 36: 2017–2026.

237. Girotti AW. Translocation as a means of disseminating lipid hydroperoxide-induced oxidative damage and effector action. *Free Radic Biol Med* 2008; 44: 956–968.

238. Wiesner P, Leidl K, Boettcher A, Schmitz G, Liebisch G. Lipid profiling of FPLC-separated lipoprotein fractions by electrospray ionization tandem mass spectrometry. *J Lipid Res* 2009; 50: 574–585.

239. Bowry VW, Stanley KK, Stocker R. High density lipoprotein is the major carrier of lipid hydroperoxides in human blood plasma from fasting donors. *Proc Natl Acad Sci U S A* 1992; 89: 10316–10320.

240. Ahotupa M, Suomela JP, Vuorimaa T, Vasankari T. Lipoprotein-specific transport of circulating lipid peroxides. *Ann Med* 2010; 42: 521–529.

241. Proudfoot JM, Barden AE, Loke WM, Croft KD, Puddey IB, Mori TA. HDL is the major lipoprotein carrier of plasma F2-isoprostanes. *J Lipid Res* 2009; 50: 716–722.

242. Christison J, Karjalainen A, Brauman J, Bygrave F, Stocker R. Rapid reduction and removal of HDL- but not LDL-associated cholesteryl ester hydroperoxides by rat liver perfused in situ. *Biochem J* 1996; 314: 739–742.

243. Panzenbock U, Stocker R. Formation of methionine sulfoxide-containing specific forms of oxidized high-density lipoproteins. *Biochim Biophys Acta* 2005; 1703: 171–181.
244. Sakai M, Miyazaki A, Hakamata H, Suginohara Y, Sakamoto YI, Morikawa W, Kobori S, Schichiri M, Horiuchi S. Reconstituted high density lipoprotein reduces the capacity of oxidatively modified low density lipoprotein to accumulate cholesteryl esters in mouse peritoneal macrophages. *Atherosclerosis* 1996; 119: 191–202.
245. Miyazaki A, Sakai M, Suginohara Y, Hakamata H, Sakamoto Y, Morikawa W, Horiuchi S. Acetylated low density lipoprotein reduces its ligand activity for the scavenger receptor after interaction with reconstituted high density lipoprotein. *J Biol Chem* 1994; 269: 5264–5269.
246. Marchesi M, Booth EA, Rossoni G, Garcia RA, Hill KR, Sirtori CR, Bisgaier CL, Lucchesi BR. Apolipoprotein A-I Milano/POPC complex attenuates post-ischemic ventricular dysfunction in the isolated rabbit heart. *Atherosclerosis* 2008; 197: 572–578.
247. Vazquez E, Sethi AA, Sampson M, Remaley AT, Chaudhry H, Cannon RO III. "Dysfunctional" high-density lipoprotein despite normal cholesterol content in sedentary obese women. *Circulation* 2009; 120: S470.
248. Klimov AN, Kozhevnikova KA, Kuzmin AA, Kuznetsov AS, Belova EV. On the ability of high density lipoproteins to remove phospholipid peroxidation products from erythrocyte membranes. *Biochemistry* 2001: 66: 300–304.
249. Ferretti G, Bacchetti T, Moroni C, Vignini A, Curatola G. Copper-induced oxidative damage on astrocytes: protective effect exerted by human high density lipoproteins. *Biochim Biophys Acta* 2003; 1635: 48–54.
250. Ferretti G, Bacchetti T, Busni D, Rabini RA, Curatola G. Protective effect of paraoxonase activity in high-density lipoproteins against erythrocyte membranes peroxidation: a comparison between healthy subjects and type 1 diabetic patients. *J Clin Endocrinol Metab* 2004; 89: 2957–2962.
251. Murphy AJ, Woollard KJ, Hoang A, Mukhamedova N, Stirzaker RA, McCormick SPA, Remaley AT, Sviridov D, Chin-Dusting J. High-density lipoprotein reduces the human monocyte inflammatory response. *Arterioscler Thromb Vasc Biol* 2008; 28: 2071–2077.
252. Murphy AJ, Suhartoyo A, Shaw J, Woollard K, Sviridov D, Chin-Dusting J. HDL and apoA-I attenuate neutrophil activation in in vitro and in vivo models of inflammation: A role for lipid rafts. *Circulation* 2009; 120 (Suppl.): S1166.
253. Terasaka N, Wang N, Yvan-Charvet L, Tall AR. High-density lipoprotein protects macrophages from oxidized low-density lipoprotein-induced apoptosis by promoting efflux of 7-ketocholesterol via ABCG1. *Proc Natl Acad Sci U S A* 2007; 104: 15093–15098.
254. Tall AR. The effects of cholesterol ester transfer protein inhibition on cholesterol efflux. *Am J Cardiol* 2009; 104: 39E–45E.
255. Vila A, Korytowski W, Girotti AW. Spontaneous transfer of phospholipid and cholesterol hydroperoxides between cell membranes and low-density lipoprotein: assessment of reaction kinetics and prooxidant effects. *Biochemistry* 2002; 41: 13705–13716.
256. Vila A, Korytowski W, Girotti AW. Spontaneous intermembrane transfer of various cholesterol-derived hydroperoxide species: kinetic studies with model membranes and cells. *Biochemistry* 2001; 40: 14715–14726.

257. Sattler W, Christison J, Stocker R. Cholesterylester hydroperoxide reducing activity associated with isolated high- and low-density lipoproteins. *Free Radic Biol Med* 1995; 18: 421–429.

258. Garner B, Witting PK, Waldeck AR, Christison JK, Raftery M, Stocker R. Oxidation of high density lipoproteins. I. Formation of methionine sulfoxide in apolipoproteins AI and AII is an early event that accompanies lipid peroxidation and can be enhanced by alpha-tocopherol. *J Biol Chem* 1998; 273: 6080–6087.

259. Garner B, Waldeck AR, Witting PK, Rye KA, Stocker R. Oxidation of high density lipoproteins. II. Evidence for direct reduction of lipid hydroperoxides by methionine residues of apolipoproteins AI and AII. *J Biol Chem* 1998; 273: 6088–6095.

260. de Souza JA, Vindis C, Negre-Salvayre A, Rye KA, Couturier M, Therond P, Chantepie S, Salvayre R, Chapman MJ, Kontush A. Small, dense HDL 3 particles attenuate apoptosis in endothelial cells: pivotal role of apolipoprotein A-I. *J Cell Mol Med* 2010; 14: 608–620.

261. Escargueil-Blanc I, Salvayre R, Vacaresse N, Jurgens G, Darblade B, Arnal JF, Parthasarathy S, Negre-Salvayre A. Mildly oxidized LDL induces activation of platelet-derived growth factor beta-receptor pathway. *Circulation* 2001; 104: 1814–1821.

262. Morrow JD. Quantification of isoprostanes as indices of oxidant stress and the risk of atherosclerosis in humans. *Arterioscler Thromb Vasc Biol* 2005; 25: 279–286.

263. Wang W, Xu H, Shi Y, Nandedkar S, Zhang H, Gao H, Feroah T, Weihrauch D, Schulte ML, Jones DW et al. Genetic deletion of apolipoprotein A-I increases airway hyperresponsiveness, inflammation, and collagen deposition in the lung. *J Lipid Res* 2010; 51: 2560–2570.

264. Bashtovyy D, Jones MK, Anantharamaiah GM, Segrest JP. Sequence conservation of apolipoprotein A-I affords novel insights into HDL structure-function. *J Lipid Res* 2011; 52: 435–450.

265. Navab M, Hama SY, Cooke CJ, Anantharamaiah GM, Chaddha M, Jin L, Subbanagounder G, Faull KF, Reddy ST, Miller NE et al. Normal high density lipoprotein inhibits three steps in the formation of mildly oxidized low density lipoprotein: step 1. *J Lipid Res* 2000; 41: 1481–1494.

266. Zhang X, Zhu X, Chen B. Inhibition of collar-induced carotid atherosclerosis by recombinant apoA-I cysteine mutants in apoE-deficient mice. *J Lipid Res* 2010; 51: 3434–3442.

267. Nguyen SD, Jeong TS, Sok DE. Apolipoprotein A-I-mimetic peptides with antioxidant actions. *Arch Biochem Biophys* 2006; 451: 34–42.

268. Boschi-Muller S, Gand A, Branlant G. The methionine sulfoxide reductases: Catalysis and substrate specificities. *Arch Biochem Biophys* 2008; 474: 266–273.

269. Zhang X, Wang J, Chen B. Effect in vitro of apolipoprotein A-V on the structure and functions of recombinant high density lipoprotein. *Biochem Biophys Res Commun* 2010; 401: 473–479.

270. Boisfer E, Stengel D, Pastier D, Laplaud PM, Dousset N, Ninio E, Kalopissis AD. Antioxidant properties of HDL in transgenic mice overexpressing human apolipoprotein A-II. *J Lipid Res* 2002; 43: 732–741.

271. Ribas V, Sanchez-Quesada JL, Anton R, Camacho M, Julve J, Escola-Gil JC, Vila L, Ordonez-Llanos J, Blanco-Vaca F. Human apolipoprotein A-II enrichment displaces paraoxonase from HDL and impairs its antioxidant properties: a new mechanism

linking HDL protein composition and antiatherogenic potential. *Circ Res* 2004; 95: 789–797.

272. Ostos MA, Conconi M, Vergnes L, Baroukh N, Ribalta J, Girona J, Caillaud JM, Ochoa A, Zakin MM. Antioxidative and antiatherosclerotic effects of human apolipoprotein A-IV in apolipoprotein E-deficient mice. *Arterioscler Thromb Vasc Biol* 2001; 21: 1023–1028.
273. Wong W-MR, Gerry AB, Putt W, Roberts JL, Weinberg RB, Humphries SE, Leake DS, Talmud PJ. Common variants of apolipoprotein A-IV differ in their ability to inhibit low density lipoprotein oxidation. *Atherosclerosis* 2007; 192: 266.
274. Davignon J. Apolipoprotein E and atherosclerosis: beyond lipid effect. *Arterioscler Thromb Vasc Biol* 2005; 25: 267–269.
275. Miyata M, Smith JD. Apolipoprotein E allele-specific antioxidant activity and effects on cytotoxicity by oxidative insults and beta-amyloid peptides. *Nat Genet* 1996; 14: 55–61.
276. Thorngate FE, Rudel LL, Walzem RL, Williams DL. Low levels of extrahepatic non-macrophage ApoE inhibit atherosclerosis without correcting hypercholesterolemia in ApoE-deficient mice. *Arterioscler Thromb Vasc Biol* 2000; 20: 1939–1945.
277. Tangirala RK, Pratico D, FitzGerald GA, Chun S, Tsukamoto K, Maugeais C, Usher DC, Pure E, Rader DJ. Reduction of isoprostanes and regression of advanced atherosclerosis by apolipoprotein E. *J Biol Chem* 2001; 276: 261–266.
278. Raffai RL, Loeb SM, Weisgraber KH. Apolipoprotein E promotes the regression of atherosclerosis independently of lowering plasma cholesterol levels. *Arterioscler Thromb Vasc Biol* 2005; 25: 436–441.
279. Navab M, Hama LS, Van LB, Fonarow GC, Cardinez CJ, Castellani LW, Brennan ML, Lusis AJ, Fogelman AM. Mildly oxidized LDL induces an increased apolipoprotein J/paraoxonase ratio. *J Clin Invest* 1997; 99: 2005–2019.
280. Trougakos IP, Lourda M, Agiostratidou G, Kletsas D, Gonos ES. Differential effects of clusterin/apolipoprotein J on cellular growth and survival. *Free Radic Biol Med* 2005; 38: 436–449.
281. Jordan-Starck TC, Witte DP, Aronow BJ, Harmony JA. Apolipoprotein J: a membrane policeman? *Curr Opin Lipidol* 1992; 1992: 75–85.
282. Dahlback B, Nielsen LB. Apolipoprotein M affecting lipid metabolism or just catching a ride with lipoproteins in the circulation? *Cell Mol Life Sci* 2009; 66: 559–564.
283. Kunitake ST, Jarvis MR, Hamilton RL, Kane JP. Binding of transition metals by apolipoprotein A-I-containing plasma lipoproteins: inhibition of oxidation of low density lipoproteins. *Proc Natl Acad Sci U S A* 1992; 89: 6993–6997.
284. Maddipati KR, Marnett LJ. Characterization of the major hydroperoxide-reducing activity of human plasma. Purification and properties of a selenium-dependent glutathione peroxidase. *J Biol Chem* 1987; 262: 17398–17403.
285. Arthur JR. The glutathione peroxidases. *Cell Mol Life Sci* 2000; 57: 1825–1835.
286. Chen N, Liu Y, Greiner CD, Holtzman JL. Physiologic concentrations of homocysteine inhibit the human plasma GSH peroxidase that reduces organic hydroperoxides. *J Lab Clin Med* 2000; 136: 58–65.
287. Shih DM, Gu L, Xia YR, Navab M, Li WF, Hama S, Castellani LW, Furlong CE, Costa LG, Fogelman AM et al. Mice lacking serum paraoxonase are susceptible to organophosphate toxicity and atherosclerosis. *Nature* 1998; 394: 284–287.

288. Tward A, Xia YR, Wang XP, Shi YS, Park C, Castellani LW, Lusis AJ, Shih DM. Decreased atherosclerotic lesion formation in human serum paraoxonase transgenic mice. *Circulation* 2002; 106: 484–490.

289. Bhattacharyya T, Nicholls SJ, Topol EJ, Zhang R, Yang X, Schmitt D, Fu X, Shao M, Brennan DM, Ellis SG et al. Relationship of paraoxonase 1 (PON1) gene polymorphisms and functional activity with systemic oxidative stress and cardiovascular risk. *JAMA* 2008; 299: 1265–1276.

290. Regieli JJ, Jukema JW, Doevendans PA, Zwinderman AH, Kastelein JJ, Grobbee DE, van der Graaf Y. Paraoxonase variants relate to 10-year risk in coronary artery disease: impact of a high-density lipoprotein-bound antioxidant in secondary prevention. *J Am Coll Cardiol* 2009; 54: 1238–1245.

291. Mackness MI, Mackness B, Durrington PN, Fogelman AM, Berliner J, Lusis AJ, Navab M, Shih D, Fonarow GC. Paraoxonase and coronary heart disease. *Curr Opin Lipidol* 1998; 9: 319–324.

292. Aviram M, Rosenblat M. Paraoxonases and cardiovascular diseases: pharmacological and nutritional influences. *Curr Opin Lipidol* 2005; 16: 393–399.

293. Guns PJ, Van Assche T, Verreth W, Fransen P, Mackness B, Mackness M, Holvoet P, Bult H. Paraoxonase 1 gene transfer lowers vascular oxidative stress and improves vasomotor function in apolipoprotein E-deficient mice with pre-existing atherosclerosis. *Br J Pharmacol* 2008; 153: 508–516.

294. Efrat M, Aviram M. Macrophage paraoxonase 1 (PON1) binding sites. *Biochem Biophys Res Commun* 2008; 376: 105–110.

295. Rozenberg O, Shiner M, Aviram M, Hayek T. Paraoxonase 1 (PON1) attenuates diabetes development in mice through its antioxidative properties. *Free Radic Biol Med* 2008; 44: 1951–1959.

296. Josse D, Ebel C, Stroebel D, Fontaine A, Borges F, Echalier A, Baud D, Renault F, Le Maire M, Chabrieres E et al. Oligomeric states of the detergent-solubilized human serum paraoxonase (PON1). *J Biol Chem* 2002; 277: 33386–33397.

297. Harel M, Aharoni A, Gaidukov L, Brumshtein B, Khersonsky O, Meged R, Dvir H, Ravelli RB, McCarthy A, Toker L et al. Structure and evolution of the serum paraoxonase family of detoxifying and anti-atherosclerotic enzymes. *Nat Struct Mol Biol* 2004; 11: 412–419.

298. James RW, Deakin SP. The importance of high-density lipoproteins for paraoxonase-1 secretion, stability, and activity. *Free Radic Biol Med* 2004; 37: 1986–1994.

299. Gaidukov L, Tawfik DS. High affinity, stability, and lactonase activity of serum paraoxonase PON1 anchored on HDL with apoA-I. *Biochemistry* 2005; 44: 11843–11854.

300. Deakin S, Moren X, James RW. Very low density lipoproteins provide a vector for secretion of paraoxonase-1 from cells. *Atherosclerosis* 2005; 179: 17–25.

301. James RW, Brulhart-Meynet M-C, Singh AK, Riederer B, Seidler U, Out R, Van Berkel TJC, Deakin S. The scavenger receptor class B, type I is a primary determinant of paraoxonase-1 association with high-density lipoproteins. *Arterioscler Thromb Vasc Biol* 2010; 30: 2121–2127.

302. Deakin SP, Bioletto S, Bochaton-Piallat ML, James RW. HDL-associated paraoxonase-1 can redistribute to cell membranes and influence sensitivity to oxidative stress. *Free Radic Biol Med* 2011; 50: 102–109.

303. Khersonsky O, Rosenblat M, Toker L, Yacobson S, Hugenmatter A, Silman I, Sussman JL, Aviram M, Tawfik DS. Directed evolution of serum paraoxonase PON3 by family shuffling and ancestor/consensus mutagenesis, and its biochemical characterization. *Biochemistry* 2009; 48: 6644–6654.

304. Meilin E, Aviram M, Hayek T. Paraoxonase 2 (PON2) decreases high glucose-induced macrophage triglycerides (TG) accumulation, via inhibition of NADPH-oxidase and DGAT1 activity: Studies in PON2-deficient mice. *Atherosclerosis* 2010; 208: 390–395.

305. Ng CJ, Hama SY, Bourquard N, Navab M, Reddy ST. Adenovirus mediated expression of human paraoxonase 2 protects against the development of atherosclerosis in apolipoprotein E-deficient mice. *Mol Genet Metab* 2006; 89: 368–373.

306. She Z-G, Zheng W, Wei Y-S, Chen H-Z, Wang A-B, Li H-L, Liu G, Zhang R, Liu J-J, Stallcup WB et al. Human paraoxonase gene cluster transgenic overexpression represses atherogenesis and promotes atherosclerotic plaque stability in apoE-null mice. *Circ Res* 2009; 104: 1160–1168.

307. Graham A, Hassall DG, Rafique S, Owen JS. Evidence for a paraoxonase-independent inhibition of low-density lipoprotein oxidation by high-density lipoprotein. *Atherosclerosis* 1997; 135: 193–204.

308. Durrington PN, Mackness B, Mackness MI. Paraoxonase and atherosclerosis. *Arterioscler Thromb Vasc Biol* 2001; 21: 473–480.

309. Karabina SA, Lehner AN, Frank E, Parthasarathy S, Santanam N. Oxidative inactivation of paraoxonase-implications in diabetes mellitus and atherosclerosis. *Biochim Biophys Acta* 2005; 1725: 213–221.

310. Aviram M, Billecke S, Sorenson R, Bisgaier C, Newton R, Rosenblat M, Erogul J, Hsu C, Dunlop C, La Du B. Paraoxonase active site required for protection against LDL oxidation involves its free sulfhydryl group and is different from that required for its arylesterase/paraoxonase activities: selective action of human paraoxonase allozymes Q and R. *Arterioscler Thromb Vasc Biol* 1998; 18: 1617–1624.

311. McPherson PA, Young IS, McEneny J. A dual role for lecithin:cholesterol acyltransferase (EC 2.3.1.43) in lipoprotein oxidation. *Free Radic Biol Med* 2007; 43: 1484–1493.

312. Ahmed Z, Ravandi A, Maguire GF, Emili A, Draganov D, La Du BN, Kuksis A, Connelly PW. Apolipoprotein A-I promotes the formation of phosphatidylcholine core aldehydes that are hydrolyzed by paraoxonase (PON-1) during high density lipoprotein oxidation with a peroxynitrite donor. *J Biol Chem* 2001; 276: 24473–24481.

313. Tselepis AD, Chapman MJ. Inflammation, bioactive lipids and atherosclerosis: potential roles of a lipoprotein-associated phospholipase A2, platelet activating factor-acetylhydrolase. *Atherosclerosis* 2002; 3 (Suppl.): 57–68.

314. Quarck R, De Geest B, Stengel D, Mertens A, Lox M, Theilmeier G, Michiels C, Raes M, Bult H, Collen D et al. Adenovirus-mediated gene transfer of human platelet-activating factor-acetylhydrolase prevents injury-induced neointima formation and reduces spontaneous atherosclerosis in apolipoprotein E-deficient mice. *Circulation* 2001; 103: 2495–2500.

315. Tsimihodimos V, Kakafika A, Tambaki AP, Bairaktari E, Chapman MJ, Elisaf M, Tselepis AD. Fenofibrate induces HDL-associated PAF-AH but attenuates enzyme activity associated with apoB-containing lipoproteins. *J Lipid Res* 2003; 44: 927–934.

316. Zalewski A, Macphee C. Role of lipoprotein-associated phospholipase A2 in atherosclerosis: Biology, epidemiology, and possible therapeutic target. *Arterioscler Thromb Vasc Biol* 2005; 25: 923–931.

317. Arakawa H, Qian J-Y, Baatar D, Karasawa K, Asada Y, Sasaguri Y, Miller ER, Witztum JL, Ueno H. Local expression of platelet-activating factor-acetylhydrolase reduces accumulation of oxidized lipoproteins and inhibits inflammation, shear stress-induced thrombosis, and neointima formation in balloon-injured carotid arteries in nonhyperlipidemic rabbits. *Circulation* 2005; 111: 3302–3309.

318. Webb NR. Getting to the core of atherosclerosis. *Nat Med* 2008; 14: 1015–1016.

319. Molina Portela MP, Raper J, Tomlinson S. An investigation into the mechanism of trypanosome lysis by human serum factors. *Mol Biochem Parasitol* 2000; 110: 273–282.

320. Miller YI, Smith A, Morgan WT, Shaklai N. Role of hemopexin in protection of low-density lipoprotein against hemoglobin-induced oxidation. *Biochemistry* 1996; 35: 13112–13117.

321. Goulinet S, Chapman MJ. Plasma LDL and HDL subspecies are heterogenous in particle content of tocopherols and oxygenated and hydrocarbon carotenoids. Relevance to oxidative resistance and atherogenesis. *Arterioscler Thromb Vasc Biol* 1997; 17: 786–796.

322. Cabana VG, Reardon CA, Feng N, Neath S, Lukens J, Getz GS. Serum paraoxonase: effect of the apolipoprotein composition of HDL and the acute phase response. *J Lipid Res* 2003; 44: 780–792.

323. Kontush A, Chantepie S, Chapman MJ. Small, dense HDL particles exert potent protection of atherogenic LDL against oxidative stress. *Arterioscler Thromb Vasc Biol* 2003; 23: 1881–1888.

324. Bergmeier C, Siekmeier R, Gross W. Distribution spectrum of paraoxonase activity in HDL fractions. *Clin Chem* 2004; 50: 2309–2315.

325. Kelso GJ, Stuart WD, Richter RJ, Furlong CE, Jordan-Starck TC, Harmony JA. Apolipoprotein J is associated with paraoxonase in human plasma. Biochemistry 1994; 33: 832–839.

326. Cheung MC, Vaisar T, Han X, Heinecke JW, Albers JJ. Phospholipid transfer protein in human plasma associates with proteins linked to immunity and inflammation. *Biochemistry* 2010; 49: 7314–7322.

327. Bisgaier CL, Sachdev OP, Megna L, Glickman RM. Distribution of apolipoprotein A-IV in human plasma. *J Lipid Res* 1985; 26: 11–25.

328. Yoshikawa M, Sakuma N, Hibino T, Sato T, Fujinami T. HDL3 exerts more powerful anti-oxidative, protective effects against copper-catalyzed LDL oxidation than HDL2. *Clin Biochem* 1997; 30: 221–225.

329. Huang JM, Huang ZX, Zhu W. Mechanism of high-density lipoprotein subfractions inhibiting copper-catalyzed oxidation of low-density lipoprotein. *Clin Biochem* 1998; 31: 537–543.

330. Shuhei N, Soderlund S, Jauhiainen M, Taskinen MR. Effect of HDL composition and particle size on the resistance of HDL to the oxidation. *Lipids Health Dis* 2010; 9: 104.

331. Sparks DL, Davidson WS, Lund-Katz S, Phillips MC. Effects of the neutral lipid content of high density lipoprotein on apolipoprotein A-I structure and particle stability. *J Biol Chem* 1995; 270: 26910–26917.

332. Curtiss LK, Bonnet DJ, Rye KA. The conformation of apolipoprotein A-I in high-density lipoproteins is influenced by core lipid composition and particle size: a surface plasmon resonance study. *Biochemistry* 2000; 39: 5712–5721.
333. Subbaiah PV, Liu M. Role of sphingomyelin in the regulation of cholesterol esterification in the plasma lipoproteins. Inhibition of lecithin-cholesterol acyltransferase reaction. *J Biol Chem* 1993; 268: 20156–20163.
334. Bolin DJ, Jonas A. Sphingomyelin inhibits the lecithin-cholesterol acyltransferase reaction with reconstituted high density lipoproteins by decreasing enzyme binding. *J Biol Chem* 1996; 271: 19152–19158.
335. Arbones-Mainar JM, Navarro MA, Carnicer R, Guillen N, Surra JC, Acin S, Guzman MA, Sarria AJ, Arnal C, Aguilera MP et al. Accelerated atherosclerosis in apolipoprotein E-deficient mice fed Western diets containing palm oil compared with extra virgin olive oils: a role for small, dense high-density lipoproteins. *Atherosclerosis* 2007; 194: 372–382.
336. Moren X, Deakin S, Liu ML, Taskinen MR, James RW. HDL subfraction distribution of paraoxonase-1 and its relevance to enzyme activity and resistance to oxidative stress. *J Lipid Res* 2008; 49: 1246–1253.
337. Gaidukov L, Viji RI, Yacobson S, Rosenblat M, Aviram M, Tawfik DS. ApoE induces serum paraoxonase PON1 activity and stability similar to ApoA-I. *Biochemistry*, 2010; 49, 532–538.
338. Rye KA, Barter PJ. Antiinflammatory actions of HDL: a new insight. *Arterioscler Thromb Vasc Biol* 2008; 28: 1890–1891.
339. Ross R. Atherosclerosis - an inflammatory disease. *N Engl J Med* 1999; 340: 115–126.
340. Barter PJ, Nicholls S, Rye KA, Anantharamaiah GM, Navab M, Fogelman AM. Antiinflammatory properties of HDL. *Circ Res* 2004; 95: 764–772.
341. Murphy AJ, Chin-Dusting JP, Sviridov D, Woollard KJ. The anti inflammatory effects of high density lipoproteins. *Curr Med Chem* 2009; 16: 667–675.
342. Wang SH, Yuan SG, Peng DQ, Zhao SP. High-density lipoprotein affects antigen presentation by interfering with lipid raft: a promising anti-atherogenic strategy. *Clin Exp Immunol* 2010; 160: 137–142.
343. Saemann MD, Poglitsch M, Kopecky C, Haidinger M, Horl WH, Weichhart T. The versatility of HDL: a crucial anti-inflammatory regulator. *Eur J Clin Invest* 2010; 40: 1131–1143.
344. Dimayuga P, Zhu J, Oguchi S, Chyu KY, Xu XO, Yano J, Shah PK, Nilsson J, Cercek B. Reconstituted HDL containing human apolipoprotein A-1 reduces VCAM-1 expression and neointima formation following periadventitial cuff-induced carotid injury in apoE null mice. *Biochem Biophys Res Commun* 1999; 264: 465–468.
345. Cockerill GW, Rye KA, Gamble JR, Vadas MA, Barter PJ. High-density lipoproteins inhibit cytokine-induced expression of endothelial cell adhesion molecules. *Arterioscler Thromb Vasc Biol* 1995; 15: 1987–1994.
346. Calabresi L, Franceschini G, Sirtori CR, De Palma A, Saresella M, Ferrante P, Taramelli D. Inhibition of VCAM-1 expression in endothelial cells by reconstituted high density lipoproteins. *Biochem Biophys Res Commun* 1997; 238: 61–65.
347. Baker PW, Rye KA, Gamble JR, Vadas MA, Barter PJ. Ability of reconstituted high density lipoproteins to inhibit cytokine-induced expression of vascular cell adhesion molecule-1 in human umbilical vein endothelial cells. *J Lipid Res* 1999; 40: 345–353.

348. Van Lenten BJ, Hama SY, de Beer FC, Stafforini DM, McIntyre TM, Prescott SM, La Du BN, Fogelman AM, Navab M. Anti-inflammatory HDL becomes pro-inflammatory during the acute phase response. Loss of protective effect of HDL against LDL oxidation in aortic wall cell cocultures. *J Clin Invest* 1995; 96: 2758–2767.

349. Scanu A, Oliviero F, Gruaz L, Sfriso P, Pozzuoli A, Frezzato F, Agostini C, Burger D, Punzi L. High-density lipoproteins downregulate CCL2 production in human fibroblast-like synoviocytes stimulated by urate crystals. *Arthritis Res Ther* 2010; 12: R23.

350. Bursill CA, Castro ML, Beattie DT, Nakhla S, van der Vorst E, Heather AK, Barter PJ, Rye K-A. High-density lipoproteins suppress chemokines and chemokine receptors in vitro and in vivo. *Arterioscler Thromb Vasc Biol* 2010; 30: 1773–1778.

351. Suzuki M, Pritchard DK, Becker L, Hoofnagle AN, Tanimura N, Bammler TK, Beyer RP, Bumgarner R, Vaisar T, de Beer MC et al. High-density lipoprotein suppresses the type I interferon response, a family of potent antiviral immunoregulators, in macrophages challenged with lipopolysaccharide. *Circulation* 2010; 122: 1919–1927.

352. Wang L, Chen WZ, Wu MP. Apolipoprotein A-I inhibits chemotaxis, adhesion, activation of THP-1 cells and improves the plasma HDL inflammatory index. *Cytokine*, 2010; 49, 194–200.

353. Yvan-Charvet L, Pagler T, Gautier E, Avagyan S, Siry R, Snoeck H, Randolph G, Tall AR. The human apoA-1 transgene suppresses myeloid proliferation and accelerated atherosclerosis in LDLr deficient mice reconstituted with ABC transporter deficient bone marrow. *Circulation* 2009; 120 (Suppl.): S1154.

354. Yvan-Charvet L, Pagler T, Gautier EL, Avagyan S, Siry RL, Han S, Welch CL, Wang N, Randolph GJ, Snoeck HW et al. ATP-binding cassette transporters and HDL suppress hematopoietic stem cell proliferation. *Science* 2010; 328: 1689–1693.

355. Okura H, Yamashita S, Ohama T, Saga A, Yamamoto-Kakuta A, Hamada Y, Sougawa N, Ohyama R, Sawa Y, Matsuyama A. HDL/apolipoprotein A-I binds to macrophage-derived progranulin and suppresses its conversion into proinflammatory granulins. *J Atheroscler Thromb* 2010; 17: 568–77.

356. Yu BL, Wang SH, Peng DQ, Zhao SP. HDL and immunomodulation: an emerging role of HDL against atherosclerosis. *Immunol Cell Biol* 2010; 88: 285–290.

357. Hyka N, Dayer JM, Modoux C, Kohno T, Edwards CK III, Roux-Lombard P, Burger D. Apolipoprotein A-I inhibits the production of interleukin-1beta and tumor necrosis factor-alpha by blocking contact-mediated activation of monocytes by T lymphocytes. *Blood* 2001; 97: 2381–2389.

358. Carpintero R, Gruaz L, Brandt KJ, Scanu A, Faille D, Combes V, Grau GE, Burger D. HDL interfere with the binding of T cell microparticles to human monocytes to inhibit pro-inflammatory cytokine production. *PLoS One* 2010; 5: e11869.

359. Gruaz L, Delucinge-Vivier C, Descombes P, Dayer JM, Burger D. Blockade of T cell contact-activation of human monocytes by high-density lipoproteins reveals a new pattern of cytokine and inflammatory genes. *PLoS One* 2010; 5: e9418.

360. Burger D, Dayer JM. High-density lipoprotein-associated apolipoprotein A-I: the missing link between infection and chronic inflammation? *Autoimmun Rev* 2002; 1: 111–117.

361. Wilhelm AJ, Zabalawi M, Grayson JM, Weant AE, Major AS, Owen J, Bharadwaj M, Walzem R, Chan L, Oka K. et al. Apolipoprotein A-I and its role in lymphocyte cholesterol homeostasis and autoimmunity. *Arterioscler Thromb Vasc Biol* 2009; 29: 843–849.

362. Wilhelm AJ, Zabalawi M, Owen JS, Shah D, Grayson JM, Major AS, Bhat S, Gibbs DP Jr, Thomas MJ, Sorci-Thomas MG. Apolipoprotein A-I modulates regulatory T cells in autoimmune LDLr-/-, ApoA-I-/- mice. *J Biol Chem* 2010; 285: 36158–36169.

363. Scanu A, Molnarfi N, Brandt KJ, Gruaz L, Dayer JM, Burger D. Stimulated T cells generate microparticles, which mimic cellular contact activation of human monocytes: differential regulation of pro- and anti-inflammatory cytokine production by high-density lipoproteins. *J Leukoc Biol* 2008; 83: 921–927.

364. Pagler TA, Wang M, Mondal M, Murphy AJ, Westerterp M, Moore KJ, Maxfield FR, Tall AR. Deletion of ABCA1 and ABCG1 impairs macrophage migration because of increased Rac1 signaling. *Circ Res* 2011; 108: 194–200.

365. Mythreye K, Satterwhite LL, Davidson WS, Goldschmidt-Clermont PJ. ApoA-I induced CD31 in bone marrow-derived vascular progenitor cells increases adhesion: implications for vascular repair. *Biochim Biophys Acta* 2008; 1781: 703–709.

366. Liao XL, Lou B, Ma J, Wu MP. Neutrophils activation can be diminished by apolipoprotein A-I. *Life Sci* 2005; 77: 325–335.

367. Fontaine E, Warwick R, Sastry P, Poullis M. Effect of foreign surface pacification with albumin, aprotinin, propofol, and high-density lipoprotein. *J Extra Corpor Technol* 2009; 41: 3–9.

368. Lapergue B, Moreno JA, Dang BQ, Coutard M, Delbosc S, Raphaeli G, Auge N, Klein I, Mazighi M, Michel JB et al. Protective effect of high-density lipoprotein-based therapy in a model of embolic stroke. *Stroke* 2010; 41: 1536–1542.

369. Sampietro T, Bigazzi F, Dal Pino B, Fusaro S, Greco F, Tuoni M, Bionda A. Increased plasma C-reactive protein in familial hypoalphalipoproteinemia: a proinflammatory condition? *Circulation* 2002; 105: 11–14.

370. Pirro M, Siepi D, Lupattelli G, Roscini AR, Schillaci G, Gemelli F, Vaudo G, Marchesi S, Pasqualini L, Mannarino E. Plasma C-reactive protein in subjects with hypo/hyperalphalipoproteinemias. *Metabolism* 2003; 52: 432–436.

371. Dullaart RP, Perton F, Kappelle PJ, de Vries R, Sluiter WJ, van Tol A. Plasma lecithin: cholesterol acyltransferase activity modifies the inverse relationship of C-reactive protein with HDL cholesterol in nondiabetic men. *Biochim Biophys Acta* 2010; 1801: 84–88.

372. Rein P, Saely CH, Beer S, Vonbank A, Drexel H. Roles of the metabolic syndrome, HDL cholesterol, and coronary atherosclerosis in subclinical inflammation. *Diabetes Care* 2010; 33: 1853–1855.

373. Zanetti M, Bosutti A, Ferreira C, Vinci P, Biolo G, Fonda M, Valente M, Cattin L, Guarnieri G, Barazzoni R. Circulating pentraxin 3 levels are higher in metabolic syndrome with subclinical atherosclerosis: evidence for association with atherogenic lipid profile. *Clin Exp Med* 2009; 9: 243–248.

374. Kent JW Jr, Comuzzie AG, Mahaney MC, Almasy L, Rainwater DL, VandeBerg JL, MacCluer JW, Blangero J. Intercellular adhesion molecule-1 concentration is genetically correlated with insulin resistance, obesity, and HDL concentration in Mexican Americans. *Diabetes* 2004; 53: 2691–2695.

375. Yamagishi S, Adachi H, Matsui T, Nakamura K, Enomoto M, Fukami A, Otsuka M, Kumagae S, Nanjo Y, Imaizumi T. Decreased high-density lipoprotein cholesterol level is an independent correlate of circulating tumor necrosis factor-alpha in a general population. *Clin Cardiol* 2009; 32: E29–E32.

376. Zhu X, Lee JY, Timmins JM, Brown JM, Boudyguina E, Mulya A, Gebre AK, Willingham MC, Hiltbold EM. Mishra N et al. Increased cellular free cholesterol in macrophage-specific Abca1 knock-out mice enhances pro-inflammatory response of macrophages. *J Biol Chem* 2008; 283: 22930–22941.

377. Cheng A, Schwartz J, Mitra P, Chait A, Kim F, Handa P. Apolipoprotein-A1 (apoA-1) diminishes lipid raft content and reduces vascular inflammation. *Circulation* 2010; 122 (Suppl.): A17092.

378. Van Linthout S, Spillmann F, Graiani G, Miteva K, Peng J, Van Craeyveld E, Meloni M, Tolle M, Escher F, Subasiguller A et al. Down-regulation of endothelial TLR4 signaling after apo A-I gene transfer contributes to improved survival in an experimental model of lipopolysaccharide-induced inflammation. *J Mol Med* 2010; 89: 151–160.

379. Yvan-Charvet L, Welch C, Pagler TA, Ranalletta M, Lamkanfi M, Han S, Ishibashi M, Li R, Wang N, Tall AR. Increased inflammatory gene expression in ABC transporter-deficient macrophages: free cholesterol accumulation, increased signaling via Toll-like receptors, and neutrophil infiltration of atherosclerotic lesions. *Circulation* 2008; 118: 1837–1847.

380. Zhu X, Owen JS, Wilson MD, Li H, Griffiths GL, Thomas MJ, Hiltbold EM, Fessler MB, Parks JS. Macrophage ABCA1 reduces MyD88-dependent Toll-like receptor trafficking to lipid rafts by reduction of lipid raft cholesterol. *J Lipid Res* 2010; 51: 3196–3206.

381. Whetzel AM, Sturek JM, Nagelin MH, Bolick DT, Gebre AK, Parks JS, Bruce AC, Skaflen MD, Hedrick CC. ABCG1 deficiency in mice promotes endothelial activation and monocyte-endothelial interactions. *Arterioscler Thromb Vasc Biol* 2010; 30: 809–817.

382. Tang C, Liu Y, Kessler PS, Vaughan AM, Oram JF. The macrophage cholesterol exporter ABCA1 functions as an anti-inflammatory receptor. *J Biol Chem* 2009; 284: 32336–32343.

383. Yin K, Deng X, Mo ZC, Zhao GJ, Jiang J, Cui LB, Tan CZ, Wen GB, Fu Y, Tang CK. Tristetraprolin-dependent posttranscriptional regulation of inflammatory cytokines mRNA expression by apolipoprotein A-I: role of ATP-binding membrane cassette transporter A1 and signal trabsducer and activator of transcription 3. *J Biol Chem* 2011; 286: 13834–13845.

384. Furnkranz A, Schober A, Bochkov VN, Bashtrykov P, Kronke G, Kadl A, Binder BR, Weber C, Leitinger N. Oxidized phospholipids trigger atherogenic inflammation in murine arteries. *Arterioscler Thromb Vasc Biol* 2005; 25: 633–638.

385. Theilmeier G, De Geest B, Van Veldhoven PP, Stengel D, Michiels C, Lox M, Landeloos M, Chapman MJ, Ninio E, Collen D et al. HDL-associated PAF-AH reduces endothelial adhesiveness in apoE-/- mice. *FASEB J* 2000; 14: 2032–2039.

386. Bancells C, Sanchez-Quesada JL, Birkelund R, Ordonez-Llanos J, Benitez S. HDL and electronegative LDL exchange anti- and pro-inflammatory properties. *J Lipid Res* 2010; 51: 2947–2956.

387. Navab M, Hama SY, Hough GP, Subbanagounder G, Reddy ST, Fogelman AM. A cell-free assay for detecting HDL that is dysfunctional in preventing the formation of or inactivating oxidized phospholipids. *J Lipid Res* 2001; 42: 1308–1317.

388. Ansell BJ, Navab M, Hama S, Kamranpour N, Fonarow G, Hough G, Rahmani S, Mottahedeh R, Dave R, Reddy ST et al. Inflammatory/antiinflammatory properties of high-density lipoprotein distinguish patients from control subjects better than high-density lipoprotein cholesterol levels and are favorably affected by simvastatin treatment. *Circulation* 2003; 108: 2751–2756.

389. Ansell BJ, Fonarow GC, Fogelman AM. The paradox of dysfunctional high-density lipoprotein. *Curr Opin Lipidol* 2007; 18: 427–434.

390. Robbesyn F, Salvayre R, Negre-Salvayre A. Dual role of oxidized LDL on the NF-kappaB signaling pathway. *Free Radic Res* 2004; 38: 541–551.

391. Xia P, Vadas MA, Rye KA, Barter PJ, Gamble JR. High density lipoproteins (HDL) interrupt the sphingosine kinase signaling pathway. A possible mechanism for protection against atherosclerosis by HDL. *J Biol Chem* 1999; 274: 33143–33147.

392. Park SH, Park JH, Kang JS, Kang YH. Involvement of transcription factors in plasma HDL protection against TNF-alpha-induced vascular cell adhesion molecule-1 expression. *Int J Biochem Cell Biol* 2003; 35: 168–182.

393. McGrath KC, Li XH, Puranik R, Liong EC, Tan JT, Dy VM, DiBartolo BA, Barter PJ, Rye KA, Heather AK. Role of 3beta-hydroxysteroid-delta 24 reductase in mediating antiinflammatory effects of high-density lipoproteins in endothelial cells. *Arterioscler Thromb Vasc Biol* 2009; 29: 877–882.

394. Patel S, Di Bartolo BA, Nakhla S, Heather AK, Mitchell TW, Jessup W, Celermajer DS, Barter PJ, Rye K-A. Anti-inflammatory effects of apolipoprotein A-I in the rabbit. *Atherosclerosis* 2011; 212: 392–397.

395. Kimura T, Tomura H, Sato K, Ito M, Matsuoka I, Im DS, Kuwabara A, Mogi C, Itoh H, Kurose H et al. Mechanism and role of high density lipoprotein-induced activation of AMP-activated protein kinase in endothelial cells. *J Biol Chem* 2010; 285: 4387–4397.

396. Norata GD, Marchesi P, Pirillo A, Uboldi P, Chiesa G, Maina V, Garlanda C, Mantovani A, Catapano AL. Long pentraxin 3, a key component of innate immunity, is modulated by high-density lipoproteins in endothelial cells. *Arterioscler Thromb Vasc Biol* 2008; 28: 925–931.

397. Norata GD, Callegari E, Marchesi M, Chiesa G, Eriksson P, Catapano AL. High-density lipoproteins induce transforming growth factor-beta 2 expression in endothelial cells. *Circulation* 2005; 111: 2805–2811.

398. Gomaraschi M, Basilico N, Sisto F, Taramelli D, Eligini S, Colli S, Sirtori CR, Franceschini G, Calabresi L. High-density lipoproteins attenuate interleukin-6 production in endothelial cells exposed to pro-inflammatory stimuli. *Biochim Biophys Acta* 2005; 1736: 136–143.

399. Coronado MT, Pozzi AO, Punchard MA, Gonzalez P, Fantidis P. Inflammation as a modulator of the HDL cholesterol-induced inteleukin-10 production by human circulating mononuclear cells. *Atherosclerosis* 2009; 202: 183–184.

400. Suc I, Meilhac O, Lajoie-Mazenc I, Vandaele J, Jurgens G, Salvayre R, Negre-Salvayre A. Activation of EGF receptor by oxidized LDL. *FASEB J* 1998; 12: 665–671.

401. Robbesyn F, Auge N, Vindis C, Cantero AV, Barbaras R, Negre-Salvayre A, Salvayre R. High-density lipoproteins prevent the oxidized low-density lipoprotein-induced epidermal growth factor receptor activation and subsequent matrix metalloproteinase-2 upregulation. *Arterioscler Thromb Vasc Biol* 2005; 25: 1206–1212.

402. Recalde D, Ostos MA, Badell E, Garcia-Otin AL, Pidoux J, Castro G, Zakin MM, Scott-Algara D. Human apolipoprotein A-IV reduces secretion of proinflammatory cytokines and atherosclerotic effects of a chronic infection mimicked by lipopolysaccharide. *Arterioscler Thromb Vasc Biol* 2004; 24: 756–761.

403. Nofer J-R, Assmann G. Atheroprotective effects of high-density lipoprotein-associated lysosphingolipids. *Trends Cardiovasc Med* 2005; 15: 265–271.

404. Shi N, Wu MP. Apolipoprotein A-I attenuates renal ischemia/reperfusion injury in rats. *J Biomed Sci* 2008; 15: 577–583.

405. Van Lenten BJ, Wagner AC, Anantharamaiah GM, Garber DW, Fishbein MC, Adhikary L, Nayak DP, Hama S, Navab M, Fogelman AM. Influenza infection promotes macrophage traffic into arteries of mice that is prevented by D-4F, an apolipoprotein A-I mimetic peptide. *Circulation* 2002; 106: 1127–1132.

406. Gomaraschi M, Baldassarre D, Amato M, Eligini S, Conca P, Sirtori CR, Franceschini G, Calabresi L. Normal vascular function despite low levels of high-density lipoprotein cholesterol in carriers of the apolipoprotein A-IMilano mutant. *Circulation* 2007; 116: 2165–2172.

407. Rosenblat M, Aviram M. Paraoxonases role in the prevention of cardiovascular diseases. *Biofactors* 2009; 35: 98–104.

408. Tolle M, Pawlak A, Schuchardt M, Kawamura A, Tietge UJ, Lorkowski S, Keul P, Assmann G, Chun J, Levkau B, et al. HDL-associated lysosphingolipids inhibit NAD(P)H oxidase-dependent monocyte chemoattractant protein-1 production. *Arterioscler Thromb Vasc Biol* 2008; 28: 1542–1548.

409. Keul P, Lucke S, von Wnuck Lipinski K, Bode C, Graler M, Heusch G, Levkau B. Sphingosine-1-phosphate receptor 3 promotes recruitment of monocyte/macrophages in inflammation and atherosclerosis. *Circ Res* 2011; 108: 314–323.

410. Baker PW, Rye KA, Gamble JR, Vadas MA, Barter PJ. Phospholipid composition of reconstituted high density lipoproteins influences their ability to inhibit endothelial cell adhesion molecule expression. *J Lipid Res* 2000; 41: 1261–1267.

411. Ashby DT, Rye KA, Clay MA, Vadas MA, Gamble JR, Barter PJ. Factors influencing the ability of HDL to inhibit expression of vascular cell adhesion molecule-1 in endothelial cells. *Arterioscler Thromb Vasc Biol* 1998; 18: 1450–1455.

412. Park CT, Wright SD. Plasma lipopolysaccharide-binding protein is found associated with a particle containing apolipoprotein A-I, phospholipid, and factor H-related proteins. *J Biol Chem* 1996; 271: 18054–18060.

413. Park CT, Wright SD. Fibrinogen is a component of a novel lipoprotein particle: factor H-related protein (FHRP)-associated lipoprotein particle (FALP). *Blood* 2000; 95: 198–204.

414. Mineo C, Deguchi H, Griffin JH, Shaul PW. Endothelial and antithrombotic actions of HDL. *Circ Res* 2006; 98: 1352–1364.

415. Tabas I. Consequences and therapeutic implications of macrophage apoptosis in atherosclerosis: the importance of lesion stage and phagocytic efficiency. *Arterioscler Thromb Vasc Biol* 2005; 25: 2255–2264.

416. Negre-Salvayre A, Dousset N, Ferretti G, Bacchetti T, Curatola G, Salvayre R. Antioxidant and cytoprotective properties of high-density lipoproteins in vascular cells. *Free Radic Biol Med* 2006; 41: 1031–1040.

417. Speidel MT, Booyse FM, Abrams A, Moore MA, Chung BH. Lipolyzed hypertriglyceridemic serum and triglyceride-rich lipoprotein cause lipid accumulation in and are cytotoxic to cultured human endothelial cells. High density lipoproteins inhibit this cytotoxicity. *Thromb Res* 1990; 58: 251–264.

418. Hamilton KK, Zhao J, Sims PJ. Interaction between apolipoproteins A-I and A-II and the membrane attack complex of complement. Affinity of the apoproteins for polymeric C9. *J Biol Chem* 1993; 268: 3632–3638.

419. Hessler JR, Robertson AL Jr, Chisolm GM III. LDL-induced cytotoxicity and its inhibition by HDL in human vascular smooth muscle and endothelial cells in culture. *Atherosclerosis* 1979; 32: 213–229.

420. Nofer JR, Levkau B, Wolinska I, Junker R, Fobker M, von Eckardstein A, Seedorf U, Assmann G. Suppression of endothelial cell apoptosis by high density lipoproteins (HDL) and HDL-associated lysosphingolipids. *J Biol Chem* 2001; 276: 34480–34485.

421. Rutti S, Ehses JA, Sibler RA, Prazak R, Rohrer L, Georgopoulos S, Meier DT, Niclauss N, Berney T, Donath MY et al. Low- and high-density lipoproteins modulate function, apoptosis, and proliferation of primary human and murine pancreatic beta-cells. *Endocrinology* 2009; 150: 4521–4530.

422. Hayashi H, Campenot RB, Vance DE, Vance JE. Apolipoprotein E-containing lipoproteins protect neurons from apoptosis via a signaling pathway involving low-density lipoprotein receptor-related protein-1. *J Neurosci* 2007; 27: 1933–1941.

423. Browne RW, Shelly WB, Bloom MS, Ocque AJ, Sandler JR, Huddleston HG, Fujimoto VY. Distributions of high-density lipoprotein particle components in human follicular fluid and sera and their associations with embryo morphology parameters during IVF. *Hum Reprod* 2008; 23: 1884–1894.

424. Brodeur MR, Brissette L, Falstrault L, Moreau R. HDL3 reduces the association and modulates the metabolism of oxidized LDL by osteoblastic cells: a protection against cell death. *J Cell Biochem* 2008; 105: 1374--1385.

425. Nofer JR, Junker R, Pulawski E, Fobker M, Levkau B, von Eckardstein A, Seedorf U, Assmann G, Walter M. High density lipoproteins induce cell cycle entry in vascular smooth muscle cells via mitogen activated protein kinase-dependent pathway. *Thromb Haemost* 2001; 85: 730–735.

426. Resink TJ, Bochkov VN, Hahn AW, Philippova MP, Buhler FR, Tkachuk VA. Low- and high-density lipoproteins as mitogenic factors for vascular smooth muscle cells: individual, additive and synergistic effects. *J Vasc Res* 1995; 32: 328–338.

427. Ko Y, Haring R, Stiebler H, Wieczorek AJ, Vetter H, Sachinidis A. High-density lipoprotein reduces epidermal growth factor-induced DNA synthesis in vascular smooth muscle cells. *Atherosclerosis* 1993; 99: 253–259.

428. Kimura T, Sato K, Malchinkhuu E, Tomura H, Tamama K, Kuwabara A, Murakami M, Okajima F. High-density lipoprotein stimulates endothelial cell migration and survival through sphingosine 1-phosphate and its receptors. *Arterioscler Thromb Vasc Biol* 2003; 23: 1283–1288.

429. Sekine Y, Demosky SJ, Stonik JA, Furuya Y, Koike H, Suzuki K, Remaley AT. High-density lipoprotein induces proliferation and migration of human prostate

androgen-independent cancer cells by an ABCA1-dependent mechanism. *Mol Cancer Res* 2010; 8: 1284–1294.

430. Sekine Y, Suzuki K, Remaley AT. HDL and sphingosine-1-phosphate activate STAT3 in prostate cancer DU145 cells via ERK1/2 and S1P receptors, and promote cell migration and invasion. *Prostate* 2011; 71: 690–699.
431. Terasaka N, Yu S, Yvan-Charvet L, Wang N, Mzhavia N, Langlois R, Pagler T, Li R, Welch CL, Goldberg IJ, Tall AR. ABCG1 and HDL protect against endothelial dysfunction in mice fed a high-cholesterol diet. *J Clin Invest* 2008; 118: 3701–3713.
432. Wilson MR, Easterbrook-Smith SB. Clusterin is a secreted mammalian chaperone. *Trends Biochem Sci* 2000; 25: 95–98.
433. Al-Jarallah A, Trigatti BL. A role for the scavenger receptor, class B type I in high density lipoprotein dependent activation of cellular signaling pathways. *Biochim Biophys Acta* 2010; 1801: 1239–1248.
434. Dimmeler S, Fleming I, Fisslthaler B, Hermann C, Busse R, Zeiher AM. Activation of nitric oxide synthase in endothelial cells by Akt-dependent phosphorylation. *Nature* 1999; 399: 601–605.
435. Mineo C, Shaul PW. HDL stimulation of endothelial nitric oxide synthase: a novel mechanism of HDL action. *Trends Cardiovasc Med* 2003; 13: 226–231.
436. Uittenbogaard A, Shaul PW, Yuhanna IS, Blair A, Smart EJ. High density lipoprotein prevents oxidized low density lipoprotein-induced inhibition of endothelial nitric-oxide synthase localization and activation in caveolae. *J Biol Chem* 2000; 275: 11278–11283.
437. Babitt J, Trigatti B, Rigotti A, Smart EJ, Anderson RG, Xu S, Krieger M. Murine SR-BI, a high density lipoprotein receptor that mediates selective lipid uptake, is N-glycosylated and fatty acylated and colocalizes with plasma membrane caveolae. *J Biol Chem* 1997; 272: 13242–13249.
438. Li XA, Guo L, Dressman JL, Asmis R, Smart EJ. A novel ligand-independent apoptotic pathway induced by scavenger receptor class B, type I and suppressed by endothelial nitric-oxide synthase and high density lipoprotein. *J Biol Chem* 2005; 280: 19087–19096.
439. Seetharam D, Mineo C, Gormley AK, Gibson LL, Vongpatanasin W, Chambliss KL, Hahner LD, Cummings ML, Kitchens RL, Marcel YL et al. High-density lipoprotein promotes endothelial cell migration and reendothelialization via scavenger receptor-B type I. *Circ Res* 2006; 98: 63–72.
440. O'Connell BJ, Denis M, Genest J. Cellular physiology of cholesterol efflux in vascular endothelial cells. *Circulation* 2004; 110: 2881–2888.
441. Radojkovic C, Genoux A, Pons V, Combes G, de Jonge H, Champagne E, Rolland C, Perret B, Collet X, Terce F et al. Stimulation of cell surface F1-ATPase activity by apolipoprotein A-I inhibits endothelial cell apoptosis and promotes proliferation. *Arterioscler Thromb Vasc Biol* 2009; 29: 1125–1130.
442. Petremand J, Bulat N, Butty AC, Poussin C, Rutti S, Au K, Ghosh S, Mooser V, Thorens B, Yang JY, Widmann C, Waeber G. Involvement of 4E-BP1 in the protection induced by HDLs on pancreatic beta-cells. *Mol Endocrinol* 2009; 23: 1572–1586.
443. Libby P, Miao P, Ordovas JM, Schaefer EJ. Lipoproteins increase growth of mitogen-stimulated arterial smooth muscle cells. *J Cell Physiol* 1985; 124: 1–8.

444. Honda HM, Wakamatsu BK, Goldhaber JI, Berliner JA, Navab M, Weiss JN. High-density lipoprotein increases intracellular calcium levels by releasing calcium from internal stores in human endothelial cells. *Atherosclerosis* 1999; 143: 299–306.

445. Grewal T, de Diego I, Kirchhoff MF, Tebar F, Heeren J, Rinninger F, Enrich C. High density lipoprotein-induced signaling of the MAPK pathway involves scavenger receptor type BI-mediated activation of Ras. *J Biol Chem* 2003; 278: 16478–16481.

446. Tamama K, Tomura H, Sato K, Malchinkhuu E, Damirin A, Kimura T, Kuwabara A, Murakami M, Okajima F. High-density lipoprotein inhibits migration of vascular smooth muscle cells through its sphingosine 1-phosphate component. *Atherosclerosis* 2005; 178: 19–23.

447. Ardans JA, Economou AP, Martinson JM Jr, Zhou M, Wahl LM. Oxidized low-density and high-density lipoproteins regulate the production of matrix metalloproteinase-1 and -9 by activated monocytes. *J Leukoc Biol* 2002; 71: 1012–1018.

448. Mason DP, Kenagy RD, Hasenstab D, Bowen-Pope DF, Seifert RA, Coats S, Hawkins SM, Clowes AW. Matrix metalloproteinase-9 overexpression enhances vascular smooth muscle cell migration and alters remodeling in the injured rat carotid artery. *Circ Res* 1999; 85: 1179–1185.

449. DeKroon RM, Mihovilovic M, Goodger ZV, Robinette JB, Sullivan PM, Saunders AM, Strittmatter WJ. ApoE genotype-specific inhibition of apoptosis. *J Lipid Res* 2003; 44: 1566–1573.

450. Kim H-J, Yoo E-K, Kim J-Y, Choi Y-K, Lee H-J, Kim J-K, Jeoung NH, Lee K-U, Park I-S, Min B-H et al. Protective role of clusterin/apolipoprotein J against neointimal hyperplasia via antiproliferative effect on vascular smooth muscle cells and cytoprotective effect on endothelial cells. *Arterioscler Thromb Vasc Biol* 2009; 29: 1558–1564.

451. Kimura T, Sato K, Kuwabara A, Tomura H, Ishiwara M, Kobayashi I, Ui M, Okajima F. Sphingosine 1-phosphate may be a major component of plasma lipoproteins responsible for the cytoprotective actions in human umbilical vein endothelial cells. *J Biol Chem* 2001; 276: 31780–31785.

452. Okajima F. Plasma lipoproteins behave as carriers of extracellular sphingosine 1-phosphate: is this an atherogenic mediator or an anti-atherogenic mediator? *Biochim Biophys Acta* 2002; 1582: 132–137.

453. Zhang B, Tomura H, Kuwabara A, Kimura T, Miura S-i, Noda K, Okajima F, Saku K. Correlation of high density lipoprotein (HDL)-associated sphingosine 1-phosphate with serum levels of HDL-cholesterol and apolipoproteins. *Atherosclerosis* 2005; 178: 199–205.

454. Kwon YG, Min JK, Kim KM, Lee DJ, Billiar TR, Kim YM. Sphingosine 1-phosphate protects human umbilical vein endothelial cells from serum-deprived apoptosis by nitric oxide production. *J Biol Chem* 2001; 276: 10627–10633.

455. Tao R, Hoover HE, Honbo N, Kalinowski M, Alano CC, Karliner JS, Raffai R. High-density lipoprotein determines adult mouse cardiomyocyte fate after hypoxia-reoxygenation through lipoprotein-associated sphingosine 1-phosphate. *Am J Physiol Heart Circ Physiol* 2010; 298: H1022–1028.

456. Frias MA, Lang U, Gerber-Wicht C, James RW. Native and reconstituted HDL protect cardiomyocytes from doxorubicin-induced apoptosis. *Cardiovasc Res* 2010; 85: 118–126.

457. Frias MA, James RW, Gerber-Wicht C, Lang U. Native and reconstituted HDL activate Stat3 in ventricular cardiomyocytes via ERK1/2: role of sphingosine-1-phosphate. *Cardiovasc Res* 2009; 82: 313–323.

458. Herzog C, Schmitz M, Levkau B, Herrgott I, Mersmann J, Larmann J, Johanning K, Winterhalter M, Chun J, Muller FU et al. Intravenous sphingosylphosphorylcholine protects ischemic and postischemic myocardial tissue in a mouse model of myocardial ischemia/reperfusion injury. *Mediators Inflamm* 2010; 2010: 425191.

459. Herrera E, Barbas C. Vitamin E: action, metabolism and perspectives. *J Physiol Biochem* 2001; 57: 43–56.

460. Nofer JR, Fobker M, Hobbel G, Voss R, Wolinska I, Tepel M, Zidek W, Junker R, Seedorf U, von Eckardstein A et al. Activation of phosphatidylinositol-specific phospholipase C by HDL-associated lysosphingolipid. Involvement in mitogenesis but not in cholesterol efflux. *Biochemistry* 2000; 39: 15199–15207.

461. Ishigami M, Swertfeger DK, Hui MS, Granholm NA, Hui DY. Apolipoprotein E inhibition of vascular smooth muscle cell proliferation but not the inhibition of migration is mediated through activation of inducible nitric oxide synthase. *Arterioscler Thromb Vasc Biol* 2000; 20: 1020–1026.

462. Ahmed Z, Babaei S, Maguire GF, Draganov D, Kuksis A, La Du BN, Connelly PW. Paraoxonase-1 reduces monocyte chemotaxis and adhesion to endothelial cells due to oxidation of palmitoyl, linoleoyl glycerophosphorylcholine. *Cardiovasc Res* 2003; 57: 225–231.

463. de Souza JA, Vindis C, Hansel B, Negre-Salvayre A, Therond P, Serrano CV Jr, Chantepie S, Salvayre R, Bruckert E, Chapman MJ et al. Metabolic syndrome features small, apolipoprotein A-I-poor, triglyceride-rich HDL3 particles with defective anti-apoptotic activity. *Atherosclerosis* 2008; 197: 84–94.

464. Norata GD, Bjork H, Hamsten A, Catapano AL, Eriksson P. High-density lipoprotein subfraction 3 decreases ADAMTS-1 expression induced by lipopolysaccharide and tumor necrosis factor-alpha in human endothelial cells. *Matrix Biol* 2004; 22: 557–560.

465. Becker AE, de Boer OJ, van Der Wal AC. The role of inflammation and infection in coronary artery disease. *Annu Rev Med* 2001; 52: 289–297.

466. Pajkrt D, Doran JE, Koster F, Lerch PG, Arnet B, van der Poll T, ten Cate JW, van Deventer SJ. Antiinflammatory effects of reconstituted high-density lipoprotein during human endotoxemia. *J Exp Med* 1996; 184: 1601–1608.

467. Levels JH, Abraham PR, van den Ende A, van Deventer SJ. Distribution and kinetics of lipoprotein-bound endotoxin. *Infect Immun* 2001; 69: 2821–2828.

468. Stoll LL, Denning GM, Weintraub NL. Potential role of endotoxin as a proinflammatory mediator of atherosclerosis. *Arterioscler Thromb Vasc Biol* 2004; 24: 2227–2236.

469. Wu A, Hinds CJ, Thiemermann C. High-density lipoproteins in sepsis and septic shock: metabolism, actions, and therapeutic applications. *Shock* 2004; 21: 210–221.

470. Levels JH, Pajkrt D, Schultz M, Hoek FJ, van Tol A, Meijers JC, van Deventer SJ. Alterations in lipoprotein homeostasis during human experimental endotoxemia and clinical sepsis. *Biochim Biophys Acta* 2007; 1771: 1429–1438.

471. Wurfel MM, Kunitake ST, Lichenstein H, Kane JP, Wright SD. Lipopolysaccharide (LPS)-binding protein is carried on lipoproteins and acts as a cofactor in the neutralization of LPS. *J Exp Med* 1994; 180: 1025–1035.

472. Levine DM, Parker TS, Donnelly TM, Walsh A, Rubin AL. In vivo protection against endotoxin by plasma high density lipoprotein. *Proc Natl Acad Sci U S A* 1993; 90: 12040–12044.
473. Xiao GL, Luo ZQ, Xiao G, Li C, Xiong XD, Yang Y, Liu HJ. High-density lipoprotein attenuates lipopolysaccharide-induced acute lung injury in mice.. *Sheng Li Xue Bao* 2008; 60: 403–408 [in Chinese].
474. Miller MA, McTernan PG, Harte AL, Silva NF, Strazzullo P, Alberti KG, Kumar S, Cappuccio FP. Ethnic and sex differences in circulating endotoxin levels: A novel marker of atherosclerotic and cardiovascular risk in a British multi-ethnic population. *Atherosclerosis* 2009; 203: 494–502.
475. Goteiner D, Craig RG, Ashmen R, Janal MN, Eskin B, Lehrman N. Endotoxin levels are associated with high-density lipoprotein, triglycerides, and troponin in patients with acute coronary syndrome and angina: possible contributions from periodontal sources. *J Periodontol* 2008; 79: 2331–2339.
476. Gao Y, Liang C, Zhao R, Lukomski S, Han R. The Scl1 of M41-type group A Streptococcus binds the high-density lipoprotein. *FEMS Microbiol Lett* 2010; 309: 55–61.
477. Hajduk SL, Moore DR, Vasudevacharya J, Siqueira H, Torri AF, Tytler EM, Esko JD. Lysis of *Trypanosoma brucei* by a toxic subspecies of human high density lipoprotein. *J Biol Chem* 1989; 264: 5210–5217.
478. Pays E, Vanhollebeke B, Vanhamme L, Paturiaux-Hanocq F, Nolan DP, Perez-Morga D. The trypanolytic factor of human serum. *Nat Rev Microbiol* 2006; 4: 477–486.
479. Thomson R, Samanovic M, Raper J. Activity of trypanosome lytic factor: a novel component of innate immunity. *Future Microbiol* 2009; 4: 789–796.
480. Samanovic M, Molina-Portela MP, Chessler AD, Burleigh BA, Raper J. Trypanosome lytic factor, an antimicrobial high-density lipoprotein, ameliorates Leishmania infection. *PLoS Pathog* 2009; 5: e1000276.
481. Cruz D, Watson AD, Miller CS, Montoya D, Ochoa MT, Sieling PA, Gutierrez MA, Navab M, Reddy ST, Witztum JL et al. Host-derived oxidized phospholipids and HDL regulate innate immunity in human leprosy. *J Clin Invest* 2008; 118: 2917–2928.
482. Johnston LD, Brown G, Gauthier D, Reece K, Kator H, Van Veld P. Apolipoprotein A-I from striped bass (Morone saxatilis) demonstrates antibacterial activity in vitro. *Comp Biochem Physiol B Biochem Mol Biol* 2008; 151: 167–175.
483. Casas AT, Hubsch AP, Rogers BC, Doran JE. Reconstituted high-density lipoprotein reduces LPS-stimulated TNF alpha. *J Surg Res* 1995; 59: 544–552.
484. Galbois A, Thabut D, Tazi KA, Rudler M, Mohammadi MS, Bonnefont-Rousselot D, Bennani H, Bezeaud A, Tellier Z, Guichard C et al. Ex vivo effects of high-density lipoprotein exposure on the lipopolysaccharide-induced inflammatory response in patients with severe cirrhosis. *Hepatology* 2009; 49: 175–184.
485. Thompson PA, Gauthier KC, Varley AW, Kitchens RL. ABCA1 promotes the efflux of bacterial LPS from macrophages and accelerates recovery from LPS-induced tolerance. *J Lipid Res* 2010; 51: 2672–2685.
486. Cai L, Ji A, de Beer FC, Tannock LR, van der Westhuyzen DR. SR-BI protects against endotoxemia in mice through its roles in glucocorticoid production and hepatic clearance. *J Clin Invest* 2008; 118: 364–375.

487. Kallio KA, Buhlin K, Jauhiainen M, Keva R, Tuomainen AM, Klinge B, Gustafsson A, Pussinen PJ. Lipopolysaccharide associates with pro-atherogenic lipoproteins in periodontitis patients. *Innate Immun* 2008; 14: 247–253.

488. Hajduk SL, Hager KM, Esko JD. Human high density lipoprotein killing of African trypanosomes. *Annu Rev Microbiol* 1994; 48: 139–162.

489. Shiflett AM, Bishop JR, Pahwa A, Hajduk SL. Human high density lipoproteins are platforms for the assembly of multi-component innate immune complexes. *J Biol Chem* 2005; 280: 32578–32585.

490. Perez-Morga D, Vanhollebeke B, Paturiaux-Hanocq F, Nolan DP, Lins L, Homble F, Vanhamme L, Tebabi P, Pays A, Poelvoorde P et al. Apolipoprotein L-I promotes trypanosome lysis by forming pores in lysosomal membranes. *Science* 2005; 309: 469–472.

491. Lecordier L, Vanhollebeke B, Poelvoorde P, Tebabi P, Paturiaux-Hanocq F, Andris F, Lins L, Pays E. C-terminal mutants of apolipoprotein L-I efficiently kill both *Trypanosoma brucei brucei* and *Trypanosoma brucei rhodesiense*. *PLoS Pathog* 2009; 5: e1000685.

492. Harrington JM, Howell S, Hajduk SL. Membrane permeabilization by trypanosome lytic factor, a cytolytic human high density lipoprotein. *J Biol Chem* 2009; 284: 13505–13512.

493. Singh IP, Baron S. Innate defences against viraemia. *Rev Med Virol* 2000; 10: 395–403.

494. Shortridge KF, Ho WK. Human serum lipoproteins as inhibitors of haemagglutination for selected togaviruses. *J Gen Virol* 1974; 23: 113–116.

495. Singh IP, Coppenhaver DH, Chopra AK, Baron S. Innate gastrointestinal immunity: characterization of broadly active viral inhibitors. *Antiviral Res* 2001; 49: 157–167.

496. Ma J, Liao XL, Lou B, Wu MP. Role of apolipoprotein A-I in protecting against endotoxin toxicity. *Acta Biochim Biophys Sin (Shanghai)* 2004; 36: 419–424.

497. Henning MF, Herlax V, Bakas L. Contribution of the C-terminal end of apolipoprotein AI to neutralization of lipopolysaccharide endotoxic effect. *Innate Immun* 2011; 17: 327–337.

498. Li Y, Dong JB, Wu MP. Human ApoA-I overexpression diminishes LPS-induced systemic inflammation and multiple organ damage in mice. *Eur J Pharmacol* 2008; 590: 417–422.

499. Tanaka N, Abe-Dohmae S, Iwamoto N, Fitzgerald ML, Yokoyama S. Helical apolipoproteins of high-density lipoprotein enhance phagocytosis by stabilizing ATP-binding cassette transporter A7. *J Lipid Res* 2010; 51: 2591–2599.

500. Srinivas RV, Birkedal B, Owens RJ, Anantharamaiah GM, Segrest JP, Compans RW. Antiviral effects of apolipoprotein A-I and its synthetic amphipathic peptide analogs. *Virology* 1990; 176: 48–57.

501. Thompson PA, Berbee JF, Rensen PC, Kitchens RL. Apolipoprotein A-II augments monocyte responses to LPS by suppressing the inhibitory activity of LPS-binding protein. *Innate Immun* 2008; 14: 365–374.

502. Lamping N, Dettmer R, Schroder NW, Pfeil D, Hallatschek W, Burger R, Schumann RR. LPS-binding protein protects mice from septic shock caused by LPS or gram-negative bacteria. *J Clin Invest* 1998; 101: 2065–2071.

503. Wurfel MM, Hailman E, Wright SD. Soluble CD14 acts as a shuttle in the neutralization of lipopolysaccharide (LPS) by LPS-binding protein and reconstituted high density lipoprotein. *J Exp Med* 1995; 181: 1743–1754.

504. Hailman E, Albers JJ, Wolfbauer G, Tu AY, Wright SD. Neutralization and transfer of lipopolysaccharide by phospholipid transfer protein. *J Biol Chem* 1996; 271: 12172–12178.

505. Berbee JF, Mooijaart SP, de Craen AJ, Havekes LM, van Heemst D, Rensen PC, Westendorp RG. Plasma apolipoprotein CI protects against mortality from infection in old age. *J Gerontol A Biol Sci Med Sci* 2008; 63: 122–126.

506. Berbee JF, van der Hoogt CC, Kleemann R, Schippers EF, Kitchens RL, van Dissel JT, Bakker-Woudenberg IA, Havekes LM, Rensen PC. Apolipoprotein CI stimulates the response to lipopolysaccharide and reduces mortality in gram-negative sepsis. *Faseb J* 2006; 20: 2162–2164.

507. Westerterp M, Berbee JFP, Pires NMM, van Mierlo GJD, Kleemann R, Romijn JA, Havekes LM, Rensen PCN. Apolipoprotein C-I is crucially involved in lipopolysaccharide-induced atherosclerosis development in apolipoprotein E knock-out mice. *Circulation* 2007; 116: 2173–2181.

508. Vanhollebeke B, Truc P, Poelvoorde P, Pays A, Joshi PP, Katti R, Jannin JG, Pays E. Human *Trypanosoma evansi* infection linked to a lack of apolipoprotein L-I. *N Engl J Med* 2006; 355: 2752–2756.

509. Molina-Portela MP, Samanovic M, Raper J. Distinct roles of apolipoprotein components within the trypanosome lytic factor complex revealed in a novel transgenic mouse model. *J Exp Med* 2008; 205: 1721–1728.

510. Wan G, Zhaorigetu S, Liu Z, Kaini R, Jiang Z, Hu CA. Apolipoprotein L1, a novel Bcl-2 homology domain 3-only lipid-binding protein, induces autophagic cell death. *J Biol Chem* 2008; 283: 21540–21549.

511. Zhaorigetu S, Wan G, Kaini R, Jiang Z, Hu CA. ApoL1, a BH3-only lipid-binding protein, induces autophagic cell death. *Autophagy* 2008; 4: 1079–1082.

512. Genovese G, Friedman DJ, Ross MD, Lecordier L, Uzureau P, Freedman BI, Bowden DW, Langefeld CD, Oleksyk TK, Uscinski Knob AL et al. Association of trypanolytic ApoL1 variants with kidney disease in African Americans. *Science* 2010; 329: 841–845.

513. Glazer NL. Variation in APOL1 gene may contribute to high rates of kidney disease in African Americans. *Circulation: Cardiovascular Genetics* 2011; 4: 98–99.

514. Pays E, Vanhollebeke B. Mutual self-defence: the trypanolytic factor story. *Microbes Infect* 2008; 10: 985–989.

515. Tytler EM, Moore DR, Pierce MA, Hager KM, Esko JD, Hajduk SL. Reconstitution of the trypanolytic factor from components of a subspecies of human high-density lipoproteins. *Mol Biochem Parasitol* 1995; 69: 9–17.

516. Vanhollebeke B, De Muylder G, Nielsen MJ, Pays A, Tebabi P, Dieu M, Raes M, Moestrup SK, Pays E. A haptoglobin-hemoglobin receptor conveys innate immunity to *Trypanosoma brucei* in humans. *Science* 2008; 320: 677–681.

517. Kieft R, Capewell P, Turner CM, Veitch NJ, MacLeod A, Hajduk S. Mechanism of *Trypanosoma brucei gambiense* (group 1) resistance to human trypanosome lytic factor. *Proc Natl Acad Sci U S A* 2010; 107: 16137–16141.

518. Gautier T, Paul C, Deckert V, Desrumaux C, Klein A, Labbe J, Le Guern N, Athias A, Monier S, Hammann A et al. Innate immune response triggered by triacyl lipid A is dependent on phospholipid transfer protein (PLTP) gene expression. *FASEB J* 2010; 24: 3544–3554.
519. Bricarello DA, Mills EJ, Petrlova J, Voss JC, Parikh AN. Ganglioside embedded in reconstituted lipoprotein binds cholera toxin with elevated affinity. *J Lipid Res* 2010; 51: 2731–2738.
520. Kivatinitz SC, Favini P, Ghidoni R. Gangliosides released by the perfused rat liver are associated to high density lipoprotein. *Biochem Int* 1992; 26: 821–829.
521. Valentino LA, Ladisch S. Localization of shed human tumor gangliosides: association with serum lipoproteins. *Cancer Res* 1992; 52: 810–814.
522. Senn HJ, Orth M, Fitzke E, Koster W, Wieland H, Gerok W. Human serum gangliosides in hypercholesterolemia, before and after extracorporeal elimination of LDL. *Atherosclerosis* 1992; 94: 109–117.
523. Ogita H, Liao J. Endothelial function and oxidative stress. *Endothelium* 2004; 11: 123–132.
524. Landmesser U, Drexler H. The clinical significance of endothelial dysfunction. *Curr Opin Cardiol* 2005; 20: 547–551.
525. Nofer JR, van der Giet M, Tolle M, Wolinska I, von Wnuck Lipinski K, Baba HA, Tietge UJ, Godecke A, Ishii I, Kleuser B et al. HDL induces NO-dependent vasorelaxation via the lysophospholipid receptor S1P3. *J Clin Invest* 2004; 113: 569–581.
526. Drew BG, Fidge NH, Gallon-Beaumier G, Kemp BE, Kingwell BA. High-density lipoprotein and apolipoprotein AI increase endothelial NO synthase activity by protein association and multisite phosphorylation. *Proc Natl Acad Sci U S A* 2004; 101: 6999–7004.
527. Beitz J, Forster W. Influence of human low density and high density lipoprotein cholesterol on the in vitro prostaglandin I2 synthetase activity. *Biochim Biophys Acta* 1980; 620: 352–355.
528. Norata GD, Callegari E, Inoue H, Catapano AL. HDL3 induces cyclooxygenase-2 expression and prostacyclin release in human endothelial cells via a p38 MAPK/CRE-dependent pathway: effects on COX-2/PGI-synthase coupling. *Arterioscler Thromb Vasc Biol* 2004; 24: 871–877.
529. Imaizumi S, Miura S, Nakamura K, Kiya Y, Uehara Y, Zhang B, Matsuo Y, Urata H, Ideishi M, Rye KA et al. Antiarrhythmogenic effect of reconstituted high-density lipoprotein against ischemia/reperfusion in rats. *J Am Coll Cardiol* 2008; 51: 1604–1612.
530. Sorrentino SA, Besler C, Rohrer L, Meyer M, Heinrich K, Bahlmann FH, Mueller M, Horvath T, Doerries C, Heinemann M et al. Endothelial-vasoprotective effects of high-density lipoprotein are impaired in patients with type 2 diabetes mellitus but are improved after extended-release niacin therapy. *Circulation* 2010; 121: 110–122.
531. Feng Y, Jacobs F, Van Craeyveld E, Brunaud C, Snoeys J, Tjwa M, Van Linthout S, De Geest B. Human ApoA-I transfer attenuates transplant arteriosclerosis via enhanced incorporation of bone marrow-derived endothelial progenitor cells. *Arterioscler Thromb Vasc Biol* 2008; 28: 278–283.
532. Feng Y, Gordts SC, Chen F, Hu Y, Van Craeyveld E, Jacobs F, Carlier V, Feng Y, Zhang Z, Xu Q et al. Topical HDL administration reduces vein graft atherosclerosis in apo E deficient mice. *Atherosclerosis* 2011; 214: 271–278.

533. Asmis R. "Topical" HDL for vein grafts: a new solution to an old problem? *Atherosclerosis* 2011; 214: 259–260.
534. Kuhn FE, Mohler ER, Satler LF, Reagan K, Lu DY, Rackley CE. Effects of high-density lipoprotein on acetylcholine-induced coronary vasoreactivity. *Am J Cardiol* 1991; 68: 1425–1430.
535. Heffernan KS, Karas RH, Kuvin JT, Jae SY, Vieira VJ, Fernhall B. Carotid artery stiffness, high-density lipoprotein cholesterol and inflammation in men with pre-hypertension. *J Hum Hypertens* 2009; 23: 590–596.
536. Ford MA, McConnell JP, Lavi S, Rihal CS, Prasad A, Sandhu GS, Hartman SJ, Lerman LO, Lerman A. Coronary artery endothelial dysfunction is positively correlated with low density lipoprotein and inversely correlated with high density lipoprotein subclass particles measured by nuclear magnetic resonance spectroscopy. *Atherosclerosis* 2009; 207: 111–115.
537. Roes SD. Alizadeh Dehnavi R, Westenberg JJ, Lamb HJ, Mertens BJ, Tamsma JT, de Roos A. Assessment of aortic pulse wave velocity and cardiac diastolic function in subjects with and without the metabolic syndrome: HDL cholesterol is independently associated with cardiovascular function. *Diabetes Care* 2008; 31: 1442–1444.
538. Feng Y, van Eck M, Van Craeyveld E, Jacobs F, Carlier V, Van Linthout S, Erdel M, Tjwa M, De Geest B. Critical role of scavenger receptor-BI-expressing bone marrow-derived endothelial progenitor cells in the attenuation of allograft vasculopathy after human apo A-I transfer. *Blood* 2009; 113: 755–764.
539. Saddar S, Yuhanna IS, Mineo C, Shaul PW. Mutations in the C-terminal transmembrane domain of scavenger receptor class B type I (SR-BI) selectively inhibit the cholesterol sensing that is required for HDL-initiated intracellular signaling. *Circulation* 2009; 120 (Suppl.): S469.
540. Zhu W, Saddar S, Seetharam D, Chambliss KL, Longoria C, Silver DL, Yuhanna IS, Shaul PW, Mineo C. The scavenger receptor class B type I adaptor protein PDZK1 maintains endothelial monolayer integrity. *Circ Res* 2008; 102: 480–487.
541. Besler C, Heinrich K, Makrides V, Riwanto M, Stein S, Verrey F, Luscher TF, Landmesser U. HDL stimulates endothelial CAT-1 expression and L-arginine uptake: A novel mechanism leading to endothelial-protective effects of HDL that is profoundly altered in patients with coronary disease. *Circulation* 2010; 122 (Suppl.): A19862.
542. Yuhanna IS, Zhu Y, Cox BE, Hahner LD, Osborne-Lawrence S, Lu P, Marcel YL, Anderson RG, Mendelsohn ME, Hobbs HH et al. High-density lipoprotein binding to scavenger receptor-BI activates endothelial nitric oxide synthase. *Nat Med* 2001; 7: 853–857.
543. Li XA, Titlow WB, Jackson BA, Giltiay N, Nikolova-Karakashian M, Uittenbogaard A, Smart EJ. High density lipoprotein binding to scavenger receptor, Class B, type I activates endothelial nitric-oxide synthase in a ceramide-dependent manner. *J Biol Chem* 2002; 277: 11058–11063.
544. Li D, Zhang Y, Ma J, Ling W, Xia M. Adenosine monophosphate activated protein kinase regulates ABCG1-mediated oxysterol efflux from endothelial cells and protects against hypercholesterolemia-induced endothelial dysfunction. *Arterioscler Thromb Vasc Biol* 2010; 30: 1354–1362.
545. Westerterp M, Koetsveld J, Yu S, Han S, Li R, Goldberg IJ, Welch CL, Tall AR. Increased atherosclerosis in mice with vascular ATP-binding cassette transporter G1 deficiency - brief report. *Arterioscler Thromb Vasc Biol* 2010; 30: 2103–2105.

546. Terasaka N, Westerterp M, Koetsveld J, Fernandez-Hernando C, Yvan-Charvet L, Wang N, Sessa WC, Tall AR. ATP-binding cassette transporter G1 and high-density lipoprotein promote endothelial NO synthesis through a decrease in the interaction of caveolin-1 and endothelial NO synthase. *Arterioscler Thromb Vasc Biol* 2010; 30: 2219–2225.

547. Van Linthout S, Spillmann F, Lorenz M, Meloni M, Jacobs F, Egorova M, Stangl V, De Geest B, Schultheiss HP, Tschope C. Vascular-protective effects of high-density lipoprotein include the downregulation of the angiotensin II type 1 receptor. *Hypertension* 2009; 53: 682–687.

548. Feng Y, Van Craeyveld E, Jacobs F, Lievens J, Snoeys J, De Geest B. Wild-type apo A-I and apo A-I(Milano) gene transfer reduce native and transplant arteriosclerosis to a similar extent. *J Mol Med* 2009; 87; 287–297.

549. Igarashi J, Michel T. S1P and eNOS regulation. *Biochim Biophys Acta* 2008; 1781: 489–495.

550. Matsuo Y, Miura S, Kawamura A, Uehara Y, Rye KA, Saku K. Newly developed reconstituted high-density lipoprotein containing sphingosine-1-phosphate induces endothelial tube formation. *Atherosclerosis* 2007; 194: 159–168.

551. Shimizu T, Nakazawa T, Cho A, Dastvan F, Shilling D, Daum G, Reidy MA. Sphingosine 1-phosphate receptor 2 negatively regulates neointimal formation in mouse arteries. *Circ Res* 2007; 101: 995–1000.

552. Gonzalez-Diez M, Rodriguez C, Badimon L, Martinez-Gonzalez J. Prostacyclin induction by high-density lipoprotein (HDL) in vascular smooth muscle cells depends on sphingosine 1-phosphate receptors: effect of simvastatin. *Thromb Haemost* 2008; 100: 119–126.

553. Gong M, Wilson M, Kelly T, Su W, Dressman J, Kincer J, Matveev SV, Guo L, Guerin T, Li XA et al. HDL-associated estradiol stimulates endothelial NO synthase and vasodilation in an SR-BI-dependent manner. *J Clin Invest* 2003; 111: 1579–1587.

554. Badeau RM, Metso J, Wahala K, Tikkanen MJ, Jauhiainen M. Human macrophage cholesterol efflux potential is enhanced by HDL-associated 17beta-estradiol fatty acyl esters. *J Steroid Biochem Mol Biol* 2009; 116: 44–49.

555. Paatela H, Mervaala E, Deb S, Wahala K, Tikkanen MJ. HDL-associated dehydroepiandrosterone fatty acyl esters: enhancement of vasodilatory effect of HDL. *Steroids* 2009; 74: 814–818.

556. Sattler K, Levkau B. Sphingosine-1-phosphate as a mediator of high-density lipoprotein effects in cardiovascular protection. *Cardiovasc Res* 2009; 82: 201–211.

557. Oravec S, Demuth K, Myara I, Hornych A. The effect of high density lipoprotein subfractions on endothelial eicosanoid secretion. *Thromb Res* 1998; 92: 65–71.

558. Furie, B., Furie, B.C. Mechanisms of thrombus formation. *N Engl J Med* 2008; 359: 938–949.

559. Ferroni P, Basili S, Davi G. Platelet activation, inflammatory mediators and hypercholesterolemia. *Curr Vasc Pharmacol* 2003; 1: 157–169.

560. Nofer JR, Brodde MF, Kehrel BE. High-density lipoproteins, platelets and the pathogenesis of atherosclerosis. *Clin Exp Pharmacol Physiol* 2010; 37: 726–735.

561. Lerch PG, Spycher MO, Doran JE. Reconstituted high density lipoprotein (rHDL) modulates platelet activity in vitro and ex vivo. *Thromb Haemost* 1998; 80: 316–320.

562. Naqvi TZ, Shah PK, Ivey PA, Molloy MD. Thomas AM, Panicker S, Ahmed A, Cercek B, Kaul S. Evidence that high-density lipoprotein cholesterol is an independent predictor of acute platelet-dependent thrombus formation. *Am J Cardiol* 1999; 84: 1011–1017.
563. Deguchi H, Pecheniuk NM, Elias DJ, Averell PM, Griffin JH. High-density lipoprotein deficiency and dyslipoproteinemia associated with venous thrombosis in men. *Circulation* 2005; 112: 893–899.
564. Doggen CJ, Smith NL, Lemaitre RN, Heckbert SR, Rosendaal FR, Psaty BM. Serum lipid levels and the risk of venous thrombosis. *Arterioscler Thromb Vasc Biol* 2004; 24: 1970–1975.
565. Eichinger S, Pecheniuk NM, Hron G, Deguchi H, Schemper M, Kyrle PA, Griffin JH. High-density lipoprotein and the risk of recurrent venous thromboembolism. *Circulation* 2007; 115: 1609–1614.
566. Jang MJ, Choi WI, Bang SM, Lee T, Kim YK, Ageno W, Oh D. Metabolic syndrome is associated with venous thromboembolism in the Korean population. *Arterioscler Thromb Vasc Biol* 2009; 29: 311–315.
567. Ageno W, Becattini C, Brighton T, Selby R, Kamphuisen PW. Cardiovascular risk factors and venous thromboembolism: a meta-analysis. *Circulation* 2008; 117: 93–102.
568. Chamberlain AM, Folsom AR, Heckbert SR, Rosamond WD, Cushman M. High-density lipoprotein cholesterol and venous thromboembolism in the Longitudinal Investigation of Thromboembolism Etiology (LITE). *Blood* 2008; 112: 2675–2680.
569. Nofer JR, Walter M, Kehrel B, Wierwille S, Tepel M, Seedorf U, Assmann G. HDL3-mediated inhibition of thrombin-induced platelet aggregation and fibrinogen binding occurs via decreased production of phosphoinositide-derived second messengers 1,2-diacylglycerol and inositol 1,4,5-tris-phosphate. *Arterioscler Thromb Vasc Biol* 1998; 18: 861–869.
570. Calabresi L, Gomaraschi M, Franceschini G. Endothelial protection by high-density lipoproteins: from bench to bedside. *Arterioscler Thromb Vasc Biol* 2003; 23: 1724–1731.
571. Higashihara M, Kinoshita M, Kume S, Teramoto T, Kurokawa K. Inhibition of platelet function by high-density lipoprotein from a patient with apolipoprotein E deficiency. *Biochem Biophys Res Commun* 1991; 181: 1331–1336.
572. Beitz J, Beitz A, Muller G, Giessler C, Mest HJ. Influence of HDL on the formation of 6-keto-PGF1 alpha and TXB2 in vitro: the importance of the source of HDL. *Eicosanoids* 1992; 5: 53–56.
573. Surya II Gorter G, Akkerman JW. Arachidonate transfer between platelets and lipoproteins. *Thromb Haemost* 1992; 68: 719–726.
574. Fujimoto Y, Tsunomori M, Muta E, Yamamoto T, Nishida H, Sakuma S, Fujita T. High density lipoprotein inhibits platelet 12-lipoxygenase activity. *Res Commun Mol Pathol Pharmacol* 1994; 85: 355–358.
575. Valiyaveettil M, Kar N, Ashraf MZ, Byzova TV, Febbraio M, Podrez EA. Oxidized high-density lipoprotein inhibits platelet activation and aggregation via scavenger receptor BI. *Blood* 2008; 111: 1962–1971.
576. Holy EW, Besler C, Landmesser U, Luscher TF, Tanner FC. High density lipoprotein (HDL) from healthy subjects, but not from patients with coronary artery disease, exerts anti-thrombotic effects on human endothelial cells. *Circulation* 2009; 120 (Suppl.): S1100.

577. Chen LY, Mehta JL. Inhibitory effect of high-density lipoprotein on platelet function is mediated by increase in nitric oxide synthase activity in platelets. *Life Sciences* 1994; 55: 1815–1821.

578. Brodde MF, Korporaal SJ, Herminghaus G, Fobker M, Van Berkel TJ, Tietge UJ, Robenek H, Van Eck M, Kehrel BE, Nofer JR. Native high-density lipoproteins inhibit platelet activation via scavenger receptor BI: Role of negatively charged phospholipids. *Atherosclerosis* 2011; 215: 374–382.

579. Imachi H, Murao K, Cao W, Tada S, Taminato T, Wong NC, Takahara J, Ishida T. Expression of human scavenger receptor B1 on and in human platelets. *Arterioscler Thromb Vasc Biol* 2003; 23: 898–904.

580. Vergeer M, Korporaal SJ, Franssen R, Meurs I, Out R, Hovingh GK, Hoekstra M, Sierts JA, Dallinga-Thie GM, Motazacker MM et al. Genetic variant of the scavenger receptor BI in humans. *N Engl J Med* 2011; 364: 136–145.

581. Korporaal SJA, Meurs I, Hauer AD, Hildebrand RB, Hoekstra M, Cate HT, Pratico D, Akkerman J-WN, Van Berkel TJC, Kuiper J et al. Deletion of the high-density lipoprotein receptor scavenger receptor BI in mice modulates thrombosis susceptibility and indirectly affects platelet function by elevation of plasma free cholesterol. *Arterioscler Thromb Vasc Biol* 2011; 31: 34–42.

582. Riddell DR, Vinogradov DV, Stannard AK, Chadwick N, Owen JS. Identification and characterization of LRP8 (apoER2) in human blood platelets. *J Lipid Res* 1999; 40: 1925–1930.

583. Ma Y, Ashraf MZ, Podrez EA. Scavenger receptor BI modulates platelet reactivity and thrombosis in dyslipidemia. *Blood* 2010; 116: 1932–1941.

584. Oslakovic C, Krisinger MJ, Andersson A, Jauhiainen M, Ehnholm C, Dahlback B. Anionic phospholipids lose their procoagulant properties when incorporated into high density lipoproteins. *J Biol Chem* 2009; 284: 5896–5904.

585. Lee M-H, Hammad SM, Semler AJ, Luttrell LM, Lopes-Virella MF, Klein RL. HDL3, but not HDL2, stimulates plasminogen activator inhibitor-1 release from adipocytes: the role of sphingosine-1-phosphate. *J Lipid Res* 2010; 51: 2619–2628.

586. Deguchi H, Fernandez JA, Hackeng TM, Banka CL, Griffin JH. Cardiolipin is a normal component of human plasma lipoproteins. *Proc Natl Acad Sci U S A* 2000; 97: 1743–1748.

587. Oslakovic C, Norstrom E, Dahlback B. Reevaluation of the role of HDL in the anticoagulant activated protein C system in humans. *J Clin Invest* 2010; 120: 1396–1399.

588. Griffin JH, Kojima K, Banka CL, Curtiss LK, Fernandez JA. High-density lipoprotein enhancement of anticoagulant activities of plasma protein S and activated protein C. *J Clin Invest* 1999; 103: 219–227.

589. Riddell DR, Graham A, Owen JS. Apolipoprotein E inhibits platelet aggregation through the L-arginine:nitric oxide pathway. Implications for vascular disease. *J Biol Chem* 1997; 272: 89–95.

590. Atta M, Crook D, Shafique F, Johnston DG, Godsland IF. Procoagulant activities of plasma factor VIIc and factor Xc are positively and independently associated with concentrations of the high-density lipoprotein apolipoprotein, apo A-II. *Thromb Haemost* 2008; 100: 391–396.

591. Karlsson H, Leanderson P, Tagesson C, Lindahl M. Lipoproteomics II: mapping of proteins in high-density lipoprotein using two-dimensional gel electrophoresis and mass spectrometry. *Proteomics* 2005; 5: 1431–1445.

592. Hoofnagle AN, Heinecke JW. Lipoproteomics: using mass spectrometry-based proteomics to explore the assembly, structure, and function of lipoproteins. *J Lipid Res* 2009; 50: 1967–1975.

593. Deguchi H, Fernandez JA, Griffin JH. Plasma cholesteryl ester transfer protein and blood coagulability. *Thromb Haemost* 2007; 98: 1160–1164.

594. Aslam R, Saeed SA, Ahmed S, Connor JD. Lipoproteins inhibit platelet aggregation and arachidonic acid metabolism in experimental hypercholesterolaemia. *Clin Exp Pharmacol Physiol* 2008; 35: 656–662.

595. Desai K, Bruckdorfer KR, Hutton RA, Owen JS. Binding of apoE-rich high density lipoprotein particles by saturable sites on human blood platelets inhibits agonist-induced platelet aggregation. *J Lipid Res* 1989; 30: 831–840.

596. Lesnik P, Vonica A, Guerin M, Moreau M, Chapman MJ. Anticoagulant activity of tissue factor pathway inhibitor in human plasma is preferentially associated with dense subspecies of LDL and HDL and with Lp(a). *Arterioscler Thromb* 1993; 13: 1066–1075.

597. Johnson JD, Henriquez RR, Tichy SE, Russell DH, McNeal CJ, Macfarlane RD. UC/MALDI-MS analysis of HDL; evidence for density-dependent post-translational modifications. *Int J Mass Spectrom* 2007; 268: 227–233.

598. Chait A, Bornfeldt KE. Diabetes and atherosclerosis: is there a role for hyperglycemia? *J Lipid Res* 2009; 50: S335–S339.

599. Chapman MJ, Le Goff W, Guerin M, Kontush A. Cholesteryl ester transfer protein: at the heart of the action of lipid-modulating therapy with statins, fibrates, niacin, and cholesteryl ester transfer protein inhibitors. *Eur Heart J* 2010; 31: 149–164.

600. Lopez-Rios L, Perez-Jimenez P, Martinez-Quintana E, Rodriguez Gonzalez G, Diaz-Chico BN, Novoa FJ, Serra-Majem L, Chirino R. Association of Taq 1B CETP polymorphism with insulin and HOMA levels in the population of the Canary Islands. *Nutr Metab Cardiovasc Dis* 2011; 21: 18–24.

601. Rohrer L, Hersberger M, von Eckardstein A. High density lipoproteins in the intersection of diabetes mellitus, inflammation and cardiovascular disease. *Curr Opin Lipidol* 2004; 15: 269–278.

602. Okin PM, Wiik BP, Hille DA, Wachtell K, Kjeldsen SE, Lindholm LA, Dahlof B, Devereux RB. Low in-treatment HDL cholesterol levels strongly predict the development of new diabetes mellitus: The LIFE Study. *Circulation* 2010; 122: A9424.

603. Fryirs M, Barter PJ, Rye KA. Cholesterol metabolism and pancreatic beta-cell function. *Curr Opin Lipidol* 2009; 20: 159–164.

604. Getz GS, Reardon CA. High-density lipoprotein function in regulating insulin secretion: possible relevance to metabolic syndrome. *Arterioscler Thromb Vasc Biol* 2010; 30: 1497–1499.

605. Fryirs MA, Barter PJ, Appavoo M, Tuch BE, Tabet F, Heather AK, Rye K-A. Effects of high-density lipoproteins on pancreatic beta-cell insulin secretion. *Arterioscler Thromb Vasc Biol* 2010; 30: 1642–1648.

606. Brunham LR, Kruit JK, Pape TD, Timmins JM, Reuwer AQ, Vasanji Z, Marsh BJ, Rodrigues B, Johnson JD, Parks JS et al. Beta-cell ABCA1 influences insulin secretion, glucose homeostasis and response to thiazolidinedione treatment. *Nat Med* 2007; 13: 340–347.

607. Salina, CA, Cruz-Bautista I, Mehta R, Villarreal-Molina MT, Perez FJ, Tusie-Luna MT, Canizales-Quinteros S. The ATP-binding cassette transporter subfamily A member 1 (ABC-A1) and type 2 diabetes: an association beyond HDL cholesterol. *Curr Diabetes Rev* 2007; 3: 264–267.

608. Vergeer M, Brunham LR, Koetsveld J, Kruit JK, Verchere CB, Kastelein JJ, Hayden MR, Stroes ES. Carriers of loss-of-function mutations in ABCA1 display pancreatic beta-cell dysfunction. *Diabetes Care* 2010; 33: 869–874.

609. Kruit JK, Brunham LR, Verchere CB, Hayden MR. HDL and LDL cholesterol significantly influence beta-cell function in type 2 diabetes mellitus. *Curr Opin Lipidol* 2010; 21: 178–185.

610. Ruan X, Li Z, Zhang Y, Yang L, Pan Y, Wang Z, Feng GS, Chen Y. Apolipoprotein A-I possesses an anti-obesity effect associated with increase of energy expenditure and upregulation of UCP1 in brown fat. *J Cell Mol Med* 2011; 15: 763–772.

611. Drew BG, Carey AL, Natoli AK, Formosa MF, Vizi D, Reddy-Luthmoodoo M, Weir JM, Barlow CK, van Hall G, Meikle PJ et al. Reconstituted high-density lipoprotein infusion modulates fatty acid metabolism in patients with type 2 diabetes mellitus. *J Lipid Res* 2011: 52: 572–581.

SECTION 2

FUNCTIONALLY DEFECTIVE HDL

HDL particles progressively lose normal biologic activities and acquire altered properties as a result of modifications in composition, structure and metabolism; such modifications typically occur in the context of high cardiovascular risk, reflecting the presence of major risk factors and, notably, dyslipidemia, the insulin resistance of Type 2 diabetes, infections, autoimmune disease and inflammation. Thus, HDL undergoes dramatic modifications in structure and composition as a result of the concerted actions of the acute phase response and inflammation [1, 2]. The intimate association between inflammation, oxidative stress, dyslipidemia and atherogenesis suggests that such HDL alterations may play a significant role in the progression of cardiovascular disease. Such altered HDL particles have been termed '**dysfunctional HDL**' [3] and HDL has been proposed to possess 'chameleon-like properties' [4, 5].

It is essential to emphasize that the presence of dysfunctional HDL has been frequently demonstrated using *ex vivo* assays, but has never been definitively proven to exist *in vivo*. Furthermore, the degree of loss of normal HDL function (*i.e.*, defective function) compared with the total absence of this function or acquirement of a novel function (*i.e.*, dysfunction) depends on the assay used to characterize HDL functionality. Indeed, HDL can be dysfunctional (with total loss of function) in cell-based or cell-free assays aimed at measuring anti-inflammatory activity [3, 6] and vasodilatory function [7–10]. On the other hand, measurements of antioxidative activity [11–15] or cholesterol efflux capacity [16–19] reveal a **deficiency in normal HDL function**, rather than a complete

High-Density Lipoproteins: Structure, Metabolism, Function, and Therapeutics, First Edition.
Anatol Kontush and M. John Chapman.

dysfunction. These considerations underscore the importance of caution when considering the very concept ofdys HDL dysfunctionality.

DEFINITIONS

Functionally deficient HDL: HDL particles that display diminished levels of their normal anti-atherogenic functionality.

HDL dysfunction: complete loss of the capacity of HDL to perform normal anti-atherogenic function, primarily anti-inflammatory and vasodilatory activities (equivalent to complete functional deficiency).

CHAPTER 8

ALTERED COMPOSITION

Compared with normal, fully functional HDL, the composition of both the protein and lipid components of HDL, *i.e.*, proteome and lipidome, can be modified in metabolic conditions associated with accelerated atherogenesis and elevated cardiovascular risk, as is typical of dyslipidemias, insulin-resistant states, inflammatory states and infectious diseases.

8.1 PROTEOME

The protein cargo of HDL is significantly altered in disease states involving dysmetabolism. Importantly, such changes in HDL-associated proteins may promote atherosclerosis, even in the absence of any decrease in plasma HDL-C concentrations [20].

A diminished HDL content of native, non-modified **apoA-I** is common in several disease states associated with elevated cardiovascular risk. Indeed, the HDL particle content of apoA-I is significantly decreased in hypercholesterolemic subjects [21]. In parallel, the HDL content of **apoE, apoC-I** and **apoC-III** is significantly elevated; **SAA** also tends to be enriched in HDL in hypercholesterolemic subjects [21]. In addition, both HDL2 and HDL display elevated content of SAA and apoC-III in patients with Metabolic Syndrome [22], reflecting chronic, low-grade inflammation [12] (see 9.3 of chapter 8). Consistent with these data, the abundance of apoC-III in HDL represents an independent predictor

High-Density Lipoproteins: Structure, Metabolism, Function, and Therapeutics, First Edition.
Anatol Kontush and M. John Chapman.

of the onset of Type 2 diabetes [23]; indeed, plasma levels of both apoC-II and apoC-III are elevated in Type 2 diabetic patients [24]. Type 2 diabetes may also involve HDL enrichment in **complement C3** protein, as observed in a subgroup of patients with the Hp 2–2 genotype [25].

Patients with coronary artery disease present multiple abnormalities in the HDL proteome, which parallel reductions in HDL-C levels and alterations in the HDL particle profile. Thus, the small, dense HDL3 of coronary artery disease patients is enriched in apoE and depleted of **apoF, PLTP** and **apoJ** [26]. A decreased content of apoF, a CETP-inhibiting protein, is suggestive of elevated CETP activity in coronary artery disease, consistent with low HDL-C levels. Importantly, all patients included in this study were taking aspirin, potentially diminishing between-group differences in inflammatory proteins. In addition, the number of included subjects was small. As a result, differences in HDL proteins between coronary artery disease patients and controls, except that in apoE, as observed in this study [26], were not detected in a previous report from the same laboratory [27]. Recently, SAA, complement C3 and **complement C9** were reported to be enriched in HDL isolated from patients with acute coronary syndrome [28].

Large, light HDL2 particles of coronary artery disease subjects contain elevated levels of apoC-III [29]. A proteomic signature composed of matrix-assisted laser desorption/ionization (MALDI)-mass spectrometry (MS) signals from apoA-I, apoC-III, Lp(a) and apoC-I accurately classified coronary artery disease and control subjects in this study. In support of these results, HDL from subjects with peripheral atherosclerosis and Metabolic Syndrome revealed an increased abundance of apoC-III and SAA4, and reduced content of apoC-I compared with controls when determined by direct proteomic analyses using surface-enhanced laser desorption/ionization (SELDI) time-of-flight (TOF) MS [30]. Furthermore, the O-glycosylated form of apoC-III is 4.7-fold elevated in coronary artery disease subjects as documented by top-down MS [31], thereby documenting HDL enrichment in apoC-III as a common denominator in cardiovascular disease states.

CRITICAL CONTRIBUTION

The HDL proteome is perturbed in patients with coronary artery disease; these alterations may involve enrichment in apoE and SAA and depletion of apoF, PLTP and apoJ [26].

The HDL proteome can be significantly altered in the acute phase. During the acute phase response, levels of apoA-I, **apoA-II** and **apoM** decrease, whereas those of **apoA-IV, apoA-V**, apoE, apoJ and **LBP** increase [1,32,33]. The HDL content of apoC-III can also be elevated by inflammation as documented in hemorrhagic fever with renal syndrome [34]. The reduction in HDL apoA-I levels in inflammatory states is related to both decreased hepatic apoA-I synthesis and to the replacement of apoA-I in HDL particles by SAA [1,2]. Indeed, SAA is able to replace apoA-I in small, dense HDL upon induction of the acute phase [35,36]; as a result, plasma levels of apoA-I decrease [37]. Elevated plasma levels of SAA are accompanied by elevated levels of lipid-free/lipid-poor apoA-I, probably because of dissociation of apoA-I from HDL [37]; poorly lipidated apoA-I is subsequently removed from the circulation by renal catabolism. In rabbits and mice, SAA can completely replace apoA-I in a subset of small, dense HDL particles, thereby functioning as a structural apolipoprotein [37,38]. As a result, SAA can account for up to 80% of total protein in dense HDL under acute inflammation; such enrichment further increases HDL protein content and density [39]. In such SAA-only HDL, 20 molecules of SAA have been estimated to replace all three molecules of apoA-I in each HDL particle, consistent with the smaller M_r of SAA relative to apoA-I.

SAA is mainly produced by the liver but also by arterial wall cells and adipocytes [40,41]. Primary murine hepatocytes secrete SAA as a monomer that subsequently associates with small, lipid-poor, apoA-I-containing HDL secreted via a separate pathway [40]. ABCA1 is required for the initial formation of HDL by SAA in mice [42]; ABCA1-mediated lipidation of SAA prolongs its residence time in the circulation and may prevent tissue deposition. The N-terminal region and the alpha-helical structure in residues 1–11 of SAA are critical to the interaction with lipid and thus to the formation of HDL [43].

The pro-atherogenic properties of SAA are intimately related to its biologic activities as a prothrombotic and pro-inflammatory mediator. SAA is present in human and murine atherosclerotic lesions and co-localizes with apoA-I, apoB and the proteoglycan, perlecan [44,45]. HDL enrichment in SAA enhances *in vitro* HDL binding to proteoglycans as a result of the presence of a proteoglycan-binding domain in the SAA molecule [45,46]. Thus, SAA may immobilize HDL particles in the arterial wall which would otherwise efflux cholesterol. In addition, SAA-enriched HDL particles are rapidly cleared from the circulation [47]. Furthermore, SAA increases monocyte and macrophage cytokine production, possibly at sites of atherosclerosis, thereby contributing to the maintenance of a local pro-inflammatory state [48]. SAA also up-regulates TF activity and TNF-alpha levels in monocytes [49] and amplifies $sPLA_2$ secretion from smooth muscle cells [50]. SAA proteins may thereby induce the recruitment of immune cells to inflammatory sites and activate enzymes that degrade the extracellular matrix. Finally, SAA-enrichment may impair the atheroprotective activities of HDL, such as its cholesterol efflux capacity and anti-inflammatory function [16,51,52].

Interestingly, normal human LDL from healthy, normolipidemic subjects contains low levels of SAA; LDL-associated SAA may serve as a risk factor for cardiac events in stable coronary artery disease [53,54]. SAA-carrying LDL represents 5%–6% of total LDL, contains increased levels of lipid and protein oxidation products and may represent circulating oxidized LDL [53]. The increased SAA content of such LDL may enhance its retention in atherosclerotic lesions; such SAA-mediated accumulation of LDL in lesions may lead to plaque instability and, ultimately, to plaque rupture [55].

Apart from its replacement by SAA, apoA-I can undergo other modifications in the circulation. Amino acid residues in apoA-I, such as tyrosine (Tyr), Met, Cys and lysine (Lys) can be selectively **modified under the action of pro-oxidants** secreted by arterial wall cells [56–58]. In human atherosclerotic lesions, proteins oxidized by hypochlorous acid (HOCl) or acrolein co-localize with extracellular apoA-I [56, 59]. Oxidized amino acid residues, including chlorotyrosines, nitrotyrosines and oxidized Met and Lys residues, are present in apoA-I isolated from plasma and from human atherosclerotic lesions [56, 57, 59, 60]. The abundance of chloro- and nitrotyrosines in apoA-I is increased in plasma of patients with cardiovascular disease [56, 61]. Furthermore, the apoA-I content of methionine sulfoxides is elevated in coronary artery disease subjects [29] and tends to be elevated in smokers [62]. Such modified forms of apoA-I display altered immune properties and can be quantitatively detected by specific antibodies [63, 64]. Furthermore, the oxidation of apoA-I may result in its fragmentation, as well as oligomerization, as occurs in aging [65].

Myeloperoxidase, a major source of reactive chlorine and nitrogen species in the arterial wall produced by activated macrophages, binds to apoA-I *in vitro* and *in vivo* and produces a distinct pattern of oxidized amino acids [66]. Specifically, the Tyr192 residue represents the major target for myeloperoxidase-catalyzed oxidation of apoA-I, whereas three other Tyr residues at positions 29, 166 and 236 are modified to a lesser extent [67]. Thus, apoA-I represents a selective target for chlorination and nitration in human atheromatous tissue catalyzed by myeloperoxidase. Furthermore, myeloperoxidase is involved in the carbamylation of the HDL protein moiety [68]. As plasma levels of myeloperoxidase are elevated in patients with coronary artery disease, correlate with the extent, and severity, of atherosclerosis and are associated with future cardiovascular risk [69–71], such a pathway may significantly contribute to pathologic oxidation of HDL proteins in atherosclerosis.

In addition, myeloperoxidase interaction with HDL may promote the binding of **ceruloplasmin** [72], a copper-containing protein capable of inducing lipid and protein oxidation via its labile copper atom [73, 74]. Interestingly, the binding affinity of myeloperoxidase to HDL markedly increases as a function of the degree of HDL modification by myeloperoxidase-derived HOCl, thereby generating a vicious cycle of myeloperoxidase-dependent oxidative modification at sites of inflammation [75]. N-chloramines generated in HDL by HOCl may underlie such high-affinity binding of the enzyme.

Protein oxidation by myeloperoxidase-generated HOCl may destabilize discoid HDL, leading to accelerated protein unfolding, dissociation and lipoprotein fusion [76]. Such alterations may reflect the reduced affinity of apoA-I for lipid, resulting from the oxidation of Met and aromatic residues in the lipid-binding faces of amphipathic alpha-helices, as well as apoA-I cross-linking into dimers and trimers on the HDL surface [76]. In contrast, extensive HDL oxidation by copper ions inhibits protein dissociation and HDL fusion, probably as a result of lipid redistribution from the core of HDL to its surface upon lipolysis and of massive apoA-I cross-linking [77]. Indeed, reactive electrophiles generated during copper-induced lipid peroxidation covalently modify apoA-I, which represents one of the most highly modified proteins of human plasma, with the formation of carboxylated, rather than noncarboxylated adducts. The His162 residue, which resides in an apoA-I domain implicated in the activation of LCAT, represents a key target for such reactive electrophiles [78].

In addition to oxidative modification, apoA-I can be **non-enzymatically glycosylated** in the presence of elevated levels of glucose, as in patients with Type 2 diabetes [79] or in elderly patients [80]. Indeed, MALDI reveals the presence of non-enzymatically glycated apoA-I in diabetic and nephropathic patients [81]. Reactive alpha-oxoaldehyde, methylglyoxal, may account for the formation of glycated apoA-I [82], which displays a modified secondary structure and associates with lower binding affinity to phospholipid, building discoid HDL with decreased alpha-helical content [83]. Interestingly, the glycation inhibitors aminoguanidine and pyridoxamine, the insulin sensitizer metformin, and the cross-link breaker alagebrium can inhibit such modification [82].

Another pathway of the pro-atherogenic modification of apoA-I involves proteolytic degradation under the action of proteases released by arterial wall cells. This pathway is exemplified by the action of chymase released by activated mast cells, leading to apoA-I proteolysis *in vivo* [84].

8.2 LIPIDOME

Although apolipoproteins and enzymes are major determinants of altered HDL function, such biologic activities are considerably influenced by changes in lipid content. HDL core **enrichment in triglyceride** with **cholesteryl ester depletion** is the most frequent abnormality of HDL lipid composition and occurs in low HDL-C hypertriglyceridemic states associated with decreased activity of LPL and hepatic lipase, and/or decreased activity of LCAT. Under such conditions, the cholesteryl ester content of HDL is diminished and the triglyceride content is raised primarily as a consequence of elevated CETP-mediated heteroexchange of triglyceride and cholesteryl ester between VLDL and HDL [85–87]. This pathway largely accounts for HDL depletion in cholesteryl ester and enrichment in triglyceride in hypertriglyceridemia typical of Type 2 diabetes and metabolic syndrome [12, 14, 22], and is accompanied by a chronic inflammatory response. A similar elevation in HDL-triglycerides, a decrease in HDL-C and an increase in inflammatory markers are observed in patients with coronary heart disease [88], as well as in the post-prandial phase [89].

CRITICAL CONTRIBUTION

Low HDL-C hypertriglyceridemia features the most frequent abnormality of the HDL lipidome, which involves core **enrichment in triglyceride** paralleled by **cholesteryl ester depletion** [87].

Interestingly, all of these metabolic alterations are frequently observed in the acute phase and during inflammation [1, 37]. Under acutephase conditions involving an enlarged pool of triglyceride-rich lipoproteins (see 9.3 of chapter 8), triglyceride typically replaces cholesteryl ester in the HDL core, resulting in a low cholesteryl ester/triglyceride ratio and a decrease in plasma HDL-C levels, another feature of the acute phase response [1]. As a result, human acute-phase HDL obtained from patients following bypass surgery is enriched in triglyceride and depleted of cholesteryl ester [90]. Similarly, HDL from subjects with malaria displays an elevated content of triglyceride [91]. Acute-phase HDL also contains elevated levels of **non-esterified fatty acids, lysophosphatidylcholine** and **isoprostanes** compared with normal HDL; in addition, cholesteryl ester levels are decreased [90]. Furthermore, HDL3 from subjects with myocardial infarction are enriched in triglyceride and depleted of **phospholipid** [92]. Induction of the acute phase response in monkeys increases plasma triglyceride levels and triglyceride content in total HDL and in all HDL subfractions, and also decreases HDL-C and HDL-cholesteryl ester; these effects are paralleled by cytokine-induced decreases in the activities of hepatic lipase, LPL and LCAT and increases in the activity of CETP [37, 93, 94]. In addition, acute-phase HDL obtained from hamsters displays reduced cholesteryl ester content and elevated phospholipid and **free cholesterol** content compared with normal HDL [19]. As a consequence of decreased LCAT activity, an increased abundance of free cholesterol in HDL is frequently observed in inflammatory states. HDL free cholesterol is also elevated in genetic LCAT deficiency [95]. In contrast, an altered phospholipid content of HDL is a less consistent finding [1]. The reduction of HDL phospholipids reported in some studies [39, 92] may reflect the elevated activity of $sPLA_2$, as frequently observed in the acute phase [96–98].

The composition of HDL can also be abnormal in other forms of dyslipidemia. In familial hypercholesterolemia, the HDL levels of triglyceride and the triglyceride/cholesteryl ester ratio are elevated [99, 100]. HDL enrichment in triglyceride is associated with accelerated cholesteryl ester transfer from LDL and HDL to triglyceride-rich lipoproteins in the plasma of patients with familial hypercholesterolemia, an observation that has been linked to abnormal properties of plasma VLDL1 [99, 101, 102]. In contrast, marked attenuation of cholesteryl ester transfer from LDL and HDL to triglyceride-rich lipoproteins, as observed in the plasma of subjects with a heterozygous CETP deficiency, results in reduced HDL content of triglyceride, elevated content of cholesteryl ester and a moderate shift in the HDL particle profile to be of a larger size [103].

In addition to core lipids, the surface lipids of HDL can also be modified in metabolic disease. Thus, low HDL-C subjects display diminished HDL content of lysophosphatidylcholine and **sphingomyelin**. Furthermore, triglyceride molecules are, in part, redistributed to the surface monolayer of such triglyceride-enriched particles [104]. In addition, patients with isolated arterial hypertension display reduced content of HDL phospholipid and altered phospholipid composition characterized by diminished content of phosphatidylcholine and elevated content of lysophosphatidylcholine, sphingomyelin and **phosphatidylethanolamine** [105].

The fatty acid composition of HDL can also be affected as shown in patients with triple vessel coronary heart disease whose HDL displays a low content of PUFAs [88]. Both an elevated sphingomyelin/phosphatidylcholine ratio and a decreased content of PUFAs can be of pathologic significance, as such modifications augment HDL surface rigidity and can thereby decrease the cholesterol efflux capacity [106], LCAT activity [107] and the antioxidative capacity of HDL [108]. These pathways can be implicated in the anti-atherogenic effects of omega-6 PUFAs [109] and of inhibitors of sphingomyelin biosynthesis [110].

In contrast, the HDL content of S1P does not appear to be modified in metabolic disease, as suggested by normal levels of HDL-associated S1P in subjects with metabolic syndrome [111]. Consistent with these data, S1P is not deficient in the plasma of hypercholesterolemic rabbits compared with control animals [112].

Finally, the HDL lipid moiety can be oxidized *in vivo* with the formation of biologically active **oxidized lipids**. For instance, HDL oxidation by HOCl produces 2-chlorohexadecanal, a chlorinated fatty aldehyde formed upon the oxidative cleavage of plasmalogen, which exerts inhibitory actions on eNOS [113]. In addition, the HDL content of thiobarbituric acid-reactive substances, a measure of aldehydic end products of lipid peroxidation, is raised in subjects with malaria [91]. However, it should be mentioned that HDL does not represent a major source of oxidized lipids compared with other lipoproteins [114, 115], a finding consistent with the capacity of HDL to inactivate such lipids.

METHODOLOGIC NOTE

Modern proteomic and lipidomic techniques allow the identification of more than 50 individual proteins and more than 200 individual lipids in HDL, in both normal and disease states. Furthermore, covalent modifications of proteins and lipids by oxidation and glycation can be quantitatively assessed using this approach.

8.3 ENZYMATIC ACTIVITIES

HDL-associated enzymes, including PON1, PAF-AH and LCAT can become dysfunctional and/or depleted under inflammatory conditions [116, 117], in metabolic diseases involving low HDL levels (Type 2 diabetes, Metabolic Syndrome) [12, 14, 15, 22] and in premature coronary heart disease [6]. Induction of the acute phase response is associated with decreased **PON1** activity, probably as a result of the displacement and replacement of PON1 by SAA [116, 117]. Furthermore, decreased PON1 activity may be caused by enzyme inactivation as a result of oxidation [118], glycation [119, 120] and/or homocysteinylation [121].

Indeed, PON1 activity falls with increasing oxidative stress, as observed in SR-BI deficient mice [122]. HOCl generation by myeloperoxidase may represent one of the pathways involved in the oxidative inactivation of PON1. Consistent with this possibility, serum concentrations of myeloperoxidase constitute an independent, negative determinant of PON1 activity [123]. In addition, LOOHs can inhibit PON1 via reaction with the free sulfhydryl Cys residues of the enzyme [124]. Oxidative stress can also impair PON1 binding to HDL [125], resulting in subnormal circulating levels of the enzyme. Furthermore, serum concentrations of PON1 are decreased to subnormal in subjects with Metabolic Syndrome [126] and patients with Type 1 and Type 2 diabetes [127–129] who feature insulin resistance, chronic inflammation and oxidative stress [130]. The contribution of glycation to the loss of PON1 activity is supported by the capacity of spermine, a glycation inhibitor, to improve PON1 activity in diabetic rats [131]. In addition, serum PON1 activity is diminished concomitant with PON1 glycosylation in patients with liver disease [132].

CLINICAL SUMMARY

Low HDL-C levels reflect both depletion of HDL particles in cholesterol, a key lipid component comprising up to 20% of total HDL particle mass, and a reduction in the number of circulating HDL particles. The depletion of HDL cholesterol is typically paralleled by triglyceride enrichment and results in the shift from large, cholesteryl ester-rich HDL2 particles to small, cholesteryl ester-poor HDL3. The altered HDL content of other functional components may also parallel low HDL-C levels, thereby further contributing to atherogenesis.

Serum PON1 activity can be impaired by smoking or by cigarette smoke [128, 133–135]. As PON1 is exclusively associated with HDL [136], such effects reflect alterations of the HDL-associated enzyme; as a corollary, smoking cessation improves PON1 activity [137]. PON1 activity is decreased although its concentration is elevated in HIV patients, along with reduced HDL-C and apoA-I levels [138], indicative of the reduction of the specific activity of the enzyme. In this study, PON1 activity was associated with the CD4+ T lymphocyte count, suggesting inflammation-related inactivation. Serum PON1 activity is similarly diminished in *Helicobacter pylori*-infected subjects along with a decrease in HDL-C and an increase in inflammation and oxidative stress [139], suggesting the involvement of related mechanisms. Consistent with these data, low PON1 activity is related to elevated CRP levels in Type 2 diabetes [140].

Thiolactonase activity may represent the physiologic activity of PON1; it is relevant, in this regard, that serum homocysteine thiolactonase activity is reduced, while levels of homocysteine thiolactone are elevated in coronary artery disease patients [141]. The oxidative modification of PON1 may well account for the loss of thiolactonase activity, which is even more sensitive to copper-catalyzed oxidation compared with arylesterase activity [142].

HDL-associated **PAF-AH** ($LpPLA_2$) activity, expressed as a percentage of total serum PAF-AH activity, is lower in familial hypercholesterolemia [143, 144], Metabolic Syndrome [145] and Type 2 diabetes [146] compared with controls. PAF-AH activity is reduced by exposure to cigarette smoke extract, which causes the modification of thiol groups in the enzyme [147]. In contrast, LDL-associated PAF-AH activity and/or levels are elevated in subjects with homozygous familial hypercholesterolemia [144] and in patients with Type 2 diabetes [148], indicating major redistribution of PAF-AH activity from apoA-I- to apoB-containing lipoproteins [144]. As LDL-associated PAF-AH predominates in the circulation and appears to represent a pro-atherogenic form of the enzyme, such an effect can contribute to elevated cardiovascular risk related to increased plasma levels of PAF-AH ($LpPLA_2$). In contrast, HDL-associated $LpPLA_2$ may substantially contribute to the anti-atherogenic activities of HDL [149].

As a result, total serum PAF-AH activity is elevated in patients with metabolic syndrome [150] and Type 1 diabetes [151] and is related to the severity of coronary artery disease [152]. PAF-AH activity is higher (+11%) in coronary artery disease cases than in controls and is positively correlated with urinary excretion concentrations of 8-epi-prostaglandin-2-alpha, a product of PAF-AH activity [153]. In a similar fashion, elevated levels of PAF-AH are paralleled by increased lysophosphatidylcholine content in LDL in Type 2 diabetes [148]. In addition, levels of PAF-AH and oxidized LDL are significantly correlated in extracts from carotid endarterectomy atheroma [154]; the ratio of PAF-AH to oxidized LDL is higher in atheromatous tissue (277:1) than in normal artery tissue (135:1) and plasma (13:1). Immunohistochemical experiments indicate that oxidized LDL and PAF-AH exist largely in an overlapping distribution in atherosclerotic plaques [154].

Finally, plasma PAF-AH is elevated in cholesterol-fed pigs; in parallel, the enzyme is redistributed from HDL to LDL [155]. Concomitantly, plasma lysophosphatidylcholine and oxidized LDL are elevated, with oxidized LDL inducing the expression of PAF-AH in macrophages. In addition both oxidized and native lipoproteins are also capable of regulating the catalytic activity of $sPLA_2$, with oxidized lipoproteins stimulating, and native lipoproteins inhibiting, the activity [156]. Together, these data suggest a mechanism of PAF-AH regulation by oxidized LDL whereby oxidized LDL induces PAF-AH resulting in the accumulation of lysophosphatidylcholine that then promotes the inflammatory reaction [155].

Finally, plasma PAF-AH activity and mass are positively, and independently, associated with smoking, as well as with exposure to second-hand smoke [157]. In contrast, HDL-associated PAF-AH activity and mass are inversely, and independently, associated with smoking. Smoking may, therefore, induce the production of the enzyme by PAF and related smoking-induced mediators, attenuating anti-atherogenic HDL-associated PAF-AH as a result of the reduction of HDL-C [157].

The activity of **LCAT**, another HDL-associated enzyme, can be diminished under inflammatory conditions [1, 95]. LCAT activity and protein are also diminished in diabetic patients with [25] or without [158] the Hp 2–2 genotype. In addition, LCAT activity is reduced by smoking and/or by exposure to cigarette smoke *in vitro* [159–163]. In contrast, coronary artery disease does not seem to impair LCAT. Thus, in apparently healthy men and women who developed coronary artery disease and matched controls that remained free of coronary artery disease during a six-year follow-up, LCAT levels did not differ between cases and controls, but were higher in women than in men [164]. Stratification into LCAT quartiles revealed a positive association with plasma LDL-C and triglyceride levels in the unexpected absence of an association with HDL-C.

Oxidation of apoA-I may contribute to the reduction of LCAT activity in inflammatory states as a result of diminished LCAT activation [165]. Specifically, the oxidation of Met148 disrupts the structure of the central loop of apoA-I which overlaps the LCAT activation domain, thereby impairing LCAT activation [166]. Mechanisms of enzyme inactivation by smoking presumably involve the oxidation of amino acid residues, primarily free thiol groups, which are critical for LCAT activity [159–161]. Similar to the inactivation of PON1, glycation may play a role in the loss of LCAT activity in Type 2 diabetes. Indeed, glycosylated haemoglobin (HbA1c) represents a strong, independent, negative predictor of LCAT activity in Type 2 diabetes [167]. Furthermore, spermine, a glycation inhibitor, restores LCAT activity in diabetic rats [131].

IMPORTANT READING

Heinecke JW. The HDL proteome: a marker-and perhaps mediator-of coronary artery disease. *J Lipid Res* 2009; 50: S167–S171.

Hoofnagle AN, Vaisar T, Mitra P, Chait A. HDL lipids and insulin resistance. *Curr Diab Rep* 2010; 10: 78–86.

REFERENCES

1. Khovidhunkit W, Kim MS, Memon RA, Shigenaga JK, Moser AH, Feingold KR, Grunfeld C. Effects of infection and inflammation on lipid and lipoprotein metabolism: mechanisms and consequences to the host. *J Lipid Res* 2004; 45: 1169–1196.
2. Esteve E, Ricart W, Fernandez-Real JM. Dyslipidemia and inflammation: an evolutionary conserved mechanism. *Clin Nutr* 2005; 24: 16–31.
3. Navab M, Hama SY, Hough GP, Subbanagounder G, Reddy ST, Fogelman AM. A cell-free assay for detecting HDL that is dysfunctional in preventing the formation of or inactivating oxidized phospholipids. *J Lipid Res* 2001; 42: 1308–1317.
4. Navab M, Berliner JA, Watson AD, Hama SY, Territo MC, Lusis AJ, Shih DM, Van Lenten BJ, Frank JS, Demer LL et al. The Yin and Yang of oxidation in the

development of the fatty streak. A review based on the 1994 George Lyman Duff Memorial Lecture. *Arterioscler Thromb Vasc Biol* 1996; 16: 831–842.

5. Van Lenten BJ, Navab M, Shih D, Fogelman AM, Lusis AJ. The role of high-density lipoproteins in oxidation and inflammation. *Trends Cardiovasc Med* 2001; 11: 155–161.
6. Ansell BJ, Navab M, Hama S, Kamranpour N, Fonarow G, Hough G, Rahmani S, Mottahedeh R, Dave R, Reddy ST et al. Inflammatory/antiinflammatory properties of high-density lipoprotein distinguish patients from control subjects better than high-density lipoprotein cholesterol levels and are favorably affected by simvastatin treatment. *Circulation* 2003; 108: 2751–2756.
7. Persegol L, Verges B, Foissac M, Gambert P, Duvillard L. Inability of HDL from type 2 diabetic patients to counteract the inhibitory effect of oxidised LDL on endothelium-dependent vasorelaxation. *Diabetologia* 2006; 49: 1380–1386.
8. Persegol L, Verges B, Gambert P, Duvillard L. Inability of HDL from abdominally obese subjects to counteract the inhibitory effect of oxidized LDL on vasorelaxation. *J Lipid Res* 2007; 48: 1396–1401.
9. Persegol L, Foissac M, Lagrost L, Athias A, Gambert P, Verges B, Duvillard L. HDL particles from type 1 diabetic patients are unable to reverse the inhibitory effect of oxidised LDL on endothelium-dependent vasorelaxation. *Diabetologia* 2007; 50: 2384–2387.
10. Sorrentino SA, Besler C, Rohrer L, Meyer M, Heinrich K, Bahlmann FH, Mueller M, Horvath T, Doerries C, Heinemann M et al. Endothelial-vasoprotective effects of high-density lipoprotein are impaired in patients with type 2 diabetes mellitus but are improved after extended-release niacin therapy. *Circulation* 2010; 121: 110–122.
11. Kontush A, Chantepie S, Chapman MJ. Small, dense HDL particles exert potent protection of atherogenic LDL against oxidative stress. *Arterioscler Thromb Vasc Biol* 2003; 23: 1881–1888.
12. Hansel B, Giral P, Nobecourt E, Chantepie S, Bruckert E, Chapman MJ, Kontush A. Metabolic syndrome is associated with elevated oxidative stress and dysfunctional dense high-density lipoprotein particles displaying impaired antioxidative activity. *J Clin Endocrinol Metab* 2004; 89: 4963–4971.
13. Kontush A, de Faria EC, Chantepie S, Chapman MJ. Antioxidative activity of HDL particle subspecies is impaired in hyperalphalipoproteinemia: relevance of enzymatic and physicochemical properties. *Arterioscler Thromb Vasc Biol* 2004; 24: 526–533.
14. Nobecourt E, Jacqueminet S, Hansel B, Chantepie S, Grimaldi A, Chapman MJ, Kontush A. Defective antioxidative activity of small, dense HDL particles in type 2 diabetes: Relationship to elevated oxidative stress and hyperglycemia. *Diabetologia* 2005; 48: 529–538.
15. Kontush A, de Faria EC, Chantepie S, Chapman MJ. A normotriglyceridemic, low HDL-cholesterol phenotype is characterised by elevated oxidative stress and HDL particles with attenuated antioxidative activity. *Atherosclerosis* 2005; 182: 277–285.
16. Banka CL, Yuan T, de Beer MC, Kindy M, Curtiss LK, de Beer FC. Serum amyloid A (SAA): influence on HDL-mediated cellular cholesterol efflux. *J Lipid Res* 1995; 36: 1058–1065.
17. Cavallero E, Brites F, Delfly B, Nicolaiew N, Decossin C, De Geitere C, Fruchart JC, Wikinski R, Jacotot B, Castro G. Abnormal reverse cholesterol transport in controlled type II diabetic patients. Studies on fasting and postprandial LpA-I particles. *Arterioscler Thromb Vasc Biol* 1995; 15: 2130–2135.

18. Brites FD, Bonavita CD, De Geitere C, Cloes M, Delfly B, Yael MJ, Fruchart J, Wikinski RW, Castro GR. Alterations in the main steps of reverse cholesterol transport in male patients with primary hypertriglyceridemia and low HDL-cholesterol levels. *Atherosclerosis* 2000; 152: 181–192.
19. Khovidhunkit W, Shigenaga JK, Moser AH, Feingold KR, Grunfeld C. Cholesterol efflux by acute-phase high density lipoprotein: role of lecithin: cholesterol acyltransferase. *J Lipid Res* 2001; 42: 967–975.
20. Heinecke JW. The HDL proteome: a marker-and perhaps mediator-of coronary artery disease. *J Lipid Res* 2009; 50: S167–S171.
21. Heller M, Schlappritzi E, Stalder D, Nuoffer JM, Haeberli A. Compositional protein analysis of high density lipoproteins in hypercholesterolemia by shotgun LC-MS/MS and probabilistic peptide scoring. *Mol Cell Proteomics* 2007; 6: 1059–1072.
22. Park KH, Shin DG, Kim JR, Hong JH, Cho KH. The functional and compositional properties of lipoproteins are altered in patients with metabolic syndrome with increased cholesteryl ester transfer protein activity. *Int J Mol Med* 2010; 25: 129–136.
23. Onat A, Hergenc G, Ayhan E, Ugur M, Kaya H, Tuncer M, Can G. Serum apolipoprotein C-III in high-density lipoprotein: a key diabetogenic risk factor in Turks. *Diabet Med* 2009; 26: 981–988.
24. Beliard S, Nogueira JP, Maraninchi M, Lairon D, Nicolay A, Giral P, Portugal H, Vialettes B, Valero R. Parallel increase of plasma apoproteins C-II and C-III in Type 2 diabetic patients. *Diabet Med* 2009; 26: 736–739.
25. Farbstein D, Kalet-Litman S, Pollak M, Blum S, Asaf R, Schwartz A, Asleh R, Perry N, Barkay I, Levy A. Decreased LCAT protein and increased activated complement fragments in the dysfunctional HDL of individuals with diabetes and the Hp 2-2 genotype. *Circulation* 2009; 120 (Suppl.); S525–c-S526.
26. Green PS, Vaisar T, Pennathur S, Kulstad JJ, Moore AB, Marcovina S, Brunzell J, Knopp RH, Zhao XQ, Heinecke JW. Combined statin and niacin therapy remodels the high-density lipoprotein proteome. *Circulation* 2008; 118: 1259–1267.
27. Vaisar T, Pennathur S, Green PS, Gharib SA, Hoofnagle AN, Cheung MC, Byun J, Vuletic S, Kassim S, Singh P et al. Shotgun proteomics implicates protease inhibition and complement activation in the antiinflammatory properties of HDL. *J Clin Invest* 2007; 117: 746–756.
28. Riwanto M, Besler C, Rohrer L, Roschitzki B, von Eckardstein A, Luescher TF, Landmesser U. Proteomic analysis of high density lipoprotein (HDL) in patients with an acute coronary syndrome reveals a markedly changed HDL-proteome as compared to healthy subjects and frequent oxidative post-translational modifications. *Circulation* 2010; 122 (Suppl.): A21471.
29. Vaisar T, Mayer P, Nilsson E, Zhao XQ, Knopp R, Prazen BJ. HDL in humans with cardiovascular disease exhibits a proteomic signature. *Clin Chim Acta* 2010; 411: 972–979.
30. Davidsson P, Hulthe J, Fagerberg B, Camejo G. Proteomics of apolipoproteins and associated proteins from plasma high-density lipoproteins. *Arterioscler Thromb Vasc Biol* 2010; 30: 156–163.
31. Mazur MT, Cardasis HL, Spellman DS, Liaw A, Yates NA, Hendrickson RC. Quantitative analysis of intact apolipoproteins in human HDL by top-down differential mass spectrometry. *Proc Natl Acad Sci U S A* 2010; 107: 7728–7733.

32. Khovidhunkit W, Duchateau PN, Medzihradszky KF, Moser AH, Naya-Vigne J, Shigenaga JK, Kane JP, Grunfeld C, Feingold KR. Apolipoproteins A-IV and A-V are acute-phase proteins in mouse HDL. *Atherosclerosis* 2004; 176: 37–44.

33. Feingold KR, Shigenaga JK, Chui LG, Moser A, Khovidhunkit W, Grunfeld C. Infection and inflammation decrease apolipoprotein M expression. *Atherosclerosis* 2008; 199: 19–26.

34. Kim J, Park HH, Choi I, Kim YO, Cho KH. Severely modified lipoprotein properties without a change in cholesteryl ester transfer protein activity in patients with acute renal failure secondary to Hantaan virus infection. *BMB Rep* 2010; 43: 535–540.

35. Parks JS, Rudel LL. Alteration of high density lipoprotein subfraction distribution with induction of serum amyloid A protein (SAA) in the nonhuman primate. *J Lipid Res* 1985; 26: 82–91.

36. Coetzee GA, Strachan AF, van der Westhuyzen DR, Hoppe HC, Jeenah MS, de Beer FC. Serum amyloid A-containing human high density lipoprotein 3. Density, size, and apolipoprotein composition. *J Biol Chem* 1986; 261: 9644–9651.

37. Cabana VG, Lukens JR, Rice KS, Hawkins TJ, Getz GS. HDL content and composition in acute phase response in three species: triglyceride enrichment of HDL a factor in its decrease. *J Lipid Res* 1996; 37: 2662–2674.

38. Cabana VG, Reardon CA, Wei B, Lukens JR, Getz GS. SAA-only HDL formed during the acute phase response in apoA-I+/+ and apoA-I-/- mice. *J Lipid Res* 1999; 40: 1090–1103.

39. Cabana VG, Siegel JN, Sabesin SM. Effects of the acute phase response on the concentration and density distribution of plasma lipids and apolipoproteins. *J Lipid Res* 1989; 30: 39–49.

40. Hoffman JS, Benditt EP. Secretion of serum amyloid protein and assembly of serum amyloid protein-rich high density lipoprotein in primary mouse hepatocyte culture. *J Biol Chem* 1982; 257: 10518–10522.

41. Malle E, Steinmetz A, Raynes JG. Serum amyloid A (SAA): an acute phase protein and apolipoprotein. *Atherosclerosis* 1993; 102: 131–146.

42. Hu W, Abe-Dohmae S, Tsujita M, Iwamoto N, Ogikubo O, Otsuka T, Kumon Y, Yokoyama S. Biogenesis of HDL by SAA is dependent on ABCA1 in the liver *in vivo*. *J Lipid Res* 2008; 49: 386–393.

43. Ohta S, Tanaka M, Sakakura K, Kawakami T, Aimoto S, Saito H. Defining lipid-binding regions of human serum amyloid A using its fragment peptides. *Chem Phys Lipids* 2009; 162: 62–68.

44. Yamada T, Kakihara T, Kamishima T, Fukuda T, Kawai T. Both acute phase and constitutive serum amyloid A are present in atherosclerotic lesions. *Pathol Int* 1996; 46: 797–800.

45. O'Brien KD, McDonald TO, Kunjathoor V, Eng K, Knopp EA, Lewis K, Lopez R, Kirk EA, Chait A, Wight TN et al. Serum amyloid A and lipoprotein retention in murine models of atherosclerosis. *Arterioscler Thromb Vasc Biol* 2005; 25: 785–790.

46. Lewis KE, Kirk EA, McDonald TO, Wang S, Wight TN, O'Brien KD, Chait A. Increase in serum amyloid A evoked by dietary cholesterol is associated with increased atherosclerosis in mice. *Circulation* 2004; 110: 540–545.

47. Hoffman JS, Benditt EP. Plasma clearance kinetics of the amyloid-related high density lipoprotein apoprotein, serum amyloid protein (apoSAA), in the mouse. Evidence for rapid apoSAA clearance. *J Clin Invest* 1983; 71: 926–934.
48. Song C, Hsu K, Yamen E, Yan W, Fock J, Witting PK, Geczy CL, Freedman SB. Serum amyloid A induction of cytokines in monocytes/macrophages and lymphocytes. *Atherosclerosis* 2009; 207: 374–383.
49. Song C, Shen Y, Yamen E, Hsu K, Yan W, Witting PK, Geczy CL, Freedman SB. Serum amyloid A may potentiate prothrombotic and proinflammatory events in acute coronary syndromes. *Atherosclerosis* 2009; 202: 596–604.
50. Sullivan CP, Seidl SE, Rich CB, Raymondjean M, Schreiber BM. Secretory phospholipase A2, group IIA is a novel serum amyloid A target gene: activation of smooth muscle cell expression by an interleukin-1 receptor-independent mechanism. *J Biol Chem* 2010; 285: 565–575.
51. Van Lenten BJ, Hama SY, de Beer FC, Stafforini DM, McIntyre TM, Prescott SM, La Du BN, Fogelman AM, Navab M. Anti-inflammatory HDL becomes proinflammatory during the acute phase response. Loss of protective effect of HDL against LDL oxidation in aortic wall cell cocultures. *J Clin Invest* 1995; 96: 2758–2767.
52. Kontush A, Chapman MJ. Functionally defective HDL: A new therapeutic target at the crossroads of dyslipidemia, inflammation and atherosclerosis. *Pharmacol Rev* 2006; 3: 342–374.
53. Ogasawara K, Mashiba S, Wada Y, Sahara M, Uchida K, Aizawa T, Kodama T. A serum amyloid A and LDL complex as a new prognostic marker in stable coronary artery disease. *Atherosclerosis* 2004; 174: 349–356.
54. Kotani K, Satoh N, Kato Y, Araki R, Koyama K, Okajima T, Tanabe M, Oishi M, Yamakage H, Yamada K et al. A novel oxidized low-density lipoprotein marker, serum amyloid A-LDL, is associated with obesity and the metabolic syndrome. *Atherosclerosis* 2009; 204: 526–531.
55. Johnson BD, Kip KE, Marroquin OC, Ridker PM, Kelsey SF, Shaw LJ, Pepine CJ, Sharaf B, Bairey Merz CN, Sopko G et al. Serum amyloid A as a predictor of coronary artery disease and cardiovascular outcome in women: The National Heart, Lung, and Blood Institute-Sponsored Women's Ischemia Syndrome Evaluation (WISE). *Circulation* 2004; 109: 726–732.
56. Bergt C, Pennathur S, Fu X, Byun J, O'Brien K, McDonald TO, Singh P, Anantharamaiah GM, Chait A, Brunzell J et al. The myeloperoxidase product hypochlorous acid oxidizes HDL in the human artery wall and impairs ABCA1-dependent cholesterol transport. *Proc Natl Acad Sci U S A* 2004; 101: 13032–13037.
57. Zheng L, Nukuna B, Brennan ML, Sun M, Goormastic M, Settle M, Schmitt D, Fu X, Thomson L, Fox PL et al. Apolipoprotein A-I is a selective target for myeloperoxidase-catalyzed oxidation and functional impairment in subjects with cardiovascular disease. *J Clin Invest* 2004; 114: 529–541.
58. Nicholls SJ, Zheng L, Hazen SL. Formation of dysfunctional high-density lipoprotein by myeloperoxidase. *Trends Cardiovasc Med* 2005; 15: 212–219.
59. Shao B, O'Brien KD, McDonald TO, Fu X, Oram JF, Uchida K, Heinecke JW. Acrolein modifies apolipoprotein A-I in the human artery wall. *Ann N Y Acad Sci* 2005; 1043: 396–403.

60. Panzenbock U, Stocker R. Formation of methionine sulfoxide-containing specific forms of oxidized high-density lipoproteins. *Biochim Biophys Acta* 2005; 1703: 171–181.

61. Pennathur S, Bergt C, Shao B, Byun J, Kassim SY, Singh P, Green PS, McDonald TO, Brunzell J, Chait A et al. Human atherosclerotic intima and blood of patients with established coronary artery disease contain high density lipoprotein damaged by reactive nitrogen species. *J Biol Chem* 2004; 279: 42977–42983.

62. Pankhurst G, Wang XL, Wilcken DE, Baernthaler G, Panzenbock U, Raftery M, Stocker R. Characterization of specifically oxidized apolipoproteins in mildly oxidized high density lipoprotein. *J Lipid Res* 2003; 44: 349–355.

63. Wang XS, Shao B, Oda MN, Heinecke JW, Mahler S, Stocker R. A sensitive and specific ELISA detects methionine sulfoxide-containing apolipoprotein A-I in HDL. *J Lipid Res* 2009; 50: 586–594.

64. Bakillah A. Nitrated apolipoprotein A-I, a potential new cardiovascular marker, is markedly increased in low high-density lipoprotein cholesterol subjects. *Clin Chem Lab Med* 2009; 47: 60–69.

65. Park KH, Shin DG, Kim JR, Cho KH. Senescence-related truncation and multimerization of apolipoprotein A-I in high-density lipoprotein with an elevated level of advanced glycated end products and cholesteryl ester transfer activity. *J Gerontol A Biol Sci Med Sci* 2010; 65: 600–610.

66. McMillen TS, Heinecke JW, LeBoeuf RC. Expression of human myeloperoxidase by macrophages promotes atherosclerosis in mice. *Circulation* 2005; 111: 2798–2804.

67. Zheng L, Settle M, Brubaker G, Schmitt D, Hazen SL, Smith JD, Kinter M. Localization of nitration and chlorination sites on apolipoprotein A-I catalyzed by myeloperoxidase in human atheroma and associated oxidative impairment in ABCA1-dependent cholesterol efflux from macrophages. *J Biol Chem* 2005; 280: 38–47.

68. Holzer M, Gauster M, Pfeifer T, Wadsack C, Fauler G, Stiegler P, Koefeler H, Beubler E, Schuligoi R, Heinemann A et al. Protein carbamylation renders high-density lipoprotein dysfunctional. *Antioxid Redox Signal* 2011; 15: 2337–2346.

69. Zhang R, Brennan ML, Fu X, Aviles RJ, Pearce GL, Penn MS, Topol EJ, Sprecher DL, Hazen SL. Association between myeloperoxidase levels and risk of coronary artery disease. *JAMA* 2001; 286: 2136–2142.

70. Meuwese MC, Stroes ES, Hazen SL, van Miert JN, Kuivenhoven JA, Schaub RG, Wareham NJ, Luben R, Kastelein JJ, Khaw KT et al. Serum myeloperoxidase levels are associated with the future risk of coronary artery disease in apparently healthy individuals: the EPIC-Norfolk Prospective Population Study. *J Am Coll Cardiol* 2007; 50: 159–165.

71. Duzguncinar O, Yavuz B, Hazirolan T, Deniz A, Tokgozoglu SL, Akata D, Demirpence E. Plasma myeloperoxidase is related to the severity of coronary artery disease. *Acta Cardiol* 2008; 63: 147–152.

72. Sokolov AV, Ageeva KV, Cherkalina OS, Pulina MO, Zakharova ET, Prozorovskii VN, Aksenov DV, Vasilyev VB, Panasenko OM. Identification and properties of complexes formed by myeloperoxidase with lipoproteins and ceruloplasmin. *Chem Phys Lipids* 2010; 163: 347–355.

73. Ehrenwald E, Chisolm GM, Fox PL. Intact human ceruloplasmin oxidatively modifies low density lipoprotein. *J Clin Invest* 1994; 93: 1493–1501.

74. Ehrenwald E, Fox PL. Role of endogenous ceruloplasmin in low density lipoprotein oxidation by human U937 monocytic cells. *J Clin Invest* 1996; 97: 884–890.
75. Marsche G, Furtmuller PG, Obinger C, Sattler W, Malle E. Hypochlorite-modified high-density lipoprotein acts as a sink for myeloperoxidase *in vitro*. *Cardiovasc Res* 2008; 79: 187–194.
76. Jayaraman S, Gantz DL, Gursky O. Effects of protein oxidation on the structure and stability of model discoidal high-density lipoproteins. *Biochemistry* 2008; 47: 3875–3882.
77. Gao X, Jayaraman S, Gursky O. Mild oxidation promotes and advanced oxidation impairs remodeling of human high-density lipoprotein *in vitro*. *J Mol Biol* 2008; 376: 997–1007.
78. Szapacs ME, Kim HY, Porter NA, Liebler DC. Identification of proteins adducted by lipid peroxidation products in plasma and modifications of apolipoprotein A1 with a novel biotinylated phospholipid probe. *J Proteome Res* 2008; 7: 4237–4246.
79. Fievet C, Igau B, Bresson R, Drouin P, Fruchart JC. Non-enzymatic glycosylation of apolipoprotein A-I and its functional consequences. *Diabete Metab* 1995; 21: 95–98.
80. Park KH, Cho KH. High-Density Lipoprotein (HDL) from elderly and reconstituted HDL containing glycated apolipoproteins A-I share proatherosclerotic and prosenescent properties with increased cholesterol influx. *J Gerontol A Biol Sci Med Sci* 2011; 66: 511–520.
81. Lapolla A, Brioschi M, Banfi C, Tremoli E, Cosma C, Bonfante L, Cristoni S, Seraglia R, Traldi P. Nonenzymatically glycated lipoprotein ApoA-I in plasma of diabetic and nephropathic patients. *Ann N Y Acad Sci* 2008; 1126: 295–299.
82. Nobecourt E, Zeng J, Davies MJ, Brown BE, Yadav S, Barter PJ, Rye KA. Effects of cross-link breakers, glycation inhibitors and insulin sensitisers on HDL function and the non-enzymatic glycation of apolipoprotein A-I. *Diabetologia* 2008; 51: 1008–1017.
83. Park KH, Jang W, Kim KY, Kim JR, Cho KH. Fructated apolipoprotein A-I showed severe structural modification and loss of beneficial functions in lipid-free and lipid-bound state with acceleration of atherosclerosis and senescence. *Biochem Biophys Res Commun* 2010; 392: 295–300.
84. Lee-Rueckert M, Silvennoinen R, Rotllan N, Judstrom I, Blanco-Vaca F, Metso J, Jauhiainen M, Kovanen PT, Escola-Gil JC. Mast cell activation *in vivo* impairs the macrophage reverse cholesterol transport pathway in the mouse. *Arterioscler Thromb Vasc Biol* 2011; 31: 520–527.
85. de Grooth GJ, Klerkx AH, Stroes ES, Stalenhoef AF, Kastelein JJ, Kuivenhoven JA. A review of CETP and its relation to atherosclerosis. *J Lipid Res* 2004; 45: 1967–1974.
86. Le Goff W, Guerin M, Chapman MJ. Pharmacological modulation of cholesteryl ester transfer protein, a new therapeutic target in atherogenic dyslipidemia. *Pharmacol Ther* 2004; 101: 17–38.
87. Lamarche B, Rashid S, Lewis GF. HDL metabolism in hypertriglyceridemic states: an overview. *Clin Chim Acta* 1999; 286: 145–161.
88. Papathanasiou A, Kostara C, Cung MT, Seferiadis K, Elisaf M, Bairaktari E, Goudevenos IA. Analysis of the composition of plasma lipoproteins in patients with extensive coronary heart disease using 1H NMR spectroscopy. *Hellenic J Cardiol* 2008; 49: 72–78.

89. Schaefer EJ, McNamara JR, Asztalos BF, Tayler T, Daly JA, Gleason JL, Seman LJ, Ferrari A, Rubenstein JJ. Effects of atorvastatin versus other statins on fasting and postprandial C-reactive protein and lipoprotein-associated phospholipase A2 in patients with coronary heart disease versus control subjects. *Am J Cardiol* 2005; 95: 1025–1032.

90. Pruzanski W, Stefanski E, de Beer FC, de Beer MC, Ravandi A, Kuksis A. Comparative analysis of lipid composition of normal and acute-phase high density lipoproteins. *J Lipid Res* 2000; 41: 1035–1047.

91. Sibmooh N, Yamanont P, Krudsood S, Leowattana W, Brittenham G, Looareesuwan S, Udomsangpetch R. Increased fluidity and oxidation of malarial lipoproteins: relation with severity and induction of endothelial expression of adhesion molecules. *Lipids Health Dis* 2004; 3: 15.

92. Clifton PM, Mackinnon AM, Barter PJ. Effects of serum amyloid A protein (SAA) on composition, size, and density of high density lipoproteins in subjects with myocardial infarction. *J Lipid Res* 1985; 26: 1389–1398.

93. Auerbach BJ, Parks JS. Lipoprotein abnormalities associated with lipopolysaccharide-induced lecithin: cholesterol acyltransferase and lipase deficiency. *J Biol Chem* 1989; 264: 10264–10270.

94. Ettinger WH, Miller LD, Albers JJ, Smith TK, Parks JS. Lipopolysaccharide and tumor necrosis factor cause a fall in plasma concentration of lecithin: cholesterol acyltransferase in cynomolgus monkeys. *J Lipid Res* 1990; 31: 1099–1107.

95. Jonas A. Lecithin cholesterol acyltransferase. *Biochim Biophys Acta* 2000; 1529: 245–256.

96. Crowl RM, Stoller TJ, Conroy RR, Stoner CR. Induction of phospholipase A2 gene expression in human hepatoma cells by mediators of the acute phase response. *J Biol Chem* 1991; 266: 2647–2651.

97. Pruzanski W, Vadas P, Browning J. Secretory non-pancreatic group II phospholipase A2: role in physiologic and inflammatory processes. *J Lipid Mediat* 1993; 8: 161–167.

98. Tietge UJ, Maugeais C, Lund-Katz S, Grass D, de Beer FC, Rader DJ. Human secretory phospholipase A2 mediates decreased plasma levels of HDL cholesterol and apoA-I in response to inflammation in human apoA-I transgenic mice. *Arterioscler Thromb Vasc Biol* 2002; 22: 1213–1218.

99. Bagdade JD, Ritter MC, Subbaiah PV. Accelerated cholesteryl ester transfer in plasma of patients with hypercholesterolemia. *J Clin Invest* 1991; 87: 1259–1265.

100. Frenais R, Ouguerram K, Maugeais C, Marchini JS, Benlian P, Bard JM, Magot T, Krempf M. Apolipoprotein A-I kinetics in heterozygous familial hypercholesterolemia: a stable isotope study. *J Lipid Res* 1991; 40: 1506–1511.

101. Guerin M, Dolphin PJ, Talussot C, Gardette J, Berthezene F, Chapman MJ. Pravastatin modulates cholesteryl ester transfer from HDL to apoB-containing lipoproteins and lipoprotein subspecies profile in familial hypercholesterolemia. *Arterioscler Thromb Vasc Biol* 1995; 15: 1359–1368.

102. Guerin M, Lassel TS, Le Goff W, Farnier M, Chapman MJ. Action of atorvastatin in combined hyperlipidemia: preferential reduction of cholesteryl ester transfer from HDL to VLDL1 particles. *Arterioscler Thromb Vasc Biol* 2000; 20: 189–197.

103. Koizumi J, Inazu A, Yagi K, Koizumi I, Uno Y, Kajinami K, Miyamoto S, Moulin P, Tall AR, Mabuchi H et al. Serum lipoprotein lipid concentration and composition in homozygous and heterozygous patients with cholesteryl ester transfer protein deficiency. *Atherosclerosis* 1991; 90: 189–196.

104. Yetukuri L, Soderlund S, Koivuniemi A, Seppanen-Laakso T, Niemela PS, Hyvonen M, Taskinen M-R, Vattulainen I, Jauhiainen M, Oresic M. Composition and lipid spatial distribution of HDL particles in subjects with low and high HDL-cholesterol. *J Lipid Res* 2010; 51: 2341–2351.

105. Ozerova IN, Perova NV, Shchel'tsyna NV, Mamedov MN. Parameters of high-density lipoproteins in patients with arterial hypertension in combination with other components of metabolic syndrome. *Bull Exp Biol Med* 2007; 143: 320–322.

106. Davidson WS, Gillotte KL, Lund-Katz S, Johnson WJ, Rothblat GH, Phillips MC. The effect of high density lipoprotein phospholipid acyl chain composition on the efflux of cellular free cholesterol. *J Biol Chem* 1995; 270: 5882–5890.

107. Subbaiah PV, Liu M. Role of sphingomyelin in the regulation of cholesterol esterification in the plasma lipoproteins. Inhibition of lecithin-cholesterol acyltransferase reaction. *J Biol Chem* 1993; 268: 20156–20163.

108. Zerrad-Saadi A, Therond P, Chantepie S, Couturier M, Rye K-A, Chapman MJ, Kontush A. HDL3-mediated inactivation of LDL-associated phospholipid hydroperoxides is determined by the redox status of apolipoprotein A-I and HDL particle surface lipid rigidity: Relevance to inflammation and atherogenesis. *Arterioscler Thromb Vasc Biol* 2009; 29: 2169–2175.

109. Harris WS, Mozaffarian D, Rimm E, Kris-Etherton P, Rudel LL, Appel LJ, Engler MM, Engler MB, Sacks F. Omega-6 fatty acids and risk for cardiovascular disease: a science advisory from the American Heart Association Nutrition Subcommittee of the Council on Nutrition, Physical Activity, and Metabolism; Council on Cardiovascular Nursing; and Council on Epidemiology and Prevention. *Circulation* 2009; 119: 902–907.

110. Tabas I. Sphingolipids and atherosclerosis: A mechanistic connection? A therapeutic opportunity? *Circulation* 2004; 110: 3400–3401.

111. de Souza JA, Vindis C, Hansel B, Negre-Salvayre A, Therond P, Serrano CV Jr, Chantepie S, Salvayre R, Bruckert E, Chapman MJ et al. Metabolic syndrome features small, apolipoprotein A-I-poor, triglyceride-rich HDL3 particles with defective anti-apoptotic activity. *Atherosclerosis* 2008; 197: 84–94.

112. Son DJ, Lee HW, Shin HW, Lee JJ, Yoo HS, Kim TJ, Yun YP, Hong JT. Enhanced release of sphingosine-1-phosphate from hypercholesterolemic platelets: role in development of hypercholesterolemic atherosclerosis. *Prostaglandins Leukot Essent Fatty Acids* 2008; 78: 383–390.

113. Marsche G, Heller R, Fauler G, Kovacevic A, Nuszkowski A, Graier W, Sattler W, Malle E. 2-Chlorohexadecanal derived from hypochlorite-modified high-density lipoprotein-associated plasmalogen is a natural inhibitor of endothelial nitric oxide biosynthesis. *Arterioscler Thromb Vasc Biol* 2004; 24: 2302–2306.

114. Stocker R. Lipoprotein oxidation: mechanistic aspects, methodological approaches and clinical relevance. *Curr Opin Lipidol* 1994; 5: 422–433.

115. Stocker R, Keaney JF Jr. Role of oxidative modifications in atherosclerosis. *Physiol Rev* 2004; 84: 1381–1478.

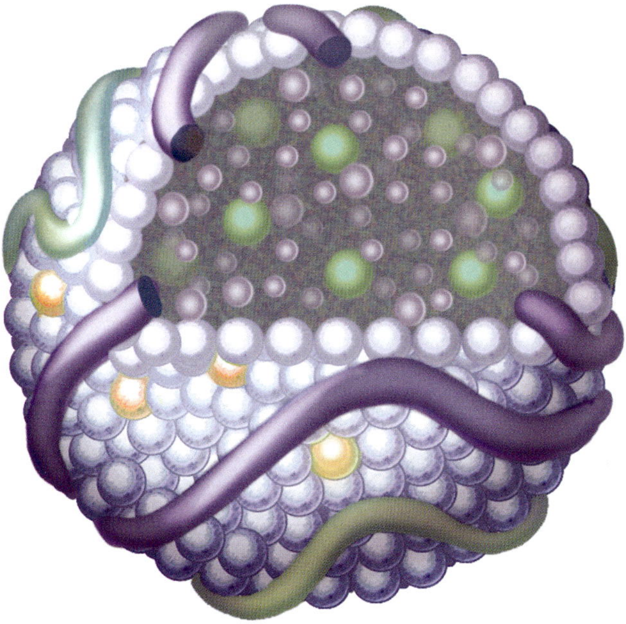

Figure 1.1 Plurimolecular, quasi-spherical, pseudomicellar HDL particle with major structural components. See page 2 for full caption.

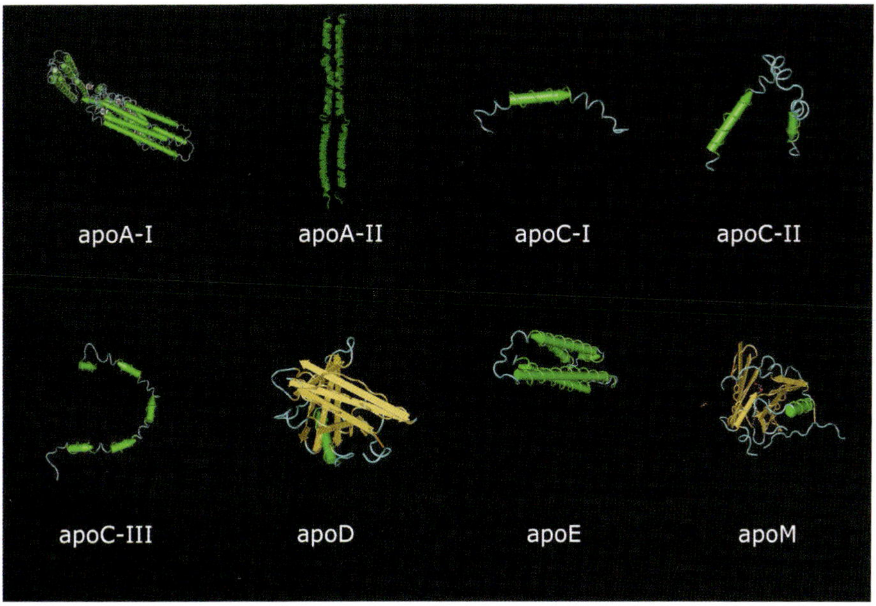

Figure 1.2 Experimentally-obtained structures of major HDL apolipoproteins. See page 10 for full caption.

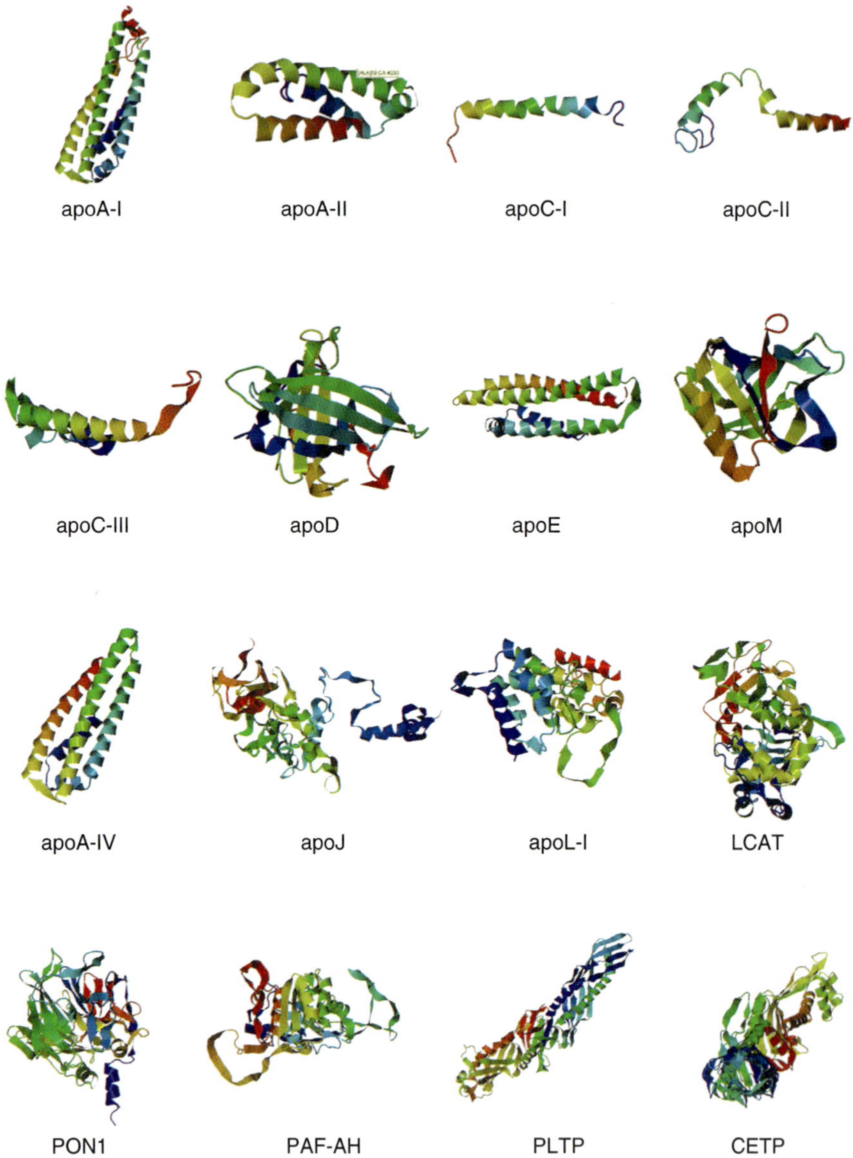

Figure 1.3 Model structures of major HDL proteins. See page 11 for full caption.

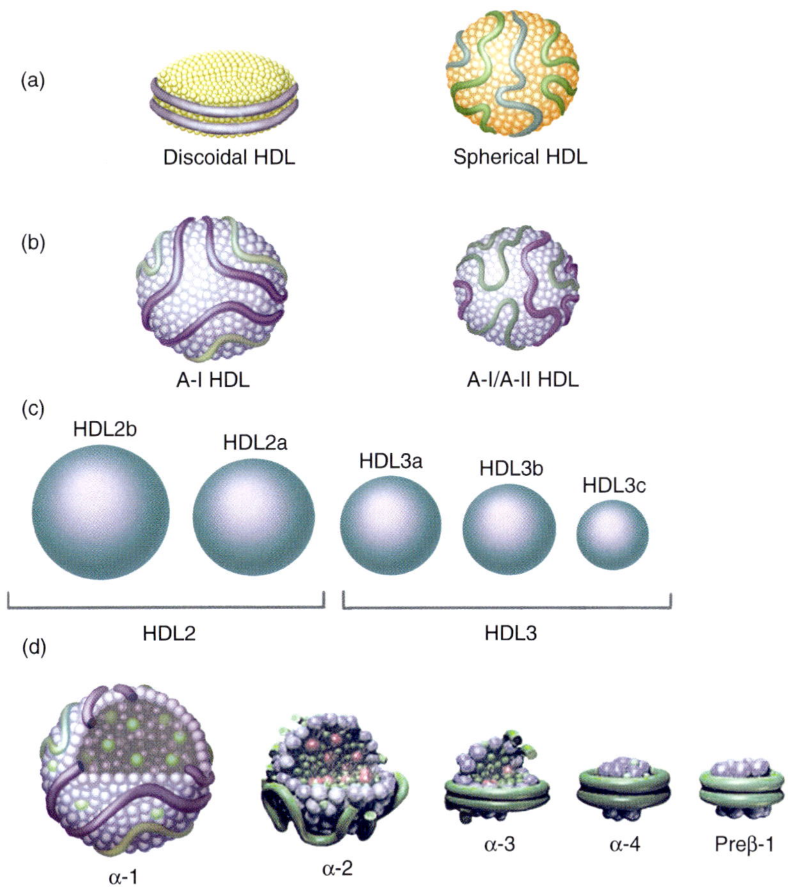

Figure 2.1 Heterogeneity of HDL particles. See page 40 for full caption.

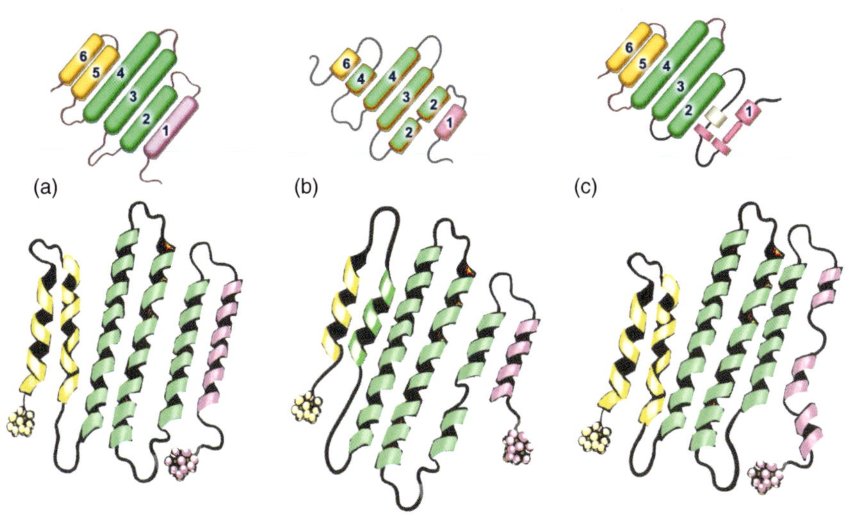

Figure 3.1 Structure of lipid-free apoA-I (modified from Reference [3]). See page 60 for full caption.

Figure 3.2 Double belt, hairpin and looped belt models of discoid HDL (modified from Reference [3]). See page 64 for full caption.

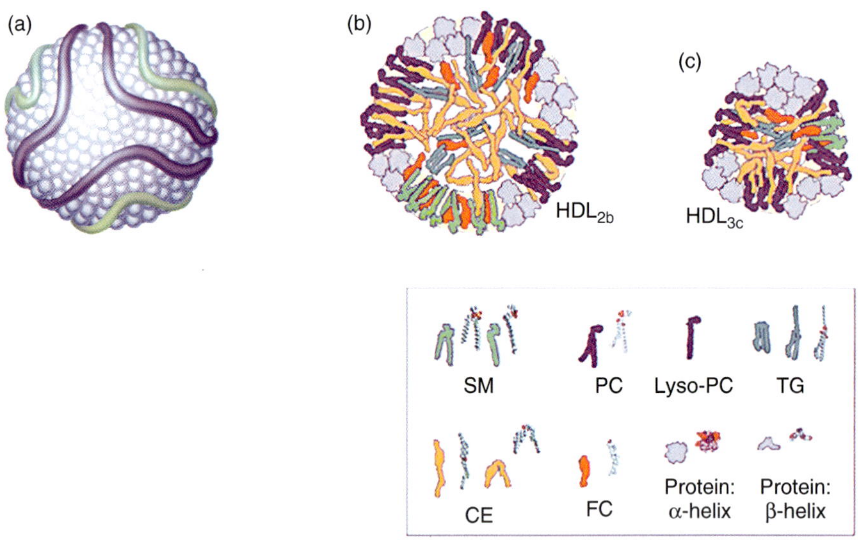

Figure 3.3 Structural organization of spherical HDL. See page 68 for full caption.

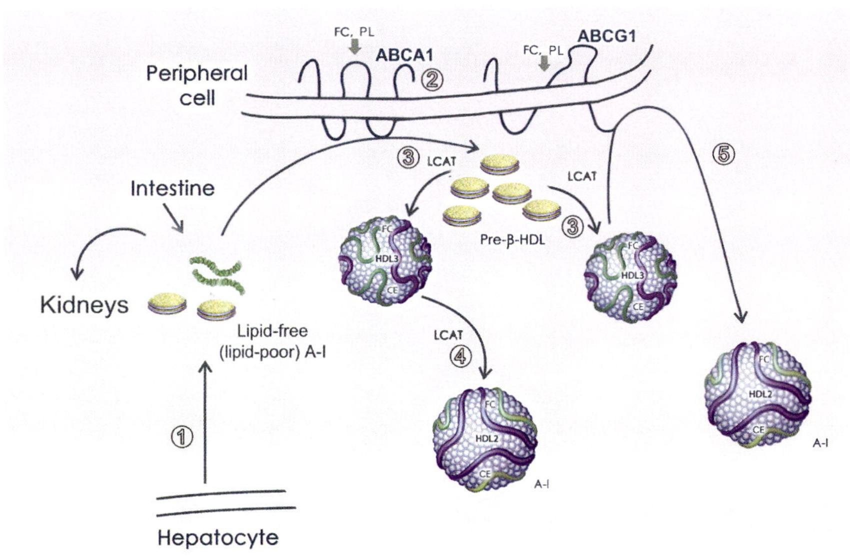

Figure 4.1 Major pathways involved in HDL formation. See page 75 for full caption.

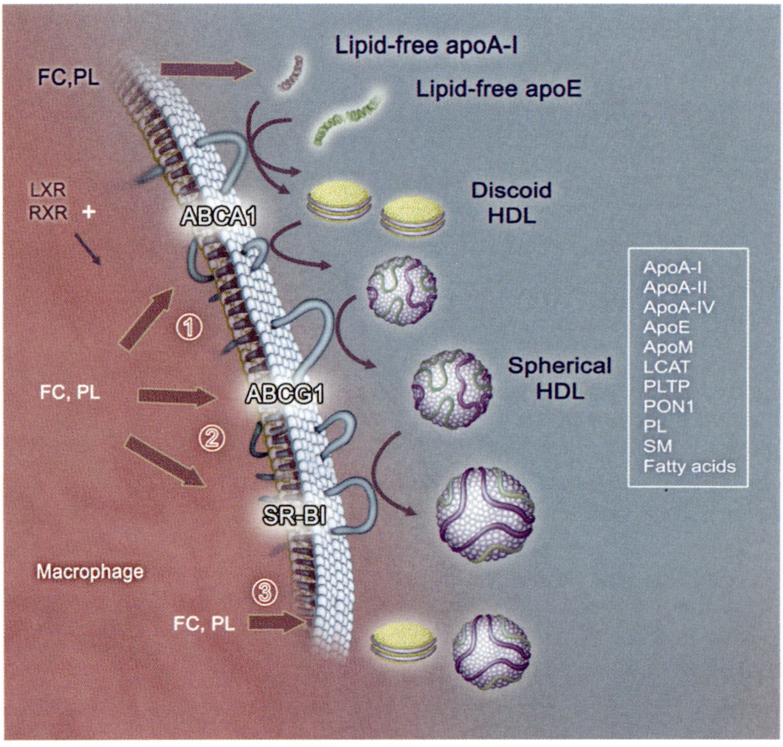

Figure 7.2 Major pathways involved in HDL-mediated cholesterol efflux. See page 201 for full caption.

116. Navab M, Hama LS, Van LB, Fonarow GC, Cardinez CJ, Castellani LW, Brennan ML, Lusis AJ, Fogelman AM. Mildly oxidized LDL induces an increased apolipoprotein J/paraoxonase ratio. *J Clin Invest* 1997; 99: 2005–2019.

117. Van Lenten BJ, Wagner AC, Nayak DP, Hama S, Navab M, Fogelman AM. High-density lipoprotein loses its anti-inflammatory properties during acute influenza A infection. *Circulation* 2001; 103: 2283–2288.

118. Jaouad L, Milochevitch C, Khalil A. PON1 paraoxonase activity is reduced during HDL oxidation and is an indicator of HDL antioxidant capacity. *Free Radic Res* 2003; 37: 77–83.

119. Hedrick CC, Thorpe SR, Fu MX, Harper CM, Yoo J, Kim SM, Wong H, Peters AL. Glycation impairs high-density lipoprotein function. *Diabetologia* 2000; 43: 312–320.

120. Ferretti G, Bacchetti T, Marchionni C, Caldarelli L, Curatola G. Effect of glycation of high density lipoproteins on their physicochemical properties and on paraoxonase activity. *Acta Diabetol* 2001; 38: 163–169.

121. Ferretti G, Bacchetti T, Masciangelo S, Bicchiega V. Effect of homocysteinylation on high density lipoprotein physico-chemical properties. *Chem Phys Lipids* 2010; 163: 228–235.

122. Van Eck M, Hoekstra M, Hildebrand RB, Yaong Y, Stengel D, Kruijt JK, Sattler W, Tietge UJF, Ninio E, Van Berkel TJC et al. Increased oxidative stress in scavenger receptor BI knockout mice with dysfunctional HDL. *Arterioscler Thromb Vasc Biol* 2007; 27: 2413–2419.

123. Jornayvaz FR, Brulhart-Meynet MC, James RW. Myeloperoxidase and paraoxonase-1 in type 2 diabetic patients. *Nutr Metab Cardiovasc Dis* 2009; 19: 613–619.

124. Tavori H, Aviram M, Khatib S, Musa R, Mannheim D, Karmeli R, Vaya J. Human carotid lesion linoleic acid hydroperoxide inhibits paraoxonase 1 (PON1) activity via reaction with PON1 free sulfhydryl cysteine 284. *Free Radic Biol Med* 2011; 50: 148–156.

125. Fuhrman B, Volkova N, Aviram M. Pomegranate juice polyphenols increase recombinant paraoxonase-1 binding to high-density lipoprotein: studies *in vitro* and in diabetic patients. *Nutrition* 2010; 26: 359–366.

126. Blatter Garin MC, Kalix B, Morabia A., James, R.W. Small, dense lipoprotein particles and reduced paraoxonase-1 in patients with the metabolic syndrome. *J Clin Endocrinol Metab* 2005; 90: 2264–2269.

127. Boemi M, Leviev I, Sirolla C, Pieri C, Marra M, James RW. Serum paraoxonase is reduced in type 1 diabetic patients compared to non-diabetic, first degree relatives; influence on the ability of HDL to protect LDL from oxidation. *Atherosclerosis* 2001; 155: 229–235.

128. Costa LG, Vitalone A, Cole TB, Furlong CE. Modulation of paraoxonase (PON1) activity. *Biochem Pharmacol* 2005; 69: 541–550.

129. Connelly PW, Zinman B, Maguire GF, Mamakeesick M, Harris SB, Hegele RA, Retnakaran R, Hanley AJG. Association of the novel cardiovascular risk factors paraoxonase 1 and cystatin C in type 2 diabetes. *J Lipid Res* 2009; 50: 1216–1222.

130. Ridker PM, Brown NJ, Vaughan DE, Harrison DG, Mehta JL. Established and emerging plasma biomarkers in the prediction of first atherothrombotic events. *Circulation* 2004; 109: IV–6–19.

131. Jafarnejad A, Bathaie SZ, Nakhjavani M, Hassan MZ. Effect of spermine on lipid profile and HDL functionality in the streptozotocin-induced diabetic rat model. *Life Sci* 2008; 82: 301–307.

132. Marsillach J, Aragones G, Mackness B, Mackness M, Rull A, Beltran-Debon R, Pedro-Botet J, Alonso-Villaverde C, Joven J, Camps J. Decreased paraoxonase-1 activity is associated with alterations of high-density lipoprotein particles in chronic liver impairment. *Lipids Health Dis* 2010; 9: 46.

133. Nishio E, Watanabe Y. Cigarette smoke extract inhibits plasma paraoxonase activity by modification of the enzyme's free thiols. *Biochem Biophys Res Commun* 1997; 236: 289–293.

134. James RW, Leviev I, Righetti A. Smoking is associated with reduced serum paraoxonase activity and concentration in patients with coronary artery disease. *Circulation* 2000; 101: 2252–2257.

135. Boemi M, Sirolla C, Testa R, Cenerelli S, Fumelli P, James RW. Smoking is associated with reduced serum levels of the antioxidant enzyme, paraoxonase, in Type 2 diabetic patients. *Diabet Med* 2004; 21: 423–427.

136. Durrington PN, Mackness B, Mackness MI. Paraoxonase and atherosclerosis. *Arterioscler Thromb Vasc Biol* 2001; 21: 473–480.

137. Erguder IB, Erguder T, Ozkan C, Bozkurt N, Soylu K, Devrim E, Durak I. Short-term effects of smoking cessation on blood antioxidant parameters and paraoxonase activity in healthy asymptomatic long-term cigarette smokers. *Inhal Toxicol* 2006; 18: 575–579.

138. Parra S, Alonso-Villaverde C, Coll B, Ferre N, Marsillach J, Aragones G, Mackness M, Mackness B, Masana L, Joven J et al. Serum paraoxonase-1 activity and concentration are influenced by human immunodeficiency virus infection. *Atherosclerosis* 2007; 194: 175–181.

139. Aslan M, Nazligul Y, Horoz M, Bolukbas C, Bolukbas FF, Gur M, Celik H, Erel O. Serum paraoxonase-1 activity in *Helicobacter pylori* infected subjects. *Atherosclerosis* 2008; 196: 270–274.

140. Dullaart RP, de Vries R, Sluiter WJ, Voorbij HA. High plasma C-reactive protein (CRP) is related to low paraoxonase-I (PON-I) activity independently of high leptin and low adiponectin in type 2 diabetes mellitus. *Clin Endocrinol* 2009; 70: 221–226.

141. Koubaa N, Nakbi A, Hammami S, Attia N, Mehri S, Ben Hamda K, Ben Farhat M, Miled A, Hammami M. Association of homocysteine thiolactonase activity and PON1 polymorphisms with the severity of acute coronary syndrome. *Clin Biochem* 2009; 42: 771–776.

142. Nguyen SD, Hung ND, Cheon-Ho P, Ree KM, Dai-Eun S. Oxidative inactivation of lactonase activity of purified human paraoxonase 1 (PON1). *Biochim Biophys Acta* 2009; 1790: 155–160.

143. Karabina SA, Elisaf M, Bairaktari E, Tzallas C, Siamopoulos KC, Tselepis AD. Increased activity of platelet-activating factor acetylhydrolase in low-density lipoprotein subfractions induces enhanced lysophosphatidylcholine production during oxidation in patients with heterozygous familial hypercholesterolaemia. *Eur J Clin Invest* 1997; 27: 595–602.

144. Tsimihodimos V, Karabina SA, Tambaki AP, Bairaktari E, Miltiadous G, Goudevenos JA, Cariolou MA, Chapman MJ, Tselepis AD, Elisaf M. Altered distribution of platelet-activating factor- acetylhydrolase activity between LDL and HDL as a function of the severity of hypercholesterolemia. *J Lipid Res* 2002; 43: 256–263.

145. Lagos KG, Filippatos TD, Tsimihodimos V, Gazi IF, Rizos C, Tselepis AD, Mikhailidis DP, Elisaf MS. Alterations in the high density lipoprotein phenotype and HDL-associated enzymes in subjects with metabolic syndrome. *Lipids* 2009; 44: 9–16.

146. Mitsutake R, Miura S, Zhang B, Saku K. HDL-associated factors provide additional prognostic information for coronary artery disease as determined by multi-detector row computed tomography. *Int J Cardiol* 2010; 143: 72–78.

147. Miyaura S, Eguchi H, Johnston JM. Effect of a cigarette smoke extract on the metabolism of the proinflammatory autacoid, platelet-activating factor. *Circ Res* 1992; 70: 341–347.

148. Iwase M, Sonoki K, Sasaki N, Ohdo S, Higuchi S, Hattori H, Iida M. Lysophosphatidylcholine contents in plasma LDL in patients with type 2 diabetes mellitus: relation with lipoprotein-associated phospholipase A2 and effects of simvastatin treatment. *Atherosclerosis* 2008; 196: 931–936.

149. Tellis CC, Tselepis AD. The role of lipoprotein-associated phospholipase A2 in atherosclerosis may depend on its lipoprotein carrier in plasma. *Biochim Biophys Acta* 2009; 1791: 327–338.

150. Tsimikas S, Willeit J, Knoflach M, Mayr M, Egger G, Notdurfter M, Witztum JL, Wiedermann CJ, Xu Q, Kiechl S. Lipoprotein-associated phospholipase A2 activity, ferritin levels, metabolic syndrome, and 10-year cardiovascular and non-cardiovascular mortality: results from the Bruneck study. *Eur Heart J* 2009; 30: 107–115.

151. Gomes MB, Cobas RA, Nunes E, Nery M, Castro-Faria-Neto HC, Tibirica E. Serum platelet-activating factor acetylhydrolase activity: a novel potential inflammatory marker in type 1 diabetes. *Prostaglandins Other Lipid Mediat* 2008; 87: 42–46.

152. Schmidt EB, Koenig W, Khuseyinova N, Christensen JH. Lipoprotein-associated phospholipase A2 concentrations in plasma are associated with the extent of coronary artery disease and correlate to adipose tissue levels of marine n-3 fatty acids. *Atherosclerosis* 2008; 196: 420–424.

153. Kim JY, Hyun YJ, Jang Y, Lee BK, Chae JS, Kim SE, Yeo HY, Jeong TS, Jeon DW, Lee JH. Lipoprotein-associated phospholipase A2 activity is associated with coronary artery disease and markers of oxidative stress: a case-control study. *Am J Clin Nutr* 2008; 88: 630–637.

154. Vickers KC, Maguire CT, Wolfert R, Burns AR, Reardon M, Geis R, Holvoet P, Morrisett JD. Relationship of lipoprotein-associated phospholipase A2 and oxidized low density lipoprotein in carotid atherosclerosis. *J Lipid Res* 2009; 50: 1735–1743.

155. De Keyzer D, Karabina S-A, Wei W, Geeraert B, Stengel D, Marsillach J, Camps J, Holvoet P, Ninio E. Increased PAFAH and oxidized lipids are associated with inflammation and atherosclerosis in hypercholesterolemic pigs. *Arterioscler Thromb Vasc Biol* 2009; 29: 2041–2046.

156. Korotaeva AA, Samoilova EV, Pirkova AA, Ameliushkina VA, Prokazova NV, Tkachuk VA, Chazov EI. Opposite effects of native and oxidized lipoproteins on the activity of secretory phospholipase A(2) group IIA. *Prostaglandins Other Lipid Mediat* 2009; 90: 37–41.

157. Tselepis AD, Panagiotakos DB, Pitsavos C, Tellis CC, Chrysohoou C, Stefanadis C. Smoking induces lipoprotein-associated phospholipase A2 in cardiovascular disease free adults: the ATTICA Study. *Atherosclerosis* 2009; 206: 303–308.

158. Nakhjavani M, Asgharani F, Khalilzadeh O, Esteghamati A, Ghaneei A, Morteza A, Anvari M. Oxidized low-density lipoprotein is negatively correlated with lecithin-cholesterol acyltransferase activity in type 2 diabetes mellitus. *Am J Med Sci* 2011; 341: 92–95.

159. McCall MR, van den Berg JJ, Kuypers FA, Tribble DL, Krauss RM, Knoff LJ, Forte TM. Modification of LCAT activity and HDL structure. New links between cigarette smoke and coronary heart disease risk. *Arterioscler Thromb* 1994; 14: 248–253.

160. Chen C, Loo G. Inhibition of lecithin: cholesterol acyltransferase activity in human blood plasma by cigarette smoke extract and reactive aldehydes. *J Biochem Toxicol* 1995; 10: 121–128.

161. Bielicki JK, Forte TM, McCall MR. Gas-phase cigarette smoke inhibits plasma lecithin-cholesterol acyltransferase activity by modification of the enzyme's free thiols. *Biochim Biophys Acta* 1995; 1258: 35–40.

162. Freeman DJ, Caslake MJ, Griffin BA, Hinnie J, Tan CE, Watson TD, Packard CJ, Shepherd J. The effect of smoking on post-heparin lipoprotein and hepatic lipase, cholesteryl ester transfer protein and lecithin:cholesterol acyl transferase activities in human plasma. *Eur J Clin Invest* 1998; 28: 584–591.

163. Imamura H, Teshima K, Miyamoto N, Shirota T. Cigarette smoking, high-density lipoprotein cholesterol subfractions, and lecithin: cholesterol acyltransferase in young women. *Metabolism* 2002; 51: 1313–1316.

164. Holleboom AG, Kuivenhoven JA, Vergeer M, Hovingh GK, van Miert JN, Wareham NJ, Kastelein JJ, Khaw KT, Boekholdt SM. Plasma levels of lecithin:cholesterol acyltransferase and risk of future coronary artery disease in apparently healthy men and women: a prospective case-control analysis nested in the EPIC-Norfolk population study. *J Lipid Res* 2010; 51: 416–421.

165. Wu Z, Wagner MA, Zheng L, Parks JS, Shy JM III, Smith JD, Gogonea V, Hazen SL. The refined structure of nascent HDL reveals a key functional domain for particle maturation and dysfunction. *Nat Struct Mol Biol* 2007; 14: 861–868.

166. Shao B, Cavigiolio G, Brot N, Oda MN, Heinecke JW. Methionine oxidation impairs reverse cholesterol transport by apolipoprotein A-I. *Proc Natl Acad Sci U S A* 2008; 105: 12224–12229.

167. Nakhjavani M, Esteghamati A, Esfahanian F, Ghanei A, Rashidi A, Hashemi S. HbA1c negatively correlates with LCAT activity in type 2 diabetes. *Diabetes Res Clin Pract* 2008; 81: 38–41.

CHAPTER 9

ABNORMAL METABOLISM

Modifications of the HDL proteome and lipidome in metabolic diseases are intimately linked to alterations in HDL metabolism. Dyslipidemia, insulin resistance, obesity, inflammation and infection all feature altered lipoprotein metabolism involving perturbed metabolism of HDL as a key component. The intimate metabolic links between atherogenic dyslipidemia, insulin resistance and inflammation hardly allow the proper separation of metabolic pathways underlying these conditions (Table 9.1). In this section, we will, therefore, group together the metabolic alterations that most typically occur, but are not exclusively present, in a given disease condition; concomitant occurrence of such alterations, which can vary in their expression under different conditions, should always be kept in mind.

9.1 DYSLIPIDEMIAS

HDL metabolism is substantially altered in dyslipidemic states, including hypertriglyceridemia, hypercholesterolemia, mixed dyslipidemia, and hypo- and hyperalphalipoproteinemia (Figs 9.1 and 9.2). **Hypertriglyceridemia** involving elevated levels of VLDL and VLDL remnants is typically characterized by decreased concentrations of HDL-C and an increased HDL content of triglyceride, resulting from the supranormal action of CETP [1]. Paralleled by elevated levels of proatherogenic small, dense LDL and potentially of Lp(a), these metabolic alterations

High-Density Lipoproteins: Structure, Metabolism, Function, and Therapeutics, First Edition.
Anatol Kontush and M. John Chapman.

TABLE 9.1 Major Pathways of Altered HDL Metabolism in Atherogenic Dyslipidemia, Insulin Resistance and Inflammation

Pathway	Disease Condition	Effects on HDL Metabolism	Actions on HDL Levels
Elevated CETP activity	Dyslipidemia Insulin resistance Inflammation	Accelerated CE transfer from HDL to TG-rich lipoproteins and reciprocal TG transfer	Subnormal HDL-C levels HDL enrichment in TG Decreased levels of large HDL Reduced mean HDL size
Accelerated lipolysis of TG-enriched HDL	Dyslipidemia Insulin resistance Inflammation	ApoA-I shedding from HDL particles, accelerated renal clearance of apoA-I and small HDL	Subnormal levels of apoA-I, HDL-C and HDL particles
Reduced LPL activity	Dyslipidemia Insulin resistance Inflammation	Diminished generation of surface constituents of TG-rich lipoproteins and their reduced transfer to the HDL pool	Subnormal HDL-C levels
Defective function or diminished expression of ABCA1	Dyslipidemia Insulin resistance Inflammation	Deficient formation of nascent HDL	Subnormal levels of HDL-C and large HDL
Reduced LCAT activity	Dyslipidemia	Reduced esterification of cholesterol in HDL and defective HDL maturation	Subnormal levels of HDL-C and large HDL
Reduced adiponectin production	Insulin resistance Inflammation	Accelerated HDL catabolism, elevated CETP activity	Subnormal HDL-C levels
Elevated PLTP activity	Insulin resistance Inflammation	Accelerated HDL remodeling	Subnormal HDL-C levels

(*continued*)

TABLE 9.1 (*Continued*)

Pathway	Disease Condition	Effects on HDL Metabolism	Actions on HDL Levels
Elevated hepatic production of SAA	Inflammation	Replacement of apoA-I by SAA in HDL	Subnormal levels of apoA-I and HDL-C
Elevated activities of $sPLA_2$ and EL	Inflammation	Hydrolysis of HDL PL	Subnormal levels of HDL-C, HDL PL and large HDL

CE, cholesteryl ester; EL, endothelial lipase; PL, phospholipid; TG, triglyceride.

underlie the phenotype of **atherogenic dyslipidemia**. Under these conditions, elevated CETP activity results in accelerated cholesteryl ester transfer from HDL to triglyceride-rich lipoproteins and in reciprocal triglyceride transfer, producing triglyceride-enriched HDL and subnormal HDL-C levels [1]. Such metabolic dysregulation is reflected in the well-established inverse correlation between plasma levels of triglyceride and HDL-C [2]. The atherogenic low HDL-C dyslipidemia associated with hypertriglyceridemia may also involve **hypercholesterolemia** and elevated LDL-C levels, resulting in **mixed hyperlipidemia**, a metabolic condition associated with markedly elevated cardiovascular risk [3].

Mechanisms leading to reduced plasma HDL-C levels and HDL particle number in hypertriglyceridemic states primarily involve the rapid removal of small HDL particles from the circulation, which results from the intravascular lipolysis of triglyceride-enriched HDL [4] (Fig. 9.2). Indeed, such HDLs are intrinsically more unstable in the circulation and contain loosely bound apoA-I which is rapidly removed by the kidneys. As a result, lipolysis of triglyceride-enriched HDL lowers HDL particle numbers by causing apoA-I to be shed from HDL particles and cleared from the circulation. Dysfunctional LPL or reduced LPL activity also contribute to the lowering of HDL levels by reducing the availability of surface constituents of triglyceride-rich lipoproteins that sequester to the plasma pool of nascent HDL particles [4] (Fig. 9.2). Increased VLDL apoB production, which results in elevated plasma triglyceride concentrations, thereby exerts a major accelerating effect on the catabolism of HDL apoA-I [5, 6]. Thus, the cholesteryl ester/triglyceride ratio represents a critical factor in determining HDL particle stability and plasma residence time; HDLs characterized by a low cholesteryl ester/triglyceride ratio are less stable than normal particles [7–10]. Importantly, accelerated apoA-I catabolism in hypertriglyceridemic states can, in part, be compensated by an elevation in apoA-I production, as suggested by the strong positive correlation between fractional catabolic rate and the production rate of apoA-I [6].

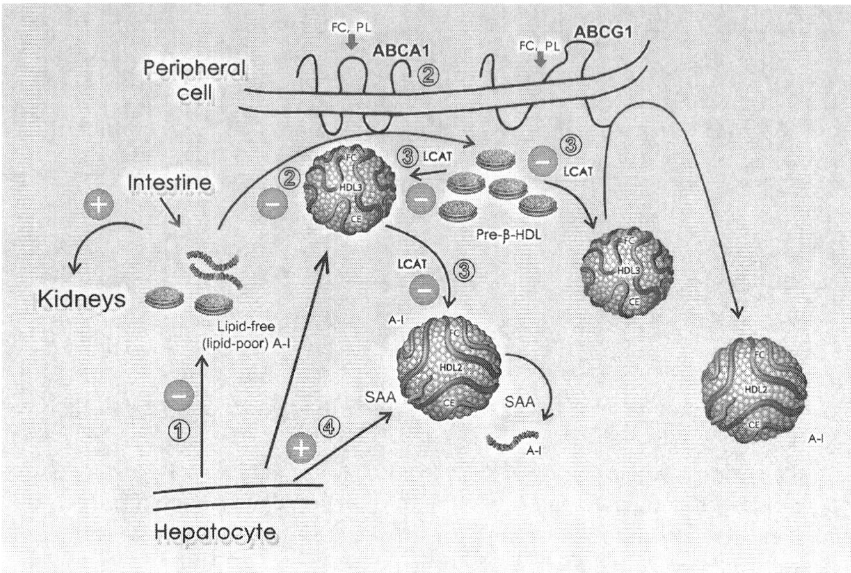

Figure 9.1 Major pathways involved in altered HDL formation in dyslipidemia, inflammation, insulin resistance and cardiovascular disease. Both (1) reduced hepatic production of apoA-I and (2) defective apoA-I lipidation upon interaction with ABCA1 secondary to ABCA1 and/or apoA-I deficiency result in defective formation of large HDL, accelerated renal clearance of apoA-I and subnormal HDL-C levels. (3) Defective HDL maturation can also be caused by diminished cholesterol esterification by LCAT. All these pathways result in an increased proportion of small HDL particles. (4) SAA produced by the liver displaces apoA-I in HDL under inflammatory conditions, with the loss of apoA-I by renal catabolism. CE, cholesteryl ester; FC, free cholesterol; PL, phospholipid.

In hypercholesterolemia, abnormalities in HDL metabolism are primarily observed as moderate decreases in plasma apoA-I and HDL-C levels [11, 12]. Indeed, subnormal plasma levels of apoA-I in familial hypercholesterolemia are related to an increased fractional catabolic rate of apoA-I in both homozygous [11] and heterozygous [12] forms of the disease. In homozygous familial hypercholesterolemia, the deleterious influence of altered apoA-I metabolism on HDL-C levels is further aggravated by decreased rates of apoA-I production [11] (Fig. 9.1). Moreover, elevated CETP activity as a result of increased numbers of apoB-containing lipoproteins—mainly LDL—also contributes to the depletion of cholesteryl ester from the plasma HDL pool in hypercholesterolemic subjects [13, 14] (Fig. 9.2). The HDL-C-lowering effects of CETP can, in part, be counteracted by apoF, apoC-I or endogenous CETP inhibitors. Indeed, plasma levels of apoF are elevated in hypercholesterolemia and apoF expression can be induced by dietary lipids [15]. Such an apoF-mediated decrease in CETP activity does not, however, appear to occur in hypertriglyceridemia which features inverse relationships between plasma triglyceride and apoF levels [16].

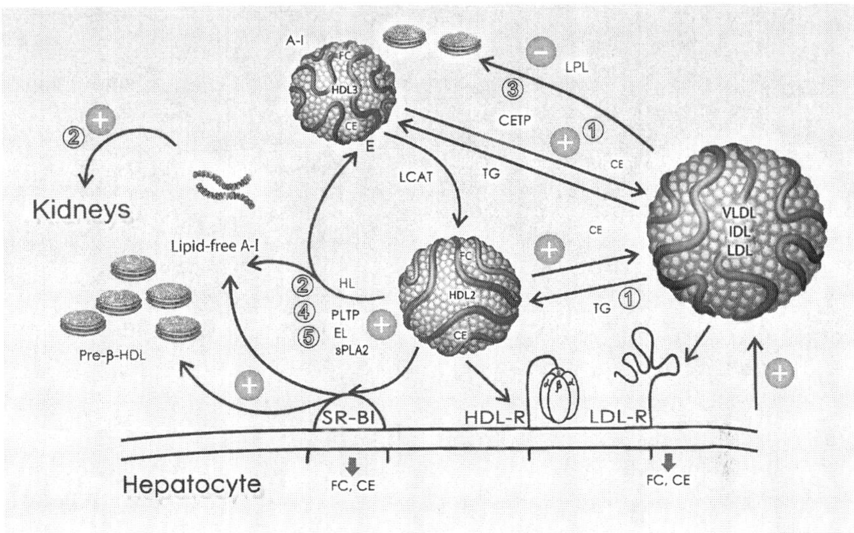

Figure 9.2 Major pathways involved in altered HDL remodeling in dyslipidemia, inflammation, insulin resistance and cardiovascular disease. (1) Elevated levels of VLDL particles resulting from increased hepatic production and/or insufficient triglyceride hydrolysis by LPL are associated with elevated CETP activity and accelerated cholesteryl ester transfer from HDL to triglyceride-rich lipoproteins and reciprocal triglyceride transfer, producing triglyceride-enriched HDL and subnormal HDL-C levels. (2) Enhanced intravascular lipolysis of triglyceride-enriched HDL particles by hepatic lipase, accelerated apoA-I shedding from HDL and enhanced catabolism ensue. (3) Defective VLDL hydrolysis by LPL reduces the transfer of surface constituents of triglyceride-rich lipoproteins to the plasma HDL pool. (4) HDL remodeling by PLTP can also be enhanced. (5) Hydrolysis of HDL phospholipid by endothelial lipase and $sPLA_2$ can be accelerated. CE, cholesteryl ester; EL, endothelial lipase; FC, free cholesterol; HDL-R, holoparticle HDL receptor; HL, hepatic lipase; LDL-R, LDL receptor; TG, triglyceride.

HDL heterogeneity and particle profile can directly reflect abnormalities in HDL metabolism [17]. Thus, plasma concentrations of the large, cholesterol-rich HDL2 subclass decrease in parallel with a decrease in HDL-C in atherogenic dyslipidemia, while concentrations of small, lipid-poor particles frequently remain unaffected [18], resulting in an elevated proportion of **small HDL**. In terms of the HDL proteome, the relative apoA-I contents of HDL2b and HDL2a are decreased, while those of pre-beta-1, pre-beta-2, HDL3a, HDL3b and HDL3c are increased in hypercholesterolemia, hypertriglyceridemia and mixed hyperlipidemia [19–22]. Furthermore, plasma concentrations of small pre-beta HDL are elevated in hypercholesterolemia, hypertriglyceridemia, LCAT deficiency and coronary heart disease, but not in CETP deficiency [19, 23, 24]. Consistent with these observations, concentrations of apoB-100 are positively correlated with pre-beta-1 HDL levels but negatively correlated with those of HDL2a and HDL2b

[25]. Furthermore, the HDL profile is shifted toward smaller particles with an increase in the LDL-C/HDL-C ratio [26]. Finally, plasma levels of small, 70-kDa HDL are elevated in mixed hyperlipidemia [27]. On the other hand, plasma levels of both LpA-I and LpA-I:A-II are reduced in hypertriglyceridemia as a consequence of the accelerated catabolism of triglyceride-enriched LpA-I and LpA-I:A-II [28, 29]. Such a shift from large to small HDL reflects delayed HDL maturation, with perturbed formation of large, cholesterol-rich particles and a diminished flux of cholesterol through the RCT pathway.

Abnormalities in HDL subfraction distribution in hypercholesterolemia also include reduced levels of apoA-I in large HDL2b and HDL2a and elevated levels of apoA-I in small pre-beta-1, pre-beta-2, HDL3b and HDL3c subfractions [19, 20]. HDL from subjects with familial hypercholesterolemia are further characterized by an elevated content of apoE in association with increased plasma levels of apoE-enriched large HDL1, which represents only a minor fraction of total HDL in normolipidemic individuals [30]. Indeed, low HDL-C subjects display an elevated apoE concentration in the plasma along with small HDL particle size [31].

OPEN QUESTION

How are alterations in HDL particle heterogeneity related to those in HDL composition (proteome, lipidome)?

Other causes of **low HDL-C dyslipidemia** include deficiency of apoA-I [32], reduced activity of LCAT [33, 34], elevated activity of hepatic lipase [35, 36], reduced activity of LPL and dysfunctionality of ABCA1 [37]. ApoA-I mutations lead to the inhibition of discrete steps in HDL biogenesis and remodeling, such as the formation of nascent HDL upon interaction of apoA-I with ABCA1, HDL maturation by LCAT, or HDL remodeling by PLTP [38] (Fig. 9.1). **ApoA-I deficiency** can arise from DNA deletions, rearrangements, or nonsense or frameshift mutations within the APOA1 gene resulting in deficient apoA-I secretion [39]. Homozygous familial apoA-I deficiency features undetectable levels of apoA-I and the complete absence of apoA-I-containing HDL particles in plasma, resulting in reduced RCT rates and tissue accumulation of cholesterol in the form of xanthomas [39]. Very low HDL-C levels in such subjects are accounted for by the presence of apoA-II, apoE- and apoA-IV-containing, small HDL particles, whereas large, lipid-rich HDLs are absent. For example, a nonsense mutation at codon 2 of apoA-I, Q[2]X, is associated with decreased levels of small alpha-3 migrating apoA-II particles and alpha-migrating apoA-IV- and apoE-containing HDL [40]. Such familial low HDL-C dyslipidemia appears to represent a rare, low HDL-C phenotype characterized by the presence of reduced concentrations of small pre-beta HDL particles [41]. In subjects with a heterozygous apoA-I deficiency, apoA-I-containing HDL is present in the circulation at intermediate levels compared with normolipidemic controls [39].

The **apoA-I Milano** variant, R173C, is well known to be associated with low HDL-C concentrations that are not, however, accompanied by elevated cardiovascular risk [42, 43] (see chapter 5.1). In the circulation, apoA-I Milano forms disulfide-linked dimers that are rapidly catabolized, resulting in low HDL levels [44, 45]. As a result, subjects with apoA-I Milano display elevated proportions of small HDL particles [46]. The mutation impacts on the C-terminal domain of apoA-I, which adopts a conformation differing from that of wild-type apoA-I [47, 48]. Both the loss of the Arg residue and its replacement by a Cys contribute to the altered properties of apoA-I Milano. The Arg residue is potentially involved in an intrahelical salt bridge with E169 that is disrupted by the loss of the positively-charged Arg and repelled by the Cys, thereby destabilizing the helix bundle domain in the apoA-I molecule and modifying its lipid binding characteristics [49]. In turn, the presence of Cys leads to the formation of dimers. The deletion of E235 in the C-terminus of apoA-I, as occurs in apoA-I Nichinan, represents another example of apoA-I deficiency, which, in this case, results from defective lipid binding secondary to the impaired helix-forming capacity of the C-terminal residues [50].

CRITICAL CONTRIBUTION

Identification of normal cardiovascular status in carriers of apoA-I Milano in association with low HDL-C and the prevalence of small HDL particles [42, 43].

Homozygous **LCAT deficiency** also leads to low HDL-C levels, corneal opacities and renal disease. In this form of hypoalphalipoproteinemia, cholesterol efflux from cholesterol-loaded macrophages is normal, while the maturation of small pre-beta HDL to large alpha-HDL is defective, resulting in accelerated catabolism of small HDL, which may lead to renal disease [51, 52]. Metabolically, LCAT deficiency also features the appearance of lipoprotein X (LpX), a highly heterogeneous, unique lipoprotein characterized by a bi-lamellar vesicle-like structure containing phospholipid, free cholesterol, apoA-I and albumin [53]. Patients with LCAT deficiency possess both pre-beta-1 and alpha-4 HDL in their plasma [39].

Genetic **ABCA1 deficiency** results in Tangier disease, a rare form of hypoalphalipoproteinemia which may enhance cardiovascular risk (see 6.1 of chapter 6). In this disorder, deficient cholesterol efflux and poor lipidation of pre-beta HDL lead to the accelerated catabolism of small LpA-I and the absence of large HDL [54]. Therefore, small HDL particles predominate in the HDL particle profile in this disorder [55]. Importantly, large nascent HDL particles (>10 nm diameter), assembled by hepatic ABCA1, generate a PI3K-mediated autocrine signal that attenuates VLDL maturation and triglyceride secretion [56]; anomalies in this pathway may underlie elevated plasma triglyceride concentrations that often occur in subjects with compromised ABCA1 function.

HDL-C levels can be reduced secondary to **elevated hepatic lipase activity** resulting in the accelerated hydrolysis of HDL components and enhanced HDL catabolism [57, 58] (Fig. 9.2). Remodeling of triglyceride-enriched HDL by hepatic lipase may result in enhanced binding, internalization and degradation of HDL, contributing to rapid renal clearance and the overall lowering of plasma HDL-C [59]. **LPL deficiency** also leads to low levels of HDL-C carried by pre-beta-1 and alpha-4 HDL particles, whereas large alpha and very large alpha-1 HDL are lacking [39]. The reduced production of surface remnants from triglyceride-rich lipoproteins in both fasting and post-prandial states may mechanistically account for this observation (Fig. 9.2).

Low HDL-C dyslipidemia frequently features elevated oxidative stress and chronic low-grade inflammation, as observed in chronic **kidney disease** and end-stage renal disease [60, 61], and are characterized by elevated triglyceride levels, diminished LCAT activity and normal PLTP and CETP activity [62]. Impaired renal clearance of apoB-containing apolipoproteins and inefficient HDL maturation can underlie the pro-atherogenic lipid profile in kidney disease [60]. Elevated oxidative stress, typical of renal disease, may also modulate HDL metabolism. Thus, advanced products of protein oxidation may inhibit the plasma clearance of HDL cholesteryl ester by SR-BI, potentially contributing to abnormal HDL composition and elevated cardiovascular risk [63].

Hyperhomocysteinemia is another metabolic disorder causing low HDL-C dyslipidemia. Hypomethylation associated with high plasma homocysteine can decrease hepatic synthesis of phosphatidylcholine, thereby inhibiting HDL particle assembly in the liver [64]. As a result, circulating levels of homocysteine and apoA-I are negatively correlated [65].

Conversely, a deficiency in CETP or hepatic lipase is associated with elevated levels of HDL-C, leading to hyperalphalipoproteinemia. Furthermore, hyperalphalipoproteinemia can arise from the increased production of apoA-I [66]. **CETP deficiency** reduces the exchange of triglyceride and cholesteryl ester between HDL and triglyceride-rich lipoproteins and elevates HDL-C as a result of cholesteryl ester retention. Familial CETP deficiency is associated with the accumulation of large, cholesteryl ester-rich HDL2 particles [66] and particularly with buoyant, apoE-containing HDL1; levels of small particles are affected to a minor degree [55, 67]. CETP deficiency also features a unique lipoprotein particle containing apoA-I, apoA-II and apoE, whose catabolism is delayed [68, 69].

Hepatic lipase deficiency results in the reduced hydrolysis of HDL triglyceride and phospholipid, leading to decelerated HDL catabolism and increased HDL levels. Large HDLs predominate in familial hepatic lipase deficiency [70]. The 2D HDL profile in this disorder is characterized by elevated alpha-1 and pre-beta-1, as well as by decreased alpha-2 HDL [39].

Finally, the stimulating action of peripheral insulin on LPL may explain the elevated levels of HDL-C in Type 1 diabetes [71], which also reveals a predominance of large HDL particles [72].

9.2 INSULIN-RESISTANT STATES

Atherogenic, low HDL-C dyslipidemia associated with hypertriglyceridemia is characteristic of insulin-resistant states associated with elevated cardiovascular risk, such as **Type 2 diabetes** and **Metabolic Syndrome** [5]. Therefore, alterations of HDL metabolism observed in atherogenic dyslipidemia (see chapter 9.1) are also present in insulin resistance, with hypertriglyceridemia, accelerated CETP activity and chronic low-grade inflammation as major contributors to enhanced HDL catabolism (Fig. 9.2).

Factors that specifically affect HDL metabolism in insulin-resistant states primarily involve **elevated plasma levels of insulin and glucose**, as well as an increased mass of adipose tissue. Plasma insulin may further accelerate HDL catabolism, acting through elevated VLDL production [6]. In addition, high plasma concentrations of glucose may hamper HDL-C biogenesis by decreasing ABCA1 expression [73]. Indeed, poor glycemic control appears to be an independent risk factor for low HDL-C in Type 2 diabetes [74].

Obesity, measured as BMI, represents the strongest determinant of decreased HDL-C levels [75]. Increased body fat storage adversely impacts on concentrations of HDL-C, displaying associations with both the degree of obesity and the distribution of adipose tissue. More specifically, intra-abdominal **visceral fat deposition** represents an important negative correlate of HDL-C [76, 77], reflecting mechanistic links between triglyceride and HDL metabolism. Mechanistically, HDL metabolism can be altered under the action of pro-inflammatory mediators secreted by the adipose tissue, as suggested by negative correlation between HDL-C levels and macrophage content in the abdominal subcutaneous adipose tissue [78].

Other pathways can also contribute to low HDL-C levels in insulin resistance. Insulin, acting through a Foxa2-mediated mechanism, is able to inhibit the expression of apoM, a key determinant of plasma pre-beta HDL and alpha-HDL concentrations, thereby contributing to a HDL-C decrease in insulin-resistant states [79]. Indeed, plasma levels of apoM are positively associated with those of HDL-C, apoA-I and apoA-II and negatively associated with fractional catabolic rates of HDL apoA-I and apoA-II [80]. Furthermore, **low plasma levels of apoM** represent an independent predictor of the fractional catabolic rate of HDL apoA-I in overweight, obese, insulin-resistant men in multivariate regression analysis [80].

The **reduction in the plasma adiponectin level** observed in Type 2 diabetes may constitute another mechanism involved in the diminution of HDL-C. Adiponectin can delay both VLDL secretion and HDL catabolism, maintaining apoA-I in the circulation [6].

Visfatin, a recently identified adipocytokine with insulin mimetic properties, may also be involved in HDL metabolism in insulin resistance, as suggested by the positive correlation between visfatin levels and HDL-C in obese adolescents [81]. Finally, HDL remodeling can be accelerated by PLTP whose activity is increased in Type 2 diabetes [82]. Importantly, **elevated PLTP activity** can also result in elevated levels of pro-atherogenic apoB-containing lipoproteins, thereby exacerbating atherosclerosis [83].

Sustained weight loss via exercise, diet or gastric surgery is an effective means of reversing the decrease in HDL-C levels in both obese and non-obese subjects [76, 84–87]. Importantly, weight loss achieved through exercise is more effective at raising HDL-C levels than dieting [76]. An improvement in insulin resistance and adiponectin levels are related to increases in HDL-C, as shown in severely obese women during surgically-induced weight loss, consistent with the important role of adiponectin in HDL-C regulation [88]. Furthermore, increased LCAT and LPL activity, as well as decreased CETP and PLTP activity, may further contribute to raising HDL levels following weight loss [75, 89] accompanied by decreases in pre-beta-1 HDL levels [90].

As a reflection of altered metabolism, HDL particle profiles are modified by insulin resistance. Thus, in the insulin-resistant states of Metabolic Syndrome and Type 2 diabetes, circulating levels of large, cholesterol-rich HDL are reduced in parallel with a decrease in HDL-C [91–94] (Fig. 9.2). In obesity and insulin resistance, plasma concentrations of large HDL particles decrease in along with HDL-C [95–98]. Furthermore, levels of alpha-1 HDL are lower and levels of alpha-2 and alpha-3 HDL are higher in obese subjects compared with lean controls [99].

In contrast, levels of small, dense, cholesterol-poor HDL3 particles and their content of apoA-I are rarely affected by insulin resistance [100–102]. On the other hand, data on plasma levels of small pre-beta HDL are controversial, revealing an inconsistency between the measurements of small HDL by different methodologies. Thus, pre-beta-1 HDL is reported to be elevated in insulin resistance [90, 103], potentially reflecting elevated production from triglyceride-rich lipoproteins by LPL, as the latter can be induced by insulin [103]. In addition, CETP-mediated generation of pre-beta HDL from alpha HDL and accelerated HDL remodeling by PLTP may contribute to the elevation of pre-beta particles [103, 104] (Fig. 9.2). Of environmental factors, overweight and high visceral adipose tissue are strongly associated with increases in small HDL, independent of other factors [105, 106]. In contrast, an almost 50% reduction of circulating levels of pre-beta-1 HDL has also been reported in Type 2 diabetes, with the percentage of pre-beta-1 HDL at about half of the normal value (6.1%) of total plasma apoA-I [107]. As pre-beta HDL formation is positively regulated by apoM, low apoM levels may provide a mechanism of explaining why pre-beta HDL concentrations are not necessarily increased in Type 2 diabetes, despite elevated PLTP activity [104].

In summary, Metabolic Syndrome, Type 2 diabetes, obesity and insulin resistance are all characterized by the **prevalence of small, dense HDL** in the HDL particle profile, indicative of both the impaired conversion from small to large HDL, in addition to accelerated turnover and remodeling of large HDL particles [24, 76]. A decrease in large, light HDL2b thus constitutes a common feature of insulin-resistant states; a novel microfluidic chip method has, therefore, been developed to specifically measure HDL2b in human plasma [108]. Using this approach, reductions in HDL2b and in the HDL2b-to-total HDL ratio can be demonstrated in patients with Type 2 diabetes with [108] and without [109] non-alcoholic fatty liver disease.

Intriguingly, low HDL-C levels may be considered, not only as a consequence, but also as a cause of Type 2 diabetes. Indeed, HDL possesses the capacity of improving the functionality of beta-cells, by acting via the removal of excess cholesterol and by exerting cellular protection from apoptosis [110, 111], as well as the capacity to increase adiponectin expression from adipocytes [112] (see 7.8 of chapter 7). Furthermore, low levels of HDL-C precede the development of Type 2 diabetes in epidemiologic studies [113].

9.3 INFLAMMATORY STATES

A decrease in circulating HDL-C levels, an increase in triglyceride levels and HDL enrichment in SAA at the expense of apoA-I are typical components of **inflammatory states** and the **acute phase reaction** [114, 115]. Thus, **HDL-C levels are decreased** at the time of acute ischemic stroke, behaving as a negative acute phase reactant [116]. In systemic lupus erythematosus, plasma levels of apoA-I are reduced, whereas those of triglyceride and apoE are elevated [117]. In Takayasu's arteritis, an idiopathic chronic inflammatory disease that causes occlusion of large arteries, low HDL-C levels associated with disease activity are observed [118, 119]. Such alterations in HDL metabolism may be of clinical value; indeed, low plasma levels of apoA-I are independently related to 30-day mortality in human sepsis [120]. Furthermore, serum HDL-C levels measured at admission are markedly reduced in patients with left-sided infective endocarditis and can predict clinical outcome [121].

Under inflammatory conditions, HDL particles transport less HDL-C but more triglyceride and are thereby rapidly catabolized and removed from the circulation. Therefore, the concentration of HDL particles in serum is decreased, as observed, for example, in patients after kidney transplantation [122]. Controversially, however, plasma levels of SAA are not associated with the fractional catabolic rate of HDL apoA-I, suggesting that SAA enrichment *per se* does not accelerate apoA-I catabolism [6].

OPEN QUESTION

What role does SAA play in HDL metabolism under pro-inflammatory conditions?

Metabolically, multiple steps in HDL metabolism are affected by inflammation. Specifically, pro-atherogenic cytokines, such as interferon-gamma and IL-1-beta, as well as inflammatory proteins, such as CRP, inhibit expression of ABCA1 [123] (Fig. 9.1). Furthermore, **activities of $sPLA_2$ and endothelial lipase are elevated**, whereas **those of LCAT, CETP and hepatic lipase are**

decreased during the inflammatory response (Fig. 9.2). **$sPLA_2$** enzymes belong to a family of pro-atherogenic enzymes that hydrolyze lipoprotein phospholipids, increasing their affinity for intimal proteoglycans. These enzymes also contribute to cholesterol loading of macrophages by non-scavenger receptor-mediated pathways and activate inflammatory pathways [124]. Elevated $sPLA_2$ activity reduces HDL-C levels, increases pro-inflammatory bioactive lipids and may thereby accelerate atherosclerosis in transgenic mice [125].

Endothelial lipase is a unique member of the triglyceride lipase gene family with a preference for phospholipids rather than triglycerides [126]; increased endothelial lipase activity is also linked to the underlying pro-inflammatory state and may contribute to a reduction in HDL-C. Secretory $sPLA_2$ may play a key role in HDL remodeling during the acute phase, as suggested by disproportionately high increases in $sPLA_2$ activity compared with changes in the activities of CETP or LCAT, which are rather commensurate to the reduction in the level of HDL-C [127]. As a consequence, induction of the acute phase response in humans often leads to the reduction of HDL phospholipid [128]. The combined action of $sPLA_2$, CETP and SAA markedly remodels HDL with the generation of lipid-poor apoA-I [127, 129], further contributing to a decrease in HDL-C.

Induction of the inflammatory response therefore underlies altered HDL metabolism, as further exemplified by the correlation between decreases in the apoA-I/HDL-C ratio with increased platelet activation [120]. Interestingly, in rheumatoid arthritis, apoA-I and HDL-C levels are decreased in plasma and increased in synovial fluid, suggesting infiltration of HDL particles in the inflamed joint with inhibition of the local production of pro-inflammatory cytokines [130]. Subsequent sequestration of apoA-I in the inflamed tissue may, in part, account for the reduction of circulating HDL and elevated cardiovascular risk in this disease.

Consistent with the key role of inflammation, low dose glucocorticoid therapy in rheumatoid arthritis induces increases in HDL-C without affecting LDL-C or triglyceride [131]. Similarly, levels of several biomarkers of inflammation are decreased, whereas those of HDL-C and apoA-I are elevated during anti-TNF therapy for ankylosing spondylitis [132]. In addition, HDL composition is improved by such treatment with the disappearance of SAA from HDL.

HDL heterogeneity is markedly affected by the inflammatory response. Thus, in systemic lupus erythematosus, the proportion of large HDL2b is diminished, whereas that of small HDL3b and HDL3c is elevated [117]. In addition, concentrations of large HDL particles are subnormal in patients with lupus compared with controls; levels of large HDL are inversely associated with disease activity and insulin resistance [133].

Low-grade inflammation is also associated with **smaller HDL particles** in healthy subjects [134]. The decrease in the HDL-C/apoA-I ratio seems to be a good marker of HDL subclass distribution into smaller size particles [122]. In contrast, concentrations of small HDL particles appear to be selectively reduced in patients with rheumatoid arthritis [135].

9.4 INFECTIOUS DISEASES

Alterations in HDL metabolism under conditions of infection are largely reminiscent of those detected under inflammatory conditions and in the acute phase (see 9.3 of chapter 9). Thus, influenza infection is associated with **decreased levels of HDL-C** in humans [136]; in rodents, endotoxin injection increases plasma triglyceride levels through increased hepatic secretion and/or delayed clearance [137].

Furthermore, HDL-C levels are decreased in subjects with *Helicobacter pylori* infection [138]. In addition, plasma levels of HDL-C and apoA-I are lower in patients in the acute phase of acute viral hepatitis compared with control subjects, whereas levels of triglycerides are higher [139]. Moreover, plasma levels of HDL-C and apoA-I are lower in the active phase of the disease than during recovery. Acute viral liver damage can contribute to the observed effects in this disease. Consistent with these data, serum levels of HDL-C and apoA-I are inversely correlated with liver reserve and disease severity in cirrhotic patients with severe sepsis [140].

Changes in HDL metabolism following infection further include **reduced plasma LCAT and CETP activities**, as well as **elevated plasma PLTP activity** paralleled by increased levels of CRP and LBP, as observed during clinical sepsis [141]. Such modifications result in HDL enrichment in triglycerides with an elevated content of lipid peroxidation products and increased lipid core fluidity, as seen in patients with malaria, suggesting oxidative modification [142].

The HDL particle profile can also be modified by infection. Thus, subjects with untreated HIV infection display reduced concentrations of HDL-C (−26%) and diminished HDL particle numbers (−21%), with reductions in both large (−50%) and small (−20%) HDL particles compared with uninfected subjects [143]. Low levels of small and large HDL thereby identify HIV-infected patients with an increased risk of cardiovascular disease independently of other cardiovascular risk factors [144]. Furthermore, pre-beta-1 HDL is decreased, whereas the mass and activity of both LCAT and CETP are elevated in HIV-positive subjects [145]. HIV infection is, therefore, associated with a modified HDL metabolism that redirects cholesterol to apoB-containing lipoproteins. Another pathway specifically involved in the impairment of normal HDL metabolism in HIV-infected subjects includes the enhanced degradation of ABCA1 by the HIV-1 Nef protein, with a subsequent decrease in cholesterol efflux from macrophages [146].

CRITICAL CONTRIBUTION

Interaction of metabolic pathways, such as those involving hypertriglyceridemia, accelerated lipolysis of triglyceride-enriched HDL, altered CETP activity and elevated hepatic production of SAA, which lead to low HDL-C levels in atherogenic dyslipidemia, insulin resistance, chronic inflammation and infection [4, 114].

9.5 CARDIOVASCULAR DISEASE

HDL metabolism in patients with established **atherosclerosis and cardiovascular disease** features alterations typical for subjects with major cardiovascular risk factors (dyslipidemia, insulin resistance, inflammation; see 9.1, 9.2 and 9.3 of chapter 9). First, **HDL-C levels are** often, but not always, **subnormal** and **SAA levels are elevated** in coronary heart disease patients [147, 148]. Thus, following **myocardial infarction** with ST-elevation, SAA in the total lipoprotein fraction reaches its peak at 48h, whereas HDL-C progressively decreases with a minimum at discharge (−26%) [149]. Following infarction, HDL becomes progressively enriched in SAA in patients with an elevated inflammatory response.

> **MAJOR CONTROVERSY**
>
> Can alterations in HDL metabolism occur in the absence of changes in circulating HDL-C levels? Can HDL metabolism and cholesterol flux via the RCT pathway be enhanced when HDL-C levels are diminished?

Other metabolic alterations, including **decreased LCAT and LPL activities** together with **elevated PLTP activity**, may also occur in patients with cardiovascular disease [82, 150] (Fig. 9.2). In contrast, CETP activity is elevated in asymptomatic subjects with an elevated cardiovascular risk [129] but is diminished during episodes of unstable angina pectoris or myocardial infarction as a result of reduced CETP expression in leukocytes [151]. In addition, ABCA1 protein concentrations in arterial tissues are lower (−45%) in patients with atherosclerotic plaques vs. controls, as determined by enzyme-linked immunosorbent assay (ELISA), suggesting impaired HDL biogenesis [152].

As observed in subjects with major cardiovascular risk factors, **small, dense HDL** also prevails in patients with established cardiovascular disease. Thus, in male participants in the Framingham Offspring Study, subjects with coronary heart disease displayed higher levels of small pre-beta-1 and alpha-3 particles and lower levels of large alpha-1, pre-alpha-1 and pre-alpha-3 particles than subjects without [153]. Similarly, subjects with cardiovascular events possessed higher levels of small pre-beta-1 and alpha-3 HDL and lower levels of large alpha-1, alpha-2, pre-alpha-1 and pre-alpha-2 HDL than subjects without such events in the VA-HIT study [154]. Furthermore, plasma pre-beta-1 and pre-beta-3 HDL concentrations were higher and alpha-1, alpha-2, pre-alpha-1 and pre-alpha-2 HDL, as well as HDL-C and apoA-I concentrations were lower in post-menopausal women with established coronary heart disease than in coronary heart disease-free post-menopausal women [155]. After adjustment for conventional cardiovascular risk factors, plasma levels of pre-beta-1 HDL were positively associated and alpha-2 HDL particle concentrations were inversely associated with the degree of coronary artery disease. Remarkably, plasma triglyceride, LDL-C and HDL-C levels were not associated with the degree of coronary atherosclerosis in this study [155]. Interestingly, plasma levels of large alpha-1 HDL were inversely associated with the presence of vulnerable atherosclerotic plaques in patients undergoing coronary angiography [156].

Similarly, levels of pre-beta-1 HDL are elevated (+31%) in coronary artery disease patients compared with controls [157]. In addition, pre-beta-1 HDL concentrations are increased (+39%) in unstable angina compared with stable coronary artery disease and represent a strong predictor of unstable angina. Notably, HDL-C levels and LCAT activity did not differ between the groups in this study, suggesting enhanced formation of pre-beta HDL as the underlying mechanism [157]. In contrast, high pre-beta-1 HDL concentrations are associated with ischemic heart disease in parallel with low LCAT levels in another study, indicative of the accelerated catabolism of pre-beta particles [150].

In addition, coronary artery disease and acute ischemic stroke are both associated with the predominance of smaller HDL particles, as determined by gradient gel electrophoresis [158, 159]. The increase in small HDL and the decrease in HDL of intermediate size, as measured by NMR, are also associated with coronary artery disease severity in men admitted for diagnostic coronary arteriography [160]. Small HDL also prevails in peripheral arterial disease [161]. Finally, coronary artery disease patients equally display elevated levels of lipid-poor apoA-I [162].

Consistent with these data, reduced mean HDL particle size is associated with an adverse cardiovascular risk profile in apparently healthy men and women [163]. Reduced mean HDL size is also paralleled by increased coronary heart disease risk, but this relationship is largely explained by traditional risk factors. In addition, mean HDL particle size determined by both gradient gel electrophoresis and NMR spectroscopy reveals an inverse relationship with coronary heart disease risk [164]. It should, however, be kept in mind that mean HDL size, as measured by these approaches, represents an operational term that does not directly reflect alterations in HDL composition and function.

CLINICAL SUMMARY

Dyslipidemia, insulin resistance, obesity, inflammation, infection and the presence of cardiovascular disease share multiple pathways leading to the altered metabolism of HDL that may result in low HDL-C levels and contribute to cardiovascular risk.

9.6 POST-PRANDIAL STATE

Atherosclerosis has been long considered as a post-prandial phenomenon, particularly as a result of the prolonged exposure of the vasculature to elevated levels of chylomicrons, VLDL and their remnants over the post-prandial period. Indeed, substantial evidence attests to the atherogenicity of triglyceride-rich lipoproteins and to their presence in human atherosclerotic plaques [165]. **Post-prandial hyperlipemia** is further distinguished by the **elevated CETP activity** and accelerated net transfer of cholesteryl ester from HDL to triglyceride-rich lipoproteins, thereby favoring the formation and accumulation of cholesteryl ester-enriched atherogenic remnant particles and cholesteryl ester-depleted, triglyceride-enriched HDL [166] (Fig. 9.2). HDL metabolism in the postprandial state is, therefore, similar to that in hypertriglyceridemia. Furthermore, hypertriglyceridemia and **low HDL-C** in the post-prandial period are accompanied by leukocyte activation in both healthy and Type 2 diabetic subjects, indicative of a **pro-inflammatory state** [167–169].

Compared with fasting levels, HDL-C is reduced up to 3 to 5 hours after the last meal in the general population [170]. Plasma HDL-C concentrations are also decreased after fat-rich meals in other populations, such as in post-menopausal women with or without Type 2 diabetes [171]. Higher fasting, but not post-prandial, CETP concentrations are associated with a larger post-prandial decrease in HDL-C, independent of the post-prandial increase in triglycerides [172, 173]. However, changes in HDL-C levels are minimal in response to normal food intake. Indeed, the maximum decrease after normal food and fluid intake from fasting levels was 0.1 mmol/l for HDL-C in 33,391 individuals from the Copenhagen General Population Study and in 9319 individuals aged between 20 and 93 years of age from the Copenhagen City Heart Study [170]. Importantly, such non-fasting lipid profiles are capable of predicting an increased risk of cardiovascular events.

The post-prandial fall in HDL-C levels is correlated with **visceral obesity**, as observed in subjects with Type 2 diabetes, consistent with the central role of body fat [169]. Thus, HDL-C levels strongly decrease as the cholesteryl ester transfer rate increases after fat loading in obese patients with Metabolic Syndrome [174]. In parallel, CETP mass increases and adiponectin decreases, pointing towards a potential role for intra-abdominal fat. Serum PLTP activity is also increased after fat loading and then decreases to nearly basal levels; serum PLTP may be involved in the clearance of post-prandial lipoproteins [82].

Post-prandial HDLs are also modified by **hepatic lipase** as they are is enriched in triglycerides and therefore represent an optimal substrate for the enzyme (Fig. 9.2). Furthermore, such HDL is better at displacing hepatic lipase than is fasting HDL as a result of its diminished content of apoE [175].

HDL particle profiles do not reveal pronounced alterations in the post-prandial phase. For example, pre-beta-1 HDL concentrations do not change post-prandially in healthy controls [103]. Equally, plasma apoA-I and LCAT concentrations do not show large post-prandial variation.

9.7 SMOKING

Smoking is also associated with **low circulating levels of HDL-C**. Interestingly, levels of both large HDL2 and small HDL3, as well as the total HDL particle concentrations, can be negatively affected by smoking [176–180]. The smoking of even one cigarette can acutely reduce HDL-C by up to -10% [176]. Passive smoking also reduces HDL-C in a dose-related manner [181, 182]. Fine particulate matter present in cigarette smoke may make a major contribution to the HDL-reducing effect, as suggested by data obtained with welding fumes; the activation of pro-inflammatory pathways appears to be involved [183]. As a corollary, smoking cessation increases plasma levels of HDL-C [184–188].

Low HDL-C in smokers is most probably related to deficient HDL maturation by LCAT as a result of **reduced LCAT activity** [189–193]. In contrast, the contribution of CETP appears less likely, as data on the influence of smoking on CETP activity are controversial [192, 194].

IMPORTANT READING

Rashid S, Watanabe T, Sakaue T, Lewis GF. Mechanisms of HDL lowering in insulin resistant, hypertriglyceridemic states: the combined effect of HDL triglyceride enrichment and elevated hepatic lipase activity. *Clin Biochem* 2003; 36: 421–429.

Khovidhunkit W, Kim MS, Memon RA, Shigenaga JK, Moser AH, Feingold KR, Grunfeld C. Effects of infection and inflammation on lipid and lipoprotein metabolism: mechanisms and consequences to the host. *J Lipid Res* 2004; 45: 1169–1196.

REFERENCES

1. Chapman MJ, Le Goff W, Guerin M, Kontush A. Cholesteryl ester transfer protein: at the heart of the action of lipid-modulating therapy with statins, fibrates, niacin, and cholesteryl ester transfer protein inhibitors. *Eur Heart J* 2010; 31: 149–164.
2. Le Goff W, Guerin M, Chapman MJ. Pharmacological modulation of cholesteryl ester transfer protein, a new therapeutic target in atherogenic dyslipidemia. *Pharmacol Ther* 2004; 101: 17–38.
3. Schaefer EJ, Asztalos BF. Increasing high-density lipoprotein cholesterol, inhibition of cholesteryl ester transfer protein, and heart disease risk reduction. *Am J Cardiol* 2007; 100: S25–S31.
4. Lamarche B, Rashid S, Lewis GF. HDL metabolism in hypertriglyceridemic states: an overview. *Clin Chim Acta* 1999; 286: 145–161.
5. Ginsberg HN. New perspectives on atherogenesis: role of abnormal triglyceride-rich lipoprotein metabolism. *Circulation* 2002; 106: 2137–2142.
6. Chan DC, Barrett PH, Ooi EM, Ji J, Chan DT, Watts GF. Very low density lipoprotein metabolism and plasma adiponectin as predictors of high-density lipoprotein apolipoprotein A-I kinetics in obese and nonobese men. *J Clin Endocrinol Metab* 2009; 94: 989–997.

7. Sparks DL, Davidson WS, Lund-Katz S, Phillips MC. Effects of the neutral lipid content of high density lipoprotein on apolipoprotein A-I structure and particle stability. *J Biol Chem* 1995; 270: 26910–26917.
8. Rashid S, Barrett PH, Uffelman KD, Watanabe T, Adeli K, Lewis GF. Lipolytically modified triglyceride-enriched HDLs are rapidly cleared from the circulation. *Arterioscler Thromb Vasc Biol* 2002; 22: 483–487.
9. Rashid S, Watanabe T, Sakaue T, Lewis GF. Mechanisms of HDL lowering in insulin resistant, hypertriglyceridemic states: the combined effect of HDL triglyceride enrichment and elevated hepatic lipase activity. *Clin Biochem* 2003; 36: 421–429.
10. Borggreve SE, De Vries R, Dullaart RP. Alterations in high-density lipoprotein metabolism and reverse cholesterol transport in insulin resistance and type 2 diabetes mellitus: role of lipolytic enzymes, lecithin:cholesterol acyltransferase and lipid transfer proteins. *Eur J Clin Invest* 2003; 33: 1051–1069.
11. Schaefer JR, Rader DJ, Ikewaki K, Fairwell T, Zech LA, Kindt MR, Davignon J, Gregg RE, Brewer HB Jr. *In vivo* metabolism of apolipoprotein A-I in a patient with homozygous familial hypercholesterolemia. *Arterioscler Thromb* 1992; 12: 843–848.
12. Frenais R, Ouguerram K, Maugeais C, Marchini JS, Benlian P, Bard JM, Magot T, Krempf M. Apolipoprotein A-I kinetics in heterozygous familial hypercholesterolemia: a stable isotope study. *J Lipid Res* 1999; 40: 1506–1511.
13. Guerin M, Dolphin PJ, Talussot C, Gardette J, Berthezene F, Chapman MJ. Pravastatin modulates cholesteryl ester transfer from HDL to apoB-containing lipoproteins and lipoprotein subspecies profile in familial hypercholesterolemia. *Arterioscler Thromb Vasc Biol* 1995; 15: 1359–1368.
14. Guerin M, Lassel TS, Le Goff W, Farnier M, Chapman MJ. Action of atorvastatin in combined hyperlipidemia: preferential reduction of cholesteryl ester transfer from HDL to VLDL1 particles. *Arterioscler Thromb Vasc Biol* 2000; 20: 189–197.
15. Izem L, Morton RE. Molecular cloning of hamster lipid transfer inhibitor protein (apolipoprotein F) and regulation of its expression by hyperlipidemia. *J Lipid Res* 2009; 50: 676–684.
16. Morton RE, Gnizak HM, Greene DJ, Cho KH, Paromov VM. Lipid transfer inhibitor protein (apolipoprotein F) concentration in normolipidemic and hyperlipidemic subjects. *J Lipid Res* 2008; 49: 127–135.
17. Tian L, Fu M. The relationship between high density lipoprotein subclass profile and plasma lipids concentrations. *Lipids Health Dis* 2010; 9: 118.
18. Kontush A, Chapman MJ. Functionally defective HDL: A new therapeutic target at the crossroads of dyslipidemia, inflammation and atherosclerosis. *Pharmacol Rev* 2006; 3: 342–374.
19. Ishida BY, Frolich J, Fielding CJ. Prebeta-migrating high density lipoprotein: quantitation in normal and hyperlipidemic plasma by solid phase radioimmunoassay following electrophoretic transfer. *J Lipid Res* 1987; 28: 778–786.
20. Xu Y, Fu M. Alterations of HDL subclasses in hyperlipidemia. *Clin Chim Acta* 2003; 332: 95–102.
21. Yang Y, Yan B, Fu M, Xu Y, Tian Y. Relationship between plasma lipid concentrations and HDL subclasses. *Clin Chim Acta* 2005; 354: 49–58.

22. Medina-Urrutia A, Juarez-Rojas JG, Martinez-Alvarado R, Jorge-Galarza E, Posadas-Sanchez R, Cardoso-Saldana G, Caracas-Portilla N, Mendoza-Perez E, Posadas-Romero C. High-density lipoprotein subclasses distribution and composition in Mexican adolescents with low HDL cholesterol and/or high triglyceride concentrations, and its association with insulin and C-reactive protein. *Atherosclerosis* 2008; 201: 392–397.

23. Miida T, Inano K, Yamaguchi T, Tsuda T, Okada M. LpA-I levels do not reflect pre beta1-HDL levels in human plasma. *Atherosclerosis* 1997; 133: 221–226.

24. Tsimihodimos V, Gazi I, Filippatos T, Kostapanos M, Lagos K, Kostara C, Tellis CC, Elisaf M, Tselepis AD. Plasma triglyceride levels and body mass index values are the most important determinants of prebeta-1 HDL concentrations in patients with various types of primary dyslipidemia. *Atherosclerosis* 2010; 208: 506–511.

25. Tian L, Wu X, Fu M, Qin Y, Xu Y, Jia L. Relationship between plasma apolipoproteinB concentrations, apolipoproteinB/apolipoproeinA-I and HDL subclasses distribution. *Clin Chim Acta* 2008; 388: 148–155.

26. Tian L, Liu Y, Qin Y, Long S, Xu Y, Fu M. Association of the low-density lipoprotein cholesterol/high-density lipoprotein cholesterol ratio and concentrations of plasma lipids with high-density lipoprotein subclass distribution in the Chinese population. *Lipids Health Dis* 2010; 9: 69.

27. Atmeh RG, Robenek H. Measurement of small high density lipoprotein subclass by an improved immunoblotting technique. *J Lipid Res* 1996; 37: 2461–2469.

28. Brinton EA, Eisenberg S, Breslow JL. Increased apo A-I and apo A-II fractional catabolic rate in patients with low high density lipoprotein-cholesterol levels with or without hypertriglyceridemia. *J Clin Invest* 1991; 87: 536–544.

29. Horowitz BS, Goldberg IJ, Merab J, Vanni TM, Ramakrishnan R, Ginsberg HN. Increased plasma and renal clearance of an exchangeable pool of apolipoprotein A-I in subjects with low levels of high density lipoprotein cholesterol. *J Clin Invest* 1993; 91: 1743–1752.

30. Keidar S, Ostlund RE Jr, Schonfeld G. Apolipoprotein E-rich HDL in patients with homozygous familial hypercholesterolemia. *Atherosclerosis* 1990; 84: 155–163.

31. Soderlund S, Watanabe H, Ehnholm C, Jauhiainen M, Taskinen MR. Increased apolipoprotein E level and reduced high-density lipoprotein mean particle size associate with low high-density lipoprotein cholesterol and features of metabolic syndrome. *Metabolism* 2010; 59: 1502–1509.

32. Ng DS, Vezina C, Wolever TS, Kuksis A, Hegele RA, Connelly PW. Apolipoprotein A-I deficiency. Biochemical and metabolic characteristics. *Arterioscler Thromb Vasc Biol* 1995; 15: 2157–2164.

33. Kuivenhoven JA, Pritchard H, Hill J, Frohlich J, Assmann G, Kastelein J. The molecular pathology of lecithin:cholesterol acyltransferase (LCAT) deficiency syndromes. *J Lipid Res* 1997; 38: 191–205.

34. Hovingh GK, Hutten BA, Holleboom AG, Petersen W, Rol P, Stalenhoef A, Zwinderman AH, de Groot E, Kastelein JJ, Kuivenhoven JA. Compromised LCAT function is associated with increased atherosclerosis. *Circulation* 2005; 112: 879–884.

35. Tato F, Vega GL, Grundy SM. Bimodal distribution of cholesteryl ester transfer protein activities in normotriglyceridemic men with low HDL cholesterol concentrations. *Arterioscler Thromb Vasc Biol* 1995; 15: 446–451.

36. Blades B, Vega GL, Grundy SM. Activities of lipoprotein lipase and hepatic triglyceride lipase in postheparin plasma of patients with low concentrations of HDL cholesterol. *Arterioscler Thromb* 1993; 13: 1227–1235.

37. Oram JF. Tangier disease and ABCA1. *Biochim Biophys Acta* 2000; 1529: 321–330.

38. Zannis VI, Koukos G, Drosatos K, Vezeridis A, Zanni EE, Kypreos KE, Chroni A. Discrete roles of apoA-I and apoE in the biogenesis of HDL species: lessons learned from gene transfer studies in different mouse models. *Ann Med* 2008; 40(Suppl. 1): 14–28.

39. Schaefer EJ, Santos RD, Asztalos BF. Marked HDL deficiency and premature coronary heart disease. *Curr Opin Lipidol* 2010; 21: 289–297.

40. Santos RD, Schaefer EJ, Asztalos BF, Polisecki E, Wang J, Hegele RA, Martinez LRC, Miname MH, Rochitte CE, Da Luz PL et al. Characterization of high density lipoprotein particles in familial apolipoprotein A-I deficiency. *J Lipid Res* 2008; 49: 349–357.

41. Soderlund S, Soro-Paavonen A, Ehnholm C, Jauhiainen M, Taskinen MR. Hypertriglyceridemia is associated with prebeta-HDL concentrations in subjects with familial low HDL. *J Lipid Res* 2005; 46: 1643–1651.

42. Sirtori CR, Calabresi L, Franceschini G, Baldassarre D, Amato M, Johansson J, Salvetti M, Monteduro C, Zulli R, Muiesan ML et al. Cardiovascular status of carriers of the apolipoprotein A-I Milano mutant: the Limone sul Garda study. *Circulation* 2001; 103: 1949–1954.

43. Chiesa G, Sirtori CR. Apolipoprotein A-I Milano: current perspectives. *Curr Opin Lipidol* 2003; 14: 159–163.

44. Roma P, Gregg RE, Meng MS, Ronan R, Zech LA, Franceschini G, Sirtori CR, Brewer HB Jr. *In vivo* metabolism of a mutant form of apolipoprotein A-I, apo A-I Milano, associated with familial hypoalphalipoproteinemia. *J Clin Invest* 1993; 91: 1445–1452.

45. Perez-Mendez O, Bruckert E, Franceschini G, Duhal N, Lacroix B, Bonte JP, Sirtori C, Fruchart JC, Turpin G, Luc G. Metabolism of apolipoproteins AI and AII in subjects carrying similar apoAI mutations, apoAI Milano and apoAI Paris. *Atherosclerosis* 2000; 148: 317–325.

46. Cheung MC, Nichols AV, Blanche PJ, Gong EL, Franceschini G, Sirtori CR. Characterization of A-I-containing lipoproteins in subjects with A-I Milano variant. *Biochim Biophys Acta* 1988; 960: 73–82.

47. Calabresi L, Tedeschi G, Treu C, Ronchi S, Galbiati D, Airoldi S, Sirtori CR, Marcel Y, Franceschini G. Limited proteolysis of a disulfide-linked apoA-I dimer in reconstituted HDL. *J Lipid Res* 2001; 42: 935–942.

48. Bhat S, Sorci-Thomas MG, Calabresi L, Samuel MP, Thomas MJ. Conformation of dimeric apolipoprotein A-I Milano on recombinant lipoprotein particles. *Biochemistry* 2010; 49: 5213–5224.

49. Alexander ET, Tanaka M, Kono M, Saito H, Rader DJ, Phillips MC. Structural and functional consequences of the Milano mutation (R173C) in human apolipoprotein A-I. *J Lipid Res* 2009; 50: 1409–1419.

50. Tanaka T, Tanaka M, Sugiura M, Kawakami T, Aimoto S, Saito H. Deletion of single amino acid E235 affects the structure and lipid interaction of human apolipoprotein A-I C-terminal peptides. *Chem Pharm Bull* 2009; 57: 499–503.

51. Rader DJ, Ikewaki K, Duverger N, Schmidt H, Pritchard H, Frohlich J, Clerc M, Dumon MF, Fairwell T, Zech L et al. Markedly accelerated catabolism of apolipoprotein A-II (ApoA-II) and high density lipoproteins containing ApoA-II in classic lecithin: cholesterol acyltransferase deficiency and fish-eye disease. *J Clin Invest* 1994; 93: 321–330.

52. Calabresi L, Pisciotta L, Costantin A, Frigerio I, Eberini I, Alessandrini P, Arca M, Bittolo Bon G, Boscutti G, Busnach G et al. The molecular basis of lecithin:cholesterol acyltransferase deficiency syndromes: A comprehensive study of molecular and biochemical findings in 13 unrelated Italian families. *Arterioscler Thromb Vasc Biol* 2005; 25: 1972–1978.

53. Nishiwaki M, Ikewaki K, Bader G, Nazih H, Hannuksela M, Remaley AT, Shamburek RD, Brewer HB Jr. Human lecithin:cholesterol acyltransferase deficiency: *in vivo* kinetics of low-density lipoprotein and lipoprotein-X. *Arterioscler Thromb Vasc Biol* 2006; 26: 1370–1375.

54. Schaefer EJ, Blum CB, Levy RI, Jenkins LL, Alaupovic P, Foster DM, Brewer HB Jr. Metabolism of high-density lipoprotein apolipoproteins in Tangier disease. *N Engl J Med* 1978; 299: 905–910.

55. Tsai MY, Li N, Sharrett AR, Shea S, Jacobs DR Jr, Tracy R, Arnett D, Arends V, Post W. Associations of genetic variants in ATP-binding cassette A1 and cholesteryl ester transfer protein and differences in lipoprotein subclasses in the multi-ethnic study of atherosclerosis. *Clin Chem* 2009; 55: 481–488.

56. Chung S, Gebre AK, Seo J, Shelness GS, Parks JS. A novel role for ABCA1-generated large pre-beta migrating nascent HDL in the regulation of hepatic VLDL triglyceride secretion. *J Lipid Res* 2010; 51: 729–742.

57. Katzel LI, Coon PJ, Busby MJ, Gottlieb SO, Krauss RM, Goldberg AP. Reduced HDL2 cholesterol subspecies and elevated postheparin hepatic lipase activity in older men with abdominal obesity and asymptomatic myocardial ischemia. *Arterioscler Thromb* 1992; 12: 814–823.

58. Bersot TP, Vega GL, Grundy SM, Palaoglu KE, Atagunduz P, Ozbayrakci S, Gokdemir O, Mahley RW. Elevated hepatic lipase activity and low levels of high density lipoprotein in a normotriglyceridemic, nonobese Turkish population. *J Lipid Res* 1999; 40: 432–438.

59. Xiao C, Watanabe T, Zhang Y, Trigatti B, Szeto L, Connelly PW, Marcovina S, Vaisar T, Heinecke JW, Lewis GF. Enhanced cellular uptake of remnant high-density lipoprotein particles: a mechanism for high-density lipoprotein lowering in insulin resistance and hypertriglyceridemia. *Circ Res* 2008; 103: 159–166.

60. Vaziri ND, Navab M, Fogelman AM. HDL metabolism and activity in chronic kidney disease. *Nat Rev Nephrol* 2010; 6: 287–296.

61. Vaziri ND. Lipotoxicity and impaired high density lipoprotein-mediated reverse cholesterol transport in chronic kidney disease. *J Ren Nutr* 2010; 20: S35–S43.

62. Pahl MV, Ni Z, Sepassi L, Moradi H, Vaziri ND. Plasma phospholipid transfer protein, cholesteryl ester transfer protein and lecithin:cholesterol acyltransferase in end-stage renal disease (ESRD). *Nephrol Dial Transplant* 2009; 24: 2541–2546.

63. Marsche G, Frank S, Hrzenjak A, Holzer M, Dirnberger S, Wadsack C, Scharnagl H, Stojakovic T, Heinemann A, Oettl K. Plasma-advanced oxidation protein products are potent high-density lipoprotein receptor antagonists *in vivo*. *Circ Res* 2009; 104: 750–757.

64. Obeid R, Herrmann W. Homocysteine and lipids: S-adenosyl methionine as a key intermediate. *FEBS Lett* 2009; 583: 1215–1225.

65. Gueant-Rodriguez RM, Spada R, Moreno-Garcia M, Anello G, Bosco P, Lagrost L, Romano A, Elia M, Gueant JL. Homocysteine is a determinant of ApoA-I and both are associated with ankle brachial index, in an Ambulatory Elderly Population. *Atherosclerosis* 2011; 214: 480–485.

66. Yamashita S, Maruyama T, Hirano K, Sakai N, Nakajima N, Matsuzawa Y. Molecular mechanisms, lipoprotein abnormalities and atherogenicity of hyperalphalipoproteinemia. *Atherosclerosis* 2000; 152: 271–285.

67. Asztalos BF, Horvath KV, Kajinami K, Nartsupha C, Cox CE, Batista M, Schaefer EJ, Inazu A, Mabuchi H. Apolipoprotein composition of HDL in cholesteryl ester transfer protein deficiency. *J Lipid Res* 2004; 45: 448–455.

68. Ikewaki K, Rader DJ, Sakamoto T, Nishiwaki M, Wakimoto N, Schaefer JR, Ishikawa T, Fairwell T, Zech LA, Nakamura H et al. Delayed catabolism of high density lipoprotein apolipoproteins A-I and A-II in human cholesteryl ester transfer protein deficiency. *J Clin Invest* 1993; 92: 1650–1658.

69. Ikewaki K, Nishiwaki M, Sakamoto T, Ishikawa T, Fairwell T, Zech LA, Nagano M, Nakamura H, Brewer HB Jr, Rader DJ. Increased catabolic rate of low density lipoproteins in humans with cholesteryl ester transfer protein deficiency. *J Clin Invest* 1995; 96: 1573–1581.

70. Cohen JC, Vega GL, Grundy SM. Hepatic lipase: new insights from genetic and metabolic studies. *Curr Opin Lipidol* 1999; 10: 259–267.

71. Merrin PK, Baynes C, Henderson A, Richmond B, Elkeles R. The effects of gender and type of diabetes on HDL-cholesterol. *Diabete Metab* 1994; 20: 20–24.

72. Colhoun HM, Otvos JD, Rubens MB, Taskinen MR, Underwood SR, Fuller JH. Lipoprotein subclasses and particle sizes and their relationship with coronary artery calcification in men and women with and without type 1 diabetes. *Diabetes* 2002; 51: 1949–1956.

73. Gao F, Yan T, Zhao Y, Yin F, Hu C. A possible mechanism linking hyperglycemia and reduced high-density lipoprotein cholesterol levels in diabetes. *J Huazhong Univ Sci Technolog Med Sci* 2010; 30: 318–321.

74. Gatti A, Maranghi M, Bacci S, Carallo C, Gnasso A, Mandosi E, Fallarino M, Morano S, Trischitta V, Filetti S. Poor glycemic control is an independent risk factor for low HDL cholesterol in patients with type 2 diabetes. *Diabetes Care* 2009; 32: 1550–1552.

75. Weissglas-Volkov D, Pajukanta P. Genetic causes of high and low serum HDL-cholesterol. *J Lipid Res* 2010; 51: 2032–2057.

76. Rashid S, Genest J. Effect of obesity on high-density lipoprotein metabolism. *Obesity* 2007; 15: 2875–2888.

77. Wildman RP, Janssen I, Khan UI, Thurston R, Barinas-Mitchell E, El Khoudary SR, Everson-Rose SA, Kazlauskaite R, Matthews KA, Sutton-Tyrrell K. Subcutaneous adipose tissue in relation to subclinical atherosclerosis and cardiometabolic risk factors in midlife women. *Am J Clin Nutr* 2011; 93: 719–726.

78. He J, Xu X, Francisco A, Ferrante A, Krakoff J. Markers of adipose tissue macrophage content are negatively associated with serum HDL-C concentrations. *Atherosclerosis* 2011; 215: 243–246.

79. Wolfrum C, Howell JJ, Ndungo E, Stoffel M. Foxa2 activity increases plasma high density lipoprotein levels by regulating apolipoprotein M. *J Biol Chem* 2008; 283: 16940–16949.
80. Ooi EMM, Watts GF, Chan DC, Nielsen LB, Plomgaard P, Dahlbäck B, Barrett PHR. Association of apolipoprotein M with high-density lipoprotein kinetics in overweight-obese men. *Atherosclerosis* 2010; 210: 326–330.
81. Jin H, Jiang B, Tang J, Lu W, Wang W, Zhou L, Shang W, Li F, Ma Q, Yang Y, Chen M. Serum visfatin concentrations in obese adolescents and its correlation with age and high-density lipoprotein cholesterol. *Diabetes Res Clin Pract* 2008; 79: 412–418.
82. Schlitt A, Schwaab B, Fingscheidt K, Lackner KJ, Heine GH, Vogt A, Buerke M, Maegdefessel L, Raaz U, Werdan K et al. Serum phospholipid transfer protein activity after a high fat meal in patients with insulin-treated type 2 diabetes. *Lipids* 2010; 45: 129–135.
83. Masson D, Deckert V, Gautier T, Klein A, Desrumaux C, Viglietta C, Pais de Barros J-P, Le Guern N, Grober J et al. Worsening of diet-induced atherosclerosis in a new model of transgenic rabbit expressing the human plasma phospholipid transfer protein. *Arterioscler Thromb Vasc Biol* 2011; 31: 766–774.
84. Yatsuya H, Jeffery RW, Erickson DJ, Welsh EM, Flood AP, Jaeb MA, Laqua PS, Mitchell NR, Langer SL, Levy RL. Sex-specific HDL cholesterol changes with weight loss and their association with anthropometric variables: the LIFE study. *Obesity* 2011; 19: 429–435.
85. Oda E, Kawai R. Changes in body weight are significantly associated with changes in fasting plasma glucose and HDL cholesterol in Japanese men without abdominal obesity (waist circumference <85cm). *Acta Diabetol* 2011; 48: 143–148.
86. Jamal M, Wegner R, Heitshusen D, Liao J, Samuel I. Resolution of hyperlipidemia follows surgical weight loss in patients undergoing Roux-en-Y gastric bypass surgery: a 6-year analysis of data. *Surg Obes Relat Dis* 2011; 7: 473–479.
87. Zhang F, Strain GW, Lei W, Dakin GF, Gagner M, Pomp A. Changes in lipid profiles in morbidly obese patients after laparoscopic sleeve gastrectomy (LSG). *Obes Surg* 2011; 21: 305–309.
88. Geloneze B, Pereira JA, Pareja JC, Lima MM, Lazarin MA, Souza IC, Tambascia MA, Chaim E, Muscelli E. Overcoming metabolic syndrome in severe obesity: adiponectin as a marker of insulin sensitivity and HDL-cholesterol improvements after gastric bypass. *Arq Bras Endocrinol Metabol* 2009; 53: 293–300.
89. Laimer MW, Engl J, Tschoner A, Kaser S, Ritsch A, Tatarczyk T, Rauchenzauner M, Weiss H, Aigner F, Patsch JR et al. Effects of weight loss on lipid transfer proteins in morbidly obese women. *Lipids* 2009; 44: 1125–1130.
90. Asztalos BF, Swarbrick MM, Schaefer EJ, Dallal GE, Horvath KV, Ai M, Stanhope KL, Austrheim-Smith I, Wolfe BM, Ali M et al. Effects of weight loss, induced by gastric bypass surgery, on HDL remodeling in obese women. *J Lipid Res* 2010; 51: 2405–2412.
91. Syvanne M, Ahola M, Lahdenpera S, Kahri J, Kuusi T, Virtanen KS, Taskinen MR. High density lipoprotein subfractions in non-insulin-dependent diabetes mellitus and coronary artery disease. *J Lipid Res* 1995; 36: 573–582.
92. Blatter Garin MC, Kalix B, Morabia A, James RW. Small, dense lipoprotein particles and reduced paraoxonase-1 in patients with the metabolic syndrome. *J Clin Endocrinol Metab* 2005; 90: 2264–2269.

93. Lagos KG, Filippatos TD, Tsimihodimos V, Gazi IF, Rizos C, Tselepis AD, Mikhailidis DP, Elisaf MS. Alterations in the high density lipoprotein phenotype and HDL-associated enzymes in subjects with metabolic syndrome. *Lipids* 2009; 44: 9–16.

94. Garcia-Sanchez C, Torres-Tamayo M, Juarez-Meavepena M, Lopez-Osorio C, Toledo-Ibelles P, Monter-Garrido M, Cruz-Robles D, Carreon-Torres E, Vargas-Alarcon G, Perez-Mendez O. Lipid plasma concentrations of HDL subclasses determined by enzymatic staining on polyacrylamide electrophoresis gels in children with metabolic syndrome. *Clin Chim Acta* 2011; 412: 292–298.

95. Garvey WT, Kwon S, Zheng D, Shaughnessy S, Wallace P, Hutto A, Pugh K, Jenkins AJ, Klein RL, Liao Y. Effects of insulin resistance and type 2 diabetes on lipoprotein subclass particle size and concentration determined by nuclear magnetic resonance. *Diabetes* 2003; 52: 453–462.

96. Festa A, Williams K, Hanley AJG, Otvos JD, Goff DC, Wagenknecht LE, Haffner SM. Nuclear magnetic resonance lipoprotein abnormalities in prediabetic subjects in the Insulin Resistance Atherosclerosis Study. *Circulation* 2005; 111: 3465–3472.

97. Goff DC Jr, D'Agostino RB Jr, Haffner SM, Otvos JD. Insulin resistance and adiposity influence lipoprotein size and subclass concentrations. Results from the Insulin Resistance Atherosclerosis Study. *Metabolism* 2005; 54: 264–270.

98. Okazaki M, Usui S, Ishigami M, Sakai N, Nakamura T, Matsuzawa Y, Yamashita S. Identification of unique lipoprotein subclasses for visceral obesity by component analysis of cholesterol profile in high-performance liquid chromatography. *Arterioscler Thromb Vasc Biol* 2005; 25: 578–584.

99. Sasahara T, Yamashita T, Sviridov D, Fidge N, Nestel P. Altered properties of high density lipoprotein subfractions in obese subjects. *J Lipid Res* 1997; 38: 600–611.

100. Hansel B, Giral P, Nobecourt E, Chantepie S, Bruckert E, Chapman MJ, Kontush A. Metabolic syndrome is associated with elevated oxidative stress and dysfunctional dense high-density lipoprotein particles displaying impaired antioxidative activity. *J Clin Endocrinol Metab* 2004; 89: 4963–4971.

101. Nobecourt E, Jacqueminet S, Hansel B, Chantepie S, Grimaldi A, Chapman MJ, Kontush A. Defective antioxidative activity of small, dense HDL particles in type 2 diabetes: Relationship to elevated oxidative stress and hyperglycemia. *Diabetologia* 2005; 48: 529–538.

102. Kontush A, de Faria EC, Chantepie S, Chapman MJ. A normotriglyceridemic, low HDL-cholesterol phenotype is characterised by elevated oxidative stress and HDL particles with attenuated antioxidative activity. *Atherosclerosis* 2005; 182: 277–285.

103. Hirayama S, Ito T, Miyazaki O, Kamimura T, Hanyu O, Seino U, Ito S, Aizawa Y, Miida T. Prebeta1-HDL is elevated in the fasting state, but markedly reduced postprandially in poorly controlled type 2 diabetic patients. *Clin Chim Acta* 2009; 401: 57–62.

104. Plomgaard P, Dullaart RP, de Vries R, Groen AK, Dahlback B, Nielsen LB. Apolipoprotein M predicts pre-beta-HDL formation: studies in type 2 diabetic and nondiabetic subjects. *J Intern Med* 2009; 266: 258–267.

105. Pietilainen KH, Soderlund S, Rissanen A, Nakanishi S, Jauhiainen M, Taskinen MR, Kaprio J. HDL subspecies in young adult twins: heritability and impact of overweight. *Obesity* 2009; 17: 1208–1214.

106. Sam S, Haffner S, Davidson MH, D'Agostino RB Sr, Feinstein S, Kondos G, Perez A, Mazzone T. Relationship of abdominal visceral and subcutaneous adipose tissue with lipoprotein particle number and size in type 2 diabetes. *Diabetes* 2008; 57: 2022–2027.

107. Pan J, Shilian P, Ishida B, Wu X, Kane JP, Malloy MJ, Charles MA. Effect of niacin on prebeta-1 high-density lipoprotein levels in diabetes. *Metabolism* 2011; 60: 292–297.

108. Ma YH, Zhao L, Xian XD, Yang D, Huang W, Wang YH, Mueller O, Chang E, Konigshofer Y, Van-Cleve M et al. A case-control study on the relationship between HDL2b and non-alcoholic fatty liver disease in Chinese type 2 diabetic patients. *Clin Chem Lab Med* 2009; 47: 1067–1072.

109. Xian X, Ma Y, Yang DD, Huang W, Wang Y, Mueller O, Chang E, Konigshofer Y, Van-Cleve M, Yang J et al. Reduced high-density lipoprotein 2b in non-obese type 2 diabetic patients analysed by a microfluidic chip method in a case-control study. *Biomarkers* 2009; 14: 619–623.

110. Drew BG, Duffy SJ, Forbes JM, Chin-Dusting J, Kaye DM, Kemp BE, Sviridov D, Steinberg GR, Kingwell BA. High density lipoprotein modulates glucose metabolism by multiple mechanisms. *Circulation* 2009; 120 (Suppl.): S468–c-S469.

111. Fryirs MA, Barter PJ, Appavoo M, Tuch BE, Tabet F, Heather AK, Rye K-A. Effects of high-density lipoproteins on pancreatic beta-cell insulin secretion. *Arterioscler Thromb Vasc Biol* 2010; 30: 1642–1648.

112. Van Linthout S, Foryst-Ludwig A, Spillmann F, Peng J, Feng Y, Meloni M, Van Craeyveld E, Kintscher U, Schultheiss H-P, De Geest B et al. Impact of HDL on adipose tissue metabolism and adiponectin expression. *Atherosclerosis* 2010; 210: 438–444.

113. Rohrer L, Hersberger M, von Eckardstein A. High density lipoproteins in the intersection of diabetes mellitus, inflammation and cardiovascular disease. *Curr Opin Lipidol* 2004; 15: 269–278.

114. Khovidhunkit W, Kim MS, Memon RA, Shigenaga JK, Moser AH, Feingold KR, Grunfeld C. Effects of infection and inflammation on lipid and lipoprotein metabolism: mechanisms and consequences to the host. *J Lipid Res* 2004; 45: 1169–1196.

115. Esteve E, Ricart W, Fernandez-Real JM. Dyslipidemia and inflammation: an evolutionary conserved mechanism. *Clin Nutr* 2005; 24: 16–31.

116. Russman AN, Schultz LR, Zaman IF, Rehman MF, Silver B, Mitsias P, Nerenz DR. A significant temporal and quantitative relationship exists between high-density lipoprotein levels and acute ischemic stroke presentation. *J Neurol Sci* 2009; 27: 53–56.

117. Juarez-Rojas J, Medina-Urrutia A, Posadas-Sanchez R, Jorge-Galarza E, Mendoza-Perez E, Caracas-Portilla N, Cardoso-Saldana G, Munoz-Gallegos G, Posadas-Romero C. High-density lipoproteins are abnormal in young women with uncomplicated systemic lupus erythematosus. *Lupus* 2008; 17: 981–987.

118. de Carvalho JF, Bonfa E, Bezerra MC, Pereira RM. High frequency of lipoprotein risk levels for cardiovascular disease in Takayasu arteritis. *Clin Rheumatol* 2009; 28: 801–805.

119. Espinola-Zavaleta N, Soto-Lopez ME, Carreon-Torres E, Gamboa R, Mejia AM, Marquez-Velasco R, Bojalil R, Huesca-Gomez C, Reyes PA, Perez-Mendez O.

Altered flow-mediated vasodilatation, low paraoxonase-1 activity, and abnormal high-density lipoprotein subclass distribution in Takayasu's arteritis. *Circ J* 2009; 73: 760–766.

120. Barlage S, Gnewuch C, Liebisch G, Wolf Z, Audebert FX, Gluck T, Frohlich D, Kramer BK, Rothe G, Schmitz G. Changes in HDL-associated apolipoproteins relate to mortality in human sepsis and correlate to monocyte and platelet activation. *Intensive Care Med* 2009; 35: 1877–1885.

121. Kahveci G, Bayrak F, Mutlu B, Gurel YE, Karaahmet T, Tigen K, Basaran Y. Clinical significance of high-density lipoprotein cholesterol in left-sided infective endocarditis. *Am J Cardiol* 2008; 101: 1170–1173.

122. Kimak E, Halabis M, Baranowicz-Gaszczyk I. Relationships between serum lipid, lipoprotein, triglyceride-rich lipoprotein, and high-density lipoprotein particle concentrations in post-renal transplant patients. *J Zhejiang Univ Sci B* 2010; 11: 249–257.

123. Yin K, Liao DF, Tang CK. ATP-binding membrane cassette transporter A1 (ABCA1): a possible link between inflammation and reverse cholesterol transport. *Mol Med* 2010; 16: 438–449.

124. Burke JE, Dennis EA. Phospholipase A2 structure/function, mechanism, and signaling. *J Lipid Res* 2009; 50 (Suppl.): S237–S242.

125. Leite JO, Vaishnav U, Puglisi M, Fraser H, Trias J, Fernandez ML. A-002 (Varespladib), a phospholipase A2 inhibitor, reduces atherosclerosis in guinea pigs. *BMC Cardiovasc Disord* 2009 9: 7.

126. Keller PM, Rust T, Murphy DJ, Matico R, Trill JJ, Krawiec JA, Jurewicz A, Jaye M, Harpel M, Thrall S et al. A high-throughput screen for endothelial lipase using HDL as substrate. *J Biomol Screen* 2008; 13: 468–475.

127. Jahangiri A, de Beer MC, Noffsinger V, Tannock LR, Ramaiah C, Webb NR, van der Westhuyzen DR, de Beer FC. HDL remodeling during the acute phase response. *Arterioscler Thromb Vasc Biol* 2009; 29: 261–267.

128. Jahangiri A. High-density lipoprotein and the acute phase response. *Curr Opin Endocrinol Diabetes Obes* 2010; 17: 156–160.

129. Zeller M, Masson D, Farnier M, Lorgis L, Deckert V, Pais de Barros JP, Desrumaux C, Sicard P, Grober J, Blache D et al. High serum cholesteryl ester transfer rates and small high-density lipoproteins are associated with young age in patients with acute myocardial infarction. *J Am Coll Cardiol* 2007; 50: 1948–1955.

130. Oliviero F, Sfriso P, Baldo G, Dayer JM, Giunco S, Scanu A, Bernardi D, Ramonda R, Plebani M, Punzi L. Apolipoprotein A-I and cholesterol in synovial fluid of patients with rheumatoid arthritis, psoriatic arthritis and osteoarthritis. *Clin Exp Rheumatol* 2009; 27: 79–83.

131. Garcia-Gomez C, Nolla JM, Valverde J, Narvaez J, Corbella E, Pinto X. High HDL-cholesterol in women with rheumatoid arthritis on low-dose glucocorticoid therapy. *Eur J Clin Invest* 2008; 38: 686–692.

132. van Eijk IC, de Vries MK, Levels JH, Peters MJ, Huizer EE, Dijkmans BA, van der Horst-Bruinsma IE, Hazenberg BP, van de Stadt RJ, Wolbink GJ et al. Improvement of lipid profile is accompanied by atheroprotective alterations in high-density lipoprotein composition upon tumor necrosis factor blockade: a prospective cohort study in ankylosing spondylitis. *Arthritis Rheum* 2009; 60: 1324–1330.

133. Chung CP, Oeser A, Raggi P, Solus JF, Avalos I, Linton MF, Fazio S, Stein CM. Lipoprotein subclasses and particle size determined by nuclear magnetic resonance spectroscopy in systemic lupus erythematosus. *Clin Rheumatol* 2008; 27: 1227–1233.

134. Vekic J, Jelic-Ivanovic Z, Spasojevic-Kalimanovska V, Memon L, Zeljkovic A, Bogavac-Stanojevic N, Spasic S. High serum uric acid and low-grade inflammation are associated with smaller LDL and HDL particles. *Atherosclerosis* 2009; 203: 236–242.

135. Chung CP, Oeser A, Raggi P, Sokka T, Pincus T, Solus JF, Linton MF, Fazio S, Stein CM. Lipoprotein subclasses determined by nuclear magnetic resonance spectroscopy and coronary atherosclerosis in patients with rheumatoid arthritis. *J Rheumatol* 2010; 37: 1633–1638.

136. Marchesi S, Lupattelli G, Lombardini R, Sensini A, Siepi D, Mannarino M, Vaudo G, Mannarino E. Acute inflammatory state during influenza infection and endothelial function. *Atherosclerosis* 2005; 178: 345–350.

137. Feingold KR, Staprans I, Memon RA, Moser AH, Shigenaga JK, Doerrler W, Dinarello CA, Grunfeld C. Endotoxin rapidly induces changes in lipid metabolism that produce hypertriglyceridemia: low doses stimulate hepatic triglyceride production while high doses inhibit clearance. *J Lipid Res* 1992; 33: 1765–1776.

138. Jia EZ, Zhao FJ, Hao B, Zhu TB, Wang LS, Chen B, Cao KJ, Huang J, Ma WZ, Yang ZJ et al. *Helicobacter pylori* infection is associated with decreased serum levels of high density lipoprotein, but not with the severity of coronary atherosclerosis. *Lipids Health Dis* 2009; 8: 59.

139. Luo L, Pu X, Wang Y, Xu N. Impaired plasma lipid profiles in acute hepatitis. *Lipids Health Dis* 2010; 9: 5.

140. Tsai MH, Peng YS, Chen YC, Lien JM, Tian YC, Fang JT, Weng HH, Chen PC, Yang CW, Wu CS. Low serum concentration of apolipoprotein A-I is an indicator of poor prognosis in cirrhotic patients with severe sepsis. *J Hepatol* 2009; 50: 906–915.

141. Levels JH, Pajkrt D, Schultz M, Hoek FJ, van Tol A, Meijers JC, van Deventer SJ. Alterations in lipoprotein homeostasis during human experimental endotoxemia and clinical sepsis. *Biochim Biophys Acta* 2007; 1771: 1429–1438.

142. Sibmooh N, Yamanont P, Krudsood S, Leowattana W, Brittenham G, Looareesuwan S, Udomsangpetch R. Increased fluidity and oxidation of malarial lipoproteins: relation with severity and induction of endothelial expression of adhesion molecules. *Lipids Health Dis* 2004; 3: 15.

143. Baker J, Ayenew W, Quick H, Hullsiek KH, Tracy R, Henry K, Duprez D, Neaton JD. High-density lipoprotein particles and markers of inflammation and thrombotic activity in patients with untreated HIV infection. *J Infect Dis* 2010; 201: 285–292.

144. Duprez DA, Kuller LH, Tracy R, Otvos J, Cooper DA, Hoy J, Neuhaus J, Paton NI, Friis-Moller N, Lampe F et al. Lipoprotein particle subclasses, cardiovascular disease and HIV infection. *Atherosclerosis* 2009; 207: 524–529.

145. Rose H, Hoy J, Woolley I, Tchoua U, Bukrinsky M, Dart A, Sviridov D. HIV infection and high density lipoprotein metabolism. *Atherosclerosis* 2008; 199: 79–86.

146. Mujawar Z, Tamehiro N, Grant A, Sviridov D, Bukrinsky M, Fitzgerald ML. Mutation of the ATP cassette binding transporter A1 (ABCA1) C-terminus disrupts HIV-1 Nef binding but does not block the Nef enhancement of ABCA1 protein degradation. *Biochemistry* 2010; 49: 8338–8349.

147. Sposito AC, Carvalho LS, Cintra RM, Araujo AL, Ono AH, Andrade JM, Coelho OR, Quinaglia e Silva JC. Rebound inflammatory response during the acute phase of myocardial infarction after simvastatin withdrawal. *Atherosclerosis* 2009; 207: 191–194.

148. Pitt B, Loscalzo J, Ycas J, Raichlen JS. Lipid levels after acute coronary syndromes. *J Am Coll Cardiol* 2008; 51: 1440–1445.

149. Gomaraschi M, Sinagra G, Serdoz LV, Pitzorno C, Fonda M, Cattin L, Calabresi L, Franceschini G. The plasma concentration of Lpa-I:A-II particles as a predictor of the inflammatory response in patients with ST-elevation myocardial infarction. *Atherosclerosis* 2009; 202: 304–311.

150. Sethi AA, Sampson M, Warnick R, Muniz N, Vaisman B, Nordestgaard BG, Tybjaerg-Hansen A, Remaley AT. High pre-beta1 HDL concentrations and low lecithin: cholesterol acyltransferase activities are strong positive risk markers for ischemic heart disease and independent of HDL-cholesterol. *Clin Chem* 2010; 56: 1128–1137.

151. Ye D, Kraaijeveld AO, Grauss RW, Willems SM, van Vark-van der Zee LC, de Jager SC, Jauhiainen M, Kuivenhoven JA, Dallinga-Thie GM, Atsma DE et al. Reduced leucocyte cholesteryl ester transfer protein expression in acute coronary syndromes. *J Intern Med* 2008; 264: 571–585.

152. Paul V, Meyer HHD, Leidl K, Soumian S, Albrecht C. A novel enzyme immunoassay specific for ABCA1 protein quantification in human tissues and cells. *J Lipid Res* 2008; 49: 2259–2267.

153. Asztalos BF, Cupples LA, Demissie S, Horvath KV, Cox CE, Batista MC, Schaefer EJ. High-density lipoprotein subpopulation profile and coronary heart disease prevalence in male participants of the Framingham Offspring Study. *Arterioscler Thromb Vasc Biol* 2004; 24: 2181–2187.

154. Asztalos BF, Collins D, Cupples LA, Demissie S, Horvath KV, Bloomfield HE, Robins SJ, Schaefer EJ. Value of high-density lipoprotein (HDL) subpopulations in predicting recurrent cardiovascular events in the Veterans Affairs HDL Intervention Trial. *Arterioscler Thromb Vasc Biol* 2005; 25: 2185–91.

155. Lamon-Fava S, Herrington DM, Reboussin DM, Sherman M, Horvath KV, Cupples LA, White C, Demissie S, Schaefer EJ, Asztalos BF. Plasma levels of HDL subpopulations and remnant lipoproteins predict the extent of angiographically-defined coronary artery disease in postmenopausal women. *Arterioscler Thromb Vasc Biol* 2008; 28: 575–579.

156. Rinehart S, Vazquez G, Joshi P, Krivitsky E, Murrieta L, Qian Z, Asztalos B, Schaefer E, Voros S. Impaired HDL metabolism and remodeling is associated with more vulnerable Coronary plaque phenotype as assessed by quantitative contrast-enhanced CT coronary angiography. *Circulation* 2010; 122 (Suppl.): A21108.

157. Tashiro J, Miyazaki O, Nakamura Y, Miyazaki A, Fukamachi I, Bujo H, Saito Y. Plasma pre beta1-HDL level is elevated in unstable angina pectoris. *Atherosclerosis* 2009; 204: 595–600.

158. Zeljkovic A, Spasojevic-Kalimanovska V, Vekic J, Jelic-Ivanovic Z, Topic A, Bogavac-Stanojevic N, Spasic S, Vujovic A, Kalimanovska-Ostric D. Does simultaneous determination of LDL and HDL particle size improve prediction of coronary artery disease risk? *Clin Exp Med* 2008; 8: 109–116.

159. Zeljkovic A, Vekic J, Spasojevic-Kalimanovska V, Jelic-Ivanovic Z, Bogavac-Stanojevic N, Gulan B, Spasic S. LDL and HDL subclasses in acute ischemic stroke: Prediction of risk and short-term mortality. *Atherosclerosis* 2010; 210: 548–554.
160. Freedman DS, Otvos JD, Jeyarajah EJ, Barboriak JJ, Anderson AJ, Walker JA. Relation of lipoprotein subclasses as measured by proton nuclear magnetic resonance spectroscopy to coronary artery disease. *Arterioscler Thromb Vasc Biol* 1998; 18: 1046–1053.
161. Mowat BF, Skinner ER, Wilson HM, Leng GC, Fowkes FG, Horrobin D. Alterations in plasma lipids, lipoproteins and high density lipoprotein subfractions in peripheral arterial disease. *Atherosclerosis* 1997; 131: 161–166.
162. Suzuki M, Wada H, Maeda S, Saito K, Minatoguchi S, Saito K, Seishima M. Increased plasma lipid-poor apolipoprotein A-I in patients with coronary artery disease. *Clin Chem* 2005; 51: 132–137.
163. Arsenault BJ, Lemieux I, Despres JP, Gagnon P, Wareham NJ, Stroes ES, Kastelein JJ, Khaw KT, Boekholdt SM. HDL particle size and the risk of coronary heart disease in apparently healthy men and women: the EPIC-Norfolk prospective population study. *Atherosclerosis* 2009; 206: 276–281.
164. Arsenault BJ, Lemieux I, Despres JP, Wareham NJ, Stroes ES, Kastelein JJ, Khaw KT, Boekholdt SM. Comparison between gradient gel electrophoresis and nuclear magnetic resonance spectroscopy in estimating coronary heart disease risk associated with LDL and HDL particle size. *Clin Chem* 2010; 56: 789–798.
165. Frank JS, Fogelman AM. Ultrastructure of the intima in WHHL and cholesterol-fed rabbit aortas prepared by ultra-rapid freezing and freeze-etching. *J Lipid Res* 1989; 30: 967–978.
166. Guerin M, Egger P, Soudant C, Le Goff W, van Tol A, Dupuis R, Chapman MJ. Cholesteryl ester flux from HDL to VLDL-1 is preferentially enhanced in type IIB hyperlipidemia in the postprandial state. *J Lipid Res* 2002; 43: 1652–1660.
167. van Oostrom AJ, Sijmonsma TP, Verseyden C, Jansen EH, de Koning EJ, Rabelink TJ, Castro Cabezas M. Postprandial recruitment of neutrophils may contribute to endothelial dysfunction. *J Lipid Res* 2003; 44: 576–583.
168. Alipour A, van Oostrom AJ, Izraeljan A, Verseyden C, Collins JM, Frayn KN, Plokker TW, Elte JW, Castro Cabezas M. Leukocyte activation by triglyceride-rich lipoproteins. *Arterioscler Thromb Vasc Biol* 2008; 28: 792–797.
169. Coutinho ER, Macedo GM, Campos FS, Bandeira FA. Changes in HDL cholesterol and in the inflammatory markers of atherogenesis after an oral fat load in type-2 diabetic patients and normal individuals. *Metab Syndr Relat Disord* 2008; 6: 153–157.
170. Langsted A, Freiberg JJ, Nordestgaard BG. Fasting and nonfasting lipid levels: Influence of normal food intake on lipids, lipoproteins, apolipoproteins, and cardiovascular risk prediction. *Circulation* 2008; 118: 2047–2056.
171. Alssema M, El-Harchaoui K, Schindhelm RK, Diamant M, Nijpels G, Kostense PJ, Teerlink T, Heine RJ, Dallinga-Thie GM, Kuivenhoven JA et al. Fasting cholesteryl ester transfer protein concentration is independently associated with the postprandial decrease in high-density lipoprotein cholesterol concentration after fat-rich meals: the Hoorn prandial study. *Metabolism* 2010; 59: 854–860.
172. Lassel TS, Guerin M, Auboiron S, Chapman MJ, Guy-Grand B. Preferential cholesteryl ester acceptors among triglyceride-rich lipoproteins during alimentary lipemia in normolipidemic subjects. *Arterioscler Thromb Vasc Biol* 1998; 18: 65–74.

173. Lassel TS, Guerin M, Auboiron S, Guy-Grand B, Chapman MJ. Evidence for a cholesteryl ester donor activity of LDL particles during alimentary lipemia in normolipidemic subjects. *Atherosclerosis* 1999; 147: 41–48.

174. Hajer GR, Dallinga-Thie GM, van Vark-van der Zee LC, Olijhoek JK, Visseren FL. Lipid-lowering therapy does not affect the postprandial drop in high density lipoprotein-cholesterol (HDL-c) plasma levels in obese men with metabolic syndrome: a randomized double blind crossover trial. *Clin Endocrinol* 2008; 69: 870–877.

175. Young EK, Chatterjee C, Sparks DL. HDL-ApoE content regulates the displacement of hepatic lipase from cell surface proteoglycans. *Am J Pathol* 2009; 175: 448–457.

176. Gnasso A, Haberbosch W, Schettler G, Schmitz G, Augustin J. Acute influence of smoking on plasma lipoproteins. *Klin Wochenschr* 1984; 62: 36–42.

177. Nesje LA, Mjos OD. Plasma HDL cholesterol and the subclasses HDL2 and HDL3 in smokers and non-smokers. *Artery* 1985; 13: 7–18.

178. Swank AM, Fell RD. Effects of acute smoking and exercise on high-density lipoprotein-cholesterol and subfractions in black female smokers. *Metabolism* 1990; 39: 343–348.

179. Batic-Mujanovic O, Zildzic M, Beganlic A, Kusljugic Z. The effect of cigarette smoking on HDL-cholesterol level. *Med Arh* 2006; 60: 90–92 [in Croatian].

180. Gossett LK, Johnson HM, Piper ME, Fiore MC, Baker TB, Stein JH. Smoking intensity and lipoprotein abnormalities in active smokers. *J Clin Lipidol* 2010; 3: 372–378.

181. Kuniyo H, Yuko Y, Hidetaka S, Miyagawa S, Nakadate T. Passive smoking is associated with lower serum HDL-C levels in school children. *Pediatr Int* 2010; 25: 252–256.

182. El-Hodhod MA, Hamdy AM, Ahmed MB, Youssef SR, Aly SM. Effect of passive smoking on blood lymphocyte apoptosis in children. *Eur J Clin Invest* 2011; 41: 387–392.

183. Rice MB, Cavallari J, Fang S, Christiani D. Acute decrease in HDL cholesterol associated with exposure to welding fumes. *J Occup Environ Med* 2011; 53: 17–21.

184. Rabkin SW. Effect of cigarette smoking cessation on risk factors for coronary atherosclerosis. A control clinical trial. *Atherosclerosis* 1984; 53: 173–184.

185. Moffatt RJ, Stamford BA, Biggerstaff KD. Influence of worksite environmental tobacco smoke on serum lipoprotein profiles of female nonsmokers. *Metabolism* 1995; 44: 1536–1539.

186. Maeda K, Noguchi Y, Fukui T. The effects of cessation from cigarette smoking on the lipid and lipoprotein profiles: a meta-analysis. *Prev Med* 2003; 37: 283–290.

187. Tamura U, Tanaka T, Okamura T, Kadowaki T, Yamato H, Tanaka H, Nakamura M, Okayama A, Ueshima H, Yamagata Z. Changes in weight, cardiovascular risk factors and estimated risk of coronary heart disease following smoking cessation in Japanese male workers: HIPOP-OHP study. *J Atheroscler Thromb* 2010; 17: 12–20.

188. Gepner AD, Piper ME, Johnson HM, Fiore MC, Baker TB, Stein JH. Effects of smoking and smoking cessation on lipids and lipoproteins: Outcomes from a randomized clinical trial. *Am Heart J* 2010; 161: 145–151.

189. McCall MR, van den Berg JJ, Kuypers FA, Tribble DL, Krauss RM, Knoff LJ, Forte TM. Modification of LCAT activity and HDL structure. New links between cigarette smoke and coronary heart disease risk. *Arterioscler Thromb* 1994; 14: 248–253.

190. Chen C, Loo G. Inhibition of lecithin: cholesterol acyltransferase activity in human blood plasma by cigarette smoke extract and reactive aldehydes. *J Biochem Toxicol* 1995; 10: 121–128.

191. Bielicki JK, Forte TM, McCall MR. Gas-phase cigarette smoke inhibits plasma lecithin-cholesterol acyltransferase activity by modification of the enzyme's free thiols. *Biochim Biophys Acta* 1995; 1258: 35–40.

192. Freeman DJ, Caslake MJ, Griffin BA, Hinnie J, Tan CE, Watson TD, Packard CJ, Shepherd J. The effect of smoking on post-heparin lipoprotein and hepatic lipase, cholesteryl ester transfer protein and lecithin:cholesterol acyl transferase activities in human plasma. *Eur J Clin Invest* 1998; 28: 584–591.

193. Imamura H, Teshima K, Miyamoto N, Shirota T. Cigarette smoking, high-density lipoprotein cholesterol subfractions, and lecithin: cholesterol acyltransferase in young women. *Metabolism* 2002; 51: 1313–1316.

194. Zaratin AC, Quintao EC, Sposito AC, Nunes VS, Lottenberg AM, Morton RE, de Faria E.C. Smoking prevents the intravascular remodeling of high-density lipoprotein particles: implications for reverse cholesterol transport. *Metabolism* 2004; 53: 858–862.

CHAPTER 10

IMPAIRED BIOLOGIC ACTIVITIES

All the major atheroprotective activities of HDL, including cholesterol efflux capacity, as well as anti-inflammatory, antioxidative, cytoprotective, vasodilatory, anti-infectious and antithrombotic activities can become deficient under conditions favoring the accelerated development of cardiovascular disease. Such pro-atherogenic conditions primarily involve atherogenic dyslipidemia, insulin resistance, inflammation and infection; metabolic alterations consistent with accelerated atherogenesis are also characteristic of the post-prandial phase and smoking. As pathways implicated in altered lipoprotein metabolism in atherogenic dyslipidemia, insulin resistance, inflammation and infection are intimately linked (see Chapter 9), clear distinctions between HDL modifications under these pro-atherogenic conditions are hardly possible; impaired biologic activities of HDL therefore markedly overlap between them.

Importantly, the clinical relevance of such impaired anti-atherogenic activities of HDL in cardiometabolic diseases remains largely indeterminate. Indeed, the very concept of HDL dysfunction has been developed using *ex vivo* assays; it is unclear whether HDL particles can also be dysfunctional in the setting of a living organism. Furthermore, such *ex vivo* data are predominantly phenomenologic and frequently document functional deficiency of isolated HDL particles in the absence of mechanistic insight.

High-Density Lipoproteins: Structure, Metabolism, Function, and Therapeutics, First Edition.
Anatol Kontush and M. John Chapman.

10.1 CHOLESTEROL EFFLUX CAPACITY

Dyslipidemia

The capacity of plasma HDL particles to remove cholesterol from peripheral cells can be reduced as a result of diminished circulating levels of HDL particles (i.e., diminished HDL quantity). The link between the reduction in plasma HDL levels, the impairment of normal clearance of cholesterol from the arterial wall and the acceleration of atherosclerosis was proposed by Miller and Miller in 1975 [1] and confirmed in subsequent studies. Indeed, the cholesterol efflux capacity of serum can be diminished in **atherogenic dyslipidemia** characterized by low levels of HDL-C. When low and high levels of HDL-C are defined as HDL-C lower or equal to a tenth percentile and HDL-C higher than or equal to the ninetieth age- and sex-specific percentile in a local population, respectively, sera from low HDL-C subjects display reduced cholesterol acceptor capacity compared with sera from subjects with high HDL-C levels [2]. In support of these data, cholesterol efflux from macrophages to plasma is reduced in hamsters fed with a cholesterol-enriched diet who display an elevated total cholesterol/HDL-C ratio [3].

HDL particles may also reveal impaired intrinsic capacity to efflux cholesterol independently of plasma HDL levels, indicative of the defective specific activity per HDL particle (i.e., diminished HDL quality). Indeed, in primary **hypertriglyceridemia** associated with low HDL-C levels, HDL particles display impaired intrinsic cholesterol efflux capacity from hepatoma Fu5AH cells. As a consequence, the cholesterol efflux capacity of serum is also reduced [4]. The diminished capacity of HDL to accept cellular cholesterol may result in the impairment of the whole RCT pathway in low HDL-C subjects. Indeed, neutral sterol excretion is significantly decreased in patients with familial hypoalphalipoproteinemia (4.0 ± 2.4 mg/kg/day) compared with normolipidemic controls (8.9 ± 6.3 mg/kg/day) [5]. Across the groups, a strong positive correlation between HDL-C levels and fecal neutral sterol excretion is found ($r = 0.53$; $p = 0.01$) [5].

MAJOR CONTROVERSY

Is the defective cholesterol efflux capacity of dyslipidemic human serum related to quantitative (low concentrations of HDL) or qualitative (intrinsic alterations of HDL particles) deficiency of HDL, or both [6, 7]?

Monogenetic forms of low HDL-C dyslipidemia can also be characterized by the presence of HDL with defective intrinsic cholesterol efflux capacity. Indeed, HDL containing apoA-I Oslo carrying the R160L substitution and the apoA-I mutant carrying the P165R substitution, two naturally occurring **apoA-I variants** associated with low HDL-C levels, are less effective in promoting cholesterol efflux from smooth muscle cells compared with normal HDL on a per particle basis [8]. Both mutants display a reduced ability to activate LCAT. Furthermore, cholesterol efflux from human fibroblasts and murine peritoneal macrophages mediated by apoA-I Nichinan, a naturally occurring human apoA-I variant with a deletion of glutamic acid at codon 235 associated with low HDL-C levels, is reduced relative to normal apoA-I [9, 10]. Such impaired cholesterol efflux capacity of apoA-I Nichinan via ABCA1 is accounted for by the disruption of the C-terminal helix which results in decreased lipid binding and reduced stability of apoA-I [11]. In addition, familial low HDL-C deficiency is characterized by reduced HDL3- and apoA-I-mediated cellular cholesterol efflux in the absence of abnormalities in cellular HDL3 binding [12]. However, not all mutations in apoA-I that cause hypoalphalipoproteinemia lead to decreased cholesterol efflux capacity. The most prominent example is provided by apoA-I Milano (see 5.1 of chapter 5 and 6.2 of chapter 6), which displays potent cholesterol efflux properties [13, 14].

The capacity of HDL particles to extract cholesterol from peripheral cells may also be impaired when hypoalphalipoproteinemia results from alterations in cellular HDL receptors, primarily **ABCA1** (Fig. 10.1). At a cellular level, defective interaction of HDL with ABCA1 may impair lipid fluxes to HDL. Indeed, defects in ABCA1, which deleteriously impact on the efflux, may cause both abnormal interaction of the receptor with HDL and impaired cellular lipid translocation. Thus, individuals from families with ABCA1 mutations display lower levels of HDL-C, higher intima-media thickness and lower cholesterol efflux from fibroblasts compared with matched controls [15]. Furthermore, macrophages from a homozygous patient with Tangier disease show virtually no apoA-I-mediated cholesterol efflux [16] (see 6.1 of chapter 6). In subjects with Tangier disease caused by the ABCA1 mutation W590S that resides in the extracellular domain and impairs apoA-I-dependent lipid efflux, ATP-dependent translocation of cholesterol and phospholipid to HDL are impaired [17]. In contrast, apoA-I binding is normal, resulting in retarded dissociation of apoA-I from ABCA1. Other mutations of the ABCA1 gene resulting in an amino acid substitution and a truncated protein or in an in-frame deletion, abrogate ABCA1-mediated cholesterol efflux from fibroblasts secondary to increased intracellular degradation of the protein and its retention in the endoplasmic reticulum [18, 19]. Importantly, cholesterol efflux from peripheral tissues *in vivo* is impaired in carriers of mutations in ABCA1 or APOA1 [20].

Similarly, deletion of **ABCG1** in mice results in severe pulmonary lipidosis and chronic inflammatory processes in lungs, consistent with impaired lipid efflux through ABCG1 which is highly expressed in pulmonary tissue [21]. Finally, cholesterol efflux from macrophages expressing the mutant P297S form of **SR-BI** is reduced compared with wild-type cells [22].

Another example involves steatosis which is associated with both dyslipidemia and decreased hepatic ABCA1 protein expression, potentially leading to impaired lipid efflux and accelerated lipid storage in hepatocytes [23]; this effect can be caused by an excess of unsaturated fatty acids in the liver. Next, chronic renal

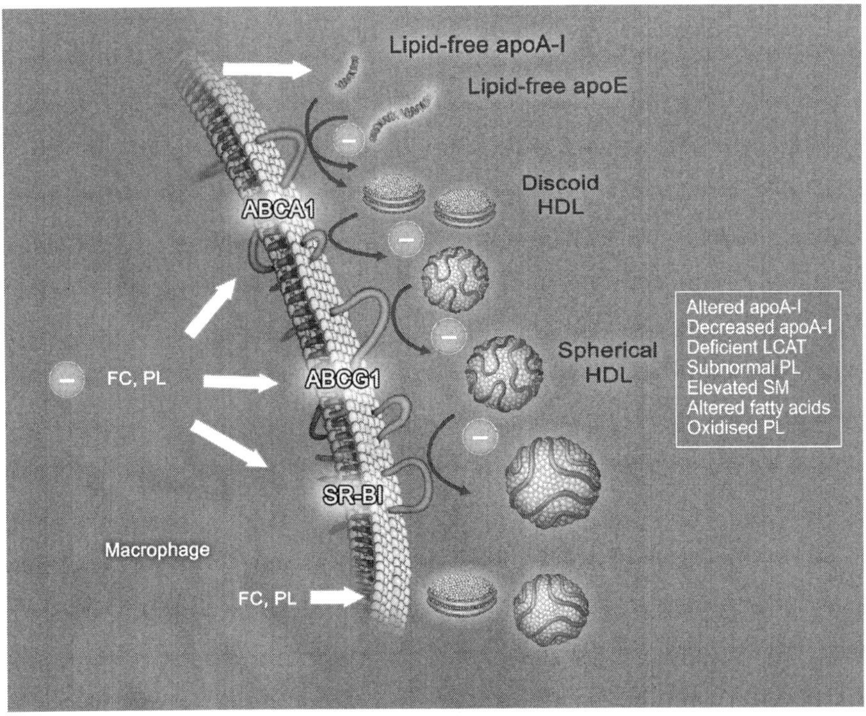

Figure 10.1 Major pathways involved in altered cholesterol efflux capacity of HDL in dyslipidemia, inflammation, insulin resistance and cardiovascular disease. All major receptor-mediated pathways of cholesterol efflux can be affected, specifically those involving ABCA1, ABCG1 and SR-BI. Compositional and structural alterations in HDL, including diminished HDL content of apoA-I, modification of apoA-I by oxidation and/or glycation, deficient LCAT activity, subnormal content of phospholipids, decreased fluidity of HDL surface lipids secondary to increased sphingomyelin/phosphatidylcholine ratio and/or altered fatty acid composition of phospholipids, may underlie such functional deficiency. Inset lists HDL components potentially involved in the altered cholesterol efflux capacity of HDL. FC, free cholesterol; PL, phospholipid; SM, sphingomyelin.

failure causes low HDL-C dyslipidemia, inflammation, oxidative stress and accelerated atherosclerosis; the ability of HDL to efflux cholesterol from lipid-loaded macrophages is diminished in patients with renal failure compared with the control group [24]. In addition, mild renal dysfunction induced by uninephrectomy in apoE -/- mice down-regulates ABCA1, blunts cholesterol efflux from macrophages, increases macrophage cholesterol and activates NFκB [25]. As a consequence, cholesterol influx to peripheral cells can be up-regulated in chronic renal failure secondary to the activation of the scavenger receptor class A type I- and LOX-1-mediated pathways and decreased LCAT activity [26].

Altered profiles of HDL subpopulations and, more specifically, a reduction in small particles, appears to underlie the deficient capacity of dyslipidemic serum to induce cholesterol efflux. Consistent with this notion, levels of **pre-beta HDL** are strongly correlated with cholesterol efflux from THP-1 macrophages to serum [2]. Furthermore, levels of pre-beta HDL are decreased in the low HDL-C vs. high-HDL-C group. In addition, positive correlations exist between ABCA1-dependent cholesterol efflux and serum pre-beta HDL concentrations in subjects with hypertriglyceridemia [27].

Data on the cholesterol efflux properties of HDL in dyslipidemic states are, however, conflicting. Normal cholesterol efflux capacity of serum from hypertriglyceridemic subjects has been reported using Fu5AH cells [28]. Similarly, cellular cholesterol efflux to plasma in subjects with or without Metabolic Syndrome does not differ, despite low HDL-C levels in Metabolic Syndrome [29]. Furthermore, sera from patients with type IV hypertriglyceridemia exhibit a major increase in the ABCA1-dependent cholesterol efflux [27]. Next, HDL from hypertriglyceridemic patients with coronary artery disease and low HDL-C levels possesses a normal capacity to extract cholesterol from smooth muscle cells [30]. Consistent with these results, triglyceride-enriched HDL are not deficient in cholesterol efflux properties from cholesterol-loaded J774 macrophages [31]. Differences in specific receptors involved in cholesterol efflux in a given cell type, as well as differences in patient phenotypes may, in part, account for such discrepancy in the results. In addition, cholesterol efflux to serum or plasma is determined, not only by levels of HDL, but also by concentrations of apoB-containing lipoproteins, as well as by the activities of lipid transfer proteins and/or enzymes. Consistent with such a high level of complexity, an elevated cholesterol esterification rate and increased PLTP activity and apoE levels may contribute to the maintenance of normal cholesterol efflux in Metabolic Syndrome [29]. The role of PLTP in cholesterol efflux is supported by studies in mice over-expressing human PLTP in the presence of human apoA-I[32].

CRITICAL CONTRIBUTION

Impairment of RCT in hypoalphalipoproteinemia compared with normolipidemic controls [5], potentially reflecting defects in both circulating levels and the intrinsic cholesterol efflux capacity of HDL.

Role of HDL Components. In terms of HDL components, **apoA-I** plays a major role in lipid efflux, consistent with the deleterious role of apoA-I mutations. ApoA-I ensures HDL interactions with cellular receptors, which is a key step in efflux. Impaired interaction of HDL particles with cellular receptors, primarily with ABCA1, as a result of alterations in apoA-I may form the basis for deficient cholesterol efflux capacity of HDL.

Abnormal lipid composition of HDL resulting from altered lipoprotein metabolism may also impair the cholesterol efflux properties of isolated HDL. This is demonstrated by the diminished capacity of large, cholesteryl ester-enriched HDL2 obtained from subjects with homozygous CETP deficiency to accept cholesterol from lipid-loaded mouse peritoneal macrophages [33]. In contrast, normalization of the lipid composition of such HDL, as a result of the transfer of excess cholesteryl ester to SR-BI-over-expressing cells, improves HDL cholesterol efflux capacity [34].

Enrichment of HDL particles in triglyceride and depletion of phospholipid may also render them deficient in their capacity to efflux cellular cholesterol. Indeed, HDL particles from cholesterol efflux-defective low HDL-C subjects are characterized by the replacement of cholesteryl ester by triglyceride and reduced content of phospholipid compared with the high HDL-C group [2]. In agreement with these results, the normal cholesterol efflux capacity of serum from hypertriglyceridemic subjects in Fu5AH cells can be related to normal contents of HDL phospholipid, a key determinant of HDL-mediated efflux [28]. Next, triglyceride-enriched HDL produced *in vitro* by co-incubation of normal HDL with CETP are weak activators of cholesterol esterification by LCAT and poor donors of cholesteryl ester to HepG2 cells [31]. Consistent with these findings, the capacity of HDL to deliver cholesteryl ester to hepatic cells through interaction with SR-BI diminishes as a result of HDL enrichment in triglyceride [35]. Furthermore, the presence of triglyceride-rich lipoproteins may have deleterious consequences for HDL-mediated cellular cholesterol efflux as demonstrated by pre-incubation of lipid-loaded macrophages with triglyceride-rich lipoproteins [36]. Together, these data reveal that alterations in the HDL lipidome can impair cellular cholesterol efflux mediated by surface receptors, such as ABCA1 and SR-BI. Furthermore, such alterations are potentially capable of modifying passive diffusion of cellular cholesterol to HDL, a pathway that may contribute significantly to the net cellular cholesterol efflux [37]; the role of the HDL lipidome for passive diffusion of cholesterol from cells remains however to be demonstrated.

Insulin Resistance

Insulin resistance is typically accompanied by subnormal levels of HDL-C (see 9.2 of chapter 9); as is the case for other forms of low HDL-C dyslipidemia, the cholesterol efflux capacity can also be deficient in insulin-resistant states. Indeed, the ability of serum to induce cellular cholesterol efflux is decreased in patients with **Type 2 diabetes** compared with controls [38]. In this study, cholesterol efflux to serum was positively related to HDL-C levels, suggesting an underlying mechanism common to other forms of low HDL-C dyslipidemia and involving quantitative HDL deficiency. Moreover, LpA-I particles from patients with well-controlled Type 2 diabetes exhibited a decreased intrinsic capacity to induce cholesterol efflux from adipose cells [39], consistent with qualitative HDL alterations. In addition, HDL obtained from diabetic individuals with the haptoglobin 2-2 genotype who had a three- to fivefold increased risk of cardiovascular disease compared with diabetic patients without the 2-2 genotype, displayed reduced cholesterol efflux capacity compared with controls [40].

Insulin resistance may diminish cellular cholesterol efflux via several pathways. Thus, the capacity of the serum to induce **ABCA1**-mediated cholesterol efflux is decreased in diabetic patients with microalbuminuria or proteinuria, whereas **SR-BI-mediated cholesterol efflux** is impaired in diabetic patients independently of albuminuria [41] (Fig. 10.1). In a similar fashion, sera from overweight/obese subjects exhibit moderately decreased SR-BI-mediated cholesterol efflux capacities in accordance with reduced concentrations of HDL-C [42]. As a consequence, weight loss induced by bariatric surgery beneficially influences both HDL levels and SR-BI-mediated cholesterol efflux to plasma in obese subjects [43]; elevated concentrations of large, light HDL2 account for the latter effect. Consistent with these data, cellular cholesterol efflux via ABCG1 is also improved by the surgery [43]. Studies in mice suggest that cholesterol efflux via SR-BI can be diminished in diabetes, despite the elevated expression of SR-BI in macrophages; this apparent paradox is resolved by the demonstration of a shift in SR-BI activity from cholesterol efflux to cholesterol influx, mediated by HDL [44].

Interestingly, ABCA1-mediated cholesterol efflux to sera from overweight/obese subjects is increased over the post-prandial period [42], as also observed in subjects displaying mixed hyperlipidemia [45]. Moreover, post-prandial lipemia enhances SR-BI and **ABCG1-dependent efflux** to large HDL2 particles in mixed hyperlipidemia. In contrast, the capacity of HDL to deliver cholesteryl ester to hepatic cells is attenuated, indicative of the reduction of direct RCT to the liver in the post-prandial state [45].

Inefficient HDL maturation represents another mechanism potentially involved in the impairment of cholesterol efflux capacity in insulin resistance. This pathway may be affected as a result of diminished **LCAT** binding to and/or activation on HDL. For example, LCAT binding to HDL appears to be diminished in diabetic carriers of the haptoglobin 2-2 polymorphism compared with diabetics without the 2-2 genotype [46]. Interaction between the Hp 2-2 protein and apoA-I, which hinders LCAT association with HDL and results in the loss of the ability of HDL to promote cholesterol efflux from macrophages, may account for this observation [46]. Deficient LCAT activation may also reflect the reduced HDL content of apoA-I, which is a feature of hypertriglyceridemic insulin-resistant states [47, 48]. In addition, the capacity of HDL particles to efflux cellular cholesterol may be altered as a consequence of diminished plasma levels of anionic peptide factor in Type 2 diabetes [49, 50].

Elevated circulating levels of insulin and leptin, an adipose tissue-derived hormone, may contribute to the impairment of cholesterol efflux from macrophages in insulin-resistant states. Indeed, HDL-mediated cholesterol efflux is suppressed by both insulin [51] and leptin [52]. Insulin inhibits cholesterol efflux from macrophages to HDL as a result of the decreased expression of neutral cholesteryl ester hydrolase and ABCG1 [51]. In addition, insulin resistance may also reduce ABCA1 protein levels and cholesterol export activity, as shown in peritoneal macrophages from diabetic, non-obese mice [53]. The reduced ABCA1 level in macrophages is associated with increased cholesterol content, consistent with impaired ABCA1-mediated cholesterol export. On the other hand, leptin accelerates cholesteryl ester accumulation in human monocyte-derived macrophages by increasing ACAT1 expression via the JAK2 and PI3K pathways [52].

Similar to other dyslipidemic states, abnormal metabolism of HDL resulting in an **altered profile and composition of HDL subpopulations** most probably accounts for alterations in the cholesterol efflux capacity of HDL in insulin-resistant states. In multiple regression analyses, plasma **pre-beta HDL** and HDL phospholipid levels are independent determinants of ABCA1- and SR-BI-dependent cholesterol efflux, respectively [42]. Furthermore, elevated cholesterol efflux via ABCA1 in the post-prandial state is partly related to elevated pre-beta HDL concentrations. Next, pre-beta HDL concentrations are significantly decreased and serum advanced glycation end products are significantly elevated along with diminished cholesterol efflux in Type 2 diabetic patients with microalbuminuria or proteinuria; levels of advanced glycation end products and pre-beta HDL are independent determinants of ABCA1-mediated cholesterol efflux [41].

The deficiency of the cholesterol efflux capacity of HDL in Type 2 diabetes is, however, far from being firmly established. Indeed, cholesterol efflux from fibroblasts to plasma in hypertriglyceridemic patients with Type 2 diabetes is increased, rather than decreased, compared with normotriglyceridemic diabetic and control plasma [54]. Multiple linear regression analysis demonstrates that cellular cholesterol efflux to plasma is positively related to pre-beta HDL formation, PLTP activity and cholesterol esterification rate, but not to the diabetic state. Unaltered pre-beta HDL formation in diabetic hypertriglyceridemia can, therefore, contribute to the maintenance of cholesterol efflux [54]. Contributions from apoB-containing lipoproteins and plasma lipid transfer proteins to the net cholesterol acceptor capacity of serum may underlie the discrepancy between the results obtained with isolated HDL compared with whole serum samples.

Role of HDL Components. The adverse impact of the **glycation** of HDL components on cholesterol efflux is supported by *in vitro* studies that have demonstrated that the cholesterol efflux capacity of apoA-I may be impaired as a consequence of non-enzymatic glycosylation [55–57]. In a similar fashion, glycation of HDL by incubation with 3-deoxyglucosone increases carboxymethyllysine content and reduces cholesterol efflux from THP-1-derived macrophages by -12% [58]. On the other hand, no relationship between serum advanced glycation end products and cholesterol efflux to serum is observed in patients with Type 2 diabetes or normoglycemic controls [38], suggesting that levels of apoA-I glycation attained *in vivo* may not be sufficient to impair the efflux. *In vitro* data suggest that the net transport of cholesterol through the RCT pathway may be adversely affected by apoA-I glycation via decreased LCAT activation and impaired lipid binding secondary to modifications of the Arg, Lys and tryptophan (Trp) residues [59]. Interestingly, inhibitors of glycation, metformin or aminoguanidine are capable of restoring cholesterol efflux mediated by HDL via decreased modification of Lys residues in apoA-I [58].

Finally, glycemia may adversely act, not only on HDL, but also on peripheral cells involved in cholesterol efflux, as suggested by decreased ABCA1 gene expression in leukocytes from hyperglycemic men [60].

Oxidative stress is a key metabolic element of insulin resistance [61]. As a corollary, **oxidative modifications** represent another factor involved in the impairment of HDL cholesterol efflux capacity in this pathology. **Myeloperoxidase** binds to HDL *in vivo* and causes oxidative modification of the HDL protein moiety with major functional implications [62, 63] (see 8.10). Chlorination, nitration and carbamylation of apoA-I catalyzed by the enzyme may contribute to the functional deficiency of HDL [64]. Indeed, *in vitro* oxidation of apoA-I by myeloperoxidase results in the selective inhibition of ABCA1-dependent cholesterol efflux from macrophages and loss of the capacity of HDL to activate LCAT [65, 66]; modification of Trp residues of apoA-I appears to be primarily responsible for this effect [67], whereas the oxidation of Tyr192 and Tyr166 residues also appears to be involved [68, 69]. In parallel, the lipid-binding capacity of apoA-I is progressively impaired [67, 69], as is the ability of apoA-I to be liberated in a lipid-free form from HDL [70]. On the other hand, Met residues of apoA-I may protect the protein from modification by myeloperoxidase [67].

Similarly, **malondialdehyde** damages apoA-I by a pathway that generates Lys adducts at specific sites on the protein [71]; such damage may facilitate the formation of macrophage foam cells by impairing cholesterol efflux via the ABCA1 pathway, potentially hindering the interaction of apoA-I with lipids and ABCA1. Intriguingly, other carbonyls resulting from lipid peroxidation and carbohydrate oxidation, such as hydroxynonenal, glycolaldehyde, glyoxal and methylglyoxal, do not impair the capacity of apoA-I to induce ABCA1-mediated cholesterol efflux [71].

CRITICAL CONTRIBUTION

Key role of myeloperoxidase-catalyzed oxidation of apoA-I in the impairment of the cholesterol efflux capacity of HDL [63, 65, 66, 68, 69].

Altered interaction of apoA-I with ABCA1 as a result of chlorination of apoA-I by myeloperoxidase appears to form a basis for the inhibitory effect of this enzyme on cholesterol efflux which can be mediated by defective activation of the JAK2 signaling pathway [72]. As a consequence of such alterations, myeloperoxidase injection reduces the net RCT rate in a mouse model [73]. Similarly, both *in vitro* HDL oxidation by copper ions and HDL modification by acrolein decreases HDL-mediated cholesterol efflux from cultured cells [74, 75]. The deleterious effect of HDL oxidation on cholesterol efflux may involve inhibition of the expression of ABCA1 [76]. Oxidized forms of cholesterol, including 7-ketocholesterol, may account for impaired cholesterol efflux from macrophage-derived foam cells mediated by apoA-I; such oxysterols act through alterations in cell membrane properties [77, 78]. In a similar fashion, oxidation of HDL lipids may underlie impairment in the cholesterol efflux properties of HDL by cigarette smoke [79, 80].

Another aspect of the action of oxidative stress on cellular cholesterol efflux concerns specific **oxidized phospholipids,** which serve as high affinity ligands for the scavenger receptors CD36 and SR-BI. Acting through this pathway, oxidized phospholipids are capable of preventing binding of HDL to SR-BI and can thereby interfere with SR-BI-mediated selective uptake of HDL-derived cholesteryl esters in hepatocytes [81].

It appears that HDL oxidation needs to be extensive in order to inhibit cellular cholesterol efflux as such a modification strongly destabilizes the particle structure [82]. In contrast, mild oxidation by copper ions or HOCl moderately destabilizes HDL and accelerates protein dissociation, cholesterol insertion and HDL fusion, thereby facilitating efflux [82]. Such moderate destabilization of HDL by low concentrations of HOCl may enhance the formation of pre-beta-migrating particles, thereby improving HDL function as a primary cholesterol acceptor [83]. In contrast, at high concentrations of HOCl, the formation of pre-beta-migrating particles is inhibited.

Oxidative stress may also impact arterial wall cells, resulting in the down-regulation of the expression of cholesterol transporters. For example, oxidative modification of HDL decreases cholesterol efflux from human umbilical vascular endothelial cells by inhibiting the expression of ABCAl [76]. Consistent with the notion of elevated oxidative stress in atherosclerotic plaques, ABCA1 expression and apoA-I binding are reduced in cultured smooth muscle cells obtained from human atherosclerotic lesions [84].

METHODOLOGIC NOTE

A wide array of cell lines are currently being used for *in vitro* measurements of cellular cholesterol efflux [THP-1, J774 and RAW 264.7 macrophages, Fu5AH hepatoma cells, Chinese hamster ovary (CHO) cells over-expressing specific HDL receptors]. Whatever the cell type, such measurements are characterized by high inter-assay variability. To minimize the variability, a pooled serum control should be used to normalize cholesterol efflux values across assays [6, 7].

Inflammation

The intrinsic cholesterol efflux capacity of HDL can be considerably impaired during inflammation and the acute phase response. Inflammation appears to predominantly impact cellular **cholesterol efflux via ABCA1** [85]. As cellular cholesterol efflux is largely mediated by apoA-I-containing HDL [86], **apoA-I replacement by SAA** can significantly impact on efflux [87]. Indeed, enrichment of HDL with SAA (up to high SAA contents of 86% of total HDL protein) results in enhanced HDL binding to, decreased cholesterol efflux capacity from, and increased selective cholesteryl ester uptake by macrophages [88, 89]. Importantly, as SAA is predominantly present in small, dense HDL3 particles, it selectively impairs the cholesterol efflux properties of HDL3. In contrast, less pronounced enrichment of HDL with SAA *in vivo* (up to 27% of total HDL protein) does not influence cholesterol efflux, but enhances HDL binding to macrophages [88]. Consistent with this conclusion, ABCA1- and ABCG1-dependent cholesterol efflux during the acute phase response following cardiac surgery is preserved along with relatively modest increases in SAA [90]. Nevertheless, SAA, in a lipid-free or lipid-poor form, may still promote cholesterol efflux by **ABCA1**, functioning as a cholesterol acceptor, as well as by increasing the availability of cellular cholesterol [91]; lipid-free apoA-I is, however, the preferred substrate for ABCA1. Furthermore, acute-phase mouse HDL containing SAA exhibits a modest, but significant, +17% increase in ABCG1-dependent efflux [92]; phospholipid enrichment, and not the presence of SAA itself, is responsible for this alteration. In contrast, acute-phase human HDL, which is not phospholipid-enriched, is somewhat less effective in mediating **ABCG1**-dependent efflux compared with normal human HDL [92].

The presence of SAA increases both HDL affinity to, and selective cholesteryl ester uptake by, macrophages, but reduces affinity to, and cholesteryl ester uptake by, hepatocytes [93–95]. Furthermore, SAA efficiently promotes cholesterol efflux from hepatoma cells, increasing the ability of acute phase HDL to interact with cellular **SR-BI** [91, 96]. The interaction of SAA-enriched acute-phase HDL with SR-BI generates progressively smaller HDL particles with the smallest particles containing only apoA-I, thereby accelerating HDL catabolism [97]. During inflammation, the number of binding sites for HDL-bound SAA increases on macrophages and decreases on hepatocytes. In addition, macrophage expression of ABCA1 is diminished [98]. Interestingly, the increased binding affinity of acute-phase HDL for macrophages is dependent on cell surface heparan sulfate, which is capable of HDL remodeling at low pH with the formation of small HDL particles possessing elevated cholesterol efflux capacity [99].

MAJOR CONTROVERSY

How does HDL enrichment in SAA during the acute-phase response act on individual steps of the RCT pathway? Does HDL enrichment in SAA accelerate or decelerate RCT as a whole?

Oxidative modifications of apoA-I as a result of a highly pro-oxidative environment typical of inflammatory states can also contribute to deficient cholesterol efflux, as described for insulin resistance. In addition, proteolytic modifications of apoA-I and other HDL apolipoproteins that occur *in vivo*, may reduce cholesterol efflux from macrophage-derived foam cells [100, 101]. Another compositional change relevant for cholesterol efflux capacity of acute-phase HDL involves enrichment in apoE [87], which may accelerate HDL catabolism via the apoB/apoE receptors. **Decreased phospholipid contents** in inflammatory HDL constitute an additional factor that contributes to deficient HDL cholesterol efflux properties, as suggested by studies in patients with periodontitis [102]. Similarly, cholesterol efflux toward plasma and HDL is significantly decreased in patients with acute sepsis [73]. Together, these changes may lead to a significant shift in HDL-mediated cholesterol transport from hepatocytes towards macrophages under acute phase conditions [93]. Consistent with this conclusion, over-expression of SAA reduces overall RCT from macrophages in the absence of systemic inflammation [73].

Similarly, acute-phase HDL from hamsters displays diminished cholesterol efflux capacity from J774 macrophages, elevated cholesterol influx capacity and decreased LCAT activity [103]. *In vitro* inactivation of LCAT in control HDL results in similar effects on cholesterol transport, identifying **LCAT** as another key player in the abnormal cholesterol transport properties of HDL particles in inflammatory states [103].

Diminished apoA-I lipidation following interaction with ABCA1 may underlie low circulating levels of HDL-C and reduced cholesterol efflux capacity of HDL in inflammatory states (Fig. 10.1). Indeed, inflammation down-regulates ABCA1 expression via an interferon-gamma-mediated pathway by inhibiting LXR alpha in a JAK/STAT-dependent manner [104]. CRP may also play a role in this effect as it inhibits cholesterol efflux from human foam cells derived from THP-1 macrophages and peripheral blood mononuclear cells via down-regulation of ABCA1 and ABCG1; increased oxidative stress and ERK1/2 activation are implicated in this pathway [105]. In a similar fashion, TNF-alpha can decrease both cholesterol efflux to apoA-I and circulating HDL-C levels by attenuating the expression of intestinal ABCA1 [106]; the latter mechanism may significantly contribute to the pathology of Crohn's disease, a chronic inflammatory condition of the gastrointestinal tract driven by TNF-alpha. In contrast, TNF-alpha can induce ABCA1 in macrophages [107].

In addition to cholesterol efflux, inflammation may impact on other steps of RCT, albeit to a lesser extent [85]. Thus, in a mouse model of sub-acute endotoxemia, macrophage expression of ABCA1 is reduced along with down-regulation of hepatic expression of ABCG5, ABCG8, and ABCB11 biliary transporters [73, 87]. Lipolytic enzymes that produce lysophospholipids and free fatty acids induced by inflammation can account for these effects. For example, group X $sPLA_2$ modulates gene expression in macrophages by generating lipolytic products that suppress LXR activation and ABCA1 expression [108]. As a result of such concerted changes in HDL metabolism, cholesterol efflux from macrophages to apoA-I and serum is reduced and RCT from macrophages to feces is diminished [73, 85, 87].

Cardiovascular Disease

Patients with cardiovascular disease typically present with dyslipidemia, oxidative stress and chronic inflammation, all of which may deleteriously impact on the cholesterol efflux properties of HDL (see 10.1 of chapter 10 above). As a consequence, apoB-depleted sera from patients with **angiographically-confirmed coronary artery disease** display impaired capacity to efflux cholesterol from cyclic AMP-treated J774 macrophages [6, 7]. Interestingly, in this study HDL-C levels represented a highly significant determinant of serum efflux capacity but only accounted for 26% of the observed variation, consistent with the role of other factors for cholesterol efflux to whole serum. The difference between coronary artery disease patients and controls was attenuated, but remained significant, after including HDL-C as an additional covariate. Importantly, in the regression analysis, serum cholesterol efflux capacity constituted a strong independent predictor of coronary artery disease status both before, and after, the inclusion of HDL-C levels. These data reinforce the concept that measurement of HDL quality, in addition to HDL quantity, may have a use in predicting cardiovascular risk [6, 7].

OPEN QUESTION

Can the cholesterol efflux capacity of HDL predict the development of atherosclerotic disease?

These results are supported by the finding of reduced cholesterol efflux from macrophages to apoA-I and HDL2 in normolipidemic subjects displaying significant stenosis on coronary angiography compared with individuals without significant stenosis [109]. Another condition facilitating the development of cardiovascular disease, notably plasma **homocysteinemia**, is also characterized by the impaired capacity of serum to induce cholesterol efflux from lipid-loaded macrophages compared with healthy controls [110]. Importantly, cholesterol efflux from macrophages is positively correlated with the particle concentration of small HDL particles, but not with circulating plasma levels of HDL-C, emphasizing the role of small HDL in the efflux.

In addition to the quantity of small HDL particles, their quality can also be altered in patients with cardiovascular disease. Specifically, the capacity to accept exogenous lipids can be diminished, as indicated by the diminished transfer rates of free cholesterol and cholesteryl esters to HDL in heart transplant patients [111].

An interesting mechanism of the impairment of cholesterol efflux may be operative in transthyretin-related amyloidoses which confer elevated cardiovascular risk. A fraction of plasma transthyretin that circulates on HDL is bound to apoA-I [112]. Such HDL-bound transthyretin is able to cleave the C-terminus of both lipid-free and lipid-bound apoA-I with the formation of discoid HDL particles displaying reduced cholesterol efflux capacity and high propensity to form amyloid-like aggregates [112].

Pathophysiologic Relevance

The impaired intrinsic cholesterol efflux capacity of HDL during inflammation may serve to redirect cholesterol to immune cells and to sites of injury and inflammation [93, 113]. Advantageous under inflammatory conditions from the point of view of immediate survival, such inefficient cholesterol efflux may become detrimental in the long-term in the context of chronic diseases, primarily of atherosclerosis. The reduced efficiency of cholesterol flux through the RCT pathway may account for the epidemiologic link between subnormal HDL-C levels and increased incidence of cardiovascular disease [114–117]; the attenuated cholesterol efflux capacity of HDL, as occurs in dyslipidemia, insulin resistance and inflammation, may contribute to accelerated atherogenesis and elevated cardiovascular risk.

IMPORTANT READING

Zhao Y, Van Berkel TJ, Van Eck M. Relative roles of various efflux pathways in net cholesterol efflux from macrophage foam cells in atherosclerotic lesions. *Curr Opin Lipidol* 2010; 21: 441–453.

10.2 ANTIOXIDATIVE ACTIVITY

Dyslipidemia

HDL particles are deficient in antioxidative activity in dyslipidemic states involving low HDL-C levels, often in association with insulin resistance [118–120]. Thus, the capacity of small, dense HDL to protect LDL from oxidative damage induced by free radicals is impaired (by up to −43%) in subjects with a normotriglyceridemic, normocholesterolemic, normoglycemic **low HDL-C phenotype** compared with normolipidemic controls [119]. In a similar fashion, small, dense HDL subpopulations are deficient in their antioxidative activity in **atherogenic dyslipidemia** of Metabolic Syndrome (by up to −23%) [118] and well-controlled Type 2 diabetes (by up to −47%) [120] that features marked hypertriglyceridemia. The intrinsic antioxidative activity of HDL particles is also reduced in subjects with **hyperalphalipoproteinemia** associated with low activity of hepatic lipase [121]. In each of these studies, antioxidative activity of HDL is deficient, both on a unit particle mass and on a particle number basis, thereby suggesting that the intrinsic properties of HDL particles, rather than HDL-C levels *per se*, are determinants of antioxidative deficiency of HDL. Furthermore, HDL-mediated protection of LDL from oxidation by copper ions is compromised in post-menopausal compared with premenopausal women; this effect is paralleled by decreased plasma levels of HDL-C [122]. Finally, a marked reduction in the antioxidative activity of HDL accompanied by low HDL-C concentrations is a feature of end-stage renal disease [123].

It is noteworthy that, despite the fact that antioxidative deficiency of HDL has been extensively documented in atherogenic low HDL-C dyslipidemias using *in vitro* assays, direct proof of its occurrence *in vivo* is still lacking. Indirect evidence includes an association between elevated levels of plasma HDL-C and reduced levels of lipid peroxidation products after a single intravenous injection of a large dose of human HDL3 (200 mg protein) in hypercholesterolemic rabbits [124], suggesting compromised antioxidative activity of autologous rabbit HDL under conditions of hypercholesterolemia.

OPEN QUESTION

Do circulating HDL particles display deficient antioxidative activity *in vivo* under metabolic conditions associated with elevated cardiovascular risk?

Role of HDL Components. The antioxidative deficiency of HDL in low HDL-C dyslipidemias is accompanied by altered compositional properties of HDL particles [118–121, 125]. First, the functionally deficient HDLs are **enriched in triglyceride** and **cholesteryl ester depleted**; such compositional alterations are observed in normotriglyceridemic, low HDL-C phenotypes, Metabolic Syndrome and Type 2 diabetes, as well as in post-menopausal women [118–122], potentially reflecting elevated CETP activity. Interestingly, HDL enrichment in triglyceride correlates with the diminished antioxidative activity and/or elevated oxidability of HDL, consistent with its functional role.

Second, dyslipidemic HDL with diminished antioxidative activity may feature a **decreased content of apoA-I** [47, 125]. In contrast, activities of HDL-associated enzymes (PON1, PAF-AH, LCAT) per unit of HDL mass typically remain unaffected by the presence of dyslipidemia [119, 121, 122]. On the other hand, concentrations and activities of such enzymes in the circulation are typically decreased as a result of low HDL levels, as exemplified by significant reductions in circulating concentrations of GSPx and LCAT, as well as of activities of GSPx and PON1 in patients with end-stage renal disease [123].

The functional significance of the compositional changes in HDL can be inferred from the two-step model of HDL-mediated protection of LDL from oxidative damage by one-electron oxidants [125]. According to this model, such protective activity can be impaired as a result of deficiency in both the transfer of oxidized lipids from LDL to HDL and the inactivation of oxidized lipids within HDL particles [125]. Inactivation of LDL-derived oxidized lipids is mediated by apoA-I and HDL-associated enzymes, which may account for the reduction of LOOHs and for the hydrolysis of short-chain oxidized phospholipids, respectively. The decreased content of apoA-I may, therefore, lead to the inefficient inactivation of LDL-derived LOOHs by Met residues of apoA-I, with subsequent accumulation of these primary products of lipid peroxidation in HDL, together with secondary products, such as short-chain oxidized phospholipids, aldehydes and modified proteins (Fig. 10.2). On the other hand, hydrolytic enzymes are unlikely to contribute to the defective antioxidative activity of HDL in dyslipidemic states not associated with insulin resistance (see below).

Mechanistically, the ineffective reduction of LOOHs by Met residues of apoA-I may also underlie the correlation between triglyceride enrichment of HDL particles and the impairment of antioxidative activity [118–122]. Indeed, the

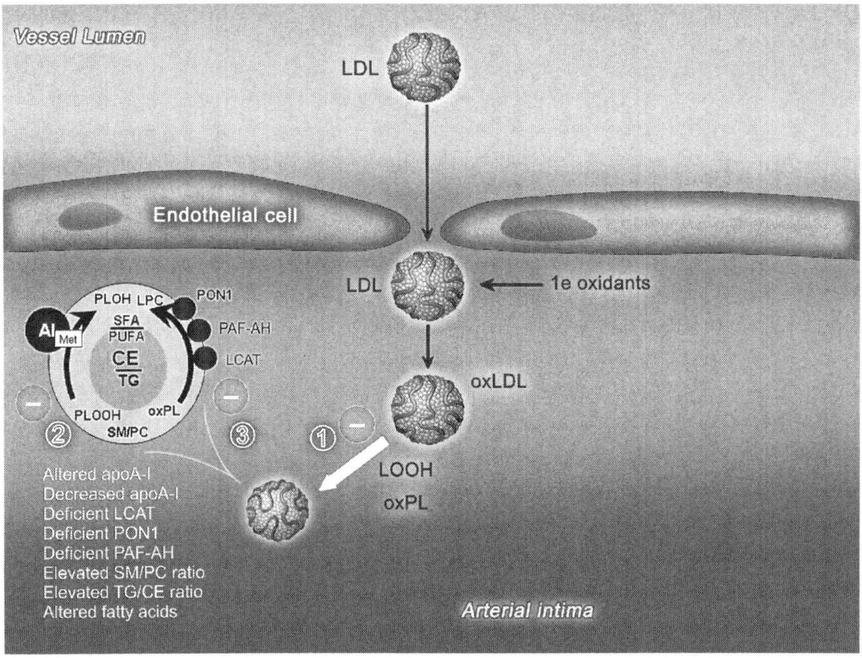

Figure 10.2 Major pathways involved in altered antioxidative activity of HDL in dyslipidemia, inflammation, insulin resistance and cardiovascular disease. The capacity of HDL to protect LDL from oxidation induced by one-electron (1e) oxidants can be impaired, involving both the transfer of oxidized lipids from oxidized LDL (oxLDL) to HDL and their subsequent inactivation by apoA-I and by HDL-associated enzymes. (1) The efficacy of the transfer of oxidized phospholipids (oxPL) to HDL can be altered secondary to the decrease in the fluidity of HDL surface lipids, resulting from increased sphingomyelin (SM)/phosphatidylcholine (PC) ratio and/or altered fatty acid composition of phospholipids, such as an elevated ratio of saturated fatty acids to PUFAs. (2) Decreased HDL content of apoA-I and modification of apoA-I by oxidation and/or glycation may result in deficient reduction of PLOOH to redox-inactive phospholipid hydroxides (PLOHs); the capacity of apoA-I to react with PLOOH may also be influenced by apoA-I conformation which can be altered by elevated triglyceride (TG)/cholesteryl ester (CE) ratios in the HDL lipid core. (3) Deficient activities of HDL enzymes may underlie the impaired hydrolysis of short-chain oxPL with the formation of lysophosphatidylcholine (LPC). Inset lists HDL components potentially involved in the altered antioxidative activity of HDL.

replacement of cholesteryl ester by triglyceride in the HDL lipid core considerably alters the conformation of the central and C-terminal domains of apoA-I [126, 127], thereby potentially modifying the accessibility of Met residues to LOOH. In addition, the replacement of cholesteryl ester by triglyceride in spherical rHDL decreases the conformational stability of apoA-I [126], resulting in triglyceride-containing particles that are unstable and which readily lose apoA-I.

Insulin Resistance

The presence of antioxidatively-deficient HDL particles is a prominent feature of insulin-resistant states, such as Metabolic Syndrome and Type 2 diabetes, which are frequently characterized by subnormal circulating concentrations of HDL-C. Indeed, small, dense HDL3 particles display impaired protection of LDL from oxidative damage in **Metabolic Syndrome** and well-controlled **Type 2 diabetes** [118, 120] (see above). Large, light HDL2 also displays a decreased capacity to protect LDL against oxidation mediated by THP-1 macrophages in poorly-controlled Type 2 diabetes [128]. Interestingly, such a deficiency of large, light, diabetic HDL2 to protect LDL from *in vitro* oxidation is not necessarily accompanied by decreases in plasma levels of HDL-C [129]. Next, the total HDL fraction obtained from subjects with Type 2 diabetes inactivates 20% less membrane hydroperoxides than HDL from control subjects [130]. Similarly, a diminished capacity of total HDL to remove LOOHs from erythrocyte membranes is a feature of poorly equilibrated **Type 1 diabetes** [131]. In addition, the antioxidative activity of HDL is significantly impaired in Type 2 diabetic patients with microalbuminuria or proteinuria along with an elevated HDL content of lipid peroxides [38, 132]. Finally, subjects with thalassemia, a disorder of iron metabolism associated with insulin resistance, possess HDL particles with a diminished capacity to protect LDL from oxidative stress [133].

The impaired antioxidative activity of small, dense HDL in Metabolic Syndrome and Type 2 diabetes is correlated with hypertriglyceridemia, hyperinsulinemia and insulin resistance, thereby suggesting that abnormalities in both lipid and glucose metabolism underlie the antioxidative deficiency of HDL particles [118, 120]. In addition, biomarkers of HDL antioxidative functionality are associated with systemic levels of oxidative stress assessed by 8-isoprostanes that are three- to fourfold elevated, implicating the role of oxidative modification.

CRITICAL CONTRIBUTION

Specific deficiency in the antioxidative activity of small, dense HDL3 particles paralleled by enhanced systemic oxidative stress in metabolic diseases associated with elevated cardiovascular risk, such as in Type 2 diabetes and Metabolic Syndrome [118, 134].

Role of HDL Components. Abnormal content and functionality of **apoA-I** may be central to the functional deficiency of HDL in insulin-resistant states and involve the combined impact of accelerated catabolism, oxidative modification, abnormal conformation and non-enzymatic glycation. Accelerated removal of apoA-I from the circulation results in diminished content of apoA-I per particle in HDL. Elevated oxidative stress in the arterial intima may lead to the oxidation of Met residues of apoA-I by one- and two-electron oxidants with the formation of methionine sulfoxides. ApoA-I contains three Met residues at positions 86, 112 and 148; oxidation of all three of them causes partial unfolding of lipid-free apoA-I, decreases its thermal stability and increases its propensity to aggregation [135]. In addition, Met oxidation inhibits the native self-association of apoA-I to form dimers and tetramers and can induce the formation of amyloid fibrils [135].

The oxidation of Met residues of apoA-I may represent a key pathway in the inefficient inactivation of LOOH by HDL (Fig. 10.2). The pathophysiologic importance of this reaction is suggested by the observation that significant levels of methionine sulfoxides (approximately 1 to 3% of total Met residues) are present on apoA-I in plasma from healthy normolipidemic subjects [136, 137]. All three Met residues of apoA-I can be present in an oxidized form in the circulation. In healthy subjects, the extent of oxidation is Met112 $\approx$ Met 148 $\gg$ Met86 [137]. Furthermore, levels of apoA-I-associated methionine sulfoxides are elevated in patients with Type 1 diabetes, consistent with increased levels of lipid peroxidation products in diabetic plasma [137]. Met112 sulfoxide is particularly elevated (about fivefold, from 1 to 5% of total Met112 in apoA-I), whereas the other two sulfoxides are increased moderately (about 1.5-fold). In addition, plasma levels of methionine sulfoxides are increased in other conditions related to elevated oxidative stress and cardiovascular risk, such as in subjects with the eNOS a/b genotype and in smokers [136].

Methionine sulfoxides can be converted back to methionines by methionine sulfoxide reductases, which play a critical role in the cellular antioxidant response [138]. As methionine sulfoxide reductase activity is, however, absent from plasma, formation of methionine sulfoxides in apoA-I may serve as a biomarker of exposure of HDL to LOOHs and/or as an integrative biomarker of oxidative stress in the vascular compartment.

Abnormal conformation of apoA-I secondary to HDL enrichment in triglyceride represents another pathway that may contribute to the antioxidative deficiency of HDL in Type 2 diabetes (Fig. 10.2); mechanistically, this pathway is similar to the alterations observed in dyslipidemic states. Non-enzymatic **glycation**, a post-translational modification resulting in glycoxidative modification of apoA-I, may also contribute to the loss of the antioxidative capacity of HDL [38]. Indeed, fructose-mediated glycation of apoA-I impairs the antioxidative activity of HDL [139]. Furthermore, lipid-free apoA-I isolated from subjects with Type 2 diabetes inhibits ROS generation by macrophages less efficiently compared with apoA-I from normolipidemic controls [140].

Altered enzymatic activities represent another potential determinant of the antioxidative deficiency of HDL in insulin resistance, potentially resulting in impaired hydrolysis of short-chain oxidized phospholipids (Fig. 10.2). Indeed, both **PON1** and **PAF-AH** activities are consistently lower in all HDL subfractions from patients with Type 2 diabetes compared with matched normolipidemic control subjects [120]. Next, HDL-associated PON1 activity is significantly decreased in Type 2 diabetic patients with microalbuminuria or proteinuria and is positively associated with the antioxidative activity of HDL [132]; multiple regression identifies urinary protein as an independent, negative determinant of HDL-PON1 activity. In addition, serum PON and lactonase activities are diminished in Type 2 diabetic patients compared with controls [141]. Serum concentrations and HDL-associated activity of PON1 are also attenuated in poorly equilibrated Type 1 diabetes [131, 142]. The role of PON1 is further consistent with another study that found no difference in HDL antioxidative activity and serum PON1 activity between Type 2 diabetic patients with glycemic control and healthy subjects [143]. Finally, deficiency in the antioxidative activity of HDL in Type 2 diabetes can also be paralleled by decreases in **LCAT** activity [46].

As is the case for apoA-I, **non-enzymatic glycation** may also contribute to the deficient functionality of HDL-associated enzymes. Glycation of PON1 is elevated in subjects with Type 2 diabetes compared with controls (17% vs. 7.5%, respectively) [130]. *In vitro* glycation of PON1 reduces the ability of HDL to inactivate membrane hydroperoxides by −50%, whereas glycoxidation reduces it by −80% [130]. In addition, PON1 activity is negatively correlated with the ability of HDL from normolipidemic controls to inactivate membrane hydroperoxides. Similarly, the antioxidative capacity of HDL and PON1 activity are negatively correlated with serum concentrations of advanced glycation end products that are elevated in diabetic patients with microalbuminuria or proteinuria [38]. Consistent with these data, the in vitro inactivation of HDL-associated enzymes, such as PON1 or LCAT, by glycation and/or oxidation is associated with a decreased capacity of HDL to protect LDL from oxidative stress [56, 144–146].

The role of enzymes in HDL antioxidative deficiency is also supported by data obtained in obese leptin-deficient (ob/ob), LDL-receptor -/- mice that possess dysfunctional HDL displaying not only decreased PON1 and LCAT activities, but also elevated levels of antibodies against oxidized LDL [147]. The contribution of the loss of PON1 activity to the defective capacity of HDL to inactivate LOOHs, however, remains questionable as LOOH does not represent a physiologic substrate for the enzyme [148, 149]. The observation that antioxidative dysfunction of HDL can be accompanied by normal PON1 activity made in patients with renal disease or in dyslipidemic subjects adds further doubt to the role of PON1 [125, 150]. A causative role of PON1 and other HDL-associated enzymes in deficient antioxidative functionality of HDL therefore remains to be demonstrated.

MAJOR CONTROVERSY

Do altered activities of HDL-associated hydrolytic enzymes contribute to the deficient capacity of HDL to protect LDL from oxidative damage by one-electron oxidants?

Other protein components that can be involved in the loss of the antioxidative activity of HDL in Type 2 diabetes include **complement C3** protein and **SAA**, which can accumulate in HDL and exert a deleterious action on HDL functionality [46, 151]. In addition, HDL-associated **apoJ** is decreased in subjects with reduced insulin sensitivity, elevated BMI and an unfavorable lipid profile [152]; the potential importance of this finding for HDL functionality, however, remains unclear. Finally, the antioxidative deficiency of HDL may be associated with an elevated content of pro-oxidants and products of oxidative modification, as evidenced by increased iron content and elevated levels of lipid peroxidation products in HDL from subjects with thalassemia [133].

Inflammation

Historically, the finding of the strongly diminished antioxidative and anti-inflammatory activities of HDL, which may even be transformed into pro-oxidative and pro-inflammatory properties, under the conditions of the **acute inflammatory response** has been at the origin of the very concept of HDL dysfunction originally developed by Fogelman et al. [153]. Their pioneering studies revealed that HDL from humans and rabbits during induction of the acute phase lose the ability to protect LDL against oxidative modification by artery wall cells in co-culture, concomitant with decreases in PON1 and PAF-AH activities [154, 155]. Furthermore, HDL from mice that are genetically predisposed to diet-induced atherosclerosis does not protect LDL against oxidation in co-cultures of artery wall cells when the mice are injected with pro-inflammatory oxidized phospholipids or fed an atherogenic diet that induces low-grade chronic inflammation [156, 157]. In addition, antioxidative deficiency of HDL is observed in both apoE -/- [157] and apoA-II transgenic [158, 159] mice displaying pro-inflammatory metabolic alterations.

The major mechanism underlying functional HDL deficiency during the acute inflammatory response probably involves the replacement of apoA-I by acute phase proteins, primarily **SAA** [151] (Fig. 10.2). As in the case of the replacement of cholesteryl ester by triglyceride, the replacement of apoA-I by SAA may cause deficient capacities of HDL to inactivate LOOHs, resulting in their accumulation in LDL and in arterial wall cells. The replacement of apoA-I by SAA may be also operative under conditions of chronic inflammatory response [151]. Indeed, the content of apoA-I in HDL tends to be reduced in metabolic diseases associated with chronic inflammation [118–120]. Remarkably, SAA is predominantly associated with small, dense HDL particles [160, 161], suggesting their impaired anti-inflammatory function under pro-inflammatory conditions.

The loss of antioxidative activity of HDL following acute phase induction occurs concomitantly with decreases in HDL-associated **PON1** and **PAF-AH** activities [154–157, 162] (Fig. 10.2). Impaired hydrolysis of short-chain oxidized phospholipids by HDL-associated enzymes may, therefore, also contribute to the deficient antioxidative activity of HDL in inflammatory states. In addition, the induction of hypertriglyceridemia may result in HDL enrichment in monosaturated fatty acids relative to PUFAs [163], thereby increasing the rigidity of HDL surface lipids; such compositional alterations may decrease the capacity of HDL to acquire and inactivate pro-inflammatory oxidized lipids [125].

Infection

It remains largely indeterminate as to whether infection impacts upon the antioxidative properties of HDL particles. HDL isolated from mice infected with influenza A virus does not protect LDL against oxidation in co-cultures of artery wall cells [162]. PON1 activity is diminished in HIV-positive patients in comparison with healthy controls. In parallel, the transfer of free cholesterol and triglycerides from a lipid nano-emulsion to HDL is decreased, suggesting that the capacity of HDL to both acquire and inactivate oxidized lipids may be defective during infection [164].

Cardiovascular Disease

Curiously, deficiency in the antioxidative activity of HDL in established cardiovascular disease remains less well studied compared with subjects with cardiovascular risk factors, such as dyslipidemia or insulin resistance. The capacity of HDL to protect LDL from oxidation is decreased in patients with Type 2 diabetes presenting cardiovascular disease, as well as in diabetic subjects with prior coronary artery bypass surgery [165]. Controversially, however, similar antioxidative deficiency of HDL is also observed in patients with Type 2 diabetes who do not display cardiovascular disease [165]. In addition, the capacity of HDL to protect LDL from oxidation induced by γ-radiolysis of water is attenuated in elderly compared with young subjects [166], consistent with the role of **age** as a major cardiovascular risk factor.

From a mechanistic point of view, the potential deficiency of the antioxidative activity of HDL in cardiovascular disease should be related to modifications of the both HDL proteome and lipidome. Indeed, circulating levels and activity of **PON1** are reduced in patients with cardiovascular disease and essential hypertension [167, 168], accompanied occasionally by elevated plasma levels of LOOHs [169]. Furthermore, homocysteine thiolactonase activity is reduced in diabetic subjects with prior coronary artery bypass surgery [165]. Such a reduction in the activities of PON1 may reflect the increased exposure of the enzyme to oxidants generated in the pro-oxidant environment of atherosclerotic plaques, primarily to HOCl; on the contrary, vitamin C may minimize the inhibitory effect of HOCl on PON1 [170].

Alterations at the level of the HDL lipidome that can impact the antioxidative properties of HDL may involve enrichment in **sphingomyelin** and in **saturated fatty acids**, both resulting in elevated rigidity of the phospholipid surface monolayer of HDL [171] (Fig. 10.2). Such structural anomalies may cause a deficiency in the capacity of HDL to acquire LDL-derived oxidized lipids and to protect LDL from free-radical-induced oxidative damage [171]. The deleterious impact of sphingomyelin and saturated fatty acids on the antioxidative activity of HDL may contribute to positive relationships between plasma levels of sphingomyelin [172] and saturated fatty acids [173] with cardiovascular risk. It is interesting, in this regard, that the deletion of sphingomyelin synthase 2 in macrophages and in the liver improves the cholesterol efflux capacity of HDL, whereas liver-specific transfer of the enzyme reverses the effect [174, 175]. As a consequence, deletion of sphingomyelin synthase 2 in macrophages reduces atherosclerosis in mice [174]. Furthermore, inhibitors of sphingomyelin biosynthesis possess anti-atherogenic properties [176].

Pathophysiologic Relevance

According to the response-to-retention hypothesis of atherosclerosis, oxidation of lipoproteins, primarily LDL, in the arterial wall is a key element in atherogenesis [177–181]. An important corollary of this hypothesis is that deficient LDL protection from oxidation may accelerate atherogenesis [182, 183].

Intriguingly, accumulation of LDL in the arterial intima has been recently proposed to represent a physiologic response to infection designed to protect susceptible segments of the arterial wall from microbial invasion [184]. Within this hypothesis, intrinsic antimicrobial activities of deposited lipids, the formation of fibrin-based matrices within the intima, vasoconstriction and mobilization of smooth muscle cells can all contribute to combating such an infectious menace. Furthermore, oxidized lipids resulting from the oxidation of lipoproteins close to sites of infection and inflammation can be disseminated through the circulatory system to promote the penetration and retention of lipoproteins in susceptible segments of the intimal tissue [184]. Oxidized lipids in the intima, such as oxidized phospholipids and oxysterols, can also act as a signal for macrophages and lymphocytes to trigger the immune response [184]. In a broader concept, common oxidation-specific epitopes formed on lipids and proteins may represent a major target of innate immunity, recognized by a variety of pattern recognition receptors [185]. By analogy with microbial pathogen-associated molecular patterns, host-derived, oxidation-specific epitopes can be considered to represent danger-associated molecular patterns [185]. The loss of antioxidative activity of HDL occurring during the acute phase response may, therefore, represent part of a defence mechanism elicited for promoting lipid peroxidation against bacteria and triggering inflammatory pathways [186].

Consistent with this concept, the impairment of the antioxidative activity of small, dense HDL in dyslipidemias involving low HDL-C levels [118–120] is associated with elevated oxidative stress, a newly recognized cardiovascular risk factor [187–189]. Indeed, dyslipidemic subjects presenting with atherogenic low HDL-C levels are characterized by both the deficient antioxidative activity of small, dense HDL and elevated systemic oxidative stress assessed as plasma levels of 8-isoprostanes, products of non-enzymatic oxidation of arachidonic acid [190, 191]. Furthermore, HDL antioxidative activity and plasma 8-isoprostanes are negatively correlated [118, 120]. Next, in subjects with controlled Type 2 diabetes, plasma 8-isoprostanes negatively correlate with HDL-C levels [120], whereas in subjects with a normotriglyceridemic low HDL-C phenotype, plasma 8-isoprostanes positively correlate with an elevated ratio of total cholesterol/HDL-C, thereby reflecting an excess of atherogenic non-HDL-C relative to anti-atherogenic HDL-C levels [119]. The elevation of plasma 8-isoprostanes in subjects with low HDL-C dyslipidemias is consistent with the elevation of F2-alpha-isoprostanes in apoA-I deficient mice, emphasizing the link between oxidative stress and HDL deficiency [192]. Inefficient protection of LDL by HDL against oxidative stress in the subendothelial space of the arterial wall may result in the delayed removal of oxidized lipids from LDL for elimination from the circulation through the liver mediated by hepatic SR-BI [193], further enhancing local oxidative stress and inflammation.

The impaired antioxidative activity of small, dense HDL particles in atherogenic dyslipidemia is intimately linked to the presence of individual cardiovascular risk factors, including hypertriglyceridemia, hyperglycemia, hyperinsulinemia, insulin resistance and a disequilibrium between circulating levels of atherogenic apoB-containing lipoproteins and anti-atherogenic HDL in favor of the former [118–120]. All these factors are characterized by their significant association with elevated systemic oxidative stress [194]. Such data strongly suggest, then, that a deficiency of antioxidative properties of HDL is associated with elevated oxidative stress independently of its underlying source and may result in attenuated atheroprotection by HDL.

IMPORTANT READING

Miller YI, Choi SH, Wiesner P, Fang L, Harkewicz R, Hartvigsen K, Boullier A, Gonen A, Diehl CJ, Que X et al. Oxidation-specific epitopes are danger-associated molecular patterns recognized by pattern recognition receptors of innate immunity. *Circ Res* 2011; 108: 235–248.

Kontush A, Chapman MJ. Antiatherogenic function of HDL particle subpopulations: Focus on antioxidative activities. *Curr Opin Lipidol* 2010; 21: 312–318.

10.3 ANTI-INFLAMMATORY ACTIVITY

Dyslipidemia

The potent anti-inflammatory activity of HDL can be deficient in **dyslipidemic states** (Fig. 10.3). Thus, HDL from subjects with familial hypercholesterolemia and premature coronary artery disease displays a reduced capacity to inhibit ICAM-1 expression in human umbilical vein endothelial cells compared with controls [195]. HDL from renal patients on hemodialysis and continuous ambulatory

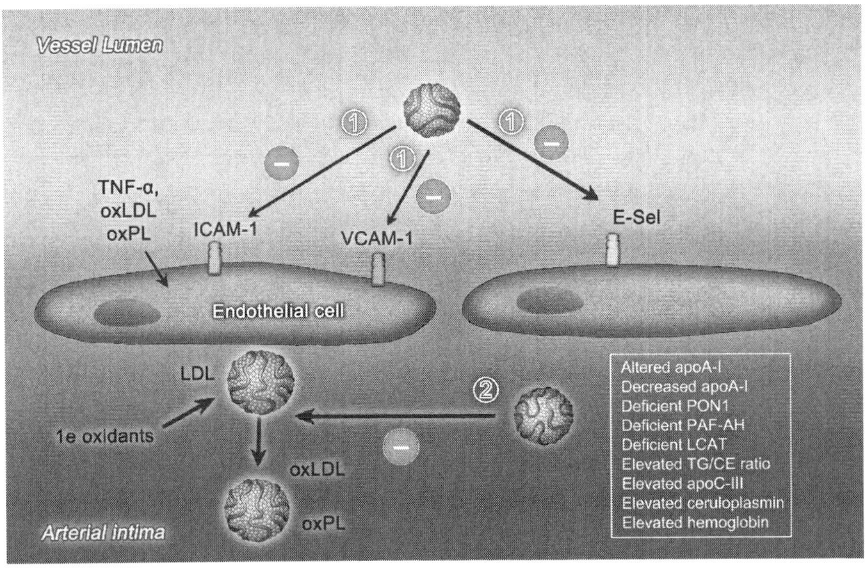

Figure 10.3 Major pathways involved in altered anti-inflammatory activity of HDL in dyslipidemia, inflammation, insulin resistance and cardiovascular disease. (1) The capacity of HDL to decrease the expression of adhesion molecules [ICAM-1, VCAM-1, E-selectin (E-Sel)] on endothelial cells activated by TNF-alpha, oxidized LDL (oxLDL) and oxidized phospholipid (oxPL) can be impaired in dyslipidemia, insulin resistance, inflammation and cardiovascular disease. (2) Equally, the capacity of HDL to inhibit the accumulation of pro-inflammatory oxidized phospholipids in LDL can be attenuated. Compositional and structural alterations in HDL, including diminished HDL content of apoA-I, modification of apoA-I by oxidation and/or glycation, deficient PON1, PAF-AH and LCAT activities, depletion of cholesteryl ester (CE) and enrichment in triglyceride (TG), apoC-III, ceruloplasmin and hemoglobin, may contribute to such functional deficiency. Inset lists HDL components potentially involved in altered anti-inflammatory activity of HDL.

peritoneal dialysis characterized by the presence of dyslipidemia is less effective than control HDL in reducing VCAM-1 expression in endothelial cells [24]. HDL particles from hemodialysis patients with end-stage renal disease are less anti-inflammatory than those of control subjects in terms of inhibiting monocyte chemotactic activity in cultured human aortic endothelial cells [196, 197]. The anti-inflammatory activity of HDL is equally diminished in patients with obstructive sleep apnea subjected to repetitive cycles of hypoxia/re-oxygenation [198]; such patients frequently exhibit Metabolic Syndrome.

Alterations in the lipidome of HDL particles parallel their functional deficiency in dyslipidemia. Compared with controls, subjects with familial hypercholesterolemia display a reduced content of phospholipids in their HDL3 accompanied by elevated serum ICAM-1 levels [195]. Moreover, HDL3 isolated from subjects with familial hypercholesterolemia and premature coronary artery disease contains increased levels of triglycerides compared with subjects with familial hypercholesterolemia but without coronary artery disease. In addition, the anti-inflammatory capacity of HDL3 particles is negatively correlated with their content of triglyceride [195].

The induction of dyslipidemia may also result in the inhibition of the anti-inflammatory function of HDL. Indeed, infusion of 20% Intralipid (15 mg/kg/h), which acutely increases serum triglyceride levels from 97 to 283 mg/dl 60 min post-infusion in healthy males, alters HDL-mediated decrease in the expression of adhesion molecules in endothelial cells in vitro [199]. Specifically, when HDL is added to TNF-alpha-stimulated human coronary artery endothelial cells and VCAM-1 and ICAM-1 expression is determined by flow cytometry, HDL post-Intralipid is significantly less anti-inflammatory compared with saline-injected controls (e.g., %VCAM-1 expression, 73% post-Intralipid vs. 54% post-saline at 8μM HDL apoA-I) [199]. In parallel, infusion of 20% Intralipid (15 mg/kg/h) induces major changes in HDL composition. Indeed, HDL triglyceride content acutely increases (5.3 vs. 3.8 wt %), whereas HDL apoA-I tends to decrease (30.5 vs. 33.9 wt %). In contrast, no change in plasma levels of triglyceride hydroperoxides, HDL-C, VCAM-1, IL-6 or Met-oxidized apoA-I is observed. There is a significant correlation between HDL triglyceride content and VCAM-1 expression mediated by HDL ($r = 0.70$, $p = 0.005$), as well as between plasma triglyceride levels and both VCAM-1 ($r = 0.71$, $p < 0.005$) and ICAM-1 ($r = 0.80$, $p < 0.005$) expression. These data show that acute **hypertriglyceridemia**, simulating the dyslipidemic post-prandial state in the absence of hyperglycemia, results in the accumulation of triglyceride-rich HDL with impaired anti-inflammatory capacity. Mechanistically, both the diminished content of **apoA-I** and the elevated content of **triglyceride** may alter HDL functionality through quantitative and qualitative changes in apoA-I; alterations in the composition of the surface lipid monolayer may also contribute to the effect [199].

It remains indeterminate as to whether deficient anti-inflammatory activity of HDL is selectively associated with a subset of HDL particles as suggested by studies with rHDL [200, 201]. Deficiency of the anti-inflammatory activity of **small, dense HDL3** is a feature of patients with familial hypercholesterolemia and coronary artery disease [195]. The specific association of the deficient antioxidative activity with small, dense HDL [118–121], together with the direct mechanistic link between anti-inflammatory and antioxidative activities via the inactivation of pro-inflammatory oxidized phospholipids [117, 125, 202], suggests that small, dense HDL may represent a major subset of the HDL plasma pool that is responsible for both potent anti-inflammatory activity under normal conditions and for deficient activity under pro-atherogenic, pro-inflammatory conditions. Consistent with this hypothesis, low concentrations of small HDL particles are associated with disease activity and coronary calcification in patients with rheumatoid arthritis [203]. This conclusion regarding the role of small HDL is further supported by the corrected anti-inflammatory properties of HDL from PLTP-deficient mice that are characterized by the prevalence of small, lipid-poor HDL particles [204].

Intriguingly, small, dense HDL3c represents the major HDL subfraction in newborns [205], consistent with an elevated need for protection against infection in early life.

Insulin Resistance

The anti-inflammatory function of HDL in insulin-resistant states remains poorly investigated. Dysfunctional, pro-inflammatory HDL is present in 50% of South Asian immigrants to the USA who are characterized by reduced insulin sensitivity and an elevated risk of coronary artery disease compared with Caucasians [206]. On logistic regression analysis, the prevalence of sub-clinical coronary artery disease using carotid intima-media thickness is associated with the presence of dysfunctional HDL.

ApoA-I glycation may lead to the loss of normal HDL function in insulin resistance. Indeed, glycated lipid-free apoA-I isolated from subjects with Type 2 diabetes does not inhibit pro-inflammatory adhesion molecule expression or neutrophil infiltration [207]. The reduced anti-inflammatory properties of non-enzymatically glycated apoA-I can be attributed to its reduced ability to inhibit cellular NFκB activation and ROS formation. In a similar fashion, non-enzymatic glycation of apoA-I *in vitro* abrogates its capacity to reduce endothelial expression of VCAM-1 and ICAM-1 and weakens the inhibition of neutrophil infiltration in a rabbit model of acute vascular inflammation induced by inserting a non-occlusive, peri-arterial carotid collar [207].

Extensive **apoA-I oxidation** by myeloperoxidase may also contribute to the deficient anti-inflammatory function of HDL [83]; underlying mechanisms may involve diminished HDL binding to SR-BI and the concurrent acquisition of specific binding to two other receptors, CD36 and scavenger receptor class A type I [62]. Furthermore, modification by myeloperoxidase is capable of conferring to HDL a pro-inflammatory activity, as demonstrated by NFκB activation and surface VCAM-1 expression on aortic endothelial cells. Such pro-atherogenic properties of meyloperoxidase-modified HDL can, in part, be compensated for by their capacity to induce cellular heme oxygenase, a protective pathway against inflammation [208].

Another pathway potentially operative in Type 2 diabetes involves **apoC-III** whose plasma levels and HDL content are frequently elevated in this disease [209, 210]. Indeed, apoC-III transcription can be induced by glucose, representing a mechanism linking hyperglycemia, hypertriglyceridemia and cardiovascular disease in Type 2 diabetes [211]. Remarkably, apoC-III-rich HDL does not reduce the adhesion of monocytes to vascular endothelial cells, whereas HDL without apoC-III decreases monocyte adhesion, suggesting that apoC-III in HDL counteracts anti-inflammatory HDL functions [212]. ApoC-III alone activates vascular endothelial cells through the activation of NFκB and induces the recruitment of monocytes to vascular endothelial cells. Furthermore, apoC-III may activate ERK and NFκB, thereby inducing pro-inflammatory adipokine expression *in vitro* and *in vivo* [213]. Specific interaction of apoC-III with cell surface proteoglycans may contribute to the observed effects [214, 215].

Inflammation

The anti-inflammatory activity of HDL becomes deficient and even transforms into *in vitro* pro-inflammatory action in the **acute phase response** [117, 202]. In contrast to functional HDL, pro-inflammatory dysfunctional HDL is unable to inactivate pro-inflammatory oxidized phospholipids produced during LDL oxidation by arterial wall cells and to prevent monocyte migration induced by oxidized LDL [117, 202] (Fig. 10.3).

CRITICAL CONTRIBUTION

The concept of HDL dysfunction derived from the transformation of anti-inflammatory activity of HDL in aortic wall cell co-cultures into pro-inflammatory activity during the acute phase response in humans and rabbits [153, 216].

Consistent with these data, the anti-inflammatory properties of HDL are impaired in **chronic inflammatory diseases**. For example, HDL from patients with active Crohn's disease and patients in remission is characterized by a reduced ability to attenuate the production of pro-inflammatory oxidized phosphatidylcholines from 1-palmitoyl-2-arachidonoyl-phosphatidylcholine compared with controls [217]. Next, anti-inflammatory effects of HDL on endothelial VCAM-1 expression, superoxide production and monocyte adhesion are diminished in antiphospholipid syndrome [218]. The presence of pro-inflammatory HDL is also associated with the increased prevalence of carotid plaque, higher intima-media thickness and accelerated subclinical atherosclerosis in systemic lupus erythematosus [219, 220]. In patients with rheumatoid arthritis, a significant relationship is observed between disease activity and inflammatory activity of HDL [221]. In a similar fashion, pro-inflammatory HDLs are present in patients infected with HIV-1 [222].

A less straightforward example of the chronic inflammatory state involves air pollution, which has been associated with significant adverse health effects leading to increased morbidity and mortality. Ultrafine particles ($<0.18\ \mu m$) present in the air enhance early atherosclerosis partly as a result of their ability to promote tissue inflammation and oxidative stress [223]. Remarkably, exposure to ultrafine particles results in an alteration of the anti-inflammatory function of HDL [223].

Mechanisms underlying deficient anti-inflammatory activity of HDL involve impaired cellular efflux of pro-inflammatory lipids and their defective inactivation (Fig. 10.3). The role of **lipid efflux** is suggested by the correlation between the formation of pro-inflammatory HDL and the impairment of HDL-mediated cholesterol efflux [224]. Another piece of evidence in support of this pathway is provided by data from mice. Specifically, ABCG1 deficiency in aortic endothelial cells decreases cholesterol efflux to HDL and activates intracellular STAT3 signaling, thereby enhancing monocyte-endothelial interactions and vascular inflammation [225]. Furthermore, aortic endothelial cells from ABCG1-deficient mice display a high expression of ICAM-1 and E-selectin [225].

Impaired **inactivation of pro-inflammatory oxidized phospholipids** is in agreement with the depletion of apoA-I and hydrolytic enzymes in HDL during the acute phase response (Fig. 10.3). Indeed, the formation of pro-inflammatory HDL correlates with decreases in the activities of various HDL-associated enzymes, such as **PON1, PAF-AH** and **LCAT**, which are replaced by acute-phase proteins, such as **SAA** and **ceruloplasmin**; the content of SAA and ceruloplasmin in HDL increases during an inflammatory response [117, 202]. Copper-containing ceruloplasmin may provide a source of transition metals for oxidative reactions, thereby accounting for the enhancement of LDL oxidative modification by acute-phase HDL [117, 202]. Indeed, purified ceruloplasmin can induce LDL oxidation because of its content of a loosely bound copper atom [226, 227]; the *in vitro* enrichment of HDL with ceruloplasmin abrogates the ability of HDL to inhibit LDL modification in aortic wall cell co-cultures [153].

METHODOLOGIC NOTE

Cell-free assays to determine anti-inflammatory activity [228, 229] largely reflect the capacity of HDL to inactivate oxidized phospholipids and are therefore mechanistically similar to the assay of the capacity of HDL to protect LDL from oxidative damage by one-electron oxidants [230].

Chronic inflammatory diseases also feature remodeling of the HDL proteome. For example, **SAA** concentrations are elevated in plasma of patients with Crohn's disease in remission relative to controls [217], whereas **PON1** activity is decreased in antiphospholipid syndrome [218]. In addition, patients with systemic lupus erythematosus display reduced PON1 activity and elevated titres of autoantibodies against HDL and apoA-I accompanied by increased circulating levels of NO metabolites, VCAM-1 and ICAM-1 [231]. Similarly, anti-HDL and anti-apoA-I antibodies are elevated in primary antiphospholipid syndrome compared with controls [232]. Anti-HDL antibodies may hamper the anti-inflammatory and antioxidative actions of HDL, favoring low-grade inflammation and enhanced oxidative stress. Indeed, the presence of anti-apoA-I IgG is associated with elevated levels of oxidized LDL in acute coronary syndrome, emphasizing the role of humoral autoimmunity as a mediator of inflammation and coronary atherogenesis [233].

Alterations in the HDL proteome in chronic inflammatory disease may expand to other components. Indeed, compared with patients with anti-inflammatory HDL, patients with rheumatoid arthritis possessing pro-inflammatory HDL display an elevated content of haptoglobin, hemoglobin and myeloperoxidase, and reduced LCAT activity [221].

Finally, the capacity of HDL3 to inhibit TNF-alpha-mediated induction of adhesion molecules and MCP-1 in human endothelial cells can be inhibited by HDL treatment with 15-lipoxygenase, consistent with the pro-inflammatory role of **oxidized lipids** transported by HDL [234].

Cardiovascular Disease

Pioneering data on the functional deficiency of HDL in cardiovascular disease were obtained by Benjamin Ansell and colleagues who documented that HDL from patients with **coronary heart disease** and normal or elevated HDL-C levels is pro-inflammatory in both cell culture and cell-free fluorescent assays [229]. In support of these data, studies from the same laboratory reveal that HDL from mice that are genetically predisposed to diet-induced atherosclerosis completely loses its anti-inflammatory properties and becomes pro-inflammatory following pro-atherogenic interventions [156, 157, 162]. Specifically, pro-inflammatory HDL can be detected in atherosclerotic apoE -/- mice [157]. In addition, transgenic mice over-expressing apoA-II feature dyslipidemia, possess pro-inflammatory HDL and develop atherosclerosis on a chow diet [158, 159]. The existence of dysfunctional HDL *in vivo* is further consistent with the abundance of apoA-I in carotid plaques from symptomatic subjects where it co-localizes with key inflammatory mediators, suggesting insufficient protection against atherosclerosis-mediated inflammation [235].

MAJOR CONTROVERSY

Are HDL particles functionally deficient and do they, in part, retain their normal anti-inflammatory function, or do they completely lose their functions under pro-inflammatory and pro-atherogenic conditions?

High plasma levels of **homocysteine** represent an independent risk factor for cardiovascular disease. Thus, HDL from hyperhomocysteinemic subjects possesses reduced anti-inflammatory activity in TNF-alpha-activated endothelial cells in terms of the attenuation of the suppressive effect on IL-6 growth-related oncogene-alpha release [110].

Interestingly, HDL PON1 activity is not decreased in coronary heart disease patients possessing pro-inflammatory HDL [229], indicating that factors other than PON1 determine anti-inflammatory HDL dysfunction *in vitro*. In contrast, serum PON1 activity is reduced in hyperhomocysteinemic subjects [110]. In a study by Ansell et al. plasma triglyceride levels were markedly elevated (+76%) in coronary heart disease patients compared with controls [229], suggesting that HDL enrichment in triglyceride may have significantly contributed to the formation of dysfunctional HDL, as observed in atherogenic low HDL-C dyslipidemias [118–121].

Furthermore, mass spectrometric analyses of pro-inflammatory HDL in mouse models of atherosclerosis identify the presence of **hemoglobin alpha and beta chains** with no apparent post-translational modification [236]. In such HDL, hemoglobin is predominantly in the oxyhemoglobin form with distinct physical and chemical properties. Specifically, oxyhemoglobin-containing pro-inflammatory HDL potently consumes nitric oxide and contracts arterial vessels *ex vivo*. Hemoglobin is also found differentially associated with HDL from coronary heart disease patients compared with healthy controls [236]. Mechanistically, hemoglobin can be tethered to HDL by the haptoglobin 2-2 protein which binds to apoA-I [46].

Moreover, not only haemoglobin, but also its scavenger proteins, haptoglobin and hemopexin, are increased in apoA-I-containing HDL, both in mouse models of hyperlipidemia and in coronary heart disease patients when compared with wild-type mice and healthy donors, respectively [237]. The association of hemoglobin, haptoglobin and hemopexin proteins with HDL positively correlates with inflammatory properties of HDL and systemic inflammation in coronary heart disease patients [237]. Interestingly, HDL from haptoglobin-null mice under atherogenic conditions does not accumulate hemoglobin and is anti-inflammatory, suggesting that haptoglobin is required for the association of hemoglobin with HDL [237]. Hemoglobin-haptoglobin complexes, therefore, may, in part, regulate the anti/pro-inflammatory properties of HDL.

Pathophysiologic Relevance

As part of the innate immune system, HDL appears to have evolved to increase inflammation in the presence of an acute phase response but to inhibit inflammation under normal circumstances. Indeed, HDL increases the production of IL-10, an anti-inflammatory cytokine, by monocytes in the basal state but decreases it in the presence of IL-1-beta, *i.e.*, under inflammatory conditions [238]. The pro-inflammatory and pro-oxidative rearrangement of HDL particles and formation of LDL-derived oxidized phospholipids have been hypothesized to form part of an evolutionary conserved mechanism of non-specific innate immunity aimed at protecting against infection [117, 202]. Such an **innate inflammatory response** may include subnormal levels of HDL-C, increased HDL-triglyceride content and altered HDL apolipoprotein composition, all of which impair cholesterol efflux capacity, as well as antioxidative and anti-inflammatory activities of HDL particles. These modifications in HDL may be aimed at redirecting cholesterol from the liver to immune cells, particularly macrophages, during infection in order to enhance cellular functions [93].

Such a response to acute infection or injury can be advantageous in the short term, but may become maladaptive in the long term. A sustained response that is not able to repair the injury, such as an emerging atherosclerotic plaque (which can be considered as a local inflammation [239]), may lead to chronic alteration in plasma lipid levels; such an adaptation may become harmful, accelerating the formation of atherosclerotic lesions [240]. The presence of anti-inflammatory-deficient HDL can facilitate, or even trigger, the accumulation of LDL-derived pro-inflammatory, pro-atherogenic oxidized phospholipids [229]. A failure of HDL to down-regulate inflammatory processes under normal circumstances may thereby result in the development of cardiovascular disease. This mechanism is consistent with the hypothesis that accelerated development of atherosclerosis in old age is related to increased inflammation and concomitant endothelial dysfunction during early life [241–243]. Within this concept, classical lipid changes associated with Metabolic Syndrome (low HDL-C and elevated triglyceride levels) are envisioned as a highly conserved evolutionary response aimed at tissue repair [240].

OPEN QUESTION

Do multiple pathways lead to the loss of the anti-inflammatory function of HDL specific to individual metabolic diseases or do they operate concomitantly?

In support of this hypothesis, all major alterations in HDL composition that lead to attenuated anti-atherogenic activities (depletion in apoA-I, PON1 and LCAT, and an increase in triglyceride and SAA) are observed during inflammation and in the acute phase response [163, 240]. It is interesting, in this regard, that circulating endotoxin levels are associated with atherosclerosis [244]. Increased levels of endotoxin may, therefore, underlie major abnormalities present in metabolic diseases associated with elevated cardiovascular risk, representing a missing link among metabolic abnormalities clustering in Metabolic Syndrome [244]. Impairment of the HDL capacity to bind LPS represents an interesting mechanism potentially contributing to elevated levels of endotoxin in metabolic diseases, which remains to be studied (see 10.6 of chapter 10).

IMPORTANT READING

Navab M, Reddy ST, Van Lenten BJ, Anantharamaiah GM, Fogelman AM. The role of dysfunctional HDL in atherosclerosis. *J Lipid Res* 2009; 50: S145–S149.

Navab M, Reddy ST, Van Lenten BJ, Fogelman AM. HDL and cardiovascular disease: atherogenic and atheroprotective mechanisms. *Nat Rev Cardiol* 2011; 8: 222–232.

10.4 CYTOPROTECTIVE ACTIVITY

Data on the potential deficiency of the cytoprotective properties of HDL are scarce. The potent **anti-apoptotic activity** of small HDL3c particles is reduced (−35%) in **Metabolic Syndrome** patients relative to normolipidemic controls [47]. Loss of the vasculoprotective properties of HDL in coronary artery disease is suggested by the inverse correlation between circulating levels of endothelial progenitor cells and HDL-C [245].

Mechanisms underlying defective cytoprotection by HDL may be related to concomitant alterations in the properties of HDL and of arterial wall cells. The attenuated anti-apoptotic activity of HDL3c in Metabolic Syndrome is correlated to abdominal obesity, atherogenic dyslipidemia and systemic oxidative stress, suggesting abnormal physicochemical properties of HDL, secondary to metabolic alterations [47]. Altered physicochemical properties of HDL3c in Metabolic Syndrome relevant to defective cytoprotective activity involve the depletion of apoA-I and cholesteryl esters, as well as enrichment in triglycerides [47].

Depletion of apoA-I, a pivotal HDL component accounting for the protection of endothelial cells from apoptosis, can critically affect HDL-mediated cytoprotection. Specifically, **SR-BI**-mediated interaction of apoA-I-depleted HDL with endothelial cells can be impaired. Moreover, apoA-I functionality can be undermined by post-translational modifications, primarily by **oxidation** and/or **glycation**. Indeed, apoA-I modified by myeloperoxidase is capable of decreasing the anti-apoptotic activity of HDL as a result of diminished HDL binding to SR-BI and elevated binding to CD36 and scavenger receptor class A type I [62]. Furthermore, glycoxidation can convert HDL into cytotoxin [246], although HDL is more resistant to such modification than LDL [247]. Thus, HDL modified by incubation with glucose *in vitro* induces apoptosis in human aortic endothelial cells [246]. Such glycoxidized HDL induces cell shrinkage, membrane blebbing and nuclear fragmentation in endothelial cells; these effects are dose-dependent relative to the glucose used for the modification of HDL. Increases in caspase-3 activity, expression of caspase-9, release of cytochrome c into the cytosol and expression of Bax and Bad follow, consistent with mitochondrial dysfunction as a key underlying pathway [246].

Core lipid components of HDL (cholesteryl ester, triglyceride) may, theoretically, contribute to defective cytoprotective activity via the altered conformation of apoA-I and impaired HDL binding to the cell surface. The involvement of S1P appears less probable as HDL contains normal levels of this bioactive lipid in patients with Metabolic Syndrome [47]. In contrast, HDL appears to display a reduced content of S1P in patients with coronary artery disease [248].

Deficiency of other HDL components, such as **PON1**, may also result in defective cytoprotection. Interestingly, PON1 deficiency in mice is associated with reduced SR-BI expression in macrophages and decreased HDL binding to the cells [249, 250]. These effects appear to be mediated by lysophosphatidylcholine, which can be generated as a result of PON1 action on oxidized phospholipids via the activation of ERK1/2 and PI3K and result in the loss of HDL-mediated cytoprotection against apoptosis. In contrast, cholesterol efflux from macrophages remains unaffected by the deletion of PON1. Such a loss of vasculoprotection may contribute to accelerated atherosclerosis observed in PON1-deficient mice [249].

Other pathways potentially contributing to the deficient cytoprotection by HDL involve defective interaction between HDL and abnormal cellular receptors, which may provoke insufficient cellular efflux of cytotoxic lipids; HDL particles *per se* may be normal in this case. Such a scenario accounts for the induction of apoptosis in macrophages from mice deficient in ABCG1, which is involved in the efflux of cytotoxic oxysterols [251]. Finally, the deficient antioxidative activity of small HDL3c, together with inefficient inhibition of intracellular ROS production, may exacerbate the defective cytoprotective properties of small HDL3 in Metabolic Syndrome [47].

IMPORTANT READING

Ferretti G, Bacchetti T, Negre-Salvayre A, Salvayre R, Dousset N., Curatola, G. Structural modifications of HDL and functional consequences. *Atherosclerosis* 2006; 184: 1–7.

10.5 VASODILATORY ACTIVITY

Data on the deficient vasodilatory activity of HDL in subjects under increased cardiovascular risk or in those with established disease are rapidly accumulating. The pioneering studies of Laurence Duvillard and colleagues have established that the ability of normolipidemic HDL to counteract the inhibition of endothelium-dependent relaxation induced by oxidized LDL in rabbit aorta rings is lost in patients with **Type 1** and **Type 2 diabetes**, as well as in dyslipidemic **abdominally obese** subjects [252–254]. Consistent with these findings, the beneficial effects on endothelial NO production, superoxide generation, NADPH oxidase activity and endothelium-dependent vasodilation in aortic rings, as well as on endothelial progenitor cell-mediated endothelial repair in a mouse model, are impaired in HDL from patients with Type 2 diabetes [255]. Moreover, HDL from patients with **stable coronary artery disease** and **acute coronary syndrome** is largely unable to stimulate endothelial NO production [256]. In parallel, intracellular signaling induced by HDL via endothelial Akt, eNOS-Ser1177 phosphorylation and eNOS-Thr495 dephosphorylation is abrogated and the endothelial-binding capacity of HDL is reduced. Furthermore, endothelial expression and membrane translocation of the cationic amino acid transporter-1 is impaired in the presence of HDL from patients with coronary disease compared with controls, whereas the activation of endothelial protein kinase C beta 2 is enhanced [257, 258]. Finally, HDL from subjects with **antiphospholipid syndrome**, which is characterized by increased thrombogenicity and pregnancy morbidity, is unable to stimulate endothelial NO production and, rather, inhibits it by −20% [218].

CRITICAL CONTRIBUTION

Loss of HDL capacity to counteract inhibition of endothelium-dependent relaxation induced by oxidized LDL in rabbit aorta rings in patients with Type 1 and Type 2 diabetes, as well as in dyslipidemic abdominally obese subjects [252–254].

Such defective vasodilatory activity of HDL may originate from multiple mechanisms (Fig. 10.4). First, HDL binding to endothelial cells can be diminished, as observed in patients with Type 2 diabetes [255]. This deficiency may reflect the decreased HDL content of **apoA-I** and/or its altered conformation as a result of changes in the core lipid composition of HDL. In support of this mechanism, the HDL content of apoA-I is reduced (−12%) and the triglyceride/cholesteryl ester ratio elevated (+36%) in abdominally obese subjects; the vasodilatory activity of HDL is positively correlated with HDL content of apoA-I and negatively with the triglyceride/cholesteryl ester ratio [253]. Furthermore, the HDL content of **triglyceride** is significantly elevated in patients with Type 2 diabetes compared with control subjects and is highly inversely correlated with the vasodilatory activity of HDL [252].

Post-translational modifications of apoA-I by **glycation** and/or **oxidation** may also play a role in defective HDL functionality. Indeed, the HDL content of fructosamine is elevated (+32%) in patients with Type 1 diabetes compared with normolipidemic controls in the absence of differences in HDL total lipid composition, mean size and PON1 activity [254]. In contrast, the HDL content of fructosamine is not correlated with the vasodilatory activity of HDL. Mechanistically, defective cellular binding of HDL containing modified apoA-I may result in the inefficient efflux of toxic **oxidized lipids** that induce vasoconstriction. The deleterious role of oxidized lipids and/or proteins is supported by the finding of elevated lipid peroxidation in HDL in Type 2 diabetes which is accompanied by an increase in HDL-associated myeloperoxidase [255].

Alterations in HDL-associated enzymes may also be related to the defective capacity of HDL to induce vasorelaxation. Indeed, **PON1** activity is diminished in patients with stable coronary artery disease and acute coronary syndrome (−62% and —70% vs. controls, respectively), whereas PON1 content is increased, suggesting inactivation of the enzyme [256]. The role of PON1 is highlighted by the observation that inhibition of PON1 activity prevents stimulation of Akt/eNOS phosphorylation, endothelial NO production and NO-dependent vasodilation by normolipidemic control HDL [256]. In addition, PON1 activity is significantly reduced in subjects with antiphospholipid syndrome compared with controls, and is inversely correlated with carotid intima-media thickness and pulse wave velocity [218]. Remarkably, both carotid intima-media thickness and pulse wave velocity are elevated in antiphospholipid syndrome compared with the control group, whereas flow-mediated dilatation is reduced.

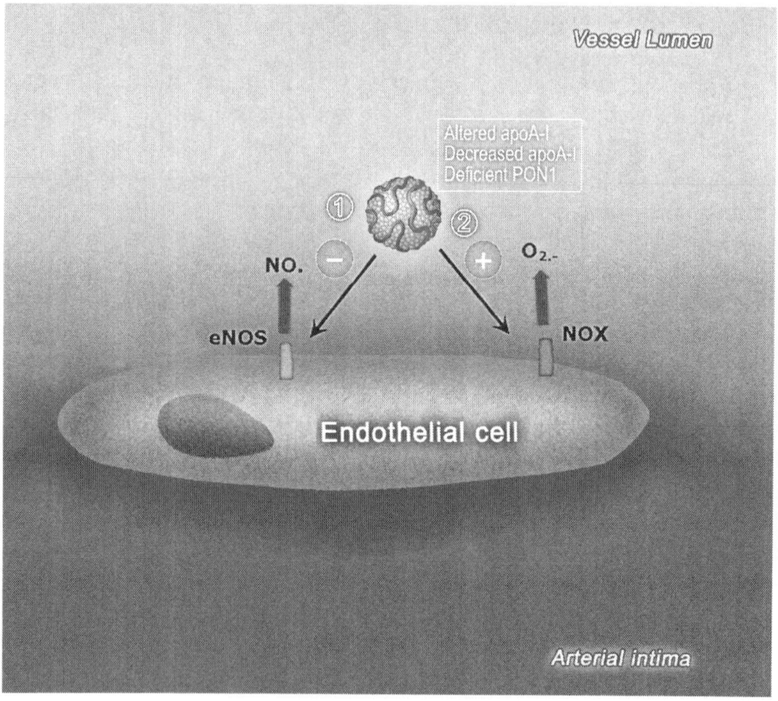

Figure 10.4 Major pathways involved in altered vasodilatory activity of HDL in dyslipidemia, inflammation, insulin resistance and cardiovascular disease. (1) The capacity of HDL to enhance NO production by eNOS in endothelial cells can be impaired in insulin resistance, chronic inflammation and cardiovascular disease. (2) HDL-mediated inhibition of cellular superoxide production by NAADPH oxidase (NOX) can be also attenuated. Compositional and structural alterations in HDL, including diminished HDL content of apoA-I, modification of apoA-I by oxidation and/or glycation, and deficient PON1 activity, may contribute to such functional deficiency. Inset lists HDL components potentially involved in altered anti-inflammatory activity of HDL.

10.6 ANTI-INFECTIOUS ACTIVITY

It remains unknown as to whether anti-infectious activities of HDL can be altered *in vivo*; indirect evidence suggests that this may be the case. Indeed, LPS is preferentially associated with apoB-containing lipoproteins in periodontitis, consistent with the decreased capacity of HDL to bind and inactivate endotoxin [259].

In addition, SR-BI-null mice display an uncontrollable inflammatory cytokine response and a markedly higher lethality rate than control mice in response to LPS, despite elevated plasma HDL-C levels [260]. These data are consistent with loss of the capacity of HDL to facilitate endotoxin inactivation. Plasma from control and SR-BI-null mice, however, exhibits a similar ability to neutralize

LPS, whereas SR-BI-null mice show decreased plasma clearance of LPS through the liver and hepatocytes compared with normal mice [260]. The essential role of SR-BI in facilitating LPS hepatic clearance therefore appears to better explain the elevated sensitivity of SR-BI-null mice to LPS compared with the presence of anti-infectiously dysfunctional HDL.

It is worth mentioning that human HDL is the target of **serum opacity factor**, a virulence determinant of *Streptococcus pyogenes* that turns serum opaque [261]. Serum opacity factor interacts with HDL in human blood by binding to apoA-I and apoA-II and causing the release of HDL lipid cargo, which coalesces to form lipid droplets, resulting in opacification. Recombinant serum opacity factor catalyzes the transfer of nearly all neutral lipids of up to 400,000 HDL particles into a single, large cholesteryl ester-rich, apoE-containing microemulsion, while releasing lipid-free apoA-I [262, 263]. Desorption of lipid-free labile apoA-I is the rate-limiting step for HDL opacification. Such disruption of HDL may attenuate its anti-infectious activity and contribute to the pathogenesis of group A streptococcal infections [261].

10.7 ANTITHROMBOTIC ACTIVITY

There is virtually no data available on the deficiency of the antithrombotic function of HDL in cardiometabolic disease. Recently, HDL from patients with **coronary artery disease** and **acute coronary syndrome** were reported to be unable to reduce thrombin-induced TF expression in endothelial cells and to increase TFPI expression and tPA release [264]. Furthermore, HDL from coronary artery disease and acute coronary syndrome patients, but not from healthy controls, enhances PAI-1 expression, thereby becoming prothrombotic. Mechanistically, such dysfunctionality of HDL particles can be related to their inability to stimulate **NO production** in endothelial cells, as demonstrated in the experiments with NO donors and inhibitors of NO formation [264]. Pathways underlying the observed effects may, therefore, resemble those implicated in the defective vasodilatory action of HDL, which involves stimulation of NO production as a key element.

Data obtained *in vitro* suggest that **oxidative modification** of HDL particles may play a role in the loss of their antithrombotic activity. Indeed, oxidized HDL3, but not native HDL3, increases PAI-1 expression in, and antigen release from, endothelial cells [265]. Molecular mechanisms underlying this phenomenon involve ERK 1/2 and p38 MAPK activation and stabilization of PAI-1 mRNA. Furthermore, HDL oxidized by HOCl raises intracellular calcium concentrations and induces surface expression of P-selectin in human platelets, accompanied by platelet aggregation [266]. The observed effects are dose-dependent and can be blocked by an antibody directed against the lipoprotein-binding domain of the platelet CD36 receptor. Finally, HOCl-oxidized HDL triggers platelet aggregation, acting via specific binding to CD36 [267].

CLINICAL SUMMARY

Complete, or partial deficiency, of multiple anti-atherogenic activities of HDL can be detected *ex vivo* in many clinical conditions associated with elevated cardiovascular risk; however, it remains to be shown whether such deficiency occurs *in vivo* and contributes to accelerated atherogenesis in humans.

REFERENCES

1. Miller GJ, Miller NE. Plasma-high-density-lipoprotein concentration and development of ischaemic heart-disease. *Lancet* 1975; 1: 16–19.
2. Nakanishi S, Vikstedt R, Soderlund S, Lee-Rueckert M, Hiukka A, Ehnholm C, Muilu M, Metso J, Naukkarinen J, Palotie L et al. Serum, but not monocyte macrophage foam cells derived from low HDL-C subjects, displays reduced cholesterol efflux capacity. *J Lipid Res* 2009; 50: 183–192.
3. Treguier M, Briand F, Boubacar A, Andre A, Magot T, Nguyen P, Krempf M, Sulpice T, Ouguerram K. Diet-induced dyslipidemia impairs reverse cholesterol transport in hamsters. *Eur J Clin Invest* 2011; 41: 92–928.
4. Brites FD, Bonavita CD, De Geitere C, Cloes M, Delfly B, Yael MJ, Fruchart J, Wikinsk, RW, Castro GR. Alterations in the main steps of reverse cholesterol transport in male patients with primary hypertriglyceridemia and low HDL-cholesterol levels. *Atherosclerosis* 2000; 152: 181–192.
5. Harchaoui KE, Franssen R, Hovingh GK, Bisoendial RJ, Stellaard F, Kuipers F, Kastelein JJP, Kuivenhoven JA, Stroes ESG, Groen AK. Reduced fecal sterol excretion in subjects with familial hypoalphalipoproteinemia. *Atherosclerosis* 2009; 207: 614–616.
6. Khera AV, Rodrigues A, de la Llera-Moya M, Rothblat GH, Rader DJ. Serum cholesterol efflux capacity, a measure of HDL-C quality, varies according to coronary artery disease status independently of HDL-C quantity. *Circulation* 2009; 120 (Suppl.): S469.
7. Khera AV, Cuchel M, de la Llera-Moya M, Rodrigues A, Burke MF, Jafri K, French BC, Phillips JA, Mucksavage ML, Wilensky RL et al. Cholesterol efflux capacity, high-density lipoprotein function, and atherosclerosis. *N Engl J Med* 2011; 364: 127–135.
8. Daum U, Leren TP, Langer C, Chirazi A, Cullen P, Pritchard PH, Assmann G, von Eckardstein A. Multiple dysfunctions of two apolipoprotein A-I variants, apoA-I(R160L)Oslo and apoA-I(P165R), that are associated with hypoalphalipoproteinemia in heterozygous carriers. *J Lipid Res* 1999; 40: 486–494.
9. Han H, Sasaki J, Matsunaga A, Hakamata H, Huang W, Ageta M, Taguchi T, Koga T, Kugi M, Horiuchi S, Arakawa K. A novel mutant, ApoA-I Nichinan (Glu235–> 0), is associated with low HDL cholesterol levels and decreased cholesterol efflux from cells. *Arterioscler Thromb Vasc Biol* 1999; 19: 1447–1455.
10. Huang W, Sasaki J, Matsunaga A, Han H, Li W, Koga T, Kugi M, Ando S, Arakawa K. A single amino acid deletion in the carboxy terminal of apolipoprotein A-I

impairs lipid binding and cellular interaction. *Arterioscler Thromb Vasc Biol* 2000; 20: 210–216.

11. Kono M, Tanaka T, Tanaka M, Vedhachalam C, Chetty PS, Nguyen D, Dhanasekaran P, Lund-Katz S, Phillips MC, Saito H. Disruption of the C-terminal helix by single amino acid deletion is directly responsible for impaired cholesterol efflux ability of apolipoprotein A-I Nichinan. *J Lipid Res* 2010; 51: 809–818.
12. Marcil M, Yu L, Krimbou L, Boucher B, Oram JF, Cohn JS, Genest J Jr. Cellular cholesterol transport and efflux in fibroblasts are abnormal in subjects with familial HDL deficiency. *Arterioscler Thromb Vasc Biol* 1999; 19: 159–169.
13. Franceschini G, Calabresi L, Chiesa G, Parolini C, Sirtori CR, Canavesi M, Bernini F. Increased cholesterol efflux potential of sera from ApoA-I Milano carriers and transgenic mice. *Arterioscler Thromb Vasc Biol* 1999; 19: 1257–1262.
14. Chiesa G, Sirtori CR. Apolipoprotein A-I Milano: current perspectives. *Curr Opin Lipidol* 2003; 14: 159–163.
15. van Dam, MJ, de Groot E, Clee SM, Hovingh GK, Roelants R, Brooks-Wilson A, Zwinderman AH, Smit AJ, Smelt AHM, Groen AK. Association between increased arterial-wall thickness and impairment in ABCA1-driven cholesterol efflux: an observational study. *Lancet* 2002; 359: 37–42.
16. Uehara Y, Tsuboi Y, Zhang B, Miura S, Baba Y, Higuchi MA, Yamada T, Rye KA, Saku K. POPC/apoA-I discs as a potent lipoprotein modulator in Tangier disease. *Atherosclerosis* 2008; 197: 283–289.
17. Nagao K, Zhao Y, Takahashi K, Kimura Y, Ueda K. Sodium taurocholate-dependent lipid efflux by ABCA1: effects of W590S mutation on lipid translocation and apolipoprotein A-I dissociation. *J Lipid Res* 2009; 50: 1165–1172.
18. Pisciotta L, Bocchi L, Candini C, Sallo R, Zanotti I, Fasano T, Chakrapani A, Bates T, Bonardi R, Cantafora A et al. Severe HDL deficiency due to novel defects in the ABCA1 transporter. *J Intern Med* 2009; 265: 359–372.
19. Cameron J, Ranheim T, Halvorsen B, Kulseth MA, Leren TP, Berge KE. Tangier disease caused by compound heterozygosity for ABCA1 mutations R282X and Y1532C. *Atherosclerosis* 2010; 209: 163–166.
20. Holleboom AG, Franssen R, Jakulj L, Decaris J, Vergeer M, Koetsveld J, Hellerstein MK, Kastelein JJ, Groen AK, Turner SM et al. *In vivo* tissue cholesterol efflux is impaired in carriers of mutations in APOA1 and ABCA1. *Circulation* 2010; 122 (Suppl.): A17881.
21. Baldan A, Gomes AV, Ping P, Edwards PA. Loss of ABCG1 results in chronic pulmonary inflammation. *J Immunol* 2008; 180: 3560–3568.
22. Vergeer M, Korporaal SJ, Franssen R, Meurs I, Out R, Hovingh GK, Hoekstra M, Sierts JA, Dallinga-Thie GM, Motazacker MM et al. Genetic variant of the scavenger receptor BI in humans. *N Engl J Med* 2011; 364: 136–145.
23. Yang Y, Jiang Y, Wang Y, An W. Suppression of ABCA1 by unsaturated fatty acids leads to lipid accumulation in HepG2 cells. *Biochimie* 2010; 92: 958–963.
24. Jurek A, Turyna B, Kubit P, Klein A. The ability of HDL to inhibit VCAM-1 expression and oxidized LDL uptake is impaired in renal patients. *Clin Biochem* 2008; 41: 1015–1018.
25. Zuo Y, Yancey P, Castro I, Khan W, Motojima M, Ichikawa I, Fogo AB, Linton MF, Fazio S, Kon V. Renal dysfunction potentiates foam cell formation by repressing ABCA1. *Arterioscler Thromb Vasc Biol* 2009; 29: 1277–1282.

26. Moradi H, Yuan J, Ni Z, Norris K, Vaziri ND. Reverse cholesterol transport pathway in experimental chronic renal failure. *Am J Nephrol* 2009; 30: 147–154.

27. Attia N, Ramaharo A, Paul JL, Cambillau M, Beaune P, Grynberg A, Simon A, Fournier N. Enhanced removal of cholesterol from macrophage foam cells to serum from type IV hypertriglyceridemic subjects. *Atherosclerosis* 2008; 198: 49–56.

28. Fournier N, Atger V, Cogny A, Vedie B, Giral P, Simon A, Moatti N, Paul JL. Analysis of the relationship between triglyceridemia and HDL-phospholipid concentrations: consequences on the efflux capacity of serum in the Fu5AH system. *Atherosclerosis* 2001; 157: 315–323.

29. Dullaart RP, Groen AK, Dallinga-Thie GM, de Vries R, Sluiter WJ, van Tol A. Fibroblast cholesterol efflux to plasma from metabolic syndrome subjects is not defective despite low high-density lipoprotein cholesterol. *Eur J Endocrinol* 2008; 158: 53–60.

30. Uint L, Sposito A, Brandizzi LI, Yoshida VM, Maranhao RC, Luz PL. Cellular cholesterol efflux mediated by HDL isolated from subjects with low HDL levels and coronary artery disease. *Arq Bras Cardiol* 2003; 81: 39–41, 35–38.

31. Skeggs JW, Morton RE. LDL and HDL enriched in triglyceride promote abnormal cholesterol transport. *J Lipid Res* 2002; 43: 1264–1274.

32. Moerland M, Samyn H, van Gent T, Jauhiainen M, Metso J, van Haperen R, Grosveld F, van Tol A, de Crom R. Atherogenic, enlarged, and dysfunctional HDL in human PLTP/apoA-I double transgenic mice. *J Lipid Res* 2007; 48: 2622–2631.

33. Ishigami M, Yamashita S, Sakai N, Arai T, Hirano K, Hiraoka H, Kameda-Takemura K, Matsuzawa Y. Large and cholesteryl ester-rich high-density lipoproteins in cholesteryl ester transfer protein (CETP) deficiency can not protect macrophages from cholesterol accumulation induced by acetylated low-density lipoproteins. *J Biochem* 1994; 116: 257–262.

34. Kinoshita M, Fujita M, Usui S, Maeda Y, Kudo M, Hirota D, Suda T, Taki M, Okazaki M, Teramoto T. Scavenger receptor type BI potentiates reverse cholesterol transport system by removing cholesterol ester from HDL. *Atherosclerosis* 2004; 173: 197–202.

35. Greene DJ, Skeggs JW, Morton RE. Elevated triglyceride content diminishes the capacity of high density lipoprotein to deliver cholesteryl esters via the scavenger receptor class B type I (SR-BI). *J Biol Chem* 2001; 276: 4804–4811.

36. Palmer AM, Murphy N, Graham A. Triglyceride-rich lipoproteins inhibit cholesterol efflux to apolipoprotein (apo) A1 from human macrophage foam cells. *Atherosclerosis* 2004; 173: 27–38.

37. Adorni MP, Zimetti F, Billheimer JT, Wang N, Rader DJ, Phillips MC, Rothblat GH. The roles of different pathways in the release of cholesterol from macrophages. *J Lipid Res* 2007; 48: 2453–2462.

38. Zhou H, Tan KC, Shiu SW, Wong Y. Increased serum advanced glycation end products are associated with impairment in HDL antioxidative capacity in diabetic nephropathy. *Nephrol Dial Transplant* 2008; 23: 927–933.

39. Cavallero E, Brites F, Delfly B, Nicolaiew N, Decossin C, De Geitere C, Fruchart JC, Wikinski R, Jacotot B, Castro G. Abnormal reverse cholesterol transport in controlled type II diabetic patients. Studies on fasting and postprandial LpA-I particles. *Arterioscler Thromb Vasc Biol* 1995; 15: 2130–2135.

40. Asleh R, Blum S, Kalet-Litman S, Alshiek J, Miller-Lotan R, Asaf R, Rock W, Aviram M, Milman U, Shapira C et al. Correction of HDL dysfunction in individuals with diabetes and the haptoglobin 2-2 genotype. *Diabetes* 2008; 57: 2794–2800.
41. Zhou H, Tan KC, Shiu SW, Wong Y. Cellular cholesterol efflux to serum is impaired in diabetic nephropathy. *Diabetes Metab Res Rev* 2008; 24: 617–623.
42. Attia N, Fournier N, Vedie B, Cambillau M, Beaune P, Ziegler O, Grynberg A, Paul J-L, Guerci B. Impact of android overweight or obesity and insulin resistance on basal and postprandial SR-BI and ABCA1-mediated serum cholesterol efflux capacities. *Atherosclerosis* 2010; 209: 422–429.
43. Aron-Wisnewsky J, Julia Z, Poitou C, Bouillot JL, Basdevant A, Chapman MJ, Clement K, Guerin M. Effect of bariatric surgery-induced weight loss on SR-BI, ABCG1-, and ABCA1-mediated cellular cholesterol efflux in obese women. *J Clin Endocrinol Metab* 2011; 96: 1151–1159.
44. Gantman A, Fuhrman B, Aviram M, Hayek T. High glucose stimulates macrophage SR-BI expression and induces a switch in its activity from cholesterol efflux to cholesterol influx. *Biochem Biophys Res Commun* 2010; 391: 523–528.
45. Julia Z, Duchene E, Fournier N, Bellanger N, Chapman MJ, Le Goff W, Guerin M. Postprandial lipemia enhances the capacity of large HDL2 particles to mediate free cholesterol efflux via SR-BI and ABCG1 pathways in type IIB hyperlipidemia. *J Lipid Res* 2010; 51: 3350–3358.
46. Farbstein D, Kalet-Litman S, Pollak M, Blum S, Asaf R, Schwartz A, Asleh R, Perry N, Barkay I, Levy A. Decreased LCAT protein and increased activated complement fragments in the dysfunctional HDL of individuals with diabetes and the Hp 2-2 genotype. *Circulation* 2009; 120 (Suppl.): S525–c–S526.
47. de Souza JA, Vindis C, Hansel B, Negre-Salvayre A, Therond P, Serrano CV Jr, Chantepie S, Salvayre R, Bruckert E, Chapman MJ et al. Metabolic syndrome features small, apolipoprotein A-I-poor, triglyceride-rich HDL3 particles with defective anti-apoptotic activity. *Atherosclerosis* 2008; 197: 84–94.
48. Kontush A, Chapman MJ. Why is HDL functionally deficient in type 2 diabetes ? *Curr Diab Rep* 2008; 8: 51–59.
49. Attia N, Domingo N, Lorec AM, Nakbi A, Hammami S, Ben Hamda K, Portugal H, Lairon D, Hammami M, Chanussot F. Reverse modulation of the HDL anionic peptide factor and phospholipid transfer protein activity in coronary artery disease and type 2 diabetes mellitus. *Clin Biochem* 2009; 42: 845–851.
50. Attia N, Lorec AM, Domingo N, Nakbi A, Hammami S, Ben Hamda K, Chaaba R, Masmoudi AS, Moulin P, Portugal H et al. High density lipoprotein-anionic peptide factor effect on reverse cholesterol transport in type 2 diabetic patients with and without coronary artery disease. *Clin Biochem* 2010; 43: 1079–1084.
51. Yamashita M, Tamasawa N, Matsuki K, Tanabe J, Murakami H, Matsui J, Suda T. Insulin suppresses HDL-mediated cholesterol efflux from macrophages through inhibition of neutral cholesteryl ester hydrolase and ATP-binding cassette transporter G1 expressions. *J Atheroscler Thromb* 2010; 17: 1183–1189.
52. Hongo S, Watanabe T, Arita S, Kanome T, Kageyama H, Shioda S, Miyazaki A. Leptin modulates ACAT1 expression and cholesterol efflux from human macrophages. *Am J Physiol Endocrinol Metab* 2009; 297: E474–482.
53. Tang C, Kanter JE, Bornfeldt KE, Leboeuf RC, Oram JF. Diabetes reduces the cholesterol exporter ABCA1 in mouse macrophages and kidneys. *J Lipid Res* 2010; 51: 1719–1728.

54. de Vries R, Groen AK, Perton FG, Dallinga-Thie GM, van Wijland MJ, Dikkeschei LD, Wolffenbuttel BH, van Tol A, Dullaart RP. Increased cholesterol efflux from cultured fibroblasts to plasma from hypertriglyceridemic type 2 diabetic patients: roles of pre beta-HDL, phospholipid transfer protein and cholesterol esterification. *Atherosclerosis* 2008; 196: 733–741.
55. Fievet C, Igau B, Bresson R, Drouin P, Fruchart JC. Non-enzymatic glycosylation of apolipoprotein A-I and its functional consequences. *Diabete Metab* 1995; 21: 95–98.
56. Ferretti G, Bacchetti T, Negre-Salvayre A, Salvayre R, Dousset N, Curatola G. Structural modifications of HDL and functional consequences. *Atherosclerosis* 2006; 184: 1–7.
57. Park KH, Cho KH. High-density lipoprotein (HDL) from elderly and reconstituted HDL containing glycated apolipoproteins A-I share proatherosclerotic and prosenescent properties with increased cholesterol influx. *J Gerontol A Biol Sci Med Sci* 2011; 66: 511–520.
58. Matsuki K, Tamasawa N, Yamashita M, Tanabe J, Murakami H, Matsui J, Imaizumi T, Satoh K, Suda T. Metformin restores impaired HDL-mediated cholesterol efflux due to glycation. *Atherosclerosis* 2009; 206: 434–438.
59. Nobecourt E, Davies MJ, Brown BE, Curtiss LK, Bonnet DJ, Charlton F, Januszewski AS, Jenkins AJ, Barter PJ, Rye KA. The impact of glycation on apolipoprotein A-I structure and its ability to activate lecithin:cholesterol acyltransferase. *Diabetologia* 2007; 50: 643–653.
60. Patel DC, Pavitt D, Albrecht C, Dhanjil S, Hannen R, Valabhji J, Higgins CF, Johnston DG. Hyperglycaemic patients have decreased ATP-binding cassette transporter-A1 (ABCA-1) gene expression: A novel disease mechanism. *Atherosclerosis* 2008; 199: 231–232.
61. Simmons RA. Developmental origins of diabetes: the role of oxidative stress. *Free Radic Biol Med* 2006; 40: 917–922.
62. Undurti A, Huang Y, Lupica JA, Smith JD, DiDonato JA, Hazen SL. Modification of high density lipoprotein by myeloperoxidase generates a pro-inflammatory particle. *J Biol Chem* 2009; 284: 30825–30835.
63. Nicholls SJ, Hazen SL. Myeloperoxidase and cardiovascular disease. *Arterioscler Thromb Vasc Biol* 2005; 25: 1102–1111.
64. Holzer M, Gauster M, Pfeifer T, Wadsack C, Fauler G, Stiegler P, Koefeler H, Beubler E, Schuligoi R, Heinemann A et al. Protein carbamylation renders high-density lipoprotein dysfunctional. *Antioxid Redox Signal* 2011; 14: 2337–2346.
65. Bergt C, Pennathur S, Fu X, Byun J, O'Brien K, McDonald TO, Singh P, Anantharamaiah GM, Chait A, Brunzell J et al. The myeloperoxidase product hypochlorous acid oxidizes HDL in the human artery wall and impairs ABCA1-dependent cholesterol transport. *Proc Natl Acad Sci U S A* 2004; 101: 13032–13037.
66. Zheng L, Nukuna B, Brennan ML, Sun M, Goormastic M, Settle M, Schmitt D, Fu X, Thomson L, Fox PL, Ischiropoulos H, Smith JD, Kinter M, Hazen SL. Apolipoprotein A-I is a selective target for myeloperoxidase-catalyzed oxidation and functional impairment in subjects with cardiovascular disease. *J Clin Invest* 2004; 114: 529–541.
67. Peng D-Q, Brubaker G, Wu Z, Zheng L, Willard B, Kinter M, Hazen SL, Smith JD. Apolipoprotein A-I tryptophan substitution leads to resistance to myeloperoxidase-mediated loss of function. *Arterioscler Thromb Vasc Biol* 2008; 28: 2063–2070.

68. Shao B, Bergt C, Fu X, Green P, Voss JC, Oda MN, Oram JF, Heinecke JW. Tyrosine 192 in apolipoprotein A-I is the major site of nitration and chlorination by myeloperoxidase, but only chlorination markedly impairs ABCA1-dependent cholesterol transport. *J Biol Chem* 2005; 280: 5983–5993.
69. Zheng L, Settle M, Brubaker G, Schmitt D, Hazen SL, Smith JD, Kinter M. Localization of nitration and chlorination sites on apolipoprotein A-I catalyzed by myeloperoxidase in human atheroma and associated oxidative impairment in ABCA1-dependent cholesterol efflux from macrophages. *J Biol Chem* 2005; 280: 38–47.
70. Cavigiolio G, Geier EG, Shao B, Heinecke JW, Oda MN. Exchange of apolipoprotein A-I between lipid-associated and lipid-free states: a potential target for oxidative generation of dysfunctional high density lipoproteins. *J Biol Chem* 2010; 285: 18847–18857.
71. Shao B, Pennathur S, Pagani I, Oda MN, Witztum JL, Oram JF, Heinecke JW. Modifying apolipoprotein A-I by malondialdehyde, but not by an array of other reactive carbonyls, blocks cholesterol efflux by the ABCA1 pathway. *J Biol Chem* 2010; 285: 18473–18484.
72. Shao B, Tang C, Heinecke JW, Oram JF. Oxidation of apolipoprotein A-I by myeloperoxidase impairs the initial interactions with ABCA1 required for signaling and cholesterol export. *J Lipid Res* 2010; 51: 1849–1858.
73. Annema W, Nijstad N, Tolle M, de Boer JF, Buijs RVC, Heeringa P, van der Giet M, Tietge UJF. Myeloperoxidase and serum amyloid A contribute to impaired *in vivo* reverse cholesterol transport during the acute phase response but not group IIA secretory phospholipase A2. *J Lipid Res* 2010; 51: 743–754.
74. Rifici VA, Khachadurian AK. Effects of dietary vitamin C and E supplementation on the copper mediated oxidation of HDL and on HDL mediated cholesterol efflux. *Atherosclerosis* 1996; 127: 19–26.
75. Shao B, Fu X, McDonald TO, Green PS, Uchida K, O'Brien KD, Oram JF, Heinecke JW. Acrolein impairs ATP binding cassette transporter A1-dependent cholesterol export from cells through site-specific modification of apolipoprotein A-I. *J Biol Chem* 2005; 280: 36386–36396.
76. Li GX, Liu P, Xu L, Miao F, Liu YF, Zhao H, Zhang ZW. Effects of high-density lipoprotein on the cholesterol efflux from endothelial cells. *Nan Fang Yi Ke Da Xue Xue Bao* 2009; 29: 2191–2193 [in Chinese].
77. Gelissen IC, Rye KA, Brown AJ, Dean RT, Jessup W. Oxysterol efflux from macrophage foam cells: the essential role of acceptor phospholipid. *J Lipid Res* 1999; 40: 1636–1646.
78. Gaus K, Dean RT, Kritharides L, Jessup W. Inhibition of cholesterol efflux by 7-ketocholesterol: comparison between cells, plasma membrane vesicles, and liposomes as cholesterol donors. *Biochemistry* 2001; 40: 13002–13014.
79. Frei B, Forte TM, Ames BN, Cross CE. Gas phase oxidants of cigarette smoke induce lipid peroxidation and changes in lipoprotein properties in human blood plasma. Protective effects of ascorbic acid. *Biochem J* 1991; 277: 133–138.
80. Ueyama K, Yokode M, Arai H, Nagano Y, Li ZX, Cho M, Kita T. Cholesterol efflux effect of high density lipoprotein is impaired by whole cigarette smoke extracts through lipid peroxidation. *Free Radic Biol Med* 1998; 24: 182–190.

81. Ashraf MZ, Kar NS, Chen X, Choi J, Salomon RG, Febbraio M, Podrez EA. Specific oxidized phospholipids inhibit scavenger receptor BI-mediated selective uptake of cholesteryl esters. *J Biol Chem* 2008; 283: 10408–10414.

82. Guha M, Gao X, Jayaraman S, Gursky O. Correlation of structural stability with functional remodeling of high-density lipoproteins: the importance of being disordered. *Biochemistry* 2008; 47: 11393–11397.

83. Pirillo A, Uboldi P, Catapano AL. Dual effect of hypochlorite in the modification of high density lipoproteins. *Biochem Biophys Res Commun* 2010; 403: 447–451.

84. Choi HY, Rahmani M, Wong BW, Allahverdian S, McManus BM, Pickering JG, Chan T, Francis GA. ATP-binding cassette transporter A1 expression and apolipoprotein A-I binding are impaired in intima-type arterial smooth muscle cells. *Circulation* 2009; 119: 3223–3231.

85. Malik P, Berisha SZ, Santore J, Agatisa-Boyle C, Brubaker G, Smith JD. Zymosan mediated inflammation impairs *in vivo* reverse cholesterol transport. *J Lipid Res* 2011; 52: 951–957.

86. Ohta T, Nakamura R, Ikeda Y, Shinohara M, Miyazaki A, Horiuchi S, Matsuda I. Differential effect of subspecies of lipoprotein containing apolipoprotein A-I on cholesterol efflux from cholesterol-loaded macrophages: functional correlation with lecithin: cholesterol acyltransferase. *Biochim Biophys Acta* 1992; 1165: 119–128.

87. McGillicuddy FC, de la Llera Moya M, Hinkle CC, Joshi MR, Chiquoine EH, Billheimer JT, Rothblat GH, Reilly MP. Inflammation impairs reverse cholesterol transport *in vivo*. *Circulation* 2009; 119: 1135–1145.

88. Banka CL, Yuan T, de Beer MC, Kindy M, Curtiss LK, de Beer FC. Serum amyloid A (SAA): influence on HDL-mediated cellular cholesterol efflux. *J Lipid Res* 1995; 36: 1058–1065.

89. Artl A, Marsche G, Lestavel S, Sattler W, Malle E. Role of serum amyloid A during metabolism of acute-phase HDL by macrophages. *Arterioscler Thromb Vasc Biol* 2000; 20: 763–772.

90. Jahangiri A, de Beer MC, Noffsinger V, Tannock LR, Ramaiah C, Webb NR, van der Westhuyzen DR, de Beer FC. HDL remodeling during the acute phase response. *Arterioscler Thromb Vasc Biol* 2009; 29: 261–267.

91. van der Westhuyzen DR, de Beer FC, Webb NR. HDL cholesterol transport during inflammation. *Curr Opin Lipidol* 2007; 18: 147–151.

92. de Beer MC, Ji A, Jahangiri A, Vaughan AM, de Beer FC, van der Westhuyzen DR, Webb NR. ATP binding cassette G1-dependent cholesterol efflux during inflammation. *J Lipid Res* 2011; 52: 345–353.

93. Kisilevsky R, Subrahmanyan L. Serum amyloid A changes high density lipoprotein's cellular affinity. A clue to serum amyloid A's principal function. *Lab Invest* 1992; 66: 778–785.

94. Artl A, Marsche G, Pussinen P, Knipping G, Sattler W, Malle E. Impaired capacity of acute-phase high density lipoprotein particles to deliver cholesteryl ester to the human HUH-7 hepatoma cell line. *Int J Biochem Cell Biol* 2002; 34: 370–381.

95. Cai L, de Beer MC, de Beer FC, van der Westhuyzen DR. Serum amyloid A is a ligand for scavenger receptor class B type I and inhibits high density lipoprotein binding and selective lipid uptake. *J Biol Chem* 2005; 280: 2954–2961.

96. van der Westhuyzen DR, Cai L, de Beer MC, de Beer FC. Serum amyloid A promotes cholesterol efflux mediated by scavenger receptor B-I. *J Biol Chem* 2005; 280: 35890–35895.
97. de Beer MC, Webb NR, Whitaker NL, Wroblewski JM, Jahangiri A, van der Westhuyzen DR, de Beer FC. SR-BI selective lipid uptake: Subsequent metabolism of acute phase HDL. *Arterioscler Thromb Vasc Biol* 2009; 29: 1298–1303.
98. Baranova I, Vishnyakova T, Bocharov A, Chen Z, Remaley AT, Stonik J, Eggerman TL, Patterson AP. Lipopolysaccharide down regulates both scavenger receptor B1 and ATP binding cassette transporter A1 in RAW cells. *Infect Immun* 2002; 70: 2995–3003.
99. Tam SP, Kisilevsky R, Ancsin JB. Acute-phase-HDL remodeling by heparan sulfate generates a novel lipoprotein with exceptional cholesterol efflux activity from macrophages. *PLoS One* 2008; 3: e3867.
100. Judström I, Jukkola H, Metso J, Jauhiainen M, Kovanen PT, Lee-Rueckert M. Mast cell-dependent proteolytic modification of HDL particles during anaphylactic shock in the mouse reduces their ability to induce cholesterol efflux from macrophage foam cells ex vivo. *Atherosclerosis* 2010; 208: 148–154.
101. Lee-Rueckert M, Silvennoinen R, Rotllan N, Judstrom I, Blanco-Vaca F, Metso J, Jauhiainen M, Kovanen PT, Escola-Gil JC. Mast cell activation *in vivo* impairs the macrophage reverse cholesterol transport pathway in the mouse. *Arterioscler Thromb Vasc Biol* 2011; 31: 520–527.
102. Pussinen PJ, Jauhiainen M, Vilkuna-Rautiainen T, Sundvall J, Vesanen M, Mattila K, Palosuo T, Alfthan G, Asikainen S. Periodontitis decreases the antiatherogenic potency of high density lipoprotein. *J Lipid Res* 2004; *45*: 139–147.
103. Khovidhunkit W, Shigenaga JK, Moser AH, Feingold KR, Grunfeld C. Cholesterol efflux by acute-phase high density lipoprotein: role of lecithin: cholesterol acyltransferase. *J Lipid Res* 2001; *42*: 967–975.
104. Hao XR, Cao DL, Hu YW, Li XX, Liu XH, Xiao J, Liao DF, Xiang J, Tang CK. IFN-gamma down-regulates ABCA1 expression by inhibiting LXRalpha in a JAK/STAT signaling pathway-dependent manner. *Atherosclerosis* 2009; *203*: 417–428.
105. Wang X, Liao D, Bharadwaj U, Li M, Yao Q, Chen C. C-reactive protein inhibits cholesterol efflux from human macrophage-derived foam cells. *Arterioscler Thromb Vasc Biol* 2008; *28*: 519–526.
106. Field FJ, Watt K, Mathur SN. TNF-alpha decreases ABCA1 expression and attenuates HDL cholesterol efflux in the human intestinal cell line Caco-2. *J Lipid Res* 2010; *51*: 1407–1415.
107. Edgel KA, LeBoeuf RC, Oram JF. Tumor necrosis factor-alpha and lymphotoxin-alpha increase macrophage ABCA1 by gene expression and protein stabilization via different receptors. *Atherosclerosis* 2010; *209*: 387–392.
108. Shridas P, Bailey WM, Gizard F, Oslund RC, Gelb MH, Bruemmer D, Webb NR. Group X secretory phospholipase A2 negatively regulates ABCA1 and ABCG1 expression and cholesterol efflux in macrophages. *Arterioscler Thromb Vasc Biol* 2010; *30*: 2014–2021.
109. Linsel-Nitschke P, Jansen H, Aherrarhou Z, Belz S, Mayer B, Lieb W, Huber F, Kremer W, Kalbitzer HR, Erdmann J et al. Macrophage cholesterol efflux correlates with lipoprotein subclass distribution and risk of obstructive coronary artery disease in patients undergoing coronary angiography. *Lipids Health Dis* 2009; 8: 14.

110. Holven KB, Aukrust P, Retterstol K, Otterdal K, Bjerkeli V, Ose L, Nenseter MS, Halvorsen B. The antiatherogenic function of HDL is impaired in hyperhomocysteinemic subjects. *J Nutr* 2008; 138: 2070–2075.

111. Puk CG, Bocchi EA, Lo Prete AC, Ferreira SM, Stolf NA, Maranhao RC. Transfer of cholesterol and other lipids from a lipid nanoemulsion to high-density lipoprotein in heart transplant patients. *J Heart Lung Transplant* 2009; 28: 1075–1080.

112. Liz MA, Gomes CM, Saraiva MJ, Sousa MM. ApoA-I cleaved by transthyretin has reduced ability to promote cholesterol efflux and increased amyloidogenicity. *J Lipid Res* 2007; 48: 2385–2395.

113. Levels JH, Pajkrt D, Schultz M, Hoek FJ, van Tol A, Meijers JC, van Deventer SJ. Alterations in lipoprotein homeostasis during human experimental endotoxemia and clinical sepsis. *Biochim Biophys Acta* 2007; 1771: 1429–1438.

114. Nofer JR, Kehrel B, Fobker M, Levkau B, Assmann G, von Eckardstein A. HDL and arteriosclerosis: beyond reverse cholesterol transport. *Atherosclerosis* 2002; 161: 1–16.

115. Assmann G, Nofer JR. Atheroprotective effects of high-density lipoproteins. *Annu Rev Med* 2003; 54: 321–341.

116. Assmann G, Gotto AM Jr. HDL cholesterol and protective factors in atherosclerosis. *Circulation* 2004; 109: III–8–14.

117. Navab M, Ananthramaiah GM, Reddy ST, Van Lenten BJ, Ansell BJ, Fonarow GC, Vahabzadeh K, Hama S, Hough G, Kamranpour N et al. The oxidation hypothesis of atherogenesis: the role of oxidized phospholipids and HDL. *J Lipid Res* 2004; 45: 993–1007.

118. Hansel B, Giral P, Nobecourt E, Chantepie S, Bruckert E, Chapman MJ, Kontush A. Metabolic syndrome is associated with elevated oxidative stress and dysfunctional dense high-density lipoprotein particles displaying impaired antioxidative activity. *J Clin Endocrinol Metab* 2004; 89: 4963–4971.

119. Kontush A, de Faria EC, Chantepie S, Chapman MJ. A normotriglyceridemic, low HDL-cholesterol phenotype is characterised by elevated oxidative stress and HDL particles with attenuated antioxidative activity. *Atherosclerosis* 2005; 182: 277–285.

120. Nobecourt E, Jacqueminet S, Hansel B, Chantepie S, Grimaldi A, Chapman MJ, Kontush A. Defective antioxidative activity of small, dense HDL particles in type 2 diabetes: Relationship to elevated oxidative stress and hyperglycemia. *Diabetologia* 2005; 48: 529–538.

121. Kontush A, de Faria EC, Chantepie S, Chapman MJ. Antioxidative activity of HDL particle subspecies is impaired in hyperalphalipoproteinemia: relevance of enzymatic and physicochemical properties. *Arterioscler Thromb Vasc Biol* 2004; 24: 526–533.

122. Zago V, Sanguinetti S, Brites F, Berg G, Verona J, Basilio F, Wikinski R, Schreier L. Impaired high density lipoprotein antioxidant activity in healthy postmenopausal women. *Atherosclerosis* 2004; 177: 203–210.

123. Moradi H, Pahl MV, Elahimehr R, Vaziri ND. Impaired antioxidant activity of high-density lipoprotein in chronic kidney disease. *Transl Res* 2009; 153: 77–85.

124. Klimov AN, Gurevich VS, Nikiforova AA, Shatilina LV, Kuzmin AA, Plavinsky SL, Teryukova NP. Antioxidative activity of high density lipoproteins *in vivo*. *Atherosclerosis* 1993; 100: 13–18.

125. Kontush A, Chapman MJ. Antiatherogenic function of HDL particle subpopulations: focus on antioxidative activities. *Curr Opin Lipidol* 2010; 21: 312–318.

126. Sparks DL, Davidson WS, Lund-Katz S, Phillips MC. Effects of the neutral lipid content of high density lipoprotein on apolipoprotein A-I structure and particle stability. *J Biol Chem* 1995; 270: 26910–26917.

127. Curtiss LK, Bonnet DJ, Rye KA. The conformation of apolipoprotein A-I in high-density lipoproteins is influenced by core lipid composition and particle size: a surface plasmon resonance study. *Biochemistry* 2000; 39: 5712–5721.

128. Gowri MS, Van der Westhuyzen DR, Bridges SR, Anderson JW. Decreased protection by HDL from poorly controlled type 2 diabetic subjects against LDL oxidation may be due to the abnormal composition of HDL. *Arterioscler Thromb Vasc Biol* 1999; 19: 2226–2233.

129. Iborra RT, Ribeiro IC, Neves MQ, Charf AM, Lottenberg SA, Negrao CE, Nakandakare ER, Passarelli M. Aerobic exercise training improves the role of high-density lipoprotein antioxidant and reduces plasma lipid peroxidation in type 2 diabetes mellitus. *Scand J Med Sci Sports* 2008; 18: 742–750.

130. Mastorikou M, Mackness B, Liu Y, Mackness M. Glycation of paraoxonase-1 inhibits its activity and impairs the ability of high-density lipoprotein to metabolize membrane lipid hydroperoxides. *Diabet Med* 2008; 25: 1049–1055.

131. Ferretti G, Bacchetti T, Busni D, Rabini RA, Curatola G. Protective effect of paraoxonase activity in high-density lipoproteins against erythrocyte membranes peroxidation: a comparison between healthy subjects and type 1 diabetic patients. *J Clin Endocrinol Metab* 2004; 89: 2957–2962.

132. Li C, Gu Q. Protective effect of paraoxonase 1 of high-density lipoprotein in type 2 diabetic patients with nephropathy. *Nephrology* 2009; 14: 514–520.

133. Unchern S, Laohareungpanya N, Sanvarinda Y, Pattanapanyasat K, Tanratana P, Chantharaksri U, Sibmooh N. Oxidative modification and poor protective activity of HDL on LDL oxidation in thalassemia. *Lipids* 2010; 45: 627–633.

134. Kontush A, Chapman MJ. Functionally defective HDL: A new therapeutic target at the crossroads of dyslipidemia, inflammation and atherosclerosis. *Pharmacol Rev* 2006; 3: 342–374.

135. Wong YQ, Binger KJ, Howlett GJ, Griffin MD. Methionine oxidation induces amyloid fibril formation by full-length apolipoprotein A-I. *Proc Natl Acad Sci U S A* 2010; 107: 1977–1982.

136. Pankhurst G, Wang XL, Wilcken DE, Baernthaler G, Panzenbock U, Raftery M, Stocker R. Characterization of specifically oxidized apolipoproteins in mildly oxidized high density lipoprotein. *J Lipid Res* 2003; 44: 349–355.

137. Brock JW, Jenkins AJ, Lyons TJ, Klein RL, Yim E, Lopes-Virella M, Carter RE, Thorpe SR, Baynes JW. Increased methionine sulfoxide content of apoA-I in type 1 diabetes. *J Lipid Res* 2008; 49: 847–855.

138. Liu F, Hindupur J, Nguyen JL, Ruf KJ, Zhu J, Schieler JL, Bonham CC, Wood KV, Davisson VJ, Rochet JC. Methionine sulfoxide reductase A protects dopaminergic cells from Parkinson's disease-related insults. *Free Radic Biol Med* 2008; 45: 242–255.

139. Park KH, Jang W, Kim KY, Kim JR, Cho KH. Fructated apolipoprotein A-I showed severe structural modification and loss of beneficial functions in lipid-free and lipid-bound state with acceleration of atherosclerosis and senescence. *Biochem Biophys Res Commun* 2010; 392: 295–300.

140. Tabet F, Lambert G, Cuesta Torres LF, Hou L, Sotirchos I, Touyz RM, Jenkins AJ, Barter PJ, Rye K-A. Lipid-free apolipoprotein A-I and discoidal reconstituted high-density lipoprotein differentially inhibit glucose-induced oxidative stress in human macrophages. *Arterioscler Thromb Vasc Biol* 2011; 31: 119–200.

141. Sonoki K, Iwase M, Sasaki N, Ohdo S, Higuchi S, Matsuyama N, Iida M. Relations of lysophosphatidylcholine in low-density lipoprotein with serum lipoprotein-associated phospholipase A2, paraoxonase and homocysteine thiolactonase activities in patients with type 2 diabetes mellitus. *Diabetes Res Clin Pract* 2009; 86: 117–123.

142. Boemi M, Leviev I, Sirolla C, Pieri C, Marra M, James RW. Serum paraoxonase is reduced in type 1 diabetic patients compared to non-diabetic, first degree relatives; influence on the ability of HDL to protect LDL from oxidation. *Atherosclerosis* 2001; 155: 229–235.

143. Sanguinetti SM, Brites FD, Fasulo V, Verona J, Elbert A, Wikinski RL, Schreier LE. HDL oxidability and its protective effect against LDL oxidation in Type 2 diabetic patients. *Diabetes Nutr Metab* 2001; 14: 27–36.

144. Hedrick CC, Thorpe SR, Fu MX, Harper CM, Yoo J, Kim SM, Wong H, Peters AL. Glycation impairs high-density lipoprotein function. *Diabetologia* 2000; 43: 312–320.

145. Ferretti G, Bacchetti T, Marchionni C, Caldarelli L, Curatola G. Effect of glycation of high density lipoproteins on their physicochemical properties and on paraoxonase activity. *Acta Diabetol* 2001; 38: 163–169.

146. Jaouad L, Milochevitch C, Khalil A. PON1 paraoxonase activity is reduced during HDL oxidation and is an indicator of HDL antioxidant capacity. *Free Radic Res* 2003; 37: 77–83.

147. Mertens A, Verhamme P, Bielicki JK, Phillips MC, Quarck R, Verreth W, Stengel D, Ninio E, Navab M, Mackness B et al. Increased low-density lipoprotein oxidation and impaired high-density lipoprotein antioxidant defense are associated with increased macrophage homing and atherosclerosis in dyslipidemic obese mice: LCAT gene transfer decreases atherosclerosis. *Circulation* 2003; 107: 1640–1646.

148. Jakubowski H. Calcium-dependent human serum homocysteine thiolactone hydrolase. A protective mechanism against protein N-homocysteinylation. *J Biol Chem* 2000; 275: 3957–3962.

149. Draganov DI, Teiber JF, Speelman A, Osawa Y, Sunahara R, La Du BN. Human paraoxonases (PON1, PON2, and PON3) are lactonases with overlapping and distinct substrate specificities. *J Lipid Res* 2005; 46: 1239–1247.

150. Hasselwander O, McEneny J, McMaster D, Fogarty DG, Nicholls DP, Maxwell AP, Young IS. HDL composition and HDL antioxidant capacity in patients on regular haemodialysis. *Atherosclerosis* 1999; 143: 125–133.

151. Van Lenten BJ, Navab M, Shih D, Fogelman AM, Lusis AJ. The role of high-density lipoproteins in oxidation and inflammation. *Trends Cardiovasc Med* 2001; 11: 155–161.

152. Hoofnagle AN, Wu M, Gosmanova AK, Becker JO, Wijsman EM, Brunzell JD, Kahn SE, Knopp RH, Lyons TJ, Heinecke JW. Low clusterin levels in high-density lipoprotein associate with insulin resistance, obesity, and dyslipoproteinemia. *Arterioscler Thromb Vasc Biol* 2010; 30: 2528–2534.

153. Van Lenten BJ, Hama SY, de Beer FC, Stafforini DM, McIntyre TM, Prescott SM, La Du BN, Fogelman AM, Navab M. Anti-inflammatory HDL becomes pro-inflammatory during the acute phase response. Loss of protective effect of HDL against LDL oxidation in aortic wall cell cocultures. *J Clin Invest* 1995; 96: 2758–2767.

154. Watson AD, Berliner JA, Hama SY, La Du BN, Faull KF, Fogelman AM, Navab M. Protective effect of high density lipoprotein associated paraoxonase. Inhibition of the biological activity of minimally oxidized low density lipoprotein. *J Clin Invest* 1995; 96: 2882–2891.

155. Watson AD, Navab M, Hama SY, Sevanian A, Prescott SM, Stafforini DM, McIntyre TM, Du BN, Fogelman AM, Berliner JA. Effect of platelet activating factor-acetylhydrolase on the formation and action of minimally oxidized low density lipoprotein. *J Clin Invest* 1995; 95: 774–782.

156. Shih DM, Gu L, Hama S, Xia YR, Navab M, Fogelman AM, Lusis AJ. Genetic-dietary regulation of serum paraoxonase expression and its role in atherogenesis in a mouse model. *J Clin Invest* 1996; 97: 1630–1639.

157. Navab M, Hama LS, Van LB, Fonarow GC, Cardinez CJ, Castellani LW, Brennan ML, Lusis AJ, Fogelman AM. Mildly oxidized LDL induces an increased apolipoprotein J/paraoxonase ratio. *J Clin Invest* 1997; 99: 2005–2019.

158. Warden CH, Hedrick CC, Qiao JH, Castellani LW, Lusis AJ. Atherosclerosis in transgenic mice overexpressing apolipoprotein A-II. *Science* 1993; 261: 469–472.

159. Castellani LW, Navab M, Van Lenten BJ, Hedrick CC, Hama SY, Goto AM, Fogelman AM, Lusis AJ. Overexpression of apolipoprotein AII in transgenic mice converts high density lipoproteins to proinflammatory particles. *J Clin Invest* 1997; 100: 464–474.

160. Cabana VG, Lukens JR, Rice KS, Hawkins TJ, Getz GS. HDL content and composition in acute phase response in three species: triglyceride enrichment of HDL a factor in its decrease. *J Lipid Res* 1996; 37: 2662–2674.

161. Cabana VG, Reardon CA, Wei B, Lukens JR, Getz GS. SAA-only HDL formed during the acute phase response in apoA-I+/+ and apoA-I-/- mice. *J Lipid Res* 1999; 40: 1090–1103.

162. Van Lenten BJ, Wagner AC, Nayak DP, Hama S, Navab M, Fogelman AM. High-density lipoprotein loses its anti-inflammatory properties during acute influenza A infection. *Circulation* 2001; 103: 2283–2288.

163. Khovidhunkit W, Kim MS, Memon RA, Shigenaga JK, Moser AH, Feingold KR, Grunfeld C. Effects of infection and inflammation on lipid and lipoprotein metabolism: mechanisms and consequences to the host. *J Lipid Res* 2004; 45: 1169–1196.

164. Daminelli EN, Spada C, Treitinger A, Oliveira TV, Latrilha Mda C, Maranhao RC. Alterations in lipid transfer to high-density lipoprotein (HDL) and activity of paraoxonase-1 in HIV+ patients. *Rev Inst Med Trop Sao Paulo* 2008; 50: 223–227.

165. Lakshman MR, Gottipati CS, Narasimhan SJ, Munoz J, Marmillot P, Nylen ES. Inverse correlation of serum paraoxonase and homocysteine thiolactonase activities and antioxidant capacity of high-density lipoprotein with the severity of cardiovascular disease in persons with type 2 diabetes mellitus. *Metabolism* 2006; 55: 1201–1206.

166. Jaouad L, de Guise C, Berrougui H, Cloutier M, Isabelle M, Fulop T, Payette H, Khalil A. Age-related decrease in high-density lipoproteins antioxidant activity is due to an alteration in the PON1's free sulfhydryl groups. *Atherosclerosis* 2006; 185: 191–200.
167. James RW. A long and winding road: defining the biological role and clinical importance of paraoxonases. *Clin Chem Lab Med* 2006; 44: 1052–1059.
168. Chen X, Wu Y, Liu L, Su Y, Peng Y, Jiang L, Liu X, Huang D. Relationship between high density lipoprotein antioxidant activity and carotid arterial intima-media thickness in patients with essential hypertension. *Clin Exp Hypertens* 2010; 32: 13–20.
169. Yildiz A, Gur M, Demirbag R, Yilmaz R, Akyol S, Aslan M, Erel O. Paraoxonase and arylesterase activities in untreated dipper and non-dipper hypertensive patients. *Clin Biochem* 2008; 41: 779–784.
170. Kunes JP, Cordero-Koning KS, Lee LH, Lynch SM. Vitamin C attenuates hypochlorite-mediated loss of paraoxonase-1 activity from human plasma. *Nutr Res* 2009; 29: 114–122.
171. Zerrad-Saadi A, Therond P, Chantepie S, Couturier M, Rye K-A, Chapman MJ, Kontush A. HDL3-mediated inactivation of LDL-associated phospholipid hydroperoxides is determined by the redox status of apolipoprotein A-I and HDL particle surface lipid rigidity: Relevance to inflammation and atherogenesis. *Arterioscler Thromb Vasc Biol* 2009; 29: 2169–2175.
172. Yeboah J, McNamara C, Jiang XC, Tabas I, Herrington DM, Burke GL, Shea S. Association of plasma sphingomyelin levels and incident coronary heart disease events in an adult population: multi-ethnic study of atherosclerosis. *Arterioscler Thromb Vasc Biol* 2010; 30: 628–633.
173. Clarke R, Shipley M, Armitage J, Collins R, Harris W. Plasma phospholipid fatty acids and CHD in older men: Whitehall study of London civil servants. *Br J Nutr* 2009; 102: 279–284.
174. Liu J, Huan C, Chakraborty M, Zhang H, Lu D, Kuo M-S, Cao G, Jiang X-C. Macrophage sphingomyelin synthase 2 deficiency decreases atherosclerosis in mice. *Circ Res* 2009; 105: 295–303.
175. Liu J, Zhang H, Li Z, Hailemariam TK, Chakraborty M, Jiang K, Qiu D, Bui HH, Peake DA, Kuo M-S et al. Sphingomyelin synthase 2 is one of the determinants for plasma and liver sphingomyelin levels in mice. *Arterioscler Thromb Vasc Biol* 2009; 29: 850–856.
176. Park T-S, Panek RL, Mueller SB, Hanselman JC, Rosebury WS, Robertson AW, Kindt EK, Homan R, Karathanasis SK, Rekhter MD. Inhibition of sphingomyelin synthesis reduces atherogenesis in apolipoprotein E-knockout mice. *Circulation* 2004; 110: 3465–3471.
177. Williams KJ, Tabas I. The response-to-retention hypothesis of early atherogenesis. *Arterioscler Thromb Vasc Biol* 1995; 15: 551–561.
178. Tabas I, Williams KJ, Boren J. Subendothelial lipoprotein retention as the initiating process in atherosclerosis: update and therapeutic implications. *Circulation* 2007; 116: 1832–1844.
179. Carew TE, Pittman RC, Marchand ER, Steinberg D. Measurement *in vivo* of irreversible degradation of low density lipoprotein in the rabbit aorta. Predominance of intimal degradation. *Arteriosclerosis* 1984; 4: 214–224.

180. Schwenke DC, Carew TE. Initiation of atherosclerotic lesions in cholesterol-fed rabbits. I. Focal increases in arterial LDL concentration precede development of fatty streak lesions. *Arteriosclerosis* 1989; 9: 895–907.

181. Schwenke DC, Carew TE. Initiation of atherosclerotic lesions in cholesterol-fed rabbits. II. Selective retention of LDL vs. selective increases in LDL permeability in susceptible sites of arteries. *Arteriosclerosis* 1989; 9: 908–918.

182. Steinberg D, Parthasarathy S, Carew TE, Khoo JC, Witztum JL. Beyond cholesterol. *N Engl J Med* 1989; 320: 915–924.

183. Chisolm GM, Steinberg D. The oxidative modification hypothesis of atherogenesis: an overview. *Free Radic Biol Med* 2000; 28: 1815–1826.

184. Le Brazidec H. Lipoproteins and protection of the arterial wall against infection: the "response to the threat of infection" hypothesis. *Atherosclerosis* 2007; 195: e21–e31.

185. Miller YI, Choi SH, Wiesner P, Fang L, Harkewicz R, Hartvigsen K, Boullier A, Gonen A, Diehl CJ, Que X et al. Oxidation-specific epitopes are danger-associated molecular patterns recognized by pattern recognition receptors of innate immunity. *Circ Res* 2011; 108: 235–248.

186. Burger D, Dayer JM. High-density lipoprotein-associated apolipoprotein A-I: the missing link between infection and chronic inflammation ? *Autoimmun Rev* 2002; 1: 111–117.

187. Schwedhelm E, Bartling A, Lenzen H, Tsikas D, Maas R, Brummer J, Gutzki FM, Berger J, Frolich JC, Boger RH. Urinary 8-iso-prostaglandin F2alpha as a risk marker in patients with coronary heart disease: a matched case-control study. *Circulation* 2004; 109: 843–848.

188. Meisinger C, Baumert J, Khuseyinova N, Loewel H, Koenig W. Plasma oxidized low-density lipoprotein, a strong predictor for acute coronary heart disease events in apparently healthy, middle-aged men from the general population. *Circulation* 2005; 112: 651–657.

189. Shishehbor MH, Zhang R, Medina H, Brennan M-L, Brennan DM, Ellis SG, Topol EJ, Hazen SL. Systemic elevations of free radical oxidation products of arachidonic acid are associated with angiographic evidence of coronary artery disease. *Free Radic Biol Med* 2006; 41: 1678–1683.

190. Davi G, Ciabattoni G, Consoli A, Mezzetti A, Falco A, Santarone S, Pennese E, Vitacolonna E, Bucciarelli T, Costantini F et al. *In vivo* formation of 8-iso-prostaglandin F2alpha and platelet activation in diabetes mellitus: effects of improved metabolic control and vitamin E supplementation. *Circulation* 1999; 99: 224–229.

191. Devaraj S, Hirany SV, Burk RF, Jialal I. Divergence between LDL oxidative susceptibility and urinary F(2)-isoprostanes as measures of oxidative stress in type 2 diabetes. *Clin Chem* 2001; 47: 1974–1979.

192. Moore RE, Kawashiri MA, Kitajima K, Secreto A, Millar JS, Pratico D, Rader DJ. Apolipoprotein A-I deficiency results in markedly increased atherosclerosis in mice lacking the LDL receptor. *Arterioscler Thromb Vasc Biol* 2003; 23: 1914–1920.

193. Christison J, Karjalainen A, Brauman J, Bygrave F, Stocker R. Rapid reduction and removal of HDL- but not LDL-associated cholesteryl ester hydroperoxides by rat liver perfused *in situ*. *Biochem J* 1996; 314: 739–742.

194. Morrow JD. Quantification of isoprostanes as indices of oxidant stress and the risk of atherosclerosis in humans. *Arterioscler Thromb Vasc Biol* 2005; 25: 279–286.

195. Balstad TR, Holven KB, Ottestad IO, Otterdal K, Halvorsen B, Myhre AM, Ose L, Nenseter MS. Altered composition of HDL3 in FH subjects causing a HDL subfraction with less atheroprotective function. *Clin Chim Acta* 2005; 35:, 171–178.
196. Vaziri ND, Moradi H, Pahl MV, Fogelman AM, Navab M. In vitro stimulation of HDL anti-inflammatory activity and inhibition of LDL pro-inflammatory activity in the plasma of patients with end-stage renal disease by an apoA-1 mimetic peptide. *Kidney Int* 2009; 76: 437–444.
197. Vaziri ND. Lipotoxicity and impaired high density lipoprotein-mediated reverse cholesterol transport in chronic kidney disease. *J Ren Nutr* 2010; 20: S35–S43.
198. Tan KC, Chow WS, Lam JC, Lam B, Wong WK, Tam S, Ip MS. HDL dysfunction in obstructive sleep apnea. *Atherosclerosis* 2006; 184: 377–382.
199. Patel S, Puranik R, Nakhla S, Lundman P, Stocker R, Wang XS, Lambert G, Rye KA, Barter PJ, Nicholls SJ et al. Acute hypertriglyceridaemia in humans increases the triglyceride content and decreases the anti-inflammatory capacity of high density lipoproteins. *Atherosclerosis* 2009; 204: 424–428.
200. Nanjee MN, Doran JE, Lerch PG, Miller NE. Acute effects of intravenous infusion of ApoA1/phosphatidylcholine discs on plasma lipoproteins in humans. *Arterioscler Thromb Vasc Biol* 1999; 19: 979–989.
201. Nicholls SJ, Dusting GJ, Cutri B, Bao S, Drummond GR, Rye KA, Barter PJ. Reconstituted high-density lipoproteins inhibit the acute pro-oxidant and proinflammatory vascular changes induced by a periarterial collar in normocholesterolemic rabbits. *Circulation* 2005; 111: 1543–1550.
202. Navab M, Berliner JA, Subbanagounder G, Hama S, Lusis AJ, Castellani LW, Reddy S, Shih D, Shi W, Watson AD et al. HDL and the inflammatory response induced by LDL-derived oxidized phospholipids. *Arterioscler Thromb Vasc Biol* 2001; 21: 481–488.
203. Chung CP, Oeser A, Raggi P, Sokka T, Pincus T, Solus JF, Linton MF, Fazio S, Stein CM. Lipoprotein subclasses determined by nuclear magnetic resonance spectroscopy and coronary atherosclerosis in patients with rheumatoid arthritis. *J Rheumatol* 2010; 37: 1633–1638.
204. Yan D, Navab M, Bruce C, Fogelman AM, Jiang XC. PLTP deficiency improves the anti-inflammatory properties of HDL and reduces the ability of LDL to induce monocyte chemotactic activity. *J Lipid Res* 2004; 45: 1852–1858.
205. Kherkeulidze P, Johansson J, Carlson LA. High density lipoprotein particle size distribution in cord blood. *Acta Paediatr Scand* 1991; 80: 770–779.
206. Dodani S, Kaur R, Reddy S, Reed GL, Navab M, George V. Can dysfunctional HDL explain high coronary artery disease risk in South Asians ? *Int J Cardiol* 2008; 129: 125–132.
207. Nobecourt E, Tabet F, Lambert G, Puranik R, Bao S, Yan L, Davies MJ, Brown BE, Jenkins AJ, Dusting GJ et al. Nonenzymatic glycation impairs the antiinflammatory properties of apolipoprotein A-I. *Arterioscler Thromb Vasc Biol* 2010; 30: 766–772.
208. Rossmann C, Rauh A, Hammer A, Windischhofer W, Zirkl S, Sattler W, Malle E. Hypochlorite-modified high-density lipoprotein promotes induction of HO-1 in endothelial cells via activation of p42/44 MAPK and zinc finger transcription factor Egr-1. *Arch Biochem Biophys* 2011; 509: 16–25.
209. Beliard S, Nogueira JP, Maraninchi M, Lairon D, Nicolay A, Giral P, Portugal H, Vialettes B, Valero R. Parallel increase of plasma apoproteins C-II and C-III in Type 2 diabetic patients. *Diabet Med* 2009; 26: 736–739.

210. Onat A, Hergenc G, Ayhan E, Ugur M, Kaya H, Tuncer M, Can G. Serum apolipoprotein C-III in high-density lipoprotein: a key diabetogenic risk factor in Turks. *Diabet Med* 2009; 26: 981–988.

211. Caron S, Verrijken A, Mertens I, Samanez CH, Mautino G, Haas JT, Duran-Sandoval D, Prawitt J, Francque S, Vallez E et al. Transcriptional activation of apolipoprotein CIII expression by glucose may contribute to diabetic dyslipidemia. *Arterioscler Thromb Vasc Biol* 2011; 31: 513–519.

212. Kawakami A, Yoshida M. Apolipoprotein CIII links dyslipidemia with atherosclerosis. *J Atheroscler Thromb* 2009; 16: 6–11.

213. Abe Y, Kawakami A, Osaka M, Uematsu S, Akira S, Shimokado K, Sacks FM, Yoshida M. Apolipoprotein CIII induces monocyte chemoattractant protein-1 and interleukin 6 expression via Toll-like receptor 2 pathway in mouse adipocytes. *Arterioscler Thromb Vasc Biol* 2010; 30: 2242–2248.

214. Gustafsson, M., Boren, J. Mechanism of lipoprotein retention by the extracellular matrix. *Curr Opin Lipidol* 2004; 15: 505–514.

215. Hiukka A, Stahlman M, Pettersson C, Levin M, Adiels M, Teneberg S, Leinonen ES, Hulten LM, Wiklund O, Oresic M et al. ApoCIII-enriched LDL in type 2 diabetes displays altered lipid composition, increased susceptibility for sphingomyelinase, and increased binding to biglycan. *Diabetes* 2009; 58: 2018–2026.

216. Navab M, Berliner JA, Watson AD, Hama SY, Territo MC, Lusis AJ, Shih DM, Van Lenten BJ, Frank JS, Demer LL et al. The Yin and Yang of oxidation in the development of the fatty streak. A review based on the 1994 George Lyman Duff Memorial Lecture. *Arterioscler Thromb Vasc Biol* 1996; 16: 831–842.

217. van Leuven SI, Hezemans R, Levels JH, Snoek S, Stokkers PC, Hovingh GK, Kastelein JJ, Stroes ES, de Groot E, Hommes DW. Enhanced atherogenesis and altered high density lipoprotein in patients with Crohn's disease. *J Lipid Res* 2007; 48: 2640–2646.

218. Charakida M, Besler C, Batuca JR, Sangle S, Marques S, Sousa M, Wang G, Tousoulis D, Delgado Alves J, Loukogeorgakis SP et al. Vascular abnormalities, paraoxonase activity, and dysfunctional HDL in primary antiphospholipid syndrome. *JAMA* 2009; 302: 1210–1217.

219. McMahon M, Grossman J, Skaggs B, Fitzgerald J, Sahakian L, Ragavendra N, Charles-Schoeman C, Watson K, Wong WK, Volkmann E et al. Dysfunctional proinflammatory high-density lipoproteins confer increased risk of atherosclerosis in women with systemic lupus erythematosus. *Arthritis Rheum* 2009; 60: 2428–2437.

220. Volkmann ER, Grossman JM, Sahakian LJ, Skaggs BJ, FitzGerald J, Ragavendra N, Charles-Schoeman C, Chen W, Gorn A, Karpouzas G et al. Low physical activity is associated with proinflammatory high-density lipoprotein and increased subclinical atherosclerosis in women with systemic lupus erythematosus. *Arthritis Care Res* 2010; 62: 258–265.

221. Charles-Schoeman C, Watanabe J, Lee YY, Furst DE, Amjadi S, Elashoff D, Park G, McMahon M, Paulus HE, Fogelman AM et al. Abnormal function of high-density lipoprotein is associated with poor disease control and an altered protein cargo in rheumatoid arthritis. *Arthritis Rheum* 2009; 60: 2870–2879.

222. Kelesidis T, Yang OO, Currier JS, Navab K, Fogelman AM, Navab M. HIV-1 infected patients with suppressed plasma viremia on treatment have pro-inflammatory HDL. *Lipids Health Dis* 2011; 10: 35.

223. Araujo JA, Nel AE. Particulate matter and atherosclerosis: role of particle size, composition and oxidative stress. *Part Fibre Toxicol* 2009; 6: 24.

224. Navab M, Reddy ST, Van Lenten BJ, Anantharamaiah GM, Fogelman AM. The role of dysfunctional HDL in atherosclerosis. *J Lipid Res* 2009; 50: S145–S149.

225. Whetzel AM, Sturek JM, Nagelin MH, Bolick DT, Gebre AK, Parks JS, Bruce AC, Skaflen MD, Hedrick CC. ABCG1 deficiency in mice promotes endothelial activation and monocyte-endothelial interactions. *Arterioscler Thromb Vasc Biol* 2010; 30: 809–817.

226. Ehrenwald E, Chisolm GM, Fox PL. Intact human ceruloplasmin oxidatively modifies low density lipoprotein. *J Clin Invest* 1994; 93: 1493–1501.

227. Ehrenwald E, Fox PL. Role of endogenous ceruloplasmin in low density lipoprotein oxidation by human U937 monocytic cells. *J Clin Invest* 1996; 97: 884–890.

228. Navab M, Hama SY, Hough GP, Subbanagounder G, Reddy ST, Fogelman AM. A cell-free assay for detecting HDL that is dysfunctional in preventing the formation of or inactivating oxidized phospholipids. *J Lipid Res* 2001; 42: 1308–1317.

229. Ansell BJ, Navab M, Hama S, Kamranpour N, Fonarow G, Hough G, Rahmani S, Mottahedeh R, Dave R, Reddy ST et al. Inflammatory/antiinflammatory properties of high-density lipoprotein distinguish patients from control subjects better than high-density lipoprotein cholesterol levels and are favorably affected by simvastatin treatment. *Circulation* 2003; 108: 2751–2756.

230. Kontush A, Chantepie S, Chapman MJ. Small, dense HDL particles exert potent protection of atherogenic LDL against oxidative stress. *Arterioscler Thromb Vasc Biol* 2003; 23: 1881–1888.

231. Batuca JR, Ames PR, Amaral M, Favas C, Isenberg DA, Delgado Alves J. Anti-atherogenic and anti-inflammatory properties of high-density lipoprotein are affected by specific antibodies in systemic lupus erythematosus. *Rheumatology (Oxford)* 2009; 48: 26–31.

232. Ames PR, Matsuura E, Batuca JR, Ciampa A, Lopez LL, Ferrara F, Iannaccone L, Alves JD. High-density lipoprotein inversely relates to its specific autoantibody favoring oxidation in thrombotic primary antiphospholipid syndrome. *Lupus* 2010; 19: 711–716.

233. Vuilleumier N, Charbonney E, Fontao L, Alvarez M, Turck N, Sanchez JC, Burkhard PR, Mensi N, Righini M, Reber G et al. Anti-(apolipoprotein A-1) IgGs are associated with high levels of oxidized low-density lipoprotein in acute coronary syndrome. *Clin Sci* 2008; 115: 25–33.

234. Pirillo A, Uboldi P, Bolego C, Kuhn H, Catapano AL. The 15-lipoxygenase-modified high density lipoproteins 3 fail to inhibit the TNF-alpha-induced inflammatory response in human endothelial cells. *J Immunol* 2008; 181: 2821–2830.

235. Patel S, Chung SH, White G, Bao S, Celermajer DS. The "atheroprotective" mediators apolipoproteinA-I and Foxp3 are over-abundant in unstable carotid plaques. *Int J Cardiol* 2010; 145: 183–187.

236. Watanabe J, Chou KJ, Liao JC, Miao Y, Meng HH, Ge H, Grijalva V, Hama S, Kozak K, Buga G et al. Differential association of hemoglobin with proinflammatory high density lipoproteins in atherogenic/hyperlipidemic mice. A novel biomarker of atherosclerosis. *J Biol Chem* 2007; 282: 23698–23707.

237. Watanabe J, Grijalva V, Hama S, Barbour K, Berger FG, Navab M, Fogelman AM, Reddy ST. Hemoglobin and its scavenger protein haptoglobin associate with apoA-1-containing particles and influence the inflammatory properties and function of high density lipoprotein. *J Biol Chem* 2009; 284: 18292–18301.

238. Coronado MT, Pozzi AO, Punchard MA, Gonzalez P, Fantidis P. Inflammation as a modulator of the HDL cholesterol-induced inteleukin-10 production by human circulating mononuclear cells. *Atherosclerosis* 2009; 202: 183–184.

239. Libby P. Inflammation in atherosclerosis. *Nature* 2002; 420: 868–874.

240. Esteve E, Ricart W, Fernandez-Real JM. Dyslipidemia and inflammation: an evolutionary conserved mechanism. *Clin Nutr* 2005; 24: 16–31.

241. Finch CE, Crimmins EM. Inflammatory exposure and historical changes in human life-spans. *Science* 2004; 305: 1736–1739.

242. Charakida M, Donald AE, Terese M, Leary S, Halcox JP, Ness A, Smith GD, Golding J, Friberg P, Klein NJ et al. for the AST. Endothelial dysfunction in childhood infection. *Circulation* 2005; 111: 1660–1665.

243. Napoli C, Pignalosa O, de Nigris F, Sica V. Childhood infection and endothelial dysfunction: A potential link in atherosclerosis ? *Circulation* 2005; 111: 1568–1570.

244. Manco M. Endotoxin as a missed link among all the metabolic abnormalities in the metabolic syndrome. *Atherosclerosis* 2009; 206: 36.

245. Petoumenos V, Nickenig G, Werner N. High-density lipoprotein exerts vasculoprotection via endothelial progenitor cells. *J Cell Mol Med* 2009; 13: 4623–4635.

246. Matsunaga T, Iguchi K, Nakajima T, Koyama I, Miyazaki T, Inoue I, Kawai S, Katayama S, Hirano K, Hokari S et al. Glycated high-density lipoprotein induces apoptosis of endothelial cells via a mitochondrial dysfunction. *Biochem Biophys Res Commun* 2001; 287: 714–720.

247. Alomar Y, Negre-Salvayre A, Levade T, Valdiguie P, Salvayre R. Oxidized HDL are much less cytotoxic to lymphoblastoid cells than oxidized LDL. *Biochim Biophys Acta* 1992; 1128: 163–166.

248. Sattler KJ, Elbasan S, Keul P, Elter-Schulz M, Bode C, Graler MH, Brocker-Preuss M, Budde T, Erbel R et al Sphingosine 1-phosphate levels in plasma and HDL are altered in coronary artery disease. *Basic Res Cardiol* 2010; 105: 821–832.

249. Fuhrman B, Gantman A, Aviram M. Paraoxonase 1 (PON1) deficiency in mice is associated with reduced expression of macrophage SR-BI and consequently the loss of HDL cytoprotection against apoptosis. *Atherosclerosis* 2010; 211: 61–68.

250. Gamliel-Lazarovich A, Gantman A, Shiner M, Coleman R, Aviram M, Keidar S. Paraoxonase 1 deficiency in mice is associated with reduced steroid biosynthesis: Effects on HDL binding, cholesteryl ester accumulation and scavenger receptor type BI expression. *Atherosclerosis* 2010; 211: 130–135.

251. Tarling EJ, Bojanic DD, Tangirala RK, Wang X, Lovgren-Sandblom A, Lusis AJ, Bjorkhem I, Edwards PA. Impaired development of atherosclerosis in ABCG1-/- apoE-/- mice: Identification of specific oxysterols that both accumulate in ABCG1-/- apoE-/- tissues and induce apoptosis. *Arterioscler Thromb Vasc Biol* 2010; 30: 1174–1180.

252. Persegol, L., Verges, B., Foissac, M., Gambert, P., Duvillard, L. Inability of HDL from type 2 diabetic patients to counteract the inhibitory effect of oxidised LDL on endothelium-dependent vasorelaxation. *Diabetologia* 2006; 49: 1380–1386.

253. Persegol L, Verges B, Gambert P, Duvillard L. Inability of HDL from abdominally obese subjects to counteract the inhibitory effect of oxidized LDL on vasorelaxation. *J Lipid Res* 2007; 48: 1396–1401.
254. Persegol L, Foissac M, Lagrost L, Athias A, Gambert P, Verges B, Duvillard L. HDL particles from type 1 diabetic patients are unable to reverse the inhibitory effect of oxidised LDL on endothelium-dependent vasorelaxation. *Diabetologia* 2007; 50: 2384–2387.
255. Sorrentino SA, Besler C, Rohrer L, Meyer M, Heinrich K, Bahlmann FH, Mueller M, Horvath T, Doerries C, Heinemann M et al. Endothelial-vasoprotective effects of high-density lipoprotein are impaired in patients with type 2 diabetes mellitus but are improved after extended-release niacin therapy. *Circulation* 2010; 121: 110–122.
256. Besler C, Heinrich K, Rohrer L, Yonekawa K, Doerries C, Manes C, Muller M, Templin C, Corti R, Maier W et al. Mechanisms leading to impaired endothelial-vasoprotective properties of high density lipoprotein (HDL) in patients with coronary disease (CAD). *Circulation* 2009; 120 (Suppl.): S1009.
257. Besler C, Heinrich K, Makrides V, Riwanto M, Stein S, Verrey F, Luscher TF, Landmesser U. HDL stimulates endothelial CAT-1 expression and L-arginine uptake: A novel mechanism leading to endothelial-protective effects of HDL that is profoundly altered in patients with coronary disease. *Circulation* 2010; 122 (Suppl.): A19862.
258. Besler, C., Heinrich, K., Rohrer, L., Riwanto, M., Doerries, C., Shih, D.M., Yonekawa, K., Stein, S., Schaefer, N., Manes, C., et al. High-density lipoprotein leads to stimulation of endothelial protein kinase C-beta 2 and inhibition of eNOS-activating pathways in patients with coronary disease: Role of paraoxonase-1 inactivation. *Circulation* 2010; 122 (Suppl.): A19687.
259. Kallio KA, Buhlin K, Jauhiainen M, Keva R, Tuomainen AM, Klinge B, Gustafsson A, Pussinen PJ. Lipopolysaccharide associates with pro-atherogenic lipoproteins in periodontitis patients. *Innate Immun* 2008; 14: 247–253.
260. Cai L, Ji A, de Beer FC, Tannock LR, van der Westhuyzen DR. SR-BI protects against endotoxemia in mice through its roles in glucocorticoid production and hepatic clearance. *J Clin Invest* 2008; 118: 364–375.
261. Courtney HS, Zhang YM, Frank MW, Rock CO. Serum opacity factor, a streptococcal virulence factor that binds to apolipoproteins A-I and A-II and disrupts high density lipoprotein structure. *J Biol Chem* 2006; 281: 5515–5521.
262. Han M, Gillard BK, Courtney HS, Ward K, Rosales C, Khant H, Ludtke SJ, Pownall HJ. Disruption of human plasma high-density lipoproteins by streptococcal serum opacity factor requires labile apolipoprotein A-I. *Biochemistry* 2009; 48: 1481–1487.
263. Gillard BK, Rosales C, Pillai BK, Lin HY, Courtney HS, Pownall HJ. Streptococcal serum opacity factor increases the rate of hepatocyte uptake of human plasma high-density lipoprotein cholesterol. *Biochemistry* 2010; 49: 9866–9873.
264. Holy EW, Besler C, Landmesser U, Luscher TF, Tanner FC. High density lipoprotein (HDL) from healthy subjects, but not from patients with coronary artery disease, exerts anti-thrombotic effects on human endothelial cells. *Circulation* 2009; 120 (Suppl.): S1100.
265. Norata GD, Banfi C, Pirillo A, Tremoli E, Hamsten A, Catapano AL, Eriksson P. Oxidised-HDL3 induces the expression of PAI-1 in human endothelial cells. Role of p38MAPK activation and mRNA stabilization. *Br J Haematol* 2004; 127: 97–104.

266. Assinger A, Schmid W, Eder S, Schmid D, Koller E, Volf I. Oxidation by hypochlorite converts protective HDL into a potent platelet agonist. *FEBS Lett* 2008; 582: 778–784.

267. Assinger A, Koller F, Schmid W, Zellner M, Babeluk R, Koller E, Volf I. Specific binding of hypochlorite-oxidized HDL to platelet CD36 triggers proinflammatory and procoagulant effects. *Atherosclerosis* 2010; 212: 153–160.

SECTION 3

THERAPEUTIC NORMALIZATION OF SUBNORMAL LEVELS AND DEFECTIVE BIOLOGIC ACTIVITIES OF HDL

Circulating levels of HDL-C are firmly established as a strong, independent, negative risk factor for premature cardiovascular disease [1, 2] (see Chapter 5). Furthermore, HDL particles possess multiple atheroprotective functions that are progressively depleted in cardiometabolic diseases, such as in low HDL-C dyslipidemias (see Section 2). Indeed, subjects in primary prevention at elevated cardiovascular risk, as well as patients in secondary prevention with symptomatic coronary atherosclerosis, possess HDL particles whose anti-atherogenic properties are impaired [3]. These considerations have led to a concept based on epidemiologic studies in which HDL constitutes a key target for the prevention and therapy of coronary heart disease, and which postulates that raising HDL-C should attenuate atherogenesis and thereby decrease cardiovascular risk [4–9]. Furthermore, the association between low HDL-C levels and the functional deficiency of HDL particles suggests that the impaired anti-atherogenic activity of HDL may be corrected by therapeutic approaches targeted to raise HDL-C and apoA-I levels and concomitantly to normalize HDL structure, composition and metabolism [4–9].

This concept becomes particularly attractive in light of the **high residual cardiovascular risk** of approximately 60%–70% that persists in subjects treated with statins and other lipid-lowering drugs. Indeed, despite achieving treatment targets for LDL-C, blood pressure and glycemia in accordance with current standards of care, patients with dyslipidemia remain at high residual risk of vascular

High-Density Lipoproteins: Structure, Metabolism, Function, and Therapeutics, First Edition.
Anatol Kontush and M. John Chapman.

events [10]. These observations become even more relevant in view of the underutilization of HDL-targeting drugs, such as niacin and fibrates for the treatment of low HDL-C in clinical practice [11]. The concept of pharmacotherapeutic **HDL-C raising** does not, however, take into account the fact that low HDL-C levels are frequently accompanied by other pro-atherogenic conditions, such as elevated levels of triglyceride-rich lipoproteins. In addition, available data on cardiovascular risk in patients with genetic forms of low HDL-C dyslipidemias are frequently inconsistent with the cardioprotective role of HDL-C raising. Indeed, no increase in cardiovascular risk is observed in several genetically-determined conditions associated with low HDL-C levels [12].

OPEN QUESTION

To what extent can HDL raising decrease residual cardiovascular risk in statin-treated patients?

The ultimate goal of developing therapeutic strategies that modulate HDL metabolism is not, however, an increase in HDL-C levels as such, but rather a decrease in cardiovascular risk. As low HDL-C levels do not always reflect a functional deficiency of HDL particles [3], such a decrease in risk may, theoretically, be achieved through an enhancement of HDL function in the absence of elevation in HDL-C levels [13]. Indeed, HDL-C is an integrative, but non-functional, measure of HDL particle size, composition and number [14]. Therefore, HDL-raising therapies should ideally target normalization of both HDL and apoA-I levels (i.e., particle numbers) and HDL functionality. As a corollary, measurement of HDL-C should be combined with the evaluation of HDL function when assessing the benefit of HDL-targeted therapies. *In vitro* assays focused on surrogate markers of HDL functionality *in vivo* need to be applied in studies of the therapeutic efficacy of HDL-raising agents [15–17].

The importance of such comprehensive assessment is further highlighted by the fact that the value of HDL-C as a surrogate end-point in predicting the reduced progression of atherosclerosis or a decrease in clinical events may be highly dependent on the mechanism of HDL-C raising [18]. The ambiguity of HDL-C as a therapeutic target is also consistent with the frequently observed dissociation between increases in HDL-C concentration and acceleration of RCT, especially if an increase in HDL-C levels results from the inhibition of cholesterol flux through HDL, such as in SR-BI deficiency [19].

According to the response-to retention hypothesis of atherosclerosis, HDL-targeting strategies should primarily promote the removal of extracellular lipid pools and intracellular atherogenic components of lipoproteins retained in the arterial intima and should lend to a diminished local pro-atherogenic response with plaque stabilization and/or regression [20]. Such considerations suggest that improvement of both the cholesterol efflux capacity together with anti-inflammatory/antioxidative activities of HDL particles should represent a major target. Specifically, the net flux of cholesterol from the arterial intima to the bile through the RCT pathway should be accelerated by an ideal HDL-targeting therapy along with decrease in the local inflammatory response and oxidative stress in the arterial wall. Development of such therapy is, however, complicated by the fact that the optimal point for intervention in the RCT pathway remains unknown [13]. Direct demonstration of the causative link between therapeutic correction of HDL functionality and delayed atherogenesis is, therefore, still outstanding.

MAJOR CONTROVERSY

Should normalization of diminished HDL-C levels, correction of defective anti-atherogenic HDL function, or both, be at the focus of therapeutic approaches targeting plasma HDL?

The key criterion by which to assess the value of the HDL-raising concept involves its capacity in affording clinical benefit by decreasing cardiovascular events and mortality. Several large-scale, randomized clinical trials, including the Armed Forces Regression Study (AFREGS) trial [21], the BIP trial [22], the VA-HIT trial [23, 24], the HDL-Atherosclerosis Treatment Study (HATS) [25], the Arterial Biology for the Investigation of the Treatment Effects of Reducing Cholesterol (ARBITER) 2 trial [26] and Arterial Biology for the Investigation of the Treatment Effects of Reducing Cholesterol 6-HDL and LDL Treatment Strategies in Atherosclerosis (ARBITER 6-HALTS) trial [27] have demonstrated that the therapeutic raising of HDL-C levels may result in slowed progression of atherosclerosis and reduced cardiovascular risk. In these studies, however, HDL-C raising was accompanied by significant decreases in circulating levels of triglyceride-rich lipoproteins, as well as by other potentially anti-atherogenic effects, such as LDL-C reduction. Direct evidence for the cardioprotective role of HDL-C raising, independently of changes in other lipoprotein classes, is, therefore, lacking in humans.

The relationship between the decrease in cardiovascular risk and an increase in HDL-C was analyzed across a wide range of populations in a number of *post-hoc* studies [28]. The rationale for such studies is provided by the correlation between decreased cardiovascular risk and increased HDL-C concentrations observed over recent decades at a population level. Indeed, the estimated net risk for cardiovascular disease in the USA decreased markedly from 1976–1980 to 1988–1994 and further improved slightly from 1988–1994 to 1999–2004, particularly in women and middle-aged persons. In parallel, HDL-C levels showed a moderate increase from 50 to 51 to 52 mg/dl [29]. Furthermore, there was a decrease in the incidence of dyslipidemia with an increase in HDL-C and a decrease in triglyceride levels over a 10-year period of follow-up in the Framingham Heart Study [30]. Specifically, mean HDL-C levels were significantly increased with multivariate-adjusted means of 44.4 and 46.6 mg/dl in men and 56.9 and 60.1 mg/dl in women (p value for trend <0.001 in both sexes). Similar trends have been observed in the UK, where HDL-C, BMI and physical activity levels rose between 1978 and 2003 after adjustment for age, whereas cigarette smoking, systolic blood pressure and non-HDL-C decreased [31]. In parallel, the age-adjusted hazard of myocardial infarction decreased by −3.8% (95% CI 2.6%–5.0%) per annum over the same period in men, which corresponds to a −62% decline over 25 years [31]. The fall in cigarette smoking explained the greatest part of the decline in the incidence of myocardial infarction (−23%), followed by changes in blood pressure (−13%), HDL-C (+12%) and non-HDL-C (−10%). Favorable changes in HDL-C levels therefore appear to have contributed to a considerable reduction in the incidence of cardiovascular disease in the UK [31]. Consistent with these data, sequential changes in HDL-C predicted future coronary heart disease events in a cohort of 6928 subjects with at least two measurements of HDL-C separated by an average of 2.6 years [32]. After multiple adjustment, a +10 mg/dl increase in HDL-C between lipid measurements was associated with a −7% (95% CI 3%–10%) lower risk of events in this study. In contrast, neither initial triglyceride levels nor a change in triglyceride predicted subsequent coronary events.

The independent effect of HDL-C raising on cardiovascular risk after lipid therapy with an average follow-up of eight years was estimated after adjustment for changes in LDL-C, triglyceride and pre-treatment blood lipid levels using data from the Framingham Offspring Study from 1975 to 2003 [33]. Potential confounders (e.g., smoking status, weight and the use of beta-blockers) were included in the model. The analyses revealed that the change in HDL-C level was a strong independent risk factor for cardiovascular events (hazard ratio, 0.79 per +5 mg/dl increase; 95% CI, 0.67–0.93). Overall, a +1% increase in HDL-C level was associated with a −2% drop in cardiovascular risk, consistent with epidemiologic data (see 5.1 of Chapter 5). This relationship remained stable across a wide range of patient subgroups and did not appear to be associated with a specific drug class. Importantly, the impact of raising HDL-C depended on pre-treatment LDL-C levels, increasing with decreasing baseline LDL-C. This study, despite being limited by the low number of cardiovascular events ($n = 79$) and residual confounding, provides evidence to support the hypothesis that raising HDL-C levels is associated with a reduction in cardiovascular risk [33].

The association between changes in HDL-C concentrations and the risk of subsequent coronary heart disease was also investigated in a prospective cohort of 4501 male physicians enrolled in the Physician's Health Study I whose HDL-C was measured in 1982 and again approximately 14 years later [34]. The Cox proportional hazards model was used to examine the association between changes in HDL-C and incident coronary heart disease. Hazard ratios were adjusted for age, baseline HDL-C, diabetes mellitus, hypertension, non-HDL-C and history of cholesterol medication. The data revealed that, compared with subjects with a decrease in HDL-C levels, multivariable-adjusted hazard ratios for coronary heart disease were 0.66 (95% CI 0.40–1.09) in subjects with no change, 0.56 (95% CI 0.35–0.89) for subjects with an increase of +2.5 to +12.5 mg/dl and 0.43 (95% CI 0.23–0.83) for subjects with an HDL-C increase of $\geq +12.5$ mg/dl [34]. These data are further supported by the analysis of determinants of cardiovascular risk in patients with stable coronary disease enrolled in the AFREGS study to receive gemfibrozil, niacin and cholestyramine in combination or matching placebos for 30 months [35]. When such patients were divided into three groups based on their therapeutic response, specifically no HDL-C increase, mild HDL-C increase and large HDL-C increase, a progressive decrease in cardiovascular events was noted across these groups (30.4%, 19.4% and 3.2%, respectively). Proportional hazards modeling also demonstrated that increasing HDL predicted a lower hazard of cardiovascular events, even after adjusting for changes in LDL-C. For every +1% increase in HDL-C achieved, a −2% decrease in events was recognized [35].

In another study, the effects of LDL-C lowering alone and of LDL-C lowering combined with HDL-C raising on coronary stenosis progression and major cardiovascular events were compared using individual patient data from 445 subjects who participated in three large-scale, double-blinded, randomized, placebo-controlled trials [Familial Atherosclerosis Treatment Study (FATS), HATS, and AFREGS] on coronary stenosis progression by quantitative coronary angiography [36]. Calculation showed that each −10% decrease in LDL-C or +10% increase in HDL-C was significantly associated with a −0.3% decrease in mean proximal coronary diameter stenosis over three years. Furthermore, each −10% decrease in LDL-C or +10% increase in HDL-C was associated with a −11% or a −22% decrease in the risk of cardiovascular event, respectively. Significantly, more rapid progression of coronary stenosis was independently associated with an increased risk of cardiovascular events, thereby validating measurements of coronary diameter as a surrogate for cardiovascular risk [36]. In a similar fashion, evaluation of the six-year progression of carotid intima-media thickness in 1809 young adults aged 32 ± 5 years revealed that of the all risk factors, only a change in HDL-C was significantly, and negatively, associated with a change in intima-media thickness in a stepwise multivariate regression analysis [37].

Such a burden of evidence in support of the beneficial impact of HDL-C raising on cardiovascular disease is counteracted by results of a meta-analysis of the association between treatment-induced change in HDL-C and cardiovascular risk [38]. When the results of 108 randomized trials of lipid-modifying interventions involving 299,310 participants at risk of cardiovascular events were adjusted for changes in LDL-C, no association between treatment-induced change in HDL-C and risk ratios for coronary heart disease deaths, coronary heart disease events, or total deaths was observed. With all trials included, a change in HDL-C explained almost no variability (<1%) in any of the outcomes, whereas a change in LDL-C explained 32% of the variability in coronary heart disease event risk [38]. Furthermore, a meta-analysis of randomized trials of cholesterol-lowering treatments in cardiovascular disease reporting on stroke found no significant association between stroke reduction and changes in HDL-C levels [39]. As for each meta-analysis, however, these studies are limited by the absence of access to individual patient data. In addition, trials included in the analyses are highly heterogeneous in terms of the intervention employed, with statin trials clearly prevailing which may, therefore, have overestimated the role of LDL-C reduction in the decrease in cardiovascular risk. Furthermore, the variation in changes in HDL-C levels over the included trials is relatively small, being less pronounced than, for example, the wide variation in changes in HDL-C levels from −37 to +35 mg/dl observed in the Framingham Offspring Study, which, in addition, had access to individual patient data [33].

OPEN QUESTION

Are increases in HDL-C concentrations related to decreases in cardiovascular risk in prospective epidemiologic studies?

The heterogeneity in the relationship between the elevation in HDL-C levels relative to cardiovascular risk reinforces the concept that the mechanistic basis of elevation in HDL-C can be critically important [40]. Indeed, genetic studies suggest that certain genetic mechanisms that lead to an elevation in HDL-C do not lower cardiovascular risk [41]; cardiovascular risk cannot thus be expected to decrease following all HDL-C-raising interventions. A final answer as to whether HDL-raising agents can decrease cardiovascular risk and, if so, which of them, will only be provided by outcome studies.

Approaches to the elevation of HDL-C concentrations attempt to improve HDL metabolism and RCT at two distinct levels: by enhancing HDL formation and by normalizing intravascular HDL remodeling. **Enhanced HDL formation** is the focus of therapeutic strategies that upregulate apoA-I synthesis in hepatocytes and accelerate the efflux of cholesterol and phospholipid from peripheral cells, mediated by ABCA1. Treatment with apoA-I, apoA-I mimetics and rHDL represent the most known interventions of this class. On the other hand, accelerated lipidation of apoA-I, attenuated activity of CETP resulting in diminished heteroexchange of cholesteryl ester and triglyceride between HDL and triglyceride-rich lipoproteins, and the inhibition of HDL uptake by the liver and/or kidney mediated by SR-BI and/or hitherto unidentified receptor(s) for HDL holoparticles provide a framework for approaches aimed at **improving intravascular HDL remodeling**. CETP inhibitors represent well known examples of this class of agents. Not infrequently, the same drug can impact both the formation and remodeling of HDL, with fibrates, statins and niacin as examples. Importantly, interventions aimed at enhanced HDL formation and improved HDL remodeling differ in their effects on HDL particle profiles. Specifically, apoA-I mimetics and rHDL preferentially raise small, cholesteryl ester-poor HDL, whereas CETP inhibitors, niacin and statins rather preferentially increase circulating levels of large, cholesteryl ester-rich, mature HDL particles. Intriguingly, data are accumulating to support the view that raising functional small HDL particles may be more important in terms of cardiovascular protection than generating large, cholesterol-rich HDL [42, 43].

CHAPTER 11

ENHANCEMENT OF HDL FORMATION AND NORMALIZATION OF INTRAVASCULAR HDL REMODELING

11.1 APOLIPOPROTEINS

The apolipoprotein moiety of HDL is intimately involved in the multiple atheroprotective activities of this lipoprotein and therefore represents the most obvious therapeutic target [44]. Among HDL apolipoproteins, **apoA-I** appears to be the most promising given its capacity to delay atherogenesis and also in view of its wide spectrum of biologic activities (Table 7), hence the hypothesis of apoA-I raising instead of HDL-C raising [45].

It is of clinical importance that lipidation of apoA-I induces the formation of **small pre-beta HDL** particles, thereby initiating the metabolic HDL cycle (Fig. 11.1 and Table 11.1). Direct intravenous administration of apoA-I is a straightforward approach that beneficially acts on atherosclerotic plaques in animal models via reduced inflammation and enhanced stability [46]; however, rapid renal filtration is a major obstacle to such treatment [47]. In addition, apoA-I is a large protein that requires venous administration and is both difficult and expensive to manufacture in endotoxin-free form. In contrast, the selective increase in circulating levels of apoA-I through **gene therapy** represents a plausible strategy to enhance HDL formation, increase HDL levels and inhibit

High-Density Lipoproteins: Structure, Metabolism, Function, and Therapeutics, First Edition.
Anatol Kontush and M. John Chapman.

TABLE 11.1 Therapeutic Agents that Enhance HDL Formation

Agents	Administration Method	Mechanisms of Action	Effects on HDL Levels	Effects on HDL Function	Limitations
Apolipoproteins	Intravenous injection Gene transfer	Enhanced nascent HDL formation	Elevation of apoA-I and HDL-C levels (i/v: transient) Increase in small HDL levels	Enhancement of cholesterol efflux capacity	Transient effects (i/v) Hypertri-glyceridemia Expensive to manufacture
Reconstituted HDL	Intravenous injection	Enhanced nascent HDL formation	Transient elevation of apoA-I and HDL-C levels Increase in small HDL levels	Enhancement of major HDL functions	Transient effects Expensive to manufacture
Apolipoprotein-mimetic peptides	Intravenous injection Oral treatment	Enhanced nascent HDL formation	Transient elevation of apoA-I and HDL-C levels Increase in small HDL levels	Enhancement of major HDL functions	Transient effects Hypertri-glyceridemia

PPAR alpha agonists	Oral treatment	Activation of apoA-I, apoA-II and ABCA1 expression Enhancement of apoA-I and apoA-II catabolism	Elevation of apoA-I, apoA-II and HDL-C levels Increases in small and medium HDL levels	Enhancement of cholesterol efflux capacity and anti-inflammatory activity	Myopathy Gastrointestinal effects
LXR agonists	Oral treatment	Upregulation of ABCA1 and ABCG1 expression	Elevation of apoA-I and HDL-C levels	Unknown	Increased CETP activity and non-HDL-C levels
Statins	Oral treatment	Upregulation of apoA-I expression	Elevation of apoA-I and HDL-C levels. Preferential increase in large HDL	Enhancement of anti-inflammatory activity	Myopathy Hepatotoxicity

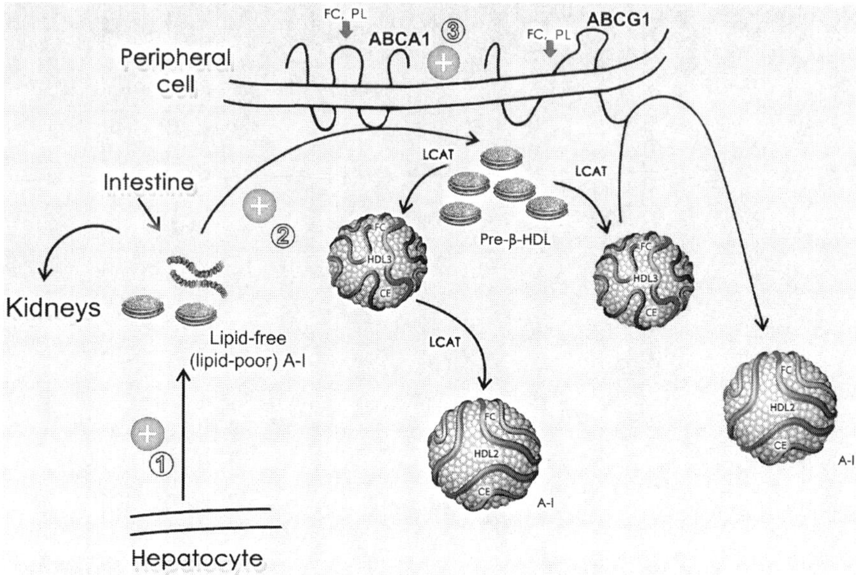

Figure 11.1 Major pathways involved in enhanced HDL formation. (1) Accelerated apoA-I production in hepatocytes (PPAR alpha agonists, PPAR gamma agonists, statins, RVX-208, pioglitazone, estrogens, niacin). (2) Accelerated lipidation of apoA-I as a result of an increase in the plasma apoA-I pool (apolipoproteins, rHDL, apoA-I mimetics). (3) Accelerated efflux of cholesterol from peripheral cells and enhanced lipidation of apoA-I as a result of the induction of ABCA1 (PPAR alpha agonists, LXR agonists, niacin). CE, cholesteryl ester; FC, free cholesterol; PL, phospholipid.

atherosclerosis. Indeed, HDL-C levels can be elevated and early atherogenesis can be inhibited by human apoA-I gene transfer to atherosclerosis-susceptible mice [48–52] as well as to cholesterol-fed rabbits [53]. Interestingly, local expression of human apoA-I in macrophages is sufficient to protect against atherogenesis without affecting plasma apoA-I and HDL-C levels [54]. In a similar fashion, expression of the human APOA1/APOC3/APOA4 gene cluster also reduces atherogenesis in mice [55, 56].

Enhanced **cholesterol removal** from macrophages via the RCT pathway underlies the atheroprotective effects of apoA-I gene transfer [56] (Table 11.1). In addition, long-term over-expression of apoA-I remodels the lesions to a more stable phenotype, consistent with the **enhanced anti-inflammatory and antioxidative activities** of HDL [57]. Remarkably, such improvements in the cardioprotective properties of HDL are paralleled by elevated levels of small pre-beta-HDL particles [52] consistent with their potent atheroprotective effects [42, 43]. Interestingly, vector-based gene transfer of wild-type human apoA-I and apoA-I Milano into LDL-receptor-deficient or apoE-deficient mice shows similar attenuation of atherosclerosis progression [13, 58, 59], despite the enhanced cardioprotective properties of apoA-I Milano in humans [60] (see 5.1 of chapter 5). The more atherogenic lipid profile in apoA-I Milano-expressing mice involving both elevated triglyceride and low HDL-C levels, together with similar capacities of wild-type apoA-I and apoA-I Milano to induce cholesterol efflux, may account for this observation [61]. In contrast, the expression of human apoA-II appears to exaggerate atherosclerosis in mouse models, potentially acting via inhibition of the RCT pathway and induction of hypertriglyceridemia [62–65]; the degree of apoA-II expression is, however, critical for its effects on atherogenesis.

ApoA-I gene delivery by viral vectors is, therefore, a promising strategy to raise circulating levels of both apoA-I and HDL and to reduce atherosclerosis. This approach may be particularly relevant in inherited apoA-I deficiency, an orphan disorder characterized by HDL-C deficiency and premature atherosclerosis [66]. Sustained gene expression in animals and humans can be attained using adeno-associated viral vectors. Indeed, delivery of human apoA-I using an adeno-associated viral serotype 8 vector results in a significant, and sustained, increase in apoA-I and HDL plasma levels for up to 16 weeks in apoA-I deficient mice [67]. HDL particles generated by this approach are fully functional in terms of cholesterol removal from cholesterol-loaded macrophages [67]. In a similar fashion, systemic delivery of a self-complementary adeno-associated viral serotype 8 vector provides a means for efficient and prolonged liver expression of human apoA-I in apoA-I- deficient mice, thereby correcting lipid abnormalities associated with murine apoA-I deficiency [68]. Together, data obtained using apoA-I transgenic animals show that, in contrast to the lack of unambiguous data on the anti-atherogenic role of HDL in humans, there is strong evidence for a direct causal role of HDL in slowing atherogenesis in animal models. The development of gene transfer technologies with a sufficiently high therapeutic index may, therefore, pave the road for selective and effective HDL-raising therapeutic intervention based on enhanced production of apoA-I [18].

Modification of apoA-I to avoid its rapid elimination through renal filtration represents an alternative to gene therapy. Using protein-engineering technology, apoA-I can be fused to the trimerization domain of human tetranectin in order to obtain a high-mass recombinant **trimeric apoA-I** [47]. The recombinant fusion protein is stable and can be highly expressed; upon purification and intravenous injection into mice, it exhibits prolonged plasma retention time compared with wild-type apoA-I. Moreover, trimeric apoA-I is biologically active in terms of the promotion of cholesterol efflux, stimulation of LCAT activity and the reduction of atherosclerosis progression in cholesterol-fed, LDL receptor-deficient mice [47]. Treatment with trimeric apoA-I may also beneficially affect the composition of atherosclerotic plaques [69]. Administration of recombinant high-mass apoA-I analogues with retarded clearance may, therefore, constitute a novel therapeutic approach to stabilize plaques and/or to reduce atherosclerosis, which is under development [70].

11.2 RECONSTITUTED HDL

DEFINITION

Reconstituted HDL: HDL particles prepared *in vitro* from purified human apoA-I and phospholipid as the essential components, potentially with other proteins and/or lipids.

Effects on HDL Levels and Metabolism

Reconstituted HDL (rHDL), sometimes termed synthetic HDL, typically consists of **apoA-I** and **phospholipid**, but may also include other proteins and/or lipids [71]. rHDL may provide an innovative approach to the management of cardiovascular disease by its ability to rapidly raise circulating HDL levels upon intravenous injection and to acutely impact atherosclerotic plaque [72] (Fig. 11.1 and Table 11.1). Raising HDL by infusing rHDL can be viewed as a replacement-type therapy because rHDL from an exogenous source is administrated to a patient [70]. Such HDL-replacement therapy represents a promising treatment strategy involving the acute administration of HDL to rapidly stabilize patients at imminent risk for future development of a myocardial infarction who frequently display diffuse atherosclerosis, or in those who present an acute coronary syndrome or unstable angina [70].

Reconstituted HDL therapy can have a surprisingly rapid effect in reducing plaque volume and in inducing egress of cholesterol-loaded cells from the plaque. As a result of the potent atheroprotective properties of **apoA-I Milano** (see 5.1 of chapter 5), rHDL containing this variant offers an especially promising approach in treating cardiovascular disease [73], particularly under acute conditions at the diseased site, namely the vulnerable, unstable atherosclerotic plaque [74]. Consistent with this notion, only five, weekly infusions of **ETC-216** (a complex of apoA-I Milano with phosphatidylcholine) at 15 or 45 mg/kg induced a −4.2% reduction in plaque volume relative to the baseline in patients with acute coronary syndromes, as assessed by intravascular ultrasound [75]. These promising findings outperform typical reductions in plaque volume established after statin therapy, one of the most efficient approaches available at present for a reduction in the progression of atherosclerosis—i.e., −2.7% and −1.9% reduction after 11 months of intensive therapy with 20 mg of rosuvastatin and 40 mg of atorvastatin, respectively [76].

In support of these data, short-term infusions of another rHDL preparation, **CSL-111**, which consists of wild-type human apoA-I and soybean phosphatidylcholine at a molar ratio of 1:150, significantly impacted on coronary atherosclerosis in the randomized, placebo-controlled Effect of rHDL on Atherosclerosis-Safety and Efficacy (ERASE) trial performed in patients with coronary artery disease [81]. Specifically, four, weekly infusions of 40 mg/kg of CSL-111 resulted in a decrease in atheroma volume (−3.4%) compared with placebo (−1.6%; p = 0.48 between groups, $p < 0.001$ vs. baseline for CSL-111). Furthermore, mean changes in plaque characterization index by intravascular ultrasound and mean changes in coronary score on quantitative coronary angiography were significantly different between the groups ($p = 0.01$ and 0.03, respectively). Further development of this promising approach in humans is, however, hampered by high intervention costs which result from the large amounts (up to 5 g) of apoA-I required per injection [70]. Another limitation involves potential hepatotoxicity as a result of residual cholate used for rHDL assembly [70]. Indeed, administration of CSL-111 at 40 mg/kg was associated with mild, self-limiting transaminase elevation, while the higher dosage of 80 mg/kg was discontinued early because of liver function test abnormalities [81].

CRITICAL CONTRIBUTIONS

Rapid regression of atherosclerotic plaques following only five, weekly intravenous infusions of ETC-216, a complex of apoA-I Milano with phosphatidylcholine, at 15 or 45 mg/kg in patients with acute coronary syndromes [75, 77].

Acute changes in lipid content and inflammatory status of atherosclerotic plaques in patients with peripheral vascular disease following a single infusion of CSL-111, a complex of wild-type human apoA-I with phosphatidylcholine, at 80 mg/kg [78].

Selective elevation in circulating concentrations of small HDL particles and a trend to reduced atheroma volume following seven, weekly injections of de-lipidated HDL in patients with acute coronary syndrome undergoing cardiac catheterization [79, 80].

A rationale for such studies of rHDL in humans has been provided by earlier results documenting a potent reduction of atherosclerosis by apoA-I Milano-containing rHDL in atherosclerotic rabbits [82, 83] and mice [84, 85]. The effects of rHDL on experimental atherogenesis are concentration-dependent. Indeed, in cholesterol-fed rabbits undergoing perivascular injury at both carotid arteries, lower doses of ETC-216 delay progression of carotid atherosclerosis, whereas higher doses of up to 150 mg/kg every four days induce significant regression of the disease (up to −6.8%) with only two administrations at the highest dose [86]. Significantly, the administration of rHDL containing apoA-I or apoA-I Milano results in the accumulation of **small HDL** particles [72] (Table 11.1) which also predominate in subjects with apoA-I Milano [73], consistent with the hypothesis regarding the potent cardioprotective properties of small, apoA-I-containing HDL [42].

Selective elevation in circulating concentrations of small HDL particles forms the basis for an alternative approach to raising plasma HDL levels that involves re-infusion of HDL after **selective *in vitro* de-lipidation** [79, 87]. This innovative approach, which involves treating plasma with sevoflourante and n-butanol, converts large, lipid-rich HDL to small, lipid-poor particles and specifically increases pre-beta-1-HDL concentrations (up to +79% of total plasma apoA-I). Moreover, selective de-lipidation markedly increases (sevenfold) the capacity of plasma to stimulate cholesterol efflux from monocytes via ABCA1, but decreases efflux via SR-BI. Treatment with de-lipidated plasma tends to reduce (−6.9%) diet-induced aortic atherosclerosis in African Green monkeys, as evaluated by intravascular ultrasound [79]. Furthermore, total atheroma volume tends to be reduced by seven, weekly injections of selectively de-lipidated HDL in patients with acute coronary syndrome undergoing cardiac catheterization compared with control plasma apheresis treatment [80]. All re-infusion sessions were tolerated well by all patients involved in this study, suggesting the feasibility of this approach as a novel adjunct therapy for acute coronary syndrome [80].

METHODOLOGIC NOTE

Administration of rHDL and apoA-I mimetic peptides involves intravenous infusion, requires large amounts of biologic material and is technically and economically demanding.

Effects on HDL Functionality

Remarkably, rHDL prepared *in vitro* displays all of the major anti-atherogenic activities documented for native HDL, notably cholesterol efflux capacity, together with anti-inflammatory, antioxidative, cytoprotective, vasodilatory, anti-infectious, antithrombotic and antidiabetic activities (see chapter 7) . The regression of atherosclerosis induced by rHDL is probably most related to accelerated **cholesterol efflux** from the arterial wall with enhanced RCT to the liver [75]. The key role of RCT is highlighted by the capacity of rHDL to acutely induce changes in the lipid content of atherosclerotic plaques. Indeed, intravenous infusion of a single dose of CSL-111 (80 mg/kg over 4 hours) leads to acute changes in plaque characteristics in patients with peripheral vascular disease compared with placebo, with a reduction in lipid content (−62%), macrophage size (−58%) and measures of inflammation (VCAM-1 expression; −28% vs. −50%) [78]. The level of HDL-C rises by +20% after infusion of rHDL and the capacity of apoB-depleted plasma to efflux cholesterol increases in parallel, suggesting enhancement of RCT as an underlying mechanism. These findings are corroborated by data obtained in patients with Type 2 diabetes receiving rHDL infusion at 80 mg/kg who featured an increased capacity of plasma (+by 40% to +60%) to efflux cholesterol from THP-1 macrophages by 1 h up to 72 h post-infusion [88].

Other anti-atherosclerotic activities of rHDL may also contribute to plaque regression in humans, consistent with the potent anti-inflammatory and vasodilatory properties of apoA-I Milano used as the basis for ETC-216 [89]. Thus, in a randomized, cross-over, placebo-controlled study in patients with Type 2 diabetes, rHDL infusion (80 mg/kg of apoA-I) improved the **anti-inflammatory properties** of isolated HDL at 4 and 72 h post-infusion, along with increases in plasma HDL-C concentration (by up to +25%) [88]. Peripheral blood monocyte CD11b expression and neutrophil adhesion to a fibrinogen matrix were reduced 72 h post-rHDL compared with placebo [88]. Furthermore, levels of TNF-alpha and monocyte activation (as CD11b expression) were decreased by a single CSL-111 infusion in patients with peripheral vascular disease [78].

Earlier data reveal that infusion of rHDL leads to the inhibition of adhesion molecule expression, attenuation of endotoxin-induced release of pro-inflammatory cytokines, reduced ROS generation and stabilized vulnerable plaque in animal models [90–94]. These results are consistent with findings in mice which show that rHDL containing wild-type apoA-I and infused into apoE-deficient animals (at a dose of 120 mg/kg of body weight) develops anti-inflammatory and **antioxidative activity** [95]. Specifically, rHDL reduces the lipid content of atherosclerotic plaques and the accumulation of macrophages in the lesion area, increases serum PON1 activity and decreases circulating levels of lipid peroxidation products. Interestingly, point apoA-I mutants V156K and R173C are even more potent than wild-type apoA-I in their anti-atherogenic effects, consistent with their lower susceptibility to chlorination by myeloperoxidase [95]. Similarly, enhanced lesion regression activity paralleled by improved anti-inflammatory and antioxidative capacities are features of another point mutant of apoA-I, V156E that naturally occurs in Oita, Japan [96].

The existence of potent antioxidative activity in rHDL particles is further supported by their capacity to inhibit superoxide production and NADPH oxidase activation in leukocytes [97]; altered assembly of NADPH oxidase subunits in lipid rafts, as a result of cholesterol efflux from cell membranes, and lipid raft disruption probably underlies this effect. In addition, rHDL injection favorably affects the distribution of antioxidative enzymes, particularly PAF-AH, between HDL and other lipoproteins in the circulation, causing a shift to HDL particles [98].

Next, rHDL can display significant **cytoprotective effects**, as illustrated by improved post-infarction survival of myocytes in rats, resulting in the attenuation of left ventricular remodeling after rHDL therapy [99]. In addition, rHDL increases circulating levels of endothelial progenitor cells in patients with Type 2 diabetes [100]. The **anti-infectious properties** of rHDL have been documented as its capacity to neutralize the toxicity of LPS in mice [101]. The potent **vasodilatory activity** of rHDL has been directly demonstrated in humans. Notably, rHDL infusion (80 mg/kg over 4 hours) increases plasma HDL-C 1.7-fold and enhances acetylcholine-induced increase in forearm blood flow in hypercholesterolemic subjects [102]. Moreover, rHDL infusion improves flow-mediated dilatation of the brachial artery (from to 2.7% to 4.5%), acting by increasing NO bioavailability. Infusion of rHDL also increases plasma HDL-C 3.2-fold and completely restores endothelial dysfunction through improved basal and stimulated NO bioactivity in subjects with familial hypoalphalipoproteinemia [103]. In contrast, a single infusion of rHDL at 80 mg/kg does not improve vascular function in patients with acute coronary syndromes, despite increasing plasma HDL-C and decreasing LDL-C [104].

An additional mechanism implicated in the anti-atherogenic action of rHDL involves inhibition of elevated platelet reactivity, as shown in patients with Type 2 diabetes. Indeed, rHDL infusion (as CSL-111 at 80 mg/kg for 4 hours) increases plasma HDL-C levels 1.4-fold and results in a >50% reduction in the *ex vivo* platelet aggregation response to multiple agonists [105]. *In vitro* studies in platelets from healthy individuals reveal that the **antithrombotic effects** of rHDL are time- and dose-dependent and result in the attenuation of platelet function and a reduction in thrombus formation (−50%) under flow. These effects can, in part, be accounted for by the phospholipid component of rHDL which enhances cholesterol efflux from platelets and reduces lipid raft assembly [105].

In addition to a potent impact on atherosclerosis, rHDL can correct glucose metabolism, consistent with the **antidiabetic activity** of native HDL. Thus, in a double-blind, placebo-controlled crossover study, rHDL infusion (80 mg/kg over 4h) acutely decreased plasma glucose and increased plasma insulin and HOMA index in patients with Type 2 diabetes [106]. Acetyl-CoA carboxylase beta phosphorylation in skeletal muscle biopsies was elevated 1.7-fold after rHDL infusion, indicative of the activation of the AMPK pathway. Furthermore, rHDL enhanced glucose uptake in primary human skeletal muscle cell cultures derived from patients with Type 2 diabetes, acting via the stimulation of ABCA1 with subsequent activation of the calcium/calmodulin-dependent protein kinase kinase and AMPK [106].

Role of HDL Components

With respect to the role of rHDL components in rHDL-mediated atheroprotection, **apoA-I** undoubtedly accounts for the majority of the observed effects, consistent with its pivotal role in the biologic activities of native HDL. The inclusion of minor apolipoproteins may also lead to profound biologic effects; thus, rHDL containing apoA-V lowers plasma triglyceride levels in a murine model [107]. The role of the **phospholipid** moiety appears to be modulatory, although bioactive lipids, such as S1P, when present, may contribute significantly to some effects of rHDL. Consistent with the role of distinct phospholipid species, the capacity of rHDL infusion to reduce atheroma size in cholesterol-fed rabbits depends, to a major degree, on the phospholipid composition of the particle [13, 108].

To further enhance the anti-inflammatory/antioxidative activities of rHDL particles, the latter can be formulated to integrate atheroprotective enzymes, such as **PON1**, which can hydrolyze pro-inflammatory lactones and/or oxidized phospholipids. As human PON1 exhibits limited stability, engineered, recombinant PON1 variants designed for higher reactivity, solubility, stability and bacterial expression, are promising candidates for treatment [109]. The lifespan of such engineered forms of PON1 can be improved relative to the wild-type enzyme. Furthermore, no adverse immune responses or other toxic effects are observed in mice subjected to repetitive administration of rHDL containing engineered enzyme [109].

Another interesting aspect of the use of rHDL particles in cardiovascular medicine involves their utilization as a targeted carrier of **biologically active agents**. In terms of modern technology, HDL represents an endogenous nanoparticle that transports lipids throughout the body, interacts with multiple cellular receptors and can be readily and flexibly reconstituted *in vitro*. Such an interpretation opens the way for the potential utilization of rHDL as a vehicle for the transport of contrasting agents, dyes, drugs and other molecules to targeted organs [110]. Importantly, the endogenous nature of rHDL particles renders them biocompatible and biodegradable and avoids recognition by the reticulo-endothelial system [111]. Indeed, HDL can be employed for the targeting and imaging of atherosclerotic plaques as it can deliver contrast-generating materials, fluorophores and nanocrystals to atherosclerotic lesions, allowing their imaging by magnetic resonance, fluorescent techniques and computed tomography [111–114]. Such rHDL particles containing **imaging agents**, such as gadolinium, rhodamine, gold, iron oxide, quantum dot nanocrystals or carboxyfluoresceine-labeled apolipoprotein-derived peptides, are macrophage-specific and can, therefore, provide information on this cell type in a non-invasive manner [112, 115, 116]. As intraplaque macrophage density is directly related to plaque vulnerability, use of macrophage-targeted rHDL may allow for the superior detection of vulnerable plaques [116]. This approach can also be used to study the metabolism of HDL particles and their interactions with arterial wall cells *in vivo* [117].

Similarly, rHDL can be used as a carrier for the delivery of lipophilic **cardiovascular drug**s, as exemplified by co-enzyme Q10 [118], curcumin [119] and Tanshinone IIA, a diterpenoid naphthoquinone found in the traditional Chinese medicine, Tanshen (*Salvia miltiorrhiza*), and widely used in treatment of cardiovascular disorders [120]. Furthermore, lipophilic peptides with significant anti-hepatitis B virus activity, such as nosiheptide, can be specifically delivered to hepatocytes when complexed to rHDL [121].

In a similar way, HDL can be employed as a multimodal nanoparticle platform for tumor targeting and imaging via nonspecific accumulation and specific binding to angiogenically activated blood vessels [122]. For imaging purposes, rHDL is prepared with amphiphilic gadolinium chelates and fluorescent dyes. For deep tissue imaging, a stable analog of bacteriochlorophyll can be transported by rHDL to tumor cells via SR-BI [123]. To target angiogenic endothelial cells, rHDL is functionalized with alpha v beta 3-integrin-specific Arg-Glycine (Gly)-Asp (RGD) peptides [122]. Reconstituted HDL may also target epidermal growth factor receptor on tumor cells [124]. Paclitaxel, a mitotic inhibitor used in cancer chemotherapy, can be efficiently delivered to tumor cells in a complex with rHDL via a selective uptake mechanism mediated by SR-BI [125, 126].

11.3 APOLIPOPROTEIN-MIMETIC PEPTIDES

DEFINITION

ApoA-I mimetic peptides, also termed apoA-I mimetics, are small, amphipathic, helical peptides that structurally resemble apoA-I and display similar biologic activities.

Effects on HDL Levels and Metabolism

Oral or intravenous administration of **small amphipathic helical peptides** that mimic HDL apolipoproteins represents another promising strategy to raise circulating HDL levels and to attenuate atherosclerosis [127–129] (Fig. 11.1 and Table 11.1). The use of small peptides can overcome difficulties related to the administration and manufacturing of full-length apoA-I [130]. Apolipoprotein mimetics were synthesized in the 1980s and were initially designed to mimic the amphipathic sequences of apoA-I [131]. Future development has led to apolipoprotein-mimetic peptides that frequently exhibit no sequence homology to a parent apolipoprotein, primarily to apoA-I, but rather contain structured analogs of the class A amphipathic helical motif that is present in apoA-I lipid binding domains and is responsible for solubilizing phospholipid. These analogs are similar in their lipid-associating structural motifs and interact with phospholipid to form peptide-lipid complexes [132].

Apolipoprotein-mimetic peptides typically include those **derived from apoA-I** [127] but also from **apoE** [133] and **apoJ** [134]. Notably, covalent binding of an apoE fragment, which is critical for binding to the LDL receptor, to an apoA-I mimetic endows the resulting peptide with potent cholesterol-lowering capacity and further increases its anti-atherogenic activity [133]. The major value of apoA-I-mimetic peptides is represented by their potent capacity to reduce atherosclerosis in experimental animals, including mice, rabbits and monkeys [135–138].

Effects on HDL Functionality

The potent anti-atherosclerotic properties of apoA-I mimetics are intimately related to their beneficial impact on HDL metabolism and biologic activities. Indeed, apoA-I-mimetic peptides activate cholesterol efflux from macrophages, increase PON1 activity, enhance the anti-inflammatory activity of HDL, increase endothelial production of NO, decrease endothelial production of superoxide, improve vasodilation, induce vascular heme oxygenase and superoxide dismutase, and inhibit endotoxin-induced inflammatory responses [135–137, 139–143].

Importantly, apoA-I mimetics are capable of activating all the major steps of RCT. Thus, synthetic ATI-5261 peptide stimulates ABCA1 **cholesterol efflux** with a molar potency similar to that of apoA-I and reduces atherosclerosis in apoE -/- mice fed high-fat diet [144]. Remarkably, a single intraperitoneal injection of lipid-free ATI-5261 at a dose of 20 mg/kg increases RCT in mice twofold. Mobilization of peripheral lipid is also observed in rats, where intravenous injection of ATI-5261 transiently raises plasma cholesterol concentrations. The half-life of ATI-5261 is approximately 10h in rats, suggesting the peptide binds to HDL and assumes its kinetic properties [144].

The bi-helical amphipathic peptide, **5A**, potently effluxes cholesterol from cells via ABCA1 [145]. Peptide 5A complexed to POPC, but not free 5A, also promotes cholesterol efflux by ABCG1 and activates RCT *in vivo*. When incubated with human serum, the 5A-POPC complex primarily binds to HDL and promotes the transfer of cholesterol from LDL to HDL [145]. The complex also modifies the HDL proteome, displacing proteins from HDL. Twenty-four hours after intravenous injection of 5A-POPC (30 mg/kg) into apoE-null mice, both the cholesterol (1.8-fold) and phospholipid (2.2-fold) content of HDL significantly increases [145]. As a consequence, when complexed with various phospholipids, peptide 5A reduces aortic plaque surface area by −29% to −53% in apoE-null mice.

Similarly, the **D-4F** peptide stimulates cholesterol efflux from mouse macrophages in a time- and dose-dependent manner [146]. Concomitantly, D-4F increases intracellular levels of cyclic adenosine monophosphate (AMP) and activates ABCA1 expression, promoting cholesterol efflux from macrophages through a pathway involving cyclic AMP, ABCA1 and protein kinase A [146]. In addition, the 4F peptide stimulates cholesterol efflux from monocyte-derived macrophages, leading to cholesterol depletion and disruption of lipid rafts [147].

In contrast, a dual-domain, cationic, apoE-mimetic peptide, **Ac-hE18A-NH2**, composed of the receptor binding domain (LRKLRKRLLR) of apoE covalently linked to the class A amphipathic peptide 18A, increases cholesterol efflux from THP-1 cells by an ABCA1-independent mechanism. Similar to apoE, the apoE mimetic peptide is recycled by macrophages and hepatocytes, thus enabling the potentiation and prolongation of its anti-atherogenic effects [148, 149]. Single-domain cationic peptides based on the lytic class L peptide 18L can also reduce atherosclerosis in apoE-null mice when orally administered [150].

Cholesterol esterification by LCAT can also be accelerated by apoA-I mimetics. Indeed, the D-4F peptide (20 mg/kg body weight once daily) **increases LCAT activity** and HDL-C levels, as well as apoA-I concentrations, in apoE-null mice at 72 hours post initial dosing [151]. In contrast, rHDL particles containing D-4F do not activate LCAT *in vitro*, suggesting that LCAT activation *in vivo* mediated by D-4F is related to its HDL-raising properties rather than to direct action on the enzyme. Finally, apoA-I mimetics can **activate cholesterol delivery to the liver**. Thus, D-4F-containing particles are 20 times more efficient in delivering cholesterol to hepatocytes *in vitro* via SR-BI compared with apoA-I-containing rHDL [152].

Significantly, apoA-I mimetics include the rapid formation of **small, lipid-poor pre-beta-HDL** both *in vivo*, when administered to animal models, and *in vitro* when added to human plasma [141, 153] (Table 11.1), consistent with a key anti-atherosclerotic role for such particles [42, 43]. Thus, incubation of human plasma with ATI-5261 dose-dependently increases small pre-beta-1 HDL, predominantly at the expense of large alpha-1 and alpha-2 particles. Enhanced pre-beta-1 HDL formation in human plasma by an apoA-I mimetic is observed after as little time as one minute [154]. The formation of pre-beta-1 HDL occurs via a highly specific mechanism, potentially involving peptide targeting of protein-protein interactions to displace apoA-I from distinct HDL particle species, thereby expanding the pool of lipid-poor apoA-I to exert anti-atherosclerotic effects [155]. The 2.5-fold increase in pre-beta HDL concentrations in the conditioned media of HepG2 cells induced by the apoE mimetic peptide Ac-hE18A-NH2 may occur by a similar mechanism [148].

In addition to the stimulation of cholesterol efflux, apoA-I mimetics inhibit the inflammatory response and attenuate oxidative stress. Such key biologic activity of the peptides may contribute to atheroprotection. Indeed, D-4F or L-4F at a dose of 10 mg/kg/d decreased aortic atherosclerosis in cholesterol-fed rabbits, along with the enhancement of the **anti-inflammatory properties** of HDL and a reduction of SAA levels [156]. Interestingly, the atherosclerotic lesion area did not correlate with HDL-C levels, but did, however, correlate with the anti-inflammatory activity of HDL, as well as with SAA levels, in this study. Furthermore, elevated plasma levels of triglyceride were inversely correlated with the level of atherosclerosis [156], consistent with the role of triglyceride in deficient HDL functionality. In addition, D-4F can prevent atherosclerosis development in diabetic apoE-null mice injected with streptozotocin [157]; this effect is associated with a reduction in hepatic arachidonic acid and oxidized fatty acid levels.

The anti-inflammatory activity of apoA-I mimetics has been further demonstrated in a rabbit model of carotid artery inflammation. In this experimental setting, apoA-I alone, as well as rHDL containing apoA-I or 5A, reduced increases in VCAM-1 and ICAM-1, in Nox4 catalytic subunits of the NADPH oxidase and in neutrophil infiltration into the arterial wall [158]. In human coronary artery endothelial cells, both 5A- and A-I-containing rHDLs inhibit TNF-alpha-induced ICAM-1 and VCAM-1 expression, as well as the NFκB signaling cascade. The effects of the 5A-rHDL are ABCA1-dependent, suggesting cellular lipid efflux is a key factor in the anti-inflammatory effects [158] (see 7.3 of chapter 7).

Similarly, the apoE mimetic peptide Ac-hE18A-NH2 dramatically lowers plasma cholesterol and LOOHs and enhances PON activity in dyslipidemic animal models [133]. Moreover, Ac-hE18A-NH2 inhibits LPS-induced VCAM-1 expression, reduces monocyte adhesion in human endothelial cells and attenuates secretion of both IL-6 and monocyte chemoattractant protein-1 from THP-1 macrophages [148].

The anti-inflammatory actions of apoA-I mimetics are broad and can be observed under several clinical conditions. For example, D-4F, in combination with pravastatin, reduces disease activity in collagen-induced arthritis in rats along with favorable changes in cytokines and chemokines, as well as with improvement in the anti-inflammatory properties of HDL [159]. Next, 4F modulates the function of monocyte-derived macrophages by regulating the expression of key cell-surface receptors. Similar to apoA-I, 4F induces profound functional changes in monocyte-derived macrophages, possibly by driving their differentiation to an anti-inflammatory phenotype [147]. Cholesterol efflux from macrophages, which leads to cholesterol depletion and disruption of lipid rafts, may mechanistically account for these effects. The protective effects of 4F may also include a decrease in sepsis-induced mortality. Indeed, 4F confers protection in rats undergoing cecal ligation and puncture injury; elevated plasma levels of HDL-C in 4F-treated rats are associated with an improvement in cardiac output and reduced mortality [160]. D-4F can, therefore, act via the transfer of LPS to the HDL fraction, resulting in endotoxin neutralization [161]. In addition, D-4F reduces renal inflammation in LDL receptor-null mice fed a Western diet [162]. D-4F is also capable of reducing pulmonary inflammation and airway resistance in an experimental murine model of asthma, acting by decreasing oxidative stress [163]. Similarly, the L-4F peptide reverses uremia-induced up-regulation of pro-atherogenic pathways in the aorta in a rat model of chronic kidney disease [164].

The potent anti-inflammatory properties of the 4F peptide are mainly caused by binding of oxidized lipids with an affinity which is four-to-six orders of magnitude higher than that of lipid-free apoA-I [165]. Thus, injecting L-4F into apoE-null mice results in a reduction in plasma levels of several hydroxy fatty acids [166], consistent with potent **antioxidative activity**. Furthermore, D-4F decreases oxidized phospholipid content in the kidneys of LDL receptor-null mice [162]. ApoA-I mimetics can also induce changes in the HDL proteome which beneficially impact HDL function. For example, the treatment of apoE-null mice with an apoA-I mimetic peptide results in the dissociation of pro-inflammatory hemoglobin-haptoglobin complexes from HDL and thereby improves HDL anti-inflammatory properties [167].

The anti-inflammatory properties of individual apoA-I mimetic peptides appear to depend on subtle differences in the configuration of the hydrophobic face of the peptides, which determines their ability to sequester pro-inflammatory oxidized lipids [168, 169]. Specifically, apoA-I mimetics that interact strongly with lipid acyl chains do not display anti-atherogenic and anti-inflammatory properties. In contrast, peptides that associate close to the lipid head group (and hence do not interact strongly with the lipid acyl chain region) are anti-atherogenic and anti-inflammatory [132].

Modification of apoA-I mimetic peptides to include a proline-punctuated alpha-helical repeat, or acetylation or biotinylation, may modulate their binding to lipoproteins and thus enhance their anti-atherogenic properties [170]. ApoA-I mimetic peptides that better mimic the punctuated alpha-helical repeats of full-length apoA-I can be more anti-inflammatory. Specifically, tandem helix apoA-I mimetic peptides, 4F-proline-4F (37mer), 4F-alanine-4F (37mer) and 4F-KVEPLRA-4F (the human apoA-I 4/5 interhelical sequence, 43mer), are more potent than monomeric 4F (18mer) in terms of their effects on HDL remodeling and cholesterol efflux, but not on LDL oxidation [171].

Promising results in terms of the inhibition of inflammatory responses in animal studies opened the way for human trials of apoA-I mimetics. The first such study revealed that a single dose of D-4F is safe and well tolerated and may improve the anti-inflammatory properties of HDL [172]. Specifically, when patients with coronary heart disease or equivalent risk received a single dose of 30, 100, 300 or 500 mg of unformulated D-4F under fasting conditions, *ex vivo* HDL-mediated inhibition of monocyte chemotactic activity induced by LDL in cultures of human aortic endothelial cells was significantly improved at 4 h at the 300 mg dose and at 2 h at the 500 mg dose compared with placebo. No significant trend in any safety parameter was observed. D-4F was detectable in plasma at all doses with maximal levels between 30 min and 2 h but its bioavailability was low [172]. In clear contrast, L-4F (also known as APL180) administered as 30 mg daily intravenous infusions for 7 days or 30 mg subcutaneous injections for 28 days in patients with coronary heart disease did not improve the anti-inflammatory activity of HDL and serum PON1 activity, despite achieving plasma levels previously shown to improve these biomarkers in animal models [173]. Paradoxically, an increase in CRP levels occurred (+49%) after seven intravenous infusions of L-4F. The peptide was well tolerated at all doses tested in this trial.

Beneficial actions of apoA-I mimetics on oxidative stress and inflammation may result in an **improvement of insulin sensitivity**. Indeed, L-4F limits weight gain, decreases visceral and subcutaneous fat, reduces plasma IL-1-beta and IL-6 levels and prevents insulin resistance in ob/ob mice, resulting in diminished circulating levels of glucose and insulin [174]. In this model, L-4F acts via elevation of aortic levels of heme oxygenase-1, a reduction of superoxide production, an increase in serum adiponectin levels and the induction of phosphorylated AMPK and phosphorylated Akt [175].

Improved arterial vasoreactivity is another important action of apoA-I mimetics. In a rat model of diabetes, D-4F increased arterial concentrations of heme oxygenase-1 and superoxide dismutase, decreased superoxide levels, reduced levels of circulating endothelial cells, reduced endothelial cell fragmentation and restored arterial vasoreactivity [143]. In a mouse model of systemic sclerosis, D-4F functioned to improve vasodilation and angiogenic potential while reducing myocardial inflammation and oxidative stress [176]. D-4F also induced heme oxygenase-1 in a mouse model of heart transplant-associated atherosclerosis [177].

ApoA-I mimetics also possess **antithrombotic activities**. Thus, injection of L-4F into apoE-null and LDL receptor-null mice reduced platelet aggregation in response to agonists and prolonged bleeding time [178]. Incubating L-4F with apoE-null platelet-rich plasma also resulted in decreased platelet aggregation and activation *in vitro*. The mechanisms of the action of the peptide may involve the sequestration of oxidized lipids that cause platelet hyper-reactivity [178].

The adverse effects of apoA-I mimetics primarily involve the **induction of hypertriglyceridemia** as a result of the inhibition of LPL [153] (Table 11.1); accumulation of ápoA-I mimetics on VLDL as a result of the saturation of HDL particles with these ligands may underlie this effect. Another drawback of apoA-I mimetics includes **transient kinetics of HDL-C elevation** as a result of their rapid removal from the circulation via kidneys [172]. Furthermore, most apoA-I mimetic peptides are **instable in the digestive system**. Indeed, typical apoA-I mimetics and apoA-I itself consist of L-amino acids that are rapidly degraded in the digestive system. In contrast, the **D-4F** peptide consists of D-amino acids, is not digested by mammalian enzymes and can be orally administrated [127]; the latter represents an important advantage in clinical development. Administered subcutaneously, D-4F or L-4F are equally efficacious, but only D-4F is efficacious orally because of the digestion of L-4F by gut proteases. Oral supplementation is also possible for small zwitterionic tetrapeptides that are too small to form an amphipathic helix but which associate with HDL and display potent anti-atherosclerotic, antioxidative and anti-inflammatory activities in apoE -/- mice [179]. A major limitation of the use of orally supplemented apoA-I mimetics consisting of D-amino acids involves their **low bioavailability** which does not exceed 1% for D-4F [70].

CLINICAL SUMMARY

ApoA-I-mimetic peptides display multiple atheroprotective activities, including the activation of cholesterol efflux from macrophages, the enhancement of anti-inflammatory and antioxidative activities of HDL, an improvement of vasodilation and insulin sensitivity, and a reduction in platelet aggregation. Limitations of apoA-I mimetics include the induction of hypertriglyceridemia, transient kinetics of HDL-C elevation, instability in the digestive system and low bioavailability.

The L-forms of apoA-I mimetics can, however, still be administered orally if they are given together with niclosamide, a chlorinated salicylanilide that has been used for more than thirty years to treat parasitic infections. Niclosamide administered orally in temporal proximity to oral L-4F, but not niclosamide alone, results in a significant improvement in the anti-inflammatory capacity of HDL in apoE-null mice [180]; plasma pre-beta-HDL levels and PON1 activity are also elevated. Furthermore, oral administration of niclosamide and L-4F, together with pravastatin to apoE-null mice, inhibits atherosclerosis compared with baseline, indicating lesion regression. In contrast, lesion progression is observed in mice receiving only niclosamide and pravastatin or L-4F and pravastatin [180]. The mechanism of the protective action of niclosamide on L-4F involves its tight association with the peptide, conserving the alpha-helical structure of L-4F and rendering it resistant to trypsin digestion. The combination of niclosamide with apolipoprotein-mimetic peptides therefore appears to be a promising method for their oral delivery [180].

Intriguingly, each biologic activity of apoA-I mimetics seems to require its own optimal structure. When bi-helical apoA-I mimetic peptides are investigated for their structure-function relationships, mean hydrophobicity, charge, size of hydrophobic face and the angle of the link between the helices appear as the major factors determining the efficiency and specificity of their cholesterol efflux properties [181]. Remarkably, the peptide with optimal parameters is more effective and specific toward cholesterol efflux than human apoA-I. The charge and size of the hydrophobic face are also the major factors affecting anti-inflammatory properties, and the presence of Cys and His residues is the main factor determining antioxidant activities [181].

Interestingly, the capacities of the peptides to promote cholesterol efflux, anti-inflammatory activity and antioxidant protection do not reveal a significant correlation. Therefore, none of the peptides is equally effective in all the anti-atherogenic functions tested. These original data show that rationalizing the design of apoA-I mimetic peptides may improve their therapeutic value [181].

MAJOR CONTROVERSY

Can transient HDL elevation as a result of the administration of rHDL or apoA-I mimetic peptides lead to the sustained inhibition of atherogenesis and translate into clinical benefit?

11.4 STATINS

Statins are inhibitors of HMG-CoA reductase whose major effect is to efficaciously decrease plasma levels of apoB-containing lipoproteins, primarily LDL, IDL, VLDL and VLDL remnants. Statins do not, however, exert significant effects on Lp(a).

Effects on HDL Levels and Metabolism

Statins induce a minor increase in HDL-C levels (by +5% to +10%) [182, 183]. Individual statins are not identical in their potency to increase HDL-C, with rosuvastatin and pitavastatin being the most potent compounds [184, 185]. Indeed, administration of pitavastatin at a dose of 1 or 2 mg/day persistently increased HDL-C by +12.5% and apoA-I levels by +9.5% in hypercholesterolemic patients with low HDL-C after 12 months of treatment in a multicenter study [186]. In addition, rosuvastatin treatment at a dose of 10 and 20 mg/day for 12 weeks dose-dependently increased HDL-C levels (by +3.4% and +5.3%, respectively) in patients with primary dyslipidemia without evidence of cardiovascular disease [187]. Higher doses of rosuvastatin can raise HDL-C concentrations even more potently as exemplified by a +11% increase in this parameter observed in hypertriglyceridemc subjects [188]. The differential effects of individual statins on HDL are further highlighted by the finding that increases in HDL-C are positively related to statin dose with rosuvastatin and simvastatin, but inversely related to dose with atorvastatin [183]. The HDL-C-raising effect of statins is under genetic control, as shown in coronary heart disease patients receiving fluvastatin [189] and pravastatin [190].

The effects of statins on HDL-C levels reflect the **stimulation of apoA-I production** and the **reduction of CETP activity** by these drugs [191–193] (Fig. 11.1 and Table 11.1). Indeed, meta-analysis of 32,258 dyslipidemic patients included in 37 randomized studies using rosuvastatin, atorvastatin and simvastatin reveals that there is no apparent relationship between a reduction in LDL-C and an increase in HDL-C [183], indicative of distinct mechanisms of action of statins on these surrogate markers of cardiovascular risk. Thus, statins modulate LpA-I and LpA-I:A-II metabolism, resulting in accelerated synthesis and delayed catabolism of apoA-I. Rosuvastatin increases hepatic apoA-I, but not apoA-II, mRNA expression, thereby selectively enhancing the production of apoA-I-containing HDL particles [194]. In contrast, HDL3 protein or HDL3 cholesteryl ester uptake by Hep G2 cells is not affected by rosuvastatin. Furthermore, the percentage increase in apoA-I levels induced by statins is virtually identical to that of HDL-C [183]. Stimulation of HDL synthesis by statins is also consistent with the elevation of small HDL3 (+8.1%) by atorvastatin 20 mg daily in patients with hypercholesterolemia after three months of treatment [195].

In addition to the up-regulation of apoA-I production, statins potently reduce plasma CETP activity [196–198] (Table 11.2). In hypercholesterolemia, HDL-C is raised by up to +10% by statins; the effect is greater in patients with low baseline HDL-C levels and is typically proportional to reductions in apoB-containing lipoproteins, indicative of the reduced rate of CETP-mediated cholesterol transfer from HDL to apoB-containing particles [183, 199]. In further support of this mechanism, the HDL-raising effect of atorvastatin in dyslipidemic APOE*3-Leiden mice requires a reduction in CETP activity [200]. Such diminished CETP activity may result in delayed apoA-I catabolism through enhanced lipidation, thereby translating into elevated HDL-C levels [193] (Fig. 11.2).

The beneficial effects of statins on HDL metabolism may also involve **elevated LCAT activity** [201, 202] and **decreased PLTP activity** [198, 203] (Fig. 11.2). Indeed, a high dose (80 mg) of atorvastatin decreases the capacity of plasma to generate pre-beta HDL particles in Type 2 diabetic patients, acting via the lowering of plasma PLTP activity and triglyceride concentrations [203]. As a result, HDL maturation by LCAT is enhanced, whereas HDL remodeling and degradation by PLTP is diminished. Furthermore, atorvastatin is capable of **reducing LPL and endothelial lipase expression** in THP-1 macrophages by reducing the activation of LXR alpha and NFκB, respectively [204] (Fig. 11.2). In addition, pitavastatin can reduce endothelial lipase expression *in vitro* and *in vivo via* the inhibition of RhoA activity [205]; the potent HDL-raising properties of this statin may be, in part, mediated by this mechanism. Finally, statins potently decrease mass and activity of $LpPLA_2$ [198].

TABLE 11.2 Therapeutic Agents that Normalize Intravascular HDL Remodeling

Agents	Administration Method	Potential Mechanisms of Action	Effects on HDL Levels	Effects on HDL Function	Shortcomings
CETP inhibitors	Oral treatment	Inhibition of CE transfer from HDL to TG-rich lipoproteins and reciprocal TG transfer	Elevation of apoA-I, apoA-II, apoE, HDL-C and HDL CE levels Preferential increase in large HDL Decrease in HDL TG	Enhancement of cholesterol efflux capacity Other effects specific to individual agents	Hypertensive effect (torcetrapib) No effect on RCT in humans
Nicotinic acid	Oral treatment	Inhibition of hormone-sensitive TG lipase in adipose tissue and DGAT2 in hepatocytes Stimulation of cholesterol efflux	Elevation of apoA-I, HDL-C and HDL CE levels Preferential increase in large HDL Decrease in HDL TG and Lp(a)	Enhancement of cholesterol efflux capacity and vasodilatory activity	Poor tolerability and compliance owing to cutaneous flushing

PPAR alpha agonists	Oral treatment	Enhancement of apoA-I production. Reduction of CETP activity secondary to decrease in plasma TG levels	Elevation of apoA-II and HDL-C levels, weaker effects on apoA-I levels Increases in small and medium HDL	Enhancement of cholesterol efflux capacity and anti-inflammatory activity	Myopathy Gastrointestinal effects
Statins	Oral treatment	Reduction of CETP activity secondary to decrease in plasma levels of VLDL and LDL	Elevation of apoA-I and HDL-C levels	Enhancement of anti-inflammatory activity	Myopathy Hepatotoxicity
Lipase inhibitors	Oral treatment	Inhibition of HDL PL hydrolysis by $sPLA_2$, EL and HL	Elevation of HDL-C levels in animal models	Unknown	Unknown
SR-BI inhibitors	Oral treatment	Inhibition of HDL CE uptake via SR-BI	Elevation of apoA-I and HDL-C levels	Unknown	Unknown

CE, cholesteryl ester; EL, endothelial lipase; HL, hepatic lipase; PL, phospholipid; TG, triglyceride.

The actions of statins on the HDL particle profile involve the preferential **increase in HDL particles of large and medium size and of alpha-mobility** [206–209]. For example, pitavastatin increases HDL-C and HDL2-C by +6.0% and +9.0%, respectively, in hypercholesterolemic patients in the absence of changes in HDL3-C [210]. Furthermore, levels of pre-beta-1 HDL and CETP mass were significantly decreased by the statin [210]. Interestingly, the effect of pitavastatin on HDL-C may be enhanced by treatment duration. Thus, HDL-C levels were raised by +14.0% and +24.9% after 12 weeks and 104 weeks, respectively, in hypercholesterolemic low HDL-C subjects treated with this drug at 2 mg/day [211]. Simvastatin (20–40 mg) increased plasma levels of HDL-C by +7% to +11%, HDL2-C by +9% to +25% and HDL3-C by +3% to +13% over a period of 10 years in hypercholesterolemic patients [212]. Fluvastatin increased plasma levels of HDL-C by +8%, HDL2-C by +24% and HDL3-C by +3% without influencing apoA-I in hypercholesterolemic patients [213]. Rosuvastatin elevated cholesterol concentration in large HDL particles (by +11% at 10 mg/day and by +22.0% at 20 mg/day), with smaller effects on intermediate HDL-C levels and no effect on those of small HDL-C in patients with primary dyslipidemia [187].

Effects on HDL Functionality

Importantly, statins exert a number of pleiotropic effects that include **anti-inflammatory** and **antioxidative activities** [214]. Antioxidative actions of statins may involve an increase in the activity of HDL-associated enzymes, as demonstrated for PON1 in hyperlipidemic subjects and patients with Type 2 diabetes treated with atorvastatin [195, 215, 216] and simvastatin [217]. In addition, pitavastatin induces PON1 expression through the activation of the p44/42 MAPK signaling cascade in hepatoma cells, suggesting beneficial effects on HDL functionality [218]. Furthermore, atorvastatin decreases antibodies to oxidized LDL and lipid peroxidation products in patients with hypercholesterolemia and Type 2 diabetes, respectively [202, 216]. Rosuvastatin (10 mg daily) also reduces $LpPLA_2$ activity and mass in patients with Type IIA dyslipidemia, decreasing plasma levels of all LDL subfractions, particularly those of dense LDL [219]. Beneficial effects of statins on HDL functionality therefore appear to be mediated, in part, by a decrease in systemic oxidative stress [195, 220]. Consistent with this mechanism, the treatment of coronary heart disease patients with simvastatin at 40 mg/d for four weeks potently enhances the anti-inflammatory function of HDL [221]. Furthermore, statins may beneficially act on the endothelium, as documented by the **improvement of endothelial function** in individuals with isolated low HDL-C treated with pravastatin [222]. Underlying mechanisms may involve an increase in circulating progenitor cells [222], as well as an enhancement of HDL-induced eNOS activation via the induction of endothelial SR-BI expression [223, 224].

Adverse effects of statin treatment involve myopathy and hepatotoxicity. Unexpectedly, statins can decrease ABCA1 and ABCG1 expression in, and

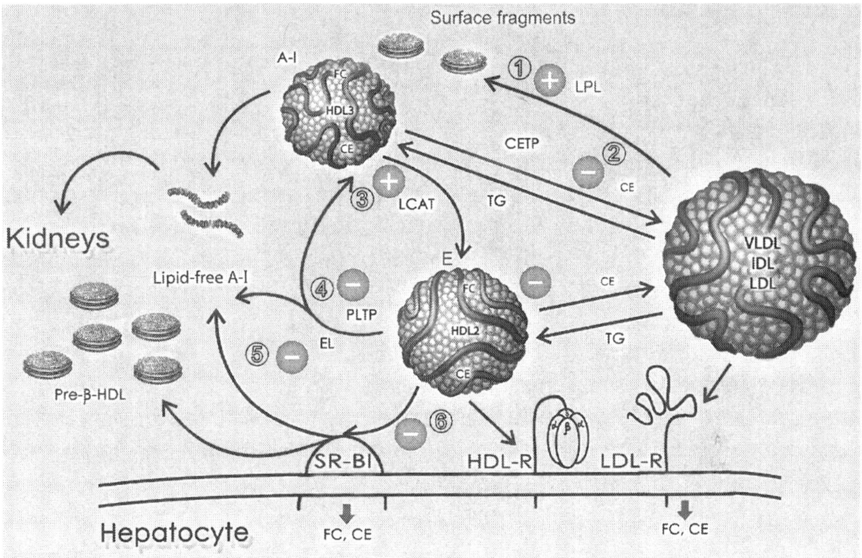

Figure 11.2 Major pathways involved in the normalization of intravascular HDL remodeling. (1) Stimulation of LPL, which can be related to the inhibition of apoC-III expression, with induction of the transfer of surface constituents of triglyceride-rich lipoproteins to the plasma HDL pool [PPAR alpha agonists (e.g., fibrates), LPL activators]. (2) Decreased CETP activity with elevation of HDL cholesteryl ester (CETP inhibitors, PPAR alpha agonists, statins, niacin). (3) Induction of cholesterol esterification by LCAT (recombinant LCAT, statins). (4) Diminished HDL remodeling by PLTP activity (statins). (5) Reduced hydrolysis of HDL phospholipid by endothelial lipase (statins) and $sPLA_2$ (varespladib). (6) Inhibition of HDL remodeling by SR-BI. CE, cholesteryl ester; EL, endothelial lipase; FC, free cholesterol; HDL-R, holoparticle HDL receptor; HL, hepatic lipase; LDL-R, LDL receptor; TG, triglyceride.

cholesterol efflux from, human macrophages. The underlying mechanism may involve impaired LXR signaling via the inhibition of oxysterol production [225]. Consistent with this notion, the addition of LXR agonists reverses the suppressing effect of statins on the ABC transporter expression [226]. Such apparent controversy is, however, resolved by the observation that supplementation of human macrophages with cholesterol reverses the statin-mediated down-regulation of ABC transporter expression [226]. In agreement with this result, the depletion of cellular cholesterol tends to accentuate the suppressive effect of statins. Furthermore, four week treatment with rosuvastatin (40 mg/day) reduces the plasma capacity to stimulate cholesterol efflux in overweight subjects with Metabolic Syndrome, secondary to the reduction in apoB-containing lipoproteins and reduced activities of CETP and LCAT, and despite increased levels of HDL-C [227]. On the other hand, simvastatin treatment improves expression of mRNA for LXR alpha and ABCA1 in circulating monocytes from Type 2 diabetic patients with hyperlipidemia [228].

Clinical Trials

The HDL-raising effect of statins may be clinically relevant. Thus, the relative changes in HDL-C were inversely related to relative changes in carotid wall thickness in the randomized, double-blind Outcome of Rosuvastatin Treatment on Carotid Artery Atheroma: A Magnetic Resonance Imaging Observation (ORION) trial in patients with moderate hypercholesterolemia and carotid stenosis treated with either a low (5 mg) or high (40/80 mg) dose of rosuvastatin for 24 months [229]. In this study, LDL-C was significantly reduced from baseline in both the low- and high-dose groups (−38.2% and −59.9%, respectively), whereas HDL-C was increased in the high-dose group in the absence of significant change in carotid plaque volume on either dose [229]. Moreover, in a Study to Evaluate the Effect of Rosuvastatin on Intravascular Ultrasound-Derived Coronary Atheroma Burden (ASTEROID), rosuvastatin at a dose of 40 mg/d for 24 months regressed coronary atherosclerosis by intravascular ultrasound and quantitative coronary angiography [230]. In parallel, rosuvastatin reduced LDL-C (−53.3%) well below 70 mg/dl to 61 mg/dl and increased HDL-C by +13.8% to 48 mg/dl. The beneficial effect of this statin on coronary atherosclerosis was correlated with both a reduction in LDL-C and an increment in HDL-C [230]. In a similar fashion, elevation in HDL-C was significantly related to atherosclerotic plaque regression in patients with coronary artery disease after six months of pravastatin therapy, as revealed by multiple regression analysis [231].

The relationship between statin-induced changes in HDL-C and cardiovascular disease progression was also assessed using a *post hoc* analysis combining raw data from four prospective, randomized trials in which 1455 patients with angiographic coronary disease underwent serial intravascular ultrasonography while receiving statin treatment for 18 or 24 months [232]. During statin therapy, LDL-C levels were reduced by −23.5% and HDL-C levels increased by +7.5%, resulting in a reduction in the LDL-C/HDL-C ratio of −26.7%. These changes were accompanied by a mean +0.5% increase in percent atheroma volume and a mean decrease in total atheroma volume of −2.4 mm^3. In multivariate analysis, mean levels of LDL-C and increases in HDL-C were independent predictors of atheroma regression. The on-treatment LDL-C/HDL-C ratio <1.5 was associated with atheroma regression; substantial atheroma regression ($\geq$5% reduction in atheroma volume) was observed in patients with levels of LDL-C less than the mean during treatment and percentage increases of HDL-C greater than the mean (+7.5%) [232]. Finally, lovastatin therapy was associated with a greater risk reduction of acute major coronary events when LDL-C levels were substantially reduced and HDL-C levels were increased by $\geq$7.5%, as observed in the Air Force/Texas Coronary Atherosclerosis Prevention Study (AFCAPS/TexCAPS) study [233]. However, it remains to be established whether statins display additional benefits in the prevention of vascular events that are related to their HDL-C-raising effects beyond LDL-C lowering [234].

CRITICAL CONTRIBUTION

Low on-treatment LDL-C/HDL-C ratio (<1.5) is associated with atheroma regression in patients with angiographic coronary disease receiving statin treatment for 18 or 24 months; substantial atheroma regression ($\geq$5% reduction in atheroma volume) is observed in patients with levels of LDL-C less than the mean during treatment and percentage increases of HDL-C greater than the mean (+7.5%) [232].

11.5 PPAR ALPHA AGONISTS

PPARs alpha, delta and gamma belong to the nuclear receptor superfamily that regulates multiple target genes involved in lipid metabolism, adipogenesis, insulin sensitivity and vascular inflammation [235] (Table 11.1). PPARs represent transcription factors activated by multiple intra- and extracellular ligands, as well as by synthetic chemical ligands known as peroxisome proliferators. **Fibrates** are PPAR alpha agonists of moderate affinity that exert multiple effects on lipid and fatty acid metabolism; in parallel, fibrates weakly activate PPAR delta and gamma [16, 236].

Effects on HDL Levels

Clinically, the administration of fibrates results in sustained HDL-C raising, in contrast to rHDL and apoA-I mimetics which produce transient increases in HDL-C; on the other hand, HDL-C response to fibrate reveals high interindividual variability. The average increase in HDL-C levels provided by fibrates is less pronounced compared with other major classes of HDL-raising agents (CETP inhibitors, niacin) and equals +10% in 53 randomized controlled trials [237]. Importantly, fenofibrate is able to increase HDL-C levels across a wide range of dyslipidemic states. The HDL-C-raising effect of fenofibrate is inversely related to baseline levels of HDL-C. For example, fenofibrate (160 mg/day for 8 weeks) lowers plasma triglyceride levels and exerts a remarkable HDL-C raising activity (+22%) in low HDL-C dyslipidemic patients, with those in the lowest range of HDL-C at baseline showing maximal benefit [201]. The HDL-C raising activity of fenofibrate is well maintained up to 15 years of treatment (mean of +24.6% at 15 years) and tends to increase over time [238]. Fenofibrate also significantly reduces triglyceride and LDL-C levels in long-term treatment (−54.9 and −28.5%, respectively).

Importantly, HDL-C-raising by fibrates is markedly attenuated by Type 2 diabetes and typically does not exceed +5% in diabetic patients [16, 239]. The relatively weak HDL-C-rasing activity of fibrates in Type 2 diabetes may, in part, reflect the elevation in plasma levels of homocysteine, a cardiovascular risk factor. Indeed, homocysteine is capable of down-regulating hepatic synthesis of apoA-I [240]. PPAR alpha agonists thus increase plasma homocysteine levels [240, 241]; increases in homocysteine induced by fenofibrate were negatively correlated with increases in HDL-C and apoA-I, but positively correlated with an elevation in apoA-II levels in the FIELD trial. PPAR alpha agonists that have a more robust effect on HDL-C and apoA-I and do not raise homocysteine would therefore be desirable.

As fibrates are relatively weak PPAR alpha activators, more potent agonists have been developed. For example, CP-778875, a potent PPAR alpha agonist designed to produce greater HDL-C increases compared with fibrates, improved the plasma lipid profile in a randomized, multicenter, double-blinded, placebo-controlled study in subjects with mixed dyslipidemia and Type 2 diabetes [241]. A 2 mg/day dose of CP-778875 given for six weeks increased concentrations of HDL-C, apoA-I, HDL2-C and HDL3-C by +14%, +13%, +12% and +19%, respectively. An unusual dose-response pattern is, however, observed in that at 6 mg/day, CP-778875 increased HDL-C only by +3% and decreased HDL2 cholesterol by −24%. Fasting triglyceride levels were decreased to a similar extent (−26%) by all three doses of CP-778875.

Other known activators of PPAR alpha that raise HDL-C levels in humans are few, despite the variety of fatty acid derivatives known to activate this receptor. An interesting example involves soy phosphatidylinositol enriched in linoleic acid [242], which increases hepatic secretion of apoA-I through the activation of G-protein- and MAPK-related pathways [242, 243].

Alteration in HDL subfraction profile is a central feature of fibrate therapy which selectively increases circulating levels of **HDL particles of small- and medium-size** (Table 11.1). Indeed, fenofibrate [244, 245], bezafibrate [208, 246, 247] and gemfibrozil [248] all selectively increase small and/or medium HDL particles in patients with Type 2 diabetes and in hypertriglyceridemic subjects. For example, fenofibrate produced a shift in HDL subspecies towards smaller HDL3 in a statin-free cohort in the FIELD trial, with HDL2-C decreased by −27.5% and HDL3-C elevated by +13.0% after five years [249]. Remarkably, the increase in small HDL induced by fibrates may attain +168% in subjects with hypertriglyceridemia, reflecting LPL stimulation [247]. Such selective elevation in the levels of small HDL is consistent with the effect of fibrates on the production of apoA-I and apoA-II. HDL-C raising by fenofibrate is associated with a shift of HDL from large to small particles paralleled by the shift from LpA-I to LpA-I:A-II, suggesting that apoA-II predominantes in small HDL induced by this drug [201].

Mechanisms of Action

PPAR alpha activation mediates a spectrum of pleiotropic effects such as the stimulation of lipid oxidation, modulation of lipoprotein metabolism and the attenuation of inflammation [250]. The actions of fibrates on lipoprotein metabolism derive from the multiplicity of target genes regulated by PPAR alpha. Indeed, in apoA-I transgenic mice, fenofibrate treatment modifies the expression of 267 genes involved in diverse molecular pathways [251]. Genes involved in phospholipid biosynthesis, lipid hydrolysis and fatty acid transport are up-regulated, whereas genes involved in fatty acid and cholesterol biosynthesis, the Krebs cycle and gluconeogenesis remain unaffected. In addition, fenofibrate modifies the expression of transcription factors and up-regulates the estrogen receptor-related gamma gene [251].

APOA1 and APOA2 are the key genes regulated by PPAR alpha; activation of the receptor therefore results in enhanced apoA-I and apoA-II production and may translate into elevated HDL-C levels (Fig. 11.1 and Table 11.1). Thus, fibrates **increase circulating levels of apoA-I and**, more potently, **apoA-II** by enhancing their production in the liver [252]. For example, **fenofibrate** (200 mg/day for 5 weeks) increases the production of both apoA-II (+43%) and LpA-I:A-II (+30%) in men with Metabolic Syndrome, accounting for significant increases in their corresponding plasma concentrations (+10 and +23%, respectively) [253]. In contrast, LpA-I kinetics and concentrations remain unaltered. As a result, circulating levels of LpA-I:A-II are elevated by fibrates.

Such stimulation of apoA-II and apoA-II production by PPAR alpha agonists is partially counteracted by enhanced fractional catabolism of these apolipoproteins. For example, the potent and specific PPAR alpha agonist LY518674 does not change circulating levels of apoA-I as a result of the accelerated turnover, which entirely counterbalances the increase in apoA-I production [254]. In parallel, however, plasma levels of apoA-II are elevated by the agonist as a result of a marked increase in apoA-II production accompanied by a smaller acceleration of catabolism.

An important aspect of the action of fibrates on HDL metabolism involves **enhanced expression of ABCA1** and accelerated biogenesis of HDL in hepatocytes, which occur in a PPAR-dependent manner [255, 256] (Fig. 11.1 and Table 11.1). Such stimulation of ABCA1 expression by PPAR alpha is regulated co-operatively with that mediated by **LXR** [257]. As a result, the plasma capacity to **promote ABCA1-mediated cholesterol efflux** is elevated following fibrate treatment, whereas SR-BI-mediated efflux remains unaffected [201]. Improved RCT may follow, as demonstrated with the novel oxybenzylglycine PPAR alpha-selective agonist BMS-711939 which induces fecal cholesterol excretion in PPAR alpha-humanized mice; such excretion is further increased upon co-treatment with a LXR agonist [258]. In addition, BMS-687453 and BMS-711939 robustly increase concentrations of apoA-I and HDL-C in human apoA-I transgenic mice and lower concentrations of LDL-C and triglyceride in fat-fed hamsters [258]. Fenofibrate is also capable of accelerating RCT in mice, as shown in human apoA-I-transgenic animals concomitant with increases in plasma HDL-C, HDL phospholipid and apoA-I levels and activation of PLTP [259]. In contrast, **gemfibrozil**, a weaker PPAR alpha agonist [260] that does not modify any of these parameters, does not influence RCT [259].

The potent activation of apoA-I production by PPAR alpha agonists is accompanied by a pronounced **decrease in plasma levels of triglyceride-rich lipoproteins** (Table 11.2). Indeed, fibrates can reduce circulating levels of triglyceride by up to −48% [237]. In addition, a potent, and highly specific, PPAR alpha agonist, LY518674, at a low dose of 100 μg/day, reduces VLDL-C (−38%) and triglyceride (−23%) levels in subjects with Metabolic Syndrome [254]. Such hypotriglyceridemic action of PPAR alpha agonists may significantly contribute to the elevation of HDL-C levels through the inhibiton of CETP-mediated heteroexchange of cholesteryl ester and triglyceride between triglyceride-rich lipoproteins and HDL [193, 261, 262] (Fig. 11.2).

The effect of fibrates on the metabolism of triglyceride-rich lipoproteins is primarily caused by the PPAR alpha-dependent **stimulation of LPL** and of a marked **reduction in apoC-III expression** [263] (Fig. 11.2). Indeed, fibrates induce both LPL, which provides release of surface fragments from triglyceride-rich lipoproteins and their transfer to HDL during lipolysis, and hepatic lipase, which facilitates the conversion of large to small HDL. As a result, fibrate action favors the normalization of HDL core lipid composition, decreasing HDL-triglyceride and increasing cholesteryl ester, an effect that is reflected in an elevation in concentrations of HDL-C. Consistent with this mechanism, fenofibrate increases the HDL-C/apoA-I ratio and LCAT levels, but decreases plasma apoC-III [201]. In addition, fibrates may decrease circulating triglyceride levels by enhancing cellular fatty acid uptake and transformation to acyl-CoA derivatives, and by increasing peroxisomal and mitochondrial beta-oxidation, thereby leading to attenuated synthesis of fatty acids and triglycerides [16].

As a consequence of LPL activation and down-regulation of apoC-III expression, fibrates act to **reduce VLDL apoB-100** through enhanced fractional catabolism, with additional inhibitory effects on VLDL apoB-100 production [252]. Fenofibrate preferentially reduces circulating concentrations of large VLDL. Similarly, the potent PPAR alpha agonist LY518674 reduces VLDL apoB-100 levels, an effect attributable to the enhanced fractional catabolism of VLDL apoB-100 with no change in its production [254]. On the other hand, the lipid-lowering effects of bezafibrate appear to be primarily mediated by the inhibition of the hepatic production of apoB-containing lipoproteins [264]. As a result, bezafibrate decreases plasma levels of both cholesterol and triglyceride, while increasing levels of HDL-C in LDL receptor-null mice. In addition, the formation of fatty streak lesions in the aortic sinus is attenuated, primarily as a result of a reduction in the ratio of non-HDL-C to HDL-C [264].

LDL apoB-100 levels typically decrease in response to fibrates as a result of **accelerated fractional catabolism of LDL**, with the exception of patients with high-to-very high plasma triglyceride levels (>400 mg/dl) [252]. In contrast, the kinetics of IDL and LDL apoB-100 are weakly affected by LY518674 treatment, consistent with enhanced lipolysis of plasma triglyceride [254]. The relatively weak LDL-C lowering activity of fibrates may be related to their ability to induce the expression of proprotein convertase subtilisin kexin type 9 (PCSK9), a key regulator of the intracellular degradation of the LDL receptor, whose gain-of-function mutations cause familial hypercholesterolemia. Thus, fenofibrate (200 mg per day) significantly increases circulating PCSK9 levels (+25%) compared with baseline in humans [265]. Such fenofibrate-induced increases in serum PCSK9 are highly correlated with changes in HDL-C, triglyceride and LDL-C levels. In contrast, another study reported a minor decrease in plasma PCSK9 concentrations (−8.5%) after six weeks of treatment with co-micronized fenofibrate (200 mg/day) [266].

Interestingly, the indirect inhibitory action of PPAR alpha agonists on plasma CETP activity as a result of reduced triglyceride levels can be, in part, counteracted by **enhanced CETP expression**. Thus, in mice expressing apoC-III and/or CETP, ciprofibrate stimulates CETP gene expression and modifies cholesterol flow through the RCT pathway, stimulating cholesteryl ester transfer from HDL to LDL and cholesterol removal from plasma through LDL [267]. In parallel, ciprofibrate increases plasma post-heparin LPL activity and decreases hepatic lipase activity in CETP transgenic mice. In addition, PPAR alpha agonists increase CETP mRNA levels, protein and activity in a human CETP transgenic animal model harboring the natural flanking regions. In these animals, both fenofibrate and a specific synthetic PPAR alpha agonist, LY970, elevate human CETP mRNA expression in the liver, as well as serum CETP protein and activity [268]. In hamsters, endogenous liver CETP mRNA levels and serum CETP activity are dose-dependently up-regulated by fenofibrate. Furthermore, WY14643, another PPAR alpha agonist, significantly increases CETP mRNA and activity [268]. Consistent with these results obtained in animal models, fenofibrate increases circulating CETP levels in dyslipidemic humans [201], although net CETP-mediated cholesteryl ester transfer from HDL to VLDL and LDL is typically reduced because of a reduction in apoB-containing particle numbers [16, 262].

Effects on HDL Functionality

PPAR alpha is expressed in endothelial and smooth muscle cells, monocytes and monocyte-derived macrophages where it can negatively regulate vascular inflammation via decreased NFκB and activator protein-1 signaling [263]. These pathways underlie the **anti-inflammatory activity** of PPAR alpha agonists which can reduce inflammation and oxidative stress and thereby contribute to atheroprotection. Thus, treatment of obese, glucose-tolerant men with fenofibrate for 12 weeks decreased circulating levels of pro-inflammatory cytokines (TNF-alpha, IL-6, IL-1 beta) and adhesion molecules (VCAM-1, ICAM-1), concomitant with a reduction in arterial pulse wave velocity by −17.4% [269]. Next, microcoated fenofibrate (160 mg/day for 12 weeks) increased PON1 concentration and activity by +14.6% and +10.1%, respectively, in dyslipidemic patients with low HDL-C levels [270]. Changes in PON1 concentration and activity were positively correlated with changes in HDL-C and negatively correlated with changes in triglyceride levels. Interestingly, HDL-raising properties accompanied by anti-inflammatory activities are a common denominator of agonists of three PPAR nuclear receptors, alpha, gamma and delta [271, 272].

Another aspect of the anti-inflammatory action of fibrates involves the redistribution of PAF-AH from apoB-containing lipoproteins to HDL in dyslipidemic patients, thereby lowering the pro-inflammatory potential of the enzyme [273]. Indeed, fenofibrate preferentially reduces $LpPLA_2$ activity and mass associated with VLDL, IDL and dense LDL in patients with Type IV dyslipidemia [219]. In parallel, HDL-associated $LpPLA_2$ activity and mass are increased, particularly in small HDL3c particles.

Clinical Trials

Despite such significant anti-atherogenic actions, **clinical trials** of cardiovascular risk reduction by fibrates have provided mixed results. Thus, treatment with gemfibrozil in the primary prevention Helsinki Heart Study and in the secondary prevention VA-HIT trial, as well as treatment with clofibrate in the primary prevention WHO trial, revealed beneficial effects [274]. In contrast, results of the secondary prevention BIP trial with bezafibrate and of the combined primary and secondary prevention, multinational, randomized, controlled FIELD trial with fenofibrate are controversial [274]. Indeed, fenofibrate did not significantly reduce the risk of the primary outcome of coronary events in the FIELD study in patients with Type 2 diabetes [239]. Furthermore, the frequency of major adverse effects, including pancreatitis and pulmonary embolism, was elevated by fenofibrate. Nonetheless, the drug reduced total cardiovascular events, mainly as a result of fewer non-fatal myocardial infarctions and revascularizations. The high drop-in rate for statin therapy in patients allocated placebo may have masked a moderately larger treatment benefit [239].

OPEN QUESTIONS

Does apoA-II elevation induced by fibrates contribute to atheroprotection?
Does homocysteine elevation compromise the clinical efficacy of fibrates?
To what degree does the enhancement of the anti-atherogenic activities of HDL contribute to clinical benefit under fibrate treatment?

Post hoc analyses of fibrate studies reveal that patients with features of Metabolic Syndrome, particularly those who are overweight and those who exhibit high plasma triglyceride and low HDL-C levels, display a disproportionately large reduction in cardiovascular events [274]. In the FIELD study performed in patients with Type 2 diabetes, fenofibrate significantly reduced cardiovascular events in those with low HDL-C or hypertension. The largest effect of fenofibrate in reducing cardiovascular risk was observed in subjects with marked dyslipidemia in whom a −27% relative risk reduction was observed. The benefits of fenofibrate are, therefore, likely to be greater when Metabolic Syndrome features are present [275]. Consistent with this result, a meta-analysis of five large trials assessing the impact of fibrates on cardiovascular endpoints indicates that there is a significant difference in the magnitude of fibrate effect across dyslipidemic subgroups [276]. Specifically, a greater effect size is found in patients with high triglyceride levels or atherogenic lipid phenotype where fibrates are estimated to reduce the cardiovascular risk by −28% or −30%. In contrast, the risk reduction is only −6% in patients with non-atherogenic dyslipidemia. Another meta-analysis of six long-term randomized controlled trials of fibrates in the prevention of cardiovascular events in a total of 11,590 patients with Type 2 diabetes reveals that long-term use of fibrates in these patients significantly reduces the risk of non-fatal myocardial infarction by −21%, but has no significant effect on mortality or on other adverse cardiovascular outcomes [277].

Importantly, the degree of HDL-C response to fibrates is independently associated with subsequent long-term mortality. When the risk of death at 16 years of follow-up was assessed among patients with coronary heart disease allocated to bezafibrate or placebo in the BIP trial, multivariate analysis demonstrated that patients allocated to bezafibrate therapy experience a −11% reduction in the risk of long-term mortality compared with placebo-allocated patients [278]. Mortality reduction among bezafibrate-allocated patients was largely related to a −22% reduction in the risk of death in patients with an upper-tertile HDL-C response to therapy (>8 mg/dl) [278]. Furthermore, the risk for a major cardiac event during a median 7.9-year follow-up period in 3020 patients with coronary artery disease enrolled in the BIP trial was related to changes in lipid levels during the study [279]. Multivariate analysis demonstrated that the benefit of HDL-C elevation was most pronounced in patients with low baseline LDL-C (≤129 mg/dl; −29% risk reduction per 5 mg/dl increment in HDL-C). A similar relationship was shown for risk reduction-associated triglyceride decrements, suggesting that the degrees of HDL-C and triglyceride response are inversely related to baseline LDL-C levels [279].

The significant cardiovascular event reduction observed with gemfibrozil in the VA-HIT trial cannot, however, be fully explained by the increment in HDL-C (+6%) [23, 24]. In the gemfibrozil arm of the trial, subjects with recurrent cardiovascular events displayed higher pre-beta-1 and lower alpha-1 and alpha-2 HDL levels than those without such events [280]. Levels of small, lipid-poor, pre-beta-1 HDL were a positive predictor of future cardiovascular events, whereas levels of large, lipid-rich alpha-1 and alpha-2 HDL were negative predictors. Levels of alpha-2 particles were superior to HDL-C in cardiovascular risk prediction after adjustment for established risk factors. Controversially, however, gemfibrozil treatment was associated with −3% to −6% decreases in the levels of both pre-beta-1 HDL and alpha-1 and alpha-2 HDL, as well as with increases in small alpha-3 (+3%) and pre-alpha-3 (+16%) particles. Such inconsistencies suggest that gemfibrozil-mediated improvement in cardiovascular risk may result, not only from the effects on HDL, but may also reflect a wider spectrum of metabolic effects of the drug [280].

Interestingly, plasma levels of small HDL3 were a powerful predictor of cardiovascular risk in insulin-resistant subjects in the VA-HIT trial [23, 24]. Furthermore, the atheroprotective effects of fibrates are, in part, related to increases in HDL3-C [281, 282]. These observations may reflect the potent atheroprotective activities of small HDL particles [42, 43]. Fibrates may, therefore, be useful, not only in inducing increase in circulating levels of HDL, but also in enhancing the functionality of small HDL particles; the latter hypothesis remains to be tested. Intriguingly, approaches to elevate HDL-C levels by raising concentrations of highly functional small HDL appear promising. The beneficial effects of probucol on cardiovascular disease, which selectively raises circulating levels of small HDL but reduces HDL-C, are consistent with this concept [283]; other biologic actions of probucol may, however, also contribute to the atheroprotection induced by this agent.

Common side effects of fibrate treatment involve myopathy and gastrointestinal effects. Interestingly, fenofibrate may decrease HDL-C levels in some patients. Such paradoxical HDL-C reduction represents an uncommon phenomenon and occurs mainly in patients on combined fibrate/statin therapy. Further, reductions of HDL-C appear to occur mainly in individuals with elevated or normal HDL-C (>50 mg/dL) [284]. This finding suggests that the HDL-C response to fibrates is under tight genetic control. Indeed, in participants of the Genetics of Lipid-Lowering Drugs and Diet Network (GOLDN) study undergoing a three-week fenofibrate trial, the HDL-C response was associated with the SNP rs3135506 (APOA5_S19W) in the APOA1/APOC3/APOA4/APOA5 gene cluster, with the minor allele being related to an attenuated response [285].

MAJOR CONTROVERSY

The clinical benefits of fibrates remain controversial; new clinical trials are urgently required to clarify this important issue.

11.6 NICOTINIC ACID

Nicotinic acid (niacin), a vitamin of the B complex, has been used for almost 50 years as a lipid-modulating drug. Niacin is presently the most effective commercially available agent for increasing HDL-C; the HDL-C raising effect of niacin may reach +35% at a 2 g/day dose [286] (Table 11.2). Meta-analysis of 30 randomized controlled trials reveals that, on average, niacin increases HDL-C by +16% across all doses studied. In parallel, niacin decreases plasma triglyceride by −20%, on average [237], and can also reduce LDL-C and Lp(a) levels. Importantly from a clinical viewpoint, niacin is frequently the only effective drug for the elevation of HDL-C in well-defined genetic or familial HDL-deficiency states [287].

Mechanisms of Action

The primary action of nicotinic acid is to transiently suppress lipolysis of triacylglycerol in adipose tissue via the **inhibition of the hormone-sensitive triglyceride lipase** [288–291] (Fig. 11.3 and Table 11.2). The hormone-sensitive lipase is activated by reversible phosphorylation under the influence of protein kinase A. The anti-lipolytic action of nicotinic acid involves a reduction in intracellular cyclic AMP levels in adipose tissue via a G-protein-coupled receptor that mediates the inhibition of adenylyl cyclase. **HM74** (also known as **GPR109B**), an orphan G-protein coupled receptor, has been identified as the nicotinic acid receptor in human adipose tissue [292]. HM74 appears to function as a low-affinity receptor for nicotinic acid, whereas the shorter homologous form (HM74A also termed GPR109A) represents a high-affinity receptor [293]. Niacin acts on HM74 and HM74A receptors both in adipocytes and immune cells.

The key feature of the mechanism of action of nicotinic acid on lipid metabolism involves **attenuated adipose tissue lipolysis** resulting in a reduction of circulating levels of non-esterified fatty acids (Fig. 11.3). The flux of non-esterified fatty acids to the liver constitutes the main substrate for hepatic triglyceride synthesis; such triglycerides may be either integrated into nascent VLDL particles and secreted into the circulation, or may alternatively be stored in the form of intracellular lipid droplets in the hepatocyte. Nicotinic acid is therefore distinguished as the sole pharmacologic agent that markedly lowers both plasma non-esterified fatty acid and VLDL-triglyceride levels, acting via **reduced hepatic VLDL production** (Fig. 11.3). Furthermore, kinetic studies reveal that niacin treatment enhances the catabolism of triglyceride-rich lipoproteins, as observed in patients with combined hyperlipidemia [294]. In contrast, MK-0354, a novel HM74A agonist, decreases plasma non-esterified fatty acids but fails to modify LDL-C, triglyceride and HDL-C levels [295]. This surprising result questions the role of HM74A as the sole receptor responsible for the lipid-modulating effects of niacin in humans. Furthermore, as a plasma decrease in non-esterified fatty acids is transient and is followed by a rebound increase in humans, the contribution of this pathway to the lipid-modulating activity of niacin remains to be unambiguously demonstrated [296]. Another element in niacin action on plasma lipids may involve direct **inhibition of DGAT2**, a key enzyme of triglyceride synthesis, in human hepatocytes [297] (Fig. 11.3). The inhibition of triglyceride synthesis by niacin potentially results in the accelerated intracellular hepatic degradation of apoB and decreased secretion of VLDL particles [296], but does not appear to involve an elevation in hepatic fat storage.

Whatever the exact mechanism of a niacin-mediated **reduction in plasma triglyceride levels**, this pathway appears to be central to HDL-C raising. Indeed, triglyceride levels are strongly inversely correlated with those of HDL-C [286]. Thus, nicotinic acid-mediated reduction in plasma triglyceride levels predictably leads to a marked HDL-C raising. This effect is intimately linked to the action of CETP. By attenuating CETP-mediated depletion

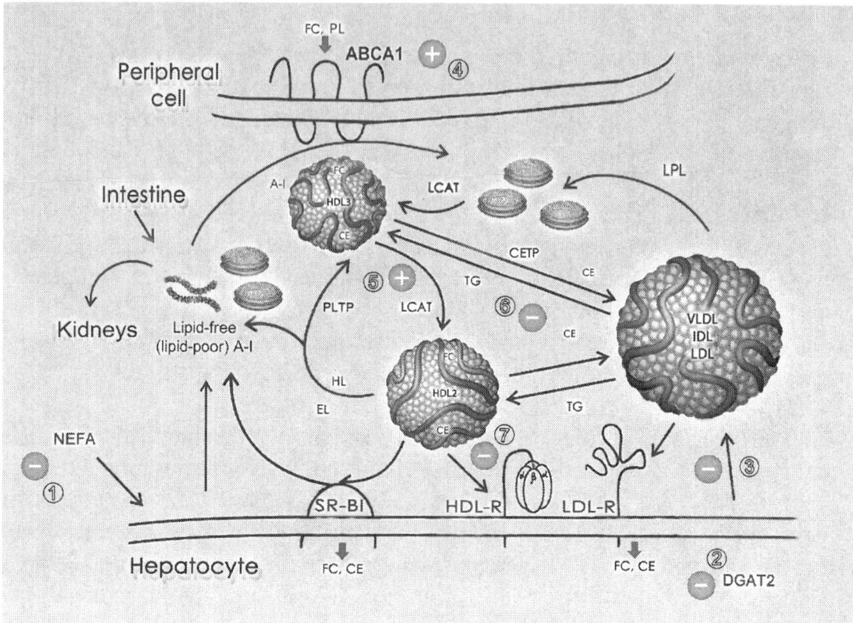

Figure 11.3 Major pathways involved in the action of nicotinic acid. (1) Transient inhibition of hormone-sensitive triglyceride lipase in adipose tissue resulting in reduced flux of non-esterified fatty acids (NEFA) to the liver. (2) Inhibition of DGAT2 in hepatocytes. (3) Decreased VLDL triglyceride secretion by the liver. (4) Accelerated efflux of cholesterol from peripheral cells mediated by ABCA1. (5) Enhanced formation of mature HDL particles, primarily of HDL2, with delayed apoA-I and HDL catabolism. (6) Diminished heteroexchange of cholesteryl ester and triglyceride between HDL and triglyceride-rich lipoproteins resulting in HDL enrichment in cholesteryl ester and depletion in triglyceride. (7) Reduced uptake of HDL2 holoparticles by the liver mediated by receptor(s) for HDL holoparticles with maintained uptake of cholesteryl ester and free cholesterol from mature HDL by SR-BI. CE, cholesteryl ester; EL, endothelial lipase; FC, free cholesterol; HDL-R, holoparticle HDL receptor; HL, hepatic lipase; LDL-R, LDL receptor; PL, phospholipid; TG, triglyceride.

of HDL-cholesteryl ester in hypertriglyceridemic states such as Metabolic Syndrome and Type 2 diabetes, the triglyceride-lowering action of nicotinic acid favors the retention of cholesteryl ester in HDL with the normalization of HDL neutral lipid content and an increase in particle size (Fig. 11.3). Prolongation of apoA-I residence time *in vivo* ensues, thus resulting in effective raising of HDL-C and apoA-I levels. This mechanism is supported by data obtained in the APOE*3Leiden transgenic mouse model expressing the human CETP transgene and displaying a human-like plasma lipid profile. In these animals, niacin markedly increases HDL-C by reducing CETP activity as a result of diminished hepatic CETP expression, an effect that is potentially secondary to a reduced hepatic cholesterol content, and decreases plasma CETP mass and (V)LDL pool size(s) [298]. In parallel, HDL apoA-I content is elevated by niacin via the reduction in the clearance of apoA-I from plasma subsequent to enhanced lipidation.

> **MAJOR CONTROVERSY**
>
> Is the therapeutic effect of niacin on cardiovascular disease related to an elevation of HDL-C levels or a reduction of triglyceride levels, or both?

Kinetic and *in vitro* studies indicate that niacin slows the hepatic catabolism of apoA-I, but not SR-BI-mediated cholesteryl ester uptake, thereby suggesting that niacin **inhibits hepatic HDL holoparticle removal** [296] (Fig. 11.3). Indeed, niacin decreases HDL uptake by the liver [299]; the inhibition of cell surface expression of the ATP synthase beta chain, a plausible HDL holoparticle receptor, may underlie this observation [300].

Another mechanism that may contribute to niacin-induced raising of HDL-C involves the **stimulation of cholesterol efflux** from peripheral cells to primary HDL acceptors via the ABCA1 membrane transporter (Fig. 11.3). Indeed, niacin activates ABCA1 via a nuclear PPAR gamma-dependent pathway [301] and may promote cholesterol efflux from adipocytes to apoA-I via ABCA1 mediated by the LXR alpha-dependent mechanism [302]. In support of this possibility, kinetic studies suggest that niacin treatment may **enhance HDL apoA-I production** in combined hyperlipidemia [294].

The multiple actions of niacin on HDL metabolism result in **elevated plasma levels of both large HDL2 and small HDL3 particles** [303, 304] (Table 11.2). In addition, niacin favorably modifies HDL composition, preferentially increasing apoA-I in the form of large HDL2-like, cholesteryl ester-rich particles [299, 304]. Decreased catabolism of HDL apoA-I by niacin explains preferential elevation in plasma concentrations of LpA-I [305]. On the other hand, niacin can decrease small pre-beta HDL as observed in Type 2 diabetic patients [306], another consequence of improved apoA-I lipidation.

Other actions of niacin on plasma lipids involve its unique capacity to effectively **lower Lp(a) levels** [193, 307] (Table 11.2). On the other hand, niacin **transiently impairs insulin sensitivity**, increasing plasma glucose and HbA1c levels [308]. Such a deterioration of insulin sensitivity in patients with impaired glucose tolerance on niacin typically occurs after short term treatment (weeks), while long term treatment (6–12 months) may significantly improve insulin sensitivity [309, 310]. Such beneficial long-term effects of niacin on insulin signaling may be related to HDL-mediated protection of pancreatic beta-cells [106, 311], as well as to **enhanced adiponectin secretion** by adipocytes [312]. Indeed, niacin up-regulates adiponectin expression in cultured adipocytes, acting via a G protein-coupled pathway [312]. As a consequence, treatment with niacin raises plasma adiponectin by +52% to +95% in patients with Metabolic Syndrome [313]. The most biologically active high-molecular weight form of adiponectin is the fraction most affected by such treatment. Furthermore, niacin exerts pleiotropic effects in adipose tissue, as indicated by decreased mean adipocyte size, increased insulin sensitivity and altered mRNA expression profile, including increased levels of mRNA for adiponectin and PPAR gamma and decreased expression of carnitine palmitoyl transferase 2, hormone sensitive lipase, nicotinic acid receptor GPR109B and fatty acid synthase [309]. Niacin-mediated reduction of serum concentrations of fetuin-A, a liver-secreted phosphoprotein and physiologic inhibitor of insulin receptor tyrosine kinase, can also contribute to the beneficial action of niacin on insulin sensitivity, as observed in individuals with Metabolic Syndrome [314]. Indeed, insulin receptor tyrosine kinase is associated with the development of insulin resistance [314].

METHODOLOGIC NOTE

Routine clinical investigation, including assessment of glucose levels, insulin sensitivity and blood pressure, should be performed when evaluating novel agents that target HDL remodeling.

Clinical Trials

Niacin exerts a clinical benefit across a wide range of cardiovascular endpoints in studies involving limited patient numbers [16]. The clinical use of niacin began with the Coronary Drug Project, which suggested favorable clinical outcomes [315]. In these studies, niacin decreased the risk of cardiovascular events when used either in monotherapy or in combination with statins. Thus, the addition of extended-release niacin to statin therapy slowed the progression of carotid atherosclerosis (measured as intima-media thickness), increased HDL-C (+21%) and decreased triglyceride (−15%) and non-HDL-C (−7%) levels after 12 months among individuals with established coronary heart disease and moderately low HDL-C in the double-blind, placebo-controlled ARBITER 2 study [26]. In the subsequent ARBITER 3 study, further regression of atherosclerosis was observed after 24 months [316]. Finally, the ARBITER 6 HDL and LDL Treatment Strategies in Atherosclerosis (HALTS) trial performed in 208 patients with coronary heart disease or a coronary heart disease risk equivalent receiving long-term (mean duration, six years) statin therapy with achieved LDL-C and HDL-C levels revealed that extended-release niacin (target dose, 2000 mg per day) displayed greater efficacy than ezetimibe (10 mg/day) in reducing both mean and maximal carotid intima-media thickness over a period of 14 months [27]. The plaque regression (as carotid intima-media thickness) of up to −6.5% achieved following niacin treatment for 24 months [316] was comparable to the potent anti-atherosclerotic effects of rHDL [75] and statins [76]. The mean HDL-C level in the niacin group markedly increased (+18.4%) in this study. In parallel, niacin therapy reduced LDL-C (−12.4%) and triglyceride (−15%). Importantly, the incidence of major cardiovascular events was lower in the niacin group than in the ezetimibe group (1% vs. 5%, $p = 0.04$) [27]. This comparative-effectiveness trial, which addressed residual cardiovascular risk persisting after statin treatment, revealed that the use of extended-release niacin caused significant regression of carotid intima-media thickness when combined with a statin [27]. The trial was terminated early on the basis of a pre-specified interim analysis; the final report, including an additional 107 subjects who completed a close-out assessment, strengthens the initial findings [317]. These data strongly support the action of niacin in attenuating atherosclerosis; the broad spectrum of effects of niacin on plasma lipids do not, however, allow a clear-cut conclusion regarding the therapeutic contribution of HDL-C raising alone [318].

Consistent with the results of the ARBITER trials, extended-release niacin (1000 mg/day) decreased carotid intima-media thickness compared with placebo (intima-media thickness change, +0.009 mm in the placebo group vs. −0.005 mm in the niacin group, $p = 0.021$ between the groups) in patients with Metabolic Syndrome after 52 weeks of treatment [310]. Furthermore, in the double-blind, randomized, placebo-controlled Oxford Niaspan study, 2 g/d modified-release niacin reduced carotid plaque area compared with placebo after 12 months when added to statin therapy in patients with low HDL-C and either Type 2 diabetes with coronary heart disease or carotid/peripheral atherosclerosis [319]. Interestingly, in both the treatment and the placebo groups, larger plaques were more prone to a reduction in size.

Finally, a meta-analysis of 11 randomized controlled trials that enrolled 2682 patients in the active group and 3934 in the control group, revealed that niacin, at doses of 1–3 g/day, significantly reduced major coronary events (relative risk reduction, −25%, 95% CI -13%, −35%), stroke (−26%, 95% CI −8%, −41%) and any cardiovascular events (−27%, 95% CI 15%, 37%) [320]. With the exception of stroke, the pooled between-group difference remained significant in sensitivity analysis. In comparison with the non-niacin group, more patients in the niacin group displayed regression of coronary atherosclerosis (+92%), whereas the proportion of patients with progression of atherosclerosis decreased by −41%. Similar effects of niacin were found on carotid intima thickness with a weighted mean difference in annual change of −17 μm/year [320]. Consistent with these data, another meta-analysis of seven randomized placebo-controlled trials that evaluated the effect of niacin on cardiovascular outcomes in coronary artery disease patients revealed that, compared with the placebo group, niacin therapy significantly reduced coronary artery revascularization (relative risk, 0.31 with 95% CI 0.15–0.63), nonfatal myocardial infarction (relative risk, 0.72; 95% CI 0.60–0.86), stroke and transient ischemic attack (relative risk, 0.76; 95% CI

0.61–0.94), as well as tended to decrease in cardiac mortality (relative risk, 0.88: 95% CI 0.77–1.01) [321].

Effects on HDL Functionality

The elevation in HDL-C levels and normalization of HDL composition and particle number by niacin are consistent with the view that niacin may normalize deficient anti-atherogenic functions of HDL particles in atherogenic dyslipidemia. Indeed, niacin treatment **increases cholesterol efflux** *in vitro* from macrophage-derived foam cells in to HDL humans [15] (Table 11.2). Furthermore, extended-release niacin treatment (1000 mg/day), largely on a statin background, **improves endothelial function** in patients with coronary artery disease and low HDL-C after 12 weeks [322]. Therapy for three months with extended-release niacin (1500 mg/d) enhances the capacity of HDL to stimulate endothelial NO production, to reduce superoxide production and to promote endothelial progenitor cell-mediated endothelial repair in patients with Type 2 diabetes [323]; endothelium-dependent flow-mediated dilatation is also improved. Similarly, extended-release niacin (1000 mg/day for 52 weeks) improves endothelial function in patients with Metabolic Syndrome [310]. In addition, short-term therapy with extended-release niacin (up to 1500 mg/day for 12 weeks) improves endothelial function (as flow-mediated vasodilation) in HIV-infected patients with low HDL-C [324].

Niacin may also exert potent **anti-inflammatory, antioxidative, antithrombotic and cytoprotective actions** that may contribute to its beneficial action on endothelial function. Indeed, niacin inhibits the TNF-alpha-induced inflammatory response in human aortic endothelial cells, decreasing NFκB activation, VCAM-1 expression and MCP-1 secretion *in vitro* [325]. Furthermore, niacin decreases endothelial ROS production and subsequent LDL oxidation induced by angiotensin II in this model. Consistent with these results, niacin treatment (1500 mg/d for 12 weeks) of patients with Type 2 diabetes reduces *in vitro* adhesion of isolated monocytes to endothelial cells after TNF-alpha stimulation [326]; attenuated endothelial expression of platelet/endothelial cell adhesion molecule-1 (PECAM-1) may underlie this finding. In addition, niacin treatment decreases the HDL content of both lipid peroxidation products and myeloperoxidase in patients with Type 2 diabetes [323].

OPEN QUESTIONS

What is the contribution of direct actions of niacin on vascular cells to its beneficial effects on cardiovascular disease?
What proportion of the clinical benefit of niacin treatment is derived from lipid-dependent vs. lipid-independent actions?

Intriguingly, evidence is accumulating that niacin can inhibit atherosclerosis via its anti-inflammatory actions, independent of changes in plasma lipid levels. Indeed, niacin reduces atherogenesis in LDL-receptor-null mice in the absence of effects on plasma lipids; diminised macrophage recruitment to atherosclerotic plaques underlies this observation [327]. The activation of macrophagic GPR109A receptor with the subsequent induction of ABCG1 expression and promotion of cellular cholesterol efflux appears to be essential for the anti-atherogenic and anti-inflammatory actions of niacin [327]. Consistent with these data, niacin inhibits the inflammatory response and protects against endothelial dysfunction, independent of changes in plasma lipid levels, in a rabbit model of acute vascular inflammation induced with a peri-arterial carotid collar [328]. Specifically, niacin treatment inhibits endothelial expression of VCAM-1, ICAM-1 and MCP-1 at 24-hour post-collar implantation. Niacin also reduces intima-media neutrophil recruitment and myeloperoxidase accumulation, enhances endothelial-dependent vasorelaxation and cyclic guanosine monophosphate production, increases subnormal vascular glutathione content, and protects against HOCl-induced endothelial dysfunction and TNF-alpha-induced vascular inflammation [328].

Niacin is also capable of attenuating TNF-alpha expression in a rat model of stroke, reducing apoptosis and improving functional outcome [329]. Moreover, CRP levels are decreased by −20% in patients with Metabolic Syndrome upon niacin treatment (1000 mg/day for 52 weeks) [310]. In addition, niacin affects platelet reactivity *in vitro*, inhibiting platelet aggregation and stimulating prostaglandin release [330]. Finally, niacin suppresses expression of the chemokines fractalkine and MCP-1 in cultured adipocytes [312].

Adverse Effects

In clinical practice, niacin preparations are distinguished by poor tolerability, poor patient compliance and suboptimal dosing, mainly because of cutaneous flushing, a process primarily mediated by prostaglandin D2 (Table 11.2). Niacin-mediated flushing is also mediated by beta-arrestin 1, an adaptor protein that participates in intracellular signalling via G-protein-coupled receptors [332]. Other adverse effects of niacin include elevations in glucose, liver enzymes and uric acid levels, which may have clinical relevance in selected patients [308]. Poor compliance as a result of flushing led to the development of extended-release niacin (Niaspan), which displays an improved compliance and safety profile [331]. High dose aspirin is typically used to ameliorate flushing; however, the discontinuation rate still remains high [333]. Some natural compounds, such as flavonoids, can decrease flushing in humans and inhibit niacin-induced plasma prostaglandin D2 and serotonin increase in rats [334, 335]; nonetheless, the effects appear to be modest.

In contrast, the selective prostaglandin D2 receptor 1 antagonist **laropiprant (MK-0524)** at 150 mg/day effectively reduces extended-release niacin-induced flushing in dyslipidemic patients after nine weeks [336]. Importantly, laropiprant does not alter the beneficial effects of niacin on the lipid profile. Furthermore, laropiprant alone has no effect on plasma lipids. Co-administration of laropiprant 30, 100 and 300 mg with extended-release niacin significantly lowers flushing symptom scores (by approximately −50% or more) at all tested doses and also significantly reduces skin blood-flow in healthy subjects [337]. The adverse effects of laropiprant are minimal and the bioavailability of niacin is unchanged [337]. A novel, accelerated extended-release niacin dosing paradigm may, therefore, be employed, which involves initiation of niacin at 1 g with subsequent advance (after four weeks) to 2 g on a background of laropiprant [336].

Moreover, when combined with extended-release niacin (1–2 g/d) in a fixed-dose tablet, laropiprant produced significantly less flushing (−65%) than niacin during the initiation (week 1) and maintenance (weeks 2–24) stages in dyslipidemic patients [338]. The proportion of patients with flushing in the niacin/laropiprant group gradually subsided to a level that approximates placebo, whilst it remained constant at approximately 30% in the niacin group. However, the proportion of patients who discontinued the study still remained high in the niacin/laropiprant group (28.3% vs.35.9% on niacin and 11.5% on placebo). The proportion of patients who discontinued the study because of the flushing was 10.2% in the niacin/laropiprant group vs. 22.2% in the niacin group. Except for flushing, niacin/laropiprant possessed a safety/tolerability profile comparable to that of niacin [338]. The ongoing clinical Treatment of HDL to Reduce the Incidence of Vascular Events (HPS2-THRIVE) trial will provide a better insight into the long-term safety of the combination of niacin with laropiprant and its efficacy in reducing cardiovascular events [339].

In addition to niacin, several other niacin receptor agonists have been developed and patented, but their clinical effects have not been described [296]. For example, 5-alkyl and aryl-pyrazole-tetrazoles have been identified as a new class of selective, small-molecule agonists of GPR109A [340, 341]. One of these molecules displays improved potency and lacks flushing at doses that effectively altered the lipid profile in rats [342]. Niacin receptor agonists may, however, theoretically result in adipocyte triglyceride accumulation, clinical adiposity and increased flushing. G protein-based ligands that activate GPR109A in a beta-arrestin-independent fashion may represent another therapeutic option as niacin mimetics [332].

CLINICAL SUMMARY

Correction of abnormal HDL remodeling by niacin potently increases HDL-C levels and may decrease both cardiovascular risk and atherogenesis in coronary heart disease patients. Flushing is a major limitation of wider niacin use that can be diminished using laropiprant, a new, selective prostaglandin D2 receptor 1 antagonist.

11.7 CETP INHIBITORS

Presently, CETP inhibitors are the most potent HDL-raising agents, with a dose-dependent HDL-C elevation of up to +100% or more for some agents [193] (Table 11.2). The therapeutic strategy to inhibit CETP and thereby raise HDL-C derives from the lipid phenotype associated with genetic deficiency of CETP, i.e., elevated HDL-C often accompanied by decreased LDL-C levels, a profile that is typically anti-atherogenic [343]. Human subjects with heterozygous CETP deficiency and HDL-C levels >60 mg/dl may exhibit a reduced risk of coronary heart disease [344, 345]. Such subjects display elevated levels of alpha-1 and subnormal levels of alpha-3 and pre-beta-1 HDL subpopulations, a profile associated with reduced cardiovascular risk [346]. Partial inhibition of CETP may, therefore, be atheroprotective. In contrast, the complete absence of CETP activity can create a potentially pro-atherogenic lipid profile involving enlarged HDL and heterogeneous LDL particles [343, 347]. The formation of such particles typical of homozygous CETP-deficient patients can be avoided by partial inhibition of CETP. Studies in rabbits, a species with naturally high levels of CETP, support the therapeutic potential of partial CETP inhibition as an approach to slowing, or even reversing, atherogenesis [4, 348]. These results are further consistent with elevated CETP activity in dyslipidemic states [193], as well as with the capacity of high circulating levels of CETP to predict accelerated atheroma progression as observed in abdominally obese patients in the Strategy To Reduce Atherosclerosis Development InVolving Administration of Rimonabant–the Intravascular Ultrasound Study (STRADIVARIUS) study [349].

> **DEFINITION**
>
> **CETP inhibitors:** small molecules or antibodies that interact with CETP and inhibit its capacity to facilitate the heterotransfer of neutral lipids between HDL and apoB-containing lipoproteins.

Based on these findings, three small-molecule inhibitors of CETP, torcetrapib, dalcetrapib (JTT-705) and anacetrapib (MK-0859), have been developed which induce an elevation in concentrations of HDL-C, while decreasing those of LDL-C and apoB (Table 11.3). A CETP vaccine that induces autoantibodies that specifically bind to endogenous CETP represents another approach in inhibiting CETP activity [348, 351, 352]. Small-molecule CETP inhibitors are chemically distinct and differ considerably in their biochemical properties, inhibitory potency and mechanisms of interaction with CETP. Indeed, torcetrapib forms a non-productive complex with HDL, whereas dalcetrapib binds to CETP irreversibly via disulfide bond formation and anacetrapib ensures reversible binding [353, 354]. Torcetrapib, dalcetrapib and anacetrapib compete with one another for binding to CETP; however, the binding site of dalcetrapib is different from that occupied by torcetrapib [355]. Each CETP inhibitor induces tight binding of CETP to HDL, indicating that these agents promote the formation of a complex between CETP and HDL, resulting in the inhibition of CETP activity [356]. The capacity of dalcetrapib to induce formation of a CETP-HDL complex is, however, low at therapeutic plasma levels of the inhibitor [357].

As a result, torcetrapib and anacetrapib inhibit CETP-mediated cholesteryl ester and triglyceride transfer with similar potencies, while dalcetrapib is a significantly less potent inhibitor in lipoprotein-deficient plasma and total plasma [356]. The half maximal inhibitory concentration (IC_{50}) values for CETP inhibition by torcetrapib are 19 and 79 nM, and by dalcetrapib, 425 nM and 10 μM, respectively [357]. Controversially, however, mathematical modeling of CETP inhibition based upon a shuttle mechanism for cholesteryl ester/triglyceride exchange between VLDL, LDL and HDL predicts that irreversible binding, such as that by dalcetrapib, should provide a more robust response compared with non-competitive inhibition provided by torcetrapib [353]. Other differences between the CETP inhibitors involve inhibition kinetics. Thus, the inhibition of CETP by both anacetrapib and torcetrapib is not time-dependent, whereas the potency of dalcetrapib significantly increases with extended pre-incubation [356]. Such time dependence of the action of dalcetrapib is consistent with S–S bond formation with Cys13 in CETP [357, 358]. Interestingly, dalcetrapib demonstrates high specificity for CETP vs. Thiol-containing enzymes, inducing a conformational change in the protein not observed with torcetrapib [355, 358].

In addition to torcetrapib, dalcetrapib and anacetrapib, other HDL-raising drugs, such as niacin, fibrates and statins can directly, or indirectly, reduce CETP activity (see chapters 11.4, 11.5 and 11.6). Other classes of CETP inhibitors include 2,3-dihydro-3,8-diphenylbenzo[1, 4]oxazines, 1,2,3,4-tetrahydroquinolines, 2-arylbenzoxazoles and benzylamino-methanones [359–362]. 2,3-Dihydro-3,8-diphenylbenzo[1, 4]oxazines represent a new class of potent CETP inhibitors. The most potent compound displays IC_{50} of 26 nM, possesses a favorable pharmacokinetic profile with good oral bioavailability in rats and increases HDL-C in human CETP transgenic mice and high-fat fed hamsters [361]. Tetrahydroquinoline A is a potent CETP inhibitor that inhibits partially purified CETP with an IC_{50} of 39 nM [359]. Moreover, pharmacophore modeling and quantitative structure-activity relationship analysis identify at least two distinct binding modes accessible to ligands within the CETP lipid-binding pocket, potentially leading to synthetic exploration of a new series of CETP inhibitors [363]. Finally, it should be noted that certain agents, such as the immunosuppressants cyclosporine A and rapamycin, can function as CETP activators [364]; these drugs cause dyslipidemia in transplant patients by increasing CETP activity via binding to site(s) distinct from the CE-binding site of CETP followed by a conformational change of the protein.

Torcetrapib

Torcetrapib became the first CETP inhibitor to enter clinical trials which revealed its marked efficacy in improving an atherogenic plasma lipid profile (Table 11.3).

Early studies demonstrated that torcetrapib dose-dependently inhibited CETP activity by −12% to −80% and increased HDL-C by +28% to +91% at 30 and 240 mg/d, respectively, in healthy young normolipidemic subjects [365]. With respect to the modulation of apolipoprotein levels, apoA-I and apoE were increased by +27% and +66%, respectively, while apoB was reduced by −26% at a torcetrapib dose of 240 mg/d [365]. Significantly, CETP inhibition by torcetrapib led to a reduction in the triglyceride content of HDL particles and increased their cholesteryl ester content [365]. In subjects with low levels of HDL-C, treatment with 120 and 240 mg of torcetrapib daily raised plasma HDL-C concentrations by +46% and +106%, respectively [366]. In parallel, torcetrapib partially corrected abnormal physicochemical properties of HDL2 and HDL3 particles in Type IIB hyperlipidemia, increasing cholesteryl ester, phospholipid and apoA-I, decreasing triglyceride and improving the cholesteryl ester/triglyceride ratio [17].

Torcetrapib treatment strongly impacts the heterogeneity of HDL particles (Table 11.3). Torcetrapib increased **large HDL2**-C (+87%) to a greater extent than small HDL3-C (+29%) and altered the distribution of cholesterol among HDL and LDL subclasses, resulting in an increased mean particle size of both HDL and LDL [367] (Fig. 11.4). In addition, torcetrapib raised plasma levels of apoA-I, increased the amount of apoA-I in alpha-1-HDL and apoA-I pool size and decreased the fractional catabolic rate of apoA-I, thereby prolonging its residence time in the circulation [367] (Fig. 11.4). In a similar fashion, torcetrapib raised plasma apoA-II concentrations by reducing the fractional catabolic rate of apoA-II without affecting its production rate [368]. Such beneficial modulation of HDL metabolism may result from normalization of apoA-I lipidation and, thus, conformation subsequent to normalized cholesteryl ester/triglyceride core lipid ratio [367]. Torcetrapib also increased LpAI:AII levels and apoA-II pool size in addition to the abundance of apoA-II in the alpha-2-migrating HDL subpopulation; concentrations of apoA-II in alpha-3-migrating HDL were, however, reduced [368].

Importantly, torcetrapib reduces plasma levels of apoB-containing, triglyceride-rich lipoproteins (Fig. 11.4). Some recent data even suggest that lowering triglyceride-rich lipoproteins by CETP inhibitors can potentially afford greater atheroprotection than HDL-C raising. Indeed, CETP inhibition reduces atherosclerosis mainly via lowering (V)LDL-C in APOE*3-Leiden mice [369]; the (V)LDL-C/HDL-C ratio is, however, high in this model, thereby down-playing the potential protective role of HDL. In non-atorvastatin-treated patients, torcetrapib-mediated CETP inhibition reduces levels of apoB-containing lipoproteins, primarily as a result of the increased clearance of VLDL apoB-100, whereas the clearance of LDL apoB-100 is not significantly affected [370]. Torcetrapib also reduces VLDL apoE pool size primarily by increasing the fractional catabolic rate of apoE while leaving the apoE content of VLDL unchanged [370]. In contrast, on a background of atorvastatin, the clearance of apoB-containing lipoproteins is accelerated as a result of the up-regulation of the LDL receptor [370]. Changes in VLDL composition, notably the enrichment in triglyceride, resulting from CETP inhibition may potentiate lipolysis of these

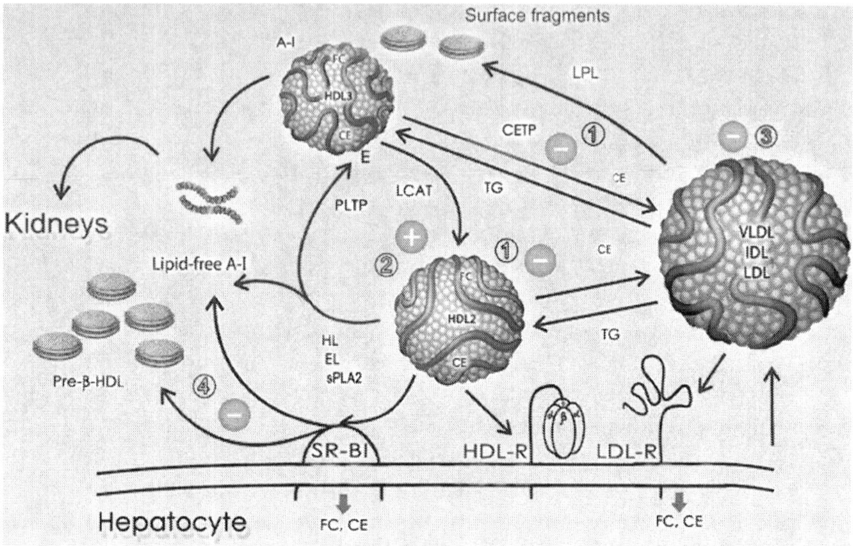

Figure 11.4 Major pathways involved in the action of CETP inhibitors. (1) Subnormal activity of CETP results in the diminished heteroexchange of cholesteryl ester and triglyceride between HDL and triglyceride-rich lipoproteins, leading to HDL enrichment in cholesteryl ester and depletion in triglyceride. (2) Preferential elevation in plasma levels of large, light, lipid-rich HDL2, with delayed apoA-I and HDL catabolism. (3) Decrease in the cholesterol load of apoB-containing lipoproteins. (4) Attenuation of the recycling of lipid-poor and lipid-free apoA-I. CE, cholesteryl ester; EL, endothelial lipase; FC, free cholesterol; HL, hepatic lipase; HDL-R, holoparticle HDL receptor; LDL-R, LDL receptor; TG, triglyceride.

particles by LPL, thereby allowing their accelerated clearance from plasma and conversion to LDL. In contrast, torcetrapib added to atorvastatin treatment does not alter VLDL apoE pool size because of increases in both the fractional catabolic rate and the production rate of VLDL apoE [371].

In addition to its action in the fasting state, torcetrapib significantly improves the post-prandial metabolism of triglyceride-rich lipoproteins, attenuating the incremental post-prandial area under the curve for VLDL1 and VLDL2 by up to −60% in subjects with mixed hyperlipidemia [372]. Concomitantly, the cholesteryl ester/triglyceride ratio in both VLDL1 and VLDL2 is significantly reduced as a result of the attenuation of post-prandial cholesteryl ester transfer to chylomicrons and VLDL1. Elevations in plasma levels of large HDL2 particles induced by CETP inhibition are also associated with enhanced catabolism of chylomicrons and their remnants [372]. The post-prandial triglyceride response to oral fat load is also decreased in CETP deficiency compared with controls [373]. Furthermore, plasma levels of remnant lipoprotein cholesterol tend to be reduced in CETP-deficient subjects.

Dalcetrapib

Another small-molecule CETP inhibitor, **dalcetrapib (JTT-705)**, at a dose of 900 mg/d, inhibits CETP by −37% and increases HDL-C by +34%, apoA-I by +15%, HDL2-C by +59% and HDL3-C by +19% in subjects with mild hyperlipidemia [374]; comparable effects are observed in patients with Type II dyslipidemia [375] and with familial hypoalphalipoproteinemia [376] (Table 11.3). In phase IIa trials, dalcetrapib, at a dose of 300, 600, and 900 mg for four weeks, raised HDL-C by up to +36% at 900 mg/d in combination with a statin and apoA-I by up to +16% (in monotherapy), and reduced CETP activity up to −38% [377]. In a double-blind trial performed in subjects with coronary heart disease or coronary heart disease risk equivalents, dalcetrapib (900 mg/day), on a background of atorvastatin (10 to 80 mg/day), for 24 weeks increased HDL-C by +33.4%, decreased CETP activity by −53.5%, increased CETP mass by +80.8% and increased apoA-I by +11.4% [378]. In patients with type II hyperlipidemia, CETP inhibition with dalcetrapib (600 mg/d for four weeks) raised plasma HDL-C by +26% and lowered triglyceride levels by −22% [379]. Dalcetrapib can also reduce concentrations of LDL-C, but only slightly and at a high dose, whereas levels of apoB remain unaffected [375]. Dalcetrapib is, therefore, a less potent CETP inhibitor compared with torcetrapib, not only *in vitro*, but also in a clinical setting. Indeed, torcetrapib at 120 mg increases HDL-C substantially more than dalcetrapib at 900 mg [366].

In rabbits, dalcetrapib at a high dietary dose of 0.75% inhibits CETP activity, increases both HDL-C concentrations and the ratio of HDL2-C to HDL3-C and, in addition, decreases the fractional esterification rate of HDL-C, again indicating preferential elevation in large HDL particles [380]. CETP-induced pre-beta HDL formation *in vitro* in human plasma is unchanged by dalcetrapib at 3 μM and enhanced at 10 μM [358]. Levels of apoE in HDL, serum PON activity and HDL-associated PAF-AH activity are elevated, whereas plasma lysophosphatidylcholine concentration is diminished. Such enhanced apoE content in HDL particles may be of special relevance to accelerated HDL particle catabolism in the liver and peripheral tissues through the LDL receptor pathway. Similarly, dalcetrapib at a high dose of 300 mg/kg daily in rabbits increased plasma total cholesterol, HDL-C, HDL2-C and HDL3-C and reduced HDL-triglyceride and CETP activity [381].

Anacetrapib

Anacetrapib (MK-0859), a second CETP inhibitor currently under clinical development, increases HDL-C to a greater extent than other molecules in this class of drugs and also markedly reduces LDL-C (Table 11.3). Interestingly, and in contrast to other CETP inhibitors, plasma levels of triglycerides were not significantly affected by anacetrapib treatment in published trials. Indeed, anacetrapib at a

TABLE 11.3 Properties of CETP Inhibitors Entered into Clinical Development

Agents	Potency	Interaction with CETP	Effects on HDL Levels	Effects on Plasma Lipids	Off-Target Effects
Torcetrapib*	High	Non-competitive inhibition	Elevation of apoA-I, apoA-II, apoE, HDL-C and HDL CE levels Preferential increase in large HDL Decrease in HDL TG content	Decrease in LDL-C, TG and apoB levels	Hypertensive effect related to increment in aldosterone levels
Dalcetrapib (JTT-705)	Low	Irreversible binding to Cys13	Elevation of apoA-I, apoE, HDL-C and HDL CE levels Preferential increase in large HDL Decrease in HDL TG content	Decrease in LDL-C and TG levels	None
Anacetrapib (MK-0859)	High	Reversible binding	Elevation of apoA-I, apoE and HDL-C levels Preferential increase in large HDL	Decrease in LDL-C, Lp(a) and apoB levels	None

*Abandoned as a result of excessive mortality in a large-scale clinical trial [350]. CE, cholesteryl ester; TG, triglyceride.

dose of 0, 10, 40, 150, or 300 mg/d for 28 days produced dose-dependent lipid-altering actions with maximal effects of +129% in HDL-C, +47% in apoA-I, −38% in LDL-C and −26% in apoB in a double-blind, randomized, placebo-controlled phase I study performed in patients with dyslipidemia [382]. Thus, LDL-C reduction provided by anacetrapib is similar to that observed with statins. The optimal dose of anacetrapib with a meal required to reach the plateau of the effects on plasma lipids in this study was 150 mg/d [382].

Another study revealed that anacetrapib 10, 40, 150, and 300 mg once daily for eight weeks produced mean percent changes from baseline to week eight for LDL-C of −16%, −27%, −40%, and −39%, respectively, and for HDL-C of +44%, +86%, +139% and +133%, respectively, in patients with primary hypercholesterolemia or mixed hyperlipidemia [354]. ApoB was reduced by up to −30% and apoA-I elevated up to +47%. Importantly, anacetrapib, both as a monotherapy and co-administrated with atorvastatin, potently reduced plasma levels of Lp(a) by up to −50% [354]. Impressively, the net result of treatment with anacetrapib on a background of atorvastatin amounted to −70% lowering of LDL-C and more than doubling of HDL-C levels [354]. In addition, anacetrapib increased plasma levels of apoE by up to +40%.

Effects on RCT

The widely accepted premise underlying the development of CETP inhibitors has long been that robust HDL-C raising should enhance **RCT** from peripheral tissues and the arterial wall to the liver and thence inhibit atherogenesis. CETP inhibitors do not, however, alter fecal excretion of cholesterol in human subjects [367] (Table 11.3), leading to ongoing discussion as to the potential of CETP inhibitors to accelerate RCT and delay atherogenesis in humans [383].

HDL-associated cholesteryl ester can be delivered to the liver either directly from HDL through selective uptake by SR-BI or indirectly via CETP-mediated transfer to apoB-containing lipoproteins (VLDL, IDL and LDL) followed by lipoprotein uptake by the hepatic LDL receptor. In humans, the CETP-mediated transfer of cholesteryl ester from HDL to apoB-containing lipoproteins represents a quantitatively significant route for the return of cholesterol from peripheral tissues to the liver [384, 385]. Importantly, the role of this **indirect reverse cholesterol transport** pathway can be increased even further when hepatic LDL receptor activity is elevated, such as under statin treatment. LDL particles constitute the major cholesteryl ester acceptor among apoB-containing lipoproteins in normolipidemic, normotriglyceridemic humans [386]. Specifically, CETP-mediated cholesteryl ester transfer from HDL is primarily targeted to LDL particles of intermediate size and density displaying an elevated affinity for the LDL receptor, which facilitates their rapid removal from plasma.

Indirect cholesterol flux from HDL to the liver mediated by apoB-containing lipoproteins appears to be attenuated rather than accelerated under CETP inhibition. However, available CETP inhibitors induce only partial CETP inhibition, indicating that a residual cholesteryl ester transfer from HDL to apoB-containing lipoproteins is maintained. Such cholesteryl ester transfer is, nonetheless, limited, as reflected by the reduction in the cholesteryl ester content of VLDL and LDL particles after torcetrapib administration [372]. Interference with this pathway by a CETP inhibitor may be even more pronounced under statin treatment when the activity of hepatic LDL receptors is high, hence the proposal to restrict CETP inhibitors to patients with low HDL-C and high triglyceride levels [283].

Reduction in cholesterol flux through the indirect pathway may, however, be compensated for by acceleration of **direct cholesterol transport** through HDL. In support of this possibility, torcetrapib treatment generates large, apoE- and cholesteryl ester-rich HDL particles that efficiently promote **cholesterol efflux** from macrophages through the ABCG1-dependent pathway [387]. Furthermore, treatment with anacetrapib (300 mg/day) for eight weeks increases HDL-C approximately twofold and improves cellular cholesterol efflux mediated by HDL [15]. This effect is associated with an increased HDL content of LCAT and apoE and is dependent on the expression of ABCG1 and ABCA1. In addition, as large HDL can efficiently efflux cellular cholesterol via the ABCG1- and SR-BI-mediated pathways [387, 388], both ABCG1- and SR-BI-mediated cholesterol efflux may be enhanced following CETP inhibition. Indeed, large HDL2 generated in subjects displaying mixed hyperlipidemia under torcetrapib treatment displays an elevated capacity to mediate cellular cholesterol efflux via both SR-BI and ABCG1 [17]. Consistent with these data, treatment of hamsters with torcetrapib raises HDL-C and increases cholesterol efflux capacity of plasma [389]. Furthermore, dalcetrapib improves cholesterol efflux from autologous labeled macrophages to plasma HDL in hamsters pretreated by daily gavage with the inhibitor (100 mg/kg BID for 7 days), along with an increase in HDL-C levels [358, 390]. In parallel, cholesterol influx from HDL to the liver can also be stimulated. By contrast, dalcetrapib (300 mg/kg/d) does not modify the cellular cholesterol efflux capacity of HDL in rabbits [381]. Thus, hepatic delivery of free cholesterol, which represents a minor route of RCT [384], may be enhanced secondary to the elevation of HDL size and lipid load.

Data from CETP-deficient subjects largely support those obtained using CETP inhibitors. Thus, homozygous CETP deficiency features the accumulation of large HDL2 particles with an elevated content of apoE and LCAT; such particles also accumulate after treatment with torcetrapib (120 mg) and display elevated cholesterol efflux capacity on THP-1 cells [391]. In addition, HDL isolated from a compound heterozygote for a known D459G variant and a novel 18-bp deletion mutation in the CETP promoter displays an elevated cholesterol efflux capacity via SR-BI [392]; the cholesteryl ester influx capacity of HDL into hepatocytes is maintained. Furthermore, serum from homozygous CETP-deficient subjects exhibits enhanced cholesterol acceptor capacity via the SR-BI dependent pathway [393]. In addition, the capacity of serum to efflux cholesterol via the ABCA1-dependent pathway is normal in subjects with homo- and heterozygous CETP deficiency. In agreement with these results, CETP over-expression in transgenic mice does not negatively affect RCT from macrophages to feces *in vivo* [394].

Moreover, the partial inhibition of CETP by torcetrapib normalizes apoA-I levels within large, alpha-migrating HDL and increases plasma concentrations of HDL apoA-I by delaying apoA-I catabolism in low HDL-C subjects [367]. Large apoA-I-containing HDL represent ideal ligands for the **selective uptake of cholesteryl ester** by SR-BI [395]; CETP inhibition may, therefore, enhance the capacity of HDL particles to deliver cholesteryl ester to the liver, thereby allowing efficient hepatic return of cholesteryl ester via the direct route of RCT. Indeed, hepatic selective uptake of HDL-cholesteryl ester is elevated following torcetrapib treatment [17]. Consistent with this conclusion, cholesteryl ester enrichment of HDL increases both the cholesteryl ester/triglyceride ratio and the selective uptake of cholesteryl ester by hepatic cells [396]. Furthermore, CETP can directly mediate the selective uptake of HDL-derived cholesteryl esters by the liver [397].

As a result, the net action of CETP inhibitors on RCT is not deleterious in CETP-expressing species. Moreover, fecal elimination of sterols is elevated by torcetrapib [389], dalcetrapib [358, 390, 398] and anacetrapib [399] in hamsters. Such effects of CETP inhibitors on RCT cannot be accounted for by their direct actions on cells participating in the RCT pathway. Indeed, treatment with torcetrapib of RAW 264.7 macrophages or HepG2 hepatocytes does not affect their intrinsic cellular capacity to efflux cholesterol to a standard acceptor [389]. Together, these data suggest that direct HDL-mediated cholesterol transport from peripheral cells to the liver can be accelerated by CETP inhibitors. The accumulation of both apoE- and apoA-I-enriched HDL, which deliver cholesterol to the liver via the LDL receptor- and SR-BI-mediated pathways, respectively, may underlie this finding. Such an acceleration of direct RCT may compensate for the deceleration of indirect cholesterol transport mediated by apoB-containing lipoproteins following CETP inhibition.

Clinical Trials

Clinical trials provide the ultimate answer regarding the efficacy, clinical benefit and safety of a drug. The first large-scale, prospective, placebo-controlled interventional trial assessing the impact of a CETP inhibitor on cardiovascular risk, Lipid Level Management to Understand its Impact in Atherosclerotic Events (ILLUMINATE) trial, was prematurely and unexpectedly terminated in December 2006 as a consequence of excess mortality in the treatment group relative to placebo [350]. In this study, more than 15,000 patients at high risk for cardiovascular disease were randomized to either torcetrapib (60 mg) or placebo on a background of atorvastatin for a median of 550 days. Despite highly favorable changes in the lipid profile in the torcetrapib arm (HDL-C elevation, +72%; LDL-C reduction, −25%), cardiovascular mortality increased by +40%, the rate of cardiovascular events by +25% and non-cardiovascular death by +100% [350]. Furthermore, the mean common carotid intima-media thickness progression was accelerated in subjects receiving torcetrapib plus atorvastatin compared with subjects receiving atorvastatin alone in the Rating Atherosclerotic Disease change by Imaging With A New CETP Inhibitor (RADIANCE) 1 and 2 studies, which examined carotid intima-media thickness in subjects with familial hypercholesterolemia and mixed dyslipidemia [400]. These data suggested adverse effects of CETP inhibition and led to heavy criticism of this therapeutic approach [401].

Indeed, some recent findings challenge the concept of CETP inhibition in lowering cardiovascular risk. First, when plasma CETP activity was measured in participants of the 1978 Framingham Heart Study who attended a routine examination in 1987–1990 and were free of cardiovascular disease, plasma CETP activity was inversely related to the incidence of cardiovascular events in multivariable analyses adjusted for standard risk factors, including HDL-C (hazard ratio for activity at or above the median, 0.72; 95% CI, 0.57–0.90, compared with below the median) [402]. Similarly, in 3256 participants of the LURIC study who were referred for coronary angiography at baseline between 1997 and 2000 with a median follow-up of 7.75 years, low plasma concentrations of CETP were associated with elevated cardiovascular and all-cause mortality [403]. Indeed, the hazard ratio for death in the lowest CETP quartile was 1.33 (CI, 1.07–1.65) compared with patients in the highest CETP quartile. Furthermore, increasing on-statin CETP mass was inversely related to coronary outcomes in the PROVE IT-TIMI 22 cohort, particularly among subjects with low LDL-C levels [404]. When analysing these results, one should, however, keep in mind that changes in CETP mass often do not parallel those in endogenous CETP activity; it is CETP activity, rather than mass, whose elevation is associated with adverse plasma lipid profile [193].

Torcetrapib treatment was, however, associated with a significant increment in aldosterone levels, with changes in serum electrolyte levels indicative of **mineralocorticoid excess**, and with an **elevation of blood pressure** (+5.4 mm Hg in systolic blood pressure) in the ILLUMINATE trial [350]. Given that every 10 mmHg increase in systolic blood pressure is associated with a +25% increase in cardiovascular disease, stroke and vascular death, this effect of torcetrapib may have negated the benefit of HDL-C elevation [13] These findings suggest an **off-target mechanism of torcetrapib toxicity** involving the activation of mineralocorticoid receptors by aldosterone with subsequent induction of hypertension [385]. Indeed, aldosterone produced in the adrenal gland activates mineralocorticoid receptors of the distal tubule of the nephron leading to increased sodium and bicarbonate readsorption in exchange for potassium excretion. Direct vascular actions of aldosterone may have further aggravated cardiovascular disease in the torcetrapib group. Data from the RADIANCE 1 and 2 studies also support mineralocorticoid-mediated, renin- and angiotensin-independent off-target toxicity in patients receiving torcetrapib as a contributing factor to an adverse outcome [400].

Moreover, pressor effects of torcetrapib and CP-532623, a closely related CETP inhibitor, were directly observed in monkeys and human subjects [405]. In humans, high- vs. low-dose CP-532623 produced greater pressor effects, despite similar maximal CETP inhibition. Inhibition of CETP was observed 48 hours post-dose, whereas blood pressure elevation was dissipated at 24 hours, temporally dissociating CETP inhibition from blood pressure changes [405]. Continuous measurements of blood pressure showed persistent elevations, whereas aldosterone changes were transient, suggesting that increases in blood pressure are not directly the result of renin-angiotensin-aldosterone system activation and may, in part, be caused by direct effects on blood vessels [405].

The aldosterone-mediated pressor effect of torcetrapib appears to be unrelated to HDL-C raising as suggested by data obtained with other CETP inhibitors. In contrast to torcetrapib, dalcetrapib and anacetrapib do not increase blood pressure in human subjects: an observation that discounts a class effect potentially related to altered steroid metabolism and enhanced aldosterone production through the elevated delivery of HDL-derived cholesterol to the adrenals [385, 406, 407]. Indeed, the incidence of adverse events was low, and similar, between placebo and dalcetrapib (300 mg and 600 mg) in human trials [377]. Adverse cardiovascular events were also similar across treatment groups, with no dose-related trends and no clinically relevant changes in blood pressure or electrocardiographic findings. Similarly, dalcetrapib showed no clinically relevant differences vs. placebo in laboratory parameters, including aldosterone, electrolytes, electrocardiograms and vital signs, including blood pressure [378]. A small (+0.05%), but significant ($p = 0.008$), increase in plasma CRP is, however, observed on dalcetrapib treatment [378]. Consistent with these data, anacetrapib does not increase blood pressure in dyslipidemic subjects, suggesting that potent CETP inhibition by itself does not increase blood pressure [382]. Similarly, anacetrapib does not affect circulating levels of CRP, aldosterone, sodium, potassium and bicarbonate [354]. The incidence of adverse events was similar for placebo and all anacetrapib treatment groups in these studies; there were no increases in systolic or diastolic blood pressure in any treatment arm [354].

The absence of a class effect of CETP inhibitors on aldosterone metabolism and blood pressure is further supported by several lines of evidence. First, experiments in animal models reveal that torcetrapib evokes an acute increase in blood pressure in all species evaluated, whereas no increase was observed with anacetrapib [408]. Second, treatment with torcetrapib is associated with increased plasma levels of aldosterone and corticosterone. Furthermore, torcetrapib releases aldosterone from adrenocortical cells *in vitro* [408]. In addition, torcetrapib, but not dalcetrapib, increases aldosterone production and cytochrome P450 subunit 11B2 mRNA levels [377]. In contrast, no increase in adrenal steroid levels is observed with anacetrapib. Interestingly, the inhibition of adrenal steroid synthesis does not inhibit the pressor response to torcetrapib, whereas adrenalectomy prevents the ability of torcetrapib to increase blood pressure in rats, indicating that the acute pressor response to torcetrapib is not mediated by adrenal steroids, but requires intact adrenal glands [408].

Another argument against the role of CETP in the hypertensive effect of torcetrapib is derived from the observation that activation of the renin-angiotensin-aldosterone system, which regulates mineralocorticoid and blood pressure responsiveness to dietary salt intake, is unrelated to plasma CETP levels in humans. Specifically, a decrease in CETP mass does not exert adverse effects on blood pressure regulation [409]. Further along this line, CETP single-nucleotide polymorphisms that are associated wth increased HDL-C levels do not result in elevated blood pressure [410]. Finally, patients with a CETP deficiency do not develop hypertension [13]. In contrast, decreasing plasma CETP activity is modestly related to increasing pulse pressure on follow-up in the Framingham community-based sample [411].

Imaging studies designed to evaluate the impact of torcetrapib on coronary atheroma with intra-vascular ultrasound in the Investigation of Lipid Level Management Using Coronary Ultrasound to Assess Reduction of Atherosclerosis by CETP Inhibition and HDL Elevation (ILLUSTRATE) trial revealed that the progression of coronary atherosclerosis was not altered by the treatment [412, 413]. Remarkably, elevated levels of HDL-C in torcetrapib-treated patients in this trial were associated with a reduction in cardiovascular events and a beneficial impact on plaque progression [350, 414], thereby suggesting that functional HDL particles were generated on drug treatment. Furthermore, an inverse relationship was observed between changes in HDL-C and the percentage atheroma volume in torcetrapib-treated patients. Specifically, the highest on-treatment HDL-C quartile was associated with significant regression of percent atheroma volume of −0.69%. In multivariable analysis, changes in HDL-C levels independently predicted the effect on atherosclerosis progression [415]. On the other hand, HDL-C elevation was not associated with carotid intima-media thickness change in the RADIANCE 1 and 2 studies [400].

OPEN QUESTION

Does HDL-C raising by CETP inhibitors translate into a reduction of cardiovascular risk?

Effects on HDL Functionality

Whereas the capacity of HDL to efflux and transport cholesterol does not appear to be compromised on torcetrapib treatment, little is known of **other atheroprotective activities** of HDL. Intriguingly, the excess of non-cardiovascular deaths observed in the torcetrapib group in the ILLUMINATE trial was primarily related to cancer (24 vs. 14 in the placebo group) and to infection (9 vs. 0) [350]. These effects appear to be specific to the drug, as meta-regression analysis of 44 randomized controlled trials corresponding to 107,773 participants reveals that increases in HDL-C up to +40% are not associated with higher non-cardiovascular death, indicative of the safety of HDL-C raising per se [416, 417]. Deleterious effects of torcetrapib on the biologic activities of HDL particles other than cholesterol efflux cannot, therefore, be excluded. Indeed, in addition to producing hypertension and baseline vasoconstriction, torcetrapib markedly inhibits acetylcholine-induced vasodilation in rabbit arteries [418]. A structurally distinct CETP inhibitor, JNJ-28545595, does not, however, affect endothelial function, despite producing similar degrees of CETP inhibition and HDL-C elevation. In contrast, torcetrapib does not impair smooth muscle function. Thus, torcetrapib impairs endothelial function *in vivo*, independently of CETP inhibition and HDL-C raising. This deleterious activity may have contributed to an increased cardiovascular risk in patients treated with torcetrapib in clinical trials [418].

Other potential drawbacks of CETP inhibition can include decreased NO production by endothelial cells secondary to reduced stimulation of eNOS by very large HDL, increased apoA-I oxidation secondary to its prolonged lifespan in the circulation and elevated secretion of endothelin-1 secondary to elevated HDL levels [419]. Another possibility is that relative decreases in plasma levels of small HDL particles possessing potent atheroprotective activities may deleteriously impact the net biologic activity of the total HDL pool following CETP inhibition. Consistent with this notion, the cardioprotective action of probucol, a unique hypolipidemic agent that raises CETP activity and reduces HDL-C, is paralleled by increases in plasma concentrations of small pre-beta-1 HDL [420]; biologic activities of such particles should be evaluated in order to assess this hypothesis. On the other hand, the anti-infectious activities of LBP or BPI, as well as the interaction of CETP with LPS, are unlikely to be affected by CETP inhibitors [421]. Similarly, the capacity of HDL to suppress macrophage TLR 4-mediated inflammatory responses, in a process partly dependent on cholesterol efflux via ABCA1 and ABCG1, is not modified by anacetrapib [15]. In addition, the capacity of HDL to protect LDL against oxidative modification and the activities of HDL-associated hydrolytic enzymes are not affected by CETP over-expression in transgenic mice [394].

Importantly, CETP inhibitors possess the potential for the improvement of all these aspects of HDL functionality. Small, triglyceride-enriched, functionally-deficient HDL accumulate in conditions associated with elevated CETP activity, as typical of metabolic diseases such as Type 2 diabetes and Metabolic Syndrome [3]. By enriching HDL in cholesteryl ester and normalizing the core lipid composition of HDL particles, CETP inhibitors may, therefore, correct their functional defects. Available data suggest that HDL functionality may be improved along with the impact of CETP inhibitors on circulating levels of HDL-C, which largely reflect the HDL content of cholesteryl ester. This conclusion may be relevant for the antioxidative and vasodilatory activities of HDL which strongly depend on the cholesteryl ester/triglyceride ratio in HDL particles [422]. Consistent with this hypothesis, dalcetrapib decreases circulating levels of oxidized LDL in familial hypoalphalipoproteinemia [376], whereas CETP inhibition *in vitro* by a monoclonal antibody renders LDL more resistant to oxidation [423], observations that could reflect improved HDL-mediated protection of LDL from oxidation *in vivo*. Furthermore, large, apoE-rich HDL produced by CETP inhibitors can be particularly effective at mediating the anti-inflammatory and anti-apoptotic effects of HDL, and in preserving normal endothelial functions, such as eNOS activity [391]. Next, endothelial function, assessed as flow-mediated dilatation, was improved in a subgroup of patients with type II hyperlipidemia displaying low baseline HDL-C levels and treated with dalcetrapib [379]. These considerations show that the concept of CETP inhibition possesses the potential to generate functional HDL particles and thus remains a viable therapeutic strategy [424].

The ultimate answer regarding the clinical utility of CETP inhibitors will be provided by ongoing multicenter, randomized, double-blind, placebo-controlled clinical trials, such as the Randomized Evaluation of the Effects of Anacetrapib Through Lipid Modification (REVEAL) trial which is to launch 2011 and enroll 30,000 patients with established vascular disease receiving a statin with follow-up for four years [425]. In this double-blind, randomized, placebo-controlled trial, patients will be randomized to 100 mg/d of anacetrapib or a matching placebo, with all participants receiving therapy with atorvastatin. The Determining the Efficacy and Tolerability of CETP Inhibition with Anacetrapib (DEFINE) study will evaluate anacetrapib (100 mg/d) in patients with coronary heart disease or at high risk for coronary heart disease [426]. The Randomised, Double-blind, Placebo-controlled Study Assessing the Effect of RO4607381 on Cardiovascular Mortality and Morbidity in Clinically Stable Patients With a Recent Acute Coronary Syndrome (Dal-OUTCOMES) study will assess the effect of dalcetrapib (600 mg/d) on cardiovascular events and mortality in patients with recent acute coronary syndrome [427]. The Dalcetrapib HDL Evaluation, Atherosclerosis and Reverse Cholesterol Transport (Dal-HEART) program also includes three surrogate end-point trials that will provide further information on the contribution of CETP to cardiovascular disease [428].

Preliminary results obtained in the DEFINE trial show that anacetrapib (100 mg/day) increased HDL-C levels by +138% relative to placebo after 18 months [429]. By 24 weeks, LDL-C levels were reduced by −39.8% vs. placebo. Through 76 weeks, no changes were noted in blood pressure, electrolyte or aldosterone levels with anacetrapib compared with placebo. Prespecified adjudicated cardiovascular events occurred in 16 patients treated with anacetrapib (2.0%) and in 21 patients on placebo (2.6%) ($p = 0.40$). The prespecified Bayesian analysis indicated that this event distribution provided a confidence of 94% that anacetrapib would not be associated with a +25% increase in cardiovascular events, as observed with torcetrapib [429]. Moreover, in the recently reported dal-PLAQUE trial dalcetrapib showed no evidence of a pathological effect related to the arterial wall over 24 months, suggesting possible beneficial vascular effects.

CLINICAL SUMMARY

Despite the failure of torcetrapib, CETP inhibitors still hold promise in terms of their ability to provide cardiovascular risk reduction in an appropriate clinical setting, particularly in low HDL-C/high triglyceride dyslipidemia.

11.8 OTHER AGENTS

LXR Agonists

LXRs are ligand-activated nuclear receptors that regulate lipid and glucose metabolism and modulate inflammation within the vascular wall [430] (Table 11.1). LXRs belong to the nuclear receptor superfamily, form heterodimers with RXRs and stimulate the expression of target genes following ligand binding [225]. Oxidized sterols represent natural ligands for LXRs. In addition, there are synthetic LXR agonists, such as T0901317 and GW3965. Therefore, LXRs operate as cholesterol sensors that protect from cholesterol overload by inhibiting intestinal cholesterol absorption, **stimulating cellular cholesterol efflux** to HDL and its transport to the liver, together with its enhanced conversion to bile acids for biliary excretion [225].

Activation of LXRs by either endogenous oxysterols or synthetic agonists induces the expression of multiple genes, including those encoding ABCA1, ABCG1, ABCG5 and ABCG8 [431]. Expression of **ABCA1** is specifically mediated by the **activation of LXR alpha** [257] . As a result, this signaling cascade increases plasma HDL-C levels and accelerates **RCT** in mice [432–434]. In parallel, trans-intestinal cholesterol transport is elevated, with intestinal ABCA1 playing an essential role in these effects [433, 434]. In contrast, macrophage LXR plays an important, but not essential, role in promoting RCT in response to an LXR agonist [435]. As functional RCT depends on biliary sterol secretion, the stimulating effect of LXR activation on RCT also requires cholesterol secretion into the bile [436]. Importantly, LXR activation promotes RCT, not only in rodents, but also in CETP-expressing species, as demonstrated by the stimulation of macrophage-to-feces RCT by the LXR agonist GW3965 in dyslipidemic hamsters [437].

In addition to the regulation of cholesterol metabolism, LXRs also exert **anti-inflammatory action** on macrophages and suppress proliferative responses in vascular cells [438]. Furthermore, LXR receptor agonism suppresses hepatic gluconeogenesis and **improves insulin sensitivity**, reflecting the ability of glucose to function as an LXR ligand [439]. As a result of such multiple anti-atherogenic actions, LXR agonists are capable of inhibiting atherosclerosis in mice. Indeed, the LXR agonist T0901317 suppresses atherogenesis and promotes regression of atherosclerotic lesions in apoE*3Leiden mice via increased expression of ABCA1, ABCG1 and apoE, decreased monocyte adhesion to the endothelium and attenuated inflammatory status [440]. However, the anti-atherogenic effect of LXR activation in rodents is not reproduced in CETP-expressing species that exhibit major adverse effects, such as elevated triglyceride and LDL-C levels. The latter observation is related to the capacity of LXR agonists to activate fatty acid synthesis via stimulation of the expression of a lipogenic transcription factor, SREBP-1c, that induces elevation of plasma triglyceride concentrations and liver steatosis [225]. Furthermore, LXR alpha can directly enhance hepatic CETP expression and plasma CETP activity, resulting in non-HDL-C increase and HDL-C decrease in CETP-expressing animals [441]. Such a **lipogenic effect** is a major deleterious feature of LXR agonists. Specific activation of LXR alpha in the absence of the effect on other LXRs is, however, capable of selectively raising HDL in the absence of the lipogenic effect. It is relevant in this regard that N,N-dimethyl-3-beta-hydroxycholenamide has recently been reported to represent the first synthetic LXR agonist that up-regulates ABCA1 and ABCG1 and attenuates plaque formation in apoE-null mice without inducing liver steatosis and hypertriglyceridemia [442].

Other drugs that may affect the LXR pathway include statins, which may impair LXR signaling by inhibiting oxysterol synthesis. In contrast, fibrates and thiazolidinediones increase LXR expression and activity [225]. Another interesting example is provided by the synthetic sphingosine analog FTY720 which is currently undergoing clinical trials as an immunomodulatory compound and acts primarily via S1P receptor activation [443]. This molecule stimulates the production of 27-hydroxycholesterol, an endogenous ligand for LXR, in human macrophages, thereby leading to the up-regulation of ABCA1 with accelerated cholesterol efflux [444]. Via this pathway, FTY720 can decrease the toxicity of free cholesterol to human macrophages, identifying a novel mechanism for therapeutic LXR activation.

PPAR Gamma Agonists

Thiazolidinediones (glitazones), a class of drugs for the treatment of Type 2 diabetes, are synthetic, high-affinity ligands for the nuclear receptor PPAR gamma, that were long believed to possess cardioprotective effects in humans and anti-atherogenic properties in animal models [445]. There are two commercially available drugs of this class, **pioglitazone** and **rosiglitazone**; the latter was, however, placed under restricted use in the USA and withdrawn from the market in Europe as a result of an increased risk of cardiovascular events.

Both pioglitazone and rosiglitazone induce elevation in HDL-C and a reduction in triglyceride levels in subjects with Type 2 diabetes and obesity [445–447]. Pioglitazone results in a more anti-atherogenic lipid profile than rosiglitazone, showing more marked effects on HDL-C and triglycerides. Furthermore, rosiglitazone may increase levels of LDL-C [446, 447], a feature that may contribute to the unfavorable safety profile of the drug. Pioglitazone treatment reduces carotid intima-media thickness progression in humans [448, 449] and prevents progression of atherosclerosis in LDL receptor-null mice [450]. Remarkably, levels of HDL-C on pioglitazone treatment inversely correlate with lesion development in mice, suggesting HDL raising as a mechanism implicated in the atheroprotection [450].

The action of thiazolidinediones on HDL involves modulation of the HDL particle profile with a **shift to small HDL3**. Thus, in a prospective open-label study, rosiglitazone (4 mg/day for 12 weeks) post-prandially reduced the proportion of large HDL2b, while increasing that of small HDL3b in Type 2 diabetes [451]. Furthermore, both pioglitazone and rosiglitazone induce preferential accumulation of small HDL3b and 3c in rabbits [452, 453].

Enhanced hepatic apoA-I production is implicated in the effects of pioglitazone on HDL metabolism [452, 453] (Fig. 11.1). In parallel, the fractional catabolic rate of apoA-I can also be increased, potentially reflecting elevated hepatic expression of SR-BI [454]. According to kinetic studies, however, pioglitazone does not affect hepatic HDL removal [272, 455]. Pioglitazone may act, in part, via the induction of PPAR alpha, a pathway that may mediate the enhanced synthesis of apoA-I and apoA-II in hepatocytes [456]. Specifically, pioglitazone increases apoA-I production by directly enhancing PPAR-response element-dependent transcription, resulting in the generation of apoA-I-containing HDL particles with improved anti-inflammatory properties [457]. In contrast, treatment with rosiglitazone **increases apoA-II production** in subjects with metabolic syndrome and low HDL-C, but has no effect on apoA-I metabolism [458].

Furthermore, PPAR gamma activators **induce cholesterol removal** from human macrophage-derived foam cells through the stimulation of the ABCA1- and ABCG1-dependent pathways [255, 459, 460]. The net effect of PPAR gamma activation on RCT is, however, less clear as it may redistribute cholesterol from fecal excretion to adipose tissue uptake [461]. The beneficial action of thiazolidinediones on HDL-C also involves a reduction in plasma CETP mass and a decrease in hepatic triglyceride content, as observed in patients with Type 2 diabetes and in apoE*3-Leiden/CETP transgenic mice [462]. Additional benefits of pioglitazone and rosiglitazone may involve an **increase in plasma adiponectin** and a reduction in inflammatory biomarkers, including CRP and TNF-alpha, in patients with Type 2 diabetes, with pioglitazone being more potent in this regard [272]. In addition, pioglitazone treatment decreases systemic levels of pro-inflammatory cytokines (TNF-alpha, IL-6, IL-1-beta) and adhesion molecules (VCAM-1, ICAM-1, E-selectin) in obese subjects [269]. ApoA-I-enriched particles produced on pioglitazone treatment display an enhanced capacity to inhibit monocyte adhesion to endothelial cells *in vitro* [457], suggesting a mechanistic basis for the **anti-inflammatory effects** of the treatment. Further consistent with its anti-inflammatory activity, rosiglitazone increases serum PON1 activity in rabbits [453].

Beneficial effects of pioglitazone on HDL-C levels and inflammation are accompanied by an improvement in endothelial function, potentially reflecting enhanced HDL functionality. Indeed, the drug enhances flow-mediated dilatation in patients with Type 2 diabetes along with increases in plasma HDL-C and adiponectin levels [463]. In addition, pulse wave velocity, a surrogate marker of arterial wall stiffness, is decreased by pioglitazone in obese subjects [269].

Lipase Inhibitors

As HDL catabolism is accelerated under the action of three lipolytic enzymes, $sPLA_2$, endothelial lipase and hepatic lipase, their inhibition can lead to increases in HDL-C levels which may, in turn, result in attenuation of atherosclerosis [13] (Table 11.3). Approaches that target hepatic lipase may, however, increase circulating levels not only of HDL, but also apoB-containing lipoproteins and do not suggest atheroprotection [13]. In contrast, inhibition of $sPLA_2$ [464] and endothelial lipase [465] results in anti-atherogenic lipid profiles and is promising.

***$sPLA_2$* Inhibitors.** Therapeutic inhibition of **$sPLA_2$** can increase HDL-C levels and delay atherosclerosis progression in animal models [464] (Fig. 11.2). A-001 is a novel inhibitor of $sPLA_2$ enzymes discovered by structure-based drug design, and **A-002 (varespladib**; 1-H-indole-3-glyoxamide) is the orally bioavailable prodrug currently in clinical development. A-001 inhibits human and mouse $sPLA_2$ group IIA, V, and X enzymes with IC_{50} values in the low nM range. A-002 (1 mg/kg) leads to high serum levels of A-001 and inhibits phospholipase A_2 activity in transgenic mice over-expressing human $sPLA_2$ group IIA on a C57BL/6J background [466].

In apoE-null mice fed a high-fat diet, A-002 (30 and 90 mg/kg twice a day for 16 weeks) reduced aortic atherosclerosis by −50% [466]. In parallel, plasma total cholesterol was decreased by one month and remained low throughout the study. In an accelerated atherosclerosis model of apoE-null mice, with angiotensin II-induced aortic lesions and aneurysms, A-002 (30 mg/kg twice a day) decreased aortic atherosclerosis by −40% and attenuated aneurysm formation [466]. Similarly, A-002 at a dose of 15 or 150 mg/kg/d (the latter to maintain serum levels of A-001 at least 10-fold over the IC_{50}) decreased *en face* atherosclerotic lesion area by −40% to −75% while increasing fibrous cap size by over +200% in apoE-/- mice on a high-fat diet [467]. In parallel, HDL-C levels were increased by +20% to +40% and plasma PON activity by +80%. When combined with pravastatin, A-002 synergistically decreased the lesion area by −50% more than A-002 alone, despite the fact that the statin alone did not impact atherosclerosis in this model [467].

A-002 is also effective against aortic cholesterol accumulation in another animal model, i.e., guinea pigs fed a high-fat diet. Specifically, A-002 given daily by oral gavage inhibited (−27%) cholesterol accumulation in the aorta compared with the placebo group [468]. Morphologic analysis of the aortic sinus revealed that the group treated with A-002 reduced atherosclerotic lesion formation by −24%. In this model, the inhibitor did not improve HDL-C concentrations, but reduced circulating levels of the inflammatory cytokines, IL-10, IL-12 and granulocyte-macrophage colony-stimulating factor. The inhibitory actions of A-002 on both HDL degradation and systemic inflammation may, therefore, underlie the beneficial effects of the inhibitor on atherosclerosis [467].

Recent data reveal that varespladib can also beneficially act on plasma lipids and inflammation in humans. Thus, in the randomized, double-blind, prospective controlled Fewer Recurrent Acute Coronary Events With Near-Term Cardiovascular Inflammation Suppression (FRANCIS) phase 2B trial, varespladib (500 mg daily) as adjunctive therapy to atorvastatin (80 mg daily) reduced mean LDL-C levels from baseline by −49.6% after eight weeks, significantly superior compared with a −43.4% reduction achieved with placebo/atorvastatin, in patients with acute coronary syndrome [469]. Respective eight-week median reductions in $sPLA_2$ group IIA levels were −82.4% and −15.6%, while hsCRP levels were lowered by −75.0% and −71.0% [469]. In addition, varespladib was capable of decreasing circulating levels of atherogenic LDL and VLDL particles on a statin background [470]. Major adverse cardiovascular events were, however, not different from placebo six months post-randomization (7.3% varespladib vs. 7.7% placebo) [469]. Similarly negative results were obtained in a study of varespladib in elective percutaneous coronary intervention. Indeed, the drug administered for 3 to 5 days before the procedure did not reduce periprocedural myonecrosis associated with the intervention [471]. Furthermore, $sPLA_2$ inhibitors were unable to beneficially act on inflammatory diseases, including rheumatoid arthritis and severe sepsis [472, 473].

The efficacy and safety of varespladib (50 to 250 mg twice daily for eight weeks) were extensively assessed in patients with stable coronary heart disease. In this study, mean $sPLA_2$ group IIA concentration fell (−86.7%) from 15.7 pmol/l to 2.1 pmol/l, in the overall active treatment group and from 15.7 pmol/l to 14.3 pmol/l in the placebo group (−4.8%; $p < 0.0001$ treatment vs. placebo) [474]. Reductions in $sPLA_2$ group IIA concentration in the varespladib groups were dose-dependent, ranging from −69.2% in the 50 mg group to −95.8% in the 500 mg group. The proportion of patients reporting treatment-emergent adverse events did not differ from placebo [474]. In support of these results, no treatment differences in elevated liver function tests on >1 occasion were observed between varespladib and placebo groups in the FRANCIS trial [469].

Endothelial Lipase Inhibitors. In contrast to $sPLA_2$, no data on the potential of selective inhibitors of **endothelial lipase** to inhibit atherogenesis is available as yet, although the synthesis of such inhibitors has been reported [475]. The therapeutic interest of selective inhibition of endothelial lipase is highlighted by the pro-atherogenic action of non-selective inhibition of endothelial lipase, together with other critical lipases in mice by a polymeric surfactant, poloxamer 407 [476]. Endothelial lipase inactivation in mice appears to increase LPS-neutralizing capacity of HDL, protecting mice against systemic inflammation induced by LPS [477, 478]. Interestingly, sulforaphane, a naturally occurring isothiocyanate present in cruciferous vegetables that possesses antioxidant and anti-inflammatory effects, inhibits endothelial lipase expression via modulation of the NFκB pathway in endothelial cells [479].

$LpPLA_2$ Inhibitors. The inhibition of yet another lipase, **$LpPLA_2$**, may diminish the production of pro-inflammatory mediators of atherosclerosis, thereby inhibiting local inflammatory reactions in plaques [464]. Therefore, phospholipase A_2 inhibitors targeted against $LpPLA_2$, such as **darapladib**, may potentially reduce cardiovascular disease [464, 480]. Indeed, selective inhibition of $LpPLA_2$ with darapladib reduces the development of advanced coronary atherosclerosis in diabetic and hypercholesterolemic swine [481]. Furthermore, darapladib markedly inhibits $LpPLA_2$ activity in plasma and atherosclerotic lesions and reduces lesion lysophosphatidylcholine content. Analysis of coronary artery gene expression shows that darapladib exerts an anti-inflammatory action, substantially reducing the expression of 24 genes associated with macrophage and T-lymphocyte function. As a consequence, darapladib treatment results in a small reduction in the necrotic core area and diminished medial destruction, leading to the formation of fewer lesions with an unstable phenotype [481].

Darapladib is also effective in humans. Thus, $LpPLA_2$ activity was inhibited by −59% with this agent (60 mg daily) after 12 months in patients with angiographically-documented coronary artery disease [482]. In the placebo-treated group, necrotic core volume increased significantly after 12 months; darapladib attenuated this increase, resulting in a significant treatment difference of $-5.2\ mm^3$. Remarkably, these intraplaque compositional changes occured without a significant treatment difference in total atheroma volume, indicative of the specific prevention of necrotic core expansion by the drug. Background therapy was comparable between the groups, with no difference in LDL-C levels at 12 months. At the same time point, there were no significant differences between groups in plaque deformability or plasma hsCRP [482]. In another study performed in coronary heart disease and coronary heart disease-risk-equivalent patients receiving atorvastatin (20 or 80 mg), darapladib (40, 80, and 160 mg) inhibited $LpPLA_2$ activity by −43%, −55% and −66% compared with placebo, and also decreased $LpPLA_2$ mass in a subset of subjects [483]. Sustained dose-dependent inhibition did not depend on the dose of atorvastatin, baseline LDL-C or baseline HDL-C levels. At 12 weeks, darapladib (160 mg) decreased IL-6 (−12.3%) and hsCRP (−13.0%) compared with placebo. The inhibition of $LpPLA_2$ produced no detrimental effects on platelet biomarkers and did not modify plasma lipids, myeloperoxidase or matrix metalloproteinase-9; no major safety concerns were noted [483].

SR-BI Inhibitors

The inhibition of HDL remodeling by SR-BI represents an additional strategy to raise HDL-C levels (Fig. 11.2 and Table 11.3). Thus, the novel compound **R-138329** transiently increased HDL-C (+41%) in normolipidemic hamsters at a dose of 100 mg/kg [484]. The compound also decreased non-HDL-C and triglyceride levels in apoE-null mice [485]. In hamsters, R-138329 also increased HDL size, delayed the clearance of HDL cholesteryl ester in mice and reduced the accumulation of HDL cholesteryl ester in the liver, whereas the clearance of the HDL protein moiety remained unaffected. These effects are mediated by R-154716, a metabolite of R-138329, which inhibits the selective uptake of HDL cholesteryl ester by hepatoma cells through SR-BI [484]. Such initial investigation of a small-molecule inhibitor of SR-BI may not, however, hold therapeutic potential, as treatment with R-138329 at 0.01% significantly increased the extent of atherosclerotic lesion formation in the aortas of apoE-null mice, despite elevated levels of HDL-C [485]. Impaired cholesterol flux via the RCT pathway as a result of SR-BI inhibition should account for this result [19]. R-138329, therefore, provides an excellent illustration of the dissociation between HDL-C levels and RCT, highlighting the limitations of HDL-C as a biomarker of cardiovascular risk.

In a similar fashion, **ITX5061**, a molecule initially characterized as a p38 MAPK inhibitor, increased HDL-C levels by +20% in hypertriglyceridemic subjects with low HDL-C levels [486]. ITX5061 also increased HDL size and moderately (+10%) augmented apoA-I levels without affecting (V)LDL-C or plasma triglyceride concentrations. Furthermore, ITX5061 decreased the fractional catabolic rate of HDL cholesteryl ester and reduced its hepatic uptake in mice, but did not affect hepatic apoA-I mRNA [486]. In transfected cells, ITX5061 inhibited SR-BI-dependent uptake of HDL-derived cholesteryl ester, consistent with the inhibition of SR-BI activity. In contrast to R-138329, however, ITX5061 reduced early atherosclerotic lesions in the aortic arch of both control and CETP-expressing LDL receptor-null mice. The reason for the distinct effects on atherosclerosis of the two SR-BI inhibitors is unclear: they may be related to the effect of ITX5061 on p38 MAPK [487] or they may reflect distinct mechanisms of action on SR-BI involving the formation of inactive HDL-SR-BI complexes [486].

Another interesting example of an SR-BI inhibitor is provided by apolipoproteins C-I, C-II and C-III, whose enrichment in HDL dose-dependently decreases SR-BI-dependent uptake of HDL cholesteryl ester by hepatocytes [488, 489]. ApoA-II also displays an inhibitory effect on the selective uptake of cholesteryl ester from HDL mediated by SR-BI [490]. Finally, compounds termed BLT-1-5 inhibit SR-BI-dependent lipid transport from HDL, but do not block HDL binding to SR-BI on the cell surface [491].

In addition to HDL-C raising, SR-BI inhibitors may possess anti-infectious activity. Indeed, as SR-BI is essential for infection with HCV, both small molecule SR-BI inhibitors and anti-SR-BI antibodies may represent promising therapeutic agents acting via the inhibition of virus entry into hepatic cells [492, 493].

Others

Other agents enhancing HDL formation primarily act by stimulation of the ABCA1-mediated pathway (Fig. 11.1). Thus, **cilostazol**, an antiplatelet agent with selective phosphodiesterase 3 inhibitory action, raises HDL-C levels in humans and ameliorates atherosclerosis in mice [494, 495]. Treatment with cilostazol increases ABCA1 and ABCG1 expression in THP-1 macrophages in a protein kinase A-independent manner. These effects translate into enhanced apoA-I- and HDL-mediated cholesterol efflux from macrophages *in vitro* and accelerated RCT from macrophages *in vivo*, and may contribute to the cardioprotective effects of the drug observed in humans [494, 495].

CLINICAL SUMMARY

Pharmacologic approaches targeting enhanced apoA-I production and HDL formation may provide rapid clinical benefits in terms of acute stabilization of atherosclerotic plaques.

Furthermore, **RVX-208** (RVX-000222), a first-in-class, orally active, small-molecule stimulator of apoA-I gene expression, under development by Resverlogix, induces apoA-I expression in HepG2 cells, leading to increased production of pre-beta- and alpha-migrating lipoprotein particles containing apoA-I [496]. Similarly, RVX-208 treatment at a dose of 15 to 60 mg/kg/day for 63 days increased serum apoA-I and HDL-C levels (by +60% and +97%, respectively) in African green monkeys. In parallel, the levels of pre-beta-1 LpA-I and alpha-1 LpA-I particles were elevated, resulting in the enhancement of ABCA1, ABCG1 and SR-BI-dependent cholesterol efflux. Treatment of healthy humans for one week with oral RVX-208 at a dose of 2 to 8 mg/kg/day increased plasma levels of apoA-I and pre-beta HDL and improved cholesterol efflux to serum [496]; however, the effects were modest and did not exceed +10%. In patients with low HDL-C levels, RVX-208 increased total HDL, as well as the alpha- and pre-beta HDL fractions [497]. In statin-treated patients with coronary artery disease, treatment with RVX-208 (50, 100, or 150 mg twice daily for 12 weeks) was associated with dose-dependent increases in apoA-I levels by up to +5.6% and HDL-C levels by up to +8.3%. In parallel, large HDL particles increased by up to +21.1% [498]. Toxicity studies in animals and phase I/II clinical trials indicated that RVX-208 was safe and well-tolerated in multiple dosing regimens; an increase in liver enzymes was, however, observed in some subjects [498].

Hepatic apoA-I synthesis and plasma HDL-C levels can be also increased by **myriocin**, a serine palmitoyltransferase inhibitor that reduces plasma levels of sphingomyelin, ceramide, S1P and glycosphingolipids [499]. These actions lead to the inhibition of ERK phosphorylation in apoE-null mice and culminate in potent inhibition of atherosclerosis. In addition, a high-throughput assay for ABCA1 upregulators using human hepatoma HepG2 cells stably transfected with ABCA1 identified four microbial compounds (pyrromycin, aclarubicin, daidzein and pratensein) as potentially potent activators of the transporter [500].

Finally, as the expression of ABCA1 is regulated by calpain-mediated proteolytic degradation, stabilization of the transporter against the degradation may result in enhanced ABCA1 expression and impoved HDL biogenesis and apoA-I lipidation. Thus, spiroquinone and diphenoquinone, products of **probucol** oxidation, reduce degradation of ABCA1, increase surface ABCA1 protein and improve the efflux of cholesterol and phospholipid from THP-1 macrophages without altering the transcription of the ABCA1 gene [501]. Accordingly, both compounds enhance ABCA1-dependent generation of HDL *in vitro* and increase hepatic ABCA1 and plasma HDL-C levels when given to rabbits. Both compounds also decrease vascular lipid deposition in cholesterol-fed rabbits without developing antioxidative activity in plasma. Such effects of spiroquinone and diphenoquinone may be related to conformational alterations of ABCA1 as a result of the high hydrophobicity of the quinones [501].

Another strategy to stimulate HDL formation involves oral administration of phospholipid. Thus, oral soy phospholipid formulations are currently under development as a treatment to increase HDL-C concentrations and to improve HDL functionality [70].

Approaches to normalize intravascular HDL remodeling include the administration of **recombinant LCAT**, which possesses markedly enhanced enzymatic activity and improved stability compared with wild-type LCAT and may induce HDL-C raising [70] (Fig. 11.2). Indeed, a single injection of recombinant LCAT to New Zealand White rabbits increased HDL-C by up to approximately fourfold and generated large HDL particles that were enriched with cholesteryl ester and apoA-I [502]. In parallel, both cholesterol efflux from peripheral tissues to plasma, with a higher plasma cholesterol esterification rate, and neutral sterol excretion into feces were elevated, indicative of enhanced RCT. As a result, the treatment of cholesterol-fed rabbits with recombinant LCAT for five months induced a sustained elevation in HDL-C (approximately threefold) and reduced atherosclerotic lesion progression in both abdominal and thoracic aortas compared with placebo [502]. Interestingly, plasma cholesteryl ester in LCAT-treated rabbits displayed a higher proportion of linoleate and lower proportion of oleate, consistent with elevated lipid phase fluidity and enhanced functionality of HDL [503]. In a similar fashion, injection of recombinant LCAT rapidly restored normal lipoprotein phenotype in LCAT-deficient mice and enhanced cholesterol efflux to isolated HDL [504]. Furthermore, the over-expression of LCAT in non-human primates led to an anti-atherogenic lipoprotein profile via the elevation of HDL-C and a reduction in apoB levels subsequent to accelerated LDL catabolism [505]. In addition, adenoviral LCAT transfer to LDL receptor-deficient rabbits increased HDL-C, induced cholesterol unloading from complex lesions and inhibited atherosclerosis progression [506]. The potential of such an approach in humans should, however, be viewed with caution in light of the recently reported dissociation between LCAT activity, RCT and atherosclerosis [507].

Other pharmacologic options to target HDL remodeling include the **activation of LPL** with an enhanced supply of surface remnants of triglyceride-rich lipoproteins for HDL generation (Fig. 11.2). Indeed, such a strategy results in elevated HDL-C levels as evidenced by lipid profiles obtained in animals treated with **ibrolipim (NO-1886)**, an LPL activator [508, 509]. The HDL-C-raising action of ibrolipim can also be related to the up-regulation of ABCA1 and ABCG1 [509, 510] and is accompanied by improvements in glucose metabolism, fatty liver and atherogenesis in animal models [511–515].

IMPORTANT READING

Tall AR, Yvan-Charvet L, Terasaka N, Pagler T, Wang N. HDL, ABC transporters, and cholesterol efflux: implications for the treatment of atherosclerosis. *Cell Metab* 2008; 7: 365–375.

Van Lenten BJ, Wagner AC, Anantharamaiah GM, Navab M, Reddy ST, Buga GM, Fogelman AM. Apolipoprotein A-I mimetic peptides. *Curr Atheroscler Rep* 2009; 11: 52–57.

Chapman MJ, Le Goff W, Guerin M, Kontush A. Cholesteryl ester transfer protein: at the heart of the action of lipid-modulating therapy with statins, fibrates, niacin, and cholesteryl ester transfer protein inhibitors. *Eur Heart J* 2010; 31: 149–164.

Khera AV, Rader DJ. Future therapeutic directions in reverse cholesterol transport. *Curr Atheroscler Rep* 2010; 12: 73–81.

REFERENCES

1. Barter PJ, Rye KA. High density lipoproteins and coronary heart disease. *Atherosclerosis* 1996; 121: 1–12.
2. Franceschini G. Epidemiologic evidence for high-density lipoprotein cholesterol as a risk factor for coronary artery disease. *Am J Cardiol* 2001; 88: 9N–13N.
3. Kontush A, Chapman MJ. Functionally defective HDL: A new therapeutic target at the crossroads of dyslipidemia, inflammation and atherosclerosis. *Pharmacol Rev* 2006; 3: 342–374.
4. Barter PJ, Brewer HB Jr, Chapman MJ, Hennekens CH, Rader DJ, Tall AR. Cholesteryl ester transfer protein: a novel target for raising HDL and inhibiting atherosclerosis. *Arterioscler Thromb Vasc Biol* 2003; 23: 160–167.
5. Ashen MD, Blumenthal RS. Clinical practice. Low HDL cholesterol levels. *N Engl J Med* 2005; 353: 1252–1260.
6. Nicholls SJ, Rye KA, Barter PJ. High-density lipoproteins as therapeutic targets. *Curr Opin Lipidol* 2005; 16: 345–349.
7. Chapman MJ. Beyond the statins: new therapeutic perspectives in cardiovascular disease prevention. *Cardiovasc Drugs Ther* 2005; 19: 135–139.
8. Duffy D, Rader DJ. Update on strategies to increase HDL quantity and function. *Nat Rev Cardiol* 2009; 6: 455–463.
9. Natarajan P, Ray KK, Cannon CP. High-density lipoprotein and coronary heart disease: current and future therapies. *J Am Coll Cardiol* 2010; 55: 1283–1299.
10. Fruchart JC, Sacks F, Hermans MP, Assmann G, Brown WV, Ceska R, Chapman MJ, Dodson PM, Fioretto P, Ginsberg HN et al. The Residual Risk Reduction Initiative: a call to action to reduce residual vascular risk in patients with dyslipidemia. *Am J Cardiol* 2008; 102: 1K–34K.
11. Toth PP, Zarotsky V, Sullivan JM, Laitinen D. Niacin and fibrate use among patients with high triglycerides and low high-density lipoprotein cholesterol. *Curr Med Res Opin* 2009; 25: 1355–1363.

12. Vergeer M, Holleboom AG, Kastelein JJP, Kuivenhoven JA. The HDL hypothesis: does high-density lipoprotein protect from atherosclerosis ? *J Lipid Res* 2010; 51: 2058–2073.

13. Joy T, Hegele RA. Is raising HDL a futile strategy for atheroprotection ? *Nat Rev Drug Discov* 2008; 7: 143–155.

14. von Eckardstein A. Implications of torcetrapib failure for the future of HDL therapy: is HDL-cholesterol the right target ? *Expert Rev Cardiovasc Ther* 2010; 8: 345–358.

15. Yvan-Charvet L, Kling J, Pagler T, Li H, Hubbard B, Fisher T, Sparrow CP, Taggart AK, Tall AR. Cholesterol efflux potential and antiinflammatory properties of high-density lipoprotein after treatment with niacin or anacetrapib. *Arterioscler Thromb Vasc Biol* 2010; 30: 1430–1438.

16. Chapman MJ, Redfern JS, McGovern ME, Giral P. Niacin and fibrates in atherogenic dyslipidemia: pharmacotherapy to reduce cardiovascular risk. *Pharmacol Ther* 2010; 126: 314–345.

17. Catalano G, Julia Z, Frisdal E, Vedie B, Fournier N, Le Goff W, Chapman MJ, Guerin M. Torcetrapib differentially modulates the biological activities of HDL2 and HDL3 particles in the reverse cholesterol transport pathway. *Arterioscler Thromb Vasc Biol* 2009; 29: 268–275.

18. Van Craeyveld E, Gordts S, Jacobs F, De Geest B. Gene therapy to improve high-density lipoprotein metabolism and function. *Curr Pharm Des* 2010; 16: 1531–1544.

19. Huby T, Doucet C, Dachet C, Ouzilleau B, Ueda Y, Afzal V, Rubin E, Chapman MJ, Lesnik P. Knockdown expression and hepatic deficiency reveal an atheroprotective role for SR-BI in liver and peripheral tissues. *J Clin Invest* 2006; 116: 2767–2776.

20. Tabas I, Williams KJ, Boren J. Subendothelial lipoprotein retention as the initiating process in atherosclerosis: update and therapeutic implications. *Circulation* 2007; 116: 1832–1844.

21. Personius BE, Brown BG, Gotto AM Jr. Effects of increasing HDL-C, lowering triglyceride and lowering LDL-C on & "fixed" atherosclerotic coronary artery disease: AFREGS (the Armed Forces Regression study). *Circulation* 1998; 98: I450–I451.

22. The-Bezafibrate-Infarction-Prevention-(BIP)-Study. Secondary prevention by raising HDL cholesterol and reducing triglycerides in patients with coronary artery disease: the Bezafibrate Infarction Prevention (BIP) study. *Circulation* 2000; 102: 21–27.

23. Rubins HB, Robins SJ, Collins D, Fye CL, Anderson JW, Elam MB, Faas FH, Linares E, Schaefer EJ, Schectman G et al. Gemfibrozil for the secondary prevention of coronary heart disease in men with low levels of high-density lipoprotein cholesterol. Veterans Affairs High-Density Lipoprotein Cholesterol Intervention Trial Study Group. *N Engl J Med* 1999; 341: 410–418.

24. Robins SJ, Collins D, Wittes JT, Papademetriou V, Deedwania PC, Schaefer EJ, McNamara JR, Kashyap ML, Hershman JM, Wexler LF et al. Relation of gemfibrozil treatment and lipid levels with major coronary events: VA-HIT: a randomized controlled trial. *JAMA* 2001; 285: 1585–1591.

25. Brown BG, Zhao XQ, Chait A, Fisher LD, Cheung MC, Morse JS, Dowdy AA, Marino EK, Bolson EL, Alaupovic P et al. (2001). Simvastatin and niacin, antioxidant vitamins, or the combination for the prevention of coronary disease. *N Engl J Med*, *345*, 1583–1592.

26. Taylor AJ, Sullenberger LE, Lee HJ, Lee JK, Grace KA. Arterial Biology for the Investigation of the Treatment Effects of Reducing Cholesterol (ARBITER) 2: a double-blind, placebo-controlled study of extended-release niacin on atherosclerosis progression in secondary prevention patients treated with statins. *Circulation* 2004; 110: 3512–3517.
27. Taylor AJ, Villines TC, Stanek EJ, Devine PJ, Griffen L, Miller M, Weissman NJ, Turco M. Extended-release niacin or ezetimibe and carotid intima-media thickness. *N Engl J Med* 2009; 361: 2113–2122.
28. Toth PP. Pharmacomodulation of high-density lipoprotein metabolism as a therapeutic intervention for atherosclerotic disease. *Curr Cardiol Rep* 2010; 12: 481–487.
29. Lopez-Jimenez F, Batsis JA, Roger VL, Brekke L, Ting HH, Somers VK. Trends in 10-year predicted risk of cardiovascular disease in the United States, 1976 to 2004. *Circ Cardiovasc Qual Outcomes* 2009; 2: 443–450.
30. Ingelsson E, Massaro JM, Sutherland P, Jacques PF, Levy D, D'Agostino RB, Vasan RS, Robins SJ. Contemporary trends in dyslipidemia in the Framingham Heart Study. *Arch Intern Med* 2009; 169: 279–286.
31. Hardoon SL, Whincup PH, Lennon LT, Wannamethee SG, Capewell S, Morris RW. How much of the recent decline in the incidence of myocardial infarction in British men can be explained by changes in cardiovascular risk factors? Evidence from a prospective population-based study. *Circulation* 2008; 117: 598–604.
32. Koro CE, Bowlin SJ, Stump TE, Sprecher DL, Tierney WM. The independent correlation between high-density lipoprotein cholesterol and subsequent major adverse coronary events. *Am Heart J* 2006; 151: 755.e1–755.e6.
33. Grover SA, Kaouache M, Joseph L, Barter P, Davignon J. Evaluating the incremental benefits of raising high-density lipoprotein cholesterol levels during lipid therapy after adjustment for the reductions in other blood lipid levels. *Arch Intern Med* 2009; 169: 1775–1780.
34. Rahilly-Tierney C, Bowman TS, Djousse L, Sesso HD, Gaziano JM. Change in high-density lipoprotein cholesterol and incident coronary heart disease in apparently healthy male physicians. *Am J Cardiol* 2008; 102: 1663–1667.
35. Devendra GP, Whitney EJ, Krasuski RA. Impact of increases in high-density lipoprotein cholesterol on cardiovascular outcomes during the Armed Forces Regression Study. *J Cardiovasc Pharmacol Ther* 2010; 15: 380–383.
36. Zhao XQ, Krasuski RA, Baer J, Whitney EJ, Neradilek B, Chait A, Marcovina S, Albers JJ, Brown BG. Effects of combination lipid therapy on coronary stenosis progression and clinical cardiovascular events in coronary disease patients with metabolic syndrome: a combined analysis of the Familial Atherosclerosis Treatment Study (FATS), the HDL-Atherosclerosis Treatment Study (HATS), and the Armed Forces Regression Study (AFREGS). *Am J Cardiol* 2009; 104: 1457–1464.
37. Koskinen J, Kahonen M, Viikari JS, Taittonen L, Laitinen T, Ronnemaa T, Lehtimaki T, Hutri-Kahonen N, Pietikainen M, Jokinen E et al. Conventional cardiovascular risk factors and metabolic syndrome in predicting carotid intima-media thickness progression in young adults: the cardiovascular risk in young Finns study. *Circulation* 2009; 120: 229–236.
38. Briel M, Ferreira-Gonzalez I, You JJ, Karanicolas PJ, Akl EA, Wu P, Blechacz B, Bassler D, Wei X, Sharman A et al. Association between change in high density lipoprotein cholesterol and cardiovascular disease morbidity and mortality: systematic review and meta-regression analysis. *BMJ* 2009; 338: b92.

39. De Caterina R, Scarano M, Marfisi R, Lucisano G, Palma F, Tatasciore A, Marchioli R. Cholesterol-lowering interventions and stroke: insights from a meta-analysis of randomized controlled trials. *J Am Coll Cardiol* 2010; 55: 198–211.

40. Khera AV, Rader DJ. Discovery and validation of new molecular targets in treating dyslipidemia: the role of human genetics. *Trends Cardiovasc Med* 2009; 19: 195–201.

41. Weissglas-Volkov D, Pajukanta P. Genetic causes of high and low serum HDL-cholesterol. *J Lipid Res* 2010; 51: 2032–2057.

42. Kontush A, Chapman MJ. Antiatherogenic small, dense HDL - guardian angel of the arterial wall ? *Nat Clin Pract Cardiovasc Med* 2006; 3: 144–153.

43. Genest J. The Yin and Yang of high-density lipoprotein cholesterol. *J Am Coll Cardiol* 2008; 51: 643–644.

44. Mackness B, Mackness M. High-density lipoprotein: why all the fuss ? *Ann Clin Biochem* 2009; 46: 5–7.

45. van der Steeg WA, Holme I, Boekholdt SM, Larsen ML, Lindahl C, Stroes ES, Tikkanen MJ, Wareham NJ, Faergeman O, Olsson AG et al. High-density lipoprotein cholesterol, high-density lipoprotein particle size, and apolipoprotein A-I: significance for cardiovascular risk: the IDEAL and EPIC-Norfolk studies. *J Am Coll Cardiol* 2008; 51: 634–642.

46. Reimers GJ, Jackson CL, Rickards J, Chan PY, Cohn JS, Rye KA, Barter PJ, Rodgers KJ. Inhibition of rupture of established atherosclerotic plaques by treatment with apolipoprotein A-I. *Cardiovasc Res* 2011; 91: 37–44.

47. Graversen JH, Laurberg JM, Andersen MH, Falk E, Nieland J, Christensen J, Etzerodt M, Thogersen HC, Moestrup SK. Trimerization of apolipoprotein A-I retards plasma clearance and preserves antiatherosclerotic properties. *J Cardiovasc Pharmacol* 2008; 51: 170–177.

48. Rubin EM, Krauss RM, Spangler EA, Verstuyft JG, Clift SM. Inhibition of early atherogenesis in transgenic mice by human apolipoprotein AI. *Nature* 1991; 353: 265–267.

49. Paszty C, Maeda N, Verstuyft J, Rubin EM. Apolipoprotein AI transgene corrects apolipoprotein E deficiency-induced atherosclerosis in mice. *J Clin Invest* 1994; 94: 899–903.

50. Plump AS, Scott CJ, Breslow JL. Human apolipoprotein A-I gene expression increases high density lipoprotein and suppresses atherosclerosis in the apolipoprotein E-deficient mouse. *Proc Natl Acad Sci U S A* 1994; 91: 9607–9611.

51. Benoit P, Emmanuel F, Caillaud JM, Bassinet L, Castro G, Gallix P, Fruchart JC, Branellec D, Denefle P, Duverger N. Somatic gene transfer of human ApoA-I inhibits atherosclerosis progression in mouse models. *Circulation* 1999; 99: 105–110.

52. Valenta DT, Bulgrien JJ, Banka CL, Curtiss LK. Overexpression of human ApoAI transgene provides long-term atheroprotection in LDL receptor-deficient mice. *Atherosclerosis* 2006; 189: 255–263.

53. Duverger N, Kruth H, Emmanuel F, Caillaud JM, Viglietta C, Castro G, Tailleux A, Fievet C, Fruchart JC, Houdebine LM et al. Inhibition of atherosclerosis development in cholesterol-fed human apolipoprotein A-I-transgenic rabbits. *Circulation* 1996; 94: 713–717.

54. Ishiguro H, Yoshida H, Major AS, Zhu T, Babaev VR, Linton MF, Fazio S. Retrovirus-mediated expression of apolipoprotein A-I in the macrophage protects against atherosclerosis in vivo. *J Biol Chem* 2001; 276: 36742–36748.
55. Vergnes L, Baroukh N, Ostos MA, Castro G, Duverger N, Nanjee MN, Najib J, Fruchart JC, Miller NE, Zakin MM et al. Expression of human apolipoprotein A-I/C-III/A-IV gene cluster in mice induces hyperlipidemia but reduces atherogenesis. *Arterioscler Thromb Vasc Biol* 2000; 20: 2267–2274.
56. Baroukh N, Ostos MA, Vergnes L, Recalde D, Staels B, Fruchart J, Ochoa A, Castro G, Zakin MM. Expression of human apolipoprotein A-I/C-III/A-IV gene cluster in mice reduces atherogenesis in response to a high fat-high cholesterol diet. *FEBS Lett* 2001; 502: 16–20.
57. Belalcazar LM, Merched A, Carr B, Oka K, Chen KH, Pastore L, Beaudet A, Chan L. Long-term stable expression of human apolipoprotein A-I mediated by helper-dependent adenovirus gene transfer inhibits atherosclerosis progression and remodels atherosclerotic plaques in a mouse model of familial hypercholesterolemia. *Circulation* 2003; 107: 2726–2732.
58. Lebherz C, Sanmiguel J, Wilson JM, Rader DJ. Gene transfer of wild-type apoA-I and apoA-I Milano reduce atherosclerosis to a similar extent. *Cardiovasc Diabetol* 2007; 6: 15.
59. Feng Y, Van Craeyveld E, Jacobs F, Lievens J, Snoeys J, De Geest B. Wild-type apo A-I and apo A-I Milano gene transfer reduce native and transplant arteriosclerosis to a similar extent. *J Mol Med* 2009; 87: 287–297.
60. Chiesa G, Sirtori CR. Apolipoprotein A-I Milano: current perspectives. *Curr Opin Lipidol* 2003; 14: 159–163.
61. Chiesa G, Stoltzfus LJ, Michelagnoli S, Bielicki JK, Santi M, Forte TM, Sirtori CR, Franceschini G, Rubin EM. Elevated triglycerides and low HDL cholesterol in transgenic mice expressing human apolipoprotein A-I Milano. *Atherosclerosis* 1998; 136: 139–146.
62. Schultz JR, Verstuyft JG, Gong EL, Nichols AV, Rubin EM. Protein composition determines the anti-atherogenic properties of HDL in transgenic mice. *Nature* 1993; 365: 762–764.
63. Warden CH, Hedrick CC, Qiao JH, Castellani LW, Lusis AJ.). Atherosclerosis in transgenic mice overexpressing apolipoprotein A-II. *Science* 1993; 261: 469–472.
64. Escola-Gil JC, Marzal-Casacuberta A, Julve-Gil J, Ishida BY, Ordonez-Llanos J, Chan L, Gonzalez-Sastre F, Blanco-Vaca F. Human apolipoprotein A-II is a pro-atherogenic molecule when it is expressed in transgenic mice at a level similar to that in humans: evidence of a potentially relevant species-specific interaction with diet. *J Lipid Res* 1998; 39: 457–462.
65. Kalopissis AD, Chambaz J. Transgenic animals with altered high-density lipoprotein composition and functions. *Curr Opin Lipidol* 2000; 11: 149–153.
66. Karathanasis SK, Norum RA, Zannis VI, Breslow JL. An inherited polymorphism in the human apolipoprotein A-I gene locus related to the development of atherosclerosis. *Nature* 1983; 301: 718–720.
67. Cimmino G, Chen W, Speidl WS, Giannarelli C, Ibanez B, Fuster V, Hajjar R, Walsh CE, Badimon JJ. Safe and sustained overexpression of functional apolipoprotein A-I/high-density lipoprotein in apolipoprotein A-I-null mice by muscular adeno-associated viral serotype 8 vector gene transfer. *J Cardiovasc Pharmacol* 2009; 54: 405–411.

68. Vaessen SF, Veldman RJ, Comijn EM, Snapper J, Sierts JA, van den Oever K, Beattie SG, Twisk J, Kuivenhoven JA. AAV gene therapy as a means to increase apolipoprotein (Apo) A-I and high-density lipoprotein-cholesterol levels: correction of murine ApoA-I deficiency. *J Gene Med* 2009; 11: 697–707.
69. Pedersen TX, Bro S, Andersen MH, Etzerodt M, Jauhiainen M, Moestrup S, Nielsen LB. Effect of treatment with human apolipoprotein A-I on atherosclerosis in uremic apolipoprotein-E deficient mice. *Atherosclerosis* 2009; 202: 372–381.
70. Remaley AT, Amar M, Sviridov D. HDL-replacement therapy: mechanism of action, types of agents and potential clinical indications. *Expert Rev Cardiovasc Ther* 2008; 6: 1203–1215.
71. Vucic E, Rosenson RS. Recombinant high-density lipoprotein formulations. *Curr Atheroscler Rep* 2011; 13: 81–87.
72. Nanjee MN, Doran JE, Lerch PG, Miller NE. Acute effects of intravenous infusion of ApoA1/phosphatidylcholine discs on plasma lipoproteins in humans. *Arterioscler Thromb Vasc Biol* 1999; 19: 979–989.
73. Sirtori CR, Calabresi L, Franceschini G. Recombinant apolipoproteins for the treatment of vascular diseases. *Atherosclerosis* 1999; 142: 29–40.
74. Newton RS, Krause BR. HDL therapy for the acute treatment of atherosclerosis. *Atheroscler* 2002; 3 (Suppl.): 31–38.
75. Nissen SE, Tsunoda T, Tuzcu EM, Schoenhagen P, Cooper CJ, Yasin M, Eaton GM, Lauer MA, Sheldon WS, Grines CL et al. Effect of recombinant ApoA-I Milano on coronary atherosclerosis in patients with acute coronary syndromes: a randomized controlled trial. *JAMA* 2003; 290: 2292–2300.
76. Hong YJ, Jeong MH, Hachinohe D, Ahmed K, Choi YH, Cho SH, Hwang SH, Ko JS, Lee MG, Park KH et al. Comparison of effects of rosuvastatin and atorvastatin on plaque regression in Korean patients with untreated intermediate coronary stenosis. *Circ J* 2011; 75: 398–406.
77. Nicholls SJ, Uno K, Kataoka Y, Nissen SE. ETC-216 for coronary artery disease. *Expert Opin Biol Ther* 2011; 11: 387–394.
78. Shaw JA, Bobik A, Murphy A, Kanellakis P, Blombery P, Mukhamedova N, Woollard K, Lyon S, Sviridov D, Dart AM. Infusion of reconstituted high-density lipoprotein leads to acute changes in human atherosclerotic plaque. *Circ Res* 2008; 103: 1084–1091.
79. Sacks FM, Rudel LL, Conner A, Akeefe H, Kostner G, Baki T, Rothblat G, de la Llera-Moya M, Asztalos B, Perlman T et al. Selective delipidation of plasma HDL enhances reverse cholesterol transport in vivo. *J Lipid Res* 2009; 50: 894–907.
80. Waksman R, Torguson R, Kent KM, Pichard AD, Suddath WO, Satler LF, Martin BD, Perlman TJ, Maltais JA, Weissman NJ et al. A first-in-man, randomized, placebo-controlled study to evaluate the safety and feasibility of autologous delipidated high-density lipoprotein plasma infusions in patients with acute coronary syndrome. *J Am Coll Cardiol* 2010; 55: 2727–2735.
81. Tardif JC, Gregoire J, L'Allier PL, Ibrahim R, Lesperance J, Heinonen TM, Kouz S, Berry C, Basser R, Lavoie MA et al. Effects of reconstituted high-density lipoprotein infusions on coronary atherosclerosis: a randomized controlled trial. *JAMA* 2007; 297: 1675–1682.
82. Ameli S, Hultgardh-Nilsson A, Cercek B, Shah PK, Forrester JS, Ageland H, Nilsson J. Recombinant apolipoprotein A-I Milano reduces intimal thickening after balloon injury in hypercholesterolemic rabbits. *Circulation* 1994; 90: 1935–1941.

83. Soma MR, Donetti E, Parolini C, Sirtori CR, Fumagalli R, Franceschini G. Recombinant apolipoprotein A-I Milano dimer inhibits carotid intimal thickening induced by perivascular manipulation in rabbits. *Circ Res* 1995; 76: 405–411.

84. Shah PK, Nilsson J, Kaul S, Fishbein MC, Ageland H, Hamsten A, Johansson J, Karpe F, Cercek B. Effects of recombinant apolipoprotein A-I Milano on aortic atherosclerosis in apolipoprotein E-deficient mice. *Circulation* 1998; 97: 780–785.

85. Shah PK, Yano J, Reyes O, Chyu KY, Kaul S, Bisgaier CL, Drake S, Cercek B. High-dose recombinant apolipoprotein A-I Milano mobilizes tissue cholesterol and rapidly reduces plaque lipid and macrophage content in apolipoprotein E-deficient mice. Potential implications for acute plaque stabilization. *Circulation* 2001; 103: 3047–3050.

86. Parolini C, Marchesi M, Lorenzon P, Castano M, Balconi E, Miragoli L, Chaabane L, Morisetti A, Lorusso V, Martin BJ et al. Dose-related effects of repeated ETC-216 (recombinant apolipoprotein A-I Milano/1-palmitoyl-2-oleoyl phosphatidylcholine complexes) administrations on rabbit lipid-rich soft plaques: in vivo assessment by intravascular ultrasound and magnetic resonance imaging. *J Am Coll Cardiol* 2008; 51: 1098–1103.

87. Sacks FM, Brewer HB, Alaupovic P, Kostner G, Asztalos B, Schaefer E, Rothblat GH, Akeefe H, Conner A, Perlman T et al. Selective plasma HDL delipidation and reinfusion: A unique new approach for acute HDL therapy in the treatment of cardiovascular disease.In: American Heart Association's Scientific Sessions. American Heart Association, New Orleans: 2004. Abstract 241.

88. Patel S, Drew BG, Nakhla S, Duffy SJ, Murphy AJ, Barter PJ, Rye KA, Chin-Dusting J, Hoang A, Sviridov D et al. Reconstituted high-density lipoprotein increases plasma high-density lipoprotein anti-inflammatory properties and cholesterol efflux capacity in patients with type 2 diabetes. *J Am Coll Cardiol* 2009; 53: 962–971.

89. Gomaraschi M, Baldassarre D, Amato M, Eligini S, Conca P, Sirtori CR, Franceschini G, Calabresi L. Normal vascular function despite low levels of high-density lipoprotein cholesterol in carriers of the apolipoprotein A-IMilano mutant. *Circulation* 2007; 116: 2165–2172.

90. Cockerill GW, Huehns TY, Weerasinghe A, Stocker C, Lerch PG, Miller NE, Haskard DO. Elevation of plasma high-density lipoprotein concentration reduces interleukin-1-induced expression of E-selectin in an in vivo model of acute inflammation. *Circulation* 2001; 103: 108–112.

91. Cockerill GW, McDonald MC, Mota-Filipe H, Cuzzocrea S, Miller NE, Thiemermann C. High density lipoproteins reduce organ injury and organ dysfunction in a rat model of hemorrhagic shock. *FASEB J* 2001; 15: 1941–1952.

92. Cuzzocrea S, Dugo L, Patel NS, Di Paola R, Cockerill GW, Genovese T, Thiemermann C. High-density lipoproteins reduce the intestinal damage associated with ischemia/reperfusion and colitis. *Shock* 2004; 21: 342–351.

93. Nicholls SJ, Dusting GJ, Cutri B, Bao S, Drummond GR, Rye KA, Barter PJ. Reconstituted high-density lipoproteins inhibit the acute pro-oxidant and proinflammatory vascular changes induced by a periarterial collar in normocholesterolemic rabbits. *Circulation* 2005; 111: 1543–1550.

94. Nicholls SJ, Cutri B, Worthley SG, Kee P, Rye KA, Bao S, Barter PJ. Impact of short-term administration of high-density lipoproteins and atorvastatin on atherosclerosis in rabbits. *Arterioscler Thromb Vasc Biol* 2005; 25: 2416–2421.

95. Cho KH, Kim JR. A reconstituted HDL containing V156K or R173C apoA-I exhibited anti-inflammatory activity in apo-E deficient mice and showed resistance to myeloperoxidase-mediated oxidation. *Exp Mol Med* 2009; 41: 417–428.

96. Cho KH. A reconstituted high density lipoprotein containing the V156E mutant of apolipoprotein A-I exhibits anti-atherosclerotic activity in Apo-E deficient mice. *J Atheroscler Thromb* 2009; 16: 217–229.

97. Peshavariya H, Dusting GJ, Di Bartolo B, Rye KA, Barter PJ, Jiang F. Reconstituted high-density lipoprotein suppresses leukocyte NADPH oxidase activation by disrupting lipid rafts. *Free Radic Res* 2009; 43: 772–782.

98. Kujiraoka T, Hattori H, Ito M, Nanjee MN, Ishihara M, Nagano M, Iwasaki T, Cooke CJ, Olszewski WL, Stepanova IP. Effects of intravenous apolipoprotein A-I/phosphatidylcholine discs on paraoxonase and platelet-activating factor acetylhydrolase in human plasma and tissue fluid. *Atherosclerosis* 2004; 176: 57–62.

99. Kiya Y, Miura S, Imaizumi S, Uehara Y, Matsuo Y, Abe S, Jimi S, Urata H, Rye KA, Saku K. Reconstituted high-density lipoprotein attenuates postinfarction left ventricular remodeling in rats. *Atherosclerosis* 2009; 203: 137–144.

100. van Oostrom O, Nieuwdorp M, Westerweel PE, Hoefer IE, Basser R, Stroes ES, Verhaar MC. Reconstituted HDL increases circulating endothelial progenitor cells in patients with type 2 diabetes. *Arterioscler Thromb Vasc Biol* 2007; 27: 1864–1865.

101. Wang Y, Zhu X, Wu G, Shen L, Chen B. Effect of lipid-bound apoA-I cysteine mutants on lipopolysaccharide-induced endotoxemia in mice. *J Lipid Res* 2008; 49: 1640–1645.

102. Spieker LE, Sudano I, Hurlimann D, Lerch PG, Lang MG, Binggeli C, Corti R, Ruschitzka F, Luscher TF, Noll G. High-density lipoprotein restores endothelial function in hypercholesterolemic men. *Circulation* 2005; 105: 1399–1402.

103. Bisoendial RJ, Hovingh GK, Levels JH, Lerch PG, Andresen I, Hayden MR, Kastelein JJ, Stroes ES. Restoration of endothelial function by increasing high-density lipoprotein in subjects with isolated low high-density lipoprotein. *Circulation* 2003; 107: 2944–2948.

104. Chenevard R, Hurlimann D, Spieker L, Bechir M, Enseleit F, Hermann M, Flammer AJ, Sudano I, Corti R, Luscher TF et al. Reconstituted HDL in acute coronary syndromes. *Cardiovasc Ther* 2011, in press.

105. Calkin AC, Drew BG, Ono A, Duffy SJ, Gordon MV, Schoenwaelder SM, Sviridov D, Cooper ME, Kingwell BA, Jackson SP. Reconstituted high-density lipoprotein attenuates platelet function in individuals with Type 2 diabetes mellitus by promoting cholesterol efflux. *Circulation* 2009; 120: 2095–2104.

106. Drew BG, Duffy SJ, Formosa MF, Natoli AK, Henstridge DC, Penfold SA, Thomas WG, Mukhamedova N, de Courten B, Forbes JM et al. High-density lipoprotein modulates glucose metabolism in patients with type 2 diabetes mellitus. *Circulation* 2009; 119: 2103–2111.

107. Shu X, Nelbach L, Weinstein MM, Burgess BL, Beckstead JA, Young SG, Ryan RO, Forte TM. Intravenous injection of apolipoprotein A-V reconstituted high-density lipoprotein decreases hypertriglyceridemia in apoav-/- mice and requires glycosylphosphatidylinositol-anchored high-density lipoprotein-binding protein 1. *Arterioscler Thromb Vasc Biol* 2010; 30: 2504–2509.

108. Adams CW, Abdulla YH, Bayliss OB, Morgan RS. Modification of aortic atheroma and fatty liver in cholesterol-fed rabbits by intravenous injection of saturated and polyunsaturated lecithins. *J Pathol Bacteriol* 1967; 94: 77–87.

109. Gaidukov L, Bar D, Yacobson S, Naftali E, Kaufman O, Tabakman R, Tawfik DS, Levy-Nissenbaum E. In vivo administration of BL-3050: highly stable engineered PON1-HDL complexes. *BMC Clin Pharmacol* 2009; 9: 18.
110. Ryan RO. Nanobiotechnology applications of reconstituted high density lipoprotein. *J Nanobiotechnology* 2010; 8: 28.
111. Skajaa T, Cormode DP, Falk E, Mulder WJM, Fisher EA, Fayad ZA. High-density lipoprotein-based contrast agents for multimodal imaging of atherosclerosis. *Arterioscler Thromb Vasc Biol* 2010; 30: 169–176.
112. Cormode DP, Frias JC, Ma Y, Chen W, Skajaa T, Briley-Saebo K, Barazza A, Williams KJ, Mulder WJ, Fayad ZA et al. HDL as a contrast agent for medical imaging. *Clin Lipidol* 2010; 4: 493–500.
113. Cormode DP, Chandrasekar R, Delshad A, Briley-Saebo KC, Calcagno C, Barazza A, Mulder WJ, Fisher EA, Fayad ZA. Comparison of synthetic high density lipoprotein (HDL) contrast agents for MR imaging of atherosclerosis. *Bioconjug Chem* 2009; 20: 937–943.
114. Luthi AJ, Patel PC, Ko CH, Mutharasan RK, Mirkin CA, Thaxton CS. Nanotechnology for synthetic high-density lipoproteins. *Trends Mol Med* 2010: 16: 553–560;
115. Cormode DP, Skajaa T, van Schooneveld MM, Koole R, Jarzyna P, Lobatto ME, Calcagno C, Barazza A, Gordon RE, Zanzonico P et al. Nanocrystal core high-density lipoproteins: a multimodality contrast agent platform. *Nano Lett* 2008; 8: 3715–3723.
116. Chen W, Vucic E, Leupold E, Mulder WJ, Cormode DP, Briley-Saebo KC, Barazza A, Fisher EA, Dathe M, Fayad ZA. Incorporation of an apoE-derived lipopeptide in high-density lipoprotein MRI contrast agents for enhanced imaging of macrophages in atherosclerosis. *Contrast Media Mol Imaging* 2008; 3: 233–242.
117. Skajaa T, Cormode DP, Jarzyna PA, Delshad A, Blachford C, Barazza A, Fisher EA, Gordon RE, Fayad ZA, Mulder WJ. The biological properties of iron oxide core high-density lipoprotein in experimental atherosclerosis. *Biomaterials* 2010; 32: 206–213.
118. Shin JY, Shin JI, Kim JS, Yang YS, Hwang Y, Yang JS, Shin D, Seo JH, Jin YS, Park YC et al. Assembly of Coenzyme Q10 nanostructure resembling nascent discoidal high density lipoprotein particle. *Biochem Biophys Res Commun* 2009; 388: 217–221.
119. Khumsupan P, Ramirez R, Khumsupan D, Narayanaswami V. Apolipoprotein E LDL receptor-binding domain-containing high-density lipoprotein: A nanovehicle to transport curcumin, an antioxidant and anti-amyloid bioflavonoid. *Biochim Biophys Acta* 2011; 1808: 352–359.
120. Zhang WL, Gu X, Bai H, Yang RH, Dong CD, Liu JP. Nanostructured lipid carriers constituted from high-density lipoprotein components for delivery of a lipophilic cardiovascular drug. *Int J Pharm* 2010; 391: 313–321.
121. Feng M, Cai Q, Shi X, Huang H, Zhou P, Guo X. Recombinant high-density lipoprotein complex as a targeting system of nosiheptide to liver cells. *J Drug Target* 2008; 16: 502–508.
122. Chen W, Jarzyna PA, van Tilborg GA, Nguyen VA, Cormode DP, Klink A, Griffioen AW, Randolph GJ, Fisher EA, Mulder WJ et al. RGD peptide functionalized and reconstituted high-density lipoprotein nanoparticles as a versatile and multimodal tumor targeting molecular imaging probe. *FASEB J* 2011; 24: 1689–1699.

123. Cao W, Ng KK, Corbin I, Zhang Z, Ding L, Chen J, Zheng G. Synthesis and evaluation of a stable bacteriochlorophyll-analog and its incorporation into high-density lipoprotein nanoparticles for tumor imaging. *Bioconjug Chem* 2009; 20: 2023–2031.

124. Zhang Z, Chen J, Ding L, Jin H, Lovell JF, Corbin IR, Cao W, Lo PC, Yang M, Tsao MS et al. HDL-mimicking peptide-lipid nanoparticles with improved tumor targeting. *Small* 2010; 6: 430–437.

125. McConathy WJ, Nair MP, Paranjape S, Mooberry L, Lacko AG. Evaluation of synthetic/reconstituted high-density lipoproteins as delivery vehicles for paclitaxel. *Anticancer Drugs* 2008; 19: 183–188.

126. Mooberry LK, Nair M, Paranjape S, McConathy WJ, Lacko AG. Receptor mediated uptake of paclitaxel from a synthetic high density lipoprotein nanocarrier. *J Drug Target* 2010; 18: 53–58.

127. Navab M, Anantharamaiah GM, Reddy ST, Hama S, Hough G, Grijalva VR, Yu N, Ansell BJ, Datta G, Garber DW et al. Apolipoprotein A-I mimetic peptides. *Arterioscler Thromb Vasc Biol* 2005; 25: 1325–1331.

128. Navab M, Shechter I, Anantharamaiah GM, Reddy ST, Van Lenten BJ, Fogelman AM. Structure and function of HDL mimetics. *Arterioscler Thromb Vasc Biol* 2010; 30: 164–168.

129. Osei-Hwedieh DO, Amar M, Sviridov D, Remaley AT. Apolipoprotein mimetic peptides: Mechanisms of action as anti-atherogenic agents. *Pharmacol Ther* 2011; 130: 83–91.

130. Navab M, Anantharamaiah GM, Fogelman AM. The effect of apolipoprotein mimetic peptides in inflammatory disorders other than atherosclerosis. *Trends Cardiovasc Med* 2008; 18: 61–66.

131. Anantharamaiah GM, Jones JL, Brouillette CG, Schmidt CF, Chung BH, Hughes TA, Bhown AS, Segrest JP. Studies of synthetic peptide analogs of the amphipathic helix. Structure of complexes with dimyristoyl phosphatidylcholine. *J Biol Chem* 1985; 260: 10248–10255.

132. Anantharamaiah GM, Mishra VK, Garber DW, Datta G, Handattu SP, Palgunachari MN, Chaddha M, Navab M, Reddy ST, Segrest JP et al. Structural requirements for antioxidative and anti-inflammatory properties of apolipoprotein A-I mimetic peptides. *J Lipid Res* 2007; 48: 1915–1923.

133. Gupta H, White CR, Handattu S, Garber DW, Datta G, Chaddha M, Dai L, Gianturco SH, Bradley WA, Anantharamaiah GM. Apolipoprotein E mimetic peptide dramatically lowers plasma cholesterol and restores endothelial function in Watanabe heritable hyperlipidemic rabbits. *Circulation* 2005; 111: 3112–3118.

134. Navab M, Anantharamaiah GM, Reddy ST, Van Lenten BJ, Wagner AC, Hama S, Hough G, Bachini E, Garber DW, Mishra VK et al. An oral apoJ peptide renders HDL antiinflammatory in mice and monkeys and dramatically reduces atherosclerosis in apolipoprotein E-null mice. *Arterioscler Thromb Vasc Biol* 2005; 25: 1932–1937.

135. Garber DW, Datta G, Chaddha M, Palgunachari MN, Hama SY, Navab M, Fogelman AM, Segrest JP, Anantharamaiah GM. A new synthetic class A amphipathic peptide analogue protects mice from diet-induced atherosclerosis. *J Lipid Res* 2001; 42: 545–552.

136. Navab M, Anantharamaiah GM, Hama S, Garber DW, Chaddha M, Hough G, Lallone R, Fogelman AM. Oral administration of an Apo A-I mimetic peptide synthesized from D-amino acids dramatically reduces atherosclerosis in mice independent of plasma cholesterol. *Circulation* 2002; 105: 290–292.

137. Li X, Chyu K-Y, Neto JRF, Yano J, Nathwani N, Ferreira C, Dimayuga PC, Cercek B, Kaul S, Shah PK. Differential effects of apolipoprotein A-I-mimetic peptide on evolving and established atherosclerosis in apolipoprotein E-null mice. *Circulation* 2004; 110: 1701–1705.

138. Busseuil, D., Shi, Y., Mecteau, M., Brand, G., Kernaleguen, A.E., Thorin, E., Latour, J.G., Rheaume, E., Tardif, J.C. Regression of aortic valve stenosis by ApoA-I mimetic peptide infusions in rabbits. *Br J Pharmacol* 2008; 154: 765–773.

139. Ou J, Ou Z, Jones DW, Holzhauer S, Hatoum OA, Ackerman AW, Weihrauch DW, Gutterman DD, Guice K, Oldham KT et al. L-4F, an apolipoprotein A-1 mimetic, dramatically improves vasodilation in hypercholesterolemia and sickle cell disease. *Circulation* 2003; 107: 2337–2341.

140. Ou Z, Ou J, Ackerman AW, Oldham KT, Pritchard KA Jr. L-4F, an apolipoprotein A-1 mimetic, restores nitric oxide and superoxide anion balance in low-density lipoprotein-treated endothelial cells. *Circulation* 2003; 107: 1520–1524.

141. Navab M, Anantharamaiah GM, Reddy ST, Hama S, Hough G, Grijalva VR, Wagner AC, Frank JS, Datta G, Garber D et al. Oral D-4F causes formation of pre-beta high-density lipoprotein and improves high-density lipoprotein-mediated cholesterol efflux and reverse cholesterol transport from macrophages in apolipoprotein E-null mice. *Circulation* 2004; 109: 3215–3220.

142. Gupta H, Dai L, Datta G, Garber DW, Grenett H, Li Y, Mishra V, Palgunachari MN, Handattu S, Gianturco SH et al. Inhibition of lipopolysaccharide-induced inflammatory responses by an apolipoprotein AI mimetic peptide. *Circ Res* 2005; 97: 236–243.

143. Kruger AL, Peterson S, Turkseven S, Kaminski PM, Zhang FF, Quan S, Wolin MS, Abraham NG. D-4F induces heme oxygenase-1 and extracellular superoxide dismutase, decreases endothelial cell sloughing, and improves vascular reactivity in rat model of diabetes. *Circulation* 2005; 111: 3126–3134.

144. Bielicki JK, Zhang H, Cortez Y, Zheng Y, Narayanaswami V, Patel A, Johansson J, Azhar S. A new HDL mimetic peptide that stimulates cellular cholesterol efflux with high efficiency greatly reduces atherosclerosis in mice. *J Lipid Res* 2010; 51: 1496–1503.

145. Amar MJ, D'Souza W, Turner S, Demosky S, Sviridov D, Stonik J, Luchoomun J, Voogt J, Hellerstein M, Sviridov D, Remaley AT. 5A apolipoprotein mimetic peptide promotes cholesterol efflux and reduces atherosclerosis in mice. *J Pharmacol Exp Ther* 2010; 334: 634–641.

146. Xie Q, Zhao SP, Li F. D-4F, an apolipoprotein A-I mimetic peptide, promotes cholesterol efflux from macrophages via ATP-binding cassette transporter A1. *Tohoku J Exp Med* 2010; 220: 223–228.

147. Smythies LE, White CR, Maheshwari A, Palgunachari MN, Anantharamaiah GM, Chaddha M, Kurundkar AR, Datta G. Apolipoprotein A-I mimetic 4F alters the function of human monocyte-derived macrophages. *Am J Physiol Cell Physiol* 2010; 298: C1538–1548.

148. Datta G, White CR, Dashti N, Chaddha M, Palgunachari MN, Gupta H, Handattu SP, Garber DW, Anantharamaiah GM. Anti-inflammatory and recycling properties of an apolipoprotein mimetic peptide, Ac-hE18A-NH(2). *Atherosclerosis* 2010; 208: 134–141.

149. Nayyar G, Handattu SP, Monroe CE, Chaddha M, Datta G, Mishra VK, Keenum TD, Palgunachari MN, Garber DW, Anantharamaiah GM. Two adjacent domains (141–150 and 151–160) of apoE covalently linked to a class A amphipathic helical peptide exhibit opposite atherogenic effects. *Atherosclerosis* 2010; 213: 449–457.

150. Handattu SP, Datta G, Epand RM, Epand RF, Palgunachari MN, Mishra VK, Monroe CE, Keenum TD, Chaddha M, Anantharamaiah GM et al. Oral administration of L-mR18L, a single domain cationic amphipathic helical peptide, inhibits lesion formation in ApoE null mice. *J Lipid Res* 2010; 51: 3491–3499.

151. Chen X, Burton C, Song X, McNamara L, Langella A, Cianetti S, Chang CH, Wang J. An apoA-I mimetic peptide increases LCAT activity in mice through increasing HDL concentration. *Int J Biol Sci* 2009; 5: 489–499.

152. Song X, Fischer P, Chen X, Burton C, Wang J. An apoA-I mimetic peptide facilitates off-loading cholesterol from HDL to liver cells through scavenger receptor BI. *Int J Biol Sci* 2009; 5: 637–646.

153. Carballo-Jane E, Chen Z, O'Neill E, Wang J, Burton C, Chang CH, Chen X, Eveland S, Frantz-Wattley B, Gagen K et al. ApoA-I mimetic peptides promote pre-beta HDL formation in vivo causing remodeling of HDL and triglyceride accumulation at higher dose. *Bioorg Med Chem* 2010; 18: 8669–8678.

154. Troutt JS, Alborn WE, Mosior MK, Dai J, Murphy AT, Beyer TP, Zhang Y, Cao G, Konrad RJ. An apolipoprotein A-I mimetic dose-dependently increases the formation of prebeta1 HDL in human plasma. *J Lipid Res* 2008; 49: 581–587.

155. Bielicki JK, Asztalos B, Johansson J, Azhar S. Abstract 1093: The novel apolipoprotein mimetic peptide ATI-5261 induces prebeta-1 HDL formation in human plasma via a highly specific mechanism involving distinct alpha-HDL subpopulations. *Circulation* 2009; 120: S445.

156. Van Lenten BJ, Wagner AC, Navab M, Anantharamaiah GM, Hama S, Reddy ST, Fogelman AM. Lipoprotein inflammatory properties and serum amyloid A levels but not cholesterol levels predict lesion area in cholesterol-fed rabbits. *J Lipid Res* 2007; 48: 2344–2353.

157. Morgantini C, Imaizumi S, Grijalva V, Navab M, Fogelman AM, Reddy ST. Apolipoprotein A-I mimetic peptides prevent atherosclerosis development and reduce plaque inflammation in a murine model of diabetes. *Diabetes* 2010; 59: 3223–3228.

158. Tabet F, Remaley AT, Segaliny AI, Millet J, Yan L, Nakhla S, Barter PJ, Rye KA, Lambert G. The 5A apolipoprotein A-I mimetic peptide displays antiinflammatory and antioxidant properties *in vitro* and in vivo. *Arterioscler Thromb Vasc Biol* 2010; 30: 246–252.

159. Charles-Schoeman C, Banquerigo ML, Hama S, Navab M, Park GS, Van Lenten BJ, Wagner AC, Fogelman AM, Brahn E. Treatment with an apolipoprotein A-1 mimetic peptide in combination with pravastatin inhibits collagen-induced arthritis. *Clin Immunol* 2008; 127: 234–244.

160. Zhang Z, Datta G, Zhang Y, Miller AP, Mochon P, Chen YF, Chatham J, Anantharamaiah GM, White CR. Apolipoprotein A-I mimetic peptide treatment inhibits

inflammatory responses and improves survival in septic rats. *Am J Physiol Heart Circ Physiol* 2009; 297: H866–873.

161. Dai L, Datta G, Zhang Z, Gupta H, Patel R, Honavar J, Modi S, Wyss JM, Palgunachari M, Anantharamaiah GM et al. The apolipoprotein A-I mimetic peptide 4F prevents defects in vascular function in endotoxemic rats. *J Lipid Res* 2010; 51: 2695–2705.
162. Buga GM, Frank JS, Mottino GA, Hakhamian A, Narasimha A, Watson AD, Yekta B, Navab M, Reddy ST, Anantharamaiah GM et al. D-4F reduces EO6 immunoreactivity, SREBP-1c mRNA levels, and renal inflammation in LDL receptor-null mice fed a Western diet. *J Lipid Res* 2008; 49: 192–205.
163. Nandedkar SD, Weihrauch D, Xu H, Shi Y, Feroah T, Hutchins W, Rickaby DA, Duzgunes N, Hillery CA, Konduri KS et al. D-4F, an apoA-1 mimetic, decreases airway hyperresponsiveness, inflammation, and oxidative stress in a murine model of asthma. *J Lipid Res* 2011; 52: 499–508.
164. Vaziri ND, Bai Y, Yuan J, Said HL, Sigala W, Ni Z. ApoA-1 mimetic peptide reverses uremia-induced upregulation of pro-atherogenic pathways in the aorta. *Am J Nephrol* 2010; 32: 201–211.
165. Van Lenten BJ, Wagner AC, Jung C-L, Ruchala P, Waring AJ, Lehrer RI, Watson AD, Hama S, Navab M, Anantharamaiah GM et al. Anti-inflammatory apoA-I-mimetic peptides bind oxidized lipids with much higher affinity than human apoA-I. *J Lipid Res* 2008; 49: 2302–2311.
166. Imaizumi S, Grijalva V, Navab M, Van Lenten BJ, Wagner AC, Anantharamiah GM, Fogelman AM, Reddy ST. L-4F differentially alters plasma levels of oxidized fatty acids resulting in more anti-inflammatory HDL in mice. *Drug Metab Lett* 2010; 4: 139–148.
167. Watanabe J, Grijalva V, Hama S, Barbour K, Berger FG, Navab M, Fogelman AM, Reddy ST. Hemoglobin and its scavenger protein haptoglobin associate with apoA-1-containing particles and influence the inflammatory properties and function of high density lipoprotein. *J Biol Chem* 2009; 284: 18292–18301.
168. Datta G, Epand RF, Epand RM, Chaddha M, Kirksey MA, Garber DW, Lund-Katz S, Phillips MC, Hama S, Navab M et al. Aromatic residue position on the nonpolar face of class A amphipathic helical peptides determines biological activity. *J Biol Chem* 2004; 279: 26509–26517.
169. Epand RM, Epand RF, Sayer BG, Datta G, Chaddha M, Anantharamaiah GM. Two homologous apolipoprotein AI mimetic peptides: Relationship between membrane interactions and biological activity. *J Biol Chem* 2004; 279: 51404–51414.
170. Wool GD, Vaisar T, Reardon CA, Getz GS. An apoA-I mimetic peptide containing a proline residue has greater in vivo HDL binding and anti-inflammatory ability than the 4F peptide. *J Lipid Res* 2009; 50: 1889–1900.
171. Wool GD, Reardon CA, Getz GS. Apolipoprotein A-I mimetic peptide helix number and helix linker influence potentially anti-atherogenic properties. *J Lipid Res* 2008; 49: 1268–1283.
172. Bloedon LT, Dunbar R, Duffy D, Pinell-Salles P, Norris R, DeGroot BJ, Movva R, Navab M, Fogelman AM, Rader DJ. Safety, pharmacokinetics, and pharmacodynamics of oral apoA-I mimetic peptide D-4F in high-risk cardiovascular patients. *J Lipid Res* 2008; 49: 1344–1352.

173. Watson CE, Weissbach N, Kjems L, Ayalasomayajula S, Zhang Y, Chang I, Navab M, Hama S, Hough G, Reddy ST et al. Treatment of patients with cardiovascular disease with L-4F, an apo-A1 mimetic, did not improve select biomarkers of HDL function. *J Lipid Res* 2011; 52: 361–373.

174. Peterson SJ, Drummond G, Kim DH, Li M, Kruger AL, Ikehara S, Abraham NG. L-4F treatment reduces adiposity, increases adiponectin levels, and improves insulin sensitivity in obese mice. *J Lipid Res* 2008; 49: 1658–1669.

175. Peterson SJ, Kim DH, Li M, Positano V, Vanella L, Rodella LF, Piccolomini F, Puri N, Gastaldelli A, Kusmic C et al. The L-4F mimetic peptide prevents insulin resistance through increased levels of HO-1, pAMPK, and pAKT in obese mice. *J Lipid Res* 2009; 50: 1293–1304.

176. Weihrauch D, Xu H, Shi Y, Wang J, Brien J, Jones DW, Kaul S, Komorowski RA, Csuka ME, Oldham KT et al. Effects of D-4F on vasodilation, oxidative stress, angiostatin, myocardial inflammation, and angiogenic potential in tight-skin mice. *Am J Physiol Heart Circ Physiol* 2007; 293: H1432–H1441.

177. Hsieh GR, Schnickel GT, Garcia C, Shefizadeh A, Fishbein MC, Ardehali A. Inflammation/oxidation in chronic rejection: apolipoprotein A-I mimetic peptide reduces chronic rejection of transplanted hearts. *Transplantation* 2007; 84: 238–243.

178. Buga GM, Navab M, Imaizumi S, Reddy ST, Yekta B, Hough G, Chanslor S, Anantharamaiah GM, Fogelman AM. L-4F alters hyperlipidemic (but not healthy) mouse plasma to reduce platelet aggregation. *Arterioscler Thromb Vasc Biol* 2010; 30: 283–289.

179. Navab M, Anantharamaiah GM, Reddy ST, Hama S, Hough G, Frank JS, Grijalva VR, Ganesh VK, Mishra VK, Palgunachari MN et al. Oral small peptides render HDL antiinflammatory in mice and monkeys and reduce atherosclerosis in apoE null mice. *Circ Res* 2005; 97: 524–532.

180. Navab M, Ruchala P, Waring AJ, Lehrer RI, Hama S, Hough G, Palgunachari MN, Anantharamaiah GM, Fogelman AM. A novel method for oral delivery of apolipoprotein mimetic peptides synthesized from all L-amino acids. *J Lipid Res* 2009; 50: 1538–1547.

181. D'Souza W, Stonik JA, Murphy A, Demosky SJ, Sethi AA, Moore XL, Chin-Dusting J, Remaley AT, Sviridov D. Structure/function relationships of apolipoprotein A-I mimetic peptides: implications for antiatherogenic activities of high-density lipoprotein. *Circ Res* 2010; 107: 217–227.

182. Chong PH, Kezele R, Franklin C. High-density lipoprotein cholesterol and the role of statins. *Circ J* 2002; 66: 1037–1044.

183. Barter PJ, Brandrup-Wognsen G, Palmer MK, Nicholls SJ. Effect of statins on HDL-C: a complex process unrelated to changes in LDL-C. Analysis of the VOYAGER Database. *J Lipid Res* 2010; 51: 1546–1553.

184. Sasaki J, Ikeda Y, Kuribayashi T, Kajiwara K, Biro S, Yamamoto K, Ageta M, Kobori S, Saikawa T, Otonari T et al. A 52-week, randomized, open-label, parallel-group comparison of the tolerability and effects of pitavastatin and atorvastatin on high-density lipoprotein cholesterol levels and glucose metabolism in Japanese patients with elevated levels of low-density lipoprotein cholesterol and glucose intolerance. *Clin Ther* 2008; 30: 1089–1101.

185. Yamashita S, Tsubakio-Yamamoto K, Ohama T, Nakagawa-Toyama Y, Nishida M. Molecular mechanisms of HDL-cholesterol elevation by statins and its effects on HDL functions. *J Atheroscler Thromb* 2010; 17: 436–451.

186. Fukutomi T, Takeda Y, Suzuki S, Ito T, Joh T, Itoh M. High density lipoprotein cholesterol and apolipoprotein A-I are persistently elevated during long-term treatment with pitavastatin, a new HMG-CoA reductase inhibitor. *Int J Cardiol* 2010; 141: 320–322.
187. Kostapanos MS, Milionis HJ, Filippatos TD, Christogiannis LG, Bairaktari ET, Tselepis AD, Elisaf MS. Dose-dependent effect of rosuvastatin treatment on HDL-subfraction phenotype in patients with primary hyperlipidemia. *J Cardiovasc Pharmacol Ther* 2009; 14: 5–13.
188. Caslake MJ, Stewart G, Day SP, Daly E, McTaggart F, Chapman MJ, Durrington P, Laggner P, Mackness M, Pears J et al. Phenotype-dependent and -independent actions of rosuvastatin on atherogenic lipoprotein subfractions in hyperlipidaemia. *Atherosclerosis* 2003; 171: 245–253.
189. Hamrefors V, Orho-Melander M, Krauss RM, Hedblad B, Almgren P, Berglund G, Melander O. A gene score of nine LDL and HDL regulating genes is associated with fluvastatin-induced cholesterol changes in women. *J Lipid Res* 2010; 51: 625–634.
190. Li J, Wang LF, Li ZQ, Pan W. Effect of R219K polymorphism of the ABCA1 gene on the lipid-lowering effect of pravastatin in Chinese patients with coronary heart disease. *Clin Exp Pharmacol Physiol* 2009; 36: 567–570.
191. Schaefer JR, Schweer H, Ikewaki K, Stracke H, Seyberth HJ, Kaffarnik H, Maisch B, Steinmetz A. Metabolic basis of high density lipoproteins and apolipoprotein A-I increase by HMG-CoA reductase inhibition in healthy subjects and a patient with coronary artery disease. *Atherosclerosis* 1999; 144: 177–184.
192. Guerin M, Lassel TS, Le Goff W, Farnier M, Chapman MJ. Action of atorvastatin in combined hyperlipidemia: preferential reduction of cholesteryl ester transfer from HDL to VLDL1 particles. *Arterioscler Thromb Vasc Biol* 2000; 20: 189–197.
193. Chapman MJ, Le Goff W, Guerin M, Kontush A. Cholesteryl ester transfer protein: at the heart of the action of lipid-modulating therapy with statins, fibrates, niacin, and cholesteryl ester transfer protein inhibitors. *Eur Heart J* 2010; 31: 149–164.
194. Qin S, Koga T, Ganji SH, Kamanna VS, Kashyap ML. Rosuvastatin selectively stimulates apolipoprotein A-I but not apolipoprotein A-II synthesis in Hep G2 cells. *Metabolism* 2008; 57: 973–979.
195. Harangi M, Seres I, Varga Z, Emri G, Szilvassy Z, Paragh G, Remenyik E. Atorvastatin effect on high-density lipoprotein-associated paraoxonase activity and oxidative DNA damage. *Eur J Clin Pharmacol* 2004; 60: 685–691.
196. Guerin M, Dolphin PJ, Talussot C, Gardette J, Berthezene F, Chapman MJ. Pravastatin modulates cholesteryl ester transfer from HDL to apoB-containing lipoproteins and lipoprotein subspecies profile in familial hypercholesterolemia. *Arterioscler Thromb Vasc Biol* 1995; 15: 1359–1368.
197. Guerin M, Lassel TS, Le Goff W, Farnier M, Chapman MJ. Action of atorvastatin in combined hyperlipidemia: preferential reduction of cholesteryl ester transfer from HDL to VLDL1 particles. *Arterioscler Thromb Vasc Biol* 2000; 20: 189–197.
198. Karalis IK, Bergheanu SC, Wolterbeek R, Dallinga-Thie GM, Hattori H, van Tol A, Liem AH, Wouter Jukema J. Effect of increasing doses of Rosuvastatin and Atorvastatin on apolipoproteins, enzymes and lipid transfer proteins involved in lipoprotein metabolism and inflammatory parameters. *Curr Med Res Opin* 2010; 26: 2301–2313.
199. McTaggart F, Jones P. Effects of statins on high-density lipoproteins: a potential contribution to cardiovascular benefit. *Cardiovasc Drugs Ther* 2008; 22: 321–338.

200. de Haan W, van der Hoogt CC, Westerterp M, Hoekstra M, Dallinga-Thie GM, Princen HM, Romijn JA, Jukema JW, Havekes LM, Rensen PC. Atorvastatin increases HDL cholesterol by reducing CETP expression in cholesterol-fed APOE*3-Leiden.CETP mice. *Atherosclerosis* 2008; 197: 57–63.

201. Franceschini G, Calabresi L, Colombo C, Favari E, Bernini F, Sirtori CR. Effects of fenofibrate and simvastatin on HDL-related biomarkers in low-HDL patients. *Atherosclerosis* 2007; 195: 385–391.

202. Harangi M, Mirdamadi HZ, Seres I, Sztanek F, Molnar M, Kassai A, Derdak Z, Illyes L, Paragh G. Atorvastatin effect on the distribution of high-density lipoprotein subfractions and human paraoxonase activity. *Transl Res* 2009; 153: 190–198.

203. Dallinga-Thie GM, van Tol A, Dullaart RP. Plasma pre beta-HDL formation is decreased by atorvastatin treatment in type 2 diabetes mellitus: Role of phospholipid transfer protein. *Biochim Biophys Acta* 2009; 1791: 714–718.

204. Qiu G, Hill JS. Atorvastatin decreases lipoprotein lipase and endothelial lipase expression in human THP-1 macrophages. *J Lipid Res* 2007; 48: 2112–2122.

205. Kojima Y, Ishida T, Sun L, Yasuda T, Toh R, Rikitake Y, Fukuda A, Kume N, Koshiyama H, Taniguchi A et al. Pitavastatin decreases the expression of endothelial lipase both *in vitro* and in vivo. *Cardiovasc Res* 2010; 87: 385–393.

206. Otvos JD, Shalaurova I, Freedman DS, Rosenson RS. Effects of pravastatin treatment on lipoprotein subclass profiles and particle size in the PLAC-I trial. *Atherosclerosis* 2002; 160: 41–48.

207. Schaefer EJ, McNamara JR, Tayler T, Daly JA, Gleason JA, Seman LJ, Ferrari A, Rubenstein JJ. Effects of atorvastatin on fasting and postprandial lipoprotein subclasses in coronary heart disease patients versus control subjects. *Am J Cardiol* 2002; 90: 689–696.

208. Kazama H, Usui S, Okazaki M, Hosoi T, Ito H, Orimo H. Effects of bezafibrate and pravastatin on remnant-like lipoprotein particles and lipoprotein subclasses in type 2 diabetes. *Diabetes Res Clin Pract* 2003; 59: 181–189.

209. Soedamah-Muthu SS, Colhoun HM, Thomason MJ, Betteridge DJ, Durrington PN, Hitman GA, Fuller JH, Julier K, Mackness MI, Neil HA. The effect of atorvastatin on serum lipids, lipoproteins and NMR spectroscopy defined lipoprotein subclasses in type 2 diabetic patients with ischaemic heart disease. *Atherosclerosis* 2003; 167: 243–255.

210. Kawano M, Nagasaka S, Yagyu H, Ishibashi S. Pitavastatin decreases plasma prebeta1-HDL concentration and might promote its disappearance rate in hypercholesterolemic patients. *J Atheroscler Thromb* 2008; 15: 41–46.

211. Teramoto T, Shimano H, Yokote K, Urashima M. Effects of pitavastatin (LIVALO Tablet) on high density lipoprotein cholesterol (HDL-C) in hypercholesterolemia. *J Atheroscler Thromb* 2009; 16: 654–661.

212. Molgaard J, Warjerstam-Elf S, Olsson AG. Efficacy and safety of simvastatin for high-risk hypercholesterolemia. *Am J Cardiol* 1999; 83: 1043–1048.

213. Broyles FE, Walden CE, Hunninghake DB, Hill-Williams D, Knopp RH. Effect of fluvastatin on intermediate density lipoprotein (remnants) and other lipoprotein levels in hypercholesterolemia. *Am J Cardiol* 1995; 76: 129A–135A.

214. Koshiyama H, Taniguchi A, Tanaka K, Kagimoto S, Fujioka Y, Hirata K, Nakamura Y, Iwakura A, Hara K, Yamamoto T et al. Effects of pitavastatin on lipid profiles and

high-sensitivity CRP in Japanese subjects with hypercholesterolemia: Kansai Investigation of Statin for Hyperlipidemic Intervention in Metabolism and Endocrinology (KISHIMEN) investigatars. *J Atheroscler Thromb* 2008; 15: 345–350.

215. Nagila A, Permpongpaiboon T, Tantrarongroj S, Porapakkham P, Chinwattana K, Deakin S, Porntadavity S.). Effect of atorvastatin on paraoxonase1 (PON1) and oxidative status. *Pharmacol Rep* 2009; 61: 892–898.

216. Abdin AA, Hassanien MA, Ibrahim EA, El-Noeman Sel D. Modulating effect of atorvastatin on paraoxonase 1 activity in type 2 diabetic Egyptian patients with or without nephropathy. *J Diabetes Complications* 2010; 24: 325–333.

217. Muacevic-Katanec D, Bradamante V, Poljicanin T, Reiner Z, Babic Z, Simeon-Rudolf V, Katanec D. Clinical study on the effect of simvastatin on paraoxonase activity. *Arzneimittelforschung* 2007; 57: 647–653.

218. Arii K, Suehiro T, Ota K, Ikeda Y, Kumon Y, Osaki F, Inoue M, Inada S, Ogami N, Takata H et al. Pitavastatin induces PON1 expression through p44/42 mitogen-activated protein kinase signaling cascade in Huh7 cells. *Atherosclerosis* 2009; 202: 439–445.

219. Saougos VG, Tambaki AP, Kalogirou M, Kostapanos M, Gazi IF, Wolfert RL, Elisaf M, Tselepis AD. Differential effect of hypolipidemic drugs on lipoprotein-associated phospholipase A2. *Arterioscler Thromb Vasc Biol* 2007; 27: 2236–2243.

220. Ceriello A, Assaloni R, Da Ros R, Maier A, Piconi L, Quagliaro L, Esposito K, Giugliano D. Effect of atorvastatin and irbesartan, alone and in combination, on postprandial endothelial dysfunction, oxidative stress, and inflammation in Type 2 diabetic patients. *Circulation* 2005; 111: 2518–2524.

221. Ansell BJ, Navab M, Hama S, Kamranpour N, Fonarow G, Hough G, Rahmani S, Mottahedeh R, Dave R, Reddy ST et al. Inflammatory/antiinflammatory properties of high-density lipoprotein distinguish patients from control subjects better than high-density lipoprotein cholesterol levels and are favorably affected by simvastatin treatment. *Circulation* 2003; 108: 2751–2756.

222. Higashi Y, Matsuoka H, Umei H, Sugano R, Fujii Y, Soga J, Kihara Y, Chayama K, Imaizumi T. Endothelial function in subjects with isolated low HDL cholesterol: role of nitric oxide and circulating progenitor cells. *Am J Physiol Endocrinol Metab* 2010; 298: E202–E209.

223. Kimura T, Mogi C, Tomura H, Kuwabara A, Im DS, Sato K, Kurose H, Murakami M, Okajima F. Induction of scavenger receptor class B type I is critical for simvastatin enhancement of high-density lipoprotein-induced anti-inflammatory actions in endothelial cells. *J Immunol* 2008; 181: 7332–7340.

224. Datar R, Kaesemeyer WH, Chandra S, Fulton DJ, Caldwell RW. Acute activation of eNOS by statins involves scavenger receptor-B1, G protein subunit Gi, phospholipase C and calcium influx. *Br J Pharmacol* 2010; 160: 1765–1772.

225. Beltowski J. Liver X receptors (LXR) as therapeutic targets in dyslipidemia. *Cardiovasc Ther* 2008; 26: 297–316.

226. Wong J, Quinn CM, Gelissen IC, Jessup W, Brown AJ. The effect of statins on ABCA1 and ABCG1 expression in human macrophages is influenced by cellular cholesterol levels and extent of differentiation. *Atherosclerosis* 2008; 196: 180–189.

227. Sviridov D, Hoang A, Ooi E, Watts G, Barrett PH, Nestel P. Indices of reverse cholesterol transport in subjects with metabolic syndrome after treatment with rosuvastatin. *Atherosclerosis* 2008; 197: 732–739.

228. Guan JZ, Tamasawa N, Murakami H, Matsui J, Tanabe J, Matsuki K, Yamashita M, Suda T. HMG-CoA reductase inhibitor, simvastatin improves reverse cholesterol transport in type 2 diabetic patients with hyperlipidemia. *J Atheroscler Thromb* 2008; 15: 20–25.

229. Underhill HR, Yuan C, Zhao XQ, Kraiss LW, Parker DL, Saam T, Chu B, Takaya N, Liu F, Polissar NL et al. Effect of rosuvastatin therapy on carotid plaque morphology and composition in moderately hypercholesterolemic patients: a high-resolution magnetic resonance imaging trial. *Am Heart J* 2008; 155: 584 e581–e588.

230. Ballantyne CM, Raichlen JS, Nicholls SJ, Erbel R, Tardif JC, Brener SJ, Cain VA, Nissen SE. Effect of rosuvastatin therapy on coronary artery stenoses assessed by quantitative coronary angiography: A Study To Evaluate the effect of Rosuvastatin On Intravascular Ultrasound-Derived coronary atheroma burden. *Circulation* 2008; 117: 2458–2466.

231. Tani S, Matsumoto M, Anazawa T, Kawamata H, Furuya S, Takahashi H, Iida K, Washio T, Kumabe N, Kobori M et al. Development of a model for prediction of coronary atherosclerotic regression: evaluation of high-density lipoprotein cholesterol level and peripheral blood monocyte count. *Heart Vessels* 2011, in press.

232. Nicholls SJ, Tuzcu EM, Sipahi I, Grasso AW, Schoenhagen P, Hu T, Wolski K, Crowe T, Desai MY, Hazen SL et al. Statins, high-density lipoprotein cholesterol, and regression of coronary atherosclerosis. *JAMA* 2007; 297: 499–508.

233. Cui Y, Watson DJ, Girman CJ, Shapiro DR, Gotto AM, Hiserote P, Clearfield MB. Effects of increasing high-density lipoprotein cholesterol and decreasing low-density lipoprotein cholesterol on the incidence of first acute coronary events (from the Air Force/Texas Coronary Atherosclerosis Prevention Study). *Am J Cardiol* 2009; 104: 829–834.

234. Kakafika A, Athyros VG, Tziomalos K, Karagiannis A, Mikhailidis DP. High density lipoprotein cholesterol and statin trials. *Curr Med Chem* 2008; 15: 2265–2270.

235. Brown JD, Plutzky J. Peroxisome proliferator-activated receptors as transcriptional nodal points and therapeutic targets. *Circulation* 2007; 115: 518–533.

236. Chapman MJ. Fibrates in 2003: therapeutic action in atherogenic dyslipidaemia and future perspectives. *Atherosclerosis* 2003; 171: 1–13.

237. Birjmohun RS, Hutten BA, Kastelein JJ, Stroes ES. Efficacy and safety of high-density lipoprotein cholesterol-increasing compounds: a meta-analysis of randomized controlled trials. *J Am Coll Cardiol* 2005; 45: 185–197.

238. Mombelli G, Banfi F, Falcioni S, Sirtori CR. Persistently increased HDL-cholesterolemia and reduced triglyceridemia in a large lipid clinic population treated with fenofibrate for 15 years or longer. *Int J Cardiol* 2009; 133: 412–414.

239. Keech A, Simes RJ, Barter P, Best J, Scott R, Taskinen MR, Forder P, Pillai A, Davis T, Glasziou P et al. Effects of long-term fenofibrate therapy on cardiovascular events in 9795 people with type 2 diabetes mellitus (the FIELD study): randomised controlled trial. *Lancet* 2005; 366: 1849–1861.

240. Taskinen M-R, Sullivan DR, Ehnholm C, Whiting M, Zannino D, Simes RJ, Keech AC, Barter PJ, for the FSI. Relationships of HDL cholesterol, apoA-I, and apoA-II with homocysteine and creatinine in patients with Type 2 diabetes treated with fenofibrate. *Arterioscler Thromb Vasc Biol* 2009; 29: 950–955.

241. Terra SG, Francone OL, Contant CF, Gao X, Lewin AJ, Nguyen TT. Efficacy and safety of a potent and selective peroxisome proliferator activated receptor alpha

agonist in subjects with dyslipidemia and type 2 diabetes mellitus. *Am J Cardiol* 2008; 102: 434–439.

242. Pandey NR, Renwick J, Misquith A, Sokoll K, Sparks DL. Linoleic acid-enriched phospholipids act through peroxisome proliferator-activated receptors alpha to stimulate hepatic apolipoprotein A-I secretion. *Biochemistry* 2008; 47: 1579–1587.
243. Pandey NR, Sparks DL. Phospholipids as cardiovascular therapeutics. *Curr Opin Investig Drugs* 2008; 9: 281–285.
244. Sasaki J, Yamamoto K, Ageta M. Effects of fenofibrate on high-density lipoprotein particle size in patients with hyperlipidemia: a randomized, double-blind, placebo-controlled, multicenter, crossover study. *Clin Ther* 2002; 24: 1614–1626.
245. Ikewaki K, Tohyama J, Nakata Y, Wakikawa T, Kido T, Mochizuki S. Fenofibrate effectively reduces remnants, and small dense LDL, and increases HDL particle number in hypertriglyceridemic men - a nuclear magnetic resonance study. *J Atheroscler Thromb* 2004; 11: 278–285.
246. Miida T, Sakai K, Ozaki K, Nakamura Y, Yamaguchi T, Tsuda T, Kashiwa T, Murakami T, Inano K, Okada M. Bezafibrate increases prebeta 1-HDL at the expense of HDL2b in hypertriglyceridemia. *Arterioscler Thromb Vasc Biol* 2000; 20: 2428–2433.
247. Ikewaki K, Noma K, Tohyama J, Kido T, Mochizuki S. Effects of bezafibrate on lipoprotein subclasses and inflammatory markers in patients with hypertriglyceridemia - a nuclear magnetic resonance study. *Int J Cardiol* 2005; 101: 441–447.
248. Kahri J, Vuorinen-Markkola H, Tilly-Kiesi M, Lahdenpera S, Taskinen MR. Effect of gemfibrozil on high density lipoprotein subspecies in non-insulin dependent diabetes mellitus. Relations to lipolytic enzymes and to the cholesteryl ester transfer protein activity. *Atherosclerosis* 1993; 102: 79–89.
249. Hiukka A, Leinonen E, Jauhiainen M, Sundvall J, Ehnholm C, Keech AC, Taskinen MR. Long-term effects of fenofibrate on VLDL and HDL subspecies in participants with type 2 diabetes mellitus. *Diabetologia* 2007; 50: 2067–2075.
250. Steiner G. Fibrates and coronary risk reduction. *Atherosclerosis* 2005; 182: 199–207.
251. Sanoudou D, Duka A, Drosatos K, Hayes KC, Zannis VI. Role of Esrrg in the fibrate-mediated regulation of lipid metabolism genes in human ApoA-I transgenic mice. *Pharmacogenomics J* 2010; 10: 165–179.
252. Shah A, Rader DJ, Millar JS. The effect of PPAR-alpha agonism on apolipoprotein metabolism in humans. *Atherosclerosis* 2010; 210: 35–40.
253. Chan DC, Watts GF, Ooi EM, Rye KA, Ji J, Johnson AG, Barrett PH. Regulatory effects of fenofibrate and atorvastatin on lipoprotein A-I and lipoprotein A-I:A-II kinetics in the metabolic syndrome. *Diabetes Care* 2009; 32: 2111–2113.
254. Millar JS, Duffy D, Gadi R, Bloedon LT, Dunbar RL, Wolfe ML, Movva R, Shah A, Fuki IV, McCoy M et al. Potent and selective PPAR-{alpha} agonist LY518674 upregulates both apoA-I production and catabolism in human subjects with the Metabolic Syndrome. *Arterioscler Thromb Vasc Biol* 2009; 29: 140–146.
255. Chinetti G, Lestavel S, Bocher V, Remaley AT, Neve B, Torra IP, Teissier E, Minnich A, Jaye M, Duverger N et al. PPAR-alpha and PPAR-gamma activators induce cholesterol removal from human macrophage foam cells through stimulation of the ABCA1 pathway. *Nat Med* 2001; 7: 53–58.

256. Hossain MA, Tsujita M, Gonzalez FJ, Yokoyama S. Effects of fibrate drugs on expression of ABCA1 and HDL biogenesis in hepatocytes. *J Cardiovasc Pharmacol* 2008; 51: 258–266.

257. Ogata M, Tsujita M, Hossain MA, Akita N, Gonzalez FJ, Staels B, Suzuki S, Fukutomi T, Kimura G, Yokoyama S. On the mechanism for PPAR agonists to enhance ABCA1 gene expression. *Atherosclerosis* 2009; 205: 413–419.

258. Mukherjee R, Locke KT, Miao B, Meyers D, Monshizadegan H, Zhang R, Search D, Grimm D, Flynn M, O'Malley KM et al. Novel peroxisome proliferator-activated receptor alpha agonists lower low-density lipoprotein and triglycerides, raise high-density lipoprotein, and synergistically increase cholesterol excretion with a liver X receptor agonist. *J Pharmacol Exp Ther* 2008; 327: 716–726.

259. Rotllan N, Llaverias G, Julve J, Jauhiainen M, Calpe-Berdiel L, Hernandez C, Simo R, Blanco-Vaca F, Escola-Gil JC. Differential effects of gemfibrozil and fenofibrate on reverse cholesterol transport from macrophages to feces in vivo. *Biochim Biophys Acta* 2011; 1811: 104–110.

260. Dullens SP, Plat J, Mensink RP. Increasing apoA-I production as a target for CHD risk reduction. *Nutr Metab Cardiovasc Dis* 2007; 17: 616–628.

261. Miyazaki T, Shimada K, Miyauchi K, Kume A, Tanimoto K, Kiyanagi T, Sumiyoshi K, Hiki M, Mokuno H, Okazaki S et al. Effects of fenofibrate on lipid profiles, cholesterol ester transfer activity, and in-stent intimal hyperplasia in patients after elective coronary stenting. *Lipids Health Dis* 2010; 9: 122.

262. Guerin M, Le Goff W, Frisdal E, Schneider S, Milosavljevic D, Bruckert E, Chapman MJ. Action of ciprofibrate in type IIb hyperlipoproteinemia: modulation of the atherogenic lipoprotein phenotype and stimulation of high-density lipoprotein-mediated cellular cholesterol efflux. *J Clin Endocrinol Metab* 2003; 88: 3738–3746.

263. Fruchart JC, Duriez P, Staels B. Peroxisome proliferator-activated receptor-alpha activators regulate genes governing lipoprotein metabolism, vascular inflammation and atherosclerosis. *Curr Opin Lipidol* 1999; 10: 245–257.

264. Inaba T, Yagyu H, Itabashi N, Tazoe F, Fujita N, Nagashima S, Okada K, Okazaki M, Furukawa Y, Ishibashi S. Cholesterol reduction and atherosclerosis inhibition by bezafibrate in low-density lipoprotein receptor knockout mice. *Hypertens Res* 2008; 31: 999–1005.

265. Troutt JS, Alborn WE, Cao G, Konrad RJ. Fenofibrate treatment increases human serum proprotein convertase subtilisin kexin type 9 levels. *J Lipid Res* 2010; 51: 345–351.

266. Lambert G, Ancellin N, Charlton F, Comas D, Pilot J, Keech A, Patel S, Sullivan DR, Cohn JS, Rye KA et al. Plasma PCSK9 concentrations correlate with LDL and total cholesterol in diabetic patients and are decreased by fenofibrate treatment. *Clin Chem* 2008; 54: 1038–1045.

267. Bighetti EJ, Patricio PR, Casquero AC, Berti JA, Oliveira HC. Ciprofibrate increases cholesteryl ester transfer protein gene expression and the indirect reverse cholesterol transport to the liver. *Lipids Health Dis* 2009; 8: 50.

268. Beyer TP, Chen Y, Porter RK, Lu D, Schmidt RJ, Mantlo NB, Konrad RJ, Cao G. Peroxisome proliferator-activated receptor alpha agonists regulate cholesterol ester transfer protein. *Lipids* 2008; 43: 611–618.

269. Ryan KE, McCance DR, Powell L, McMahon R, Trimble ER. Fenofibrate and pioglitazone improve endothelial function and reduce arterial stiffness in obese glucose tolerant men. *Atherosclerosis* 2007; 194: e123–130.

270. Phuntuwate W, Suthisisang C, Koanantakul B, Chaloeiphap P, Mackness B, Mackness M. Effect of fenofibrate therapy on paraoxonase1 status in patients with low HDL-C levels. *Atherosclerosis* 2008; 196: 122–128.

271. Barish GD, Atkins AR, Downes M, Olson P, Chong LW, Nelson M, Zou Y, Hwang H, Kang H, Curtiss L et al. PPARdelta regulates multiple proinflammatory pathways to suppress atherosclerosis. *Proc Natl Acad Sci U S A* 2008; 105: 4271–4276.

272. Vijay SK, Mishra M, Kumar H, Tripathi K. Effect of pioglitazone and rosiglitazone on mediators of endothelial dysfunction, markers of angiogenesis and inflammatory cytokines in type-2 diabetes. *Acta Diabetol* 2009; 46: 27–33.

273. Tsimihodimos V, Kakafika A, Tambaki AP, Bairaktari E, Chapman MJ, Elisaf M, Tselepis AD. Fenofibrate induces HDL-associated PAF-AH but attenuates enzyme activity associated with apoB-containing lipoproteins. *J Lipid Res* 2003; 44: 927–934.

274. Barter PJ, Rye K-A. Is there a role for fibrates in the management of dyslipidemia in the Metabolic Syndrome ? *Arterioscler Thromb Vasc Biol* 2008; 28: 39–46.

275. Scott R, O'Brien R, Fulcher G, Pardy C, D'Emden M, Tse D, Taskinen MR, Ehnholm C, Keech A. Effects of fenofibrate treatment on cardiovascular disease risk in 9,795 individuals with type 2 diabetes and various components of the metabolic syndrome: the Fenofibrate Intervention and Event Lowering in Diabetes (FIELD) study. *Diabetes Care* 2009; 32: 493–498.

276. Bruckert E, Labreuche J, Deplanque D, Touboul PJ, Amarenco P. Fibrates effect on cardiovascular risk is greater in patients with high triglyceride levels or atherogenic dyslipidemia profile: a systematic review and metanalysis. *J Cardiovasc Pharmacol* 2011; 57: 262–272.

277. Saha SA, Arora RR. Fibrates in the prevention of cardiovascular disease in patients with type 2 diabetes mellitus–a pooled meta-analysis of randomized placebo-controlled clinical trials. *Int J Cardiol* 2010; 141: 157–166.

278. Goldenberg I, Boyko V, Tennenbaum A, Tanne D, Behar S, Guetta V. Long-term benefit of high-density lipoprotein cholesterol-raising therapy with bezafibrate: 16-year mortality follow-up of the bezafibrate infarction prevention trial. *Arch Intern Med* 2009; 169: 508–514.

279. Goldenberg I, Benderly M, Sidi R, Boyko V, Tenenbaum A, Tanne D, Behar S. Relation of clinical benefit of raising high-density lipoprotein cholesterol to serum levels of low-density lipoprotein cholesterol in patients with coronary heart disease (from the Bezafibrate Infarction Prevention Trial). *Am J Cardiol* 2009; 103: 41–45.

280. Asztalos BF, Collins D, Horvath KV, Bloomfield HE, Robins SJ, Schaefer EJ. Relation of gemfibrozil treatment and high-density lipoprotein subpopulation profile with cardiovascular events in the Veterans Affairs High-Density Lipoprotein Intervention Trial. *Metabolism* 2008; 57: 77–83.

281. Ruotolo G, Ericsson CG, Tettamanti C, Karpe F, Grip L, Svane B, Nilsson J, de Faire U, Hamsten A. Treatment effects on serum lipoprotein lipids, apolipoproteins and low density lipoprotein particle size and relationships of lipoprotein variables to progression of coronary artery disease in the Bezafibrate Coronary Atherosclerosis Intervention Trial (BECAIT). *J Am Coll Cardiol* 1998; 32: 1648–1656.

282. Syvanne M, Nieminen MS, Frick MH, Kauma H, Majahalme S, Virtanen V, Kesaniemi YA, Pasternack A, Ehnholm C, Taskinen MR. Associations between

lipoproteins and the progression of coronary and vein-graft atherosclerosis in a controlled trial with gemfibrozil in men with low baseline levels of HDL cholesterol. *Circulation* 1998; 98: 1993–1999.

283. Sirtori CR, Mombelli G. CETP antagonism versus agonism in cardiovascular prevention and plaque regression. *Clin Lipidology* 2009; 4: 63–78.

284. Mombelli G, Pazzucconi F, Bondioli A, Zanaboni A, Gaito S, Calabresi L, Sirtori CR. Paradoxical decrease in high-density lipoprotein cholesterol with fenofibrate: a quite rare phenomenon indeed. *Cardiovasc Ther* 2010; 28: 153–160.

285. Liu Y, Ordovas JM, Gao G, Province M, Straka RJ, Tsai MY, Lai CQ, Zhang K, Borecki I, Hixson JE et al. Pharmacogenetic association of the APOA1/C3/A4/A5 gene cluster and lipid responses to fenofibrate: the genetics of lipid-lowering drugs and diet network study. *Pharmacogenet Genomics* 2009; 19: 161–169.

286. Chapman MJ, Assmann G, Fruchart JC, Shepherd J, Sirtori C. Raising high-density lipoprotein cholesterol with reduction of cardiovascular risk: the role of nicotinic acid - a position paper developed by the European Consensus Panel on HDL-C. *Curr Med Res Opin* 2004; 20: 1253–1268.

287. Alrasadi K, Awan Z, Alwaili K, Ruel I, Hafiane A, Krimbou L, Genest J. Comparison of treatment of severe high-density lipoprotein cholesterol deficiency in men with daily atorvastatin (20mg) versus fenofibrate (200mg) versus extended-release niacin (2g). *Am J Cardiol* 2008; 102: 1341–1347.

288. Ganji SH, Kamanna VS, Kashyap ML. Niacin and cholesterol: role in cardiovascular disease (review). *J Nutr Biochem* 2003; 14: 298–305.

289. Rosenson RS. Antiatherothrombotic effects of nicotinic acid. *Atherosclerosis* 2003; 171: 87–96.

290. Karpe F, Frayn KN. The nicotinic acid receptor–a new mechanism for an old drug. *Lancet* 2004; 363: 1892–1894.

291. Meyers CD, Kamanna VS, Kashyap ML. Niacin therapy in atherosclerosis. *Curr Opin Lipidol* 2004; 15: 659–665.

292. Tunaru S, Kero J, Schaub A, Wufka C, Blaukat A, Pfeffer K, Offermanns S. PUMA-G and HM74 are receptors for nicotinic acid and mediate its anti-lipolytic effect. *Nat Med* 2003; 9: 352–355.

293. Wise A, Foord SM, Fraser NJ, Barnes AA, Elshourbagy N, Eilert M, Ignar DM, Murdock PR, Steplewski K, Green A et al. Molecular identification of high and low affinity receptors for nicotinic acid. *J Biol Chem* 2003; 278: 9869–9874.

294. Lamon-Fava S, Diffenderfer MR, Barrett PHR, Buchsbaum A, Nyaku M, Horvath KV, Asztalos BF, Otokozawa S, Ai M, Matthan NR et al. Extended-release niacin alters the metabolism of plasma apolipoprotein (apo) A-I and apoB-containing lipoproteins. *Arterioscler Thromb Vasc Biol* 2008; 28: 1672–1678.

295. Martres P. HM74a agonists: will they be the new generation of nicotinic acid ? *Curr Top Med Chem* 2009; 9: 428–435.

296. Kamanna VS, Kashyap ML. Nicotinic Acid (niacin) receptor agonists: will they be useful therapeutic agents ? *Am J Cardiol* 2007; 100: S53–S61.

297. Ganji SH, Tavintharan S, Zhu D, Xing Y, Kamanna VS, Kashyap ML. Niacin noncompetitively inhibits DGAT2 but not DGAT1 activity in HepG2 cells. *J Lipid Res* 2004; 45: 1835–1845.

298. van der Hoorn JWA, de Haan W, Berbee JFP, Havekes LM, Jukema JW, Rensen PCN, Princen HMG. Niacin increases HDL by reducing hepatic expression and plasma levels of cholesteryl ester transfer protein in APOE*3Leiden.CETP mice. *Arterioscler Thromb Vasc Biol* 2008; 28: 2016–2022.

299. Sakai T, Kamanna VS, Kashyap ML. Niacin, but not gemfibrozil, selectively increases LP-AI, a cardioprotective subfraction of HDL, in patients with low HDL cholesterol. *Arterioscler Thromb Vasc Biol* 2001; 21: 1783–1789.

300. Zhang LH, Kamanna VS, Zhang MC, Kashyap ML. Niacin inhibits surface expression of ATP synthase beta chain in HepG2 cells: implications for raising HDL. *J Lipid Res* 2008; 49: 1195–1201.

301. Rubic T, Trottmann M, Lorenz RL. Stimulation of CD36 and the key effector of reverse cholesterol transport ATP-binding cassette A1 in monocytoid cells by niacin. *Biochem Pharmacol* 2004; 67: 411–419.

302. Wu ZH, Zhao SP. Niacin promotes cholesterol efflux through stimulation of the PPARgamma-LXRalpha-ABCA1 pathway in 3T3-L1 adipocytes. *Pharmacology* 2009; 84: 282–287.

303. McKenney JM, McCormick LS, Schaefer EJ, Black DM, Watkins ML. Effect of niacin and atorvastatin on lipoprotein subclasses in patients with atherogenic dyslipidemia. *Am J Cardiol* 2001; 88: 270–274.

304. Morgan JM, Capuzzi DM, Baksh RI, Intenzo C, Carey CM, Reese D, Walker K. Effects of extended-release niacin on lipoprotein subclass distribution. *Am J Cardiol* 2003; 91: 1432–1436.

305. Kamanna VS, Kashyap ML. Mechanism of action of niacin. *Am J Cardiol* 2008; 101: 20B–26B.

306. Pan J, Shilian P, Ishida B, Wu X, Kane JP, Malloy MJ, Charles MA. Effect of niacin on prebeta-1 high-density lipoprotein levels in diabetes. *Metabolism* 2011; 60: 292–297.

307. Vogt A, Kassner U, Hostalek U, Steinhagen-Thiessen E. Correction of low HDL cholesterol to reduce cardiovascular risk: practical considerations relating to the therapeutic use of prolonged-release nicotinic acid (Niaspan). *Int J Clin Pract* 2007; 61: 1914–1921.

308. Bays H. Safety of niacin and simvastatin combination therapy. *Am J Cardiol* 2008; 101: 3B–8B.

309. Linke A, Sonnabend M, Fasshauer M, Hollriegel R, Schuler G, Niebauer J, Stumvoll M, Bluher M. Effects of extended-release niacin on lipid profile and adipocyte biology in patients with impaired glucose tolerance. *Atherosclerosis* 2009; 205: 207–213.

310. Thoenes M, Oguchi A, Nagamia S, Vaccari CS, Hammoud R, Umpierrez GE, Khan BV. The effects of extended-release niacin on carotid intimal media thickness, endothelial function and inflammatory markers in patients with the metabolic syndrome. *Int J Clin Pract* 2007; 61: 1942–1948.

311. Drew BG, Duffy SJ, Forbes JM, Chin-Dusting J, Kaye DM, Kemp BE, Sviridov D, Steinberg GR, Kingwell BA. High density lipoprotein modulates glucose metabolism by multiple mechanisms. *Circulation* 2009; 120 (Suppl.): S468–c–S469.

312. Digby JE, McNeill E, Dyar OJ, Lam V, Greaves DR, Choudhury RP. Anti-inflammatory effects of nicotinic acid in adipocytes demonstrated by suppression of fractalkine, RANTES, and MCP-1 and upregulation of adiponectin. *Atherosclerosis* 2010; 209: 89–95.

313. Westphal S, Luley C. Preferential increase in high-molecular weight adiponectin after niacin. *Atherosclerosis* 2008; 198: 179–183.

314. Kaushik SV, Plaisance EP, Kim T, Huang EY, Mahurin AJ, Grandjean PW, Mathews ST. Extended-release niacin decreases serum fetuin-A concentrations in individuals with metabolic syndrome. *Diabetes Metab Res Rev* 2009; 25: 427–434.

315. The Coronary Drug Project Research Group. Clofibrate and niacin in coronary heart disease. *JAMA* 1975; 231: 360–381.

316. Taylor AJ, Lee HJ, Sullenberger LE. The effect of 24 months of combination statin and extended-release niacin on carotid intima-media thickness: ARBITER 3. *Curr Med Res Opin* 2006; 22: 2243–2250.

317. Villines TC, Stanek EJ, Devine PJ, Turco M, Miller M, Weissman NJ, Griffen L, Taylor AJ. The ARBITER 6-HALTS Trial (Arterial Biology for the Investigation of the Treatment Effects of Reducing Cholesterol 6-HDL and LDL Treatment Strategies in Atherosclerosis): final results and the impact of medication adherence, dose, and treatment duration. *J Am Coll Cardiol* 2010; 55: 2721–2726.

318. Blumenthal RS, Michos ED. The HALTS trial–halting atherosclerosis or halted too early ? *N Engl J Med* 2009; 361: 2178–2180.

319. Lee JM, Robson MD, Yu LM, Shirodaria CC, Cunnington C, Kylintireas I, Digby JE, Bannister T, Handa A, Wiesmann F et al. Effects of high-dose modified-release nicotinic acid on atherosclerosis and vascular function: a randomized, placebo-controlled, magnetic resonance imaging study. *J Am Coll Cardiol* 2009; 54: 1787–1794.

320. Bruckert E, Labreuche J, Amarenco P. Meta-analysis of the effect of nicotinic acid alone or in combination on cardiovascular events and atherosclerosis. *Atherosclerosis* 2010; 210: 353–361.

321. Duggal JK, Singh M, Attri N, Singh PP, Ahmed N, Pahwa S, Molnar J, Singh S, Khosla S, Arora R. Effect of niacin therapy on cardiovascular outcomes in patients with coronary artery disease. *J Cardiovasc Pharmacol Ther* 2010; 15: 158–166.

322. Warnholtz A, Wild P, Ostad MA, Elsner V, Stieber F, Schinzel R, Walter U, Peetz D, Lackner K, Blankenberg S et al. Effects of oral niacin on endothelial dysfunction in patients with coronary artery disease: results of the randomized, double-blind, placebo-controlled INEF study. *Atherosclerosis* 2009; 204: 216–221.

323. Sorrentino SA, Besler C, Rohrer L, Meyer M, Heinrich K, Bahlmann FH, Mueller M, Horvath T, Doerries C, Heinemann M et al. Endothelial-vasoprotective effects of high-density lipoprotein are impaired in patients with type 2 diabetes mellitus but are improved after extended-release niacin therapy. *Circulation* 2010; 121: 110–122.

324. Chow DC, Stein JH, Seto TB, Mitchell C, Sriratanaviriyakul N, Grandinetti A, Gerschenson M, Shiramizu B, Souza S, Shikuma C. Short-term effects of extended-release niacin on endothelial function in HIV-infected patients on stable antiretroviral therapy. *AIDS* 2010; 24: 1019–1023.

325. Ganji SH, Qin S, Zhang L, Kamanna VS, Kashyap ML. Niacin inhibits vascular oxidative stress, redox-sensitive genes, and monocyte adhesion to human aortic endothelial cells. *Atherosclerosis* 2009; 202: 68–75.

326. Tavintharan S, Woon K, Pek LT, Jauhar N, Dong X, Lim SC, Sum CF. Niacin results in reduced monocyte adhesion in patients with type 2 diabetes mellitus. *Atherosclerosis* 2011; 215: 176–179.

327. Lukasova M, Malaval C, Gille A, Kero J, Offermanns S. Nicotinic acid inhibits progression of atherosclerosis in mice through its receptor GPR109A expressed by immune cells. *J Clin Invest* 2011; 121: 1163–1173.

328. Wu BJ, Yan L, Charlton F, Witting P, Barter PJ, Rye K-A. Evidence that niacin inhibits acute vascular inflammation and improves endothelial dysfunction independent of changes in plasma lipids. *Arterioscler Thromb Vasc Biol* 2010; 30: 968–975.

329. Shehadah A, Chen J, Zacharek A, Cui Y, Ion M, Roberts C, Kapke A, Chopp M. Niaspan treatment induces neuroprotection after stroke. *Neurobiol Dis* 2010; 40: 277–283.

330. Serebruany V, Malinin A, Aradi D, Kuliczkowski W, Norgard NB, Boden WE. The *in vivo* effects of niacin on platelet biomarkers in human volunteers. *Thromb Haemost* 2010; 104: 311–317.

331. Kamanna VS, Ganji SH, Kashyap ML. Niacin: an old drug rejuvenated. *Curr Atheroscler Rep* 2009; 11: 45–51.

332. Walters RW, Shukla AK, Kovacs JJ, Violin JD, DeWire SM, Lam CM, Chen JR, Muehlbauer MJ, Whalen EJ, Lefkowitz RJ. Beta-Arrestin1 mediates nicotinic acid-induced flushing, but not its antilipolytic effect, in mice. *J Clin Invest* 2009; 119: 1312–1321.

333. Alves JD, Steinhagen-Thiessen E, Darioli R, Hostalek U, Vogt A. Influence of the timing of low-dose aspirin on tolerability of prolonged-release nicotinic acid in patients at elevated cardiovascular risk. *Curr Med Res Opin* 2008; 24: 2815–2820.

334. Kalogeromitros D, Makris M, Chliva C, Aggelides X, Kempuraj D, Theoharides TC. A quercetin containing supplement reduces niacin-induced flush in humans. *Int J Immunopathol Pharmacol* 2008; 21: 509–514.

335. Papaliodis D, Boucher W, Kempuraj D, Theoharides TC. The flavonoid luteolin inhibits niacin-induced flush. *Br J Pharmacol* 2008; 153: 1382–1387.

336. Paolini JF, Mitchel YB, Reyes R, Kher U, Lai E, Watson DJ, Norquist JM, Meehan AG, Bays HE, Davidson M et al. Effects of laropiprant on nicotinic acid-induced flushing in patients with dyslipidemia. *Am J Cardiol* 2008; 101: 625–630.

337. Lai E, De Lepeleire I, Crumley TM, Liu F, Wenning LA, Michiels N, Vets E, O'Neill G, Wagner JA, Gottesdiener K. Suppression of niacin-induced vasodilation with an antagonist to prostaglandin D2 receptor subtype 1. *Clin Pharmacol Ther* 2007; 81: 849–857.

338. Maccubbin D, Bays HE, Olsson AG, Elinoff V, Elis A, Mitchel Y, Sirah W, Betteridge A, Reyes R, Yu Q et al. Lipid-modifying efficacy and tolerability of extended-release niacin/laropiprant in patients with primary hypercholesterolaemia or mixed dyslipidaemia. *Int J Clin Pract* 2008; 62: 1959–1970.

339. Hussein AA. Nicholls SJ. Critical appraisal of laropiprant and extended-release niacin combination in the management of mixed dyslipidemias and primary hypercholesterolemia. *Ther Clin Risk Manag* 2010; 6: 183–190.

340. Shen HC. Acyl hydroxypyrazoles as novel agonists for high-affinity nicotinic acid receptor GPR109A: WO2008051403. *Expert Opin Ther Pat* 2009; 19: 1149–1155.

341. Imbriglio JE, Chang S, Liang R, Raghavan S, Schmidt D, Smenton A, Tria S, Schrader TO, Jung JK, Esser C et al. GPR109a agonists. Part 1: 5-Alkyl and 5-aryl-pyrazole-tetrazoles as agonists of the human orphan G-protein coupled receptor GPR109a. *Bioorg Med Chem Lett* 2009; 19: 2121–2124.

342. Schmidt D, Smenton A, Raghavan S, Carballo-Jane E, Lubell S, Ciecko T, Holt TG, Wolff M, Taggart A, Wilsie L et al. Pyrazole acids as niacin receptor agonists for the treatment of dyslipidemia. *Bioorg Med Chem Lett* 2009; 19: 4768–4772.

343. Le Goff W, Guerin M, Chapman MJ. Pharmacological modulation of cholesteryl ester transfer protein, a new therapeutic target in atherogenic dyslipidemia. *Pharmacol Ther* 2004; 101: 17–38.

344. Curb JD, Abbott RD, Rodriguez BL, Masaki K, Chen R, Sharp DS, Tall AR. A prospective study of HDL-C and cholesteryl ester transfer protein gene mutations and the risk of coronary heart disease in the elderly. *J Lipid Res* 2004; 45: 948–953.

345. Ridker PM, Pare G, Parker AN, Zee RYL, Miletich JP, Chasman DI. Polymorphism in the CETP gene region, HDL cholesterol, and risk of future myocardial infarction: Genomewide analysis among 18 245 initially healthy women from the Women's Genome Health Study. *Circ Cardiovasc Genet* 2009; 2: 26–33.

346. Asztalos BF, Horvath KV, Kajinami K, Nartsupha C, Cox CE, Batista M, Schaefer EJ, Inazu A, Mabuchi H. Apolipoprotein composition of HDL in cholesteryl ester transfer protein deficiency. *J Lipid Res* 2004; 45: 448–455.

347. Brewer HB Jr. Increasing HDL cholesterol levels. *N Engl J Med* 2004; 350: 1491–1494.

348. Gaofu Q, Jun L, Xiuyun Z, Wentao L, Jie W, Jingjing L. Antibody against cholesteryl ester transfer protein (CETP) elicited by a recombinant chimeric enzyme vaccine attenuated atherosclerosis in a rabbit model. *Life Sci* 2005; 77: 2690–2702.

349. Nicholls SJ, Rodes-Cabau J, Despres J-P, Brennan D, Bayturan O, Uno K, Shreevatsa A, Kastelein JJ, Gaudin C, Job B et al. High cholesteryl ester transfer protein mass predicts accelerated progression of coronary atherosclerosis in abdominally obese patients. *Circulation* 2009; 120: S427.

350. Barter PJ, Caulfield M, Eriksson M, Grundy SM, Kastelein JJ, Komajda M, Lopez-Sendon J, Mosca L, Tardif JC, Waters DD et al. Effects of torcetrapib in patients at high risk for coronary events. *N Engl J Med* 2007; 357: 2109–2122.

351. Rittershaus CW, Miller DP, Thomas LJ, Picard MD, Honan CM, Emmett CD, Pettey CL, Adari H, Hammond RA, Beattie DT et al. Vaccine-induced antibodies inhibit CETP activity in vivo and reduce aortic lesions in a rabbit model of atherosclerosis. *Arterioscler Thromb Vasc Biol* 2000; 20: 2106–2112.

352. Davidson MH, Maki K, Umporowicz D, Wheeler A, Rittershaus C, Ryan U. The safety and immunogenicity of a CETP vaccine in healthy adults. *Atherosclerosis* 2003; 169: 113–120.

353. Potter LK, Sprecher DL, Walker MC, Tobin FL. Mechanism of inhibition defines CETP activity: a mathematical model for CETP *in vitro*. *J Lipid Res* 2009; 50: 2222–2234.

354. Bloomfield D, Carlson GL, Sapre A, Tribble D, McKenney JM, Littlejohn TW III, Sisk CM, Mitchel Y, Pasternak RC. Efficacy and safety of the cholesteryl ester transfer protein inhibitor anacetrapib as monotherapy and coadministered with atorvastatin in dyslipidemic patients. *Am Heart J* 2009; 157: 352–360 e352.

355. Maugeais C, Magg C, Dernick G, Matile H, von der Mark E, Pflieger P, Huber W, Thoma R, Kakutani M, Okamoto H et al. Dalcetrapib binds to and changes the conformation of CETP in a unique manner (differing to that observed with torcetrapib). *Circulation* 2009; 120 (Suppl.): S445.

356. Ranalletta M, Bierilo KK, Chen Y, Milot D, Chen Q, Tung E, Houde C, Elowe NH, Garcia-Calvo M, Porter G et al. Biochemical characterization of cholesteryl ester transfer protein inhibitors. *J Lipid Res* 2010; 51: 2739–2752.

357. Niesor EJ, von der Marck E, Brousse M, Maugeais C. Inhibition of cholesteryl ester transfer protein (CETP): Different *in vitro* characteristics of RO4607381/JTT-705 and torcetrapib (TOR). *Atherosclerosis* 2008; 199: 231–259.

358. Niesor EJ, Magg C, Ogawa N, Okamoto H, von der Mark E, Matile H, Schmid G, Clerc RG, Chaput E, Blum-Kaelin D et al. Modulating cholesteryl ester transfer protein activity maintains efficient pre-beta-HDL formation and increases reverse cholesterol transport. *J Lipid Res* 2010; 51: 3443–3454.

359. Rano TA, Sieber-McMaster E, Pelton PD, Yang M, Demarest KT, Kuo GH. Design and synthesis of potent inhibitors of cholesteryl ester transfer protein (CETP) exploiting a 1,2,3,4-tetrahydroquinoline platform. *Bioorg Med Chem Lett* 2009; 19: 2456–2460.

360. Abu Sheikha G, Abu Khalaf R, Melhem A, Albadawi G. Design, synthesis, and biological evaluation of benzylamino-methanone based cholesteryl ester transfer protein inhibitors. *Molecules* 2010; 15: 5721–5733.

361. Wang A, Prouty CP, Pelton PD, Yong M, Demarest KT, Murray WV, Kuo GH. Synthesis and discovery of 2,3-dihydro-3,8-diphenylbenzo[1,4]oxazines as a novel class of potent cholesteryl ester transfer protein inhibitors. *Bioorg Med Chem Lett* 2010; 20: 1432–1435.

362. Kallashi F, Kim D, Kowalchick J, Park YJ, Hunt JA, Ali A, Smith CJ, Hammond ML, Pivnichny JV, Tong X et al. 2-Arylbenzoxazoles as CETP inhibitors: raising HDL-C in cynoCETP transgenic mice. *Bioorg Med Chem Lett* 2011; 21: 558–561.

363. Abu Khalaf R, Abu Sheikha G, Bustanji Y, Taha MO. Discovery of new cholesteryl ester transfer protein inhibitors via ligand-based pharmacophore modeling and QSAR analysis followed by synthetic exploration. *Eur J Med Chem* 2010; 45: 1598–1617.

364. Tory R, Sachs-Barrable K, Hill JS, Wasan KM. Cyclosporine A and Rapamycin induce *in vitro* cholesteryl ester transfer protein activity, and suppress lipoprotein lipase activity in human plasma. *Int J Pharm* 2008; 358: 219–223.

365. Clark RW, Sutfin TA, Ruggeri RB, Willauer AT, Sugarman ED, Magnus-Aryitey G, Cosgrove PG, Sand TM, Wester RT, Williams JA et al. Raising high-density lipoprotein in humans through inhibition of cholesteryl ester transfer protein: an initial multidose study of torcetrapib. *Arterioscler Thromb Vasc Biol* 2004; 24: 490–497.

366. Brousseau ME, Schaefer EJ, Wolfe ML, Bloedon LT, Digenio AG, Clark RW, Mancuso JP, Rader DJ. Effects of an inhibitor of cholesteryl ester transfer protein on HDL cholesterol. *N Engl J Med* 2004; 350: 1505–1515.

367. Brousseau ME, Diffenderfer MR, Millar JS, Nartsupha C, Asztalos BF, Welty FK, Wolfe ML, Rudling M, Bjorkhem I, Angelin B et al. Effects of cholesteryl ester transfer protein inhibition on high-density lipoprotein subspecies, apolipoprotein A-I metabolism, and fecal sterol excretion. *Arterioscler Thromb Vasc Biol* 2005; 25: 1057–1064.

368. Brousseau ME, Millar JS, Diffenderfer MR, Nartsupha C, Asztalos BF, Wolfe ML, Mancuso JP, Digenio AG, Rader DJ, Schaefer EJ. Effects of cholesteryl ester transfer protein inhibition on apolipoprotein A-II-containing HDL subspecies and apolipoprotein A-II metabolism. *J Lipid Res* 2009; 50: 1456–1462.

369. de Vries-van der Weij J, Zadelaar S, Toet K, Havekes LM, Kooistra T, Rensen PC. Human CETP aggravates atherosclerosis by increasing VLDL-cholesterol rather than by decreasing HDL-cholesterol in APOE*3-Leiden mice. *Atherosclerosis* 2009; 206: 153–158.

370. Millar JS, Brousseau ME, Diffenderfer MR, Barrett PH, Welty FK, Faruqi A, Wolfe ML, Nartsupha C, Digenio AG, Mancuso JP et al. Effects of the cholesteryl ester transfer protein inhibitor torcetrapib on apolipoprotein B100 metabolism in humans. *Arterioscler Thromb Vasc Biol* 2006; 26: 1350–1356.

371. Millar JS, Brousseau ME, Diffenderfer MR, Barrett PHR, Welty FK, Cohn JS, Wilson A, Wolfe ML, Nartsupha C, Schaefer PM et al. Effects of the cholesteryl ester transfer protein inhibitor torcetrapib on VLDL apolipoprotein E metabolism. *J Lipid Res* 2008; 49: 543–549.

372. Guerin M, Le Goff W, Duchene E, Julia Z, Nguyen T, Thuren T, Shear CL, Chapman MJ. Inhibition of CETP by torcetrapib attenuates the atherogenicity of postprandial TG-rich lipoproteins in type IIB hyperlipidemia. *Arterioscler Thromb Vasc Biol* 2008; 28: 148–154.

373. Inazu A, Nakajima K, Nakano T, Niimi M, Kawashiri MA, Nohara A, Kobayashi J, Mabuchi H. Decreased post-prandial triglyceride response and diminished remnant lipoprotein formation in cholesteryl ester transfer protein (CETP) deficiency. *Atherosclerosis* 2008; 196: 953–957.

374. de Grooth GJ, Kuivenhoven JA, Stalenhoef AF, de Graaf J, Zwinderman AH, Posma JL, van Tol A, Kastelein JJ. Efficacy and safety of a novel cholesteryl ester transfer protein inhibitor, JTT-705, in humans: a randomized phase II dose-response study. *Circulation* 2002; 105: 2159–2165.

375. Kuivenhoven JA, de Grooth GJ, Kawamura H, Klerkx AH, Wilhelm F, Trip MD, Kastelein JJP. Effectiveness of inhibition of cholesteryl ester transfer protein by JTT-705 in combination with pravastatin in Type II dyslipidemia. *Am J Cardiol* 2005; 95: 1085–1088.

376. Bisoendial RJ, Hovingh GK, El Harchaoui K, Levels JHM, Tsimikas S, Pu K, Zwinderman AE, Kuivenhoven JA, Kastelein JJP, Stroes ESG. Consequences of cholesteryl ester transfer protein inhibition in patients with familial hypoalphalipoproteinemia. *Arterioscler Thromb Vasc Biol* 2005; 25: e133–e134.

377. Stein EA, Stroes ES, Steiner G, Buckley BM, Capponi AM, Burgess T, Niesor EJ, Kallend D, Kastelein JJ. Safety and tolerability of dalcetrapib. *Am J Cardiol* 2009; 104: 82–91.

378. Stein EA, Roth EM, Rhyne JM, Burgess T, Kallend D, Robinson JG. Safety and tolerability of dalcetrapib (RO4607381/JTT-705): results from a 48-week trial. *Eur Heart J* 2010; 31: 480–488.

379. Hermann F, Enseleit F, Spieker LE, Periat D, Sudano I, Hermann M, Corti R, Noll G, Ruschitzka F, Luscher TF. Cholesterylestertransfer protein inhibition and endothelial function in type II hyperlipidemia. *Thromb Res* 2009; 123: 460–465.

380. Zhang B, Fan P, Shimoji E, Xu H, Takeuchi K, Bian C, Saku K. Inhibition of cholesteryl ester transfer protein activity by JTT-705 increases apolipoprotein E-containing high-density lipoprotein and favorably affects the function and enzyme composition of high-density lipoprotein in rabbits. *Arterioscler Thromb Vasc Biol* 2004; 24: 1910–1915.

381. Kobayashi J, Okamoto H, Otabe M, Bujo H, Saito Y. Effect of HDL, from Japanese white rabbit administered a new cholesteryl ester transfer protein inhibitor JTT-705, on cholesteryl ester accumulation induced by acetylated low density lipoprotein in J774 macrophage. *Atherosclerosis* 2002; 162: 131–135.

382. Krishna R, Anderson MS, Bergman AJ, Jin B, Fallon M, Cote J, Rosko K, Chavez-Eng C, Lutz R, Bloomfield DM et al. Effect of the cholesteryl ester transfer protein inhibitor, anacetrapib, on lipoproteins in patients with dyslipidaemia and on 24-h ambulatory blood pressure in healthy individuals: two double-blind, randomised placebo-controlled phase I studies. *Lancet* 2007; 370: 1907–1914.

383. Rader DJ. Illuminating HDL–is it still a viable therapeutic target ? *N Engl J Med* 2007; 357: 2180–2183.

384. Schwartz CC, VandenBroek JM, Cooper PS. Lipoprotein cholesteryl ester production, transfer, and output in vivo in humans. *J Lipid Res* 2004; 45: 1594–1607.

385. Kontush A, Guerin M, Chapman MJ. Spotlight on HDL-raising therapies: insights from the torcetrapib trials. *Nat Clin Pract Cardiovasc Med* 2008; 5: 329–336.

386. Guerin M, Dolphin PJ, Chapman MJ. Preferential cholesteryl ester acceptors among the LDL subspecies of subjects with familial hypercholesterolemia. *Arterioscler Thromb* 1994; 14: 679–685.

387. Yvan-Charvet L, Matsuura F, Wang N, Bamberger MJ, Nguyen T, Rinninger F, Jiang XC, Shear CL, Tall AR. Inhibition of cholesteryl ester transfer protein by torcetrapib modestly increases macrophage cholesterol efflux to HDL. *Arterioscler Thromb Vasc Biol* 2007; 27: 1132–1138.

388. Thuahnai ST, Lund-Katz S, Dhanasekaran P, de la Llera-Moya M, Connelly MA, Williams DL, Rothblat GH, Phillips MC. Scavenger receptor class B type I-mediated cholesteryl ester-selective uptake and efflux of unesterified cholesterol. Influence of high density lipoprotein size and structure. *J Biol Chem* 2004; 279: 12448–12455.

389. Tchoua U, D'Souza W, Mukhamedova N, Blum D, Niesor E, Mizrahi J, Maugeais C, Sviridov D. The effect of cholesteryl ester transfer protein overexpression and inhibition on reverse cholesterol transport. *Cardiovasc Res* 2008; 77: 732–739.

390. Niesor EJ, Okamoto H, Maugeais C, von der Mark E, Brousse M, Blum-Kaelin D, Chaput E, Kakutani M. The effects of dalcetrapib on macrophage reverse cholesterol transport in a rodent model. *Circulation* 2009; 120 (Suppl.): S468.

391. Tall AR. The effects of cholesterol ester transfer protein inhibition on cholesterol efflux. *Am J Cardiol* 2009; 104: 39E–45E.

392. Plengpanich W, Le Goff W, Poolsuk S, Julia Z, Guerin M, Khovidhunkit W. CETP deficiency due to a novel mutation in the CETP gene promoter and its effect on cholesterol efflux and selective uptake into hepatocytes. *Atherosclerosis* 2011; 216: 370–373.

393. Miwa K, Inazu A, Kawashiri M, Nohara A, Higashikata T, Kobayashi J, Koizumi J, Nakajima K, Nakano T, Niimi M et al. Cholesterol efflux from J774 macrophages and Fu5AH hepatoma cells to serum is preserved in CETP-deficient patients. *Clin Chim Acta* 2009; 402: 19–24.

394. Rotllan N, Calpe-Berdiel L, Guillaumet-Adkins A, Suren-Castillo S, Blanco-Vaca F, Escola-Gil JC. CETP activity variation in mice does not affect two major HDL antiatherogenic properties: macrophage-specific reverse cholesterol transport and LDL antioxidant protection. *Atherosclerosis* 2008; 196: 505–513.

395. Rhainds D, Brodeur M, Lapointe J, Charpentier D, Falstrault L, Brissette L. The role of human and mouse hepatic scavenger receptor class B type I (SR-BI) in the selective uptake of low-density lipoprotein-cholesteryl esters. *Biochemistry* 2003; 42: 7527–7538.

396. Greene DJ, Skeggs JW, Morton RE. Elevated triglyceride content diminishes the capacity of high density lipoprotein to deliver cholesteryl esters via the scavenger receptor class B type I (SR-BI). *J Biol Chem* 2001; 276: 4804–4811.

397. Gauthier A, Lau P, Zha X, Milne R, McPherson R. Cholesteryl ester transfer protein directly mediates selective uptake of high density lipoprotein cholesteryl esters by the liver. *Arterioscler Thromb Vasc Biol* 2005; 25: 2177–2184.

398. Okamoto H, Ogawa N, Takahashi D, Niesor EJ, Kakutani M. Dalcetrapib mediated increase in macrophage reverse cholesterol transport does not correlate with raised HDL-C in hamsters. *Circulation* 2010; 122 (Suppl.): A23016.

399. Johns DG, Gagen K, Castro-Perez J, Roddy TA, Hubbard BK. A model to predict clinical efficacy of CETP inhibition: Anacetrapib elevates HDL, lowers LDL and increases sterol excretion in the Syrian golden hamster. *Circulation* 2010; 122 (Suppl.): A18072.

400. Vergeer M, Bots ML, van Leuven SI, Basart DC, Sijbrands EJ, Evans GW, Grobbee DE, Visseren FL, Stalenhoef AF, Stroes ES et al. Cholesteryl ester transfer protein inhibitor torcetrapib and off-target toxicity: a pooled analysis of the rating atherosclerotic disease change by imaging with a new CETP inhibitor (RADIANCE) trials. *Circulation* 2008; 118: 2515–2522.

401. Sirtori CR, Mombelli G. Viability of developing CETP inhibitors. *Cardiovasc Ther* 2008; 26: 135–146.

402. Vasan RS, Pencina MJ, Robins SJ, Zachariah JP, Kaur G, D'Agostino RB, Ordovas JM. Association of circulating cholesteryl ester transfer protein activity with incidence of cardiovascular disease in the community. *Circulation* 2009; 120: 2414–2420.

403. Ritsch A, Scharnagl H, Eller P, Tancevski I, Duwensee K, Demetz E, Sandhofer A, Boehm BO, Winkelmann BR, Patsch JR et al. Cholesteryl ester transfer protein and mortality in patients undergoing coronary angiography: The Ludwigshafen Risk and Cardiovascular Health Study. *Circulation* 2010; 121: 366–374.

404. Khera AV, Wolfe ML, Cannon CP, Qin J, Rader DJ. On-statin cholesteryl ester transfer protein mass and risk of recurrent coronary events (from the pravastatin or atorvastatin evaluation and infection therapy-thrombolysis in myocardial infarction 22 [PROVE IT-TIMI 22] study). *Am J Cardiol* 2010; 106: 451–456.

405. Blasi E, Bamberger M, Knight D, Engwall M, Wolk R, Winter S, Betts A, John-Baptiste A, Keiser J. Effects of CP-532,623 and torcetrapib, cholesteryl ester transfer protein inhibitors, on arterial blood pressure. *J Cardiovasc Pharmacol* 2009; 53: 507–516.

406. Chapman MJ. Therapeutic elevation of HDL-cholesterol to prevent atherosclerosis and coronary heart disease. *Pharmacol Ther* 2006; 111: 893–908.

407. Xing Y, Cohen A, Rothblat G, Sankaranarayanan S, Weibel G, Royer L, Francone OL, Rainey WE. Aldosterone production in human adrenocortical cells is stimulated by high-density lipoprotein 2 (HDL2) through increased expression of aldosterone synthase (CYP11B2). *Endocrinology* 2011; 152: 751–763.

408. Forrest MJ, Bloomfield D, Briscoe RJ, Brown PN, Cumiskey AM, Ehrhart J, Hershey JC, Keller WJ, Ma X, McPherson HE et al. Torcetrapib-induced blood pressure elevation is independent of CETP inhibition and is accompanied by increased circulating levels of aldosterone. *Br J Pharmacol* 2008; 154: 1465–1473.

409. Krikken JA, Dallinga-Thie GM, Navis G, Dullaart RP. Renin-angiotensin-aldosterone responsiveness to low sodium and blood pressure reactivity to angiotensin-II are unrelated to cholesteryl ester transfer protein mass in healthy subjects. *Expert Opin Ther Targets* 2008; 12: 1321–1328.
410. Sofat R, Hingorani AD, Smeeth L, Humphries SE, Talmud PJ, Cooper J, Shah T, Sandhu MS, Ricketts SL, Boekholdt SM et al. Separating the mechanism-based and off-target actions of cholesteryl ester transfer protein inhibitors with CETP gene polymorphisms. *Circulation* 2010; 121: 52–62.
411. Zachariah JP, Pencina MJ, Lyass A, Kaur G, D'Agostino RB, Ordovas JM, Vasan RS. Circulating plasma cholesteryl ester transfer protein activity and blood pressure tracking in the community. *J Hypertens* 2011; 29: 863–868
412. Kastelein JJ, van Leuven SI, Burgess L, Evans GW, Kuivenhoven JA, Barter PJ, Revkin JH, Grobbee DE, Riley WA, Shear CL et al. Effect of torcetrapib on carotid atherosclerosis in familial hypercholesterolemia. *N Engl J Med* 2007; 356: 1620–1630.
413. Nissen SE, Tardif JC, Nicholls SJ, Revkin JH, Shear CL, Duggan WT, Ruzyllo W, Bachinsky WB, Lasala GP, Tuzcu EM. Effect of torcetrapib on the progression of coronary atherosclerosis. *N Engl J Med* 2007; 356: 1304–1316.
414. Nicholls SJ, Brennan DM, Wolski K, Kalidindi SR, Moon K-W, Tuzcu EM, Nissen SE. Abstract 684: Changes in levels of high density lipoprotein cholesterol predict the impact of torcetrapib on progression of coronary atherosclerosis: Insights from ILLUSTRATE. *Circulation* 2007; 116: II_127.
415. Nicholls SJ, Tuzcu EM, Brennan DM, Tardif JC, Nissen SE. Cholesteryl ester transfer protein inhibition, high-density lipoprotein raising, and progression of coronary atherosclerosis: insights from ILLUSTRATE (Investigation of Lipid Level Management Using Coronary Ultrasound to Assess Reduction of Atherosclerosis by CETP Inhibition and HDL Elevation). *Circulation* 2008; 118: 2506–2514.
416. Burillo E, Andres EM, Mateo-Gallego R, Fiddyment S, Jarauta E, Cenarro A, Civeira F. High-density lipoprotein cholesterol increase and non-cardiovascular mortality: a meta-analysis. *Heart* 2010; 96: 1345–1351.
417. Burillo E, Andres EM, Mateo-Gallego R, Fiddyment S, Jarauta E, Cenarro A, Civeira F. Increase in high-density lipoprotein cholesterol is not related with higher non-cardiovascular mortality; systematic review and meta-regression analysis. *Circulation* 2009; 120 (Suppl.): S406–c–407.
418. Connelly MA, Parry TJ, Giardino EC, Huang Z, Cheung WM, Chen C, Cools F, Van der Linde H, Gallacher DJ, Kuo GH et al. Torcetrapib produces endothelial dysfunction independent of cholesteryl ester transfer protein inhibition. *J Cardiovasc Pharmacol* 2010; 55: 459–468.
419. Duriez P. CETP inhibition. *Lancet* 2007; 370: 1882–1883.
420. Miida T, Seino U, Miyazaki O, Hanyu O, Hirayama S, Saito T, Ishikawa Y, Akamatsu S, Nakano T, Nakajima K et al. Probucol markedly reduces HDL phospholipids and elevated prebeta1-HDL without delayed conversion into alpha-migrating HDL: putative role of angiopoietin-like protein 3 in probucol-induced HDL remodeling. *Atherosclerosis* 2008; 200: 329–335.
421. Clark RW, Cunningham D, Cong Y, Subashi TA, Tkalcevic GT, Lloyd DB, Boyd JG, Chrunyk BA, Karam GA, Qiu X et al. Assessment of cholesteryl ester transfer protein inhibitors for interaction with proteins involved in the immune response to infection. *J Lipid Res* 2010; 51: 967–974.

422. Kontush A, Chapman MJ. Why is HDL functionally deficient in type 2 diabetes ? *Curr Diab Rep* 2008; 8: 51–59.

423. Sugano M, Sawada S, Tsuchida K, Makino N, Kamada M. Low density lipoproteins develop resistance to oxidative modification due to inhibition of cholesteryl ester transfer protein by a monoclonal antibody. *J Lipid Res* 2000; 41: 126–133.

424. Orloff DG. Regulatory considerations in the development of high-density lipoprotein therapies. *Am J Cardiol* 2007; 100: S10–S14.

425. Mitka M. CETP inhibition shows promise as way to reduce cardiovascular disease risk. *JAMA* 2011; 305: 136–137.

426. Cannon CP, Dansky HM, Davidson M, Gotto AM Jr, Brinton EA, Gould AL, Stepanavage M, Liu SX, Shah S, Rubino J et al. Design of the DEFINE trial: determining the EFficacy and tolerability of CETP INhibition with AnacEtrapib. *Am Heart J* 2009; 158: 513–519 e513.

427. Schwartz GG, Olsson AG, Ballantyne CM, Barter PJ, Holme IM, Kallend D, Leiter LA, Leitersdorf E, McMurray JJ, Shah PK et al. Rationale and design of the dal-OUTCOMES trial: efficacy and safety of dalcetrapib in patients with recent acute coronary syndrome. *Am Heart J* 2009; 158: 896–901 e893.

428. Robinson JG. Dalcetrapib: a review of Phase II data. *Expert Opin Investig Drugs* 2010; 19: 795–805.

429. Cannon CP, Shah S, Dansky HM, Davidson M, Brinton EA, Gotto AM, Stepanavage M, Liu SX, Gibbons P, Ashraf TB et al. Safety of anacetrapib in patients with or at high risk for coronary heart disease. *N Engl J Med* 2010; 363: 2406–2415.

430. Barish GD, Evans RM. PPARs and LXRs: atherosclerosis goes nuclear. *Trends Endocrinol Metab* 2004; 15: 158–165.

431. Baldan A, Bojanic DD, Edwards PA. The ABCs of sterol transport. *J Lipid Res* 2009; 50: S80–S85.

432. Tang W, Ma Y, Jia L, Ioannou YA, Davies JP, Yu L. Niemann-Pick C1-Like 1 is required for an LXR agonist to raise plasma HDL cholesterol in mice. *Arterioscler Thromb Vasc Biol* 2008; 28: 448–454.

433. van der Veen JN, van Dijk TH, Vrins CL, van Meer H, Havinga R, Bijsterveld K, Tietge UJ, Groen AK, Kuipers F. Activation of the liver X receptor stimulates trans-intestinal excretion of plasma cholesterol. *J Biol Chem* 2009; 284: 19211–19219.

434. Jakulj L, Vissers MN, van Roomen CP, van der Veen JN, Vrins CL, Kunne C, Stellaard F, Kastelein JJ, Groen AK. Ezetimibe stimulates faecal neutral sterol excretion depending on abcg8 function in mice. *FEBS Lett* 2010; 584: 3625–3628.

435. Yasuda T, Grillot D, Billheimer JT, Briand F, Delerive P, Huet S, Rader DJ. Tissue-specific liver X receptor activation promotes macrophage reverse cholesterol transport in vivo. *Arterioscler Thromb Vasc Biol* 2010; 30: 781–786.

436. Nijstad N, Gautier T, Briand F, Rader DJ, Tietge UJ. Biliary sterol secretion is required for functional in vivo reverse cholesterol transport. *Circulation* 2009; 120 (Suppl.): S443.

437. Briand F, Treguier M, Andre A, Grillot D, Issandou M, Ouguerram K, Sulpice T. Liver X receptor activation promotes macrophage-to-feces reverse cholesterol transport in a dyslipidemic hamster model. *J Lipid Res* 2010; 51: 763–770.

438. Shibata N, Glass CK. Regulation of macrophage function in inflammation and atherosclerosis. *J Lipid Res* 2009; 50: S277–S281.

439. Nomiyama T, Bruemmer D. Liver X receptors as therapeutic targets in metabolism and atherosclerosis. *Curr Atheroscler Rep* 2008; 10: 88–95.

440. Verschuren L, de Vries-van der Weij J, Zadelaar S, Kleemann R, Kooistra T. LXR agonist suppresses atherosclerotic lesion growth and promotes lesion regression in apoE*3Leiden mice: time course and mechanisms. *J Lipid Res* 2009; 50: 301–311.

441. Honzumi S, Shima A, Hiroshima A, Koieyama T, Ubukata N, Terasaka N. LXR alpha regulates human CETP expression *in vivo* and in transgenic mice. *Atherosclerosis* 2010; 212: 139.

442. Kratzer A, Buchebner M, Pfeifer T, Becker TM, Uray G, Miyazaki M, Miyazaki-Anzai S, Ebner B, Chandak PG, Kadam RS et al. Synthetic LXR agonist attenuates plaque formation in apoE-/- mice without inducing liver steatosis and hypertriglyceridemia. *J Lipid Res* 2009; 50: 312–326.

443. Nofer J-R, Bot M, Brodde M, Taylor PJ, Salm P, Brinkmann V, van Berkel T, Assmann G, Biessen EAL. FTY720, a synthetic sphingosine 1 phosphate analogue, inhibits development of atherosclerosis in low-density lipoprotein receptor-deficient mice. *Circulation* 2007; 115: 501–508.

444. Blom T, Back N, Mutka A-L, Bittman R, Li Z, de Lera A, Kovanen PT, Diczfalusy U, Ikonen E. FTY720 Stimulates 27-Hydroxycholesterol Production and Confers Atheroprotective Effects in Human Primary Macrophages. *Circ Res* 2010; 106: 720–729.

445. Yki-Jarvinen H. Thiazolidinediones and the liver in humans. *Curr Opin Lipidol* 2009; 20: 477–483.

446. Rizzo M, Christ ER, Rini GB, Spinas GA, Berneis K. The differential effects of thiazolidindiones on atherogenic dyslipidemia in type 2 diabetes: what is the clinical significance ? *Expert Opin Pharmacother* 2008; 9: 2295–2303.

447. Doggrell SA. Clinical trials with thiazolidinediones in subjects with Type 2 diabetes–is pioglitazone any different from rosiglitazone ? *Expert Opin Pharmacother* 2008; 9: 405–420.

448. Langenfeld MR, Forst T, Hohberg C, Kann P, Lubben G, Konrad T, Fullert SD, Sachara C, Pfutzner A. Pioglitazone decreases carotid intima-media thickness independently of glycemic control in patients with type 2 diabetes mellitus: results from a controlled randomized study. *Circulation* 2005; 111: 2525–2531.

449. Yamasaki Y, Katakami N, Furukado S, Kitagawa K, Nagatsuka K, Kashiwagi A, Daida H, Kawamori R, Kaku K. Long-term effects of pioglitazone on carotid atherosclerosis in Japanese patients with type 2 diabetes without a recent history of macrovascular morbidity. *J Atheroscler Thromb* 2010; 17: 1132–1140.

450. Nakaya H, Summers BD, Nicholson AC, Gotto AM Jr, Hajjar DP, Han J. Atherosclerosis in LDLR-knockout mice is inhibited, but not reversed, by the PPARgamma ligand pioglitazone. *Am J Pathol* 2009; 174: 2007–2014.

451. Rizzo M, Vekic J, Koulouris S, Zeljkovic A, Jelic-Ivanovic Z, Spasojevic-Kalimanovska V, Rini GB, Sakellariou D, Pastromas S, Mikhailidis DP et al. Effects of rosiglitazone on fasting and postprandial low- and high-density lipoproteins size and subclasses in type 2 diabetes. *Angiology* 2010; 61: 584–590.

452. Carreon-Torres E, Juarez-Meavepena M, Cardoso-Saldana G, Gomez CH, Franco M, Fievet C, Luc G, Juarez-Oropeza MA, Perez-Mendez O. Pioglitazone increases the fractional catabolic and production rates of high-density lipoproteins apo AI in the New Zealand White Rabbit. *Atherosclerosis* 2005; 181: 233–240.

453. Carreon-Torres E, Rendon-Sauer K, Monter-Garrido M, Toledo-Ibelles P, Gamboa R, Menjivar M, Lopez-Marure R, Luc G, Fievet C, Cruz D et al. Rosiglitazone modifies HDL structure and increases HDL-apo AI synthesis and catabolic rates. *Clin Chim Acta* 2009; 401: 37–41.

454. Ahmed RA, Murao K, Imachi H, Yu X, Li J, Wong NC, Ishida T. Human scavenger receptor class B type 1 is regulated by activators of peroxisome proliferators-activated receptor-gamma in hepatocytes. *Endocrine* 2009; 35: 233–242.

455. Qin S, Liu T, Kamanna VS, Kashyap ML. Pioglitazone stimulates apolipoprotein A-I production without affecting HDL removal in HepG2 cells: Involvement of PPAR-alpha. *Arterioscler Thromb Vasc Biol* 2007; 27: 2428–2434.

456. Qin S, Liu T, Kamanna VS, Kashyap ML. Pioglitazone stimulates apolipoprotein A-I production without affecting HDL removal in HepG2 cells: Involvement of PPAR-{alpha}. *Arterioscler Thromb Vasc Biol* 2007; 27: 2428–2434.

457. Zhang L-H, Kamanna VS, Ganji SH, Xiong X-M, Kashyap ML. Pioglitazone increases apolipoprotein A-I production by directly enhancing PPRE-dependent transcription in HepG2 cells. *J Lipid Res* 2010; 51: 2211–2222.

458. Millar JS, Ikewaki K, Bloedon LT, Wolfe ML, Szapary PO, Rader DJ. Effect of rosiglitazone on HDL metabolism in subjects with metabolic syndrome and low HDL. *J Lipid Res* 2011; 52: 136–142.

459. Chawla A, Boisvert WA, Lee CH, Laffitte BA, Barak Y, Joseph SB, Liao D, Nagy L, Edwards PA, Curtiss LK et al. A PPAR gamma-LXR-ABCA1 pathway in macrophages is involved in cholesterol efflux and atherogenesis. *Mol Cell* 2001; 7: 161–171.

460. Ozasa H, Ayaori M, Ogura M, Takiguchi S-i, Nakaya K, Terao Y, Uto-Kondo H, Yakushiji E, Iizuka M, Sasaki M et al. Pioglitazone enhances cholesterol efflux from macrophages by increasing ABCA1/ABCG1 expressions via LXR alpha/PPAR gamma pathway: Findings from *in vitro* and ex vivo studies. *Circulation* 2010; 122 (Suppl.): A15463.

461. Toh SA, Millar JS, Billheimer J, Fuki I, Naik SU, Macphee C, Walker M, Rader DJ. PPAR gamma activation redirects macrophage cholesterol from fecal excretion to adipose tissue uptake in mice via SR-BI. *Biochem Pharmacol* 2011; 81: 934–941.

462. Jonker JT, Wang Y, de Haan W, Diamant M, Rijzewijk LJ, van der Meer RW, Lamb HJ, Tamsma JT, de Roos A, Romijn JA et al. Pioglitazone decreases plasma cholesteryl ester transfer protein mass, associated with a decrease in hepatic triglyceride content, in patients with type 2 diabetes. *Diabetes Care* 2010; 33: 1625–1628.

463. Tsuchiya K, Akaza I, Yoshimoto T, Hirata Y. Pioglitazone improves endothelial function with increased adiponectin and high-density lipoprotein cholesterol levels in type 2 diabetes. *Endocr J* 2009; 56: 691–698.

464. Suckling K. Phospholipase A2s: developing drug targets for atherosclerosis. *Atherosclerosis* 2010; 212: 357–366.

465. Yasuda T, Ishida T, Rader DJ. Update on the role of endothelial lipase in high-density lipoprotein metabolism, reverse cholesterol transport, and atherosclerosis. *Circ J* 2010; 74: 2263–2270.

466. Fraser H, Hislop C, Christie RM, Rick HL, Reidy CA, Chouinard ML, Eacho PI, Gould KE, Trias J. Varespladib (A-002), a secretory phospholipase A2 inhibitor, reduces atherosclerosis and aneurysm formation in ApoE-/- mice. *J Cardiovasc Pharmacol* 2009; 53: 60–65.

467. Shaposhnik Z, Wang X, Trias J, Fraser H, Lusis AJ. The synergistic inhibition of atherogenesis in apoE-/- mice between pravastatin and the $sPLA_2$ inhibitor varespladib (A-002). *J Lipid Res* 2009; 50: 623–629.

468. Leite JO, Vaishnav U, Puglisi M, Fraser H, Trias J, Fernandez ML. A-002 (Varespladib), a phospholipase A2 inhibitor, reduces atherosclerosis in guinea pigs. *BMC Cardiovasc Disord* 2009; 9: 7.

469. Rosenson RS, Hislop C, Elliott M, Stasiv Y, Goulder M, Waters D. Effects of varespladib methyl on biomarkers and major cardiovascular events in acute coronary syndrome patients. *J Am Coll Cardiol* 2010; 56: 1079–1088.

470. Rosenson RS, Elliott M, Stasiv Y, Hislop C. Randomized trial of an inhibitor of secretory phospholipase A2 on atherogenic lipoprotein subclasses in statin-treated patients with coronary heart disease. *Eur Heart J* 2011; 32: 999–1005

471. Dzavik V, Lavi S, Thorpe K, Yip PM, Plante S, Ing D, Overgaard CB, Osten MD, Lan J, Robbins K et al. The $sPLA_2$ Inhibition to Decrease Enzyme Release After Percutaneous Coronary Intervention (SPIDER-PCI) Trial. *Circulation* 2010; 122: 2411–2418.

472. Bradley JD, Dmitrienko AA, Kivitz AJ, Gluck OS, Weaver AL, Wiesenhutter C, Myers SL, Sides GD. A randomized, double-blinded, placebo-controlled clinical trial of LY333013, a selective inhibitor of group II secretory phospholipase A2, in the treatment of rheumatoid arthritis. *J Rheumatol* 2005; 32: 417–423.

473. Zeiher BG, Steingrub J, Laterre PF, Dmitrienko A, Fukiishi Y, Abraham E. LY315920NA/S-5920, a selective inhibitor of group IIA secretory phospholipase A2, fails to improve clinical outcome for patients with severe sepsis. *Crit Care Med* 2005; 33: 1741–1748.

474. Rosenson RS, Hislop C, McConnell D, Elliott M, Stasiv Y, Wang N, Waters DD. Effects of 1-H-indole-3-glyoxamide (A-002) on concentration of secretory phospholipase A2 (PLASMA study): a phase II double-blind, randomised, placebo-controlled trial. *Lancet* 2009; 373: 649–658.

475. Goodman KB, Bury MJ, Cheung M, Cichy-Knight MA, Dowdell SE, Dunn AK, Lee D, Lieby JA, Moore ML, Scherzer DA et al. Discovery of potent, selective sulfonylfuran urea endothelial lipase inhibitors. *Bioorg Med Chem Lett* 2009; 19: 27–30.

476. Johnston TP. Poloxamer 407 as a general lipase inhibitor: its implications in lipid metabolism and atheroma formation in C57BL/6 mice. *J Pharm Pharmacol* 2010; 62: 1807–1812.

477. Hara T, Ishida T, Kojima Y, Hirata K-i. Abstract 1276: Inactivation of endothelial lipase increases anti-inflammatory HDL particles. *Circulation* 2009; 120: S469.

478. Hara T, Ishida T, Kojima Y, Tanaka H, Yasuda T, Shinohara M, Toh R, Hirata K-i. Targeted deletion of endothelial lipase increases HDL particles with anti-inflammatory properties both *in vitro* and in vivo. *J Lipid Res* 2011; 52: 57–67.

479. Kivela AM, Makinen PI, Jyrkkanen HK, Mella-Aho E, Xia Y, Kansanen E, Leinonen H, Verma IM, Yla-Herttuala S, Levonen AL. Sulforaphane inhibits endothelial lipase expression through NF-kappaB in endothelial cells. *Atherosclerosis* 2010; 213: 122–128.

480. Corson MA. Phospholipase A2 inhibitors in atherosclerosis: the race is on. *Lancet* 2009; 373: 608–610.

481. Wilensky RL, Shi Y, Mohler ER III, Hamamdzic D, Burgert ME, Li J, Postle A, Fenning RS, Bollinger JG, Hoffman BE et al. Inhibition of lipoprotein-associated phospholipase A2 reduces complex coronary atherosclerotic plaque development. *Nat Med* 2008; 14: 1059–1066.
482. Serruys PW, Garcia-Garcia HM, Buszman P, Erne P, Verheye S, Aschermann M, Duckers H, Bleie O, Dudek D, Botker HE et al. for the Integrated Biomarker and Imaging Study, I. Effects of the direct lipoprotein-associated phospholipase A2 inhibitor darapladib on human coronary atherosclerotic plaque. *Circulation* 2008; 118: 1172–1182.
483. Mohler ER III, Ballantyne CM, Davidson MH, Hanefeld M, Ruilope LM, Johnson JL, Zalewski A. The effect of darapladib on plasma lipoprotein-associated phospholipase A2 activity and cardiovascular biomarkers in patients with stable coronary heart disease or coronary heart disease risk equivalent: the results of a multicenter, randomized, double-blind, placebo-controlled study. *J Am Coll Cardiol* 2008; 51: 1632–1641.
484. Nishizawa T, Kitayama K, Wakabayashi K, Yamada M, Uchiyama M, Abe K, Ubukata N, Inaba T, Oda T, Amemiya Y. A novel compound, R-138329, increases plasma HDL cholesterol via inhibition of scavenger receptor BI-mediated selective lipid uptake. *Atherosclerosis* 2007; 194: 300–308.
485. Kitayama K, Nishizawa T, Abe K, Wakabayashi K, Oda T, Inaba T, Amemiya Y. Blockade of scavenger receptor class B type I raises high density lipoprotein cholesterol levels but exacerbates atherosclerotic lesion formation in apolipoprotein E deficient mice. *J Pharm Pharmacol* 2006; 58: 1629–1638.
486. Masson D, Koseki M, Ishibashi M, Larson CJ, Miller SG, King BD, Tall AR. Increased HDL cholesterol and apoA-I in humans and mice treated with a novel SR-BI inhibitor. *Arterioscler Thromb Vasc Biol* 2009; 29: 2054–2060.
487. Muslin AJ. MAPK signalling in cardiovascular health and disease: molecular mechanisms and therapeutic targets. *Clin Sci* 2008; 115: 203–218.
488. Huard K, Bourgeois P, Rhainds D, Falstrault L, Cohn JS, Brissette L. Apolipoproteins C-II and C-III inhibit selective uptake of low- and high-density lipoprotein cholesteryl esters in HepG2 cells. *Int J Biochem Cell Biol* 2005; 37: 1308–1318.
489. de Haan W, Out R, Berbee JF, van der Hoogt CC, van Dijk KW, van Berkel TJ, Romijn JA, Jukema JW, Havekes LM, Rensen PC. Apolipoprotein CI inhibits scavenger receptor BI and increases plasma HDL levels in vivo. *Biochem Biophys Res Commun* 2008; 377: 1294–1298.
490. Krasteva V, Brodeur MR, Tremblay FL, Falstrault L, Brissette L. Apolipoprotein C-I reduces cholesteryl esters selective uptake from LDL and HDL by binding to HepG2 cells and lipoproteins. *Biochim Biophys Acta* 2010; 1801: 42–48.
491. Nieland TJ, Shaw JT, Jaipuri FA, Duffner JL, Koehler AN, Banakos S, Zannis VI, Kirchhausen T, Krieger M. Identification of the molecular target of small molecule inhibitors of HDL receptor SR-BI activity. *Biochemistry* 2008; 47: 460–472.
492. Syder AJ, Lee H, Zeisel MB, Grove J, Soulier E, Macdonald J, Chow S, Chang J, Baumert TF, McKeating JA et al. Small molecule scavenger receptor BI antagonists are potent HCV entry inhibitors. *J Hepatol* 2010; 54: 48–55.
493. Catanese MT, Graziani R, von Hahn T, Moreau M, Huby T, Paonessa G, Santini C, Luzzago A, Rice CM, Cortese R et al. High-avidity monoclonal antibodies against the human scavenger class B type I receptor efficiently block hepatitis C virus infection in the presence of high-density lipoprotein. *J Virol* 2007; 81: 8063–8071.

494. Nakaya K, Ayaori M, Kondo H, Ogura M, Yakushiji E, Takiguchi S, Terao Y, Ozasa H, Ohsuzu F, Ikewaki K. Cilostazol enhances macrophage reverse cholesterol transport *in vitro* and in vivo. *Circulation* 2009; 120 (Suppl.): S445.

495. Nakaya K, Ayaori M, Uto-Kondo H, Hisada T, Ogura M, Yakushiji E, Takiguchi S, Terao Y, Ozasa H, Sasaki M et al. Cilostazol enhances macrophage reverse cholesterol transport *in vitro* and in vivo. *Atherosclerosis* 2010; 213: 135–141.

496. Bailey D, Jahagirdar R, Gordon A, Hafiane A, Campbell S, Chatur S, Wagner GS, Hansen HC, Chiacchia FS, Johansson J et al. RVX-208: a small molecule that increases apolipoprotein A-I and high-density lipoprotein cholesterol *in vitro* and in vivo. *J Am Coll Cardiol* 2010; 55: 2580–2589.

497. McNeill E. RVX-208, a stimulator of apolipoprotein AI gene expression for the treatment of cardiovascular diseases. *Curr Opin Investig Drugs* 2010; 11: 357–364.

498. Nicholls SJ, Gordon A, Johansson J, Wolski K, Ballantyne CM, Kastelein JJ, Taylor A, Borgman M, Nissen SE. Efficacy and safety of a novel oral inducer of apolipoprotein A-I synthesis in statin-treated patients with stable coronary artery disease: a randomized controlled trial. *J Am Coll Cardiol* 2011; 57: 1111–1119.

499. Glaros EN, Kim WS, Garner B. Myriocin-mediated up-regulation of hepatocyte apoA-I synthesis is associated with ERK inhibition. *Clin Sci* 2010; 118: 727–736.

500. Gao J, Xu Y, Yang Y, Yang Y, Zheng Z, Jiang W, Hong B, Yan X, Si S. Identification of upregulators of human ATP-binding cassette transporter A1 via high-throughput screening of a synthetic and natural compound library. *J Biomol Screen* 2008; 13: 648–656.

501. Arakawa R, Tsujita M, Iwamoto N, Ito-Ohsumi C, Lu R, Wu C-A, Shimizu K, Aotsuka T, Kanazawa H, Abe-Dohmae S et al. Pharmacological inhibition of ABCA1 degradation increases HDL biogenesis and exhibits antiatherogenesis. *J Lipid Res* 2009; 50: 2299–2305.

502. Zhou M, Sawyer J, Kelley K, Fordstrom P, Chan J, Tonn G, Carlson T, Retter M, Meininger D, Cheng J et al. Lecithin cholesterol acyltransferase promotes reverse cholesterol transport and attenuates atherosclerosis progression in New Zealand white rabbits. *Circulation* 2009; 120 (Suppl.): S1175.

503. Kontush A, Chapman MJ. Antiatherogenic function of HDL particle subpopulations: focus on antioxidative activities. *Curr Opin Lipidol*, 2010; *21*, 312–318.

504. Rousset X, Vaisman B, Auerbach B, Krause BR, Homan R, Stonik J, Csako G, Shamburek R, Remaley AT. Effect of recombinant human lecithin cholesterol acyltransferase infusion on lipoprotein metabolism in mice. *J Pharmacol Exp Ther*, 2010; *335*, 140–148.

505. Amar MJ, Shamburek RD, Vaisman B, Knapper CL, Foger B, Hoyt RF Jr, Santamarina-Fojo S, Brewer HB Jr, Remaley AT. Adenoviral expression of human lecithin-cholesterol acyltransferase in nonhuman primates leads to an antiatherogenic lipoprotein phenotype by increasing high-density lipoprotein and lowering low-density lipoprotein. *Metabolism*, 2009; *58*, 568–575.

506. Van Craeyveld E, Lievens J, Jacobs F, Feng Y, Snoeys J, De Geest B. Apolipoprotein A-I and lecithin:cholesterol acyltransferase transfer induce cholesterol unloading in complex atherosclerotic lesions. *Gene Ther*, 2009; *16*, 757–765.

507. Calabresi L, Franceschini G. Lecithin:cholesterol acyltransferase, high-density lipoproteins, and atheroprotection in humans. *Trends Cardiovasc Med*, 2010; *20*, 50–53.

508. Tsutsumi K, Inoue Y, Shima A, Iwasaki K, Kawamura M, Murase T. The novel compound NO-1886 increases lipoprotein lipase activity with resulting elevation of high density lipoprotein cholesterol, and long-term administration inhibits atherogenesis in the coronary arteries of rats with experimental atherosclerosis. *J Clin Invest* 1993; 92: 411–417.

509. Chen SG, Xiao J, Liu XH, Liu MM, Mo ZC, Yin K, Zhao GJ, Jiang J, Cui LB, Tan CZ et al. Ibrolipim increases ABCA1/G1 expression by the LXRalpha signaling pathway in THP-1 macrophage-derived foam cells. *Acta Pharmacol Sin* 2010; 31: 1343–1349.

510. Zhang C, Yin W, Liao D, Huang L, Tang C, Tsutsumi K, Wang Z, Liu Y, Li Q, Hou H et al. NO-1886 upregulates ATP binding cassette transporter A1 and inhibits diet-induced atherosclerosis in Chinese Bama minipigs. *J Lipid Res* 2006; 47: 2055–2063.

511. Kusunoki M, Tsutsumi K, Inoue Y, Hara T, Miyata T, Nakamura T, Ogawa H, Sakakibara F, Fukuzawa Y, Okabayashi N et al. Lipoprotein lipase activator NO-1886 improves fatty liver caused by high-fat feeding in streptozotocin-induced diabetic rats. *Metabolism* 2004; 53: 260–263.

512. Yin W, Yuan Z, Tsutsumi K, Xie Y, Zhang Q, Wang Z, Fu G, Long G, Yang Y. A lipoprotein lipase-promoting agent, NO-1886, improves glucose and lipid metabolism in high fat, high sucrose-fed New Zealand white rabbits. *Int J Exp Diabesity Res* 2003; 4: 27–34.

513. Yin W, Tsutsumi K, Yuan Z, Yang B. Effects of the lipoprotein lipase activator NO-1886 as a suppressor agent of atherosclerosis in aorta of mild diabetic rabbits. *Arzneimittelforschung* 2002; 52: 610–614.

514. Kusunoki M, Tsutsumi K, Hara T, Ogawa H, Nakamura T, Miyata T, Sakakibara F, Fukuzawa Y, Suga T, Kakumu S et al. A lipoprotein lipase activator, NO-1886 prevents impaired endothelium-dependent relaxation of aorta caused by exercise in aged rats. *Exp Gerontol* 2002; 37: 891–896.

515. Hara T, Kusunoki M, Tsutsumi K, Okada K, Sakamoto S, Ohnaka M, Nakamura T, Miyata T, Nakayama K, Fukatsu A et al. A lipoprotein lipase activator, NO-1886, improves endothelium-dependent relaxation of rat aorta associated with aging. *Eur J Pharmacol* 1998; 350: 75–79.

CHAPTER 12

COMBINATION THERAPIES

Administration of a statin together with an HDL-C-raising agent, such as niacin or a fibrate represents the major type of combination therapy targeting plasma HDL. Such combinations provide a straightforward example of an approach aimed at reducing the residual cardiovascular risk persisting following statin treatment. In addition, as distinct HDL-C raising agents function through complementary mechanisms, their effects may be additive, or even synergistic. The association of such medications may also be clinically relevant and constitutes an interesting area for future research.

Another advantage of combination therapies involves the potential for the use of lower doses of each agent compared with their use in monotherapy in order to obtain additive or synergistic therapeutic effects; such an approach may also lead to a reduction in adverse effects.

12.1 NIACIN AND STATINS

The first group of combination therapies targeting both LDL-C and HDL-C comprises the simultaneous administration of **niacin and a statin**. Two trials evaluating niacin combined with either a statin or a bile acid sequestrant reported marked HDL-C increases ranging from +25% to +41% with an unprecedented reduction in cardiovascular event rates ranging from −60% to −72% [1, 2]. Such effects

High-Density Lipoproteins: Structure, Metabolism, Function, and Therapeutics, First Edition.
Anatol Kontush and M. John Chapman.

can be accounted for by the additive benefit of a concomitant reduction of LDL-C and a raising of HDL-C. These impressive reduction rates correspond well to the estimated risk reduction based on the relationship between HDL-C levels and cardiovascular events in epidemiologic studies, i.e., +1 mg/dl HDL-C increase is associated with a −2% to −3% reduction in cardiovascular events [3, 4].

Risk factor modelling suggests that raising HDL-C levels using nicotinic acid in statin-treated patients is likely to yield significant incremental clinical benefits of up to −29% risk reduction in subjects with diabetes or an abnormal lipid profile [5]. Risk reductions in cardiovascular events may reach −83% for the highest doses of extended-release niacin and simvastatin [6]. Several large clinical trials of combination therapy with statin and niacin, such as the Heart Protection Study 2: Treatment of HDL to Reduce the Incidence of Vascular Events (HPS2-THRIVE) are currently in progress to address the clinical efficacy of this combination. Recently, one such trial, the Atherothrombosis Intervention in Metabolic Syndrome with Low HDL/High Triglycerides and Impact on Global Health Outcomes (AIM-HIGH) study, was however stopped early by the FDA [7]. This trial enrolled 3414 patients with a history of cardiovascular disease, low HDL-C, and high triglycerides who were all prescribed simvastatin 40 mg per day and then randomized to Niaspan in gradually increasing doses up to 2000 mg/day or placebo. Of the patients, 515 were given a second LDL-lowering drug, ezetimibe, in order to maintain LDL-C levels at the target range of 40-80 mg/dl. The trial was started in September 2005, but was stopped early due to the lack of incremental benefit on cardiovascular risk reduction in the extended-release niacin plus simvastatin treatment group over simvastatin alone. At the time the trial was stopped, the primary end point (a composite of fatal or nonfatal myocardial infarction, stroke, acute coronary syndrome, or revascularization) had occurred at an annualized rate of 5.8% in the niacin group vs 5.6% with placebo. Unexpectedly, during the 32-month follow-up period, there were 28 strokes (1.6%) in the niacin group vs 12 strokes (0.7%) in the control group, but nine strokes in the niacin group occurred after the drug had been discontinued. Therefore, it remains unclear what role, if any, niacin contributed to this imbalance in ischemic stroke [7]. Several factors, including insufficient statistical power of the study, its short duration and aggressive management of LDL-C prior to inclusion, can potentially account for the lack of clinical benefit of niacin in this trial.

Effects of the niacin/simvastatin combination on HDL metabolism involve an elevation in large HDL. Indeed, an increase in the content of apoA-I in large alpha-1 HDL to >20 mg/dl with niacin/simvastatin combination is associated with lack of progression and, in some individuals, with regression of coronary atherosclerosis [8]. However, the addition of extended-release niacin to a statin, despite a relative improvement in HDL-C and LDL-C, is not always superior to a statin alone in terms of the reduction in carotid atherosclerosis, as shown in elderly patients with clinically-evident atherosclerosis receiving 1500 mg niacin per day for 18 months [9].

Such beneficial results obtained with niacin/statin combinations are related to additional, clinically-relevant improvements in multiple lipid parameters beyond statin monotherapy. Thus, full-dose niacin/moderate-dose simvastatin combination reveals sustained beneficial actions on atherogenic apoB-containing lipoproteins that are at least comparable to those of high-dose simvastatin, while also raising HDL-C by +20% compared with a +5.9% increment on simvastatin (80 mg) alone [10]. Importantly, the niacin/simvastatin combination potently reduces plasma levels of Lp(a), an effect not observed using simvastatin alone [10]. Furthermore, plasma HDL-C levels are increased to a greater degree and triglyceride levels are lowered more with extended-release niacin/simvastatin compared with simvastatin monotherapy [11, 12]. Next, adding extended-release niacin to atorvastatin results in beneficial effects on the lipid profile in coronary heart disease patients with low HDL-C levels, with an additional elevation of HDL-C and further lowering of LDL-C compared with monotherapy [13].

The safety profile of niacin/statin combinations is good. Extended-release niacin/simvastatin combinations are generally well tolerated, with flushing being the most common adverse reaction, and there is a lack of unanticipated adverse events [11, 12, 14]. The addition of laropiprant and niacin to ongoing statin treatment improves the tolerability profile in dyslipidemic patients while producing lipid-modifying benefits comparable to those on niacin and a statin [15]. The types of adverse effects that occur with this triple combination are those typically associated with niacin alone, with the exception of flushing. Finally, a non-prescription time-release niacin formulation, Slo-Niacin©, at 1.5 g/day combined with atorvastatin 10 mg/day is also capable of improving plasma levels of lipids, apolipoproteins and inflammatory biomarkers without hepatotoxicity over three months [16].

CRITICAL CONTRIBUTION

Concomitant administration of niacin and a statin efficiently decreases LDL-C, VLDL-C and Lp(a), raises HDL-C and provides marked reduction in cardiovascular event rates [1, 2].

Interestingly, combination therapy of atorvastatin 10/20 mg and niacin 2 g for a year remodels the HDL proteome, decreasing the abundance of apoE (−41%) in HDL3 and apoE concentration in plasma [17]. Other changes in the HDL3 proteome with niacin/statin treatment include an elevation in the abundances of apoF (+85%), PLTP (+247%) and apoJ (+198%), resulting in a protein composition that more closely resembles that of HDL3 in healthy controls. Interestingly, decreased apoF content in HDL3 from patients with coronary artery disease is suggestive of elevated CETP activity and is consistent with low HDL-C levels [17]. These compositional changes may result in the enhanced functionality of HDL after treatment. Indeed, statins may enhance HDL functionality by acting synergistically with a CETP inhibitor [18] or an apoA-I mimetic peptide [19].

12.2 FIBRATES AND STATINS

Combining **a fibrate with a statin** produces a second major group of combination drugs that concomitantly target both LDL-C and HDL-C levels. This approach delivers improvements in the plasma lipid profile that are superior to monotherapy with a statin. Indeed, fenofibric acid combined with a low-dose statin results in greater changes in HDL-C (+16.8%) and triglycerides (−43.9%) than low-dose statin monotherapy (+4.7% and −18.1%, respectively), and greater reductions in LDL-C (−34.0%) than fenofibric acid monotherapy (−5.3%) in patients with mixed dyslipidemia and Type 2 diabetes [20]. Similarly, fenofibric acid combined with a moderate-dose statin results in greater changes in HDL-C (+16.3%) and triglycerides (−43.4%) than moderate-dose statin monotherapy (+8.7% and −24.2%, respectively), and greater reductions in LDL-C (−32.6%) than fenofibric acid monotherapy (−5.3%). Furthermore, combination therapy with fenofibric acid and simvastatin provides a more effective control of multiple lipid parameters than monotherapy alone in patients with mixed dyslipidemia [21]. Specifically, HDL-C is increased (+17.8% vs. +7.2%) and triglycerides decreased (−37.4% vs. −14.2%) more potently by the combination relative to simvastatin alone, respectively. In addition, a combination of fenofibrate and simvastatin improved lipid, prothrombotic and inflammatory parameters over 12 months of therapy in Type 2 diabetic patients with combined dyslipidemia [22]. Finally, a fixed-dose combination tablet containing 40/100 mg atorvastatin/fenofibrate produced a greater reduction in triglycerides than atorvastatin (40 mg) or higher-dose fenofibrate (145 mg) alone in a 12-week study in patients with dyslipidemia [23]. Treatment with the combination tablet is also associated with a greater increase in HDL-C compared with atorvastatin alone and a greater reduction in non-HDL-C compared with fenofibrate alone. Such fenofibric acid/statin combination therapies are well tolerated with a safety profile similar to each drug in monotherapy [20, 21, 23, 24]; they may however increase risks of hepato- and muscular toxicity, including that of rhabdomyolysis.

Data on the clinical efficacy of fibrate/statin combinations are, however, less persuasive than one would have expected on the basis of their effects on plasma lipids. Thus, in the Action to Control Cardiovascular Risk in Diabetes (ACCORD) trial, the combination of fenofibrate and simvastatin did not reduce the rate of fatal cardiovascular events, nonfatal myocardial infarction or nonfatal stroke, compared with simvastatin alone in 5518 patients with Type 2 diabetes treated with open-label simvastatin and randomly assigned to receive either masked fenofibrate or placebo for a mean of 4.7 years [25]. These results do not support the routine use of combination therapy with fenofibrate and simvastatin to reduce cardiovascular risk in the majority of high-risk patients with Type 2 diabetes. Prespecified subgroup analyses, however, suggests heterogeneity in the treatment effect according to sex, with a benefit for men and possible harm for women, a possible interaction according to lipid subgroup, and a possible benefit for patients with both high baseline triglycerides and a low baseline level of HDL-C. A substantial decrease in triglycerides in the placebo group (−10% vs. −26%

in the treatment arm) paralleled by an increase in HDL-C (+6% vs. +8% in the treatment arm) may, at least in part, account for such a negative outcome. Thus, the addition of fenofibrate to a statin may still benefit patients with Type 2 diabetes who have substantial dyslipidemia. This conclusion is consistent with *post hoc* subgroup analyses in the earlier Helsinki Heart Study, BIP and FIELD trials [25, 26]. Furthermore, fibrates lead to a moderate cardiovascular risk reduction in a subgroup of patients with atherogenic dyslipidemia in combination with statins in the ACCORD Lipid study [25, 27, 28]. It therefore remains to be determined whether a combination of a fibrate plus statin reduces cardiovascular risk beyond that achieved with a statin alone in diabetic patients displaying elevated triglyceride and low HDL-C levels [29].

METHODOLOGIC NOTE

As a result of safety concerns, the administration of combination therapies of a statin with an HDL-raising drug should be accompanied by the monitoring of potential hepato- and muscular toxicity.

12.3 OTHER COMBINATIONS

Instead of a statin, HDL-raising drugs can also be combined with another LDL-lowering agent, such as **ezetimibe**. Thus, co-administration of fenofibrate and ezetimibe produces complementary, and favorable, effects on LDL-C, triglycerides, HDL-C and non-HDL-C levels, and is well tolerated in patients with mixed hyperlipidemia [30]. In addition, this combination promotes similar increases in HDL2-C and HDL3-C.

Recently, beneficial effects of simvastatin/ezetimibe combination on major atherosclerotic events were reported in patients with chronic kidney disease in the randomized, placebo-controlled Study of Heart and Renal Protection (SHARP) trial [31]. A more sophisticated approach involves the use of a triple combination of a statin and ezetimibe with an HDL-raising agent. This strategy is based on the superiority of statin/ezetimibe combinations over either statin or ezetimibe monotherapy in terms of their effects on both LDL-C and HDL-C levels. For example, treatment with ezetimibe/simvastatin (10/20 mg) vs. atorvastatin (10 and 20 mg/day) results in an increment in HDL3-C levels in hypercholesterolemic patients with Type 2 diabetes [32].

Indeed, combination treatment with ezetimibe/simvastatin, plus extended-release niacin shows superior lipid-altering efficacy compared with extended-release niacin or an ezetimibe/simvastatin combination in patients with type IIa or IIb hyperlipidemia and is generally well tolerated aside from niacin-associated flushing [33]. Such a triple combination offers a broad, lipid-modulating therapy with enhanced lipid effects beyond LDL-C in these patients. Similarly, a combination of ezetimibe/simvastatin plus niacin is generally well tolerated and was superior to ezetimibe/simvastatin alone in improving several lipoprotein parameters during a 64-week trial in patients with hyperlipidemia [34]. Ezetimibe/simvastatin can also be combined with a fibrate, such as fenofibrate; this combination produces favorable effects on atherogenic lipoprotein subclasses in patients with mixed hyperlipidemia [35]. However, it remains largely indeterminate how ezetimibe impacts HDL particles; one study demonstrates decreases in SAA levels in small HDL3 after ezetimibe treatment [36].

Other combinations of lipid-modulating drugs have also been tested, such as those involving cannabinoid receptor antagonists (see chapter 13) or PPAR gamma agonists. For example, a combination of rimonabant (20 mg/day) and micronized fenofibrate (200 mg/day) may further improve metabolic parameters in overweight/obese patients with dyslipidemia compared with each monotherapy [37]. Such an improvement is particularly pronounced for HDL-C. Indeed, this combination results in an increment in HDL-C levels of +42% after three months, while HDL-C is not significantly altered in each monotherapy group. A combination of fenofibrate and rosiglitazone can also beneficially impact plasma lipid levels [38]; however, some patients may display dramatic decreases in HDL-C when given this combination as a result of an interaction between the two agents.

IMPORTANT READING

Brown BG, Stukovsky KH, Zhao XQ. Simultaneous low-density lipoprotein-C lowering and high-density lipoprotein-C elevation for optimum cardiovascular disease prevention with various drug classes, and their combinations: a meta-analysis of 23 randomized lipid trials. *Curr Opin Lipidol* 2006; 17: 631–636.

REFERENCES

1. Brown BG, Hillger L, Zhao XQ, Poulin D, Albers JJ. Types of change in coronary stenosis severity and their relative importance in overall progression and regression of coronary disease. Observations from the FATS Trial. Familial Atherosclerosis Treatment Study. *Ann N Y Acad Sci* 1995; 748: 407–417.
2. Brown BG, Zhao XQ, Chait A, Fisher LD, Cheung MC, Morse JS, Dowdy AA, Marino EK, Bolson EL, Alaupovic P et al. Simvastatin and niacin, antioxidant vitamins, or the combination for the prevention of coronary disease. *N Engl J Med* 2001; 345: 1583–1592.
3. Gordon DJ, Probstfield JL, Garrison RJ, Neaton JD, Castelli WP, Knoke JD, Jacobs DR Jr, Bangdiwala S, Tyroler HA. High-density lipoprotein cholesterol and cardiovascular disease. Four prospective American studies. *Circulation* 1989; 79: 8–15.

4. Wilson PW. High-density lipoprotein, low-density lipoprotein and coronary artery disease. *Am J Cardiol* 1990; 66: 7A–10A.
5. Belsey J, de Lusignan S, van Vlymen J, Chan T, Hague N. Reducing coronary risk by raising HDL-cholesterol: risk modelling the addition of nicotinic acid to existing therapy. *Curr Med Res Opin* 2008; 24: 2703–2709.
6. Charland SL, Malone DC. Prediction of cardiovascular event risk reduction from lipid changes associated with high potency dyslipidemia therapy. *Curr Med Res Opin* 2010; 26: 365–375.
7. *FDA Statement on the AIM-HIGH Trial.*2011; http://www.fda.gov/Drugs/DrugSafety/PostmarketDrugSafetyInformationforPatientsandProviders/ucm256841.htm.
8. Asztalos BF, Batista M, Horvath KV, Cox CE, Dallal GE, Morse JS, Brown GB, Schaefer EJ. Change in alpha1 HDL concentration predicts progression in coronary artery stenosis. *Arterioscler Thromb Vasc Biol* 2003; 23: 847–852.
9. Sibley CT, Gottlieb I, Cox C, Godoy G, Spooner AE, Bluemke DA, Lima JA. Comparative effect of statins vs niacin on MRI measured regression of carotid atherosclerosis in a randomized clinical trial: The NIA Plaque Study. *Circulation* 2009; 120 (Suppl.): S376.
10. Airan-Javia SL, Wolf RL, Wolfe ML, Tadesse M, Mohler E, Reilly MP. Atheroprotective lipoprotein effects of a niacin-simvastatin combination compared to low- and high-dose simvastatin monotherapy. *Am Heart J* 2009; 157: 687 e681–e688.
11. Sanford M, Curran MP. Niacin extended-release/simvastatin. *Drugs* 2008; 68: 2373–2386.
12. Ballantyne CM, Davidson MH, McKenney J, Keller LH, Bajorunas DR, Karas RH. Comparison of the safety and efficacy of a combination tablet of niacin extended release and simvastatin vs simvastatin monotherapy in patients with increased non-HDL cholesterol (from the SEACOAST I study). *Am J Cardiol* 2008; 101: 1428–1436.
13. Harikrishnan S, Rajeev E, Tharakan JA, Titus T, Ajit Kumar VK, Sivasankaran S, Krishnamoorthy KM, Nair K. Efficacy and safety of combination of extended release niacin and atorvastatin in patients with low levels of high density lipoprotein cholesterol. *Indian Heart J* 2008; 60: 215–222.
14. Karas RH, Kashyap ML, Knopp RH, Keller LH, Bajorunas DR, Davidson MH. Long-term safety and efficacy of a combination of niacin extended release and simvastatin in patients with dyslipidemia: the OCEANS study. *Am J Cardiovasc Drugs* 2008; 8: 69–81.
15. Shah S, Ceska R, Gil-Extremera B, Paolini JF, Giezek H, Vandormael K, Mao A, McCrary Sisk C, Maccubbin D. Efficacy and safety of extended-release niacin/laropiprant plus statin vs. doubling the dose of statin in patients with primary hypercholesterolaemia or mixed dyslipidaemia. *Int J Clin Pract* 2010; 64: 727–738.
16. Knopp RH, Retzlaff BM, Fish B, Dowdy A, Twaddell B, Nguyen T, Paramsothy P. The SLIM Study: Slo-Niacin® and atorvastatin treatment of lipoproteins and inflammatory markers in combined hyperlipidemia. *J Clin Lipidol* 2009; 3: 167–178.
17. Green PS, Vaisar T, Pennathur S, Kulstad JJ, Moore AB, Marcovina S, Brunzell J, Knopp RH, Zhao XQ, Heinecke JW. Combined statin and niacin therapy remodels the high-density lipoprotein proteome. *Circulation* 2008; 118: 1259–1267.

18. Catalano G, Julia Z, Frisdal E, Vedie B, Fournier N, Le Goff W, Chapman MJ, Guerin M. Torcetrapib differentially modulates the biological activities of HDL2 and HDL3 particles in the reverse cholesterol transport pathway. *Arterioscler Thromb Vasc Biol* 2009; 29: 268–275.
19. Navab M, Anantharamaiah GM, Hama S, Hough G, Reddy ST, Frank JS, Garber DW, Handattu S, Fogelman AM. D-4F and statins synergize to render HDL antiinflammatory in mice and monkeys and cause lesion regression in old apolipoprotein E-null mice. *Arterioscler Thromb Vasc Biol* 2005; 25: 1426–1432.
20. Jones PH, Cusi K, Davidson MH, Kelly MT, Setze CM, Thakker K, Sleep DJ, Stolzenbach JC. Efficacy and safety of fenofibric acid co-administered with low- or moderate-dose statin in patients with mixed dyslipidemia and type 2 diabetes mellitus: results of a pooled subgroup analysis from three randomized, controlled, double-blind trials. *Am J Cardiovasc Drugs* 2010; 10: 73–84.
21. Mohiuddin SM, Pepine CJ, Kelly MT, Buttler SM, Setze CM, Sleep DJ, Stolzenbach JC. Efficacy and safety of ABT-335 (fenofibric acid) in combination with simvastatin in patients with mixed dyslipidemia: a phase 3, randomized, controlled study. *Am Heart J* 2009; 157: 195–203.
22. Derosa G, Maffioli P, Salvadeo SA, Ferrari I, Gravina A, Mereu R, Palumbo I, D'Angelo A, Cicero AF. Fenofibrate, simvastatin and their combination in the management of dyslipidaemia in type 2 diabetic patients. *Curr Med Res Opin* 2009; 25: 1973–1983.
23. Davidson MH, Rooney MW, Drucker J, Eugene Griffin H, Oosman S, Beckert M. Efficacy and tolerability of atorvastatin/fenofibrate fixed-dose combination tablet compared with atorvastatin and fenofibrate monotherapies in patients with dyslipidemia: a 12-week, multicenter, double-blind, randomized, parallel-group study. *Clin Ther* 2010; 31: 2824–2838.
24. Wierzbicki AS, Morrell J, Hemsley D, McMahon Z, Crook MA, Wray R. The effect of fibrate-statin combination therapy on cardiovascular events: a retrospective cohort analysis. *Curr Med Res Opin* 2010; 26: 2141–2146.
25. Ginsberg HN, Elam MB, Lovato LC, Crouse JR III, Leiter LA, Linz P, Friedewald WT, Buse JB, Gerstein HC, Probstfield J et al. Effects of combination lipid therapy in type 2 diabetes mellitus. *N Engl J Med* 2010; 362: 1563–1574.
26. Scott R, O'Brien R, Fulcher G, Pardy C, D'Emden M, Tse D, Taskinen MR, Ehnholm C, Keech A. Effects of fenofibrate treatment on cardiovascular disease risk in 9,795 individuals with type 2 diabetes and various components of the metabolic syndrome: the Fenofibrate Intervention and Event Lowering in Diabetes (FIELD) study. *Diabetes Care* 2009; 32: 493–498.
27. Tenenbaum A, Fisman EZ. "If it ain't broke, don't fix it": a commentary on the positive-negative results of the ACCORD Lipid study. *Cardiovasc Diabetol* 2010; 9: 24.
28. Wierzbicki AS. Fibrates: no ACCORD on their use in the treatment of dyslipidaemia. *Curr Opin Lipidol* 2010; 21: 352–358.
29. Barter PJ, Rye K-A. Is there a role for fibrates in the management of dyslipidemia in the metabolic syndrome? *Arterioscler Thromb Vasc Biol* 2008; 28: 39–46.
30. Tribble DL, Farnier M, Macdonell G, Perevozskaya I, Davies MJ, Gumbiner B, Musliner TA. Effects of fenofibrate and ezetimibe, both as monotherapy and in coadministration, on cholesterol mass within lipoprotein subfractions and low-density

lipoprotein peak particle size in patients with mixed hyperlipidemia. *Metabolism* 2008; 57: 796–801.

31. Baigent C, Landray MJ, Reith C, Emberson J, Wheeler DC, Tomson C, Wanner C, Krane V, Cass A, Craig J, Neal B, Jiang L, Hooi LS, Levin A, Agodoa L, Gaziano M, Kasiske B, Walker R, Massy ZA, Feldt-Rasmussen B, Krairittichai U, Ophascharoensuk V, Fellstrom B, Holdaas H, Tesar V, Wiecek A, Grobbee D, de Zeeuw D, Gronhagen-Riska C, Dasgupta T, Lewis D, Herrington W, Mafham M, Majoni W, Wallendszus K, Grimm R, Pedersen T, Tobert J, Armitage J, Baxter A, Bray C, Chen Y, Chen Z, Hill M, Knott C, Parish S, Simpson D, Sleight P, Young A, Collins R. The effects of lowering LDL cholesterol with simvastatin plus ezetimibe in patients with chronic kidney disease (Study of Heart and Renal Protection): a randomised placebo-controlled trial. *Lancet* 2011; 377: 2181–2192.
32. Tomassini JE, Mazzone T, Goldberg RB, Guyton JR, Weinstock RS, Polis A, Jensen E, Tershakovec AM. Effect of ezetimibe/simvastatin compared with atorvastatin on lipoprotein subclasses in patients with type 2 diabetes and hypercholesterolaemia. *Diabetes Obes Metab* 2009; 11: 855–864.
33. Guyton JR, Brown BG, Fazio S, Polis A, Tomassini JE, Tershakovec AM. Lipid-altering efficacy and safety of ezetimibe/simvastatin coadministered with extended-release niacin in patients with type IIa or type IIb hyperlipidemia. *J Am Coll Cardiol* 2008; 51: 1564–1572.
34. Fazio S, Guyton JR, Polis AB, Adewale AJ, Tomassini JE, Ryan NW, Tershakovec AM. Long-term safety and efficacy of triple combination ezetimibe/simvastatin plus extended-release niacin in patients with hyperlipidemia. *Am J Cardiol* 2010; 105: 487–494.
35. Farnier M, Perevozskaya I, Taggart WV, Kush D, Mitchel YB. VAP II analysis of lipoprotein subclasses in mixed hyperlipidemic patients on treatment with ezetimibe/simvastatin and fenofibrate. *J Lipid Res* 2008; 49: 2641–2647.
36. Hirano T, Nohtomi K, Nakanishi N, Watanabe T, Hyodo T, Taira T. Ezetimibe decreases serum amyloid A levels in HDL3 in hemodialysis patients. *Clin Nephrol* 2010; 74: 282–287.
37. Florentin M, Liberopoulos EN, Filippatos TD, Kostara C, Tselepis A, Mikhailidis DP, Elisaf M. Effect of rimonabant, micronised fenofibrate and their combination on cardiometabolic risk factors in overweight/obese patients: a pilot study. *Expert Opin Pharmacother* 2008; 9: 2741–2750.
38. Schwing W, Hustak L, Taylor HC. Paradoxical severe decrease in HDL cholesterol due to rosiglitazone-fenofibrate interaction. *Endocr Pract* 2010; 16: 382–388.

CHAPTER 13

OTHER PHARMACOLOGIC APPROACHES

13.1 CANNABINOID TYPE 1 RECEPTOR ANTAGONISTS

The endogenous cannabinoid system plays a central role in the regulation of energy homeostasis, fatty acid synthesis and glucose and lipid metabolism. Overactivation of the endocannabinoid system is associated with obesity, dyslipidemia, insulin resistance and low HDL-C dyslipidemia; patients with coronary artery disease exhibit activation of the endocannabinoid system with elevated levels of blood endocannabinoids and increased expression of cannabinoid receptors type 1 (CB1) in coronary atheroma [1]. These observations have led to the concept of the inhibition of the endocannabinoid system as a viable approach in treating obesity and reducing the associated cardiovascular risk.

Rimonabant was the first selective endocannabinoid CB1 receptor antagonist to be extensively investigated in a framework of the Rimonabant in Obesity (RIO) program, which comprised four 1–2 year placebo-controlled, randomized clinical trials recruiting more than 6600 overweight/obese patients with, or without, comorbidities [2]. In these trials, rimonabant (20 mg daily) consistently reduced body weight, waist circumference, triglycerides, blood pressure, insulin resistance and CRP levels, and increased HDL-C concentrations in both non-diabetic and Type 2 diabetic overweight/obese patients [2–4]. Adiponectin levels were elevated, an effect that correlated with HDL-C elevation, while small, dense LDL-C levels were reduced in patients receiving rimonabant (20 mg) compared

High-Density Lipoproteins: Structure, Metabolism, Function, and Therapeutics, First Edition.
Anatol Kontush and M. John Chapman.

with those receiving placebo in the RIO Lipids study. Furthermore, in the RIO Diabetes study, a −0.7% reduction in HbA1c levels was observed in metformin- or sulphonylurea-treated patients with Type 2 diabetes, an effect confirmed in the randomized, double-blind, placebo-controlled Study Evaluating Rimonabant Efficacy in Drug-Naive Diabetic Patients (SERENADE) trial in drug-naive diabetic patients [5]. Almost half of the metabolic changes occurred beyond weight loss, consistent with direct peripheral effects. The positive effects observed after one year were maintained after two years. Moreover, rimonabant exhibited anti-inflammatory effects on macrophages, which may provide beneficial effects on atherogenesis [1].

Specifically, the RIO studies enrolled patients who were either obese or displayed at least one comorbidity. All patients received daily treatment with rimonabant (5 or 20 mg) or placebo for 1 year plus a hypocaloric diet (600 kcal/day deficit) and advice on increasing physical activity. After 1 year of randomized treatment, changes from baseline with 20 mg rimonabant in the nondiabetic population were as follows: body weight −6.5 kg; waist circumference −6.4 cm; HDL-C +16.4%; triglycerides −6.9%; fasting insulin −0.6 muU/ml; and, HOMA index −0.2 [6]. In the diabetic population, 20 mg rimonabant reduced HbA1c levels by −0.6%. Next, in the SERENADE trial, rimonabant monotherapy (20 mg/day) was assessed in drug-naive Type 2 diabetic patients over a six-month period. In this population, rimonabant induced improvements relative to placebo in HbA1c, weight, waist circumference, fasting plasma glucose, triglycerides and HDL-C (+10.1 vs. +3.2% in the placebo group; $p < 0.0001$) [5]. In the randomized placebo-controlled An International Study of Rimonabant in Dyslipidemia With Atherogenic Risk In Abdominally Obese Patients (ADAGIO-Lipids) trial in abdominally obese patients with atherogenic dyslipidemia, rimonabant given at 20 mg/d for 1 year produced significantly greater changes from baseline relative to placebo in HDL-C (+7.4%) and triglyceride (−18%) levels, as well as in mean LDL and HDL particle sizes, and apoA-I, apoB, HDL2, HDL3, CRP and adiponectin levels [7]. In parallel, rimonabant decreased abdominal subcutaneous adipose tissue cross-section, with a greater reduction in visceral adipose tissue, thereby reducing the ratio of visceral/subcutaneous adipose tissue. Rimonabant also reduced liver fat content, as well as systolic and diastolic blood pressure relative to placebo in this study [7].

Importantly, rimonabant inhibited the progression of atherosclerosis in humans and in animal models. Thus, in a randomized, double-blind, placebo-controlled, two-group, parallel-group Strategy To Reduce Atherosclerosis Development InVolving Administration of Rimonabant-the Intravascular Ultrasound Study (STRADIVARIUS) trial, patients with abdominal obesity and Metabolic Syndrome were randomized to receive rimonabant (20 mg daily) or matching placebo for 18 months [8]. In the treatment group, total atheroma volume decreased by −2.2 mm^3 vs. an increase of +0.88 mm^3 in the placebo group ($p = 0.03$). In the rimonabant group, HDL-C levels increased +5.8 mg/dl (+22.4%) vs. +1.8 mg/dl (+6.9%; $p < 0.001$) in the placebo group, and median triglyceride levels decreased −24.8 mg/dl (−20.5%) vs.

−8.9 mg/dl (−6.2%; $p < 0.001$) after placebo. Rimonabant-treated patients also displayed a greater decrease in hsCRP, less increase in HbA1c levels, a larger reduction in body weight and a greater decrease in waist circumference [8]. This promising result was not, however, translated into a reduced number of cardiovascular events, as observed in a large, double-blind, placebo-controlled Comprehensive Rimonabant Evaluation Study of Cardiovascular ENDpoints and Outcomes (CRESCENDO) trial involving 18,695 patients with previously manifested vascular disease or at increased cardiovascular risk [9]. At a mean follow-up of 13.8 months, the composite primary endpoint of cardiovascular death, myocardial infarction or stroke occurred in 364 (3.9%) patients assigned to rimonabant 20 mg and 375 (4.0%) assigned to placebo (hazard ratio 0.97, 95% CI 0.84–1.12) [9].

Consistent with the human data, CB1 receptor blockade also slowed atherogenesis in apoE-deficient mice [10]. Indeed, oral administration of rimonabant (8 mg/kg/day) to mice for three months significantly reduced the relative area of atherosclerotic lesions in the aorta, an effect accompanied by an increase in serum adiponectin levels, without effects on body weight or serum cholesterol levels. Rimonabant tended to increase serum HDL-C ($p = 0.05$); the relative area of atherosclerotic lesions in the aorta correlated inversely with serum HDL-C levels. The mechanism of the beneficial action of the drug on atherosclerosis involved activation of the RCT system, as suggested by up-regulation of the mRNA expression of SR-BI and ABCG1 in THP-1 macrophages [10].

The clinical development of rimonabant has, however, been hampered by its adverse safety profile. As early as 2007, a meta-analysis of published randomized controlled trials to assess the efficacy and safety of rimonabant revealed that the drug caused significantly more adverse events (odds ratio, 1.4; $p = 0.0007$), including serious adverse events (odds ratio, 1.4; $p = 0.03$), than did the placebo [11]. Furthermore, patients given rimonabant were 2.5 times more likely to discontinue the treatment because of depressive mood disorders than those who were given placebo (odds ratio, 2.5; $p = 0.01$). In addition, anxiety caused more patients to discontinue treatment in rimonabant groups compared with placebo groups (odds ratio, 3.0; $p = 0.03$) [11]. In the subsequent RIO programme, gastrointestinal side effects were the most common adverse effects observed in individual trials. However, when the RIO trials were combined, a small, but significant, increase in the incidence of depression and anxiety became apparent with rimonabant 20 mg [4]. Overall, discontinuation rates were similar across treatment groups, except discontinuation from adverse events, which occurred more frequently with 20 mg rimonabant vs. placebo [6]. Adverse events that occurred more frequently with rimonabant vs. placebo included dizziness (10.9% vs. 2.1%), nausea (8.7% vs. 3.6%), anxiety (5.8% vs. 3.6%), depressed mood (5.8% vs. 0.7%), and paresthesia (2.9% vs. 1.4%) [5]. In the STRADIVARIUS trial, psychiatric adverse effects were more common in the rimonabant group (43.4% vs. 28.4%, $p < 0.001$) [8]. In the ADAGIO-Lipids trial, gastrointestinal, nervous system, psychiatric and general adverse events were more common with rimonabant 20 mg; anxiety and depression were the most frequent adverse events [7]. As an inevitable corollary of such an unfavorable safety profile, the CRESCENDO trial of rimonabant 20 mg was prematurely discontinued because of concerns by health regulatory authorities in three countries about suicide and other serious side-effects in treated individuals [9]. Specifically, gastrointestinal (33% vs. 22%), neuropsychiatric (32% vs. 21%) and serious psychiatric side-effects (2.5% vs. 1.3%) were significantly increased with rimonabant compared with placebo. Moreover, four patients in the rimonabant group, and one in the placebo group, committed suicide [9]. The marked consistency of these findings resulted in the withdrawal of the approval of the drug by the European Medicines Agency in 2009.

At present, it is doubtful whether the benefits of rimonabant outweigh the risks [4]. Given the largely psychiatric profile of the adverse effects caused by the drug, selective development of CB1 blockers is required that are not able to cross the blood-brain barrier in order to limit the side effects of rimonabant on the central nervous system [12].

13.2 ESTROGENS

Estrogen supplementation can increase HDL-C levels by +5% to +15% [13]. For example, in the Postmenopausal Estrogen/Progestin Interventions (PEPI) trial, 28-day cycles of both unopposed estrogen and estrogen combined with progesterone significantly increased HDL-C by +2% to +9% in post-menopausal women [14]. Next, in the Heart and Estrogen/Progestin Replacement Study (HERS), post-menopausal women with coronary heart disease were randomized to placebo or conjugated equine estrogen (0.625 mg/day) plus medroxyprogesterone acetate (2.5 mg/day) [15]. At one year, patients in the estrogen/progestin group displayed an increment in HDL-C of +10% compared with the placebo group.

HDL-C raising by estrogens is paralleled by **increases in the number of circulating HDL particles**. Indeed, hormone replacement therapy with conjugated estrogens (0.625 mg/day) combined with medroxyprogesterone acetate (2.5 mg/day) raised HDL-C and HDL particle concentrations as measured by NMR and reduced LDL-C relative to placebo in post-menopausal women after one year [16]. In contrast, neither unopposed estrogen, nor estrogen with progestin, lowered LDL particle concentration.

The elevation of HDL-C levels by estrogens appears to be mediated by the **activation of hepatic apoA-I production** [17] and the **inhibition of hepatic lipase** [18]. Thus, oral contraception by estrogen plus progestin (30 μg ethinylestradiol, 75 μg gestodene) in normolipidemic young women for three months induced changes in apoA-I metabolism characterized by increases in the production rate and pool size, without an effect on the fractional catabolic rate of apoA-I [19]. An additional pathway contributing to HDL-C raising by estrogens may involve the **stimulation of cholesterol efflux** by estradiol fatty acyl esters which are known to incorporate into HDL. As a result, HDL-mediated macrophage cholesterol efflux potential can be enhanced [20]. Indeed, HDL enriched in estradiol oleate, as well as HDL obtained from premenopausal women, display superior cholesterol efflux capacity from THP-1 macrophages compared with HDL from age-matched men [20]; this effect is mediated by SR-BI and the estrogen receptor. Oral estrogen can also **induce ABCA1 expression** in peripheral cells, as observed in post-menopausal women on hormone replacement therapy [21].

In parallel to their effects on HDL-C, estrogens considerably modify the profile of circulating HDL particles. In the placebo-controlled, randomized Estrogen Replacement and Atherosclerosis (ERA) trial, conjugated equine estrogen (0.625 mg/day) alone or in combination with medroxyprogesterone acetate (2.5 mg/day) was adminstrated for 3.2 years to post-menopausal women with established coronary atherosclerosis [22]. Relative to placebo, both treatments induced elevation in HDL-C, apoA-I and large alpha-1, alpha-2 and pre-alpha-1 HDL subpopulations; such changes were accompanied by a reduction in plasma concentrations of remnant lipoprotein cholesterol. In contrast, HDL size distribution was shifted from large, light HDL2 to small, dense HDL3 particles by estrogen plus progestin in normolipidemic young women after three months [19]; differences in baseline metabolic profiles may account for the distinct results obtained in these studies.

The effects of estrogens on lipid metabolism may also vary depending upon the route of administration. Thus, oral administration of estrogen replacement therapy increased HDL-C and apoA-I levels, decreased total cholesterol and LDL-C and increased triglyceride concentrations in hysterectomized women after 12 weeks [23]. Furthermore, oral estrogen supplementation resulted in triglyceride enrichment of LDL and HDL particles. In clear contrast, transdermal administration of estrogen did not modify lipoprotein composition but improved the plasma triglyceride/HDL-C ratio. As triglyceride enrichment of HDL particles is often accompanied by deficient functionality [24], lipoprotein response to oral estrogen replacement therapy may involve the formation of functionally deficient HDL with the potential to deleteriously impact cardiovascular risk.

It is highly relevant in this regard that oral estrogen replacement therapy does not reduce cardiovascular risk despite consistent elevation in HDL-C concentration. Indeed, no benefit of such treatment in reducing coronary artery disease events was observed in the HERS study over approximately four years, with a pattern of early increased risk with oral estrogen/progestin treatment [15]. Moreover, when healthy, post-menopausal women with an intact uterus were randomized to either oral estrogen/progestin or placebo in the Women's Health Initiative trial, the Data Safety and Monitoring Board stopped the trial at a mean of 5.2 years duration in view of an increase in the rates of coronary artery disease, stroke, pulmonary embolism and invasive breast cancer in the active group [25]. Such adverse effects outweighed the reductions in the rates of colorectal cancers and hip fractures associated with estrogen/progestin therapy. Therefore, the current consensus is that oral estrogen therapy, although raising HDL-C concentrations, should not be used in the primary prevention of cardiovascular disease in post-menopausal women.

Interestingly, androgens reduce HDL-C in contrast to estrogens; a reduction in the cholesterol content of large HDL particles, subsequent to enhanced CETP activity resulting from attenuated expression of apoC-I, a CETP inhibitor, may underlie this effect [26].

13.3 OTHERS

Non-nucleoside reverse transcriptase inhibitors employed for HIV treatment constitute another class of drugs whose action involves an elevation in HDL-C levels [27]. This effect may be particularly important from a clinical viewpoint as HIV infection is associated with a significant risk of premature cardiovascular disease [28]. These drugs predominantly induce an increase in plasma levels of large HDL [29, 30] and appear to act via stimulation of apoA-I production in HIV-1-infected patients [31] and animal models [29]. Specifically, the expression and production rate of human apoA-I in the liver can be increased in mice expressing human apoA-I and treated with nevirapine or efavirenz, resulting in accelerated RCT [29]. In contrast, the direct effects of these drugs on cholesterol efflux from mouse macrophages and human monocyte-derived macrophages to HDL and apoA-I are negligible [32].

Next, **liver-selective thyromimetics** that selectively activate thyroid hormone receptor beta, increase plasma HDL-C, reduce total cholesterol through the induction of hepatic LDL receptors, promote RCT and protect from atherosclerosis through the induction of biliary sterol secretion in animal models [33–35]. The applicability of these findings to humans remains unclear.

Finally, several approaches appear to be capable of improving HDL quality in the absence of changes in HDL-C levels. Thus, **aspirin** can beneficially impact HDL quality without affecting HDL-C in mice via the stimulation of PON1 activity as a result of the induction of both PON1 and apoA-I gene expression in the liver [36]. This effect can be mediated by salicylate, a product of aspirin hydrolysis *in vivo*. Similarly, **selenium** increases serum PON1 activity and concentration in hypercholesterolemic rats in the absence of HDL-C elevation [37].

CLINICAL SUMMARY

Agents that raise HDL-C levels may differ markedly in their clinical effects.

REFERENCES

1. Sugamura K, Sugiyama S, Nozaki T, Matsuzawa Y, Izumiya Y, Miyata K, Nakayama M, Kaikita K, Obata T, Takeya M et al. Activated endocannabinoid system in coronary artery disease and antiinflammatory effects of cannabinoid 1 receptor blockade on macrophages. *Circulation* 2009; 119: 28–36.

2. Scheen AJ. CB1 receptor blockade and its impact on cardiometabolic risk factors: overview of the RIO programme with rimonabant. *J Neuroendocrinol* 2008; 20 (Suppl. 1): 139–146.

3. Wright SM, Dikkers C, Aronne LJ. Rimonabant: new data and emerging experience. *Curr Atheroscler Rep* 2008; 10: 71–78.

4. Doggrell SA. Is rimonabant efficacious and safe in the treatment of obesity? *Expert Opin Pharmacother* 2008; 9: 2727–2731.

5. Rosenstock J, Hollander P, Chevalier S, Iranmanesh A. SERENADE: the Study Evaluating Rimonabant Efficacy in Drug-naive Diabetic Patients: effects of monotherapy with rimonabant, the first selective CB1 receptor antagonist, on glycemic control, body weight, and lipid profile in drug-naive type 2 diabetes. *Diabetes Care* 2008; 31: 2169–2176.

6. Van Gaal L, Pi-Sunyer X, Despres JP, McCarthy C, Scheen A. Efficacy and safety of rimonabant for improvement of multiple cardiometabolic risk factors in overweight/obese patients: pooled 1-year data from the Rimonabant in Obesity (RIO) program. *Diabetes Care* 2008; 31 (Suppl. 2): S229–S240.

7. Despres J-P, Ross R, Boka G, Almeras N, Lemieux I for the, A.-L.I. Effect of rimonabant on the high-triglyceride/low-HDL-cholesterol dyslipidemia, intraabdominal adiposity, and liver fat: The ADAGIO-Lipids Trial. *Arterioscler Thromb Vasc Biol* 2009; 29: 416–423.

8. Nissen SE, Nicholls SJ, Wolski K, Rodes-Cabau J, Cannon CP, Deanfield JE, Despres JP, Kastelein JJ, Steinhubl SR, Kapadia S et al. Effect of rimonabant on progression of atherosclerosis in patients with abdominal obesity and coronary artery disease: the STRADIVARIUS randomized controlled trial. *JAMA* 2008; 299: 1547–1560.

9. Topol EJ, Bousser MG, Fox KA, Creager MA, Despres JP, Easton JD, Hamm CW, Montalescot G, Steg PG, Pearson TA et al. Rimonabant for prevention of cardiovascular events (CRESCENDO): a randomised, multicentre, placebo-controlled trial. *Lancet* 2010; 376: 517–523.

10. Sugamura K, Sugiyama S, Fujiwara Y, Matsubara J, Akiyama E, Maeda H, Ohba K, Matsuzawa Y, Konishi M, Nozaki T et al. Cannabinoid 1 receptor blockade reduces atherosclerosis with enhances reverse cholesterol transport. *J Atheroscler Thromb* 2010; 17: 141–147.

11. Christensen R, Kristensen PK, Bartels EM, Bliddal H, Astrup A. Efficacy and safety of the weight-loss drug rimonabant: a meta-analysis of randomised trials. *Lancet* 2007; 370: 1706–1713.

12. Taskinen MR. ADAGIO-Lipids gives promises but faces the setbacks. *Arterioscler Thromb Vasc Biol* 2009; 29: 339–340.

13. Rader DJ. Effects of nonstatin lipid drug therapy on high-density lipoprotein metabolism. *Am J Cardiol* 2003; 91: 18E–23E.

14. The Writing Group for the PEPI Trial. Effects of estrogen or estrogen/progestin regimens on heart disease risk factors in postmenopausal women. The Postmenopausal Estrogen/Progestin Interventions (PEPI) Trial. *JAMA* 1995; 273: 199–208.

15. Hulley S, Grady D, Bush T, Furberg C, Herrington D, Riggs B, Vittinghoff E. Randomized trial of estrogen plus progestin for secondary prevention of coronary heart disease in postmenopausal women. Heart and Estrogen/progestin Replacement Study (HERS) Research Group. *JAMA* 1998; 280: 605–613.

16. Hsia J, Otvos JD, Rossouw JE, Wu L, Wassertheil-Smoller S, Hendrix SL, Robinson JG, Lund B, Kuller LH. Lipoprotein particle concentrations may explain the absence of coronary protection Women's Health Initiative Hormone in the trials. *Arterioscler Thromb Vasc Biol* 2008; 28: 1666–1671.

17. Jin FY, Kamanna VS, Kashyap ML. Estradiol stimulates apolipoprotein A-I- but not A-II-containing particle synthesis and secretion by stimulating mRNA transcription rate in Hep G2 cells. *Arterioscler Thromb Vasc Biol* 1998; 18: 999–1006.

18. Jones DR, Schmidt RJ, Pickard RT, Foxworthy PS, Eacho PI. Estrogen receptor-mediated repression of human hepatic lipase gene transcription. *J Lipid Res* 2002; 43: 383–391.

19. Duvillard L, Dautin G, Florentin E, Jeannin A, Pais de Barros JP, Lagrost L, Petit JM, Gambert P, Verges B. Increased apolipoprotein AI production rate and redistribution of high-density lipoprotein size induced by estrogen plus progestin as oral contraceptive. *J Clin Endocrinol Metab* 2009; 94: 4891–4897.

20. Badeau RM, Metso J, Wahala K, Tikkanen MJ, Jauhiainen M. Human macrophage cholesterol efflux potential is enhanced by HDL-associated 17beta-estradiol fatty acyl esters. *J Steroid Biochem Mol Biol* 2009; 116: 44–49.

21. Darabi M, Rabbani M, Ani M, Zarean E, Panjehpour M, Movahedian A. Increased leukocyte ABCA1 gene expression in post-menopausal women on hormone replacement therapy. *Gynecol Endocrinol* 2011; 27: 701–705.

22. Lamon-Fava S, Herrington DM, Reboussin DM, Sherman M, Horvath K, Schaefer EJ, Asztalos BF. Changes in remnant and high-density lipoproteins associated with hormone therapy and progression of coronary artery disease in postmenopausal women. *Atherosclerosis* 2009; 205: 325–330.

23. Vrablik M, Fait T, Kovar J, Poledne R, Ceska R. Oral but not transdermal estrogen replacement therapy changes the composition of plasma lipoproteins. *Metabolism* 2008; 57: 1088–1092.

24. Kontush A, Chapman MJ. Functionally defective HDL: A new therapeutic target at the crossroads of dyslipidemia, inflammation and atherosclerosis. *Pharmacol Rev* 2006; 3: 342–374.

25. Rossouw JE, Anderson GL, Prentice RL, LaCroix AZ, Kooperberg C, Stefanick ML, Jackson RD, Beresford SA, Howard BV, Johnson KC et al. Risks and benefits of estrogen plus progestin in healthy postmenopausal women: principal results From the Women's Health Initiative randomized controlled trial. *JAMA* 2002; 288: 321–333.

26. Nantermet P, Harada S, Liu Y, Cheng S, Johnson C, Yu Y, Kimme D, Holder D, Hodor P, Phillips R et al. Gene expression analyses in cynomolgus monkeys provides mechanistic insight into high-density lipoprotein-cholesterol reduction by androgens in primates. *Endocrinology* 2008; 149: 1551–1561.

27. Meynard JL, Lacombe K, Poirier JM, Massot O, Lefebvre B, Bouchaud O, Guiard-Schmidt JB, Martinez A, Benlian P, Girard PM. Sustained increase of HDL cholesterol over a 72-week period in HIV-infected patients exposed to an antiretroviral regimen including lopinavir/ritonavir. *J Int Assoc Physicians AIDS Care (Chic Ill)* 2008; 7: 311–316.

28. Friis-Moller N, Reiss P, Sabin CA, Weber R, Monforte A, El-Sadr W, Thiebaut R, De Wit S, Kirk O, Fontas E et al. Class of antiretroviral drugs and the risk of myocardial infarction. *N Engl J Med* 2007; 356: 1723–1735.

29. Tohyama J, Billheimer JT, Fuki IV, Rothblat GH, Rader DJ, Millar JS. Effects of nevirapine and efavirenz on HDL cholesterol levels and reverse cholesterol transport in mice. *Atherosclerosis* 2009; 204: 418–423.
30. den Boer MAM, Westerterp M, de Vries-van der Weij J, Wang Y, Hu L, Espirito Santo SMS, Kooistra T, Reiss P, Romijn JA, Havekes LM et al. Ritonavir protects against the development of atherosclerosis in APOE*3-Leiden mice. *Atherosclerosis* 2010; 210: 381–387.
31. Franssen R, Sankatsing RR, Hassink E, Hutten B, Ackermans MT, Brinkman K, Oesterholt R, Arenas-Pinto A, Storfer SP, Kastelein JJ et al. Nevirapine increases high-density lipoprotein cholesterol concentration by stimulation of apolipoprotein A-I production. *Arterioscler Thromb Vasc Biol* 2009; 29: 1336–1341.
32. Mukhamedova N, Rose H, Cui HL, Grant A, Tchoua U, Dart A, Bukrinsky M, Sviridov D. Antiretroviral compounds and cholesterol efflux from macrophages. *Atherosclerosis* 2009; 206: 439–443.
33. Ness GC, Lopez D, Chambers CM, Newsome WP, Cornelius P, Long CA, Harwood HJ Jr. Effects of L-triiodothyronine and the thyromimetic L-94901 on serum lipoprotein levels and hepatic low-density lipoprotein receptor, 3-hydroxy-3-methylglutaryl coenzyme A reductase, and apo A-I gene expression. *Biochem Pharmacol* 1998; 56: 121–129.
34. Angelin B, Rudling M. Lipid lowering with thyroid hormone and thyromimetics. *Curr Opin Lipidol* 2010; 21: 499–506.
35. Tancevski I, Demetz E, Eller P, Duwensee K, Hoefer J, Heim C, Stanzl U, Wehinger A, Auer K, Karer R et al. The liver-selective thyromimetic T-0681 influences reverse cholesterol transport and atherosclerosis development in mice. *PLoS One* 2010; 5: e8722.
36. Jaichander P, Selvarajan K, Garelnabi M, Parthasarathy S. Induction of paraoxonase 1 and apolipoprotein A-I gene expression by aspirin. *J Lipid Res* 2008; 49: 2142–2148.
37. Kaur HD, Bansal MP. Studies on HDL associated enzymes under experimental hypercholesterolemia: possible modulation on selenium supplementation. *Lipids Health Dis* 2009; 8: 55.

CHAPTER 14

LIFESTYLE MODIFICATIONS

> **CLINICAL SUMMARY**
>
> A series of lifestyle modifications, including exercise, weight loss, moderate alcohol consumption and diet, may efficiently increase HDL-C levels, typically by +5% to +10%.

14.1 EXERCISE

Regular aerobic exercise raises plasma HDL-C levels by +6% to +25%, with an average of +4.6% provided by a meta-analysis [1]. Increases in HDL-C by **intense aerobic exercise** are even more pronounced and can attain +43% [2]. Regular aerobic exercise therefore represents a commonly recommended means of raising plasma HDL-C [3]. The HDL-raising effect of regular exercise is dose-dependent, increasing with weekly exercise volume [4]. Indeed, a meta-analysis reveals that exercise volume is the primary determinant of effects on HDL-C at a threshold of approximately 700–1500 kcal/week of energy expenditure through training [5]. As a corollary, lack of exercise is associated with low HDL-C concentrations, as documented in the NHANES cohort [6]. In a similar fashion, a sedentary lifestyle involving excessive television viewing is associated with low HDL-C in adolescence [7].

High-Density Lipoproteins: Structure, Metabolism, Function, and Therapeutics, First Edition.
Anatol Kontush and M. John Chapman.

Significantly, physical training is capable of raising HDL-C in both normo- and dyslipidemic subjects, including those with cardiovascular disease [8], polycystic ovary syndrome [9] and those on hemodialysis [10]. The response of HDL-C to exercise training is, however, heterogeneous and can be enhanced in subjects with low HDL-C concentrations at baseline [11], such as in those with hypertriglyceridemia and cardiovascular disease [12]. Indeed, chronic aerobic exercise increases HDL-C by approximately +9% in patients with cardiovascular disease [8], an effect that is greater than that observed in a healthy population [1, 13, 14]. Such an inverse relationship between the HDL-raising effect of exercise and baseline HDL-C levels may also contribute to greater lipoprotein responses to exercise in men compared with women [15]; however, the latter finding remains controversial [13, 14].

Increases in the level of physical activity are associated with increases in HDL-C across several ethnic groups, as documented in the ARIC study [16]. In this study, exercise raised HDL-C by 3–5 mg/dl per 180 metabolic equivalent-min/week corresponding to 1h walking or 0.5h playing basketball.

MAJOR CONTRIBUTION

Regular aerobic exercise raises plasma HDL-C levels by +6% to +25% [1]; increases in HDL-C by intense aerobic exercise can attain +43% [2]. More specifically, exercise raises HDL-C by 3–5 mg/dl per 180 metabolic equivalent-min/week corresponding to 1h walking or 0.5h playing basketball [16].

Interestingly, decreases in cardiovascular risk under the action of physical exercise are typically greater than can be expected from epidemiologic data on the relationship between the risk and HDL-C. Indeed, epidemiologic studies document a cardiovascular risk reduction of −2% to −3% for each +1 mg/dl increase in HDL-C; as exercise raises HDL-C by +1 to +2 mg/dl [1, 17], it should only decrease cardiovascular risk by −3% to −6%, much less than the reductions of up to −40% observed in epidemiologic studies [18, 19]. This discrepancy suggests that other mechanisms that are not directly related to HDL should contribute to the cardioprotective effects of physical training. As regular exercise efficiently decreases circulating triglycerides, such a lipid-lowering effect may provide an additional contribution to the cardioprotective action of physical training.

Physical exercise markedly impacts the HDL particle profile. First, exercise often exerts opposite effects on plasma levels of **large HDL2-C** and small HDL3-C, increasing the former and decreasing the latter [9, 20]. The net effect of exercise on plasma HDL-C levels can, therefore, be weaker than that on HDL2-C. Indeed, a meta-analysis demonstrates that increases in HDL2-C induced by aerobic exercise average +11% [21]. As a consequence, exercise increases the mean size of HDL, as revealed by NMR spectroscopy and electron microscopy [9, 22, 23]. Another frequently observed effect of aerobic exercise involves elevated plasma levels of **pre-beta HDL** [24–27]. Inconsistent with these data, however, pre-beta HDL levels did not differ between trained athletes and sedentary subjects [24].

The composition of HDL particles can also be modified by exercise, which impacts both the lipidome and proteome. Indeed, exercise modulates the metabolism of **apoA-I**, as suggested by the elevated plasma levels of this apolipoprotein observed after exercise in some [28–30] but not all [31, 32] studies. In addition, plasma levels of apoA-I are higher in trained subjects compared with a sedentary population [26, 33]. Importantly, the HDL particle content of apoA-I is elevated in trained athletes who also display a diminished HDL content of apoA-II, indicative of the replacement of apoA-II by apoA-I [23]. This effect is probably caused by the delayed clearance of apoA-I from the circulation as a result of a decreased fractional catabolic rate [31, 34]. In contrast, apoA-I production remains unaffected by exercise [31, 34]. Like apoA-I, the catabolism of apoA-II can be also slowed, but presumably to a lesser extent.

It is probable that a **decreased activity of CETP** underlies the HDL-C-raising effect of exercise (Fig. 14.1). Indeed, both the activity and mass of CETP are decreased in athletes compared with controls [23], whereas physical inactivity increases CETP mass with decreases in HDL-C and increases in triglycerides [35]. Moreover, intense exercise decreases CETP activity [36]. Data regarding the relationship between exercise and CETP activity are, however, conflicting, as no difference in CETP activity or mass was observed between trained and sedentary subjects [25].

Other key elements of HDL metabolism that are affected by exercise involve hydrolytic lipases. Indeed, regular training appears to **decrease hepatic lipase activity** [37, 38] (Fig. 14.1). In contrast, intense exercise increases the activity of hepatic lipase in trained subjects [39, 40]. Next, both intense and regular exercise **increase LPL activity** (Fig. 14.1), potentially contributing to the HDL-raising effects of physical training via the enhanced production of surface remnants during accelerated catabolism of triglyceride-rich lipoproteins [4, 41, 42].

Exercise significantly modulates enzymatic activities associated with HDL. Both intense and chronic exercise **increase LCAT activity** [27, 43, 44] (Fig. 14.1). Consistent with these data, LCAT activity and mass is increased in trained subjects compared with sedentary controls [23, 25, 45, 46]. Both the activity and mass of HDL-associated PON1 are also augmented in athletes [23].

Along with an elevation in HDL-C concentration, physical exercise can improve the anti-atherogenic functionality of HDL. **Cholesterol efflux** from fibroblasts and macrophages to serum is greater in trained athletes than in sedentary controls [25, 47]. In addition, increased physical activity induces cholesterol efflux from macrophages to plasma of obese women after substantial weight reduction [48]. **Elevated expression of ABCA1**, as observed in hepatic and peripheral cells, may contribute to this phenomenon [26, 27, 49]. Increased plasma levels of pre-beta HDL, a potent acceptor of cholesterol via ABCA1,

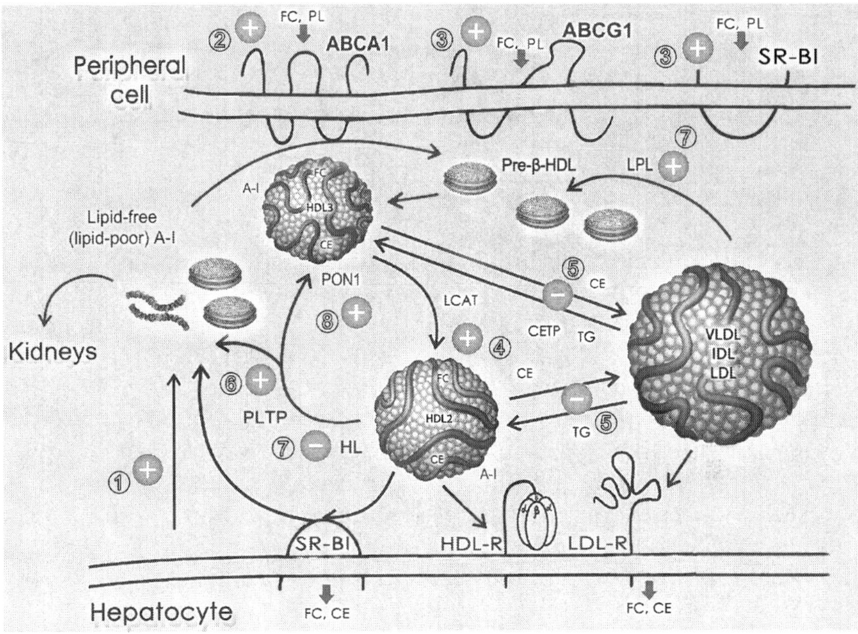

Figure 14.1 Major pathways involved in the beneficial influence of lifestyle modification on HDL. (1) Stimulation of hepatic production of HDL by blocking endocytic recycling of apoA-I (PUFAs). (2) Accelerated efflux of cholesterol from peripheral cells and enhanced lipidation of apoA-I as a result of the induction of ABCA1 (alcohol, PUFAs, polyphenols). (3) Accelerated efflux of cholesterol from peripheral cells and enhanced formation of HDL as a result of the induction of ABCG1 and SR-BI (polyphenols). (4) Induction of cholesterol esterification by LCAT (exercise, alcohol, PUFAs). (5) Decreased activity of CETP which results in the diminished heteroexchange of cholesteryl ester and triglyceride between HDL and triglyceride-rich lipoproteins (exercise, alcohol, polyphenols). (6) Accelerated HDL remodeling as a result of the induction of PLTP (alcohol). (7) Normalization of HDL remodeling by hydrolytic enzymes involving decreased activity of HL and increased activity of LPL (exercise, PUFAs). (8) Induction of HDL-associated PON1 (alcohol, polyphenols). CE, cholesteryl ester; EL, endothelial lipase; FC, free cholesterol; HDL-R, holoparticle HDL receptor; HL, hepatic lipase; LDL-R, LDL receptor; PL, phospholipid; TG, triglyceride.

following exercise and weight reduction provide additional support for the importance of this pathway [24–27, 48].

HDL from athletes display elevated *in vitro* **antioxidative activity** compared with controls [23]. Furthermore, regular aerobic training improves HDL-mediated protection of LDL against oxidative stress in patients with Type 2 diabetes [50] and Metabolic Syndrome [51]. Such functional improvement, which is characteristic for both large HDL2 and small HDL3 particles [51, 52], can be accounted for by elevated HDL content of apoA-I and increased content and activities of LCAT and PON1.

Finally, exercise may beneficially impact the anti-inflammatory activity of HDL. Indeed, elevation of HDL-C induced by lifestyle intervention in healthy adults correlates with a reduction of vascular inflammation, as assessed by vascular 18F-fluorodeoxyglucose uptake, a marker of arterial inflammation [53].

14.2 ALCOHOL

Numerous studies have documented the HDL-raising effect of alcohol consumption [54]. **Moderate alcohol consumption** of about two drinks per day (about 20–30 g alcohol) in men and one drink per day (about 10–15 g alcohol) in women increases HDL-C by approximately +10%, independent of age and BMI. The effect of alcohol on HDL-C is dose-dependent [55]; **elevated alcohol consumption** leads to further elevation of HDL-C levels [56]. Specifically, alcohol consumption of 30–40 g/day (2–3 drinks/day) or more has been shown to increase HDL-C levels by +12 to +15%. In contrast, alcohol consumption is associated with deleterious effects on plasma triglycerides whose circulating levels are elevated. The beneficial effect of alcohol on HDL levels can be of major importance for public health, as moderate alcohol consumption is associated with a decreased risk of cardiovascular disease in numerous epidemiologic studies [57–59].

The relationship between alcohol intake and cardiovascular risk is J-shaped with minimal risk associated with moderate amounts of alcohol consumption. An increase in cardiovascular risk occurs at heavy alcohol intake (over about 60 g, equivalent to approximately four drinks per day) despite further elevation in HDL-C levels compared with moderate intake. The adverse effects of heavy alcohol consumption on other cardiovascular risk factors, such as blood pressure and inflammation, appear to account for this phenomenon [60]. Interestingly, the epidemiologic association between moderate alcohol consumption and reduced cardiovascular risk can, at least in part, reflect the better social and health status of moderate alcohol drinkers [59], calling for future studies with rigorous multiple adjustment for potential confounders.

The effects of alcohol intake on HDL may differ in healthy subjects and dyslipidemic patients, as suggested by the blunted action of red wine consumption on HDL-C levels in hypercholesterolemic individuals [61]. The HDL-raising effect of alcohol can also be modulated by genetic background, primarily by genes involved in HDL metabolism, as exemplified by the higher HDL-C and apoA-I levels typical of carriers of the B1 Taq1B polymorphism in the CETP gene upon alcohol consumption compared with the B2 carriers [62, 63]. In addition, the presence of the E2 or E3 allele of the APOE gene, or the C-480T polymorphisms in the LIPC gene may augment the response of HDL-C to alcohol intake [54]. In contrast, the relationship of alcohol consumption to HDL-C appears to be independent of behavioral factors, such as smoking [64]. Controversially [55], the effects of alcohol intake on HDL levels may depend on the type of alcoholic beverage [65], with both liquor [66] and wine [60] intake reportedly being most strongly related to increased HDL-C.

Importantly, the effects of alcohol on the lipoprotein profile in general, and on HDL in particular, account for the greatest part of the association between moderate alcohol intake and reduced cardiovascular risk [67, 68]; antithrombotic, anti-inflammatory and antidiabetic actions of alcohol may contribute to the rest [68]. Indeed, levels of HDL-C alone explained approximately 50% of the observed association between alcohol and cardiovascular risk in the large-scale Health Professionals Follow-Up Study, Nurses' Health Study and Honolulu Heart Program [69, 70].

Moderate alcohol intake **raises plasma levels of small HDL3** to a greater extent than those of large HDL2; however, heavy alcohol consumption is associated with preferential elevation of HDL2 [71]. NMR data reveal that the numbers of medium and large HDL particles gradually increase with alcohol consumption, leading to larger mean HDL size [72]. Moderate alcohol intake also increases circulating levels of small, lipid-poor, pre-beta HDL [73]. Plasma concentrations of other lipoprotein subspecies, including LpA-I and LpA-I:A-II, as well as major HDL apolipoproteins, including apoA-I, apoA-II and apoE, are also elevated by alcohol, but to a variable extent [74–76].

Alcohol intake also alters the lipid composition of HDL. Indeed, the HDL **content of phospholipid and of PUFAs tends to increase**, whereas HDL triglycerides do not change [77, 78]. In addition, alcohol promotes the conversion of HDL phosphatidylcholine to phosphatidylethanol, an alteration that modulates multiple aspects of HDL functionality, such as anti-inflammatory and cytoprotective activities, related, in part, to altered binding to endothelial cells [67, 79, 80]. Specifically, phosphatidylethanol-enriched HDL particles activate intracellular signaling pathways that involve CD36 and Cla-1 and which enhance the production of vascular endothelial growth factor [80].

OPEN QUESTION

Are HDL particles formed following alcohol ingestion fully functional?

Alcohol intake affects multiple anti-atherogenic activities of HDL. First, HDL isolated after alcohol ingestion frequently [73, 81, 82], but not always [83, 84], displays an **elevated capacity to efflux cholesterol** from cells, primarily macrophages. Compositional changes in HDL, involving elevated content of phospholipid [81] and enhanced particle surface fluidity as a consequence of elevated content of PUFA [78], may facilitate cholesterol efflux. Intriguingly, such changes may also account for the enhanced antioxidative and anti-inflammatory activities of HDL particles [77, 85], suggesting the existence of a common mechanism underlying improved HDL function (see chapter7).

Several additional steps involved in RCT are affected by alcohol (Fig. 14.1). First, cellular **ABCA1** expression is elevated by alcohol intake [26, 73]; second, **LCAT** activity is increased; third, **CETP** activity is diminished following heavy alcohol consumption; fourth, **PLTP** activity is enhanced [81]; and, fifth HDL binding to hepatic SR-BI is elevated in alcohol consumers [86]. In contrast, alcohol abuse may decrease the transfer of cholesteryl ester from HDL to hepatocytes [83, 84].

Finally, an increase in the activity of the HDL-associated anti-atherogenic enzyme **PON1** is induced by alcohol [54, 87] (Fig. 14.1); such an effect may result in improved anti-inflammatory and antioxidative properties of HDL. In contrast, the plasma activity of $LpPLA_2$ was unaffected by alcohol intake in overweight subjects [88].

14.3 NUTRITIONAL FACTORS

Circulating levels of HDL are strongly influenced by nutrition. Major nutritional factors that impact on HDL-C concentrations involve dietary content of individual fats and carbohydrates; a number of other nutrients, such as vitamins and polyphenols may also modulate HDL metabolism and are capable of HDL raising, as well as of enhancing HDL functionality [3].

OPEN QUESTIONS

Does HDL-C raising induced by individual nutrients result in the enhancement of HDL functionality?
How do nutritional factors interact with genes determining HDL metabolism?

Dietary Fats

Dietary fats potently modulate plasma lipoprotein concentrations and play a key role in the development of coronary heart disease. **High-fat diets** typically **increase both HDL-C and LDL-C levels** [89], whereas **low-fat diets** may **decrease HDL-C**, apoA-I and LDL-C in normal weight, overweight and obese subjects [90–92]. However, the latter statement does not always hold true and HDL-C concentrations may remain stable when a significant reduction of fat content in the diet is accompanied by considerable weight loss [93–95]. The effects of low-fat diets on HDL metabolism may involve decelerated production and accelerated catabolism of apoA-I [96, 97], as well as reduced production and catabolism of HDL-apoA-II, as observed in obese men with Metabolic Syndrome [98]. By contrast, high-fat diets delay apoA-I catabolism [97].

In parallel to decreases in HDL-C and LDL-C, a low caloric diet is capable of reducing both the plasma cholesterol efflux capacity [99] and lactonase activity of PON1 [90]. The HDL particle profile may also be affected, with decreased formation of small pre-beta HDL [100].

Cholesterol. Individual lipid classes differ considerably in their effects on plasma HDL-C. Raising dietary cholesterol intake results in elevated plasma concentrations of total cholesterol, LDL-C and HDL-C [101]. Such **HDL-C-raising action** of dietary cholesterol is largely accounted for by elevated levels of HDL2-C, rather than those of HDL3-C [101]. Importantly, cholesterol feeding normally increases LDL-C to a greater extent than HDL-C, causing dyslipidemia and accelerating atherosclerosis [102]. A dietary source of cholesterol may be important for the net effect on the balance between LDL and HDL, as an addition of one egg per day to a normal fat diet can raise HDL-C levels without influencing LDL-C [103].

Phospholipids. Phospholipids from various sources (i.e., soybean, safflower, egg and fish roe) can exert beneficial effects on plasma lipid and lipoprotein levels which involve **HDL-C raising**. For example, dimiristoyl phosphatidylcholine (but not lecithin) raises HDL-C and apoA-I, improves HDL function and prevents the development of atherosclerotic lesions in apoE-null mice [104] and in cholesterol-fed rabbits [105]. The primary site of action of dietary phospholipids appears to be in the intestinal lumen, where they interfere with sterol absorption, although enrichment of HDL in phospholipid with subsequent enhancement of anti-atherogenic activies of HDL may also occur. Dietary phospholipids can also stimulate bile acid and cholesterol secretion [106]. Thus, dietary polyunsaturated phosphatidylcholine increases fecal sterol excretion in humans [107]. In addition, phosphatidylinositol accelerates biliary sterol excretion in rats [108].

Saturated Fatty Acids. Dietary fatty acids are even more potent regulators of plasma lipoprotein concentrations than dietary cholesterol in man. Fatty acids and cholesterol in the diet may, however, interact in their effects on plasma lipoprotein concentrations.

High intakes of most fatty acids **raise HDL-C**, with saturated fatty acids exerting the strongest effects. Indeed, an increase in the intake of saturated fatty acids relative to monounsaturated fatty acids and PUFAs or to carbohydrate increases HDL-C concentrations in both animals and humans when dietary intake of cholesterol is kept constant [109–111]. This effect can probably be explained by the delayed clearance of apoA-I from plasma. In contrast, when dietary consumption of total fat, saturated fat and cholesterol is reduced, decreased hepatic expression and diminished secretion of apoA-I by the liver ensue, resulting in HDL-C lowering [96, 112, 113]. Such HDL-C reduction resulting from decreased saturated fat intake is not associated with marked changes in HDL particle composition [114]. Saturated fat can, however, deleteriously impact anti-atherogenic functionality of HDL. Thus, the consumption of a saturated fat reduces the anti-inflammatory potential of post-prandial HDL and impairs arterial endothelial function in humans [115].

In parallel, saturated fatty acids increase LDL-C levels by decreasing LDL catabolism via the LDL receptor [102]. Importantly, such an **LDL raising** effect frequently exceeds the extent of the HDL raising, resulting in an elevated LDL-C/HDL-C ratio. A reduction in the saturated fat intake in the appropriate dietary context therefore represents an essential approach in correcting dyslipidemia and decreasing cardiovascular risk. Indeed, animal studies conclusively demonstrate that diets poor in saturated fat reduce the LDL-C/HDL-C ratio and delay atherogenesis [116]. Accordingly, dietary saturated fat intake is recommended by the National Cholesterol Education Program (NCEP) not to exceed 7% of total energy in order to reduce coronary heart disease risk [117]. Reducing saturated fat intake, primarily as a means of lowering LDL-C, is therefore a focus of dietary recommendations for the prevention and treatment of cardiovascular disease.

There are differences between individual types of saturated fatty acids with regard to their effects on LDL-C and HDL-C [118]. For example, lauric acid results in greater increases in both LDL-C and HDL-C compared with longer chain saturated fatty acids [118]. The net effect of saturated fat on lipoprotein metabolism may also depend on the dietary intervention regime. For instance, the substitution of saturated fat for carbohydrate increases HDL-C concentrations, with no effect on the total cholesterol/HDL-C ratio [118].

Monounsaturated Fatty Acids. Dietary monosaturated fatty acids exert hypocholesterolemic effects when substituted for saturated fatty acids [102, 119–121]. Accelerated clearance of LDL apoB may underlie this phenomenon [122]. The substitution of saturated fat by monounsaturated fat in the diet may also lead to minor reduction in HDL-C concentrations [102, 120, 122]. When isoenergetically substituted for saturated fat, monounsaturated fat is, however, associated with a more favorable plasma lipoprotein profile, specifically, a lower ratio of LDL-C to HDL-C. When **substituted for carbohydrate**, monosaturated fatty acids result in significant **increases in HDL-C** and reductions in the total cholesterol/HDL-C ratio [118]. The addition of monosaturated fatty acids to the diet may induce HDL raising, as documented by increased levels of HDL-C and apoA-I in children supplemented with olive oil-enriched skim milk [123] and in hypercholesterolemic patients [124]. Despite that, switching to a monounsaturated fat diet is insufficient in providing protection against the development of atherosclerosis, as shown in monkeys [125].

In clear contrast, trans isomers of monounsaturated fatty acids (primarily elaidic acid), which are formed during hydrogenation, exert deleterious effects on the plasma lipoprotein profile when compared with their cis counterparts [102, 126–129]. When added to the diet, trans-monosaturated fatty acids typically elevate LDL-C and reduce HDL-C concentrations.

PUFAs. When **substituted for carbohydrate**, PUFAs **increase HDL-C** and decrease the total cholesterol/HDL-C ratio [118]. On the other hand, the replacement of saturated fat with polyunsaturated fat decreases total cholesterol, LDL-C and HDL-C [130], consistent with the observation that all fatty acids elevate HDL-C when substituted for carbohydrates, but the effect diminishes with increasing unsaturation of the fatty acids [120].

n-3 Fatty acids found in fish oil, especially eicosapentaenoic acid, are frequently, but not always [131], capable of increasing HDL-C and apoA-I levels when added to the diet. Indeed, n-3 PUFA raises HDL-C in patients with mixed dyslipidemia [132] and hypercholesterolaemia [133]. Earlier meta-analysis of the effects of omega-3 fatty acids (at doses of 1–7 g/day for at least 2 weeks in 72 placebo-controlled trials) on HDL-C levels suggested a statistically significant increase in HDL-C vs. placebo in subjects with triglyceride concentrations of less than 177 mg/dl, although the relative HDL-C increase from baseline with omega-3 fatty acid treatment was only about +5% [134]. However, even such a modest effect is not fully supported by more recent analyses which detected only a moderate (+1.6 mg/dl) [135], or no [131], effect of omega-3 fatty acid supplementation on HDL-C. Such a discrepancy may, in part, reflect differences in the fatty acid formulations between the studies. Indeed, when purified algal docosahexaenoic acid was supplemented alone, it resulted in a mean HDL-C increment of +6.5% at a mean dose of 2092 mg/d across eight studies [136].

The HDL-raising activity of PUFAs is further consistent with epidemiologic data documenting positive associations of plasma concentrations of n-6 and n-3 PUFAs with HDL-C [137, 138]. n-6 PUFAs reportedly display stronger HDL-C-raising properties compared to their n-3 counterparts, as suggested by data obtained in apoE -/- mice [139]. PUFA supplementation may be particularly efficient in combination with plant sterols, which additionally lower LDL-C [133].

HDL-C elevation by PUFAs primarily reflects the increased cholesterol content of **large HDL2** particles [140, 141]. Supplementation with n-3 PUFAs redistributes cholesterol carried by HDL from small HDL3 to large HDL2 particles; these changes may exert opposite actions on the net level of HDL-C, potentially accounting for the null effect observed in some studies [140]. Consistent with these data, serum levels of n-6 PUFAs, specifically those of linoleic acid and arachidonic acid, are positively associated with circulating concentrations of large HDL in middle-aged men from the USA [142].

Along with changes in HDL-C, PUFAs markedly lower triglyceride concentrations [131]. Indeed, PUFA-rich diets favor the production of small, triglyceride-poor VLDL particles by the liver, thereby reducing circulating triglyceride levels. Such potentially cardioprotective effects of PUFAs are associated with the down-regulation of hepatic genes of triglyceride synthesis and may contribute to HDL raising via **reduced CETP activity** [143]; through this pathway, PUFAs may beneficially influence the chemical composition of HDL.

The results obtained in animal models suggest that mechanisms contributing to HDL raising by PUFAs may involve: (i) elevated hepatic production of nascent HDL as a result of **enhanced expression of apoA-I and ABCA1** [139]; (ii) the stimulation of hepatic secretion of spherical HDL by partial block of the endocytic recycling of apoA-I [144]; (iii) **elevated** hepatic expression and plasma activity of **LCAT** [139, 141, 145]; and, (iv) normalization of HDL remodeling by hydrolytic enzymes involving **decreased activity of hepatic lipase** and **increased activity of LPL** [141] (Fig. 14.1). It remains to be shown whether these pathways can be specifically affected by distinct PUFA subclasses.

Importantly, HDL-raising after a PUFA diet may **beneficially impact RCT**. Indeed, fish oil accelerates RCT from macrophages with increased fecal excretion in mice [143]. Augmented hepatic excretion of macrophage-derived and HDL-derived cholesterol with the participation of hepatic SR-BI, ABCG5 and ABCG8 may provide a mechanistic basis for this activity [139, 143, 145]. **Cholesterol efflux** from macrophages via ABCA1 can also be beneficially affected by a fish oil diet, presumably reflecting elevated surface fluidity of HDL particles [143]. The improved cholesterol-transporting capacity of HDL particles following a PUFA diet is further consistent with the absence of changes in cholesterol efflux from macrophages in healthy males, despite significant decreases of HDL-C when compared with a saturated fatty acid diet [146]. Furthermore, the consumption of polyunsaturated fat improves the **anti-inflammatory activity** of post-prandial HDL in humans [115]. In addition marine n-3 fatty acids may reduce plasma levels of pro-atherogenic $LpPLA_2$, as suggested by the inverse correlation between the content of eicosapentaenoic acid in adipose tissue and plasma $LpPLA_2$ [147].

Finally, PUFAs exert multiple cardioprotective effects, independently of plasma lipoproteins, which include decreases in platelet aggregation, immune response and blood pressure [102, 131, 137]. Together with actions on plasma HDL and triglyceride, such effects may form a mechanistic basis for the reduction of cardiovascular risk following PUFA supplementation. Indeed, clinical trials that replaced saturated fat with polyunsaturated fat have generally shown a reduction in cardiovascular events, although several studies showed no effects [130]. As a corollary, dietary intake of PUFAs is recommended to constitute up to 10% of total energy [117].

Carbohydrates

Dietary carbohydrates are associated with atherogenic dyslipidemia. Elevated dietary intake of carbohydrates may, therefore, significantly contribute to the prevalence of dyslipidemia. Indeed, carbohydrate intake was **inversely associated with plasma HDL-C** and positively associated with plasma triglyceride concentrations in white participants in the ARIC study [148]. In a similar fashion, increases in carbohydrate intake associated with decreases in HDL-C levels were observed in participants of the Strong Heart Family Study [149]. Consistent with these data, a representative sample survey of USA residents derived from the third NHANES cohort reveals that high glycemic load is associated with low HDL-C concentrations in multivariate analysis. Specifically, HDL-C levels are inversely related to the intake of simple carbohydrates, rather than complex carbohydrates [150]. Furthermore, foods high in complex carbohydrates and fibre are associated with a reduced risk of coronary heart disease and Type 2 diabetes [151]. These findings suggest that low glycemic load diets may exert beneficial effects on cardiovascular risk via their action on HDL-C [152].

It is important in this context that added sugars (caloric sweeteners used as ingredients in processed or prepared foods) represent an increasing and potentially modifiable component in the diet. Cross-sectional studies among USA adults ($n = 6113$) from the NHANES 1999–2006 reveal a mean of 15.8% of consumed calories from added sugars, much higher than recommended by the WHO (<10%)

or the American Heart Association (<5%) [153]. Among participants consuming less than 5%, 5% to less than 17.5%, 17.5% to less than 25% and 25% or greater of total energy as added sugars, adjusted mean HDL-C levels were, respectively, 58.7, 57.5, 53.7, 51.0, and 47.7 mg/dl ($p < 0.001$ for trend), geometric mean triglyceride levels were 105, 102, 111, 113, and 114 mg/dl ($p < 0.001$ for trend), and LDL-C levels were 116, 115, 118, 121, and 123 mg/dl ($p = 0.047$ for trend) [153]. Similar data were obtained in a cross-sectional study of 2157 US adolescents in the NHANES 1999–2004, which demonstrated an inverse correlation between added sugar intake and HDL-C concentrations [154]. Furthermore, soft drink intake was associated with low HDL-C in the Oslo Health study [155]. Elevated fructose intake, which induces hepatic lipogenesis and VLDL secretion, can account for the dysmetabolic effects observed. Dietary consumption of added sugars should, therefore, be reduced in order to increase HDL-C levels and correct dyslipidemia [153].

As an important therapeutic corollary, carbohydrate-restricted diets significantly decrease body weight and independently improve plasma HDL-C and triglycerides [156–159]. Further along this line, a low carbohydrate Mediterranean diet induced elevation of HDL-C concentrations in overweight diabetic patients after 12 months [160]. In contrast, a traditional Mediterranean diet and the 2003 American Diabetic Association diet were ineffective in this study.

Proteins

Numerous studies demonstrate that **soy protein** may improve plasma lipoprotein profile [161]. However, the beneficial effects of soy protein are largely confined to decreases in LDL-C and total cholesterol, whereas effects on HDL-C levels are inconsistent [162]. Specifically, long-term (>12 weeks) treatment and high doses appear to be required for HDL raising by soy protein supplementation [163].

Indeed, a recent meta-analysis of 30 randomized controlled trials revealed that the inclusion of modest amounts of soy protein (about 25 g/day, or 25% of the usual total daily protein intake) in the diet of adults with normocholesterolemia or mild hypercholesterolemia resulted in small decreases in total cholesterol and LDL-C. In contrast, HDL-C levels remained unchanged [162]. On the other hand, soy protein may improve HDL functionality, even in the absence of changes in HDL-C, as suggested by the elevated serum activity of PON1 observed in a double-blind, randomized trial of 50 g/day of soy protein consumed by hypercholesterolemic post-menopausal women for 10 weeks [164].

Interestingly, the active components of soy protein responsible for LDL-lowering are unlikely to include the protein component itself, but rather appear to involve soy isoflavones, which are adsorbed to soy protein [165], and/or polyunsaturated fats, fiber, vitamins and minerals [166].

Another example of the influence of dietary protein on HDL metabolism is provided by **gluten**, a composite of proteins from grass-related grains that displays HDL-lowering properties [167]. Celiac disease is characterized by a gluten-induced villous atrophy of the upper small intestine which has an active role in lipoprotein metabolism. As a result, HDL-C levels are low in this condition [168]. When a gluten-free diet is employed to treat celiac disease, elevation in HDL-C, total cholesterol and apoA-I levels is observed [169]; the HDL-raising effect of the diet may be explained by increased secretion of apoA-I by intestinal cells [169].

Minor Dietary Components

Polyphenols. Polyphenols, a group of chemical substances found in plants, possess HDL-raising properties and can improve HDL function.

Berries are a particularly rich source of polyphenols. A single-blind, randomized, placebo-controlled trial revealed that moderate amounts of berry consumed for eight weeks increased HDL-C (+5.2%) in middle-aged subjects without affecting total cholesterol and triglyceride levels [170]. Similarly, HDL-C levels can be raised by polyphenol-rich grapefruit and grapefruit juice in obese adults [171]. In addition, berry consumption inhibits platelet function and decreases blood pressure.

Anthocyanins, a subclass of polyphenolic compounds that exert beneficial effects on lipid profiles in animal models, are major components in the HDL-raising effect of berries. Indeed, berry-derived anthocyanin supplements (160 mg twice daily for 12 weeks) increased HDL-C (+13.7%) and decreased LDL-C concentrations (−13.6%) in a double-blind, randomized, placebo-controlled trial performed in dyslipidemic subjects [172].

Dark chocolate is one of the richest sources of polyphenols, such as flavonoids. When given for a week to healthy volunteers to provide 700 mg of flavonoids/day, dark chocolate raised HDL-C by +9% and reduced LDL-C by −6% [173]. Decreased platelet reactivity was also observed at the end of the study. Furthermore, high-cocoa polyphenol-rich chocolate (45 g/d for eight weeks) increased HDL-C levels by +8.6% in Type 2 diabetes patients [174].

Vegetables that are high in polyphenols include garlic, among others. Consistent with the capacity of polyphenols to elevate HDL-C, extended-released garlic powder tablets, Allicor (600 mg daily), increased HDL-C levels by +11.5% in patients with mild hypercholesterolemia in a double-blinded, placebo-controlled, randomized study when given for 12 weeks [175]. In parallel, LDL-C was decreased by up to −13.8%. Data in animal models support findings in humans, as exemplified by the fact that apple polyphenols supplemented in the diet at 0.6% increased HDL-C and decreased triglycerides in Golden Syrian hamsters maintained on a 0.1% cholesterol diet [176].

Mechanisms responsible for the HDL-raising action of polyphenols may involve **CETP inhibition** (Fig. 14.1). Indeed, in dyslipidemic subjects, both the mass and activity of plasma CETP are decreased by anthocyanins (−10.4% and −6.3%, respectively) [172]. In a similar fashion, apple polyphenols suppress plasma CETP activity, but did not influence CETP mass in hamsters[176]. Separate *in vitro* experiments provide evidence for direct action of anthocyanines [172] and apple polyphenols [176] on CETP.

In addition to raising HDL levels, polyphenols can improve the anti-atherogenic activities of HDL. Indeed, anthocyanines increase cellular **cholesterol efflux** to serum in dyslipidemic humans by +20%, a level that exceeds elevation of HDL-C concentrations of +13.7%, and is indicative of the improved HDL capacity to efflux cholesterol [172] (Fig. 14.1). Resveratrol, a major polyphenolic constituent of red wine, enhances apoA-I-mediated cholesterol efflux from J774 macrophages by **up-regulating ABCA1** and also improves HDL resistance to oxidative modification *in vitro* [177]. Caffeic acid and ferulic acid, pyrocatehol-based polyphenols and major phenolic acids of coffee, enhance cholesterol efflux from THP-1 macrophages mediated by HDL, but not by apoA-I [178]. As both compounds increase mRNA and protein levels of ABCG1 and SR-BI in macrophages without influencing the expression of ABCA1, enhanced cholesterol efflux induced by caffeic and ferulic acids can be mediated by **ABCG1** and **SR-BI** (Fig. 14.1). Consistent with this conclusion, mRNA and protein levels of ABCG1 and SR-BI, as well as cholesterol efflux to HDL, are elevated in macrophages isolated from healthy subjects after coffee consumption compared with baseline [178]. Such *in vitro* findings are suggestive of enhanced RCT from macrophages to feces as demonstrated by measurements in mouse models *in vivo* [178]. In a similar fashion, the expression of SR-BI can be elevated by isoflavones [179].

The activities of anti-atherogenic enzymes carried by HDL, primarily of **PON1**, can also be improved by polyphenols (Fig. 14.1). Indeed, chlorogenic acids, phenolics present in significant amounts in coffee, many fruits and *Ilex paraguariensis* extracts, protect PON1 in HDL from inactivation by HOCl [180]. Next, pomegranate juice, a rich source of polyphenols, up-regulates PON1 expression in hepatocytes, an effect that results in accelerated secretion and elevated activity of the enzyme [181]. Pomegranate juice polyphenols can also increase PON1 association with HDL, thereby stabilizing the enzyme [182]. Punicalagin and gallic acid, major polyphenolic components of pomegranate juice, exert similar effects, which appear to be mediated by the intracellular PPAR gamma/protein kinase A/cyclic AMP signaling cascade.

As an *in vivo* corollary of these *in vitro* data, pomegranate juice enhances HDL-associated arylesterase, PON and lactonase activities of PON1 in patients with Type 2 diabetes when given for four weeks at 50 ml/day [183]. Such enhancement of catalytic activities is accompanied by elevated levels of HDL-bound PON1, suggestive of the stabilization of the enzyme as a result of its enhanced association with HDL. In a similar fashion, quercetin increases hepatic expression of PON1, serum and liver PON1 activities, and serum homocysteine thiolactonase in rats, resulting in an improved capacity of HDL to protect LDL from oxidative stress *in vitro* [184, 185].

Another element in support of the stimulating action of polyphenols on PON1 is provided by epidemiologic data that demonstrate a positive association between a flavonoid-rich diet and arylesterase activity of PON1 in patients with Type 2 diabetes [186].

Vitamins. The effects of dietary vitamins, other than niacin (vitamin B3), on HDL-C levels in humans are weak, if there are any. Thus, supplementation with vitamin A, vitamin C, vitamin D or vitamin E does not provide consistent effects on circulating HDL concentrations across different populations [187–191]. These results contrast with data obtained in animal models or in cell culture studies. For example, vitamin D insufficiency may result in elevated concentrations of total cholesterol and HDL-C in vitamin D receptor-deficient mice [192]. On the other hand, vitamin E, in the form of alpha-tocopherol, reduces oxidized LDL-mediated stimulation of LXR alpha and decreases CD36, ABCA1 and ABCG1 expression, thereby compromising cholesterol efflux [193]. Such discrepancies highlight major metabolic differences between human and animal studies and emphasize caution when interpreting the results of *in vitro* studies.

Other Minor Components. Among other nutrients, the carotenoid astaxanthin possesses the capacity to improve atherogenic dyslipidemia via increasing HDL-C and decreasing triglyceride concentrations both in humans and in animal models [194]. In contrast, supplementation with beta-carotene does not affect HDL-C levels in humans [195, 196].

Another example of HDL-raising activity is provided by terpenoid-rich oils obtained from the seeds of *Alpinia zerumbet*, a Chinese plant of the ginger family popularly recognized as a hepatoprotector [197]. Such seeds contain 0.51% essential oils that are largely comprised of monoterpenoids, oxygenated monoterpenoids, sesquiterpenoids and oxygenated sesquiterpenoids. Interestingly, such oils are reportedly capable of potently increasing HDL-C levels.

HDL levels can be elevated by selenium [198]. Indeed, the supplementation of hepatocytes with selenium increases the expression of apoA-I and mitigates repression of apoA-I expression under conditions of oxidative stress by suppressing signaling via the NFκB pathway and enhancing signaling via PPAR alpha [199]. In parallel, however, selenium supplementation increases plasma levels of total cholesterol and LDL-C [200].

Recent interest in the health-enhancing properties of probiotics stimulated studies of their action on HDL. Thus, *Enterococcus faecium* CRL183, a probiotic microorganism, was shown to raise HDL-C levels by +43% in rabbits with induced hypercholesterolemia [201]. In contrast, nattokinase, an enzyme present in a Japanese food called natto, was unable to modulate circulating levels of lipids in a randomized, double-blind, placebo-controlled study carried out in hyperlipidemic patients. This negative result is consistent with the absence of significant influence of probiotics on plasma HDL levels in humans [202, 203]. On the other hand, a combination of nattokinase with red yeast rice increases HDL-C by +7.5% [204]; however, the latter food contains lovastatin which may underlie the HDL-C raising effect [205, 206].

IMPORTANT READING

Katcher HI, Hill AM, Lanford JL, Yoo JS, Kris-Etherton PM. Lifestyle approaches and dietary strategies to lower LDL-cholesterol and triglycerides and raise HDL-cholesterol. *Endocrinol Metab Clin North Am* 2009; 38: 45–78.

REFERENCES

1. Leon AS, Sanchez OA. Response of blood lipids to exercise training alone or combined with dietary intervention. *Med Sci Sports Exerc* 2001; 33: S502–S515.
2. Leaf DA, Parker DL, Schaad D. Changes in VO2max, physical activity, and body fat with chronic exercise: effects on plasma lipids. *Med Sci Sports Exerc* 1997; 29: 1152–1159.
3. Katcher HI, Hill AM, Lanford JL, Yoo JS, Kris-Etherton PM. Lifestyle approaches and dietary strategies to lower LDL-cholesterol and triglycerides and raise HDL-cholesterol. *Endocrinol Metab Clin North Am* 2009; 38: 45–78.
4. Leaf DA. The effect of physical exercise on reverse cholesterol transport. *Metabolism* 2003; 52: 950–957.
5. Durstine JL, Grandjean PW, Davis PG, Ferguson MA, Alderson NL, DuBose KD. Blood lipid and lipoprotein adaptations to exercise: a quantitative analysis. *Sports Med* 2001; 31: 1033–1062.
6. Healy GN, Matthews CE, Dunstan DW, Winkler EA, Owen N. Sedentary time and cardio-metabolic biomarkers in US adults: NHANES 2003-06. *Eur Heart J* 2011; 32: 590–597.
7. Martinez-Gomez D, Rey-Lopez JP, Chillon P, Gomez-Martinez S, Vicente-Rodriguez G, Martin-Matillas M, Garcia-Fuentes M, Delgado M, Moreno LA, Veiga OL et al. Excessive TV viewing and cardiovascular disease risk factors in adolescents. The AVENA cross-sectional study. *BMC Public Health* 2010; 10: 274.
8. Kelley GA, Kelley KS, Franklin B. Aerobic exercise and lipids and lipoproteins in patients with cardiovascular disease: a meta-analysis of randomized controlled trials. *J Cardiopulm Rehabil* 2006; 26: 131–139.
9. Brown AJ, Setji TL, Sanders LL, Lowry KP, Otvos JD, Kraus WE, Svetkey PL. Effects of exercise on lipoprotein particles in women with polycystic ovary syndrome. *Med Sci Sports Exerc* 2009; 41: 497–504.
10. Masuda R, Imamura H, Mizuuchi K, Miyahara K, Kumagai H, Hirakata H. Physical activity, high-density lipoprotein cholesterol subfractions and lecithin:cholesterol acyltransferase in dialysis patients. *Nephron Clin Pract* 2009; 111: c253–c259.
11. Leon AS, Gaskill SE, Rice T, Bergeron J, Gagnon J, Rao DC, Skinner JS, Wilmore JH, Bouchard C. Variability in the response of HDL cholesterol to exercise training in the HERITAGE Family Study. *Int J Sports Med* 2002; 23: 1–9.

12. Thompson PD, Rader DJ. Does exercise increase HDL cholesterol in those who need it the most? *Arterioscler Thromb Vasc Biol* 2001; 21: 1097–1098.
13. Kelley GA, Kelley KS, Tran ZV. Aerobic exercise and lipids and lipoproteins in women: a meta-analysis of randomized controlled trials. *J Womens Health* 2004; 13: 1148–1164.
14. Kelley GA, Kelley KS. Aerobic exercise and lipids and lipoproteins in men: a meta-analysis of randomized controlled trials. *J Mens Health Gend* 2006; 3: 61–70.
15. Lewis DA, Kamon E, Hodgson JL. Physiological differences between genders. Implications for sports conditioning. *Sports Med* 1986; 3: 357–369.
16. Monda KL, Ballantyne CM, North KE. Longitudinal impact of physical activity on lipid profiles in middle-aged adults: the Atherosclerosis Risk in Communities Study. *J Lipid Res* 2009; 50: 1685–1691.
17. Leon AS, Rice T, Mandel S, Despres JP, Bergeron J, Gagnon J, Rao DC, Skinner JS, Wilmore JH, Bouchard C. Blood lipid response to 20 weeks of supervised exercise in a large biracial population: the HERITAGE Family Study. *Metabolism* 2000; 49: 513–520.
18. Kohl HW III. Physical activity and cardiovascular disease: evidence for a dose response. *Med Sci Sports Exerc* 2001; 33: S472–S483.
19. Pate RR, Pratt M, Blair SN, Haskell WL, Macera CA, Bouchard C, Buchner D, Ettinger W, Heath GW, King AC et al. Physical activity and public health. A recommendation from the Centers for Disease Control and Prevention and the American College of Sports Medicine. *JAMA* 1995; 273: 402–407.
20. Nye ER, Carlson K, Kirstein P, Rossner S. Changes in high density lipoprotein subfractions and other lipoproteins by exercise. *Clin Chim Acta* 1981; 113: 51–57.
21. Kelley GA, Kelley KS. Aerobic exercise and HDL2-C: a meta-analysis of randomized controlled trials. *Atherosclerosis* 2006; 184: 207–215.
22. Kraus WE, Houmard JA, Duscha BD, Knetzger KJ, Wharton MB, McCartney JS, Bales CW, Henes S, Samsa GP, Otvos JD et al. Effects of the amount and intensity of exercise on plasma lipoproteins. *N Engl J Med* 2002; 347: 1483–1492.
23. Lee H, Park JE, Choi I, Cho KH. Enhanced functional and structural properties of high-density lipoproteins from runners and wrestlers compared to throwers and lifters. *BMB Rep* 2009; 42: 605–610.
24. Jafari M, Leaf DA, Macrae H, Kasem J, O'Conner P, Pullinger C, Malloy M, Kane JP. The effects of physical exercise on plasma prebeta-1 high-density lipoprotein. *Metabolism* 2003; 52: 437–442.
25. Olchawa B, Kingwell BA, Hoang A, Schneider L, Miyazaki O, Nestel P, Sviridov D. Physical fitness and reverse cholesterol transport. *Arterioscler Thromb Vasc Biol* 2004; 24: 1087–1091.
26. Hoang A, Tefft C, Duffy SJ, Formosa M, Henstridge DC, Kingwell BA, Sviridov D. ABCA1 expression in humans is associated with physical activity and alcohol consumption. *Atherosclerosis* 2008; 197: 197–203.
27. Khabazian BM, Ghanbari-Niaki A, Safarzadeh-Golpordesari A, Ebrahimi M, Rahbarizadeh F, Abednazari H. Endurance training enhances ABCA1 expression in rat small intestine. *Eur J Appl Physiol* 2009; 107: 351–358.
28. Kiens B, Jorgensen I, Lewis S, Jensen G, Lithell H, Vessby B, Hoe S, Schnohr P. Increased plasma HDL-cholesterol and apo A-1 in sedentary middle-aged men after physical conditioning. *Eur J Clin Invest* 1980; 10: 203–209.

29. Thompson PD, Yurgalevitch SM, Flynn MM, Zmuda JM, Spannaus-Martin D, Saritelli A, Bausserman L, Herbert PN. Effect of prolonged exercise training without weight loss on high-density lipoprotein metabolism in overweight men. *Metabolism* 1997; 46: 217–223.

30. Couillard C, Despres JP, Lamarche B, Bergeron J, Gagnon J, Leon AS, Rao DC, Skinner JS, Wilmore JH, Bouchard C. Effects of endurance exercise training on plasma HDL cholesterol levels depend on levels of triglycerides: evidence from men of the Health, Risk Factors, Exercise Training and Genetics (HERITAGE) Family Study. *Arterioscler Thromb Vasc Biol* 2001; 21: 1226–1232.

31. Thompson PD, Cullinane EM, Sady SP, Flynn MM, Bernier DN, Kantor MA, Saritelli AL, Herbert PN. Modest changes in high-density lipoprotein concentration and metabolism with prolonged exercise training. *Circulation* 1988; 78: 25–34.

32. Wood PD, Stefanick ML, Dreon DM, Frey-Hewitt B, Garay SC, Williams PT, Superko HR, Fortmann SP, Albers JJ, Vranizan KM et al. Changes in plasma lipids and lipoproteins in overweight men during weight loss through dieting as compared with exercise. *N Engl J Med* 1988; 319: 1173–1179.

33. Sasaki J, Tanabe Y, Tanaka H, Saku K, Shindo M, Arakawa K. Elevated levels of HDL2-cholesterol and apo A-I in national class Japanese male marathon runners. *Atherosclerosis* 1988; 70: 175–177.

34. Herbert PN, Bernier DN, Cullinane EM, Edelstein L, Kantor MA, Thompson PD. High-density lipoprotein metabolism in runners and sedentary men. *JAMA* 1984; 252: 1034–1037.

35. Mazzucco S, Agostini F, Mangogna A, Cattin L, Biolo G. Prolonged inactivity up-regulates cholesteryl ester transfer protein independently of body fat changes in humans. *J Clin Endocrinol Metab* 2010; 95: 2508–2512.

36. Foger B, Wohlfarter T, Ritsch A, Lechleitner M, Miller CH, Dienstl A, Patsch JR. Kinetics of lipids, apolipoproteins, and cholesteryl ester transfer protein in plasma after a bicycle marathon. *Metabolism* 1994; 43: 633–639.

37. Haskell WL. Exercise-induced changes in plasma lipids and lipoproteins. *Prev Med* 1984; 13: 23–36.

38. Thompson PD, Cullinane EM, Sady SP, Flynn MM, Chenevert CB, Herbert PN. High density lipoprotein metabolism in endurance athletes and sedentary men. *Circulation* 1991; 84: 140–152.

39. Seip RL, Moulin P, Cocke T, Tall A, Kohrt WM, Mankowitz K, Semenkovich CF, Ostlund R, Schonfeld G. Exercise training decreases plasma cholesteryl ester transfer protein. *Arterioscler Thromb* 1993; 13: 1359–1367.

40. Durstine JL, Haskell WL. Effects of exercise training on plasma lipids and lipoproteins. *Exerc Sport Sci Rev* 1994; 22: 477–521.

41. Sady SP, Thompson PD, Cullinane EM, Kantor MA, Domagala E, Herbert PN. Prolonged exercise augments plasma triglyceride clearance. *JAMA* 1986; 256: 2552–2555.

42. Kantor MA, Cullinane EM, Herbert PN, Thompson PD. Acute increase in lipoprotein lipase following prolonged exercise. *Metabolism* 1984; 33: 454–457.

43. Frey I, Baumstark MW, Berg A, Keul J. Influence of acute maximal exercise on lecithin:cholesterol acyltransferase activity in healthy adults of differing aerobic performance. *Eur J Appl Physiol Occup Physiol* 1991; 62: 31–35.

44. Urata H, Sasaki J, Tanabe Y, Kiyonaga A, Yoshida T, Tanaka H, Shindo M, Arakawa K. Effect of mild aerobic exercise on serum lipids and apolipoproteins in patients with coronary artery disease. *Jpn Heart J* 1987; 28: 27–34.
45. Tsopanakis C, Kotsarellis D, Tsopanakis A. Plasma lecithin: cholesterol acyltransferase activity in elite athletes from selected sports. *Eur J Appl Physiol Occup Physiol* 1988; 58: 262–265.
46. Lopez A, Vial R, Balart L, Arroyave G. Effect of exercise and physical fitness on serum lipids and lipoproteins. *Atherosclerosis* 1974; 20: 1–9.
47. Gupta AK, Ross EA, Myers JN, Kashyap ML. Increased reverse cholesterol transport in athletes. *Metabolism* 1993; 42: 684–690.
48. Kralova Lesna I, Suchanek P, Kovar J, Poledne R. Life style change and reverse cholesterol transport in obese women. *Physiol Res* 2009; 58 (Suppl. 1): S33–S38.
49. Ghanbari-Niaki A, Saghebjoo M, Hedayati M. A single session of circuit-resistance exercise effects on human peripheral blood lymphocyte ABCA1 expression and plasma HDL-C level. *Regul Pept* 2010; 166: 42–47.
50. Ribeiro IC, Iborra RT, Neves MQ, Lottenberg SA, Charf AM, Nunes VS, Negrao CE, Nakandakare ER, Quintao EC, Passarelli M. HDL atheroprotection by aerobic exercise training in type 2 diabetes mellitus. *Med Sci Sports Exerc* 2008; 40: 779–786.
51. Casella-Filho A, Chagas AC, Maranhao RC, Trombetta IC, Cesena FH, Silva VM, Tanus-Santos JE, Negrao CE, da Luz PL. Effect of exercise training on plasma levels and functional properties of high-density lipoprotein cholesterol in the metabolic syndrome. *Am J Cardiol* 2011; 107: 1168–1172.
52. Iborra RT, Ribeiro IC, Neves MQ, Charf AM, Lottenberg SA, Negrao CE, Nakandakare ER, Passarelli M. Aerobic exercise training improves the role of high-density lipoprotein antioxidant and reduces plasma lipid peroxidation in type 2 diabetes mellitus. *Scand J Med Sci Sports* 2008; 18: 742–750.
53. Lee SJ, On YK, Lee EJ, Choi JY, Kim BT, Lee KH. Reversal of vascular 18F-FDG uptake with plasma high-density lipoprotein elevation by atherogenic risk reduction. *J Nucl Med* 2008; 49: 1277–1282.
54. Brinton EA. Effects of ethanol intake on lipoproteins and atherosclerosis. *Curr Opin Lipidol* 2010; 21: 346–351.
55. Brien SE, Ronksley PE, Turner BJ, Mukamal KJ, Ghali WA. Effect of alcohol consumption on biological markers associated with risk of coronary heart disease: systematic review and meta-analysis of interventional studies. *BMJ* 2011; 342: d636.
56. Weissglas-Volkov D, Pajukanta P. Genetic causes of high and low serum HDL-cholesterol. *J Lipid Res* 2010; 51: 2032–2057.
57. Gaziano JM, Gaziano TA, Glynn RJ, Sesso HD, Ajani UA, Stampfer MJ, Manson JE, Hennekens CH, Buring JE. Light-to-moderate alcohol consumption and mortality in the Physicians' Health Study enrollment cohort. *J Am Coll Cardiol* 2000; 35: 96–105.
58. Sesso HD. Alcohol and cardiovascular health: recent findings. *Am J Cardiovasc Drugs* 2001; 1: 167–172.
59. Hansel B, Thomas F, Pannier B, Bean K, Kontush A, Chapman MJ, Guize L, Bruckert E. Relationship between alcohol intake, health and social status and cardiovascular risk factors in the urban Paris-Ile-De-France Cohort: is the cardioprotective action of alcohol a myth? *Eur J Clin Nutr* 2010; 64: 561–568.

60. Foerster M, Marques-Vidal P, Gmel G, Daeppen JB, Cornuz J, Hayoz D, Pecoud A, Mooser V, Waeber G, Vollenweider P et al. Alcohol drinking and cardiovascular risk in a population with high mean alcohol consumption. *Am J Cardiol* 2009; 103: 361–368.

61. Andrade AC, Cesena FH, Consolim-Colombo FM, Coimbra SR, Benjo AM, Krieger EM, Luz PL. Short-term red wine consumption promotes differential effects on plasma levels of high-density lipoprotein cholesterol, sympathetic activity, and endothelial function in hypercholesterolemic, hypertensive, and healthy subjects. *Clinics* 2009; 64: 435–442.

62. Zhou Y, Yin R, Deng Y, Li Y, Wu J. Interactions between alcohol intake and the polymorphism of rs708272 on serum high-density lipoprotein cholesterol levels in the Guangxi Hei Yi Zhuang population. *Alcohol* 2008; 42: 583–591.

63. Jensen MK, Mukamal KJ, Overvad K, Rimm EB. Alcohol consumption, TaqIB polymorphism of cholesteryl ester transfer protein, high-density lipoprotein cholesterol, and risk of coronary heart disease in men and women. *Eur Heart J* 2008; 29: 104–112.

64. Wakabayashi I. Associations of alcohol drinking and cigarette smoking with serum lipid levels in healthy middle-aged men. *Alcohol Alcohol* 2008; 43: 274–280.

65. Volcik KA, Ballantyne CM, Fuchs FD, Sharrett AR, Boerwinkle E. Relationship of alcohol consumption and type of alcoholic beverage consumed with plasma lipid levels: differences between Whites and African Americans of the ARIC study. *Ann Epidemiol* 2008; 18: 101–107.

66. de Jong HJ, de Goede J, Oude Griep LM, Geleijnse JM. Alcohol consumption and blood lipids in elderly coronary patients. *Metabolism* 2008; 57: 1286–1292.

67. Hannuksela ML, Liisanantti MK, Savolainen MJ. Effect of alcohol on lipids and lipoproteins in relation to atherosclerosis. *Crit Rev Clin Lab Sci* 2002; 39: 225–283.

68. Djousse L, Lee IM, Buring JE, Gaziano JM. Alcohol consumption and risk of cardiovascular disease and death in women: potential mediating mechanisms. *Circulation* 2009; 120: 237–244.

69. Langer RD, Criqui MH, Reed DM. Lipoproteins and blood pressure as biological pathways for effect of moderate alcohol consumption on coronary heart disease. *Circulation* 1992; 85: 910–915.

70. Mukamal KJ, Jensen MK, Gronbaek M, Stampfer MJ, Manson JE, Pischon T, Rimm EB. Drinking frequency, mediating biomarkers, and risk of myocardial infarction in women and men. *Circulation* 2005; 112: 1406–1413.

71. Gardner CD, Tribble DL, Young DR, Ahn D, Fortmann SP. Associations of HDL, HDL(2), and HDL(3) cholesterol and apolipoproteins A-I and B with lifestyle factors in healthy women and men: the Stanford Five City Project. *Prev Med* 2000; 31: 346–356.

72. Muth ND, Laughlin GA, von Muhlen D, Smith SC, Barrett-Connor E. High-density lipoprotein subclasses are a potential intermediary between alcohol intake and reduced risk of cardiovascular disease: The Rancho Bernardo Study. *Br J Nutr* 2010; 104: 1034–1042.

73. Beulens JW, Sierksma A, van Tol A, Fournier N, van Gent T, Paul JL, Hendriks HF. Moderate alcohol consumption increases cholesterol efflux mediated by ABCA1. *J Lipid Res* 2004; 45: 1716–1723.

74. Lecomte E, Herbeth B, Paille F, Steinmetz J, Artur Y, Siest G. Changes in serum apolipoprotein and lipoprotein profile induced by chronic alcohol consumption and withdrawal: determinant effect on heart disease? *Clin Chem* 1996; 42: 1666–1675.
75. Gottrand F, Beghin L, Duhal N, Lacroix B, Bonte JP, Fruchart JC, Luc G. Moderate red wine consumption in healthy volunteers reduced plasma clearance of apolipoprotein AII. *Eur J Clin Invest* 1999; 29: 387–394.
76. Senault C, Betoulle D, Luc G, Hauw P, Rigaud D, Fumeron F. Beneficial effects of a moderate consumption of red wine on cellular cholesterol efflux in young men. *Nutr Metab Cardiovasc Dis* 2000; 10: 63–69.
77. Schafer C, Parlesak A, Eckoldt J, Bode C, Bode JC, Marz W, Winkler K. Beyond HDL-cholesterol increase: phospholipid enrichment and shift from HDL3 to HDL2 in alcohol consumers. *J Lipid Res* 2007; 48: 1550–1558.
78. Perret B, Ruidavets JB, Vieu C, Jaspard B, Cambou JP, Terce F, Collet X. Alcohol consumption is associated with enrichment of high-density lipoprotein particles in polyunsaturated lipids and increased cholesterol esterification rate. *Alcohol Clin Exp Res* 2002; 26: 1134–1140.
79. Hannuksela ML, Ramet ME, Nissinen AE, Liisanantti MK, Savolainen MJ. Effects of ethanol on lipids and atherosclerosis. *Pathophysiology* 2004; 10: 93–103.
80. Liisanantti MK, Savolainen MJ. Phosphatidylethanol mediates its effects on the vascular endothelial growth factor via HDL receptor in endothelial cells. *Alcohol Clin Exp Res* 2009; 33: 283–288.
81. Makela SM, Jauhiainen M, Ala-Korpela M, Metso J, Lehto TM, Savolainen MJ, Hannuksela ML. HDL2 of heavy alcohol drinkers enhances cholesterol efflux from raw macrophages via phospholipid-rich HDL 2b particles. *Alcohol Clin Exp Res* 2008; 32: 991–1000.
82. Kralova Lesna I, Suchanek P, Stavek P, Poledne R. May an alcohol induced increase of HDL be considered as atheroprotective? *Physiol Res* 2009; 59: 407–413.
83. Rao MN, Liu QH, Marmillot P, Seeff LB, Strader DB, Lakshman MR. High-density lipoproteins from human alcoholics exhibit impaired reverse cholesterol transport function. *Metabolism* 2000; 49: 1406–1410.
84. Marmillot P, Munoz J, Patel S, Garige M, Rosse RB, Lakshman MR. Long-term ethanol consumption impairs reverse cholesterol transport function of high-density lipoproteins by depleting high-density lipoprotein sphingomyelin both in rats and in humans. *Metabolism* 2007; 56: 947–953.
85. Kontush A, Chapman MJ. Antiatherogenic function of HDL particle subpopulations: focus on antioxidative activities. *Curr Opin Lipidol* 2010; 21: 312–318.
86. Sierksma A, Vermunt SH, Lankhuizen IM, van der Gaag MS, Scheek LM, Grobbee DE, van Tol A, Hendriks HF. Effect of moderate alcohol consumption on parameters of reverse cholesterol transport in postmenopausal women. *Alcohol Clin Exp Res* 2004; 28: 662–666.
87. Costa LG, Giordano G, Furlong CE. Pharmacological and dietary modulators of paraoxonase 1 (PON1) activity and expression: the hunt goes on. *Biochem Pharmacol* 2011; 81: 337–344.
88. Beulens JW, van den Berg R, Kok FJ, Helander A, Vermunt SH, Hendriks HF. Moderate alcohol consumption and lipoprotein-associated phospholipase A2 activity. *Nutr Metab Cardiovasc Dis* 2008; 18: 539–544.

89. Knopp RH, Walden CE, Retzlaff BM, McCann BS, Dowdy AA, Albers JJ, Gey GO, Cooper MN. Long-term cholesterol-lowering effects of 4 fat-restricted diets in hypercholesterolemic and combined hyperlipidemic men. The Dietary Alternatives Study. *JAMA* 1997; 278: 1509–1515.

90. Kotani K, Sakane N, Sano Y, Tsuzaki K, Matsuoka Y, Egawa K, Yoshimura M, Horikawa C, Kitagawa Y, Kiso Y et al. Changes on the physiological lactonase activity of serum paraoxonase 1 by a diet intervention for weight loss in healthy overweight and obese women. *J Clin Biochem Nutr* 2009; 45: 329–334.

91. Jebb SA, Lovegrove JA, Griffin BA, Frost GS, Moore CS, Chatfield MD, Bluck LJ, Williams CM, Sanders TA. Effect of changing the amount and type of fat and carbohydrate on insulin sensitivity and cardiovascular risk: the RISCK (Reading, Imperial, Surrey, Cambridge, and Kings) trial. *Am J Clin Nutr* 2010; 92: 748–758.

92. Bloomer RJ, Kabir MM, Canale RE, Trepanowski JF, Marshall KE, Farney TM, Hammond KG. Effect of a 21 day Daniel Fast on metabolic and cardiovascular disease risk factors in men and women. *Lipids Health Dis* 2010; 9: 94.

93. Gould KL, Ornish D, Kirkeeide R, Brown S, Stuart Y, Buchi M, Billings J, Armstrong W, Ports T, Scherwitz L. Improved stenosis geometry by quantitative coronary arteriography after vigorous risk factor modification. *Am J Cardiol* 1992; 69: 845–853.

94. Ornish D, Brown SE, Scherwitz LW, Billings JH, Armstrong WT, Ports TA, McLanahan SM, Kirkeeide RL, Brand RJ, Gould KL. Can lifestyle changes reverse coronary heart disease? The Lifestyle Heart Trial. *Lancet* 1990; 336: 129–133.

95. Ornish D, Scherwitz LW, Billings JH, Brown SE, Gould KL, Merritt TA, Sparler S, Armstrong WT, Ports TA, Kirkeeide RL et al. Intensive lifestyle changes for reversal of coronary heart disease. *JAMA* 1998; 280: 2001–2007.

96. Brinton EA, Eisenberg S, Breslow JL. A low-fat diet decreases high density lipoprotein (HDL) cholesterol levels by decreasing HDL apolipoprotein transport rates. *J Clin Invest* 1990; 85: 144–151.

97. Hayek T, Ito Y, Azrolan N, Verdery RB, Aalto-Setala K, Walsh A, Breslow JL. Dietary fat increases high density lipoprotein (HDL) levels both by increasing the transport rates and decreasing the fractional catabolic rates of HDL cholesterol ester and apolipoprotein (Apo) A-I. Presentation of a new animal model and mechanistic studies in human Apo A-I transgenic and control mice. *J Clin Invest* 1993; 91: 1665–1671.

98. Ng TW, Chan DC, Barrett PH, Watts GF. Effect of weight loss on HDL-apoA-II kinetics in the metabolic syndrome. *Clin Sci (Lond)* 2009; 118, 79–85.

99. Vasudevan MM, Tchoua U, Gillard BK, Jones PH, Pownall HJ, Ballantyne CM. Effect of weight loss on macrophage cholesterol efflux to plasma of obese individuals with metabolic syndrome. *Circulation* 2009; 120 (Suppl.): S443.

100. Shige H, Nestel P, Sviridov D, Noakes M, Clifton P. Effect of weight reduction on the distribution of apolipoprotein A-I in high-density lipoprotein subfractions in obese non-insulin-dependent diabetic subjects. *Metabolism* 2000; 49: 1453–1459.

101. Hopkins PN. Effects of dietary cholesterol on serum cholesterol: a meta-analysis and review. *Am J Clin Nutr* 1992; 55: 1060–1070.

102. Schaefer EJ. Lipoproteins, nutrition, and heart disease. *Am J Clin Nutr* 2002; 75: 191–212.

103. Mayurasakorn K, Srisura W, Sitphahul P, Hongto PO. High-density lipoprotein cholesterol changes after continuous egg consumption in healthy adults. *J Med Assoc Thai* 2008; 91: 400–407.

104. Navab M, Hama S, Hough G, Fogelman AM. Oral synthetic phospholipid (DMPC) raises high-density lipoprotein cholesterol levels, improves high-density lipoprotein function, and markedly reduces atherosclerosis in apolipoprotein E-null mice. *Circulation* 2003; 108: 1735–1739.

105. Cho BH, Park JR, Nakamura MT, Odintsov BM, Wallig MA, Chung BH. Synthetic dimyristoylphosphatidylcholine liposomes assimilating into high-density lipoprotein promote regression of atherosclerotic lesions in cholesterol-fed rabbits. *Exp Biol Med* 2010; 235: 1194–1203.

106. Cohn JS, Wat E, Kamili A, Tandy S. Dietary phospholipids, hepatic lipid metabolism and cardiovascular disease. *Curr Opin Lipidol* 2008; 19: 257–262.

107. Greten H, Raetzer H, Stiehl A, Schettler G. The effect of polyunsaturated phosphatidylcholine on plasma lipids and fecal sterol excretion. *Atherosclerosis* 1980; 36: 81–88.

108. Shirouchi B, Nagao K, Furuya K, Inoue N, Inafuku M, Nasu M, Otsubo K, Koga S, Matsumoto H, Yanagita T. Effect of dietary phosphatidylinositol on cholesterol metabolism in Zucker (fa/fa) rats. *J Oleo Sci* 2009; 58: 111–115.

109. Brousseau ME, Schaefer EJ, Stucchi AF, Osada J, Vespa DB, Ordovas JM, Nicolosi RJ. Diets enriched in unsaturated fatty acids enhance apolipoprotein A-I catabolism but do not affect either its production or hepatic mRNA abundance in cynomolgus monkeys. *Atherosclerosis* 1995; 115: 107–119.

110. Chong KS, Nicolosi RJ, Rodger RF, Arrigo DA, Yuan RW, MacKey JJ, Georas S, Herbert PN. Effect of dietary fat saturation on plasma lipoproteins and high density lipoprotein metabolism of the rhesus monkey. *J Clin Invest* 1987; 79: 675–683.

111. Parks JS, Rudel LL. Different kinetic fates of apolipoproteins A-I and A-II from lymph chylomicra of nonhuman primates. Effect of saturated versus polyunsaturated dietary fat. *J Lipid Res* 1982; 23: 410–421.

112. Velez-Carrasco W, Lichtenstein AH, Welty FK, Li Z, Lamon-Fava S, Dolnikowski GG, Schaefer EJ. Dietary restriction of saturated fat and cholesterol decreases HDL ApoA-I secretion. *Arterioscler Thromb Vasc Biol* 1999; 19: 918–924.

113. Stucchi AF, Hennessy LK, Vespa DB, Weiner EJ, Osada J, Ordovas JM, Schaefer EJ, Nicolosi RJ. Effect of corn and coconut oil-containing diets with and without cholesterol on high density lipoprotein apoprotein A-I metabolism and hepatic apoprotein A-I mRNA levels in cebus monkeys. *Arterioscler Thromb* 1991; 11: 1719–1729.

114. Ordovas JM, Cupples LA, Corella D, Otvos JD, Osgood D, Martinez A, Lahoz C, Coltell O, Wilson PW, Schaefer EJ. Association of cholesteryl ester transfer protein-TaqIB polymorphism with variations in lipoprotein subclasses and coronary heart disease risk: the Framingham study. *Arterioscler Thromb Vasc Biol* 2000; 20: 1323–1329.

115. Nicholls SJ, Lundman P, Harmer JA, Cutri B, Griffiths KA, Rye KA, Barter PJ, Celermajer DS. Consumption of saturated fat impairs the anti-inflammatory properties of high-density lipoproteins and endothelial function. *J Am Coll Cardiol* 2006; 48: 715–720.

116. Nicolosi RJ. Dietary fat saturation effects on low-density-lipoprotein concentrations and metabolism in various animal models. *Am J Clin Nutr* 1997; 65: 1617S–1627S.

117. Executive Summary of The Third Report of The National Cholesterol Education Program (NCEP) Expert Panel on Detection, Evaluation, And Treatment of High Blood Cholesterol In Adults (Adult Treatment Panel III). *JAMA* 2001; 285: 2486–2497.

118. Schaefer EJ, Lamon-Fava S, Ausman LM, Ordovas JM, Clevidence BA, Judd JT, Goldin BR, Woods M, Gorbach S, Lichtenstein AH. Individual variability in lipoprotein cholesterol response to National Cholesterol Education Program Step 2 diets. *Am J Clin Nutr* 1997; 65: 823–830.

119. Lichtenstein AH, Ausman LM, Carrasco W, Jenner JL, Gualtieri LJ, Goldin BR, Ordovas JM, Schaefer EJ. Effects of canola, corn, and olive oils on fasting and postprandial plasma lipoproteins in humans as part of a National Cholesterol Education Program Step 2 diet. *Arterioscler Thromb* 1993; 13: 1533–1542.

120. Mensink RP, Katan MB. Effect of dietary fatty acids on serum lipids and lipoproteins. A meta-analysis of 27 trials. *Arterioscler Thromb* 1992; 12: 911–919.

121. Dreon DM, Vranizan KM, Krauss RM, Austin MA, Wood PD. The effects of polyunsaturated fat vs monounsaturated fat on plasma lipoproteins. *JAMA* 1990; 263: 2462–2466.

122. Brousseau ME, Stucchi AF, Vespa DB, Schaefer EJ, Nicolosi RJ. A diet enriched in monounsaturated fats decreases low density lipoprotein concentrations in cynomolgus monkeys by a different mechanism than does a diet enriched in polyunsaturated fats. *J Nutr* 1993; 123: 2049–2058.

123. Estevez-Gonzalez MD, Saavedra-Santana P, Lopez-Rios L, Chirino R, Cebrero-Garcia E, Pena-Quintana L, Betancor-Leon P. HDL cholesterol levels in children with mild hypercholesterolemia: Effect of consuming skim milk enriched with olive oil and modulation by the TAQ 1B polymorphism in the CETP gene. *Ann Nutr Metab* 2010; 56: 288–293.

124. Jenkins DJ, Chiavaroli L, Wong JM, Kendall C, Lewis GF, Vidgen E, Connelly PW, Leiter LA, Josse RG, Lamarche B. Adding monounsaturated fatty acids to a dietary portfolio of cholesterol-lowering foods in hypercholesterolemia. *CMAJ* 2010; 182: 1961–1967.

125. Rudel LL, Parks JS, Sawyer JK. Compared with dietary monounsaturated and saturated fat, polyunsaturated fat protects African green monkeys from coronary artery atherosclerosis. *Arterioscler Thromb Vasc Biol* 1995; 15: 2101–2110.

126. Mensink RP, Katan MB. Effect of dietary trans fatty acids on high-density and low-density lipoprotein cholesterol levels in healthy subjects. *N Engl J Med* 1990; 323: 439–445.

127. Judd JT, Clevidence BA, Muesing RA, Wittes J, Sunkin ME, Podczasy JJ. Dietary trans fatty acids: effects on plasma lipids and lipoproteins of healthy men and women. *Am J Clin Nutr* 1994; 59: 861–868.

128. Lichtenstein AH, Ausman LM, Carrasco W, Jenner JL, Ordovas JM, Schaefer EJ. Hydrogenation impairs the hypolipidemic effect of corn oil in humans. Hydrogenation, trans fatty acids, and plasma lipids. *Arterioscler Thromb* 1993; 13: 154–161.

129. Lichtenstein AH, Ausman LM, Jalbert SM, Schaefer EJ. Effects of different forms of dietary hydrogenated fats on serum lipoprotein cholesterol levels. *N Engl J Med* 1999; 340: 1933–1940.

130. Siri-Tarino PW, Sun Q, Hu FB, Krauss RM. Saturated fat, carbohydrate, and cardiovascular disease. *Am J Clin Nutr* 2010; 91: 502–509.

131. Hartweg J, Farmer AJ, Holman RR, Neil A. Potential impact of omega-3 treatment on cardiovascular disease in type 2 diabetes. *Curr Opin Lipidol* 2009; 20: 30–38.

132. Maki KC, McKenney JM, Reeves MS, Lubin BC, Dicklin MR. Effects of adding prescription omega-3 acid ethyl esters to simvastatin (20mg/day) on lipids and lipoprotein particles in men and women with mixed dyslipidemia. *Am J Cardiol* 2008; 102: 429–433.

133. Khandelwal S, Demonty I, Jeemon P, Lakshmy R, Mukherjee R, Gupta R, Snehi U, Niveditha D, Singh Y, van der Knaap HC et al. Independent and interactive effects of plant sterols and fish oil n-3 long-chain polyunsaturated fatty acids on the plasma lipid profile of mildly hyperlipidaemic Indian adults. *Br J Nutr* 2009; 102: 722–732.

134. Harris WS. n-3 Fatty acids and lipoproteins: comparison of results from human and animal studies. *Lipids* 1996; 31: 243–252.

135. Balk EM, Lichtenstein AH, Chung M, Kupelnick B, Chew P, Lau J. Effects of omega-3 fatty acids on serum markers of cardiovascular disease risk: a systematic review. *Atherosclerosis* 2006; 189: 19–30.

136. Holub BJ. Docosahexaenoic acid (DHA) and cardiovascular disease risk factors. *Prostaglandins Leukot Essent Fatty Acids* 2009; 81: 199–204.

137. Sun Q, Ma J, Campos H, Rexrode KM, Albert CM, Mozaffarian D, Hu FB. Blood concentrations of individual long-chain n-3 fatty acids and risk of nonfatal myocardial infarction. *Am J Clin Nutr* 2008; 88: 216–223.

138. Motoyama KR, Curb JD, Kadowaki T, El-Saed A, Abbott RD, Okamura T, Evans RW, Nakamura Y, Sutton-Tyrrell K, Rodriquez BL et al. Association of serum n-6 and n-3 polyunsaturated fatty acids with lipids in 3 populations of middle-aged men. *Am J Clin Nutr* 2009; 90: 49–55.

139. Zhang L, Geng Y, Xiao N, Yin M, Mao L, Ren G, Zhang C, Liu P, Lu N, An L et al. High dietary n-6/n-3 PUFA ratio promotes HDL cholesterol level, but does not suppress atherogenesis in apolipoprotein E-null mice 1. *J Atheroscler Thromb* 2009; 16: 463–471.

140. Wooten JS, Biggerstaff KD, Ben-Ezra V. Responses of LDL and HDL particle size and distribution to omega-3 fatty acid supplementation and aerobic exercise. *J Appl Physiol* 2009; 107: 794–800.

141. Soulimane-Mokhtari NA, Guermouche B, Saker M, Merzouk S, Merzouk H, Hichami A, Madani S, Khan NA, Prost J. Serum lipoprotein composition, lecithin cholesterol acyltransferase and tissue lipase activities in pregnant diabetic rats and their offspring receiving enriched n-3 PUFA diet. *Gen Physiol Biophys* 2008; 27: 3–11.

142. Choo J, Ueshima H, Curb JD, Shin C, Evans RW, El-Saed A, Kadowaki T, Okamura T, Nakata K, Otake T et al. Serum n-6 fatty acids and lipoprotein subclasses in middle-aged men: the population-based cross-sectional ERA-JUMP study. *Am J Clin Nutr* 2010; 91: 1195–1203.

143. Nishimoto T, Pellizzon MA, Aihara M, Stylianou IM, Billheimer JT, Rothblat G, Rader DJ. Fish oil promotes macrophage reverse cholesterol transport in mice. *Arterioscler Thromb Vasc Biol* 2009; 29: 1502–1508.

144. Pandey NR, Renwick J, Rabaa S, Misquith A, Kouri L, Twomey E, Sparks DL. An induction in hepatic HDL secretion associated with reduced ATPase expression. *Am J Pathol* 2009; 175: 1777–1787.

145. Sheril A, Jeyakumar SM, Jayashree T, Giridharan NV, Vajreswari A. Impact of feeding polyunsaturated fatty acids on cholesterol metabolism of dyslipidemic obese rats of WNIN/GR-Ob strain. *Atherosclerosis* 2009; 204: 136–140.

146. Kralova Lesna I, Suchanek P, Kovar J, Stavek P, Poledne R. Replacement of dietary saturated FAs by PUFAs in diet and reverse cholesterol transport. *J Lipid Res* 2008; 49: 2414–2418.

147. Schmidt EB, Koenig W, Khuseyinova N, Christensen JH. Lipoprotein-associated phospholipase A2 concentrations in plasma are associated with the extent of coronary artery disease and correlate to adipose tissue levels of marine n-3 fatty acids. *Atherosclerosis* 2008; 196: 420–424.

148. Nettleton JA, Volcik KA, Hoogeveen RC, Boerwinkle E. Carbohydrate intake modifies associations between ANGPTL4[E40K] genotype and HDL-cholesterol concentrations in white men from the Atherosclerosis Risk in Communities (ARIC) study. *Atherosclerosis* 2009; 203: 214–220.

149. Mosher MJ, Lange LA, Howard BV, Lee ET, Best LG, Fabsitz RR, Maccluer JW, North KE. Sex-specific interaction between APOE genotype and carbohydrate intake affects plasma HDL-C levels: the Strong Heart Family Study. *Genes Nutr* 2008; 3: 87–97.

150. Starc TJ, Shea S, Cohn LC, Mosca L, Gersony WM, Deckelbaum RJ. Greater dietary intake of simple carbohydrate is associated with lower concentrations of high-density-lipoprotein cholesterol in hypercholesterolemic children. *Am J Clin Nutr* 1998; 67: 1147–1154.

151. Habib K, Butt MS. Hypocholestrolemic and hypoglycemic effects of dietetic flour in rats. *Asia Pac J Clin Nutr* 2003; 12 (Suppl.): S22.

152. Culberson A, Kafai MR, Ganji V. Glycemic load is associated with HDL cholesterol but not with the other components and prevalence of metabolic syndrome in the third National Health and Nutrition Examination Survey, 1988–1994. *Int Arch Med* 2009; 104: 1034–1042.

153. Welsh JA, Sharma A, Abramson JL, Vaccarino V, Gillespie C, Vos MB. Caloric sweetener consumption and dyslipidemia among US adults. *JAMA* 2010; 303: 1490–1497.

154. Welsh JA, Sharma A, Cunningham SA, Vos MB. Consumption of added sugars and indicators of cardiovascular disease risk among US adolescents. *Circulation* 2011; 123: 249–257.

155. Hostmark AT. The Oslo Health Study: soft drink intake is associated with the metabolic syndrome. *Appl Physiol Nutr Metab* 2010; 35: 635–642.

156. Mutungi G, Ratliff J, Puglisi M, Torres-Gonzalez M, Vaishnav U, Leite JO, Quann E, Volek JS, Fernandez ML. Dietary cholesterol from eggs increases plasma HDL cholesterol in overweight men consuming a carbohydrate-restricted diet. *J Nutr* 2008; 138: 272–276.

157. LeCheminant JD, Smith BK, Westman EC, Vernon MC, Donnelly JE. Comparison of a reduced carbohydrate and reduced fat diet for LDL, HDL, and VLDL subclasses during 9-months of weight maintenance subsequent to weight loss. *Lipids Health Dis* 2010; 9: 54.

158. Can AS, Uysal C, Palaoglu KE. Short term effects of a low-carbohydrate diet in overweight and obese subjects with low HDL-C levels. *BMC Endocr Disord* 2010; 10: 18.

159. Larsen RN, Mann NJ, Maclean E, Shaw JE. The effect of high-protein, low-carbohydrate diets in the treatment of type 2 diabetes: a 12 month randomised controlled trial. *Diabetologia* 2011; 54: 731–740.
160. Elhayany A, Lustman A, Abel R, Attal-Singer J, Vinker S. A low carbohydrate Mediterranean diet improves cardiovascular risk factors and diabetes control among overweight patients with type 2 diabetes mellitus: a 1-year prospective randomized intervention study. *Diabetes Obes Metab* 2010; 12: 204–209.
161. Onuegbu AJ, Olisekodiaka JM, Onibon MO, Adesiyan AA, Igbeneghu CA. Consumption of soymilk lowers atherogenic lipid fraction in healthy individuals. *J Med Food* 2011; 14: 257–260.
162. Harland JI, Haffner TA. Systematic review, meta-analysis and regression of randomised controlled trials reporting an association between an intake of circa 25g soya protein per day and blood cholesterol. *Atherosclerosis* 2008; 200: 13–27.
163. Zhan S, Ho SC. Meta-analysis of the effects of soy protein containing isoflavones on the lipid profile. *Am J Clin Nutr* 2005; 81: 397–408.
164. Shidfar F, Ehramphosh E, Heydari I, Haghighi L, Hosseini S, Shidfar S. Effects of soy bean on serum paraoxonase 1 activity and lipoproteins in hyperlipidemic postmenopausal women. *Int J Food Sci Nutr* 2009; 60: 195–205.
165. Taku K, Umegaki K, Sato Y, Taki Y, Endoh K, Watanabe S. Soy isoflavones lower serum total and LDL cholesterol in humans: a meta-analysis of 11 randomized controlled trials. *Am J Clin Nutr* 2007; 85: 1148–1156.
166. Sacks FM, Lichtenstein A, Van Horn L, Harris W, Kris-Etherton P, Winston M. Soy protein, isoflavones, and cardiovascular health: an American Heart Association Science Advisory for professionals from the Nutrition Committee. *Circulation* 2006; 113: 1034–1044.
167. Aziz A, Xiao CW, Cockell KA, Sarwar Gilani G, Cruz-Hernandez C, Nimal Ratnayake WM. Impact of dietary protein on lipid metabolism in hamsters is source-dependent and associated with changes in hepatic gene expression. *Br J Nutr* 2008; 100: 503–511.
168. Ciampolini M, Bini S. Serum lipids in celiac children. *J Pediatr Gastroenterol Nutr* 1991; 12: 459–460.
169. Capristo E, Malandrino N, Farnetti S, Mingrone G, Leggio L, Addolorato G, Gasbarrini G. Increased serum high-density lipoprotein-cholesterol concentration in celiac disease after gluten-free diet treatment correlates with body fat stores. *J Clin Gastroenterol* 2009; 43: 946–949.
170. Erlund I, Koli R, Alfthan G, Marniemi J, Puukka P, Mustonen P, Mattila P, Jula A. Favorable effects of berry consumption on platelet function, blood pressure, and HDL cholesterol. *Am J Clin Nutr* 2008; 87: 323–331.
171. Silver HJ, Dietrich MS, Niswender KD. Effects of grapefruit, grapefruit juice and water preloads on energy balance, weight loss, body composition, and cardiometabolic risk in free-living obese adults. *Nutr Metab* 2011; 8: 8.
172. Qin Y, Xia M, Ma J, Hao Y, Liu J, Mou H, Cao L, Ling W. Anthocyanin supplementation improves serum LDL- and HDL-cholesterol concentrations associated with the inhibition of cholesteryl ester transfer protein in dyslipidemic subjects. *Am J Clin Nutr* 2009; 90: 485–492.
173. Hamed MS, Gambert S, Bliden KP, Bailon O, Singla A, Antonino MJ, Hamed F, Tantry US, Gurbel PA. Dark chocolate effect on platelet activity, C-reactive protein and lipid profile: a pilot study. *South Med J* 2008; 101: 1203–1208.

174. Mellor DD, Sathyapalan T, Kilpatrick ES, Beckett S, Atkin SL. High-cocoa polyphenol-rich chocolate improves HDL cholesterol in Type 2 diabetes patients. *Diabet Med* 2010; 27: 1318–1321.

175. Sobenin IA, Andrianova IV, Demidova ON, Gorchakova T, Orekhov AN. Lipid-lowering effects of time-released garlic powder tablets in double-blinded placebo-controlled randomized study. *J Atheroscler Thromb* 2008; 15: 334–338.

176. Lam CK, Zhang Z, Yu H, Tsang SY, Huang Y, Chen ZY. Apple polyphenols inhibit plasma CETP activity and reduce the ratio of non-HDL to HDL cholesterol. *Mol Nutr Food Res* 2008; 52: 950–958.

177. Berrougui H, Grenier G, Loued S, Drouin G, Khalil A. A new insight into resveratrol as an atheroprotective compound: Inhibition of lipid peroxidation and enhancement of cholesterol efflux. *Atherosclerosis* 2009; 207: 420–427.

178. Uto-Kondo H, Ayaori M, Ogura M, Nakaya K, Ito M, Suzuki A, Takiguchi S-i, Yakushiji E, Terao Y, Ozasa H et al. Coffee consumption enhances high-density lipoprotein-mediated cholesterol efflux in macrophages. *Circ Res* 2010; 106: 779–787.

179. Yang Y, Jiang W, Wang L, Zhang ZB, Si SY, Hong B. Characterization of the isoflavone pratensein as a novel transcriptional up-regulator of scavenger receptor class B type I in HepG2 cells. *Biol Pharm Bull* 2009; 32: 1289–1294.

180. Gugliucci A, Bastos DH. Chlorogenic acid protects paraoxonase 1 activity in high density lipoprotein from inactivation caused by physiological concentrations of hypochlorite. *Fitoterapia* 2009; 80: 138–142.

181. Khateeb J, Gantman A, Kreitenberg AJ, Aviram M, Fuhrman B. Paraoxonase 1 (PON1) expression in hepatocytes is upregulated by pomegranate polyphenols: A role for PPAR-gamma pathway. *Atherosclerosis* 2010; 208: 119–125.

182. Fuhrman B, Volkova N, Aviram M. Pomegranate juice polyphenols increase recombinant paraoxonase-1 binding to high-density lipoprotein: studies in vitro and in diabetic patients. *Nutrition* 2010; 26: 359–366.

183. Rock W, Rosenblat M, Miller-Lotan R, Levy AP, Elias M, Aviram M. Consumption of Wonderful Variety pomegranate juice and extract by diabetic patients increases paraoxonase 1 association with high-density lipoprotein and stimulates its catalytic activities. *J Agric Food Chem* 2008; 56: 8704–8713.

184. Gong M, Garige M, Varatharajalu R, Marmillot P, Gottipatti C, Leckey LC, Lakshman RM. Quercetin up-regulates paraoxonase 1 gene expression with concomitant protection against LDL oxidation. *Biochem Biophys Res Commun* 2009; 379: 1001–1004.

185. Boesch-Saadatmandi C, Niering J, Minihane AM, Wiswedel I, Gardeman A, Wolffram S, Rimbach G. Impact of apolipoprotein E genotype and dietary quercetin on paraoxonase 1 status in apoE3 and apoE4 transgenic mice. *Atherosclerosis* 2010; 211: 110–113.

186. Lixandru D, Mohora M, Coman A, Stoian I, van Gils C, Aerts P, Manuel YKB. Diet and paraoxonase 1 enzymatic activity in diabetic foot patients from Romania and Belgium: Favorable association of high flavonoid dietary intake with arylesterase activity. *Ann Nutr Metab* 2010; 56: 294–301.

187. Sutken E, Inal M, Ozdemir F. Effects of vitamin E and gemfibrozil on lipid profiles, lipid peroxidation and antioxidant status in the elderly and young hyperlipidemic subjects. *Saudi Med J* 2006; 27: 453–459.

188. Rajpathak SN, Xue X, Wassertheil-Smoller S, Van Horn L, Robinson JG, Liu S, Allison M, Martin LW, Ho GY, Rohan TE. Effect of 5 y of calcium plus vitamin D supplementation on change in circulating lipids: results from the Women's Health Initiative. *Am J Clin Nutr* 2010; 91: 894–899.
189. Maki KC, Rubin MR, Wong LG, McManus JF, Jensen CD, Lawless A. Effects of vitamin D supplementation on 25-hydroxyvitamin D, high-density lipoprotein cholesterol, and other cardiovascular disease risk markers in subjects with elevated waist circumference. *Int J Food Sci Nutr* 2011; 62: 318–327.
190. Abdollahzad H, Eghtesadi S, Nourmohammadi I, Khadem-Ansari M, Nejad-Gashti H, Esmaillzadeh A. Effect of vitamin C supplementation on oxidative stress and lipid profiles in hemodialysis patients. *Int J Vitam Nutr Res* 2009; 79: 281–287.
191. McRae MP. Vitamin C supplementation lowers serum low-density lipoprotein cholesterol and triglycerides: a meta-analysis of 13 randomized controlled trials. *J Chiropr Med* 2008; 7: 48–58.
192. Wang JH, Keisala T, Solakivi T, Minasyan A, Kalueff AV, Tuohimaa P. Serum cholesterol and expression of ApoAI, LXRbeta and SREBP2 in vitamin D receptor knock-out mice. *J Steroid Biochem Mol Biol* 2009; 113: 222–226.
193. Rode S, Rubic T, Lorenz RL. alpha-Tocopherol disturbs macrophage LXRalpha regulation of ABCA1/G1 and cholesterol handling. *Biochem Biophys Res Commun* 2008; 369: 868–872.
194. Yoshida H, Yanai H, Ito K, Tomono Y, Koikeda T, Tsukahara H, Tada N. Administration of natural astaxanthin increases serum HDL-cholesterol and adiponectin in subjects with mild hyperlipidemia. *Atherosclerosis* 2010; 209: 520–523.
195. Ribaya-Mercado JD, Ordovas JM, Russell RM. Effect of beta-carotene supplementation on the concentrations and distribution of carotenoids, vitamin E, vitamin A, and cholesterol in plasma lipoprotein and non-lipoprotein fractions in healthy older women. *J Am Coll Nutr* 1995; 14: 614–620.
196. van Poppel G, Hospers J, Buytenhek R, Princen HM. No effect of beta-carotene supplementation on plasma lipoproteins in healthy smokers. *Am J Clin Nutr* 1994; 60: 730–734.
197. Lin LY, Peng CC, Liang YJ, Yeh WT, Wang HE, Yu TH, Peng RY. Alpinia zerumbet potentially elevates high-density lipoprotein cholesterol level in hamsters. *J Agric Food Chem* 2008; 56: 4435–4443.
198. Tara F, Maamouri G, Rayman MP, Ghayour-Mobarhan M, Sahebkar A, Yazarlu O, Ouladan S, Tavallaie S, Azimi-Nezhad M, Shakeri MT et al. Selenium supplementation and the incidence of preeclampsia in pregnant Iranian women: a randomized, double-blind, placebo-controlled pilot trial. *Taiwan J Obstet Gynecol* 2010; 49: 181–187.
199. Stahle JA, Vunta H, Channa Reddy C, Sandeep Prabhu K. Regulation of expression of apolipoprotein A-I by selenium status in human liver hepatoblastoma cells. *Eur J Nutr* 2009; 48: 283–290.
200. Bleys J, Navas-Acien A, Stranges S, Menke A, Miller ER III, Guallar E. Serum selenium and serum lipids in US adults. *Am J Clin Nutr* 2008; 88: 416–423.
201. Cavallini DC, Bedani R, Bomdespacho LQ, Vendramini RC, Rossi EA. Effects of probiotic bacteria, isoflavones and simvastatin on lipid profile and atherosclerosis in cholesterol-fed rabbits: a randomized double-blind study. *Lipids Health Dis* 2009; 8: 1.

202. Ooi LG, Liong MT. Cholesterol-lowering effects of probiotics and prebiotics: a review of in vivo and in vitro findings. *Int J Mol Sci* 2010; 11: 2499–2522.
203. Greany KA, Bonorden MJ, Hamilton-Reeves JM, McMullen MH, Wangen KE, Phipps WR, Feirtag J, Thomas W, Kurzer MS. Probiotic capsules do not lower plasma lipids in young women and men. *Eur J Clin Nutr* 2008; 62: 232–237.
204. Yang NC, Chou CW, Chen CY, Hwang KL, Yang YC. Combined nattokinase with red yeast rice but not nattokinase alone has potent effects on blood lipids in human subjects with hyperlipidemia. *Asia Pac J Clin Nutr* 2009; 18: 310–317.
205. Li YG, Zhang F, Wang ZT, Hu ZB. Identification and chemical profiling of monacolins in red yeast rice using high-performance liquid chromatography with photodiode array detector and mass spectrometry. *J Pharm Biomed Anal* 2004; 35: 1101–1112.
206. Liu J, Zhang J, Shi Y, Grimsgaard S, Alraek T, Fonnebo V. Chinese red yeast rice (*Monascus purpureus*) for primary hyperlipidemia: a meta-analysis of randomized controlled trials. *Chin Med* 2006; 1: 4.

CONCLUSIONS AND PERSPECTIVES

Plasma HDL is a small, dense and protein-rich lipoprotein compared with other lipoprotein classes (chylomicrons, VLDL, IDL, LDL). HDL particles can be discoid or spherical, containing apolipoproteins and enzymes involved in lipid metabolism, in addition to acute-phase proteins, complement proteins and antiproteases. HDL particles are highly heterogeneous in their structure, composition (proteome, lipidome) and physicochemical properties; such a pronounced level of HDL heterogeneity is probably determined by interactions of apoA-I and other structural amphipathic proteins with lipids, as well as by specific protein-protein interactions.

The structure of HDL particles largely reflects the properties of their exchangeable apolipoprotein components, primarily those of apoA-I, and varies as a function of HDL composition. The basic features of the structural organization of apoA-I are, however, similar between individual HDL subpopulations, including discoid and spherical HDL. Specifically, apoA-I adopts a double belt-like conformation in both discoid and spherical particles and predominantly covers the surface of HDL compared with lipid. It remains largely unclear how the presence of proteins other than apoA-I influences the particle structure of HDL.

Intravascular HDL metabolism is under the influence of multiple receptors, lipid transfer proteins and enzymes (ABCA1, ABCG1, LCAT, CETP, PLTP, LPL, hepatic lipase, endothelial lipase, SR-BI), each of which is capable of modifying circulating levels of HDL-C. HDL metabolism is based on the dynamic nature of the apoA-I molecule, is intimately linked to the metabolism

High-Density Lipoproteins: Structure, Metabolism, Function, and Therapeutics, First Edition.
Anatol Kontush and M. John Chapman.

of apoB-containing lipoproteins and ensures the processes of cellular cholesterol efflux and, in part, RCT. Cellular cholesterol efflux occurs via several receptor-dependent and independent pathways; the relative contributions of individual cellular receptors (ABCA1, ABCG1, SR-BI) and other pathways (passive diffusion) to the efflux remains to be established for major cell types involved in atherogenesis, including monocytes, macrophages, smooth muscle cells, endothelial cells, lymphocytes and adipocytes.

Consistent with the concept of RCT, low circulating levels of cholesterol carried by HDL (i.e., HDL-C) are a significant and independent predictor of cardiovascular disease. In a similar fashion, low circulating levels of apoA-I also constitute a biomarker of elevated cardiovascular risk; plasma concentrations and/or the enzymatic activities of HDL components other than total cholesterol and apoA-I possess the potential to further improve risk assessment.

HDL-C levels are under strong control by multiple genes; the heritability of HDL-C concentrations ranges between 40% and 60% and can attain up to 80%. In contrast, the effects on the HDL-C levels of individual variants are small, each accounting for <5%, and the majority for <1%–2%. These data raise the question of how such small effects can combine to translate into the high degree of heritability.

In addition to cellular cholesterol efflux capacity, HDL particles possess other biologic activities, including antioxidative, anti-inflammatory, cytoprotective, antithrombotic, anti-infectious, vasodilatory and antidiabetic actions which may provide protection from atherosclerosis and cardiovascular disease. HDL particles may, therefore, provide a key for efficient atheroprotection in humans. However, it remains indeterminate as to which biologic activities of HDL are relevant for the protection of the arterial wall from atherogenesis. Intriguingly, cellular efflux of cholesterol and/or other lipids may mechanistically underlie multiple biologic activities of HDL, such as anti-inflammatory, antithrombotic and antidiabetic actions, with apoA-I apparently playing a pivotal role.

HDL particle subpopulations are heterogeneous in their biologic activities. Notably, small, dense, protein-rich HDL3 displays potent anti-atherogenic activities both *in vivo* and *in vitro*. It is tempting to suggest that specific biologic functions of HDL are confined to individual subpopulations of distinct proteome and/lipidome; evidence in support of this hypothesis is accumulating. The identification of such clinically relevant HDL subpopulations, followed by assessment of their plasma levels and/or biologic activities, may represent another approach to further improve our evaluation of cardiovascular risk.

The composition, particle profile and atheroprotective properties of HDL can be compromised along with subnormal circulating levels of HDL under conditions associated with accelerated development of atherosclerosis, such as in atherogenic low HDL-C dyslipidemias typical of metabolic diseases, including Metabolic Syndrome and Type 2 diabetes. Relationships between alterations in the HDL composition (proteome, lipidome), particle heterogeneity and function largely remain to be characterized. A deficiency in HDL function may result in accelerated atherosclerosis; low HDL content of functional components, altered particle profile and deficient anti-atherogenic activities of HDL may, therefore, all contribute to atherogenesis and enhance cardiovascular risk.

Functional HDL deficiency is intimately associated with alterations in HDL metabolism. Multiple pathways, such as enhanced production of SAA, altered CETP activity, oxidative modification and glycation, which operate concomitantly, may lead to the loss of the anti-atherogenic function of HDL in metabolic diseases. The formation of HDL particles with attenuated anti-atherogenic activities may be mechanistically related to HDL enrichment in SAA, apoE, triglyceride and oxidized lipids, depletion of apoA-I, LCAT, PON1, apoF, apoJ, PLTP, S1P, phospholipid and cholesteryl ester, and covalent modification of key HDL apolipoproteins, including site-specific oxidation and glycation.

Thus, dyslipidemic subjects in primary prevention at elevated cardiovascular risk, as well as patients in secondary prevention with symptomatic coronary atherosclerosis who are treated by a statin, may possess HDL particles whose anti-atherogenic properties are impaired. Raising HDL-C accompanied by the normalization of HDL function may, therefore, attenuate atherogenesis and thereby decrease the residual cardiovascular risk in such subjects. The therapeutic normalization of HDL metabolism in terms of both the quantity (circulating levels) and quality (anti-atherogenic function) of HDL particles may represent a novel approach in attenuating atherosclerosis and reducing cardiovascular risk. The induction of selective increase in the circulating concentrations of HDL particle subpopulations possessing potent anti-atherogenic activity is particularly promising. Other clinical applications of HDL particles may involve the use of artificial HDL in drug delivery and imaging studies of atherosclerosis.

HDL metabolism can be normalized to different degrees by agents enhancing HDL formation and/or improving remodeling (apolipoproteins, rHDL, apoA-I mimetics, CETP inhibitors, niacin, PPAR alpha and gamma agonists, LXR agonists, lipase inhibitors and others), which can be administrated acutely or chronically. Clinical evidence is accumulating that HDL-based therapies can be effective against cardiovascular disease; such evidence primarily includes data on the acute effects of rHDL on the characteristics of atherosclerotic plaques in humans. In addition, rHDL infusion can exert antidiabetic actions on pancreatic beta-cells and antithrombotic actions on platelets, as well as protect against infection. Studies involving an array of potential candidate molecules targeting HDL on the basis of chronic administration are also ongoing. In particular, HDL-C raising in low HDL-C dyslipidemia by CETP inhibitors and niacin hold promise to translate into the reduction of cardiovascular risk persisting after statin treatment; the clinical potential of these agents is being assessed in large-scaled, randomized clinical trials.

Importantly, novel HDL-based therapies may not only delay the progression of cardiovascular disease and stabilize atherosclerotic plaques, but may even favor plaque regression. Indeed, raising functional plasma HDL levels through enhanced apoA-I production may be sufficient to favorably remodel atheromata. Moreover, a stronger effect can be achieved when high levels of functional HDL are combined with large reductions in the concentration of circulating apoB-lipoproteins. Specifically, atherosclerosis regression can be observed when plasma LDL-C concentrations are reduced and HDL functionality is enhanced along the RCT pathway and beyond.

Lifestyle interventions, primarily exercise and weight loss, as well as nutritional approaches, may also prove useful to act on plasma HDL. However, it remains largely unknown whether HDL-C raising induced by individual nutrients results in the enhancement of HDL functionality. Furthermore, interactions between nutritional factors and genes determining HDL metabolism need to be characterized.

The development of HDL-based therapeutic approaches will probably require targeted measurements of specific HDL components, HDL particle profile and/or HDL functionality as biomarkers. Clinically-relevant assays allowing high throughput screening of HDL composition, heterogeneity and/or functionality will therefore be critical for such development. Comprehensive studies of all aspects of HDL biology, together with large clinical trials, will soon provide an answer to the major remaining question in the field of HDL—are HDL-based therapies capable of further decreasing the cardiovascular risk persisting after statin therapy?

INDEX

High-Density Lipoproteins: Structure, Metabolism, Function, and Therapeutics, First Edition.
Anatol Kontush and M. John Chapman.